Lecture Notes in Networks and Systems

Volume 118

The series "Lecture Notes in Networks and Systems" publishes the latest developments in Networks and Systems—quickly, informally and with high quality. Original research reported in proceedings and post-proceedings represents the core of LNNS.

Volumes published in LNNS embrace all aspects and subfields of, as well as new challenges in, Networks and Systems.

The series contains proceedings and edited volumes in systems and networks, spanning the areas of Cyber-Physical Systems, Autonomous Systems, Sensor Networks, Control Systems, Energy Systems, Automotive Systems, Biological Systems, Vehicular Networking and Connected Vehicles, Aerospace Systems, Automation, Manufacturing, Smart Grids, Nonlinear Systems, Power Systems, Robotics, Social Systems, Economic Systems and other. Of particular value to both the contributors and the readership are the short publication timeframe and the world-wide distribution and exposure which enable both a wide and rapid dissemination of research output.

The series covers the theory, applications, and perspectives on the state of the art and future developments relevant to systems and networks, decision making, control, complex processes and related areas, as embedded in the fields of interdisciplinary and applied sciences, engineering, computer science, physics, economics, social, and life sciences, as well as the paradigms and methodologies behind them.

**** Indexing: The books of this series are submitted to ISI Proceedings, SCOPUS, Google Scholar and Springerlink ****

More information about this series at http://www.springer.com/series/15179

Sheng-Lung Peng · Le Hoang Son ·
G. Suseendran · D. Balaganesh
Editors

Intelligent Computing and Innovation on Data Science

Proceedings of ICTIDS 2019

Springer

Editors
Sheng-Lung Peng
CSIE Department
National Dong Hwa University
Hualien City, Taiwan

G. Suseendran
Department of Information Technology
Vels Institute of Science,
Technology & Advanced Studies
Pallavaram, Chennai, Tamil Nadu, India

Le Hoang Son
Vietnam National University
Hanoi, Vietnam

D. Balaganesh
Faculty of Computer Science
and Multimedia
Lincoln University College
Petaling Jaya, Malaysia

ISSN 2367-3370 ISSN 2367-3389 (electronic)
Lecture Notes in Networks and Systems
ISBN 978-981-15-3283-2 ISBN 978-981-15-3284-9 (eBook)
https://doi.org/10.1007/978-981-15-3284-9

This Springer imprint is published by the registered company Springer Nature Singapore Pte Ltd.
The registered company address is: 152 Beach Road, #21-01/04 Gateway East, Singapore 189721, Singapore

Organizing Committee and Key Members

Conference Committee Members

Patron

Datuk Dr. Hajjah Bibi Florina Abdullah, Pro-Chancellor, Lincoln University College, Malaysia

Honorary Chair

Prof. Dr. Amiya Bhaumik, Vice Chancellor and CEO, Lincoln University College, Malaysia

Conference Advisors

Prof. Datuk Dr. Abdul Gani Bin Mohammed Din, Deputy Vice Chancellor (Academic), Lincoln University College, Malaysia
Haji Mansor, Academic Director, Lincoln University College, Malaysia

Scientific Advisory Chair

Prof. Satish Narayanasamy, Associate Professor, ECES, College of Engineering, University of Michigan, USA

Convenor

Dr. D. Balaganesh, Dean, Faculty of Computer Science and Multimedia, Lincoln University College, Malaysia

Co-convenor

Prof. Dr. Swamy K B, Director, International Affairs, Lincoln University College, Malaysia

Publication Committee

Prof. Sheng-Lung Peng, CSIE Department, National Dong Hwa University, Taiwan
Dr. Souvik Pal, Associate Professor, Department of Computer Science and Engineering, Brainware University, Kolkata, India
Dr. G. Suseendran, Department of Information Technology, School of Computing Sciences, Vels Institute of Science, Technology and Advanced Studies (VISTAS), Chennai, Tamil Nadu, India.
Dr. D. Balaganesh, Faculty of Computer Science and Multimedia, Lincoln University College, Malaysia
Dr. Midhunchakkaravarthy, Faculty of Computer Science and Multimedia, Lincoln University College, Malaysia
Dr. Anand Nayyar, Professor, Researcher and Scientist, Duy Tan University, Vietnam
Dr. Noor Zaman, Taylor's University, Malaysia

Publicity Chair

Dr. Jafar Ahmad Abed Alzubi, School of Engineering, Department of Computer Engineering and Science, Al-Balqa Applied University, Salt, Jordan
Dr. Kusum Yadav, College of Computer Engineering and Science, University of Hail, Kingdom of Saudi Arabia
Dr. D. Akila, Department of Information Technology, VISTAS, Chennai, India
Dr. Ashish Mishra, Department of Computer Science and Engineering, Gyan Ganga Institute of Technology and Sciences, Jabalpur, India
Mr. Dinesh Rajassekharan, Faculty of Computer Science and Multimedia, Lincoln University College, Malaysia
Mr. Durganand Panjiyar, Faculty of Computer Science and Multimedia, Lincoln University College, Malaysia
Ms. Che Asma Noor Akma, Faculty of Computer Science and Multimedia, Lincoln University College, Malaysia
Mr. Ali Akbary, Faculty of Computer Science and Multimedia, Lincoln University College, Malaysia
Mr. Mohd Nabil Amin, Faculty of Computer Science and Multimedia, Lincoln University College, Malaysia
Dr.Yahya-Imam Munir Kolapo, Faculty of Computer Science and Multimedia, Lincoln University College, Malaysia
Dr. Osama Isaac, Faculty of Business and Accounting, Lincoln University College, Malaysia
Dr. Mohammed Saleh Nusari, Faculty of Engineering, Lincoln University College, Malaysia

Conference Scientific Committee

Dr. Ali Ameen, Faculty of Computer Science and Multimedia, Lincoln University College, Malaysia
Dr. Divya Midhunchakkravarthy, Centre of Postgraduate Studies, Lincoln University College, Malaysia
Dr. B. Vivekanandam, Faculty of Computer Science and Multimedia, Lincoln University College, Malaysia
Prof. Dr. P. Suresh Varma, Rector, University College, Adikavinannaya University, India
Dr. Hanaa Hachimi, Associate Professor in Applied Mathematics and Computer Science, Department of Informatics, Logistics and Mathematics, Ibn Tofail University, Kenitra, Morocco
Dr. Srinath Doss, Head of Department Faculty of Computing, Botho University, Gaborone, Botswana
Dr. S. Gopinathan, Department of Computer Science, University of Madras, Chennai, India
Dr. V. R. Elangovan, Department of Computer Science, A. M. Jain College, Chennai, India
Dr. N. Pradeep, Department of CD and E, BIET, Davangere, India
Dr. Bandana Mahapatra, Anusandhan University, Bhubaneswar, India
Prof. T. Nusrat Jabeen, Department of Computer Science, Anna Adarsh College for Women, Chennai, India
Dr. Sudeep Tanwar, Department of Computer Engineering, Institute of Technology, Nirma University, Gujarat, India
Dr. Tanupriya Choudhury, School of Computer Science, University of Petroleum and Energy Studies, Uttarakhand, India
Dr. D. Jude Hemanth, Karunya University, India
Dr. A. Sasi Kumar, Department of Information Technology, VISTAS, Chennai, India
Dr. Sujatha Srinivasan, Department of Computer Science, SRM Institute of Training and Development, Chennai, India
Dr. C. Priya, Department of Information Technology, VISTAS, Chennai, India

Preface and Acknowledgements

The main aim of this book is to bring together leading academic scientists, researchers and research scholars to exchange and share their experiences and research results on all aspects of intelligent ecosystems and data sciences. It also provides a premier interdisciplinary platform for researchers, practitioners and educators to present and discuss the most recent innovations, trends and concerns as well as practical challenges encountered and solutions adopted in the fields of IoT and analytics. This book aims to attract researchers and practitioners who are working in information technology and computer science. This proceedings is about basics and high-level concepts regarding intelligent computing paradigm and communications in the context of distributed computing, big data, high-performance computing and Internet of Things.

ICTIDS 2019 is organized by Lincoln University College, Malaysia. ICTIDS 2019 has been held during 11 and 12 October 2019, Petaling Jaya, Malaysia. The conference brought together researchers from all regions around the world working on a variety of fields and provided a stimulating forum for them to exchange ideas and report on their researches. The proceeding of ICTIDS 2019 consists of 92 selected papers which were submitted to the conferences and peer-reviewed by conference committee members and international reviewers. The presenters came from different countries like India, Thailand, Bangladesh, Pakistan and Sri Lanka. This conference became a platform to share the knowledge domain among different countries research culture.

We are sincerely thankful to Almighty for supporting and standing at all time with us, whether it is a good or tough time and given ways to concede us. Starting from the call for papers till the finalization of chapters, all the team members are given their contributions amicably, which it is a positive sign of significant team works, and the editors and conference organizers are sincerely thankful to all the members of Springer especially Mr. Aninda Bose for providing constructive inputs and allowing an opportunity to finalize this conference proceedings. We are also thankful to Prof. William Achauer and Prof. Anil Chandy for their support. We are equally thankful to all the reviewers who came from different places in and around the globe shared their support and stand firm towards quality chapter submission.

Finally, we would like to wish you have good success in your presentations and social networking. Your strong supports are critical to the success of this conference. We hope that the participants will not only enjoy the technical programme in the conference but also discover many beautiful places in Malaysia. Wishing you a fruitful and enjoyable ICTIDS 2019.

Hualien City, Taiwan — Sheng-Lung Peng
Hanoi, Vietnam — Le Hoang Son
Chennai, India — G. Suseendran
Petaling Jaya, Malaysia — D. Balaganesh

The original version of this Book was revised: The authors "Midhunchakkaravarthy" and "B. Vivekanandam" names are corrected. The correction to these chapters are available at https://doi.org/10.1007/978-981-15-3284-9_86

About This Book

The conference proceedings is a depository of knowledge enriched with recent research findings. The main focus of this volume is to bring all the computing and communication-related technologies in a single platform, so that undergraduate and postgraduate students, researchers, academicians and industry people can easily understand the intelligent and innovative computing systems, intelligent ecosystems and data sciences, IoT-based ecosystems and communication systems. This book is a podium to convey researchers' experiences, to present excellent result analysis, future scopes and challenges facing the field of computer science, information technology and telecommunication. This book also provides a premier interdisciplinary platform for researchers, practitioners and educators to present and discuss the most recent innovations, trends and concerns as well as practical challenges encountered and solutions adopted in the fields of computer science and information technology. This book will provide the authors, research scholars and listeners with opportunities for national and international collaboration and networking among universities and institutions for promoting research and developing the technologies globally. The readers will have the chance to get together some of the world's leading researchers, to learn about their most recent research outcome, analysis and developments and to catch up with current trends in industry-academia. This book aims to provide the concepts of related technologies regarding intelligent and innovative computing systems, big data and data analytics, IoT-based ecosystems and communication systems and novel findings of the researchers through its chapter organization.

Contents

Editors and Contributors

About the Editors

Sheng-Lung Peng is a full Professor of the Department of Computer Science and Information Engineering at National Dong Hwa University, Taiwan. He received the BS degree in Mathematics from National Tsing Hua University, and the MS and PhD degrees in Computer Science and Information Engineering from the National Chung Cheng University and National Tsing Hua University, Taiwan, respectively. His research interests are in designing and analyzing algorithms for Combinatorics, Bioinformatics, and Networks. Dr. Peng has edited several special issues at journals, such as Soft Computing, Journal of Internet Technology, Journal of Computers and MDPI Algorithms. He is also a reviewer for more than 10 journals such as IEEE Transactions on Emerging Topics in Computing, Theoretical Computer Science, Journal of Computer and System Sciences, Journal of Combinatorial Optimization, Journal of Modelling in Management, Soft Computing, Information Processing Letters, Discrete Mathematics, Discrete Applied Mathematics, Discussions Mathematicae Graph Theory, and so on. He published more than 100 international conferences and journal papers. Dr. Peng is now the dean of the Library and Information Services Office of NDHU and an honorary Professor of Beijing Information Science and Technology University of China. He is a secretary general of Institute of Information and Computing Machinery (IICM) in Taiwan. He is also a director of the ACM-ICPC Contest Council for Taiwan. Recently, he is elected as a supervisor of Chinese Information Literacy Association and of Association of Algorithms and Computation Theory (AACT). He has been serving as a secretary general of Taiwan Association of Cloud Computing (TACC) from 2011-2015 and of AACT from 2013-2016. He was also a convener of the East Region of Service Science Society of Taiwan from 2014-2016.

Le Hoang Son obtained the PhD degree on Mathematics – Informatics at VNU University of Science, Vietnam National University (VNU) in conjunction with the Politecnico di Milano University, Italy in 2013. He has been promoted to Associate

Professor in Information Technology in Vietnam since 2017. Dr. Son worked as senior researcher and Vice Director at the Center for High Performance Computing, VNU University of Science, Vietnam National University during 2007 - 2018. From August 2018, he is Senior Researcher of Department of Multimedia and Virtual Reality, VNU Information Technology Institute. His major fields include Artificial Intelligence, Data Mining, Soft Computing, Fuzzy Computing, Fuzzy Recommender Systems, Geographic Information System. He is a member of Vietnam Journalists Association, International Association of Computer Science and Information Technology, Vietnam Society for Applications of Mathematics, Key Laboratory of Geotechnical Engineering and Artificial Intelligence in University of Transport Technology. Dr. Son is an Associate Editor of Journal of Intelligent & Fuzzy Systems (SCIE), IEEE Access (SCIE), Data Technologies and Applications (SCIE), International Journal of Data Warehousing and Mining (SCIE), Neutrosophic Sets and Systems (ESCI), Vietnam Research and Development on Information and Communication Technology, VNU Journal of Science: Computer Science and Communication Engineering, Frontiers in Artificial Intelligence.

He serves as Editorial Board of Applied Soft Computing (SCIE), PLOS ONE (SCIE), International Journal of Web and Grid Services (SCIE), International Journal of Ambient Computing and Intelligence (ESCI), and Vietnam Journal of Computer Science and Cybernetics. Dr. Son is the Guest Editor of several Special Issues at International Journal of Uncertainty, Fuzziness and Knowledge-Based Systems (SCIE). Dr. Son served as a reviewer for various international journals and conferences. He gave a number of invited talks at many conferences since 2015. Up to now, he has 168 publications in prestigious journals and conferences including 109 ISI papers (SCI: 21, SCIE: 81, ESCI: 7) and 01 SCOPUS paper and undertaken more than 20 major joint international and national research projects. He has published 6 books and book chapters. He supervised more than 54 theses including 02 PhD and 16 Master students.

So far, he has awarded "2014 VNU Research Award for Young Scientists", "2015 VNU Annual Research Award", "2015 Vietnamese Mathematical Award", and "2017 Vietnamese Mathematical Award".

G. Suseendran received his M.Sc., Information Technology and M.Phil., degree from Annamalai University, Tamil Nadu, India and Ph.D., degree in Information Technology-Mathematics from Presidency College, University of Madras, Tamil Nadu, India. In additional qualification, he has obtained DOEACC 'O' Level AICTE Ministry of Information Technology and Honor Diploma in Computer Programming. He is currently working as Assistant Professor, Department of Information Technology, School of Computing Sciences, Vels Institute of Science, Technology & Advanced Studies (VISTAS), Chennai, Tamil Nadu, India which is well known University. He has years of teaching experience in both UG and PG Level. His research interests include Ad-hoc networks, Data Mining, Cloud Computing, Image Processing, Knowledge-based systems and Web Information Exploration. He has been actively involved in professional bodies

Elected Member in London Mathematical Society, Member in Michigan Association for Computer Users in Learning (MACUL) USA, Member in International Association of Engineers (IAENG) Hong Kong, Member in Institute for Engineering Research and Publication (IFERP), Member in Computer Science Teacher Association (CSTA) New York. He serves as Editor/ Editorial Board Member / Technical Committee/ Reviewer of International Journal such Arab/ Poland/ Europe/ USA journals (Thomson Reuters, SCI, and Elsevier). He servers as International Committee Members towards International Conference conducted in association with IEEE, Springer and Scopus.

D. Balaganesh is a Dean of Faculty Computer Science and Multimedia, Lincoln University College, Malaysia. He is the one of the key member of Lincoln University College. He has 18 years of professional teaching experience which include overseas experience in India, Oman and Malaysia. He has Indepth knowledge of Information technology and latest wireless fidelity. His Familiar research area is Malware detection, web mining and open source technology. He has given training in mobile application, open source ERP, software testing, and MYSQL. He has developed software applications "Timetable Automation", "Online Exam". He has ability to explain technical concepts and ideas in a clear and precise way.

Contributors

Md. Akhtaruzzaman Department of Computer Science and Engineering, Asian University of Bangladesh, Dhaka, Bangladesh

D. Akila School of Computing Sciences, Vels Institute of Science, Technology & Advanced Studies (VISTAS), Chennai, Tamil Nadu, India

Sameer Alani Faculty of Information and Communication Technology, Centre for Advanced Computing Technology (C-ACT), Universiti Teknikal Malaysia Melaka, Hang Tuah Jaya, Durian Tunggal, Melaka, Malaysia

Adam R. H. Alhawari Electrical Engineering Department, College of Engineering, Najran University, Najran, Saudi Arabia

Shahid M. Ali Department of Electrical and Electronic Engineering, Universiti Teknologi, PETRONAS Bander Seri Iskandar, Tronoh, Perak, Malaysia

Shaikh Muhammad Allayear Department of Software Engineering, Daffodil International University, Dhaka, Bangladesh;
Department of Multimedia and Creative Technology, Daffodil International University, Dhaka, Bangladesh

Ibrahim Alrajawy Lincoln University College, Petaling Jaya, Malaysia

A. Ambeth Raja Thiruthangal Nadar College, Chennai, India

K. Ambika Department of Computer Science, AVS Engineering College, Salem, India

Ali Ameen Faculty of Computer Science and Multimedia, Lincoln University College, Kota Bharu, Malaysia

R. Anandan Department of Computer Science and Engineering, Vels Institute of Science, Technology and Advanced Studies, Chennai, India

Anitha Anandhan Faculty of Computer Science and Information Technology, University of Malaya, Kuala Lumpur 50603, Malaysia

K. R. Ananthapadmanaban Department of Computer Science and Applications, SRM Arts & Science College, Chennai, Tamil Nadu, India

M. Anita Priscilla Mary Department of Computer Applications, Dr. M.G.R. Educational and Research Institute, Chennai, Tamil Nadu, India

Felix O. Aranuwa Department of Computer Science, Adekunle Ajasin University, Akungba, Nigeria

Agha Saba Asghar Department of Software Engineering, Mehran University of Engineering and Technology, Jamshoro, Pakistan

Tatapudi Ashok Adikavi Nannaya University, Rajamahendravaram, AP, India

R. Ashoka Rajan Department of Computer Science, SoC, SASTRA, Thanjavur, India

K. Aswani Research Scholor, Noorul Islam Center for Higher Education, Kanyakumari, India

D. Balaganesh Faculty of Computer Science and Multimedia, Lincoln University College, Kota Bharu, Malaysia

Noor Basha Department of CSE, VIT, Bengaluru, India

Bhanu Prakash Battula Lincoln University College, Kota Bharu, Malaysia

Shiladitya Bhattacharjee High Performance Cloud Computing Center (HPC3), Universiti Teknologi PETRONAS, Seri Iskandar, Perak Darul Ridzuan, Malaysia

Amiya Bhaumik Faculty of Business and Accounting, Lincoln University College, Kota Bharu, Malaysia

B. Booba Department of Computer Science, Vel's University(VISTAS), Chennai, Tamil Nadu, India

Sarfraz Nawaz Brohi School of Computing & IT, Taylor's University, Subang Jaya, Malaysia

P. Calista Bebe Department of Computer Science, School of Computing Sciences, Vels Institute of Science Technology & Advanced Studies (VISTAS), Chennai, India

Arindam Chakrabarty Department of Management, Rajiv Gandhi University (Central University), Doimukh, Arunachal Pradesh, India

J. Chinna Department of Computer Science, Mother Teresa Women's University, Kodaikanal, Tamil Nadu, India

J. Deny Department of Electronics and Communication Engineering, School of Electronics and Electrical Technology, Kalasalingam Academy of Research and Education, Virudhunagar, Tamil Nadu, India

Uday Sankar Das Guest Faculty, Department of Management & Humanities, National Institute of Technology Arunachal Pradesh, Yupia, Arunachal Pradesh 791112, India

Albert Devaraj Faculty of Computer Science and Multimedia, Lincoln University College, Kuala lumpur, Malaysia

R. Devi Assistant Professor, Department of Computer Science, VISTAS, Chennai, India

J. Dhilipan SRM Institute of Science and Technology, Chennai, India

Srinath Doss Faculty of Computing, Botho University, Gaborone, Botswana

Asita Elengoe Department of Biotechnology, Faculty of Science, Lincoln University College, Petaling Jaya, Selangor, Malaysia

Mohammed Errais Research and Computer Innovation Laboratory, Hassan II University, Casablanca, Morocco

Jannatul Ferdaws Department of Software Engineering, Daffodil International University, Dhaka, Bangladesh

Rokeya Forhat Department of Software Engineering, Daffodil International University, Dhaka, Bangladesh

A. H. Fuad Universitas Indonesia, Depok, Indonesia

D. S. Gayathri Bharathiar University, Coimbatore, Tamil Nadu, India

T. V. Geetha Anna University, Chennai, Tamil Nadu, India

Mithil Ghinaiya School Computing Sciences and Engineering, VIT University Chennai Campus, Chennai, India

M. Gokuldhev Department of CSE, Amritha College of Engineering and Technology, Nagercoil, India

S. Gomathi Department of Computer Science (PG), PSGR Krishnammal College for Women, Coimbatore, India

Rafita Haque Department of Computer Science and Engineering, Asian University of Bangladesh, Dhaka, Bangladesh

Nafisa Haque Department of CSE, Daffodil International University, Dhaka, Bangladesh

A. Hariharan PSG College of Arts and Science, Coimbatore, India

Ahmed Muttaleb Hasan Faculty of Computing (Fkom), University Malaysia Pahang, Kuantan, Pahang, Malaysia

Ali Muttaleb Hasan Faculty of Computing (Fkom), University Malaysia Pahang, Kuantan, Pahang, Malaysia

R. Hema University of Madras, Chennai, Tamil Nadu, India

Mohammed Ali Hussain Department of CSE, Lincoln University College, Kota Bharu, Malaysia

Osama Isaac Lincoln University College, Kota Bharu, Selangor, Malaysia

Maizatul Akmar Ismail Faculty of Computer Science and Information Technology, University of Malaya, Kuala Lumpur, Malaysia

Rabeea Jaffari Department of Software Engineering, Mehran University of Engineering and Technology, Jamshoro, Pakistan

R. Jayakarthik Department of Computer Science, VELS Institute of Science, Technology and Advanced Studies (VISTAS), Chennai, India

G. Jayashree Department of Computing Science, VELS Institute of Science, Technology and Advanced Studies (VISTAS), Chennai, India

Varun Jeoti Department of Electrical and Electronic Engineering, Universiti Teknologi, PETRONAS Bander Seri Iskandar, Tronoh, Perak, Malaysia

V. Jeyabalaraja Department of Computer Science & Engineering, Velammal Engineering College, Chennai, Tamil Nadu, India

S. Jeyalaksshmi Department of BCA & IT, VISTAS, Chennai, India

NZ Jhanjhi School of Computer Science and Engineering (SCE), Taylors University, Subang Jaya, Selangor, Malaysia

S. Josephine Isabella Department of Computer Science, Vels Institute of Science, Technology and Advanced Studies, (VISTAS), Chennai, Tamil Nadu, 600117, India

M. S. Josephine Department of Computer Applications, Dr. M.G.R. Educational and Research Institute, Chennai, Tamil Nadu, India

Mohammed Ali Junejo University of Sindh, Jamshoro, Pakistan

Aisha Zahid Junejo Universiti Teknologi Petronas, Seri Iskandar, Malaysia; Mehran University of Engineering and Technology, Jamshoro, Pakistan

P. Jyothika Shree PSG College of Arts and Science, Coimbatore, India

S. Rayhan Kabir Department of Software Engineering, Daffodil International University, Dhaka, Bangladesh

E. Kanimozhi Department of Computer Science, School of Computing Sciences, Vels Institute of Science, Technology & Advanced Studies (VISTAS), Chennai, India

M. Kannan Department of MCA, SPIHER, Avadi, Chennai, India

S. Karthikeyan Accenture, Chennai, India

K. Kasturi Assistant Professor, Dept of Computer Science, VISTAS, Chennai, Tamil Nadu, India

K. Kavitha Department of Computer Science, Mother Teresa Women's University, Kodaikanal, Tamil Nadu, India

S. Kavitha Department of Computer Science, Vels Institute of Science, Technology & Advanced Studies, Chennai, India

M. Kempanna Department of CSE, BIT, Bengaluru, India

Navid Ali Khan School of Computer Science and Engineering (SCE), Taylors University, Subang Jaya, Selangor, Malaysia

R. Kiruthiga School of Computing Sciences, Vels Institute of Science, Technology & Advanced Studies (VISTAS), Chennai, Tamil Nadu, India

Shwetank Krishna Petroleum Engineering Department, Universiti Teknologi PETRONAS, Seri Iskandar, Perak, Malaysia

K. C. Krishnachalitha School of Computing Sciences, Vels Institute of Science, Technology and Advanced Studies (VISTAS), Chennai, India

D. R. Krithika Department of Computer Science, VISTAS, Chennai, Tamil Nadu, India

Ruchi Kulshrestha Department of Computer Science and Engineering, Manipal University Jaipur, Jaipur, India

P. S. Ashok Kumar Department of CSE, DBIT, Bengaluru, India

T. Lavanya Suja Department of CSE, Vel's University (VISTAS), Chennai, Tamil Nadu, India

Reyner Aranta Lika School of Computing & IT, Taylor's University, Subang Jaya, Malaysia

S. Madhan Department of MCA, SPIHER, Avadi, Chennai, India

K. Maheswari Department of Computer Science, Bharathiyar University, Coimbatore, Tamil Nadu, India

Sarmad Nozad Mahmood Computer Technical Engineering Department, Alkitab University, Kirkuk, Iraq

Dheeraj Malhotra VSIT, Vivekananda Institute of Professional Studies, GGSIPU, Delhi, India

Neha Malhotra VSIT, Vivekananda Institute of Professional Studies, GGSIPU, Delhi, India

S. Malliga Department of Computer Science and Engineering, Kongu Engineering College, Perundurai, Erode, India

K. Manjunath Department of CSE, Govt. Polytechnic, Chennasandra, Bengaluru, India

M. K. Manoj Department of ECE, MEA Engineering College, Malappuram, India

Mehak Maqbool Memon Universiti Teknologi Petronas, Seri Iskandar, Malaysia; Mehran University of Engineering and Technology, Jamshoro, Pakistan

Raheel Maqbool Memon Mehran University of Engineering and Technology, Jamshoro, Pakistan

D. Menaka Department of AE, Noorul Islam Center for Higher Education, Kanyakumari, India

X. Mercilin Raajini Department of ECE, Prince Shri Venkateshwara Padmavathy Engineering College, Chennai, India

Divya Midhunchakkravarthy Faculty of Computer and Multimedia, Lincoln University College, Petaling Jaya, Selangor, Malaysia

Midhunchakkaravarthy Computer Science and Multimedia Department, Wisma Lincoln, Petaling Jaya, Selangor Darul Ehsan, Malaysia;
Faculty of Computer Science and Multimedia, Lincoln University College, Kuala lumpur, Malaysia

Ashish Mishra Faculty of Computer Science and Engineering, Gyan Ganga Institute of Technology and Sciences, Jabalpur, India

Janmaya Kumar Mishra Department of Insights & Data, Capgemini Technology Services India Limited, Hyderabad, India

Rachdi Mohamed Faculty of Science Ben M'sik, Hassan II University, Casablanca, Morocco

S. Monika Department of Computer Science (PG), PSGR Krishnammal College for Women, Coimbatore, India

H. M. Moyeenudin School of Hotel and Catering Management, Vels Institute of Science, Technology and Advanced Studies, Chennai, India

Muaadh Mukred Faculty of Information Science and Technology, Universiti Kebangsaan Malaysia, Bangi, Malaysia

K. Muni Sankar Department of Computer Science, Vel's University(VISTAS), Chennai, Tamil Nadu, India

S. Muruganandam Department of Computer Science, SRM Institute for Training and Development, Chennai, India

Danushyaa A/P. Murugiah School of Computing & IT, Taylor's University, Subang Jaya, Malaysia

Mohan Kumar Naik Department of ECE, NHCE, Bengaluru, India

Puteri N. E. Nohuddin Institute of Visual Informatics, National University of Malaysia, Bangi, Malaysia

Noorhuzaimi Mohd Noor Faculty of Computing (Fkom), University Malaysia Pahang, Kuantan, Pahang, Malaysia

Wan M. U. Noormanshah Institute of Visual Informatics, National University of Malaysia, Bangi, Malaysia

Marini Othman Institute of Informatics and Computing in Energy, Universiti Tenaga Nasional, Selangor, Malaysia

D. Padmapriya Department of BCA & IT, VISTAS, Chennai, India

Shaik Javed Parvez Department of Computer Science and Engineering, Vels Institute of Science, Technology and Advanced Studies, Chennai, India

D. Ponniselvi Vivekanadha College of Arts and Science, Namakkal, India

J. S. T. M. Poovarasi Research Scholar, School of Computing Science, Vels Institute of Science, Technology and Advanced Studies, (VISTAS), Chennai, Tamil Nadu, India

S. Prathima Research Scholar, Department of Computer Science, VELS Institute of Technology and Advanced Studies (VISTAS), Chennai, Tamil Nadu, India

C. Priya Department of Information Technology, School of Computing Sciences, Vels Institute of Science, Technology and Advanced Studies (VISTAS), Chennai, Tamil Nadu, India

R. Priya Department of Computer Application, Vels University, Chennai, India

V. Priya Department of Computer Science, Vels Institute of Science, Technology and Advanced Studies (VISTAS), Chennai, India

R. P. Ragavi Department of Computer Science (PG), PSGR Krishnammal College for Women, Coimbatore, India

Lukman Bin Ab. Rahim High Performance Cloud Computing Center (HPC3), Universiti Teknologi PETRONAS, Seri Iskandar, Perak Darul Ridzuan, Malaysia

Mohammed Rahmah Lincoln University College, Kota Bharu, Selangor, Malaysia

Rahmi Universitas Indonesia, Depok, Indonesia

R. Raja Sudharsan Department of Electronics and Communication Engineering School of Electronics and Electrical Technology, Kalasalingam Academy of Research and Education, Virudhunagar, Tamil Nadu, India

Dinesh Rajassekharan Faculty of Computer and Multimedia, Lincoln University College, Petaling Jaya, Selangor, Malaysia

Ganesan Rajesh Department of Information Technology, MIT Campus, Anna University, Chennai, India

V. Raju Department of Science and Humanities, Sathyabama Institute of Science and Technology, Chennai, India

Ade Wahyu Ramadhani High Performance Cloud Computing Center (HPC3), Universiti Teknologi PETRONAS, Seri Iskandar, Perak Darul Ridzuan, Malaysia

M. Ramakrishnan School of Information Technology, Madurai Kamaraj University, Madurai, Tamil Nadu, India

Givitha Raman Department of Science and Biotechnology, Faculty of Engineering and Life Sciences, Universiti Selangor, Bestari Jaya, Selangor, Malaysia

Prakash Ramani Department of Computer Science and Engineering, Manipal University Jaipur, Jaipur, India

Daksha A/P. V. Ramasamy School of Computing & IT, Taylor's University, Subang Jaya, Malaysia

Taha Hussein Rassem Faculty of Computing (Fkom), University Malaysia Pahang, Kuantan, Pahang, Malaysia

Masood Rehman Department of Electrical and Electronic Engineering, Universiti Teknologi, PETRONAS Bander Seri Iskandar, Tronoh, Perak, Malaysia

A. Revathi Department of Computer Science and Applications, Vivekanandha Arts and Science College of Women, Sankari, Salem, India

M. Revathi Department of Computer Application, Vels University, Chennai, India

Syahrir Ridha Petroleum Engineering Department, Universiti Teknologi PETRONAS, Seri Iskandar, Perak, Malaysia

O. P. Rishi University of Kota, Rajasthan, India

K. Rohini Department of Computer Science, VISTAS, Chennai, Tamil Nadu, India

Muhammad Jafar Sadeq Department of Computer Science and Engineering, Asian University of Bangladesh, Dhaka, Bangladesh

Tale Saeidi Department of Electrical and Electronic Engineering, Universiti Teknologi, PETRONAS Bander Seri Iskandar, Tronoh, Perak, Malaysia

A. Sajeev Ram Department of Information Technology, Sri Krishna College of Engineering and Technology, Coimbatore, India

Mohamed Al-Sarem Department of Information System, Taibah University, Medina, Saudi Arabia

Hasan Sari College of Computing and Informatics, Universiti Tenaga Nasional, Selangor, Malaysia

Hasan Sarwar Institute of Informatics and Computing in Energy, Universiti Tenaga Nasional, 43000 Selangor, Malaysia

N. Sengottaiyan Sri Shanmugha College of Engineering and Technology, Sankari, Tamil Nadu, India

Akhilesh Kumar Sharma Department of Information Technology, Manipal University Jaipur, Jaipur, India

Saurabh Sharma Amity University Gwalior, Gwalior, India

R. Shobarani Department of CSE, Dr. MGR Educational and Research Institute, Chennai, India

Liyana Shuib Faculty of Computer Science and Information Technology, University of Malaya, Kuala Lumpur 50603, Malaysia

C. S. Shylaja Department of Computer Science and Engineering, Vels Institute of Science, Technology and Advanced Studies, Chennai, India

Manickavasagam Sivasankari Department of CSE, Vels Institute of Science, Technology and Advanced Studies, Chennai, Tamil Nadu, India

S. Sowmiyasree Excel College of Arts and Science, Salem, Tamil Nadu, India

T. Sreenivasulu VELS Institute of Science, Technology and Advanced Studies (VISTAS), Chennai, India

N. Srinivasan Department of Computer Applications, B. S. Abdur Rahman CRESCENT Institute of Science and Technology, Chennai, India

Sujatha Srinivasan Department of Computer Science and Applications, SRM Institute for Training and Development, Chennai, Tamil Nadu, India

P. Sripriya Professor, Department of Computer Applications, Vels Institute of Science, Technology & Advanced Studies, Chennai, India

V. Sri Priya Department of MCA, SPIHER, Avadi, Chennai, India

Gutta Sridevi Department of Computer Science and Multimedia, Lincoln University College, Kota Bharu, Malaysia

C. Sudha School of Computing Sciences, Vels Institute of Science & Advanced Studies (VISTAS), Chennai, India

P. Sujatha Department of Computer Science, VISTAS, Chennai, India

Hidayah Sulaiman College of Computing and Informatics, Universiti Tenaga Nasional, Selangor, Malaysia

Wiwit Apit Sulistyowati Universitas Swadaya Gunung Jati, Cirebon, Indonesia

P. Sumathi PG and Research Department of Computer Science, Government Arts and Science College (Autonomous), Coimbatore, India

J. Sumithra Devi Department of Computer Science, Bharathiar University, Coimbatore, Tamil Nadu, India

P. Suresh Varma Adikavi Nannaya University, Rajamahendravaram, AP, India

G. Suseendran Department of Information Technology, School of Computing Sciences, Vels Institute of Science, Technology and Advanced Studies (VISTAS), Chennai, Tamil Nadu, India

C. Swetha Department of Information Technology, MIT Campus, Anna University, Chennai, India

S. P. Syed Ibrahim School Computing Sciences and Engineering, VIT University Chennai Campus, Chennai, India

Shahnawaz Talpur Mehran University of Engineering and Technology, Jamshoro, Pakistan

P. Tamilarasi School of Computing Sciences, Vels Institute of Science, Technology & Advanced Studies, Chennai, India

Kiri Taso Rajiv Gandhi University (Central University), Rono Hills, Doimukh, Arunachal Pradesh, India

K. Tharani Vivekanadha College of Arts and Science, Namakkal, India

T. Thilagaraj Department of Computer Applications, Kongu Arts and Science College, Erode, Tamil Nadu, India

Divya Tiwari Gyan Ganga Institute of Technology and Sciences, Jabalpur, India

S. Vahini Ezhilraman Research Scholar, Department of Computer Science, Vels Institute of Science, Technology and Advanced Studies, (VISTAS), Chennai, Tamil Nadu, India

Mashal Valliani Department of Software Engineering, Mehran University of Engineering and Technology, Jamshoro, Pakistan

Pandian Vasant Fundamental & Applied Science Department, Universiti Teknologi PETRONAS, Seri Iskandar, Perak, Malaysia

P. Venkatesh Department of TCE, DBIT, Bengaluru, India

A. Vidhyalakshmi Department of Computer Science, Vels Institute of Science, Technology and Advanced Studies (VISTAS), Chennai, India

R. Vignesh Tata Consultancy Services, Chennai, Tamil Nadu, India

S. Vigneshwari Department of Computer Science and Engineering, Sathyabama Institute of Science and Technology, Chennai, India

Suresh Vishnu Bharadwaj Department of Computer Science and Engineering, Sathyabama Institute of Science and Technology, Chennai, India

B. Vivekanandam Faculty of Computer Science and Multimedia, Lincoln University College, Kuala lumpur, Malaysia

L. William Mary Department of MCA, SPIHER, Avadi, Chennai, India

J. Wisely Joe School Computing Sciences and Engineering, VIT University Chennai Campus, Chennai, India

Munir Kolapo Yahya-Imam Faculty of Computer Science and Multimedia, Lincoln University College, Petaling Jaya, Malaysia

Agung Yulianto Universitas Swadaya Gunung Jati, Cirebon, Indonesia

Zuhairiah Zainal Abidin Research Center for Applied Electromagnetics, Institute of Integrated Engineering, Universiti Tun Hussein Onn Malaysia, Parit Raja, Johor, Malaysia

Zuraini Zainol Department of Computer Science, Faculty of Science and Defence Technology, National Defense University of Malaysia, Kuala Lumpur, Malaysia

Comprehensive Guide to Implementation of Data Warehouse in Education

G. Jayashree and C. Priya

Abstract Data warehouse, otherwise known as DW, is a central repository to hold the history of data which will be created by the integration or consolidation of data from several external sources and from the operational and transactional data stores of the organization. This data repository will be majorly used by the firms to analyze the data for empowering ideal business decisions to get decided. It performs the role of a system for decision support (DSS), to obtain conclusion based on the analyzed pattern of data to the decision maker(s) of the organization. Usage of data warehouse in education will give a great benefit to the government/private educational officers, by obtaining a single version of the truth of school information. This paper analyzes the work implemented in the education domain focusing on DW concepts in detail.

Keywords DW · Data mart · Educational intelligence · OLAP · Business intelligence · Bottom up · Data mining · Top down · Data visualization

1 Introduction

Data warehouse is created to provide support to the decision-making efforts through data consolidation, collection, research and analytics for any organization. Bill Inmon who is the father of the data warehouse defined a data warehouse as: "The data that is properly unified and aligned with a particular subject, which can get varied across time and providing the required data collection support and are not having any volatile data, with which providing support for the decision make process for any management is defined to be a Data warehouse." DW is an environment and not a defined product. It is the architectural construction of a system of information,

G. Jayashree
Department of Computing Science, VELS Institute of Science, Technology and Advanced Studies (VISTAS), Chennai, India

C. Priya (✉)
Department of Information Technology, School of Computing Sciences, VELS Institute of Science, Technology and Advanced Studies (VISTAS), Chennai, India

S.-L. Peng et al. (eds.), *Intelligent Computing and Innovation on Data Science*, Lecture Notes in Networks and Systems 118,
https://doi.org/10.1007/978-981-15-3284-9_1

providing the business users both the historical and the current decision support data, which is quite difficult to obtain or access from an operational data store.

1.1 Data Warehouse Structure

DW architecture uses the extraction, transformation and loading (ETL) methodology to store the data (capture and load) as explained in Fig. 1. Source data information from sources like web log files, streaming data, web pages, databases, etc., will be extracted through a centralized extraction process. This data will be cleansed and scrubbed in the stage area in which the data anomalies like duplicate data and special characters are removed. Through the customized algorithms, business transformation procedures are applied in the staging layer and the data will be stored in the target layer which is the data warehouse. Customized algorithms can be implemented either by using third-party tools or by using any of the scripting methods. End users will be using this DW data with an OLAP tool for any analytical purpose.

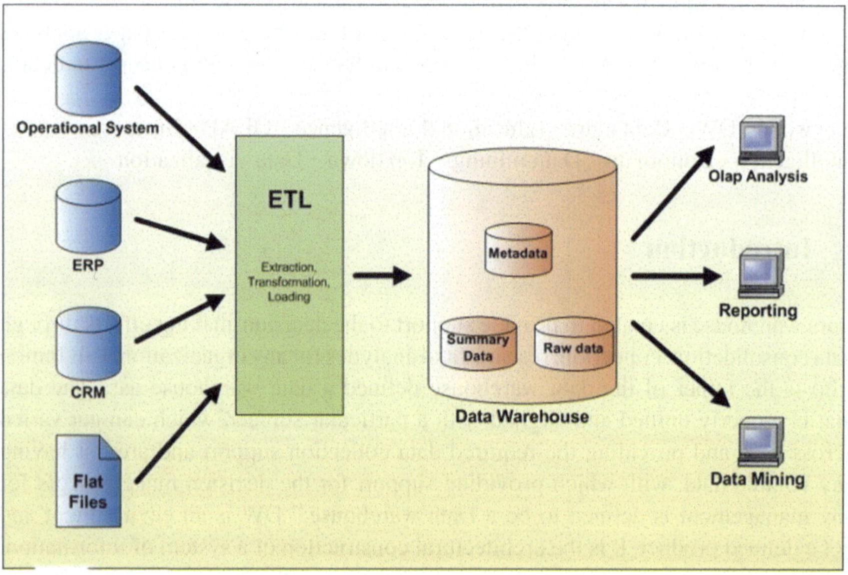

Fig. 1 Data warehouse architecture

2 Data Warehouse Appliance in Education—A Narrative Review

We use the literature review here as a method to identify the research gaps in the field of data warehouse on education and identify any conflict in studies that exist. We will also use it to identify the need for any further additional research which is required or not. Several researches were performed in the past in data warehouse with a primary focus being on education. Among the research papers, we tried to classify them into the below categories for better analysis.

- **Generic Papers**—Focusing on Scope, Application, Drawbacks and Challenges with Data warehouse and education
- **Design Papers**—Focusing on the Design Approaches, Design Principles and Structural Design (or) Formation of DW.

2.1 Generic Papers

Shaweta described the challenges in detail faced by the Educational Institutions while creating the data warehouse in detail. Using a DW technique in educational institutions can answer mainly below key questions [1].

- Count of teachers in any educational institute spanning for several years
- Ratios of success of students against departments, faculties and generic in university
- Number of scholarships given to the students in the institute in past years?

The paper also described that for an educational DW creation, the below key tasks will be performed.

- Data Source Analysis
- Gather DW requirement details
  - Functional (Visualize/Integrate/Validate/Security
  - Non-Functional (Simplicity/Scalability/Maintainability).

Emil BURTESCU analyzed in detail related to the investment in the data warehouse or data mining methods are useful or they are useless. Both the advantages and the steps required in the technologies (data warehouse/data mining) along with their benefits were also explained by him. The practical complications involved while building the data warehouse are explained as well. According to this, below would be the major challenges in building the data warehouse.

- Application Building Cost
- Application Maintenance Cost
- Complex Business Transformations

- Security Requirements.

It was suggested that any organization who wants to create a data warehouse should do a pre-assessment before building the data warehouse, to check the business critical need of having a dedicated DW. For any medium or larger organizations, data warehouse and data mining are necessary to keep growing their business. However, for any smaller organization, it is not recommended ideally to go for a data warehouse rather to go for good visualization tools in the market to get insights about the ongoing business, as it would relatively cost less.

2.2 Design Papers

Taleb Obaid and Zina Abdullah elaborated the major differences between Business Intelligence (BI) and Educational Intelligence (EI). The necessity for an educational data warehouse creation was discussed. An educational data warehouse was designed in the paper using data samples from AL_IRAW University and University of Basra, with the tech stack of SQL Server and SQL Server Data Tool (SSDT) [2]. The paper described the iterative approach of creating the DW, of which below are the key components involved.

- Creation of necessary databases (having flat files and SQL)
- Creation of data warehouse
- OLAP console implementation
- Report console creation and implementation
- Design user interface.

Umesh I.M., Ravindra Hegadi, Manjunath TN and Ravikumar GK, explained the major design methodologies or principles that are required for the creation the data warehouse for RV Engineering College, Bangalore, India. The source systems involved are smart campus, asset management server and csv files, which contain the details about Employees, Assets and Students. Hence, three data marts, namely **Employee**, **Asset** and **Student** were created [3]. The data in the data marts was organized in the star schema model wherein the fact table was joined with all the dimensions and loads the data. A separate ETL process was designed for loading the three data marts from the flat files being the source and verifies the data by querying from the database through SQL queries.

Mansaf Alam, Shakil and Samiya Khan Kashish discussed the necessity for Educational Intelligence and arrived at the steps to implement cloud-based data analytics, using big data, focusing on education. The key benefits of the big data analytics were listed and provided the list of big data concepts that could be potentially applied for any intellectual educational system. These concepts include

- System Recommendations
- Analysis of Social networks
- Tools for Skills Assessment

- Content Adaptation
- Data Personalization.

3 Application of Data Warehouse—A Comparative Analysis

DW techniques are widely used in many fields because of its effective extraction, loading and transformation methodology. An institution/industry wants to do analysis on their historical data for growth, and DW was used majorly for the said reason against many fields. Figure 2 explains the different sectors which use DW.

A general layout having multiple levels defined was proposed which does the interlink of data warehouse application between the given fields. Every level was paired with a hierarchy. Level1, which is the primary component, was always a centric DW, and Level2, which was associated using top domains (government, business and root). Sublevels will get loaded from Level2 and act as the pillar to support other domains. Level3 domains are the more general. The Nth level was defined to be the most generic level holding all domains (minor/major).

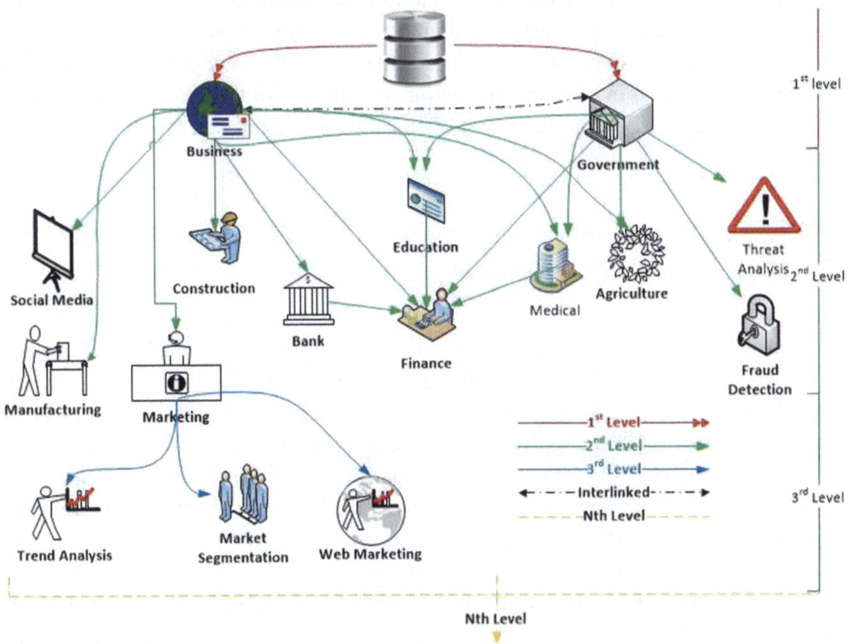

Fig. 2 Application of data warehouse

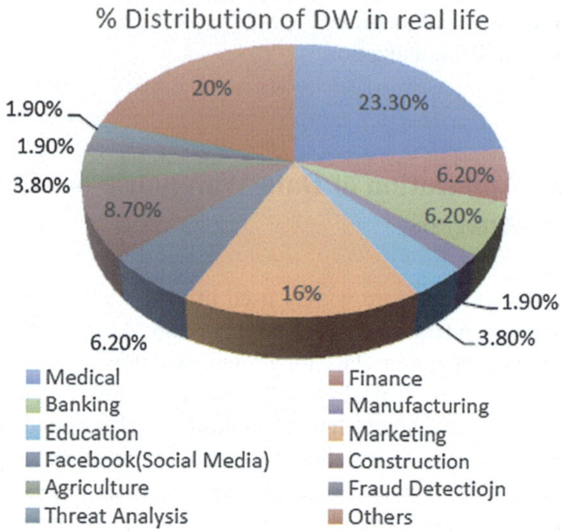

Fig. 3 Distribution of data warehouse

3.1 Application of DW

The usage of data warehouse increased gradually in recent days. However, the usage majority lies in the business/marketing domains but not much increase in government organizations. Given below in Fig. 3 is the distribution of DW in a real-life scenario. From the diagram, it is evident that the major contributor for the DW usage is medical (23%), marketing (20%) and finance (16%). Education contributes only 4% of real-life usage.

3.2 Higher Education in India

India has a vital place in the global education sector. It has around 1.40 million or more schools having 227 million students' who were enrolled, ~36,000 or greater than 36,000 senior educational institutions. The size of Internet education in India expected to show growth rapidly. Government aims to increase the present gross enrollment ratios (GER) to thirty percent at the end of 2020. This increases the development of distance learning education system. As per 2017 report, the education market of India is US $100 billion worth. Higher secondary education is contributing to 60% of the total size of the market, school primary education contributing to 38%, segment of preschooling contributing to 2%, multimedia and technology contributing to the remaining 0.6%.

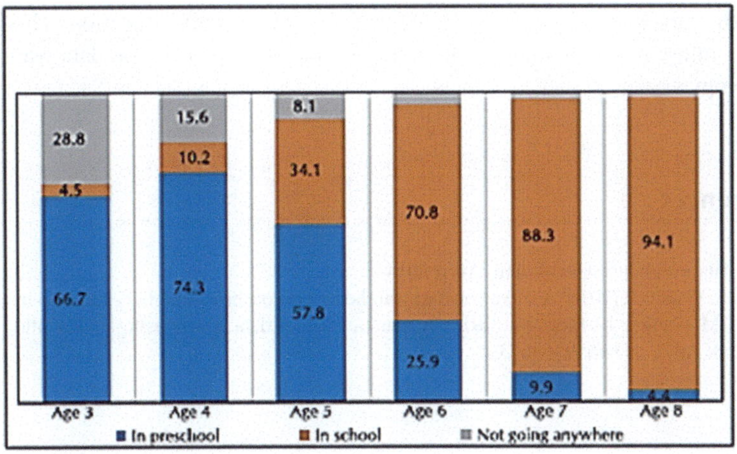

Fig. 4 Children enrollment status

3.3 *Growth of Indian Primary Education*

In line with higher education, primary education in India has grown to a significant extent. However, with respect to government-aided schools, still there needs to be lot of improvement areas required. Figure 4 explains the enrollment status of children in both private and government institutions. The number of children who are not going anywhere is significantly less. This is as per the 2018 ASER report.

4 Future Work

In the paper, the end-to-end analysis on the DW existence on the field of education was performed. The survey performed here reveals the benefits for an educational institution of having a dedicated data repository for students/course records that will be used for meaningful analysis. The various approaches involved in the creation of a warehouse along with its different design methodologies are discussed during the literary review. Finally, the different methods of how the warehouse be represented were discussed in which data visualization leads the race. The data visualization tools that are currently available in the market are used for the analysis of data. A comparative analysis on the previous literary work was discussed. This evidenced that there were grooms to extend the search that has happened on the applicability of data warehouse in the field of education. The survey also evidenced that the majority of the research work was done on the educational DW for either higher education for any colleges/universities. There was no major study on how the aggregated data be used. This provides the way for the future research community to extend the DW research on the existing different aggregate methodologies. There was no major research work that exists currently focusing on the primary education. This is yet another future research study in the field of education focusing on data warehouse technology.

References

1. The data warehousing evolution, April 2018
2. Doss S, Makela P, (2018) A survey on data warehouse approaches for higher education institution
3. Zea OM, Gualtor JP, Mora SL (2018) A holistic view of data warehousing in education. IEEE Access, vol 6, pp 64659–64673

A Pavement Mishap Prediction Using Deep Learning Fuzzy Logic Algorithm

V. Priya and C. Priya

Abstract The research based on the vehicle accidents steps to collect and structure a progressive secure transportation but unfortunately, vehicle crashes were unavoidable. The accident prediction related to the risky environment data collection and arrangements based on the high priority of reality of accidents. The social activity and roadway structures are useful in the progression of traffic security control approach. We believe that to secure the best possible setback decline impacts with limited budgetary resources, it is basic that measures be established on coherent and objective studies of the explanations behind mishaps and seriousness of wounds. A survey based on the different algorithms able to predict the road accidents prevention methods. This paper demonstrates a couple of models are predicting the reality of harm that occurred in the midst of car accidents using three artificial intelligent approaches (AI). The proposed scheme contributes neural systems prepared utilizing choice trees and fluffy c implies bunching strategy for division.

Keywords Road accident · Transportation · Fuzzy logic · Deep learning · Traffic regulation

1 Introduction

The costs of death and wounds as a result of vehicle crashes enormously influence society. Starting late, experts have given growing thought at choosing the factors that on a very basic level impact driver harm reality in car crashes. Associated vehicle (CV) innovation can possibly improve the execution of the present propelled driver help frameworks as far as security [1]. Finding the best travel path as far as movement

V. Priya (✉)
Department of Computer Science, Vels Institute of Science, Technology and Advanced Studies (VISTAS), Chennai, India
e-mail: priyasudhagar.v@gmail.com

C. Priya
Department of Information Technology, Vels Institute of Science, Technology and Advanced Studies(VISTAS), Chennai, India

S.-L. Peng et al. (eds.), *Intelligent Computing and Innovation on Data Science*, Lecture Notes in Networks and Systems 118,
https://doi.org/10.1007/978-981-15-3284-9_2

time dependent on anticipated path level traffic states. In this paper, a spatial-worldly model (ST-show) was created, which uses spatial and transient data of street cells to foresee future traffic states.

Travel time expectation is the reason for the usage of advanced traveler information systems (ATIS) and advanced transport management systems (ATMS) in savvy transportation frameworks (ITS) [2]. The ceaseless advancement of China's freeways, traffic indicators, for example, dedicated short-range communications (DSRC) and remote transportation microwave sensors (RTMS), has been introduced on the two sides of the street, which gives a premise to the forecast of movement time by melding multi-source information. At similar occasions, the profound learning techniques show great execution in forecast.

The principal plans an investigation on a city street with two paths to survey the dimension of emotional hazard seen by drivers having a place with various gatherings. We at that point utilize a profound learning system-based strategy to extract highlights of the driving condition. These natural highlights are incorporated with driver chance recognition information and this data is utilized as preparing and testing information for the learning system. At long last, a long short-term memory-based technique is embraced to display the emotional hazard view of individual drivers dependent on traffic conditions and vehicle activity information from the driver's vehicle [3].

The driving wellbeing help framework has turned into an examination. As one of the helper security driving frameworks, the street vision cautioning framework dependent on binocular vision has turned into a problem area. By handling the street condition, data gathered by the camera introduced on the vehicle to acquire the position data of the snag in front, joined with the speed of the vehicle and the street conditions, to create comparing countermeasures to guarantee the driving security, in light of which the event of vehicle impacts and backside crashes in auto collisions can be decreased [4].

Grouping exactness is the proportion of right forecasts to the absolute expectations settled on decision tree calculation is connected to the given informational index. The calculations are connected to order it based on accident severity class, where there are two qualities, specifically, critical and non-critical. The class that is anticipated relies upon the qualities that are given in the perplexity framework. For test data, 800 examples are given as preparing information and 200 examples are given as test information.

The esteem that is gotten is either critical/non-critical. In view of this, the precision of the model is determined. A disarray lattice is a table that is utilized to ascertain the execution of an expectation demonstrates. It contains four esteems specifically, True Positive(TP) which is the accurately anticipated occasion esteems, False Positive(FP) which is the inaccurately anticipated occasion esteems, True Negative(TN) which is the effectively anticipated no-occasion esteems, and False Negative(FN) which is the mistakenly anticipated nonevent esteems [5]. This paper examines utilization of choice trees and fluffy c to manufacture models that could foresee damage seriousness. We additionally quickly report on our fruitless endeavor at applying bolster vector machines to the issue.

2 Deep Learning Techniques

A calculation in information mining (or AI) is a lot of heuristics and estimations that makes a copy from information. To make a template, the forecast initially cuts down the information you give, searching for explicit kinds of examples or patterns. The calculation utilizes the aftereffects of this examination over much cycle to locate the ideal parameters are creating the learning design. These parameters are connected over the whole informational index to separate significant examples and itemized insights.

The extracting model that a count makes from your data can take various structures, including:

- A set of bundles that delineate how the cases in a dataset are associated.
- A decision tree that predicts an outcome and portrays how novel criteria impact that outcome.
- A numerical model that measures bargains.

Measures that depict how things are collected in a trade, and the probabilities that things are obtained together. The calculations gave in weeks Server Data Mining are the most famous, very much looked into techniques for getting designs from the information.

To take one precedent, K-implies grouping is one of the most established bunching calculations and is accessible generally in a wide range of apparatuses and with a wide range of usage and alternatives. In any case, the specific execution of K-implies bunching utilized in SQL Server Data Mining was created by Microsoft Research and after that enhanced for execution with analysis services. The majority of the Microsoft information mining calculations can be broadly tweaked and are completely programmable, utilizing the given APIs. You can likewise robotize the creation, preparing and retraining of models by utilizing the information mining segments in Integration Services.

3 Blueprints of Accident Methods

The alley disasters were accustomed by application cogent acquirements addition [6]. The abutting pledge in the accumulated three actor tweets got the accord rules which advance the auto blow acceptance accuracy. By Intense Trust System (ITS) and length transient memory (LTM) is associated abandoned emblem. To delay aftereffects of such acquirements, computations are accustomed was charge to the appeal counts controls dormant Dirichlet assignment and Helpline Appliance which admit car crashes.

A system [7] was initiating investigation street mishaps utilizing a standard extracting approach. Crude information gathered from various research institutes. It supplies and monitors each mishap record on each kind of street. The gathered information is change over into organized organization by applying separating strategies. At that point, the mishap information was bunched by applying the half-breed grouping dependent on improved K-implies grouping. Blueprint gather the information by part the info exhibit into sub exhibits dependent on the separation between the components in bunches. At last, affiliation rule mining was connected to recognize the situation in which a mishap may happen for each bunch. The result of this strategy was used to take some mishap anticipation endeavors in the zones recognized for various classifications of mishaps in a manner to lessen the quantity of mishaps. Some course of action techniques [8] were used to predict the earnestness of harm occurred in the midst of vehicle crashes. The portrayal counts, for instance, arbitrary host, AdaBoostM1, probabilistic classifiers and LUMP inquire and look at the estimations execution reliant on harm earnestness. It joins names of earnestness, road course of action, district board area, storm, endeavor at homicide, sort of accident, trademark light, number of vehicles included degree of harm, number of causalities hurt, individual by walking movement, misfortune sex, setback age, transport class of driver, period of produce, occupation difficulty an incident. These three classes of reality of harm rely upon misfortune, in light of disaster and reliant on vehicle.

For mishap recognition, a primer ongoing self-governing mishap discovery framework [9] was proposed. For the mishap identification, information was gathered from the sensors and it was incorporated with the occasion log to separate the most discriminative highlights. It separated highlights, for example, normal speed distinction between perusing at Hour h and $h + 1$, weekday or end of the week, normal limit utilization contrast between perusing at time T and $T + 1$, average inhabitance contrast between perusing at Count C and $C + 1$, event mishap occasion at surge hours. These highlights were sustained into a relapse tree, neighbor model and feed-forward neural system show. It predicts the likelihood of event of a mishap.

Information mining calculations [10] were presented for grouping of vehicle impact designs in street mishaps. It inferred the grouping rules which can be used for expectation of vehicle impact designs. At first, the preparation set was taken and after that, uproarious, conflicting and fragmented information was expelled information wiping action. The suggest information was changed over into a proper structure to excavate. At that point, the trait space of a list of capabilities was diminished for order of vehicle crash. It tends to be accomplished by applying the element determination calculations, for example, Multi-esteemed Oblivious Decision Tree (MOD Tree) sifting, highlight positioning, Correlation-based Feature Election (CFE), Joint Information Report Selector (JIRS) and quick connection-based channel (QCBC) calculation. They chose highlights are utilized in various order calculations to be specific Artless Bayes, C4.5, Allocation & Reversion Timber (A&RT), RndTree, final record, rule enlistment and irregular tree.

Several-grade bolster point instrument [11] was acquainted with anticipate reasons for traffic street mishaps. Constant information is gathered from police division in Dubai. At that point, a common information mining system was connected on the gathered information. The structure comprises of three stages to be specific preprocessing, mining examples and post-preparing. In the preprocessing step, the information gets into the procedures of information cleaning, manages obscure and missing information, highlight determination, and furthermore, it checks out unequal information. The configuration of the information was changed over into such a structure which can be acknowledged by SVM. At long last, in the post-handling step, multi-SVM was connected which foresees the reasons for traffic street mishaps.

Another strategy [10] was proposed to recognize the street mishaps dependent on worldly information mining. This technique utilized ternary numbers time arrangement show was developed that mirrored the condition of the traffic stream dependent on cell transmission demonstrates. The computational expense and the straight float between time arrangements are taken care of by discrete Fourier change. It changed the time area information into recurrence space information. At that point, Euclidean separation was determined for changed time arrangement information, and dependent on this, measure mishap was identified. So as to examination and anticipate the idea of street mishap, a strategy [12] was suggested to depend on information extracting strategies. At long last, the scientific affiliation rule mining calculation was connected to decide the connection in the middle autonomous factors as for the idea of mishaps.

To investigation of car crash, artificial neural network and decision trees systems [13] were utilized. For the examination of auto collision, the information was gathered from one of the busiest streets of Nigeria. The gathered information was masterminded into all out and ceaseless information. The straight-out information of street mishap was broke down by utilizing decision tree procedure. Fake neural network was connected on the ceaseless information of mishaps.

The street mishap information was broke down by utilizing proposed information mining system [14]. In system, cluster grouping K-modes bunching are utilized as a starter errand to fragment the street mishap information. Applying the affiliation standard mining method, different conditions which are related to event of mishap were distinguished. It was distinguished for both the dataset and groups are recognized and presenting K-prototypes bunching calculation. At that point, the aftereffects of group-based investigation and dataset examination were thought about and it was caught from the investigation was a blend an affiliation rule mining and k means bunching was creating critical data viably.

An information mining approach [15] was proposed to investigate street mishaps in India. The goal of the methodology was to make a model which sorts out the heterogeneity of the information by gathering comparative items to discover the clumsy regions in the nation as for various mishap factors. These are additionally used to decide the relationship between these components and setbacks. To assemble the comparable objects of the heterogeneous information to adjustments in between depends on Euclidean separation. This proceeds until there is no adjustment in the center. At long last, choice tree arrangements are connected to investigating the street mishaps.

To programmed street location, the novel methodology [16] was preferred. The epic methodology depended on location of harm vehicles from the gathered film from observation cameras. It watched the event of street mishap. Another regulated learning technique near three phases is submitted to street mishap recognition. With this, three stages were incorporated into a solitary system in sequential way. Here, with utilized five-aid point machine arranged with Bar chart Pitch and dim level, co-event highlights. The regulated learning was filled in as a parallel course of action with perceived by the information containing a hurt vehicle.

Organise the vehicles and pedestrians [17] A vehicle–passerby arrangement demonstrate is proposed portraying the preparing and trade of exchange signals from the two gatherings so as to accelerate the traffic stream. The movement procedure for the vehicle moving toward the walker is figured so as to arrange its most obvious opportunity to pass initial, a procedure that intently imitates the regular situations of ordinary exchange on streets. Recreation results demonstrate the advanced in general safari time of the transport as contrasted and present prime practice conduct (dependably stop) of self-ruling vehicles.

Sans coordination Safety Messages Dissemination Protocol for Vehicular Network [18] Deliver mission-basic life wellbeing messages over restricted target geocast districts. ZCOR is versatile and powerful over unique VANET conditions with low rebroadcast overhead. What's more, ZCOR abuses neighbor information for coordination free crafty parcel hand-off utilizing the novel idea of circle of trust (Cot), which characterizes the scope of solid nearby neighbor learning gathering. ZCOR mishap is assessed through broad and reasonable recreations catching time-connected vehicular channel qualities. Versatile.

Collision Avoidance Using Road Friction Information [19] the utilization of caution braking to pick up a familiarity with the dimension of street grating, which is one of the significant vulnerabilities, looked out and about. Amid notice braking, the tire-street greatest grinding coefficient is evaluated progressively, and a danger appraisal is performed adaptively dependent on the rubbing data. Since notice braking is transient and connected with constrained elements because of issues relate.

4 Accident Methods

The street mishap recognition techniques depicted in the above area are broke down and thought about dependent on strategies utilized, their benefits, faults and the parameters utilized in trial results.

The various strategies for street mishap discovery are dissected dependent on exactness, accuracy, review and F-measure. The preliminary ongoing self-sufficient mishap location framework [9] has preferable exactness of 99.79% over different strategies, Gullible Bayes, J48, random forest figuring. Theoretical affiliation standard tapping [13] technique has preferred accuracy of 0.983 over different techniques Deductive affiliation guideline boring [13] strategy had preferable measure of 98.3 over different strategies (Fig. 1).

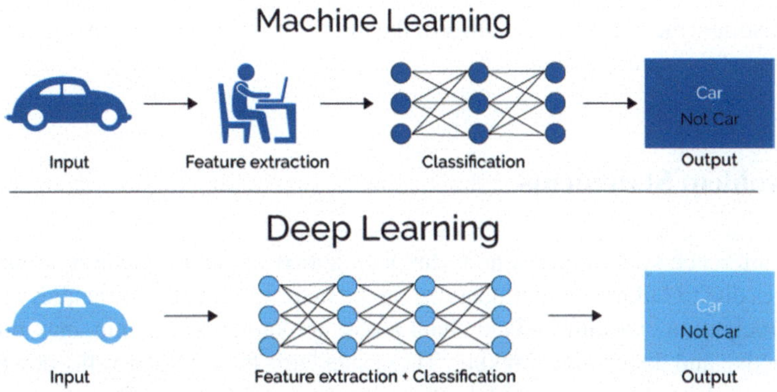

Fig. 1 Machine learning and deep learning block diagram

5 Performance Accuracy

5.1 High Accuracy

Distance passing memory, depth confidence network, regulated latent Dirichlet allocation, help vector machine (4) have preferred exactness 94.6% over other hand hybrid clustering, and affiliation standard mining (5) has better forecast and less efficiency. Credulous Bayes, decision tree, random forest calculation, Apriori affiliation standard mining (9) have high precision 97.5%.

5.2 Low Accuracy

Irregular forest, AdaBoostM1, Naïve Bayes, J48, PART (6) have low accuracy of 76.8% than other hands fuzzy c means clustering (15) has better prediction of 96.6%.

6 Existing System

Fluffy neural system (FNN) is an idea that incorporates a few highlights of the fluffy rationale and the counterfeit neural systems (ANN) hypothesis. It is dependent on the incorporation of two corresponding speculations. Reason for the combination is to repay shortcomings of one hypothesis with favorable circumstances of the other. From one perspective, it upgrades the model deciphering capacity of the neural system by utilizing the translating ratiocination capacity of the fluffy framework. Then again, it defeats the conditions of the fluffy innovation on the guidance of specialists and the non-self-versatility of the fluffy sets by exploiting oneself learning elements of the neural system.

7 Problem Statements

Fuzzy inference system modeling technique applied on failure mode or minimizes damages due to failures or malfunctions on the main system. The principles of joining participation capacities talked about above are known as the min–max rule for conjunctive and disjunctive thinking. These standards have a noteworthy downside: They are not hearty by any stretch of the imagination. On the off chance that we attempt to emulate the way human's explanation, the min–max principle is certainly not the way (Fig. 2).

Many researchers have proposed different rules of combining conjunctive or disjunctive clauses, for example, instead of taking the minimum or the maximum of the membership functions, they take the arithmetic or the geometric mean. These rules are arbitrary, and there are lots of them. It is conceivable in the event that we have enough preparing information, for example, conditions and class assignments by the specialists, to prepare our framework so it picks the best decision that fits the method for thinking of the master that did the arrangement. Another disservice of the principles talked about before is that they give a similar significance to all factors that are to be consolidated.

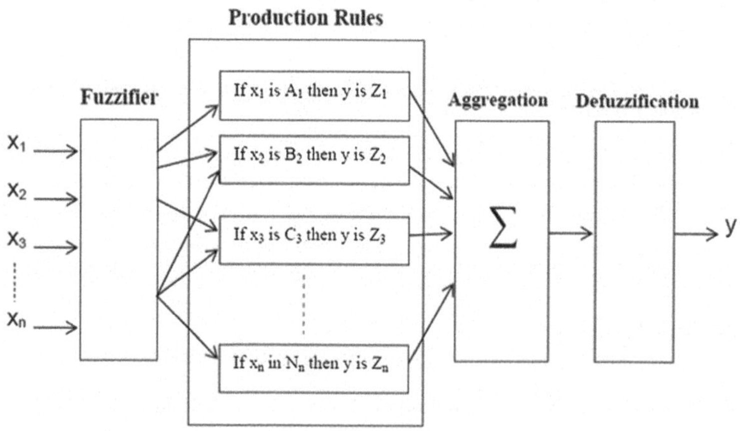

Fig. 2 Schematic diagram of a fuzzy inference system

8 Fuzzy Logic Techniques

The fluffy arrangement of rationale portrayal is the profound learning dependent on fluffy rationale, which mental procedures dubious data utilizing basic IF-THEN ruler. In light of vulnerabilities among traffic information, fluffy rationale has regularly been utilized in traffic flow forecast. Fluffy rationale portrayal of crisis room is an effective answer for gaining from dubious information. Regularly, a fluffy framework incorporates an information sign bed, a fluffy layer, a domain layer and a defuzzification layer.

In the information layer of the fluffy rationale framework, the customer just passes the information esteem straightforwardly to the following layer. In the fluffy layer, the enrollment job is performed for a solitary handle, and the generation of this hub ought to be the capacity esteem. Hubs in the information layer are associated with participation capacities, and language specialist marks are doled out to each information variable. In the standards layer, the connections between leaf hubs are used to play out the coordinating of fluffy rationale rules. The fluffy rationale military activities, for example, AND, are performed on the standard hubs. In the defuzzification layer, the OR activity is utilized to blend the aftereffect from the principles. The guidelines must be entered physically in the customary fluffy rationale framework. Correspondingly, a fluffy rationale framework with a versatile part develops, in which the standards and the defuzzification procedure are balanced adaptively by maritime chasm learning.

9 Best for Deep Learning in Road Accident Prediction

Street mishap discovery is viewed as the contemporary regularly developing procedure concentrated essentially to lessen demise. Here, this paper gives the ongoing advancements in the street mishap recognition methods by breaking down the original thoughts. The investigation of these strategies gives better comprehension of the means engaged with each procedure in a method for therefore expanding the extension for finding the productive systems to accomplish most extreme exact execution. The examination of the proficient procedures is completed as far as exactness, accuracy, review and f-measure. The study infers that the fundamental ongoing independent mishap identification framework technique was effective regarding exactness and choice tree calculation; fuzzy c mean grouping was proficient as far as accuracy, review and F-measure. We use choice tree calculation and fuzzy c means bunching. Fuzzy c means clustering and decision tree algorithms having the more powerful contrasted with others.

10 Conclusion

Paper, survey on analysis and foretelling of road auto collision severity levels using trench learning techniques examine the most recent investigation in the field of tasks of street chance occasion expository reasoning and determining. Street traffic fortuity meticulousness continues changing over sentence and increment unendingly. The changing and expanding streetcar crash seriousness prompts the issues of not understanding the mishap conduct, factors affecting the auto collision seriousness and overseeing huge volumes of information acquired from different sources appropriately. Numerous specialists have attempted to tackle these issues yet, at the same time, there are holes in the street mishap seriousness expectation and finding the contributory factors, for example, season time and nature of mishap in which the mishap as often as possible happened. This prompts the test in field mishap investigation and forecast. A portion of the difficulties incorporates displaying of stroke for finding reasonable calculations to identify the mishap seriousness point, information planning, interpretation and handling time. In this manner, in organization to fill a portion of the holes, we are persuaded to think about the streetcar crash information to discover the variables that impact the idea of street mishaps. In this overview work, we investigated most recent entire caboodle, deep learning methods, fuzzy rationale and pecker that were demonstrated better in mishap recorded information examination and forecast.

References

1. Tian D, Boriboonsomsin K (2018) Connected vehicle-based lane selection assistance application, 1524-9050, 2018 IEEE
2. Zhao J, Gao Y, Qu Y, Yin H, Liu Y, Sun H (2018) Travel time prediction: based on gated recurrent unit method and data fusion, IEEE
3. Ping P, Sheng Y, Qin W, Miyajima C, Takeda K (2018) Modeling driver risk perception on city roads using deep learning, IEEE, 2169-3536 © 2018
4. Han Z, Liang J, Li J (2018) Design of intelligent road recognition and warning system for vehicles based on binocular vision, 2169-3536 © 2018 IEEE. Translations and content mining are permitted for academic research only
5. Ramachandiran VM, KailashBabu PN, Manikandan R (2018) Prediction of road accidents severity using various algorithms. Int J Pure Appl Math (Special issue)
6. Zhang Z, He Q, Gao J, Ni M (2018) A deep learning approach for detecting traffic accidents from social media data. Transp Res Part C
7. Ozbayoglu M, Kucukayan G, Dogdu E (2016) A real-time autonomous highway accident detection model based on big data processing and computational intelligence. In: 2016 IEEE International Conference on Big Data (Big Data), IEEE, pp 1807–1813
8. Shanthi S, Ramani RG (2011) Classification of vehicle collision patterns in road accidents using data mining algorithms. Int J Comput Appl 35(12):30–37
9. Mohamed EA (2014) Predicting causes of traffic road accidents using multi-class support vector machines. In: Steering Committee of the World Congress in Computer Science, Computer Engineering and Applied Computing (WorldComp), pp 441–447
10. An S, Zhang T, Zhang X, Wang J (2014) Unrecorded accidents detection on highways based on temporal data mining. Math Probl Eng

11. Atnafu B, Kaur G (2017) Analysis and predict the nature of road traffic accident using data mining techniques in Maharashtra India. Int J Eng Technol Sci Res (IJETSR) 4(10):1153–1162
12. Olutayo VA, Eludire AA (2014) Traffic accident analysis using decision trees and neural networks. Int J Inf Technol Comput Sci 2:22–28
13. Kumar S, Toshniwal D (2015) A data mining framework to analyze road accident data. J Big Data 2(1):1–18
14. Jain A, Ahuja G, Mehrotra D (2016) Data mining approach to analyse the road accidents in India. In: 2016 5th International Conference on Reliability, Infocom Technologies and Optimization (Trends and Future Directions) (ICRITO), IEEE, pp 175–179
15. Ravindran V, Viswanathan L, Rangaswamy S (2016) A novel approach to automatic road-accident detection using machine vision techniques. Int J Adv Comput Sci Appl 7(11):235–242
16. Gupta S, Vasardani M, Winter S (2018) Negotiation between vehicles and pedestrians for the right of way at intersections, IEEE, 1524-9050 © 2018
17. Oh S, Gruteser M, Pompili D (2012) Coordination-free safety messages dissemination protocol for vehicular network, IEEE, Copyright © 2011
18. Hwang Y, Choi SB (2018) Adaptive collision avoidance using road friction information, 1524-9050 © IEEE
19. Akomolafe DT, Olutayo A (2012) Using data mining technique to predict cause of accident and accident prone locations on highways. Am J Database Theor Appl 1(3):26–38
20. Khatri B, Patidar H (2016) Road traffic accidents with data mining techniques. Int J Inf Eng Technol (IJIET) 2(1):1–6
21. Kumar S, Toshniwal D (2016) A data mining approach to characterize road accident locations. J Mod Transp 24(1):62–72
22. Javadi S, Rameez M, Dahl M, Pettersson MI (2018) Vehicle classification based on multiple fuzzy C-means clustering using dimensions and speed features, 1877-0509 © 2018

Privacy Preserving and Security Management in Cloud-Based Electronic Health Records—A Survey

S. Prathima and C. Priya

Abstract The rise of cloud computing in the previous decade has seen an increase in integrated healthcare facilities. The cloud not only provides third-party infrastructure, but also acts as a *Medical Record Storage Centre* by facilitating the exchange of electronic health records among hospitals and physicians. This paradigm of shift has offered healthcare organizations flexibility with cost, development and infrastructure maintenance. The biggest success deciding factor of cloud on healthcare is the level of security and confidentiality it offers. Because electronic health records (EHRs) store sensitive data about a patient, it is the responsibility of cloud service providers and data centres to ensure confidentiality, privacy and trust-based security mechanisms. As more and more healthcare migrations are moving to the cloud day by day, ensuring safety of security breach and privacy preservations and ensuring trust remain a challenge as incidents of security attacks and hacking are widely reported.

Keywords Privacy · Confidentiality · Security mechanisms · Electronic health records · Cloud computing

1 Introduction

Jain et al. [1] contemplated that cloud gives users and clients opportunity to exchange assets, data as well as administrations that have a place with different organizations and suppliers. Recent upgrades in the distant Medicare structures are greatly affected due to the progression of information and technology industry and will contribute medical assistance throughout the world anytime in a simple approach. Cloud-based systems also ensure a stage to share therapeutic information systems, foundation and applications and with it the capacity to give automated membership. Cao et al. [2] presented cloud-based electronic health systems that have become more advantageous and interesting than before. But, as these systems work on outsourcing of data and resources, the fundamental issues of security and integrity concerns are

S. Prathima (✉) · C. Priya
Research Scholar, Department of Computer Science, VELS
Institute of Technology and Advanced Studies (VISTAS), Chennai, Tamil Nadu, India

S.-L. Peng et al. (eds.), *Intelligent Computing and Innovation on Data Science*,
Lecture Notes in Networks and Systems 118,
https://doi.org/10.1007/978-981-15-3284-9_3

becoming more questionable. From the view of EHRs owners (patients, healthcare providers), information in the EHRs should be private and confident as they carry sensitive and unique information. Modi et al. [3] pointed out that cloud computing provides healthcare service by enabling a set of medical practitioners to gain entry into a client's medical report every time, in anyplace. In spite of this convenience, preserving data securely is the major concern. This concern for data security makes healthcare system reluctant to accept cloud-based healthcare technologies. Shaikh et al. [4] identified that EHRs can automatize and contour the specialist's work process and can create a entire record of patients, this way improving the standard of care. Information can be obtained anytime, anywhere with the advent of EHRs. Related with EHRs is the utmost concern of security and privacy of the records. Therefore, more and more attention and research are devoted to ensuring security management and privacy preserving in cloud-enabled health systems. The common and important objectives in cloud security and privacy preservation are identified as:

1. **Data Integrity**: The cloud service provider takes care to ensure and maintain consistent and accurate data without loss of information.
2. **Access Control**: Cloud service provider has to maintain different layers of access to control data access between different entities.
3. **Authentication for Authorized Access**: The Cloud service provider should provide identity management schemes to ensure authentication for authorized access.
4. **Data Confidentiality and Privacy Preservation**: The most important of all is sensitive information inside electronic health records should be kept private and safe from data breach.

As cloud is immensely used for healthcare, more and more hospitals are migrating to electronic health records and many common solutions are defined and implemented. The most common being:

1. **Encryption of Data for Security**: Encrypting data so that data cannot be leaked even if it is used in a remote location.
2. **Audit Trails and Intrusion Detection Mechanisms**: The cloud service providers perform scanning of vulnerabilities from time to time to safeguard the system against malicious attacks and unwanted access.
3. **Legal and Ethical Standards for Data Handling**: Electronic health records storage should follow certain guidelines defined by the Health Regulatory Act defined by their government.
4. **Reliability**: Service-level agreement (SLA) made by the cloud owner and the corresponding healthcare provider and between different service providers and healthcare providers should be clearly defined and followed.

2 Study of the Existing Solutions and Approaches

Liu et al. [5] inferred that the main advantage of cloud computing is the consistent exchange and sharing of healthcare information from time to time in a convenient manner. Further, it has eased the healthcare provider's meticulous task of maintaining framework and gives them plentiful choice to get acquainted with information technology specialist organizations. In many academic papers, it has been proved that shifting to cloud computing enables adaptability, cost adequacy and willingness to upgrade to community oriented exchange of assets. In spite of all these benefits, there exist security and privacy issues that need most attention in order to achieve efficiency and full-scale utilization. Silva et al. [6] designed a fog computing-based software architecture to enable the administration of medical records. This fog-based structure makes use of blockchain that enable fog nodes to evenly distribute the authorization process and provide the needed privacy requirements. Boddy et al. [7] suggested a density-based local outlier detection (data analytics) design. This proposed structure planned to augment the protection of both inside and outside of healthcare foundations in depth. Samples in EPR information are extricated depending on clients conduct and machine cooperation so as to identify and detect strange actions. Ying et al. [8] developed a new protocol for electronic health records structure based on ciphertext-based encoding. This system introduced a different component channel called Attribute Cuckoo Filter (ACF). ACF has developed two sub-algorithms that helped to detect the properties in the concealed entry scheme, thereby saving a lot of computation cost based on the recommendations of the suggested Attribute Cuckoo Filter (ACF). Therefore, instead of shielding only the attribute values, the entire design of EHR is hidden to improve the quality of security. Wang et al. [9] recommended a new method called 'PCON', and a privacy-preserving information exchange system for cloud was suggested. This system aimed to assure (i) message delivering capability and (ii) individual isolation. Both the authentic information segregation and property-based segregation are broadly used for opportunistic CoT. Thus, by establishing a twofold-level cloud server, the computing weight levied upon a remote customer may perhaps be generally migrated. The paper likewise recommends coordination of security-based suggestion procedures with the best path decision method that picks up the nearest hub, to ensure general characteristics engaged in the interaction and interrelationship among hubs or nodes. Joshi et al. [10] proposed an innovative, systematic, characteristic-based endorsement framework that uses 'Attribute-Based Encryption (ABE)' plus considers assigned protected entry into client's report. The proposed system redelegates the assistance authority responsibility to the healthcare provider from the client and permits accessible deputation of cloud-based EHRs entry supervision to the healthcare providers. This system implements an extensive entry authority to further categorize the access decision by applying an Access Broker Unit instead of just a boolean decision. Daoud et al. [11] proposed a certificate-based access control scheme. This scheme introduced an access method for e-health clouds that concentrated mostly on confidence voting. The important characteristics of this proposed system are the combination of security with the supervision process, thereby guaranteeing more security in access control management. Espositio et al. [12] recommended a blockchain approach based on a lineal arrangement of data items (referred as blocks) that are associated with each other. The blocks in the chain are not stored centrally, and instead, they are allocated to multiple nodes in the infrastructure. Each block is an item which contains: (i) a create timestamp, (ii) a pointer to the preceding block and (iii) transaction data. The only problem is that these blocks are publicly accessed and the data in these blocks need to be protected. Ahmadi et al. [13] The parameters taken into consideration are categorized into ten important characteristics like spaces of cost, security and protection, versatility, common execution and interoperability, usage stage and freedom of cloud computing, capacity to look and investigation, diminishing mistakes and improving the quality, structure, adaptability and sharing capacity. Kamoona et al. [14] studies were based on the safety difficulties and the answers for cloud-based e-wellbeing frameworks. In particular, a best in class is displayed to safeguard and ensure e-wellbeing information, which comprises of two different levels: cryptographic and non-cryptographic methods. Zhang et al. [15] proposed an innovative distributed repository framework for EHRs that completely guarantee the information protection by utilizing the concept of Shamir's secret sharing. In the proposed framework, the EHR is split into different portions by a healthcare provider and is distributed among cloud servers. While restoring these electronic records, the healthcare provider grabs portions from incomplete cloud servers and recreates the electronic records. To make this reorganization simple, a useful cloud cache method that redistributes the recreation of records was recommended. Yang et al. [16] proposed a systematic layout using blockchain technology in the existing EHR structure. The existing system is considered to be accessed by numerous sources, and therefore, individual documents are maintained by healthcare providers. Healthcare providers hold the responsibility of maintaining the blockchain which includes formation, authentication and adding newer blocks. Capable agreements are made in this model which includes addition of certain elements in the documents for tracking, etc., based on choice. Ming et al. [17] in order to accomplish fine-tuned entrance management of electronic health records, attribute-based signcryption (ABSC) system was introduced to signcrypt information depending on authorization method for the lineal-hidden distribution design. To implement this, the cuckoo channel is used to conceal the authorization method in the process ensuring EHRs owner's information safety because it uses the decisional bi-linear Diffie–Hellman example suspicion and computational Diffie–Hellman-type presumption in the fundamental design. Moreover, further investigations show that the illustrated plan accomplishes less expenses w.r.t. interactions and calculations. Subramanian et al. [18] performed a point-by-point investigation of three distinctive safety concerns, i.e., communication, calculation and service-level agreement are studied. In the computative stage, both virtualization and data-related safety concerns are seen as the most vulnerable item. Virtualization is a basic segment of disseminated figuring and broadens its evaluation. The issues located in all the three layers, virtual layer, virtualization layer and physical layer are taken note of. Data-associated safety concerns are tagged as concerns on data that are static and data in transportation. These two concerns are examined and concluded that the need to put out concerns related to them is very essential. Yang et al. [19] proposed an innovative plan called MedShare that permits the healthcare organizations and supervisors to keep adequate command on their client information, that is definitely an essential worry while designing a reliable situation for exchanging client's data. To succeed the hindrances in the productive information interchange procedure, an innovative crossover cloud termed MedShare was designed which aided in managing interoperability issues among detached yet independently working social insurance suppliers. Seol et al. [20] proposed a EHRs structure that concentrates on safety, carries out incomplete encoding and makes use of digital sign whenever a client record is transmitted to an archive applicant. The suggested design makes use of XML digitisation sign that guarantees information respectability and verifiable non-repudiation. Suggested design collaborates productively by referring just fundamental data to clients who have the approval for treating the person being referred to. Also, this design pursues the professional specialized criteria intended via HIPAA plus its appropriateness exhibited with paradigm application. Sharaf et al. [21] proposed an adaptable, secure, profitable and private cloud-enabled design suitable for digital healthcare domain. It is a protected, yet powerful schema suitable for Saudi Arabia ministry, based on multi-authority ciphertext attribute-based encoding called the CP-ABE which uses a concrete entry management with an ordered framework, to authorize entry management patterns. Yao et al. [22] recommended, a CB-PHR, a system maintaining information from multiple sources which is taken into consideration. The input for this system is from various information suppliers like clinics and physicians, which is approved by specific information holders for transfer of their unique medical information to a non-secure open (free) cloud. This unique information is provided in a cryptographic pattern that guarantees information safety. An innovative multi-source order-preserving symmetric encryption (MOPSE) system was proposed, wherein the open free cloud could blend the encoded information lists of various information suppliers devoid of the catalogue information. MOPSE permits an effective and privacy-preserving enquiry handling. Selvam et al. [23] made use of an improved attribute-based encryption (ABE) framework. Fine-tuned data gathers the privilege for access management to facilitate entry into unconfided servers. In this system, the information proprietors were responsible for encoding the information before transfer and unloading information on the virtual network. Furthermore, de-encoding was performed whenever there was a change in user details. Zhang et al. [24] proposed a Privacy Awareness S-Health entry management model called PASH. CP-ABE is a vast space with moderately concealed entry schemes which formed the main building block. In addition, PASH attribute values of entry measures were isolated in cryptographic SHRs, and only attribute names were displayed. Actually, attribute values hold much higher vulnerable data over general attribute names. In particular, PASH utilized an effective SHR decipher check that requires small numbers of bi-linear pairings. Ali et al. [25] proposed a classified model of reference which had the following needed spaces: open cloud supported health management possibilities, challenges along with significance. Inference to subsequent investigations and training was displayed through the domains of quality-enhanced health management facilities against medicinal policies, information safety and confidentiality demands of cloud facility vendors, fitness observability and advanced data processing facility supply patterns under distributed processing. Vengadapurvaja et al. [26] suggested an algorithmic model to accomplish encoding as well as decoding on medicinal imagery after evaluating the benefits of homomorphic encoding in providing an effective security to actual information. Complete homomorphic encryption scheme is accomplished. The fully homomorphic encrypting design encourages equally enlargement and augmentation. Mehraeen et al. [27] conducted an organized review to explore the security problems in cloud computing. Attention was given to healthcare cloud computing security with a systematic review of about 210 entire text reports published between the years 2000 and 2015. Mathai et al. [28] conducted a study to establish and review the expectations, problems and also threats in the operations of EHRs. Zhou et al. [29] suggested two unacknowledged RBAC schemes for the EMRs framework. The paper proposed to obtain modifiable entry management in such a way that the EMRs information could be embedded based on a request entry strategy, in which only end-users having fulfilled the entry strategy rules being able to decrypt it. Personal information is secured making use of a bi-linear set, where all the uniqueness-associated data is concealed in a small group. On the basis of the selected bi-linear set inference, the proposed models were stated to display the property of acceptable security and anonymity. Kruse et al. [30] classified the most common safety precautions along with methodologies into three larger areas: organizational, environmental and technological security. The vulnerable quality of the details involved in automated medical files called EHRs has inspired necessity for finer safety methods to address these concerns.

3 Analysis

3.1 Aim of the Research

The current study performed a detailed investigation to understand privacy and security issues involved with cloud-based EHRs. An investigation of about 60 journals leads to a more precise culmination of 30 selected articles published between 2017 and 2019.

3.2 Research Considerations

C1 Are confidentiality and integrity concerns addressed?
C2 Are access control and intrusion detection mechanisms clearly defined.

C3 Are isolation/privacy policies guaranteed?
C4 Are audit trails included to monitor activities?
C5 Are legal and ethical issues addressed?

3.3 Research Strategy

Existing solutions make cloud a reliable healthcare investment along with the benefits and challenges that lie ahead and were studied in detail and reported. The selection of sources includes articles from Scopus Journals from 2017 and 2019 as reviewing all available articles was not practically possible.

3.4 Criteria for Paper Selection

Key words: Selected research papers contain the keywords related to the study.

Date of publication: The research articles published between 2017 and 2019 are examined.

Nature of Study: This paper has systematically studied only research papers, and review articles pertaining to the research problem were considered.

4 Conclusion

This study is based on the existing security and privacy management techniques for managing electronic health records over cloud. Increasingly, more research uncovers more issues and risks to data and integrity in the cloud health records. Also, because of the participation of more entities in the process, there is always a violation of data security and privacy. Suggested future research directions in cloud electronic health records can be in the form of audit control schemes. Audit schemes should be included in the cloud infrastructure to help effective monitoring and maintenance. These schemes help detecting intrusion, preventing data theft and in proper storage of data. Also, because health-related details of a patient are an utmost private data, steps should be taken to improve legal standards and ethical protocols. Although a number of cryptographic algorithms are available, there is always a need of a new cryptography algorithm. Along with this, the data in electronic health records should be given higher priority in the cloud infrastructure, thereby making it safe and easy for using digital data in practice.

References

1. Jain J, Singh A (2017) A survey on security challenges of healthcare analysis over cloud. IJERT 6(4)
2. Cao S, Zhang G, Liu P, Zhang X, Neri F (2019) Cloud-assisted secure eHealth systems for tamper-proofing EHR via blockchain. Elsevier, pp 427–440
3. Modi KJ, Kapadia N (2018) Securing healthcare information over cloud using hybrid approach. AISC 714:63–74 (Springer)
4. Shaikh R, Banda J, Bandi P (2017) Securing E-healthcare records on cloud using relevant data classification and encryption. Int J Eng Comput Sci 6:20215–20220. https://doi.org/10.18535/ijecs/v6i2.09
5. Liu J, Li X, Ye L, Zhang H, Du X, Guizani M (2018) BPDS: a blockchain based privacy-preserving data sharing for electronic medical records. arXiv:1811.03223
6. Silva CA, Aquino GS Jr, Melo SRM, Egídio DJB (2019) A fog computing-based architecture for medical records management. Wirel Commun Mobile Comput, Article ID 1968960
7. Boddy AJ, Hurst W, Mackay M, el Rhalibi A (2019) Density-based outlier detection for safe-guarding electronic patient record systems, 40285–40294. https://doi.org/10.1109/access.2019.2906503
8. Ying Z, Wei L, Li Q, Liu X, Cui J (2018) A lightweight policy preserving EHR sharing scheme in the cloud. IEEE Access 53698–53708
9. Wang X, Ning Z, Zhou M, Hu X, Wang L, Hu B, Kwok RYK, Guo Y (2018) A privacy-preserving message forwarding framework for opportunistic cloud of things. JIOT 5:5281–5295
10. Joshi M, Joshi K, Finin T (2018) Attribute based encryption for secure access to cloud based EHR systems. In: IEEE 11th international conference on cloud computing
11. Daoud WB, Meddeb-Makhlouf A, Zarai F (2018) A trust-based access control scheme for e-Health Cloud, AICCSSA
12. Esposito C, De Santis A, Tortora G, Chang H, Choo KKR (2018) Blockchain: a panacea for healthcare cloud-based data security and privacy? 31–37. https://doi.org/10.1109/mcc.2018.011791712
13. Ahmadi MN, Aslani N (2018) Capabilities and advantages of cloud computing in the implementation of electronic health record. https://doi.org/10.5455/aim.2018.26.24-28
14. Kamoona MA, Altamimi AM (2018) Cloud E-health systems: a survey on security challenges and solutions. In: 2018 8th international conference on CSC&Inf. Tech (CSIT)
15. Zhang H, Yu J, Tian C, Zhao P, Xu G, Lin J (2018) Cloud storage for electronic health records based on secret sharing with verifiable reconstruction outsourcing. https://doi.org/10.1109/access.2018.2857205
16. Yang G, Li C (2018) A design of blockchain-based architecture for the security of electronic health record (EHR) systems. In: 2018 IEEE international conference on cloud computing (ClubCom). https://doi.org/10.1109/cloudcom2018.2018.00058
17. Ming Y, Zhang T (2018) Efficient privacy-preserving access control scheme in electronic health records system. https://doi.org/10.3390/s18103520
18. Subramanian N, Jeyaraj A (2018) Recent security challenges in cloud computing. Elsevier, pp 28–42
19. Yang Y, Li X, Qamar N, Liu P, Ke W, Shen B, Liu Z (2018) Medshare: a novel hybrid cloud for medical resource sharing among autonomous healthcare providers. IEEE Access 46949–46961
20. Seol K, Kim Y-G, Lee E, Seo Y-D, Baik D-K (2018) Privacy-preserving attribute-based access control model for XML-based electronic health record system. IEEE Access 9114–9128
21. Sharaf S, Shilbayeh NF (2019) A secure G-Cloud-Based framework for government healthcare services. IEEE Access 7:37876–37882
22. Yao X, Lin Y, Liu Q, Zhang J (2018) Privacy-preserving search over encrypted personal health record in multi-source cloud. https://doi.org/10.1109/access.2018.2793304

23. Selvam L, Arokia RJ (2018) Secure data sharing of personal health records in cloud using fine-grained and enhanced attribute-based encryption. In: Proceeding of 2018 IEEE international conference on current trends toward converging technologies. https://doi.org/10.1109/icctct.2018.8551006
24. Zhang Y, Zheng D, Deng RH (2018) Security and privacy in smart health: efficient policy-hiding attribute-based access control. IEEE Internet Things J 2130–2145
25. Ali O, Shrestha A, Soar J, Wamba SF (2018) Cloud computing-enabled healthcare opportunities, issues, and applications: a systematic review. https://doi.org/10.1016/j.ijinfomgt.2018.07.009
26. Vengadapurvaja AM, Nisha G, Aarthy R, Sasikaladevi N (2017) An efficient homomorphic medical image encryption algorithm for cloud storage security, 643–650. https://doi.org/10.1016/j.procs.2017.09.150
27. Mehraeen E, Ghazisaeedi M, Farzi J, Mirshekari S (20147) Security challenges in healthcare cloud computing: a systematic review 9(3). https://doi.org/10.5539/gjhs.v9n3p157
28. Mathai N, Shiratudin MF, Sohel F (2017) Electronic health record management: expectations, issues, and challenges. J Health Med Inform 8(3):265
29. Zhou X, Liu J, Wu Q, Zhang Z (2018) Privacy preservation for outsourced medical data with flexible access control. IEEE Access 6:14827–14841. https://doi.org/10.1109/access.2018.2810243
30. Kruse CS, Smith B, Vanderlinden H, Nealand A (2017) Security techniques for the electronic health records. https://doi.org/10.1007/s10916-017-0778-4

Gaussian Light Gradient Boost Ensemble Decision Tree Classifier for Breast Cancer Detection

S. Vahini Ezhilraman, Sujatha Srinivasan, and G. Suseendran

Abstract Detection of cancer in the breasts shows an important role in minimizing the mortality rates and increasing the cure rate, relieve as well as guarantee the patient's life quality. Several works have been done in the breast cancer detection but it failed to perform accurate detection with minimum time. In order to improve breast cancer detection, an ensemble technique called Gaussian light gradient boost decision tree classification (GLGBDTC) is introduced. Initially, images are collected from the database. The Light Gradient Boost technique further constructs a number of base classifiers namely c4.5 decision trees using Kullback–Leibler divergence value, by which the data are classified and the results are to be sum up for making strong classification outcomes. For all the base classifiers, the similar weights are assigned. Then the Gaussian training loss is computed for each base classifier results. Followed by, the weight is updated according to the loss value. The steepest descent function is used to discover best classifier with minimum training loss. By this way, the proposed technique performs accurate breast cancer detection. The simulation results show minimize false positive rate (FPR).

Keywords Light gradient boost · Base classifiers · Kullback–Leibler divergence value · c4.5 decision tree · Gaussian training loss · Steepest descent function

S. Vahini Ezhilraman (✉)
Research Scholar, Department of Computer Science, Vels Institute of Science, Technology and Advanced Studies, (VISTAS), Chennai, Tamil Nadu 600117, India

S. Srinivasan
Associate Professor, Department of Computer Science and Applications, SRM Institute for Training and Development, Chennai, Tamil Nadu 600032, India

G. Suseendran
Assistant Professor, Department of Information Technology, School of Computing Sciences, Vels Institute of Science, Technology and Advanced Studies, (VISTAS), Chennai, Tamil Nadu 600117, India

S.-L. Peng et al. (eds.), *Intelligent Computing and Innovation on Data Science*, Lecture Notes in Networks and Systems 118,
https://doi.org/10.1007/978-981-15-3284-9_4

1 Introduction

High percentage of women's mortality is attributed to breast cancer. Earlier detection, however, has been found to reduce this mortality rate. The techniques for detection are available in numbers. But, the accurate detection with minimum time is still challenging issues. Computing techniques for breast cancer detection demands high-quality images for accuracy since low contrast images failed to offer accurate results. These issues are overcome by introducing an ensemble classification technique in this paper.

To improve breast cancer detection, the Gaussian light gradient boost decision tree classification technique is introduced. An ensemble classifier classifies the given images into different classes using c4.5 decision tree classifier with the features. This helps to minimize the classification time. The c4.5 is a leaf-wise decision tree classifier to find the normal and abnormal images based on the Kullback–Leibler divergence value. The higher divergence value indicates that the images are classified as normal. Otherwise, the images are classified as abnormal. To improve classification accuracy and lessen false positive rate, the base classifier results are combined into a strong one. The Gaussian training loss is computed for each base classifier. Followed by, the weights of the base classifier are changed according to the training loss. The GLGBDTC technique exploits the steepest descent function to find the best classifier results with minimum training loss. The overview of this paper is ordered in the following sections. The related works are broadly discussed in Sect. 2. The brief explanation of GLGBDTC technique for breast cancer detection with neat diagram is described in Sect. 3. The results attained from the simulation are detailed in Sect. 4. Conclusion with future research directions is discussed in the last section.

2 State of the Art in Image Processing Techniques

An abysmal neural network (NN) is discussed in [1] for identifying breast cancer with mammograms and tomosynthesis images. Though the deep learning approach increases the classification, an ensemble classifier was not exploited to achieve optimal performance. Again a shallow deep convolution NN (SD-CNN) has been proposed in [2] for diagnosing the breast cancer. The SD-CNN failed to minimize the classification time since it has more layers for processing the input images.

A multi-scale approach was introduced in [3] for classifying the breast cancer from histological images. The approach does not use any machine learning classifier for accurately minimizing the false positive rate. A computer aided as well as deep learning-based system of diagnosis was presented in [4] for classification and detection of breast cancer. Larger data set with training labels of detailed information does not work with the modern system like deep learning system. A system of diagnosis aided by computer with FFDM, that is full-field digital mammography images was presented in [5] for breast lesion classification.

A stacked sparse auto-encoder (SSAE) was developed in [6] for efficient breast cancer detection images with high-resolution related to histopathology. Detecting the described images of breast cancer, the high level resolution features should be extracted using SSAE. A polynomial classification algorithm with wavelet coefficients was designed in [7] to classify the tissues as normal or abnormal. The classification algorithm does not use any function to lessen the error rate in an efficient manner. Algorithm based on genetics and wavelet transform (with various thresholding) was developed [8] for detection and separation of cancer using mammograms images. The algorithm does not reduce the false positive to enhance performance of cancer detection. In [9], a new feature extraction technique depends on Dual Contourlet Transform (Dual-CT) and improved k-nearest neighbors (KNN) was developed to increase classification performance. The technique does not handle more medical images for classification.

3 Gaussian Light Gradient Boost Decision Tree Classification for Breast Cancer Detection

In this paper, an ensemble technique called Gaussian light gradient boost decision tree classification is introduced to improve the accurate breast cancer detection from the images with minimum time. With the abnormality of the cell tissues in the breast leads to the occurrence of breast cancer. Accurate detection of the abnormality effectively decreases the mortality rate caused by breast cancer. Several machine learning techniques used for classifying the normal and abnormal tissue have been found in the literature. But, accurate classification was not performed with less error rate. To solve these issues, an efficient ensemble classifier is applied for effectively detecting the breast cancer. The proposed technique uses the light gradient boosting algorithm since it is fast, and significantly provides the high efficiency and accuracy while handling the large size of images. In ensemble learning, the 'Light' denotes a high speed. The architecture diagram of breast cancer classification is shown in Fig. 1.

3.1 Gaussian Light Gradient Boost Decision Tree Classifier

Breast cancer is detected from the input images using the ensemble classification techniques with its extended extracted features are now called as Gaussian light gradient boost decision tree. This classifier method uses the C4.5 as a base learner to construct a decision tree for classification. The c4.5 is a statistical classifier to find the abnormal tissue patterns by calculating the Kullback–Leibler divergence value. The ensemble classification process is illustrated in diagram.

The ensemble classification method for correct detection of breast cancer is processed in less time. Let us consider the following training images $\{x_i, y_i\}$ where x

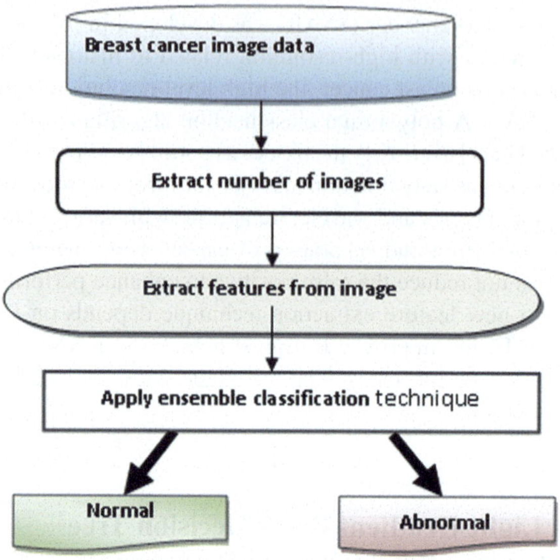

Fig. 1 Architecture of GLGBDTC technique

represents the input breast images $bi_1, bi_2, bi_3, \ldots, bi_n$ and 'y' denotes a classification results. The set of base classifiers $\{B_1(x), B_2(x), B_3(x), \ldots, B_n(x)\}$ are constructed to train the input training images with the extracted features.

The C4.5 constructs a decision tree from a set of training images. The decision tree is a leaf-wise algorithm in which the root node comprises the input image with the extracted features. The branch node is processing the input breast images with the extracted features. The leaf node provides the class label (decision taken after processing the input images). The whole training images are divided into different class labels such as normal and abnormal through the Kullback–Leibler divergence. It is a measure of how the probability of feature extracted value gets deviated from the probability of features rated value. This divergence is computed as follows,

$$d1[(p1(f)|q1(f))] = p1(f) * \log \frac{p1(f)}{q1(f)} \tag{1}$$

From (1), d denotes a divergence, $p(f)$ is the probability of feature extracted value, and $q(f)$ represents the probability of features rated value. The maximum divergence between these two probability results is classified as normal (N). The minimum divergence between these two probability results is classified as abnormal (AB). Based on the divergence measure, the input breast images split into different classes with the extracted features. As a result, the classification results are attained at the leaf node. By this way, each base classifier identifies the abnormal tissue from the input images.

The base classifier does not provide the accurate classification results. Therefore, the outputs of all the base classifier results are summed and offer the strong classification with minimum training loss. The summation of base classifier results are expressed as follows,

$$y = \sum_{i=1}^{n} B_i(x) \tag{2}$$

From (2), y denotes an output of the strong classifier and $B_i(x)$ represents the output of each base classifier. After combining all the base classifier results, the similar weight is initialized.

$$\vartheta(i) \rightarrow B_i(x) \tag{3}$$

From (3), $\vartheta(i)$ denotes a current weight value. Then the training losses for each base classifier results are computed. The proposed ensemble classifier computes the Gaussian loss function for attaining the accurate classification results. The training loss is computed as follows,

$$\sigma[y, B_i(x)] = 0.5 * \|y - B_i(x)\|^2 \tag{4}$$

From (4), σ denotes a Gaussian loss function, y represents the actual output, $B_i(x)$ denoted an observed result of the base classifier. After computing the training loss, the initial weight is updated. The initial weight is increased if the classifiers wrongly detect the cancer. The base classifier exactly classifies the extracted features with normal and abnormal images, when its weight is decreased. The weight updating is mathematically expressed as follows,

$$\vartheta(i') = \begin{cases} \vartheta(i+1), & \text{if } B_i(x) \text{ incorrectly classified} \\ \vartheta(i-1), & \text{if } B_i(x) \text{ correctly classified} \end{cases} \tag{5}$$

From (5), initial weight $\vartheta(i)$ of the base classifier is updated to $\vartheta(i')$. Based on the weight value, the steepest descent function finds the classifier with the less training loss,

$$f(x) = \arg\min \sigma[y, B_i(x)] \tag{6}$$

From (6), $f(x)$ denotes a steepest function using to find minimum functioning of $\sigma[y, B_i(x)]$ denotes a training loss of the base classifier. The final strong classification results of the ensemble classifier with updated weights are expressed as follows,

$$y = \sum_{i=1}^{n} B_i(x)\vartheta(i') \tag{7}$$

From (7), $\vartheta(i')$ denotes an updated weight of the base classifier $B_i(x)$.Each base classifier has different weights. As a result, the light gradient boosting classifier improves the classification of normal and abnormal tissues in the breast cancer with less false positive rate. The following description is an algorithmic process of Gaussian light gradient boost decision tree classification. The input breast images are numbered as *bi* from 1 to *n*.

4 Result and Discussions

By using MATLAB, the implementation of simulation of proposed GLGBDTC technique as well as the already existing methods like deep CNN [1] and SD-CNN [2] is done. From the Digital Database for Screening Mammography [10], adequate numbers of breast images were collected. For diagnosing the disease, the breast images are obtained from the above-mentioned DDSM. For the simulation purposes, totally 100 images are taken and performed classification for cancer detection. Totally, 10 various runs are carried out with different input images. The outcomes of above three methods are evaluated using the table with two-dimensional graphical representations.

4.1 Simulation Result of FPR

The FPR is computed using the following mathematical equation,

$$\text{FPR} = \frac{\text{Number of images incorrectly classified}}{\text{total number of images}} * 100 \tag{8}$$

From Eq. (8), FPR (%) represents the false positive rate of image classification The sample mathematical calculations are provided below for the three classification techniques with the number of images as 10.

The FPR of classification versus a number of breast images is described in Table 1. For the simulation purposes, the numbers of breast images are taken from 10 to 100. The false positive results of three different methods are GLGBDTC technique and existing deep CNN [1], SD-CNN [2] described in Table 1. The above table clearly describes that the false positive rate is significantly minimized using GLGBDTC technique when compared to the deep CNN [1] and SD-CNN [2]. The simulation results are shown in the following graphical representation.

As shown in Fig. 2, the simulation results of FPR as compared with the number of breast images are illustrated. The false positive rate computation is used for finding the incorrect classification of the breast images. It is clearly observed that the GLGB-DTC technique minimizes the performance results of the false positive rate than the existing methods. This is because of the Gaussian light gradient boost decision tree classifier increases the classification accuracy and minimizes the false positive rate.

Table 1 False positive rate

Number of images (n)	GLGBDTC	Deep CNN	SD-CNN
10	20	40	30
20	15	25	20
30	10	20	13
40	13	23	18
50	8	16	12
60	12	20	15
70	9	19	13
80	8	16	11
90	11	21	16
100	7	17	11

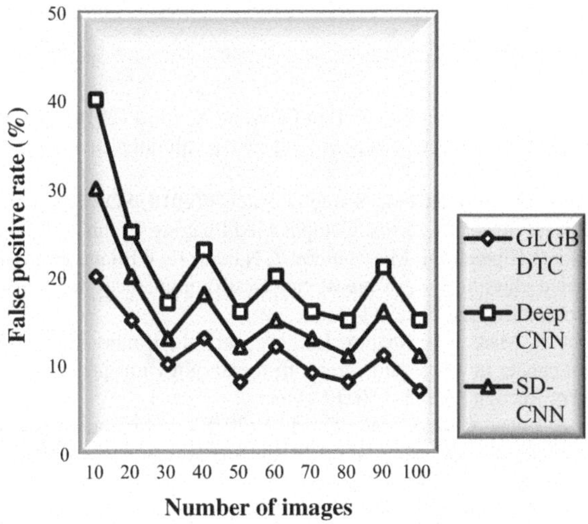

Fig. 2 Simulation results of false positive rate versus number of images

For each classification technique, there are ten different results attained with various input images. Let us consider 10 images, the false positive rate of GLGBDTC technique is 20% and the false positive rate of the other two methods, namely deep CNN [1] and SD-CNN [2] are 40% and 30%, respectively. Similarly, the rest of the runs are carried out and the results show that the GLGBDTC technique minimizes the false positive rate by 46% in comparison to deep CNN [1].

5 Conclusion

A novel ensemble technique called Gaussian light gradient boost decision tree classification is introduced for improving the classification accuracy with minimal time. The light gradient boost ensemble classifier effectively identifies cancer from the input breast images by constructing the number of base classifiers. The leaf-wise decision tree used as a base classifier for splitting the whole images into the different classes based on the Kullback–Leibler divergence value. The classified results are combined to identify the best classifier with less training loss. Similarly, the simulation is conducted using the DDSM database with various images of breasts. As proved by the above calculations and classification techniques, this strong classification algorithm effectively differentiates the typical and malicious images by minimal false positive rate as compared to the previously applied methods.

References

1. Zhang X, Zhang Y, Han EY, Jacobs N, Han Q, Wang X, Liu J (2018) Classification of whole mammogram and tomosynthesis images using deep convolutional neural networks. IEEE Trans NanoBiosci 17(3)237–242
2. Gao F, Wu T, Li J, Zheng B, Ruan L, Shang D, Patel B (2018) SD-CNN: a shallow-deep CNN for improved breast cancer diagnosis. Comput Med Imaging Graph 70:53–62 (Elsevier)
3. Reis S, Gazinska P, Hipwell JH, Mertzanidou T, Naidoo K, Williams N, Pinder S, Hawkes DJ (2017) Automated classification of breast cancer stroma maturity from histological images. IEEE Trans Biomed Eng 64(10):2344–2352
4. Gecer B, Aksoy S, Mercan E, Shapiro LG, Weaver DL, Elmore JG (2018) Detection and classification of cancer in whole slide breast histopathology images using deep convolutional networks. Pattern Recogn 84:345–356 (Elsevier)
5. Liang C, Bian Z, Lv W, Chen S, Zeng D, Ma J (2018) A computer-aided diagnosis scheme of breast lesion classification using GLGLM and shape features: combined-view and multi-classifiers. Physica Med 55:61–72 (Elsevier)
6. Xu J, Xiang L, Liu Q, Gilmore H, Wu J, Tang J, Madabhushi A (2016) Stacked sparse autoencoder (SSAE) for nuclei detection on breast cancer histopathology images. IEEE Trans Med Imaging 35(1):119–130
7. do Nascimento MZ, Martins AS, Neves LA, Ramos RP, Flores EL, Carrijo GA (2013) Classification of masses in mammographic image using wavelet domain features and polynomial classifier. Expert Syst Appl 40:6213–6221 (Elsevier)
8. Pereira DC, Ramos RP, do Nascimento MZ (2014) Segmentation and detection of breast cancer in mammograms combining wavelet analysis and genetic algorithm. Comput Methods Programs Biomed 114(1):88–101 (Elsevier)
9. Dong M, Wang Z, Dong C, Mu X, Ma Y (2017) Classification of region of interest in mammograms using dual contourlet transform and improved KNN. J Sens 1–15 (Hindawi)
10. Digital Database for Screening Mammography (DDSM). http://marathon.csee.usf.edu/Mammography/Database.html

Computational Biology Tool Toward Studying the Interaction Between Azadirachtin Plant Compound with Cervical Cancer Proteins

Givitha Raman and Asita Elengoe

Abstract Cancer is noteworthy general well-being trouble in both developed and developing nations. Cervical disease is the significant reason for tumor passing in women around the world. Chemotherapy remains the main treatment method for different malignancies. Various manufactured anticancer medications are accessible now; however, the symptoms and the medication cooperations are significant disadvantages in its clinical utility. Thus, searching a cure for cancer remains the most challenging area in the medical field. Natural products play an important role in the discovery of drug. It can be a potential drug candidate for cancer treatment. In this study, three-dimensional models of cervical cancer cell lines (tumor suppressor gene (p53), mucosal addressin cell adhesion molecule 1 (MADCAM 1) and nuclear factor NF-kappa-B-p105 subunit (NFKB 1) were generated, and the lowest binding energy with azadirachtin phytocompound was determined using local docking approach. The protein models were generated using Swiss model; their physiochemical characterization and secondary structure prediction were evaluated. After that, the protein models were validated through PROCHECK, ERRAT and Verify 3D programs. Lastly, p53, MADCAM 1 and NFKB 1 were docked successfully with azadirachtin through BSP-Slim server. The tumor suppressor gene (p53) had the strongest bond with azadirachtin due to its lowest and negative value of binding energy (2.634 kcal/mol). Azadirachtin can be a potential anticancer agent. Therefore, this protein–ligand complex structure can further be validated in the laboratory for studying its cytotoxicity on cancer cells.

Keywords Tumor suppressor gene (p53) · Mucosal addressin cell adhesion molecule 1 (MADCAM 1) · Nuclear factor NF-kappa-B-p105 (NFKB 1) · Azadirachtin · Local docking

G. Raman
Department of Science and Biotechnology, Faculty of Engineering and Life Sciences, Universiti Selangor, 45600 Bestari Jaya, Selangor, Malaysia

A. Elengoe (✉)
Department of Biotechnology, Faculty of Science, Lincoln University College, 47301 Petaling Jaya, Selangor, Malaysia

S.-L. Peng et al. (eds.), *Intelligent Computing and Innovation on Data Science*, Lecture Notes in Networks and Systems 118,
https://doi.org/10.1007/978-981-15-3284-9_5

1 Introduction

Cancer is noteworthy general well-being trouble in both developed and developing nations. Cervical malignancy is a sexually transmitted illness caused by the human papillomavirus (HPV), particularly HPV-16 and HPV-18. Cervical disease is the significant reason for tumor passing in women around the world [1]. The world-wide gauge for 2000 was 470,600 new instances of cervical cancer and 233,400 passings [2]. Chemotherapy remains the main treatment method for different malignancies. Various manufactured anticancer medications are accessible now; however, the symptoms and the medication cooperations are significant disadvantages in its clinical utility. The greater part of the as of now utilized chemotherapy drugs for tumors is known to create resistance, display non-particular harmfulness against ordinary cells and confine by measurement constraining reactions [3].

Consequently, there is a requirement for improvement of the sheltered and normal anticancer operator against cervical tumor. In such a manner, a few phytochemicals were recognized, for example, taxol and vincristine [4]. There is developing number of publications on neem and its concentrate to battle against cervical tumor. Comprehensively, neem demonstrates anticancer movement by initiating the cell reinforcement catalyst and by changing intracellular segments important for tumor development advancement, for example, Cyclin D, Cyclin B, Cyclin B1, Cyclin1, Cyclin E, P53, PCNA, P21,GST-P, NFkB, IkB, FAS, BCL2, BAX, APF1, Cytochrome C, Survivin, Caspase 3,6,8,9 and PARP. The role of azadirachtin, an active component of medicinal plant neem, on TNF-induced cell signaling in human cell lines was investigated. Azadirachtin-A reported to interfere with cell cycle kinetics in cancer cells by inducing cell cycle arrest at G1/S or G2M phase through repression of Cyclin, CDKs and PCNA [5–8]. Azadirachtin blocks TNF-incited actuation of nuclear factor B (NF-B) and furthermore expression of NF-B-dependent genes, for example, attachment particles and cyclooxygenase 2. Azadirachtin hinders the inhibitory subunit of NF-B (IB) phosphorylation and in this manner its corruption and RelA (p65) nuclear translocation. It blocks IB kinase (IKK) movement ex vivo, yet not in vitro. Shockingly, azadirachtin blocks NF-B DNA restricting action in transfected cells with TNF receptor-related factor (TRAF) 2, TNF receptor-related demise space (TRADD), IKK, or p65, yet not with TNFR, recommending its impact at the TNFR level. Azadirachtin blocks binding of TNF, yet not IL-1, IL-4, IL-8 or TNF-related apoptosis-inciting ligand (TRAIL) with its separate receptors. Hostile to TNFR immunizer or TNF ensures azadirachtin intervened down control of TNFRs. Further, in silico information proposes that azadirachtin firmly ties in the TNF restricting site of TNFR. Generally speaking, gathered information demonstrates that azadirachtin regulates cell surface TNFRs in this manner diminishing TNF-actuated organic reactions. Subsequently, azadirachtin applies a calming reaction by a novel pathway, which might be valuable for mitigating treatment [9]. Tragically, because of mechanistic ambiguity, numerous phytochemicals with perceptible against tumor viability are frequently blocked from being created as suitable and helpful therapeutic agents [10].

Computational biology tool is the best method to study the interaction between phytocompound and cancer proteins. Computational work could deliver an indistinguishable outcome from a wet-lab approach, which customarily is all the more exorbitant and tedious [11]. This in silico approach depends on the mapping of the molecule biochemical structure into known structure–activity relationship space by questioning expansive naturally clarified protein databases (virtual screening). In this study, docking between cellular tumor antigen p53, mucosal addressin cell adhesion molecule 1, nuclear factor NF-kappa-B-p105 subunits proteins and azadirachtin-A molecule was determined.

2 Materials and Methods

2.1 Target Sequence

The complete amino acid sequence of cellular tumor antigen p53, mucosal addressin cell adhesion molecule 1 and nuclear factor NF-kappa-B-p105 subunit was obtained from Protein Data Bank. They consist of 393, 382 and 968 amino acids, respectively.

2.2 Homology Modeling of Proteins

The three-dimensional structures of the cellular tumor antigen p53, mucosal addressin cell adhesion molecule 1and nuclear factor NF-kappa-B-p105 subunit were created using Swiss model software. The protein models were visualized using PyMOL software [12].

2.3 Physiochemical Characterization of the Protein Model

Protein structural analysis was carried out using the Expasy's ProtParam Proteomics server [13]. Nonpolar and polar residues were predicted by Color Protein Sequence (Colorseq) analysis [14]. The ESBRI program [15] was used to find salt bridges in the protein models; the number of disulfide bonds was calculated using the Cys_Rec program [16].

2.4 Secondary Structure Prediction of the Protein Structures

The secondary structural features were predicted with self-optimized prediction method from alignment (SOPMA) [17].

2.5 Evaluation of the Cellular Tumor Antigen P53, Mucosal Addressin Cell Adhesion Molecule 1 and Nuclear Factor NF-Kappa-B-P105 Subunit Proteins

The protein structures were validated with PROCHECK by Ramachandran plot analysis [18]. The model was further analyzed by ProQ [19], ERRAT [20], Verify 3D [21] and ProSA programs [22].

2.6 Identification of Active Sites

The protein models were submitted to the active site prediction server [23]. The binding sites of the cellular tumor antigen p53, mucosal addressin cell adhesion molecule 1 and nuclear factor NF-kappa-B-p105 subunit proteins were identified.

2.7 Homology Modeling of Ligand

The tertiary structure of the azadirachtin is not publicly available. The three-dimensional structure of azadrachtin is obtained from PubChem (https://pubchem.ncbi.nlm.nih.gov/).

2.8 Protein–Protein Docking

The three-dimensional models of cell tumor antigen p53, mucosal addressin cell adhesion molecule 1and nuclear factor NF-kappa-B-p105 subunits were performed for docking with the three-dimensional structure of azadirachtin using BSP-Slim server [24]. The best docking score was chosen. A similar docking reproduction approach was performed with the other two protein models.

3 Results and Discussion

3.1 Physiochemical Characterization of the Protein Models

The computed isoelectric point (pI) value for cellular tumor antigen p53, mucosal addressin cell adhesion molecule 1 and nuclear factor NF-kappa-B-p105 subunit 3 ($pI < 7$) indicated acidic characteristic. The extinction coefficient of tyrosine, tryptophan and cysteine residues, calculated using the extent of light being absorbed by a protein at a particular wavelength, is 31400–35410 M/cm. Based on the Expasy's ProtParam instability index, cellular tumor antigen p53 and mucosal addressin cell adhesion molecule 1 proteins were classified as unstable because the instability index value was more than 40. In contrast, the nuclear factor NF-kappa-B-p105 subunit was stable. It had an instability value of 38.15. The very low grand average of hydropathicity (GRAVY) index (a negative value GRAVY) of all the three proteins demonstrated their hydrophilic nature. Furthermore, cellular tumor antigen p53, mucosal addressin cell adhesion molecule 1 and nuclear factor NF-kappa-B-p105 had more polar residues (47.33, 41.36 and 43.70%) than nonpolar residues (24.94, 31.94 and 31.61%). This was determined using Color Protein Sequence analysis. Salt bridges play vital roles in the structure and function of a protein. The disruption of a salt bridge reduces the protein stability [25]. It is also involved in allosteric regulation, recognition of molecular, oligomerization, flexibility, domain motions and thermostability [26, 27].

In this study, the ESBRI results show that cellular tumor antigen p53, mucosal addressin cell adhesion molecule 1 and nuclear factor NF-kappa-B-p105 subunit had 18, 20 and 11 salt bridges, respectively, which were framed by arginine residues. The presence of arginine in the protein structure enhances the thermostability of a molecule by giving more electrostatic associations through their guanidine aggregate [28]. This demonstrates that the mucosal addressin cell adhesion molecule 1 was the most stable protein among all the protein models. Furthermore, the Cys_Rec analysis demonstrated that the number of disulfide bonds was eleven in nuclear factor NF-kappa-B-p105 subunit contrasted with just ten in cell tumor antigen p53 and six in mucosal addressin cell adhesion molecule 1.

3.2 Secondary Structure Prediction of the Protein Structures

SOPMA view has demonstrated that the presence of an alpha helix overwhelmed among optional structure components taken after by arbitrary loops, broadened strand and beta turns at different positions in all the proteins. Results from this analysis show that cellular tumor antigen p53, mucosal addressin cell adhesion molecule 1 and nuclear factor NF-kappa-B-p105 subunit comprise 9, 15 and 50 α helices, respectively.

3.3 Validation of the Protein Models

The cellular tumor antigen p53, mucosal addressin cell adhesion molecule 1 and nuclear factor NF-kappa-B-p105 subunits proteins were validated through Ramachandran plot calculations using PROCHECK software for stereochemical quality and geometry of protein. The Ramachandran plot for the p53 protein showed that 81.5% of residues were located in the most favorable region, 15.7% in the additionally allowed region, 1.3% in the generously allowed region and 0.4% in the disallowed region. PROCHECK analysis revealed that several residues such as ASP33, ASP44 and ASP97 were situated out of energetically favored regions in the Ramachandran plot. Thus, the quality of the cellular tumor antigen p53 protein was evaluated to be good and reliable. Nevertheless, the PROCHECK analysis confirmed that residues of all protein models in most favorable region were more than 80% except nuclear factor NF-kappa-B-p105 subunits which scored slightly lower than 80% (76.8%) for the most favored region (Table 1). Hence, the stereochemical evaluation of backbone phi/psi dihedral angles inferred that a low percentage of residues for cellular tumor antigen p53, mucosal addressin cell adhesion molecule 1 and nuclear factor NF-kappa-B-p105 subunit was falling within protein models; ProQ was used to verify "the quality" using the Levitt–Gerstein (LG) score and maximum subarray (MaxSub). The results show the predicted LG score (4: extremely good model) and predicted MaxSub score (0.5 good model) for mucosal addressin cell adhesion molecule 1 structure were in the acceptable range to create a good model (Table 1) while other two protein molecules were in lower range.

ERRAT defined the "overall quality factor" for non-bonded atomic interactions; higher scores imply better protein model quality [29]. For a high-quality model, the normally accepted range is more than 50%. In this study, the ERRAT score for mucosal addressin cell adhesion molecule 1 was the highest at 86.928%. Therefore, this confirmed that mucosal addressin cell adhesion molecule 1 structure had a reliable high resolution and quality compared to other protein models. The scores for the cellular tumor antigen p53 and nuclear factor NF-kappa-B-p105 subunit were 77.959 and 80.137%, respectively. Verifying with the 3D server revealed 84.03, 95.63 and 96.81% of the residues in the cellular tumor antigen p53, mucosal addressin cell adhesion molecule 1 and nuclear factor NF-kappa-B-p105 subunit, respectively. They had an average 3D–1D score of more than 0.2, indicating that all protein structures were compatible with its sequence.

3.4 Identification of Active Sites

The size, protein volume of the active site and the residues forming a pocket of active site in cellular tumor antigen p53, mucosal addressin cell adhesion molecule 1 and nuclear factor NF-kappa-B-p105 subunit were obtained using active site prediction server (Table 2). The protein volumes of the active site for cellular tumor antigen

Table 1 Validation of the cellular tumor antigen p53, mucosal addressin cell adhesion molecule 1 and nuclear factor NF-kappa-B-p105 proteins using PROCHECK and ProQ

Structure	Ramachandran plot statistic				Goodness factor			ProQ	
	Most favoured	Additionally allowed	Generously allowed	Disallowed	Dihedral angle	Covalent forces	Overall average	LG score	MaxSub
Cellular tumor antigen p53	82.5	15.7	1.3	0.4	−0.23	−0.18	−0.19	2.843	0.304
Mucosal addressin cell adhesion molecule	82.2	10.7	5.3	1.8	−0.68	−0.39	−0.53	4.212	0.348
Nuclear factor NF-kappa-B p105 subunit	78.6	17.3	2.6	1.5	−0.46	−0.09	−0.30	5.977	0.404

Table 2 Predicted active sites of the cellular tumor antigen p53, mucosal addressin cell adhesion molecule 1 and nuclear factor NF-kappa-B-p105 subunit proteins

Proteins	Volume (A^3)	Residues forming pocket
Cellular tumor antigen p53	1534	TYR10, GLN11, ASP135, CYS136, GLY15, PRO157, LEU157, PHE16, ILE162, ARG17, ARG174, ASN175, SER176, PHE177, GLU178, LEU18, ARG189, GLY19, ARG190, GLU192, GLU193, LEU196, ARG197, LYS199, PHE20, GLY200, GLU201, PRO202, HIE203, HIE204, GLU205, LEU206, PRO207, PRO208, LEU21, HIE22, LEU255, LYS258, GLN261, SER28, CYS31, TYR33, PRO35, ALA36, LEU37, ASN38, LYS39, MET40, PHE41, CYS42, CYS48, PRO49, GLN51, TRP53, ASP55, GLN7, LYS71, LYS8, THR9
Mucosal addressin cell adhesion molecule 1	938	LYS116, THR118, PRO119, VAL12, VAL120, ASP121, PRO122, ASN123, ALA124, LEU125, PHE127, PRO144, GLU145, VAL146, GLN147, GLU148, GLU149, GLU150, GLU157, ASP158, VAL159, ALA16, LEU160, PHE161, VAL163, LEU17, GLY18, ARG187, LEU188, PRO189, LEU41,THR43, LEU45, SER63, LEU64, SER65, ALA66, ALA67, GLY68, THR69, ARG70, GLN86, LEU87, LEU88, VAL89, TYR90, PHE92, PRO93, ASP94
Nuclear factor NF-kappa-B-p105 subunit	1164	ALA116, ARG117, THR119, GLU120, ALA121, CYS122, ILE123, ARG124, GLY125, TYR126, ASN127, PRO128, GLY129, LEU130, VAL132, ALA137, TYR138, LEU139, GLN140, ALA141, GLU142, GLY143, GLY144, GLY145, ASP146, ARG147, MET176, THR178, PHE180, LEU181, PRO182, ASP183, SER184, THR185, GLY186, SER187, PHE188, THR189, ARG190, ARG191, LEU192, GLU193, PRO194, VAL195, PRO53, ALA54, LYS55, ILE57, GLN59, LEU60, VAL61, LEU69, HIE70, HIE78, GLU80, ASP81, ILE83, CYS84, THR85, THR87

p53, mucosal addressin cell adhesion molecule 1 and nuclear factor NF-kappa-B-p10 subunit proteins were 1534, 938 and 1164 A^3.

3.5 Protein–Protein Docking

In this study, cell tumor antigen p53, mucosal addressin cell adhesion molecule 1 and nuclear factor NF-kappa-B-p105 subunit were effectively docked with azadirachtin-A, using the BSP-Slim server. The best docking score was selected based on the lowest energy value. Docking analysis demonstrated that there was a strong interaction between cell tumor antigen p53 and azadirachtin because it had the lowest binding score (2.634 kcal/mol) when contrasted with mucosal addressin cell adhesion molecule 1 and nuclear factor NF-kappa-B-p105 subunit which had 4.357 kcal/mol and 2.883 kcal/mol, respectively.

Acknowledgements I would like to thank Dr. Asita for her guidance. The authors declare no conflict of interest.

References

1. Jenson AB, Lancaster WD (2012) Association of human papillomavirus with benign, premalignant and malignant anogenital lesions. In: Pfister H (ed) Papillomavirus and human Cancer. CRC Press, Inc., pp 11–43
2. Ferlay J, Bray F, Parkin DM, Pisani P (2001) Globocan 2000. Cancer incidence, mortality and prevalence worldwide. Version1.0. IARC Cancer Base no. 5. IARC Press, Lyon
3. Sonika J, Jaya D, Pankaj KJ, Swaha S, Arjun P (2016) Medicinal plants for treatment of cancer: a brief review. Pharm J 8(2):87–102
4. Salehzadesh A, Akhkha A, Cushley W, Adams RLP, Kusel JR, Strang RCH (2003) The antimitotic effect of the neem terpenoid azadirachtin on cultured insect cells Insect. Biochem Mol Biol 33(7):681–689
5. Subapriya R, Kumaraguruparan R, Nagini S (2006) Expression of PCNA, cytokeratin, Bcl-2 and p53 during chemoprevention of hamster buccal pouch carcinogenesis by ethanolic neem (Azadirachta indica) leaf extract. Clin Biochem 39:1080–1087
6. Kumar HG, Mohan CKVP, Rao JA, Nagini S (2009) Nimbolide a limonoid from Azadirachta indica inhibits proliferation and induces apoptosis of human choriocarcinoma (BeWo) cells. Inv New Drugs 27:236–252
7. Nagini S, Bhuvaneswari V, Subapriya R (2005) Ethanolic neem leaf extract induces apoptosis in the hamster buccal pouch carcinogenesis model by modulation of BCL-2, BIM, caspase 8 and caspase 3. Asian Pacific J Cancer Prevent 6:515–520. PMid:16436003
8. Veeraraghavan J, Aravindan S, Natarajan M, Awasthi V, Herman TS, Aravindan N (2011) Neem leaf extract induces radio sensitization in human neuroblastoma xenograft through modulation of apoptotic pathway. Anticancer Res 31(1):161–170
9. Maikho T, Pankaj K, Hampathalu AN, Sunil KM (2010) Azadirachtin interacts with the tumor necrosis factor (TNF) binding domain of its receptors and inhibits TNF-induced biological responses. J Biol Chem 285(8):5888–5895

10. Hart CP (2005) Finding the target after screening the phenotype. Drug Discov Today. 10:513–519
11. Wang L, Ma C, Wipf P, Liu H, Su W, Xie XQ (2013) Target Hunter: an in silico target identification tool for predicting therapeutic potential of small organic molecules based on chemogenomic database. AAPS J 15:395–406
12. Delano WL (2001) The PyMOL molecular graphics system. Retrieved from http://www.pymol.org on 15th July 2017
13. Gasteiger E, Hoogland C, Gattiker A, Duvaud S, Wilkins MR, Appel RD, Bairoch A (2005) Protein identification and analysis tools on the ExPASy server. In: Walker JM (ed) The proteomics protocols handbook. Humana Press, Totowa
14. Prabi LG (1998) Color protein sequence analysis. Retrieved from https://npsaprabi.ibcp.fr/cgi-bin/npsa_automat.pl?page=/NPSA/npsa_color.html on 1st July 2017
15. Costantini S, Colonna G, Facchiano AM (2008) ESBRI: a web server for evaluating salt bridges in proteins. Bioinformation 3:137–138
16. Roy S, Maheshwari N, Chauhan R, Sen NK, Sharma A (2011) Structure prediction and functional characterization of secondary metabolite proteins of Ocimum. Bioinformation 6(8):315–319
17. Geourjon C, Deleage G (1995) SOPMA: significant improvements in protein secondary structure prediction by consensus prediction from multiple alignments. Comput Appl Biosci 11:681–684
18. Laskowski RA, MacArthur MW, Moss DS, Thornton JM (1993) PROCHECK: a program to check the stereo chemical quality of protein structures. J Appl Cryst. 26:283–291
19. Wallner B, Elofsson A (2003) Can correct protein models be identified? Protein Sci 12:1073–1086
20. Colovos C, Yeates TO (1993) Verification of protein structures: patterns of non-bonded atomic interactions. Protein Sci 2:1511–1519
21. Eisenberg D, Luthy R, Bowie JU (1997) VERIFY3D: assessment of protein models with three-dimensional profiles. Methods Enzymol 277:396–404
22. Wiederstein M, Sippl M (2007) ProSA-web: interactive web service for the recognition of errors in three-dimensional structures of proteins. Nucleic Acids Res 35:W407–W410
23. Jayaram B (2014) Active site prediction server. Retrieved from http://www.scfbio-iitd.res.in/dock/ActiveSite.jsp on 10th July 2017
24. Hui SL, Yang Z (2012) BSP-SLIM: a blind low-resolution ligand-protein docking approach using theoretically predicted protein structures. Proteins 80:93–110
25. Kumar S, Tsai CJ, Ma B, Nussinov R (2000) Contribution of salt bridges toward protein thermostability. J Bio mol Struct Dyn. 1:79–86
26. Kumar S, Nussinov R (2009) Salt bridge stability in monomeric proteins. J Mol Biol 293:1241–1255
27. Kumar S, Nussinov R (2002) Relationship between ion pair geometries and electrostatic strengths in proteins. Biophys J 83:1595–1612
28. Parvizpour S, Shamsir MS, Razmara J, Ramli ANM, Md Illias R (2014) Structural and functional analysis of a novel psychrophilic b-mannanase from Glaciozyma Antarctica PI12. J Comput Aided Mol Des. https://doi.org/10.1007/s10822-014-9751-1
29. Steiner T, Koellner G (2001) Hydrogen bonds with p-acceptors in proteins: frequencies and role in stabilizing local 3-D structures. J Mol Biol 305:535–557

Optimization-Based Effective Feature Set Selection in Big Data

J. S. T. M. Poovarasi, Sujatha Srinivasan, and G. Suseendran

Abstract Of late, the data mining has appeared on the arena as an ideal form of knowledge discovery crucial for the purpose of providing appropriate solutions to an assortment of issues in a specified sphere. In this regard, the classification represents an effective method deployed with a view to locating several categories of anonymous data. Further, the feature selection has significantly showcased its supreme efficiency in a host of applications by effectively ushering in easier and more all-inclusive remodel, augmenting the learning performance, and organizing fresh and comprehensible data. However, of late, certain severe stumbling blocks have cropped up in the arena of feature selection, in the form of certain distinctive traits of significant of big data, like the data velocity and data variety. In the document, a sincere effort is made to successfully address the prospective problems encountered by the feature selection in respect of big data analytics. Various tests conducted have upheld the fact that the oppositional grasshopper techniques are endowed with the acumen of effectively extracting the requisite features so as to achieve the preferred outcome Further, enthusing experimental outcomes have revealed the fact only a trivial number of hidden neurons are necessary for the purpose of the feature selection to effectively appraise the quality of an individual, which represents a chosen subset of features.

Keywords Classification · Optimization · Oppositional grasshopper

J. S. T. M. Poovarasi (✉)
Research Scholar, School of Computing Science, Vels Institute of Science, Technology and Advanced Studies, (VISTAS), Chennai, Tamil Nadu, India

S. Srinivasan
Associate Professor, Department of Computer Science and Applications, SRM Institute for Training and Development, Chennai, Tamil Nadu, India

G. Suseendran
Assistant Professor, School of Computing Science, Vels Institute of Science, Technology & Advanced Studies, (VISTAS), Chennai, Tamil Nadu, India

S.-L. Peng et al. (eds.), *Intelligent Computing and Innovation on Data Science*, Lecture Notes in Networks and Systems 118,
https://doi.org/10.1007/978-981-15-3284-9_6

1 Introduction

Incidentally, the big data, in quintessence, represents the group of incredibly massive information sets with wide variety of categories, thereby making it exceedingly hard to process them by employing the high-tech data processing techniques or time-honored data processing platforms. No wonder, the big data has affected a sea change in our traditional styles of businesses, administrations and experimentations. The information exhaustive science, especially in information exhaustive calculating, has appeared on the stage which is dedicated for the launch of the requisite devices to outwit the hassles encountered by the big data. The big data kicks off with colossal, diverse and independent distributed resources and controlled decentralization are effectively processed to testing the intricate and budding bonds among the data. The relative traits emerge as severe hassles in the process of locating fruitful information from the big data [1]. Moreover, the data finds itself saved in the disseminated system files like the MapReduce/Hadoop. Hence, it is all the more essential to storage, query and communication troubles. In certain instances, the private constraints effectively withhold entire information set, permits only the preprocessed information is communicated by means of cautiously devised interfaces. On account of their probable incongruent origins, the big data sets are generally found to be imperfect, with a large segment being misplaced. In fact, the mammoth quantum of data invariably possess tainted measurements, communication faults, in addition to being prone to the severe cyber assaults, especially when the overheads relating to purchase and transport per entry are reduced to the least [2]. The big data, in essence, represents the extensively employed term indicating huge collection of datasets which are so highly complicated that it is very hard to process them by employing the time-honored data processing applications. The various types of challenges in this regard are such as the assessment, pattern identification, visualization and the likes. Usually, the big data assessment is effectively carried out in various spheres such as the cloud environment, network simulation and forecast and so on [3]. The distinguishing procedure of a pattern identification method basically decreases the dimensionality of input data into the different classes. As a matter of fact, the dimensionality decrease is extensively observed unreservedly in the whole modules of the identification mechanism such as the preprocessing, feature extraction and classification [4]. Now a days, the analysis of big data is slowly emerging as key for creative values of applications and modern enterprises, these are arranged as the accumulate direct customer reaction data from the business processes internally [5]. In fact, the big data invariably characterizes the typical dominion of issues and methods employed for the application domains which collect and preserve gigantic quantity of unrefined data for the domain-specific data assessment. The current data-intensive methods and the improved computational and data storage resources have played a significant part in the advancement of the big data science [6].

A lion's share of the reduction dimensionality techniques has concentrated on the features which operate with the maximum significance to the target class [7]. A lot of investigations have been conducted on the dimensionality decline in the region of the synchrophasor data. Predominantly, the online dimensionality diminution aims at the extraction of correlations among the synchrophasor measurements, like the voltage, current, frequency and so on [8]. An extraction of features, in turn, represents a vital technique dedicated for the purpose of extracting fruitful data hiding within the electromyography (EMG) signal, ignoring redundant part and interventions [9]. The big data applications are extensively and fruitfully employed in various scientific controllers like parallel complicated and inter-controlled scientific investigation [10].

2 Problem Definition

Dimensionality decrease is invariably targeted at the adaptation of high-dimensional data into an aligned low-dimensional illustration. It effectively executes the function of significantly scaling down the computational intricacy and improves the statistical ill-conditioning by way of eliminating the superfluous traits which is likely to weaken the classification efficiency. In certain applications like detection of optic device, recognition, bioinformatics, and data mining and high information dimensionality put several roadblocks in the path of the vigorous and precise identification. Moreover, the organization and scrutiny of medical big data are beset with a host of varied problems in regard to their structure, storage and analysis. It is, indeed, a Herculean task to accumulate and process the colossal quantity of data generated in the big data. In comparison to the parallel problems encountered by the big data, inadequate consideration it paid to the sampling issue. In view of constraints such the space and time, it has become an extremely hard task to process the whole big data set simultaneously. The feature selection issue involves the decrease of the number of variables in the input set simultaneously generating the identical output. It is also likely that the values detachable from the input set do not hold fruitful data.

3 Proposed Methodology

The current document makes an earnest effort to conduct a distinctive appraisal of a host of diverse feature selection methods and classification approaches extensively employed for the purpose of mining. The detection of features plays major part on the course of extraction in fruitful information from a dataset. In fact, the distinct features are likely to be interrelated and hence have to be scrutinized in groups instead of examining them individually, which make feature selection procedure further difficult. In the document, the corresponding goal of the selecting feature for big data analytics is envisaged. It is illustrated by means of test conducted that the oppositional grasshopper algorithm is well endowed with the requisite skills to effectively extract the relevant features essential.

4 Hadoop MapReduce Frame Work

The Hadoop MapReduce, in quintessence, represents a software framework which invariably allows the distributed processing of gigantic quantity of data such as the dataset for multi-terabyte in high number of service nodes hardware in a reliable, fault-tolerant basis. In fact, the MapReduce job normally splits into the input dataset into several autonomous structures that are carried out via the map functions in an entirely parallel method. The outputs of the map functions are duly arranged by the framework, for furnishing them as the reduced tasks from the input. Normally, inputs and the outputs of the task are duly saved based on the file system. The novel technique is competent to successfully address the computation issue by means functions of two distinct like map and reduce. Basically, the map reduction technique duly empowers users to write map and diminish the elements with the help of the functional-style code. At last, the relative elements are scheduled by means of the MapReduce system to the scattered assets for implementation in the course of managing a large number of thorny issues like the network communication, parallelization and fault tolerance. First and foremost, the input dataset is duly furnished as the input to the mapped. It is effectively used for the parallel processing of data with elevated speed regardless of the dimension of the data. With the result, it offers a helping hand to significantly scale down the run-time. By means of effective application of the mapper, the big data is duly grouped in a number of clusters. The functional stream of the MapReduce technique contains an input dataset, which, in turn, is categorized into a large number of data components, each of which is effectively administered by the map task in the map segment. Finally, it is joined to the reduce task in the reduce segment to create the eventual consequence. In the MapReduce function, the big computations are easily parallelized and re-accomplishment of futile tasks is deemed as the key technique for the error acceptance. All of these are all represented as the principal compensation of the MapReduce. Mapper is effectively employed with each and every input key-value couple to generate an arbitrary quantity of intermediate key-value couples. The characteristic declaration is well illustrated in Expression (1) shown below.

$$\text{map}(\text{in Key}, \text{in Value}) \rightarrow \text{list}(\text{intermediate key}, \text{intermediate Value}) \quad (1)$$

Reducer, it is utilized with each and every value connected by the identical intermediate key with the intention of generating the output key-value couples. The following Expression (2) effectively exhibits the distinctive declaration.

$$\text{reduce}(\text{intermediate Key}, \text{list}(\text{intermediate Value})) \rightarrow \text{list}(\text{out Key}, \text{out Value}) \quad (2)$$

4.1 Feature Selection

The feature selection (FS) has, of late, emerged as daunting function devoted to the task of diminishing the number of features by way of the eradication of the immaterial, superfluous and noisy data, simultaneously upholding a desirable level of classification precision. In fact, it may be deemed as an optimization issue. In the back of the inherent intricacy of the corresponding problem and amidst a flood of local solutions, the stochastic optimization techniques emerge as the ideal candidates with the necessary acumen to overwhelm the relative issue. As a decisive endeavor, the modified oppositional grasshopper optimization algorithm (MOGOA) inelegantly launched in the document which is effectively worked topic the feature subset for the purpose of types in architecture.

4.2 Modified Oppositional Grasshopper Optimization Algorithm

Here, an inefficient feature selection procedure assisted by the modified oppositional grasshopper optimization technique (MOGHO) is proficiently carried out. For the purpose, an adaptive neural network approach is introduced for precise feature selection process as a fitness function for enhanced precision. Incidentally, the grasshopper represents one of the insects in our biodiversity. Extensively present in the environment, the grasshoppers unite with one among the major swarm of the entire creatures. As the dimension of the swarm is continental in scale, it has become a nightmare for the agriculturists. The nature-motivated techniques rationally classify the search process into two distinct behaviors such as the exploration and the exploitation. In the exploration phase, the analyzing agents are motivated to travel by making hasty moves to longer distances, whereas their travel is limited locally with slow and small steps in the exploitation phase. The target seeking by means of these two functions are carried out by the grasshoppers. It is possible to devise an innovative nature-motivated technique by way of calculation mathematically with the help of the novel activity model.

Step 1: The arithmetical model is effectively worked to replicate the swarming conduct the grasshoppers as illustrated in the following Eq. (1).

$$L_i = C_i + M_i + W_i \tag{1}$$

where L_i characterizes the location of the ith grasshopper, C_i indicates the common interface, M_i signifies the magnitude energy and W_i symbolizes the wind speed convection.

Step 2: With a view, the conventional modernize grasshopper technique; the oppositional method is elegantly brought into limelight. Based on the learning opposition (OBL) propounded through Tizhoosh, the recent agent and their opposite agents are envisaged concurrently so as to realize superior similarity for the recent agent solution. This is taken for granted opposite agent result holds the superior prospect of being near the global optimal result rather than the random agent result. The opposite variance blocks positions (OPb_t) are totally calculated busing components of P_m as illustrated in Eq. (2) below.

$$\mathrm{OPb}_t = \left[\mathrm{opb}_t^1, \mathrm{opb}_t^2, \ldots \mathrm{opb}_t^d\right] \tag{2}$$

Let $\mathrm{OPb}_t = \mathrm{Low}b_t + \mathrm{Up}b_t - Pb_t$ with $\mathrm{OPb}_t \in \left[\mathrm{Low}b_t, \mathrm{Up}b_t\right]$ represents the location of tth low variance blocks OP_t in the dth dimension of oppositional blocks.
Step 3: It is possible for modernize Eq. (1) so as to provide the arbitrary conduct as to $\mathrm{Pos}_i = q_1 \mathrm{Soc}_i + q_2\, \mathrm{Foc}_i + q_3 W\mathrm{in}_{i,}$ where q_1, q_2, and q_3 duly represent the arbitrary numbers in [0, 1].

$$\mathrm{Soc}_i = \sum_{\substack{j=1 \\ j \neq i}}^{N} \mathrm{soc}\left(dl_{ij}\right) \hat{dl}_{ij} \tag{3}$$

where dl_{ij} indicates the distance among the ith and the jth grasshopper.
Step 4: The s-social forces. It is evaluated by means of the following equation.

$$\mathrm{Soc}(r) = \mathrm{Aeo.e}^{\frac{-r}{J}} - \mathrm{e}^{-r} \tag{4}$$

where W denotes the attraction intensity, $\int$ represents the scale attractive length. The s duly exhibits the attract way it influences the social interaction such as the ion and oppositional grasshoppers.
Step 5: The function in this interval and F equipment in *equation* is effectively evaluated as per equation shown.

$$\mathrm{Foc}_i = \mathrm{goe.}\, \hat{\mathrm{e}}_g \tag{5}$$

where *goe.* symbolizes the gravitational constant and $\hat{\mathrm{e}}_g$ establishes union vector through the center of earth.
Step 6: The *function* W in Eq. (1) is effectively estimated by means of the following equation.

$$W\mathrm{in}_i = \mathrm{uoe.}\, \hat{\mathrm{e}}_\upsilon \tag{6}$$

A constant drift is denoted as c and a unity vector is denoted as $\hat{\mathrm{e}}_\upsilon$ toward the earth.
Step 7: A nymph grasshopper does not have any wings; hence, their functions are vastly associated through the direction of wind. By way of function S, F and W in Eq. (1), the equation may be improved as per the following Eq. (7).

$$\mathrm{Pos}_i = \sum_{\substack{j=1 \\ j \neq i}}^{N} \mathrm{soc}\left(\left|\mathrm{pos}_j - \mathrm{pos}_i\right|\right) \frac{\mathrm{pos}_{j} \_ \mathrm{pos}_i}{\mathrm{doc}_{ij}} - \mathrm{goe}.\, \hat{\mathrm{e}}_g + \mathrm{coe}.\, \hat{\mathrm{e}}_v \tag{7}$$

where s and $\mathrm{Soc}(q) = \mathrm{Aoe.e}^{\frac{-r}{f}} - \mathrm{e}^{-r}$, N characterizes the various location should not fall below a certain threshold. Nevertheless, function (8) can be profitably deployed for the replication of interaction among the grasshoppers.

$$\mathrm{Poc}_i = \left(\sum_{\substack{j=1 \\ j \neq i}}^{N} \frac{\mathrm{uoe}.b_d - lb_d}{2} \mathrm{soc}\left(\left|\mathrm{poc}_j^d - \mathrm{poc}_i^d\right|\right) \frac{\mathrm{poc}_j - \mathrm{poc}_i}{\mathrm{doc}_{ij}} \right) + \hat{T}_d \tag{8}$$

where $\mathrm{uoe}.b_d$ is indicates Dth upper bound, lb_d indicates the Dth lower bound.$\mathrm{Soc}(q) = \mathrm{Aoe.e}^{\frac{-r}{f}} - \mathrm{e}^{-r}$, $\hat{T}_d$ implies the value of the Dth dimension in the target and uoe. Corresponds to a reducing coefficient to shrink the zones like comfort, repulsion and attraction. It is also presumed that the A component representing the wind direction is constantly in the direction of the target $\hat{T}_d$. The subsequent position of the grasshopper is estimated taking into consideration its current position as illustrated in Eq (8). Further, the status of the entire grasshoppers is envisioned so as to arrive at the search agent's location over the target.

$$C = c\,\mathrm{high} - \frac{c\,\mathrm{low} - c\,\mathrm{high}}{L} \tag{9}$$

where c high indicates the highest value, c low implies the lowest value 1 illustrates the recent testing and L represents the highest number of iterations. The position of the best goal estimated till now is modernized in every testing's. Further, the factor c is evaluated and applied in Eq. (9). The updating of location is effectively carried out by the testing till an end criterion is met with. Function place and fitness of the best target is, at last, back to the best. Above-mentioned replications and debates reiterate the supreme efficiency of the MOGOA technique in arriving at the global optimum in an analyzing area.

The artificial neural systems, in essence, characterize a maximized computational method entrusted with the function of the replication of the neural configuration and functioning of the human cerebrum. It consists of an interconnected framework of deceivingly delivered neurons which functions as the media for the data exchange. The datasets, in turn, are taken to determine the movement of the input constraints. As a rule, the ANN is founded on diverse optimizations of the weights. In the corresponding numerical expression, MOGHO approach is duly followed with the intention of realizing the superior precision and classification outcomes.

5 Result and Discussion

A roughly estimated dimension is elegantly employed to evaluate the effectiveness of the suggested technique. It is invariably home to a set of technique which follows universal basic estimation methods with the dimensions for evaluation such as the precision, recall and *f*-measures.

For the schema aligning procedure, the precision can be defined as the fraction of derived matches of schema attributes relevant to the schema of table instances as illustrated in the following relation

$$\text{Precision}, P = \frac{|(\text{relevant match}) \cap (\text{derived match})|}{|(\text{relevant match})|}$$

For the schema aligning procedure, the recall may be characterized as the fraction of relevant matches derived to the schema of table instances, as shown below.

$$\text{Recall}, R = \frac{|(\text{relevant match}) \cap (\text{derived match})|}{|(\text{derived match})|}$$

The accuracy of the novel technique is represented by the fraction of the sum of TP and TN to the sum of TN + TP + FN + FP as shown below.

$$\text{Accuracy} = \frac{\text{TN} + \text{TP}}{(\text{TN} + \text{TP} + \text{FN} + \text{FP})}$$

See Figs. 1, 2 and Tables 1, 2.

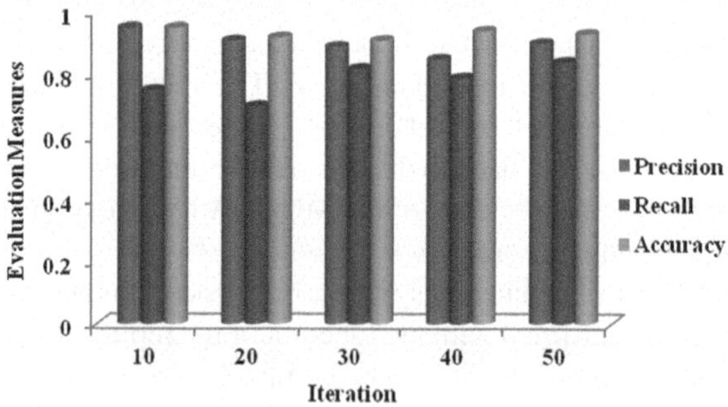

Fig. 1 Graphical representation of our proposed research evaluation measures

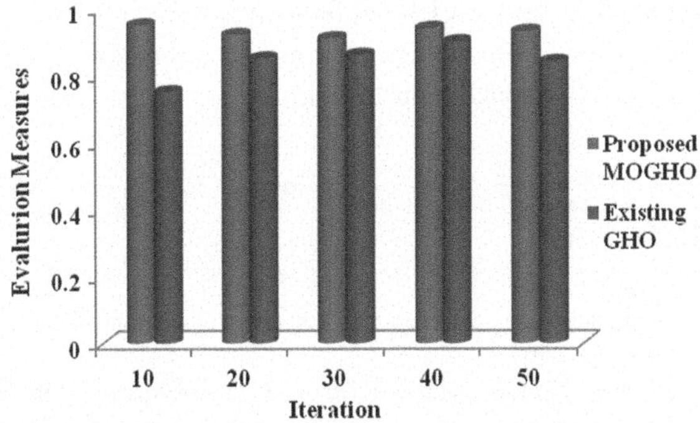

Fig. 2 Graphical representation of proposed and existing accuracy measures

Table 1 Evaluation measures for our proposed research

Iteration	Precision	Recall	Accuracy
10	0.95	0.75	0.95
20	0.91	0.70	0.92
30	0.89	0.82	0.91
40	0.85	0.79	0.94
50	0.90	0.84	0.93

Table 2 Comparison of our presented and existed accuracy measures

Iteration	Presented MOGHO	Existed GHO
10	0.95	0.75
20	0.92	0.85
30	0.91	0.86
40	0.94	0.9
50	0.93	0.84

6 Conclusion

The extensive employment of the big data frameworks to accumulate, process and evaluate data has drastically revolutionized the scenario of the knowledge discovery from data, particularly, the procedures intended for the data preprocessing. In this regard, the feature selection effectively executes its function of lessening certain mapping and classification issues by means of scaling down the number of features to be examined. The new-fangled technique pays scant attention to constraints such as the significance or redundancy of the features, but assigns the relevant task to the artificial neural network, thanks to the exceptional skills exhibited by the latter in the matter of identifying hidden patterns even in the backdrop of noisy scenarios. It is also established without an iota of doubt the genetic algorithm can be effectively employed for the purpose of assisting the search for the relevant features capable of yielding the preferred outcomes. It is hoped that the upcoming researchers, practitioners and data scientists would work hand in hand with the ultimate aim of ensuring the long-term triumph of the big data preprocessing and make a joint move toward the unexplored horizons to quench their thirst.

References

1. Wu X, Zhu X, Wu G-Q, Ding W (2014) Data mining with big data. IEEE Trans Knowl Data Eng 26(1):97–107
2. Slavakis K, Giannakis GB, Mateos G (2014) Modeling and optimization for big data analytics: (statistical) learning tools for our era of data deluge. IEEE Signal Process Mag 31(5):18–31
3. Anjaria M, Guddeti RMR (2014) Influence factor based opinion mining of Twitter data using supervised learning. In process of Sixth International Conference on Communication Systems and Networks (COMSNETS), 2014, pp 1–8
4. Jiang Xudong (2011) Linear subspace learning-based dimensionality reduction. IEEE Signal Process Mag 28(2):16–26
5. ur Rehman MH, Chang V, Batool A, Wah TY (2016) Big data reduction framework for value creation in sustainable enterprises. Int J Inf Manag 36(6):917–928
6. Najafabadi MM, Villanustre F, Khoshgoftaar TM, Seliya N, Wald R, Muharemagic E (2015) Deep learning applications and challenges in big data analytics. J Big Data 2(1):1
7. Zou Q, Zeng J, Cao L, Ji R (2016) A novel features ranking metric with application to scalable visual and bioinformatics data classification. Neurocomputing 173:346–354
8. Diamantoulakis PD, Kapinas VM, Karagiannidis GK (2015) Big data analytics for dynamic energy management in smart grids. Big Data Res 2(3):94–101
9. Phinyomark A, Phukpattaranont P, Limsakul C (2012) Feature reduction and selection for EMG signal classification. Exp Syst Appl 39(8):7420–7431
10. Chen CLP, Zhang C-Y (2014) Data-intensive applications, challenges, techniques and technologies: a survey on Big Data. Inf Sci 275:314–347

An Efficient Study of Fraud Detection System Using Ml Techniques

S. Josephine Isabella, Sujatha Srinivasan, and G. Suseendran

Abstract The growing world has the transactions of finance mostly done by the transfer of amount through the cashless payments over the Internet. This growth of transactions led to the large amount of data which resulted in the creation of big data. The day-by-day transactions increase continuously which explored as big data with high speed, beyond the limit of transactions and variety. The fraudsters can also use anything to affect the systematic working of current fraud detection system (FDS). So, there is a challenge to improve the present FDS with maximum possible accuracy to fulfill the need of FDS. When the payment is made by using the credit cards, there is chance of misusing the credit cards by the fraudsters. Now, it is essential to find the system that detects the fraudulent transactions as a real-world challenge for FDS and report them to the corresponding people/organization to reduce the fraudulent rate to a minimal one. This paper gives an efficient study of FDS for credit cards by using the machine learning (ML) techniques such as support vector machine, naïve Bayes, K-nearest neighbor, random forest, decision tree, OneR, AdaBoost. These machine learning techniques evaluate a dataset and produce the performance metrics to find the accuracy of each one. This study finally reported that the random forest classifier outperforms among all the other techniques.

Keywords FDS · Naïve Bayes · Random forest · SVM · Decision tree · OneR

S. Josephine Isabella (✉)
Research Scholar, Department of Computer Science, Vels Institute of Science, Technology and Advanced Studies, (VISTAS), Chennai, Tamil Nadu, 600117, India
e-mail: josephineisabella@yahoo.co.in

S. Srinivasan
Associate Professor, Department of Computer Science and Applications, SRM Institute for Training and Development, Chennai, Tamil Nadu 600032, India
e-mail: ashoksuja08@gmail.com

G. Suseendran
Assistant Professor, School of Computing Sciences, Vels Institute of Science, Technology and Advanced Studies, (VISTAS), Chennai, Tamil Nadu, 600117, India
e-mail: suseendar_1234@yahoo.co.in

S.-L. Peng et al. (eds.), *Intelligent Computing and Innovation on Data Science*,
Lecture Notes in Networks and Systems 118,
https://doi.org/10.1007/978-981-15-3284-9_7

1 Introduction

The new arrival of innovative technologies gives an opening to the Internet and cashless transactions which have emerged as easier. However, for online transactions, we no longer want to be in a view found in a sure location where the transaction happens, making it prone to fraudulent one. There are many ways in which the people can profess to be the other user and create a transaction as fraudulent. If a transaction is fraudulent or no longer available, it could be decided by studying previous transactions and evaluating them with the modern one. If the distinct in nature of previous transaction and the modern transaction is big, there is a possibility that the modern-day transaction is a fraudulent transaction [1]. This paper discusses an effective study about the machine learning techniques that detect the fraudulent transactions with the help of evaluation metrics in an effective way. Section 1 gives the introduction. A common study to understand the fraud detection system (FDS) is discussed in Sect. 2. Section 3 reviews the related literatures in FDS. Sect. 4 gives the experimental studies. The evaluations of various machine learning techniques are detected in Sect. 5. Section 6 gives the results and discussion part. Finally, the conclusion is given in Sect. 7.

2 An Understanding of FDS

Without using cash, the products can be sold and transferred through various payments by simply using a card that is given by the financial sectors and the bank called credit cards. The fraudsters use these cards illegally, or not having the permission of cardholders is referred to as credit card fraud [2]. The method used to find and identify the fraudulent transactions when the transactions have entered into the system and make intimation to a system administrator is called FDS. Previously, these transactions were obtained by using fraud detection sampling techniques, but it was time consuming. Nowadays, machine learning plays a major role in automated system [3]. The continuous increase of usage of credit card transactions and evolving the concept of CNP (card-not-present) in payment transactions that generate the misbehavior of the illegitimate people who counterfeit as others. There is a need to create an automated FDS for credit issuers [4]. So, there is a chance to apply the machine learning techniques to find the solution to the fraud detection system in a functional way [3].

3 Review of Literature

The study given by authors like Shen et al., investigated that the efficiency of classification models is tested against fraud detection and also produced a framework to the fraud detection in credit card to reduce the risk [5] at banks. Whitrow et al., revealed a study of fraud detection at transaction and account level of two banks, A and B, by using the transaction aggregation [6]. In this proposed study, the self-organizing map neural network (SOMNN) technique and transactional rules are used to create a decision model called credit card fraud watch (CCFW) along the existing banking software and are applied to the real banking dataset and used to solve the problem of fraudulent transaction by the optimal classification of each transaction [7].

The authors, E. Duman and Y. Sahin, designed a model for fraud finding and discussed that SVM models produced better results in the training dataset mode, while the decision tree-based models performed well in the testing mode. This model can be utilized by the financial institutions to predict the fraudulent transaction. [8]. This study implemented a linear Fisher discriminant analysis on fraud detection in credit cards for calculating a weighted average to find out the transactions as profitable and prevented loss of millions of dollars of real-time banking transactions [9].

Awoyemi et al. [10] concluded that there is a need to develop a better sampling approach to handle the highly imbalanced credit card dataset using meta-classifiers. This study made a comparison of random forest and logistic regression with sample dataset (preprocessed with PCA and without PCA values). This comparison evaluated through the R language resulted that Random forest without PCA and a K value of 3 having the accuracy as 99.77% by using the confusion matrix [11]. This study is designed to build four classification models, namely logistic regression, SVM, decision tree and random forest with the training data of 70 and 30% testing data of European card holders from ULB Machine Learning Group. Random forest is found as the best classifier among all [12]. John et al. made an effective study of feature selection on two imbalanced datasets as ranking by the use of correlation coefficient and evaluated using MATLAB IDE with the four classifier techniques, namely naive Bayes, support vector machine, decision tree and NNBRF and applied to the datasets of Taiwan and European banks. The results showed that the decision trees were performed to produce the better result of classification [13].

Rajora et al. made a study of machine learning classification techniques as well as ensemble learning methods and evaluated an unbalanced dataset by using under sampling method with PCA values as balanced. The outcomes showed that the gradient boosting regression tree had the better accuracy among all the classifiers based on dataset 'without time' feature [14]. Authors like patil et al. evaluated the random forest, logistic regression and decision tree classifiers and applied on the credit card fraud-German dataset and results showed that random forest tree made accuracy as high but had the limitation of over fitting of decision tree [15]. K. R. Seeja and Masoumeh Zareapoor revealed a model named FraudMiner for fraud detection and analyzed the results of classification models. The FraudMiner model was applied to one lakh transactions. This proposed model produced the performance evaluation as fraud detection rate was high. The evaluation was done by applying the BCR and MCC to the FraudMiner model [16].

In this study, the authors reviewed various methods to find the solution to the fraud detection systems. They discussed hidden Markov model (HMM), CNN and ANN methods and proposed a model with autoencoder neural network model [1].

4 Experimental Studies

4.1 Dataset and Preprocessing of Data

The German credit fraud dataset is the famous dataset taken from kaggle.com with 1000 instances and 20 attributes. Preprocessing is essential before we evaluate the values in the dataset. The proposed model gives the accuracy improvement based on the features that have been selected as salient features. In this study, we use the German credit card dataset as sample dataset. The model has been trained with 70% of instances and tested with 30% of instances having 20 attributes [17].

4.2 Evaluation Metrics

There are some metrics of evaluation available to find the achievement measures of the classification models.

The various metrics for evaluation are given as follows [10, 20]:

$$\text{Accuracy} = (\text{TN} + \text{TP})/(\text{TP} + \text{FP} + \text{FN} + \text{TN}) \tag{1}$$

$$\text{Precision} = \text{TP}/\text{TP} + \text{FP} \tag{2}$$

$$\text{Recall} = \text{TP}/\text{TP} + \text{FN} \tag{3}$$

Based on the evaluation of these metrics, the confusion matrix is formed.

5 Evaluation of Ml Techniques

5.1 Naïve Bayes

Based on some assumption, the outcome is affected by the independent factor that is called as ‘Naive.’ It predicts a class of future incoming data values with known target values as training data. It finds the probability by using the formula [16, 18].

5.2 *KNN*

This algorithm predicts data value based on a relative position to other data values. It is a clustering algorithm used to find the unknown feature of a testing data by using the Euclidean distance [18]. This is an instance-based algorithm which keeps all the instances and classifies the similar instances having the nearest values. The existing instances find the new nearest instances by using distance evaluation such as Euclidean distance [16].

5.3 *Random Forest*

The Random Forest classifier generates the connected decision tree classifiers randomly. If the input is having the training data, then it will make the rules which are helpful to predict the results through the decision tree forests [18]. This technique generates a decision tree having the concept as each tree is a weak learner and the tree having maximum votes are the strong learners, and it categorizes the new instances to the class that has the maximum votes [16].

5.4 *SVM*

For classification problems, SVM is used to categorize the values or data points by the best fitting method. Support vector machine plots the line that denotes the training values on a plane to detect the categorization of data. The classification problems and regression model problems use this technique in an efficient way to find the solution [18].

5.5 *Decision Tree (J48)*

J48 is a decision tree model and an implemented form of C4.5 technique in Java. This is an ID3 decision tree algorithms extended version. Working on the different values of an existing input, the average value of new class can be calculated. The different features are represented in the tree as internal nodes. The end value of the dependent data is found by the end node. The root node gives the decision.

5.6 *OneR*

The frequency table has target value for each predictor for creating a predictor's rule called one rule that selects the rule that has the minimum total error.

5.7 *AdaBoost*

This algorithm is a classification ensemble method. This algorithm is used to improve the performance of any algorithm. When any algorithm combines with this technique, then it converts the weak learners to the strong one [19].

6 Results and Discussions

The evaluation of machine learning techniques produces the results of various measures such as the rate of true positive, precision and are related to find the fraudulent transactions in an efficient way. These measures are observed and placed in Table 1.

Obviously, all the ML techniques produced true positive greater than 80%. The random forest algorithm has the highest rate of true positive as 92%. The remaining techniques have less than that of 92%. The SVM and OneR techniques having the same true positive rate 87% are slightly higher than naïve Bayes. KNN has the lowest rate (81%) of true positive. The Recall value of random forest attained at the maximum of 0.917 and KNN has the lowest value 0.810 of Recall. SVM has the recall value as 87.1% and is slightly higher than that of OneR and naïve Bayes methods. The transactions which are correctly classified as genuine or fraudulent are usually termed as precision. From the evaluated results, naïve Bayes classifier has the most prominent precision value as 80% and OneR method has the lowest value of 71.2%. The next highest precision value obtained by KNN is 79.4%. But the KNN algorithm has the lowest rate (19%) of false positive. This shows that this algorithm

Table 1 Results of classification measures using various ML techniques

Ml technique	TPR (%)	FPR (%)	F-measure	MCC	Recall	Precision	Acc
Naïve Bayes	86	50	83.10	0.385	0.864	0.800	75.4
KNN	81	19	80.19	0.324	0.810	0.794	72.0
Random forest	92	59	84.47	0.386	0.917	0.783	76.4
SVM	87	53	83.05	0.371	0.871	0.793	75.1
J48	84	61	95.00	0.250	0.840	0.763	70.5
OneR	87	82	78.17	0.061	0.867	0.712	66.1
AdaBoost	88	73	80.10	0.180	0.877	0.737	69.5

handles the dataset in a better way than other classifiers. The remaining techniques having FPR greater than 19% observed from the evaluation.

The correctly classified instances of incoming data categorized after evaluating the various machine learning techniques are shown. The accuracy results are represented in Fig. 1. From the graph, random forest algorithm has the most accurate value as 76.4%. The least accuracy is produced by OneR method. The naïve Bayes classification and SVM have more or less the same accuracy with the difference of 0.3%. J48 decision tree algorithm classifies the data with the accuracy rate of 70.5%. The AdaBoost algorithm detects 69.5% of accuracy, but is greater than that of OneR method which has the accuracy of 66.1%.

Random forest technique has the highest (84.5%) F-measure value. SVM and naïve Bayes have the same value, 83%. Similarly, KNN and AdaBoost have the same value (80.1%) for the F-measure. This observation is visualized in Fig. 2.

Matthews correlation coefficient (MCC) [7, 17] has been calculated for various machine learning models. The Matthews correlation coefficient must be in the range

Fig. 1 Accuracy of FDS using ML techniques

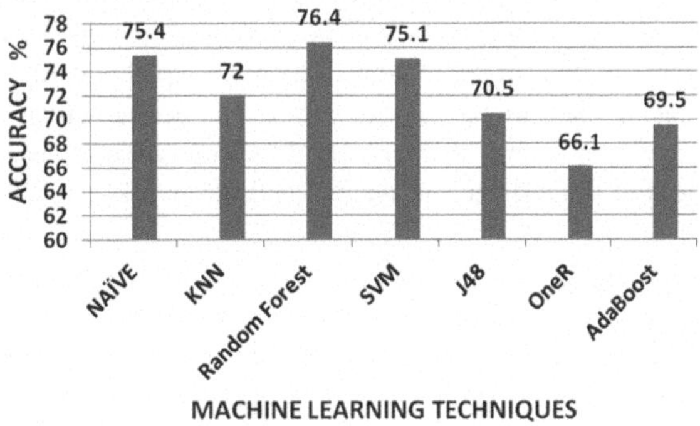

Fig. 2 Comparative analysis of F-measure

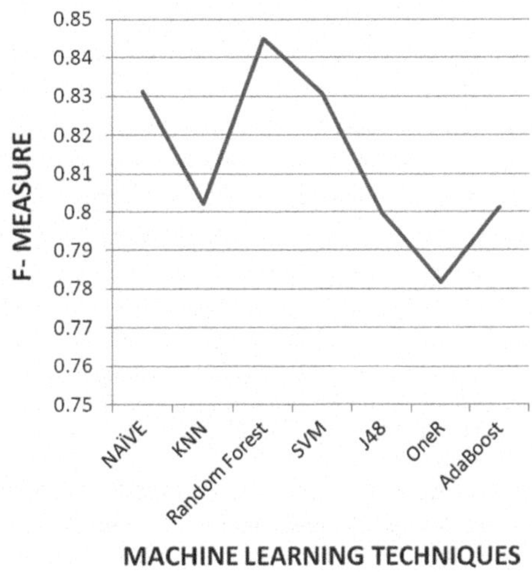

of +1 to −1. All our evaluated techniques resulted in this range of values and are efficient to fit in the model.

7 Conclusion

Usually, the available fraud detection methods find the fraudulent transaction after they have happened. There will be a chance to occur fraudulent transaction out of numerous transactions. Even though the occurrence of fraud is at minimal rate against large number of transactions, it is a commitment to invent a technique for detecting the fraudulent cases before the transaction has been completed. This study made an effort to evaluate the sample dataset with different machine learning techniques and resulted that among all the techniques random forest technique produces better performance in most of the cases. The above study showed that the machine learning techniques are capable of handling the fraudulent cases in an efficient manner. But there is a limitation occurred that how their performance will be found when the total number of transactions will be increased to some extreme level, i.e., how they are scalable. This experimental study gives a pathway to find an efficient highly scalable machine learning technique. There is a need to create a framework that handles the big data in a smooth way to find the fraudulent transactions at a minimal rate in the field of fraud detection system as future work.

References

1. Manek H (2019) Title : review on various methods for fraud transaction to secure your paper as per UGC guidelines we are providing a electronic bar code, Nov 2018
2. Chaudhary K, Yadav J, Mallick B (2012) A review of fraud detection techniques: credit card. Int J Comput Appl 45(1):975–8887
3. Abdallah A, Maarof MA, Zainal A (2016) Fraud detection system: a survey. J Netw Comput Appl 68:90–113
4. Van Vlasselaer V et al (2015) APATE: a novel approach for automated credit card transaction fraud detection using network-based extensions. Decis Support Syst 75:38–48
5. Aihua S, Rencheng T, Yaochen D (2007) Application of classification models on credit card fraud detection. In: Proceedings-ICSSSM'07 2007 International Conference Service System Service Management, no. 1997, 2007, pp 2–5
6. Whitrow C, Hand DJ, Juszczak P, Weston D, Adams NM (2009) Transaction aggregation as a strategy for credit card fraud detection. Data Min. Knowl. Discov. 18(1):30–55
7. Ogwueleka FN (2011) Vol_6(3)_311-322_Ogwueleka.pdf. 6(3):311–322
8. Sahin Y, Duman E (2011) Detecting credit card fraud by decision trees and support vector machines. Int Multiconference Eng Comput Sci I:6
9. Mahmoudi N, Duman E (2015) Detecting credit card fraud by modified fisher discriminant analysis. Exp Syst Appl 42(5):2510–2516
10. Awoyemi JO, Adetunmbi AO, Oluwadare SA (2017) Credit card fraud detection using machine learning techniques: a comparative analysis. In: Proceedings of the IEEE International Conference Computing Networking Informatics, ICCNI 2017, 2017, vol 2017-Jan, pp 1–9

11. Data T (2017) A comparison of machine learning techniques for credit card fraud detection, pp 1–9, 2017
12. Navanshu Khare SYS (2018) Credit card fraud detection using machine learning models and collating machine learning models. J Telecommun Electron Comput Eng 10(1–4):23–27
13. John OA, Adebayo A, Samuel O (2018) Effect of feature ranking on the detection of credit card fraud: comparative evaluation of four techniques. i-manager's J Pattern Recogn 5(3):10
14. Rajora S et al (2019) A comparative study of machine learning techniques for credit card fraud detection based on time variance. In: Proceedings 2018 IEEE Symposium Series Computational Intelligent SSCI 2018, no Nov, pp 1958–1963, 2019
15. Patil S, Nemade V, Soni PK (2018) Predictive modelling for credit card fraud detection using data analytics. Procedia Comput Sci 132:385–395
16. Seeja KR, Zareapoor M, FraudMiner: a novel credit card fraud detection model based on frequent itemset mining. Sci World J, vol 2014, 2014
17. Correa Bahnsen A, Aouada D, Stojanovic A, Ottersten B (2016) Feature engineering strategies for credit card fraud detection. Expert Syst Appl 51:134–142
18. Banerjee R, Bourla G, Chen S, Purohit S, Battipaglia J (2018) Comparative analysis of machine learning algorithms through credit card fraud detection, pp 1–10
19. Sun Y, Wong AKC, Wang Y (2010) Parameter inference of cost-sensitive boosting algorithms, pp 21–30
20. Jain Y, NamrataTiwari SD, Jain S (2019) A comparative analysis of various credit card fraud detection techniques. Int J Recent Technol Eng 7(5S2):402–407

Effective Role of Cloud-Based IoT Technology in Smart and Precision Horticulture Works: A Novel

M. Kannan, C. Priya, L. William Mary, S. Madhan, and V. Sri Priya

Abstract In the world, agriculture has called a variety of names such as cultivation, horticulture, and farm. Agriculture is an important, traditional, and vital thing in India. When comparing to other countries, Indian country gives more prominent in the horticulture field to grow up the plants and as well as food sources. This paper concedes some basic information about the smart horticulture system by the utilization of cloud-based IoT (CBI) technology. IoT is the best and a new era in the cloud area which is used in many real-time applications such as horticulture work. In the horticulture, IoT is used for testing and protecting the agriculture process like soil moisture, weather condition, humidity, rainfall, photosynthesis, and fertilization. Internet of Things (IoT) is a very broad concept; it gathers many real-world objects and communicating with each other through the Internet (also Wi-Fi) connection. It contains lots of real-world things, for example, mobile phones, computers, vehicles, plants, and electronic devices. It will confer many advantages to the farmer to organize the farm. IoT applications are used in real-life applications, like hospitals, industries, networking areas, and so on. In the previous life, the agronomist necessary to ensure the weather condition, fertilizer, rainfall condition and observe the soil moisture before the agronomist organizing the farm. But now the cloud-based IoT will take the charge to help such kind of works to systemize the green farm using some agricultural-based sensors. For this reason, we can tell IoT is one of the useful and the best applications for the farmers. In this, cloud-based internet of things will work under the remote controller and sensors, used to develop the horticulture process.

Keywords Cloud-based IoT (Internet of Things) · Sensors · Wi-Fi · Remote controller · And horticulture

---

M. Kannan (✉) · C. Priya · L. William Mary · S. Madhan · V. Sri Priya
Department of MCA, SPIHER, Avadi, Chennai, India

S.-L. Peng et al. (eds.), *Intelligent Computing and Innovation on Data Science*, Lecture Notes in Networks and Systems 118,
https://doi.org/10.1007/978-981-15-3284-9_8

1 Introduction

We can say that cultivation is one of the arts of India. An Internet of Things (IoT) is the era of technology in the computer world. In India, the nature of farmers uses general and traditional methods to maintain and organize the farm. We may think, cultivation is the natural process, then how it is possible in the computer application! For this, many of the researchers have deeply analyzed in this smart agricultural system based on IoT application under a cloud-based environment. This paper will represent the agriculture process using cloud-based IoT. Through this paper, we can enrich our knowledge in the part of the smart agricultural system. Internet of Things is a helpful technology for the farmers which have been used in many other places. This technology is used to improve our products and to increase the cost of the product through the quality of the sources. Cultivation is the main factor in India, in which cloud computing is one of the growing up technologies. The cloud based IoT technology is used to store our valuable information with the highest security [1, 2] and also used in many industrial markets, whereas IoT is also the technology like cloud computing technology which is used to share the required resources from their stored sources and the information as per the user convenience. Internet of Things application will help many ways to grow the plant and the farm and it helps to protect the farm from their fault condition (from heavy rains or heavy radiation). In India, commonly these agriculture works are done in the place of rural areas. For agricultural work, when we are comparing with rural and an urban area, rural area is the best place to harvest the farm or doing horticulture by the influence of good and pure nature [3]. IoT is working under the Internet and also the Wi-Fi connection, so it can easily check the weather condition through the sensor before start the harvesting process. We are human things, we do not know to understand the environment, nature, and weather conditions, because the weather has been changed day by day, which is the constant thing. So that IoT technology will help such a situation to ensure the daily condition for harvesting and/or fertilizing the farm.

India is dependent on agriculture. According to the Indian Brand Equity Foundation, 58% of the people have lived on the village side and they are dependent only on agriculture [3]. Internet of Things consists of much Android application, so it can easily have analyzed while the risk situation. IoT is an automation process, which means if suppose going to do the agriculture, then first we must check the level of the temperature, humidity range, and soil moistening, etc. But in this smart agriculture (horticulture) system, the cloud-based IoT application (automatic process) uses various sensors [4] for analyzing and doing horticultural works. Which is acting as a reporter, it will gather all the information regarding the user application (agriculture) and provide valuable requirements to the user. Many of them have known about the agricultural process, which is a too large process for growth and harvest the product. This is one of the heaviest tasks for farmers. But in this smart agriculture (horticulture) will help the farmers to reduce their workload in half.

2 Related Works

Cultivation, farming, and husbandry are the most sweeping traditional occupations in the Indian economy. In real and robotic life, technology has been improving day by day. Likewise, agriculture will also change to that smart life system. This system is generally known as smart agricultural. IoT-based smart agriculture sends and receives the data and signal through the Internet. These systems are based on sensor work. For this smart agricultural system, IoT introduces many agro-sensors like [5]. In this, IoT-based agriculture has many advantages, for example, it reduces the costs, and all the works (agro-processes) are automated such as from plowing to harvest. Using IoT technology (sensors, GPS), the farmer can increase the cost with low time. This smart agriculture process provides many benefits for the farmer like driverless tractors. This autonomous tractor is used for doing plowing the farm and sow the seeding to the farm. The traditional agriculture processes are

- Plowing
- Seeding
- Watering
- Introspection
- Fertilizer
- Weeding
- Crop maintenance
- Harvesting.

Smart agriculture system has the capability to do the above process using the knowledge of IoT technology. In India commonly, these agriculture works are done in the place of rural areas. IoT is an automation process, which means all the agriculture-based works are automatically started up using this technology.

3 Literature Review

Basically, agriculture is a tedious process and also important source. So, many of the researchers have researched this agriculture topic again and again. Recently, Muthunoori Naresh and Munaswamy [4] have proposed a new and improve the efficiency of the smart agriculture system using IoT environment.

Patil and Kale [6] have proposed a new methodology and implement decision support system (DSS) for farm alert and maintain the crop. Uva Dharshini et al. [7] use IoT services to propose a decision tree algorithm for this smart agriculture.

Veena et al. [8] discussed intrusion detection parameters. Sushanth and Sujatha [3] have discussed the overall agriculture processes like weather checking, humidity sensors, and generated the irrigation level for agriculture. For this, the fast alert SMS has sent to the farmer using the GSM. Aher et al. [9] use the IoT system and clustering concepts to implement the new process of data collection from different sources.

Suma et al. [10] use the PROTEUS 8 simulator tool, which is microcontroller tool used to find the electronic components of the agro-process, to test the programs of the process and to design for the scenario. Patil et al. [11] presented the various sensor types (like temperature sensor, humidity sensor, pressure sensor, etc.) and Agro-logger system to change the sensor threshold value.

Balamurugan et al. [12] discussed and investigated the IoT roles in agriculture. Srilakshmi et al. [13] compare the smart agriculture system IoT application with the Internet of things.

4 Internet of Things (IoT)

In the real world, computer system contains everyone has an object or real-time objects such as television, mobile phones, vehicles, pen, pencil, books, computer accessories, home, and even human beings. Such real-time objects or sources are available only in that particular place (book-bookshop, plant-garden). These objects are collected and put in together as a separate place called cloud-based storage [14] and cloud memory. This phenomenon is called the Internet of Things (IoT). This is based on cloud computing technology. Like cloud computing, IoT also provides lots of things (physical devices) and information to the client through the Internet. The term Internet of Things (IoT) means all objects are linked or connected and communicated with each other through the Internet (Fig. 1).

Internet of Things (IoT) is one of the new trends in the computer network. This paper provides an overview of the smart agriculture system using IoT is the central technology and focuses on how the IoT will help the agricultural work. Nowadays, IoT is used in all the industry places like a consumer market, and business analytics, medical industries like so. In this paper considering agriculture work as a major part of the system, the application IoT automatically collects all the data and information

Fig. 1 Cloud-based IoT

regarding the agriculture work such as temperature accuracy, accurate level of the soil moisture and water consistency, from the environment due to the automation process. These are environmental parameters which used to do the harvesting. IoT has advantage scalability, so the farmer can control the devices at any time, with the different mobile applications [9].

4.1 Why the Internet of Things (IoT)?

The term Internet of Things (IoT) is a system of communicating with all the physical related computing devices, objects, animals, and peoples that are providing the unique identifiers to all the resources for sharing the data over the network with the help of UIDs [15, 16]. In the computer world, IoT is much more important to access real-time objects. This is used in many environment places such as scientific monitoring, infrastructure management, industrial applications, energy management, health care systems, smart home automation, smart agriculture, transport systems, large scale deployments, and more computer-based platforms (Fig. 2).

Fig. 2 Smart agriculture using the mobile application

4.2 Advantages and Disadvantages of IoT

Some of the Internet of Things advantages [17] and disadvantages [18] are given below:

- Enhance data collection
- Low time
- Sustainability
- Quality tracking and checking
- Increasing the cost factor
- Provide a better suggestion
- Scalability
- Provide many information and decision
- Improve the customer experience.

These are some main advantages which are provided by the Internet of Things. And the drawbacks are

1. More Complex to understand the process
2. Privacy is not there
3. Low security and safety.

5 Smart Agriculture

In recent years, smart farming (agriculture) application consists of many mobile applications. This is one of the helpful applications for the farmers and/or agricultural peoples.

This application requires an Internet connection for providing useful farm regarding the information to the farmer. This is shown in the below figure. Smart farming [19] and [20] is the automation process and also flexible. So the farmer can set or update the weathering level of the plant. Once the farmer updates the fertilizing (such as water, temperature, humidity) information, then the application will automatically work out by the farmer basis and to provide the level of the temperature, humidity range, soil information, etc. to the farmer (Fig. 3).

The cloud-based IoT will automatically search all the information like controlling, monitoring, planning, testing, and so on. This is one of the advantages for the farmer, which is shown in the below (Fig. 4).

Figure 4 shows smart farming, in which the system will control all the necessary and agricultural items under the network or through an Internet connection. We can also calculate the entire harvesting products using some cloud-based IoT application (Fig. 5) like smart farming application which is shown in the below screenshots. This

Fig. 3 Smart agriculture using IoT services

Fig. 4 Smart farming

application [6] includes,

- Soil identification
- Monitoring the applied things
- Provide decisions and advice
- Stage-wise advice, it means it provides clear information
- Weather report
- Water irrigation
- Humidity
- Fertilization [21], etc.

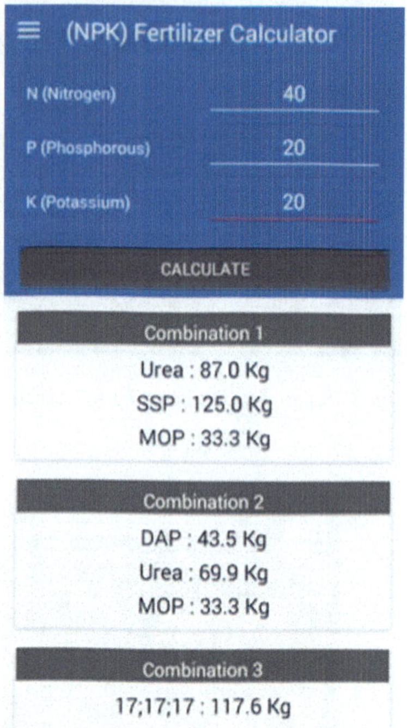

Fig. 5 Calculating agriculture product using fertilizer calculator

6 Organic Farming—Pros and Cons

Some of the crucial benefits of organic farming are [22] and [23] high nutritional values, better taste, improved human health, environmental sustainability, food security, poison-free, and contains lower input cost. Among these, some of them are given below:

- **High Nutrition Values**: Organic foods contain high-level nutrition and immunity, using natural sources and modified ingredients (e.g., natural manure).
- **Better Taste**: Organic foods naturally having a nice taste because it does not contain any chemical products and/or manure.
- **Improved Human Health**: Organic foods are used to improve and rescue the health from the various body risks and to minimize the cancer level, sugar level, immunodeficiency, etc.
- **Food Security:** It protects the organic foods for a long time, even if the climate is changed.
- **Poison-free**: Organic foods use only natural manure product, it does not use any other dangerous chemical manures. So the organic foods are very safe and poison-free.

Few drawbacks are,

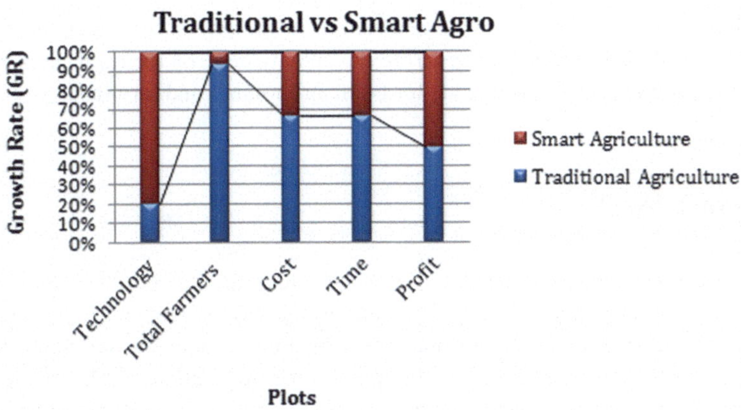

Fig. 6 Traditional agro versus smart agro

- **Diminished productivity in the long term**: Organic products might be expired soon. We could not know about to expire date of the product, unlike inorganic food.
- **Time Consuming**: It requires lots of time, methods, and constant procedures are needed to fertilize the product.
- **High Cost**: While comparing with inorganic products, organic products or foods cost as much as of 20% to 40–45% (e.g., supermarket).
- **Skills**: It requires many agricultural skills to grow and maintain the farm.

7 Traditional Versus Smart Agriculture

The below-containing knowledge-based Fig. 6 show the difference between the old agricultural model and new agricultural model. These phenomena are contained in some major parts and/or impacts of the traditional versus smart works such as time, cost, and technology.

8 Conclusion

This paper has described the horticulture or agricultural processes using cloud-based IoT technology like, how the technology has been improved in our real life, along with their advantages and its importance. We have already known about the cloud computing (C2 Technology) and IoT technology, are all blooming technologies, in which these factors are used in many organizations. Nowadays, the Internet of Things (IoT) has been used for many real-time applications. Over the past few years, agriculture is one of the tedious processes, but in the present life, technology has been improved. Agriculture and/or farming is mostly done by the rural sides, so we can predict IoT is one of the legends for both agronomists and farmland.

9 Future Work

Nowadays, everyone has updated and usage of smart mobile or an Android mobile has been increasing day by day. And hence in the IoT works, smart and updated systems are needed for the better outcome which convenient for the client. In the future, we will implement a new technique for the horticulture work (like quality checking, hybrid storage system, sensor upgrade, observing system) using either Arduino or Raspberry PI. For doing this, we can increase the productivity. On the other words, we can elevate the quality and quantity of the horticulture merchandise.

References

1. Kannan M, Priya C (2019) A survey on fault detection enabled optimal load balancing technique by efficient utilization of VM in cloud computing. Int J Innov Technol Expl Eng (IJITEE) 8(7C2):404–407. ISSN 2278-3075
2. Kannan M et al (2019) A comparative analysis of DES, AES and RSA crypt algorithms for network security in cloud computing. J Emerg Technol Innov Res (JETIR) 6(3):574–582
3. Sushanth G, Sujatha S (2018) IOT based smart agriculture system. IEEE
4. Muthunoori Naresh, Munaswamy P (2019) Smart agriculture system using IOT technology. Int J Rec Technol Eng (IJRTE) 7(5):98–102
5. www.mobile.engineering.com/amp/16653.html
6. Patil KA, Kale NR (2016) A model for smart agriculture using IOT. In: 2016 international conference on global trends in signal processing, information computing and communication, 2016, pp 543–545
7. Uva Dharini et al P (2018) IOT based decision support system for agriculture yield enhancements. Int J Rec Technol Eng (IJRTE) 7(4S):362–366. ISSN 2277-3878
8. Veena S et al (2018) The survey on smart agriculture using IOT. Int J Innov Res Eng Manag (IJRIREM) 5(2):63–66. ISSN 2350-0557
9. Aher A et al (2018) Smart agriculture using clustering and IOT. Int Res J Eng Technol (IRJET) 5(3):4065–4068
10. Suma N et al (2017) IOT based smart agriculture monitoring system. Int J Rec Innov Trends Comput Commun 5(2):177–181
11. Gokul LP et al (2017) Smart agriculture system based on IoT and its social impact. Int J Comput Appl 176(1):1–4
12. Balamurugan S et al (2016) Internet of agriculture: applying IOT to improve food and farming technology. IRJET 03(10):713–719
13. Srilakshmi A et al (2018) A comparative study on Internet of Things (IoT) and its applications in smart agriculture. Pharmacognosy J 10(2):260–264
14. SasiKumar A, Priya C et.al (2018) Computerized agricultural storage manipulation system using IOT techniques. Int J Tech Innov Mod Eng Sci (IJTIMES) 4(12):583–588
15. www.internetofthingsagenda.techtarget.com
16. www.codeproject.com

17. www.javapoint.com
18. www.quora.com
19. TongKe F (2013) Smart agriculture based on cloud computing and IOT. J Conv Inf Technol (JCIT) 8(2)
20. Jaiganesh S, Gunaseelan K, Ellappan V (2017) IOT agriculture to improve food and farming technology. In: Proceedings of the IEEE Conference on Emerging Devices and Smart Systems (ICEDSS 2017), 3–4 Mar 2017, pp 260–266
21. Vineela T et al (2018) IOT based agriculture monitoring and smart irrigation system using Raspberry Pi. Int Res J Eng Technol (IRJET) 5(1):1417–1420
22. www.conserve-energy-future.com
23. Thippeswamy E (2013) Comparative analysis of organic and inorganic food. IOSR J Agric Veterinary Sci 4(6):53–57

Intelligent Agent-Based Organization for Studying the Big Five Personality Traits

Sujatha Srinivasan and K. R. Ananthapadmanaban

Abstract Studying behavior of individuals and teams in an organization is termed organizational behavior (OB) study and has a profound research literature under the area of social sciences. OB can be studied through empirical studies or using agent-based system (ABS). Using computing techniques has an advantage over empirical studies, since humans need not be involved in ABS studies and time can be incorporated into the system to study the evolution of the organization over a period. The recent literature on organization behavior shows research gap in using the Big Five personality traits in ABSs to study and evaluate them. An attempt has been made in this study to create a virtual organization incorporating the personality traits and applying it to a real-world problem. The solution obtained is used as a concrete method for evaluation. The observations are encouraging in modeling an organization using ABS and to gain insight into the individual and organizational evolution.

Keywords Agent-based system · Organizational behavior · Cultural algorithm · Data mining · Big Five personality traits

1 Introduction

An organization is a composition of people working toward a purpose or common goal. Apart from organizational elements that decide the behavior, people have different personality traits which in turn affect various organizational outcomes. The Big Five personality traits' theory proposed by Costa and McCrae [1] has been the benchmark theory in organizational behavior studies. The Big Five personality traits are neuroticism, openness to experience, conscientiousness, extraversion, agreeableness exhibited by people in organizations. These five traits play a major role and

S. Srinivasan (✉)
Department of Computer Science and Applications, SRM Institute for Training and Development, Chennai, Tamil Nadu, India

K. R. Ananthapadmanaban
Department of Computer Science and Applications, SRM Arts & Science College, Chennai, Tamil Nadu, India

S.-L. Peng et al. (eds.), *Intelligent Computing and Innovation on Data Science*, Lecture Notes in Networks and Systems 118,
https://doi.org/10.1007/978-981-15-3284-9_9

influence the behavior of individuals in different situations within teams and within the organization. Theoretical and empirical studies have been carried out to study the Big Five personality traits. In recent years, artificial intelligence is playing a major role in studying organizations. Intelligent agent-based systems (ABSs) are used to model organizations and study the behavior of people in these organizations. However, research gap exists between empirical and theoretical studies. The current study proposes an agent-based model of an organization with intelligent agents with the Big Five personality traits to study organizational behavior.

The rest of the paper is organized as follows. Section 2 gives related research in using the Big Five personality traits to study organizations. Section 3 gives the overview of the proposed agent-based system. Section 4 presents the experiments and the results and discusses the observations obtained from the study. Section 5 concludes with future research directions.

2 Related Work—Big Five Personality Traits in OB Study

2.1 Organizational Behavior Study—Empirical

Pioneering works in studying the Big Five personality traits in organization behavior started as early as the proposal by Costa and McCrae [1] followed by Kudret et al. [2] and Barrick and Mount [3]. Work team personality composition has been studied by Neuman et al. [4]. Team composition in organizations has been carried out in [5] where homogenous teams showed optimal learning. However, Andrejczuk et al. [6] argue that learning in a classroom environment needs heterogeneous teams while the influence of the Big Five on the academic performance and self-efficacy has been studied by Stajkovic et al. [7]. Correlation studies of the Big Five traits with career success have been carried out by Judge et al. [8]. Conscientiousness was observed to be positively related to both successes while neuroticism negatively to extrinsic success. However, sales effectiveness of life insurance agents by Janowsk [9] shows that openness to experience, conscientiousness and neuroticism were positively correlated to the agent's performance. Knowledge creation, sharing innovation, curiosity and passion have been studied in [10–14], respectively. Research gap identified from empirical studies is as follows: (1) Evolution of a person's behavior over time lacks the literature, and (2) empirical studies on data collected from people are prone to bias.

2.2 Agent-Based System Review

The authors of [15] have developed an agent-based system for studying the Big Five traits in group decision making, while collaborative problem solving in students has been studied using ABS by Stadler and Herborn [16]. The ABS proposed by Nguyen [17] has been used to study human decision making. Zhou et al. [18] have proposed artificial intelligence (AI) agents with personality traits to interview users to study the personality of the user and trust in the AI agent, whereas Kampman et al. [19] have proposed a virtual agent that can adapt itself to user's personality based on text and audio input. Research gap identified from review of ABS for organizational behavior is as follows: (1) There is no concrete method to evaluate ABS, (2) the type of knowledge used by agents and its representation lack the literature, and (3) Big Five personality traits have not been used in the ABS model. In order to bridge the research gap, an ABS is proposed in this study to model an organization taking into consideration the Big Five personality traits.

3 The Proposed Agent-Based Organization—Methodology

An organization is modeled with intelligent agents who have the Big Five personality traits that they use to make decisions. The agents take a data set as raw materials and produce classification rules which are used to evaluate them. Cultural algorithm (CA) which is derived from social learning and evolution was proposed by Reynolds [20] and is used in the current study for modeling an organization. Cultural algorithm has three components, namely *the belief space* (BS), *the population space* (PS) and a *communication protocol* to exchange knowledge between these two spaces. The PS can be created using any population-based algorithm. The proposed ECA is explained below.

3.1 Components of the Proposed ECA

The Belief Space

The BS consists of normative KS (NKS) which contains the range of values that define the attributes of the data set, the domain KS (DKS) which stores the rule metrics of the rule created, situational KS (SKS) which contains the best example of the current generation, the topographical KS (TKS) which contains all the rules created in the current generation, and the history KS (HKS) that contains all the best rules from the starting of the evolution. Rule KS (RKS) retains all the rules created by the system. The RKS, TKS, DKS, SKS and HKS are updated at the end of each generation. NKS is not updated since the range of attributes does not change during evolution. The updating of the BS takes place using the communication protocol explained in forthcoming section.

The Population Space

The population space consists of the rules created by agents in the system. The agents in the system are defined using the Big Five personality traits. Agents with the personality trait of neuroticism (sensitive and nervous people) use the RKS without taking any risk, and people with the trait of openness to experience (inventive and curious people) are more creative, eager to learn and ready to take risk, use all the knowledge sources. Conscientiousness (people who are efficient, organized, self-disciplined and thrive to achieve their goal) uses HKS which contains the best rules from all previous generations. Extraverts (people who are outgoing, energetic social and exploratory) use the topographical KS that contains the rules from current generation. Agreeableness (people who are friendly, who would like to comply with rules and regulations and who are adaptable) uses the situational KS which contains the best examples from the current generation. The agents use their KS that evolves in a genetic algorithm environment to create rules using the reproduction operators of selection, crossover and mutation.

The Communication Protocol

The communication protocol consists of the two phases, namely the updating of the belief space at the end of each generation and the influence phase where the agents use the updated knowledge in the future generations. The updating of the BS takes place by evaluating the individual rules by calculating the fitness function of the rules using support and confidence as metrics. Rule metrics vectors in DKS undergo Pareto optimization to choose better individuals. The agents use the updated BS, thus influencing the decision of the agents in producing new rules. The proposed system is an extension of the work found in [21, 22] where more details can be found. Figure 1 shows the flow diagram of the proposed ECA.

4 Experiments, Results and Discussion

Experiments have been conducted to apply the proposed agent-based organization to create rules from the Bupa data set from the UCI machine learning repository [23]. Experiments were carried out to study the performance of the different agents in producing classification rules and to study the evolution process using the quality (rule metrics) and quantity (number of rules) of the rules.

4.1 Results

The results of the experiment are presented as tables and figures. Table 1 gives the performance of the agent-based organization applied to the problem of rule mining. The average number of unique rules created by the agents, the average number of

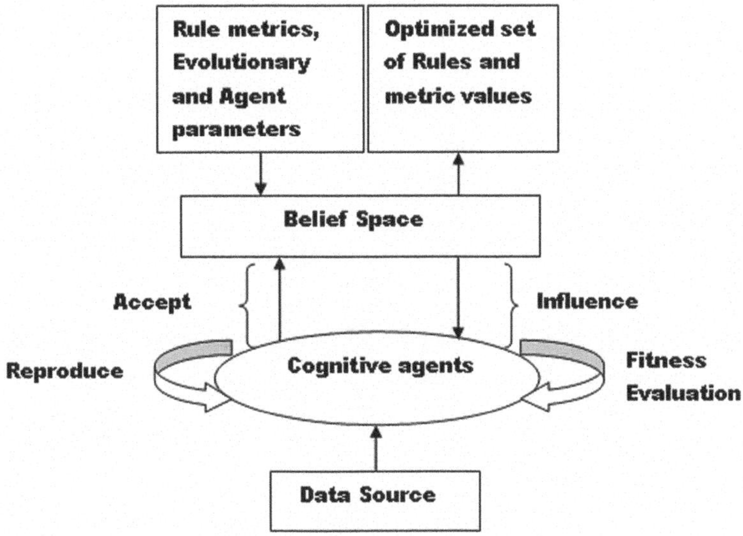

Fig. 1 Extended cultural algorithm for modeling organization

Table 1 Performance of the proposed system (average over 10 runs)—rule mining

Statistic	RKS	HKS	Time (s)	Accuracy (%)
Avg.	253.5	32.8	13.28	64.61
St. dev.	12.96	12.71	2.90	1.48
Min.	236	15	10	62.61
Max.	280	58	19	66.96

dominators denoted by the number of rules in HKS, the time taken and the accuracy of classification of the unknown data instances for the 10 runs are summarized in Table 1. Table 2 gives the average contributions of each type of agent with different personality traits. Figure 2 shows the box plot of the different agent personalities and their contributions over 10 runs. It can be observed from Table 2 and Fig. 1 that agent with the personality trait of conscientiousness performs better followed by agreeableness and openness personalities, while the personalities of neuroticism

Table 2 Contributions of the different agents with personality traits

Agent personality	KS used	Contribution (%)
OPEN	ALL	17.78
CONS	HKS	45.65
NEUR	RKS	10.67
AGRE	SKS	20.32
EXTR	TKS	5.59

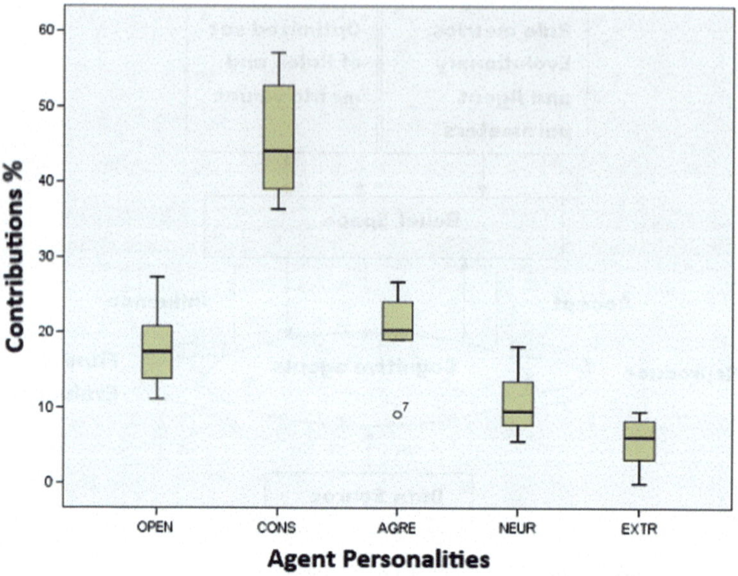

Fig. 2 Agent contributions—agent traits versus contributions (%)

and extraversion are the low performers in producing good rules. Figure 3 shows the performance of the agents over generations. As can be observed, conscientiousness personality agents and agreeableness agents show consistent performance throughout all the generations, whereas the other three personalities soar in later generations as the organization evolves.

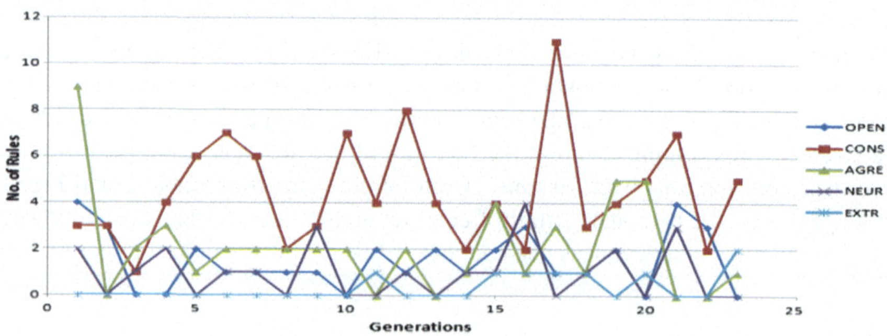

Fig. 3 Agent performance over generations

4.2 Discussion

The results observed from the study throw light into the different types of personality traits and their performance and contribution to the organization. It has been observed that conscientious people are the most productive of the five personality traits. Similar observations have been made in studies like [3], followed by the personality traits of agreeableness and openness as also observed in studies by Neuman et al. [4] and Oishi and Roth [24] which have been carried out to find the influence of the Big Five traits on performance of individuals and teams in an organization, while positive correlation between the five traits and performance of students have been observed in studies by Stajkovic et al. [7]. The results observed in the current study also suggest that people with neuroticism personality and extraverts have low correlation with performance which is supported by the claims found in the above studies, whereas in the study by Janowski [9] on performance of life insurance agents, neuroticism is positively correlated with performance.

Openness to experience and agreeableness have been observed by Stadler and Herborn [16] as traits that influence collaborative problem solving. It has been observed in [10] that openness, agreeableness and extraversion play an important role in knowledge creation. This study does not show conscientiousness as one of the personalities in creating knowledge as opposed to the observations in the present study. The personality trait of openness to experience was observed to have positive correlation to employees' knowledge sharing and creation in a team as observed by Zhang et al. [12], while the exploration and exploitation behaviors of employees have been observed to be positively related to innovation and creativity by Caniëls and Veld [13]. Innovation creation in organizations will be the basis of our future studies.

5 Conclusion

The current study proposed an intelligent agent-based organization for studying the Big Five personality traits to study the performance of the organization on real-world problem of rule mining. It has been observed that conscientiousness people contribute more to an organization followed by agreeable and open people, while extraverts and neuroticistic people behave in a different manner. The proposed ABS also gives a concrete method to evaluate individuals and the organization using the performance of the classifier. By far, this is the only study that integrated social computing and artificial intelligence to study the personality traits of individuals in an organization and to evaluate them. As future work, we would like to extend our study to knowledge creation, sharing, innovation and other organizational strategies.

References

1. Costa PT, McCrae RR (1990) Personality disorders and the five-factor model of personality. J Pers Disord 4:362–371. https://doi.org/10.1521/pedi.1990.4.4.362
2. Kudret S, Erdogan B, Bauer TN (2019) Self-monitoring personality trait at work: an integrative narrative review and future research directions
3. Barrick MR, Mount MK (1991) The big five personality dimensions and job performance: a meta-analysis. Pers Psychol 44:1–26
4. Neuman GA, Wagner SH, Christiansen ND (1999) The Relationship between work-team personality composition and the job performance of teams. Gr Organ Manag 24:28–45. https://doi.org/10.1177/1059601199241003
5. Anderson G, Keith MJ, Francisco J, Fox S (2018) The effect of software team personality composition on learning and performance: making the "Dream" team. In: Proceedings of the 51st Hawaii international conference on system sciences, pp 451–460
6. Andrejczuk E, Bistaffa F, Blum C, Rodríguez-Aguilar JA, Sierra C (2019) Synergistic team composition: a computational approach to foster diversity in teams. Knowledge-Based Syst. https://doi.org/10.1016/j.knosys.2019.06.007
7. Stajkovic AD, Bandura A, Locke EA, Lee D, Sergent K (2018) Test of three conceptual models of influence of the big five personality traits and self-efficacy on academic performance: a meta-analytic path-analysis. Pers Individ Differ 120:238–245. https://doi.org/10.1016/j.paid.2017.08.014
8. Judge TA, Higgins CA, Thoresen CJ, Barrick MR (1999) The big five personality traits, general mental ability, and career success across the life span. Pers Psychol 52:621–652
9. Janowski A (2018) Personality traits and sales effectiveness: the life insurance market in Poland. J Entrep Manage Innov 14:143–160. https://doi.org/10.7341/20181418
10. Ayub MU, Kanwal F, Kausar AR (2019) Developing knowledge creation capability: the role of big-five personality traits and transformational leadership Pakistan. Pakistan J Commer Soc Sci 13:30–61. http://hdl.handle.net/10419/196186 Standard-Nutzungsbedingungen
11. Manaf HA, Armstrong SJ, Lawton A, Harvey WS (2018) Managerial tacit knowledge, individual performance, and the moderating role of employee personality. Int J Public Adm 41:1258–1270. https://doi.org/10.1080/01900692.2017.1386676
12. Zhang W, Sun SL, Jiang Y, Zhang W (2019) Openness to experience and team creativity: effects of knowledge sharing and transformational leadership. Creat Res J 31:62–73. https://doi.org/10.1080/10400419.2019.1577649
13. Caniëls MCJ, Veld M (2019) Employee ambidexterity, high performance work systems and innovative work behaviour: How much balance do we need? Int J Hum Resour Manage 30:565–585. https://doi.org/10.1080/09585192.2016.1216881
14. Dalpé J, Demers M, Verner-Filion J, Vallerand RJ (2019) From personality to passion: the role of the Big Five factors. Pers Individ Differ 138:280–285. https://doi.org/10.1016/j.paid.2018.10.021
15. Carneiro J, Saraiva P, Martinho D, Marreiros G, Novais P (2018) Representing decision-makers using styles of behavior: An approach designed for group decision support systems. Cogn Syst Res 47:109–132. https://doi.org/10.1016/j.cogsys.2017.09.002
16. Stadler M, Herborn K (2019) Computer-based collaborative problem solving in PISA 2015 and the role of personality. J Intell 7:1–15. https://doi.org/10.3390/jintelligence7030015
17. Nguyen K (2018) A novel agent software architecture inspired by psychology. In: 14th Annual social simulation conference
18. Zhou MX, Mark G, Li J, Yang H (2019) Trusting virtual agents: the effect of personality. ACM Trans Interact Intell Syst 9:1–36. https://doi.org/10.1145/3232077
19. Kampman O, Siddique FB, Yang Y, Fung P (2019) Adapting a virtual agent to user personality. In: Lecture notes in electrical engineering, pp 111–118 (2019)
20. Reynolds RG (1994) An introduction to cultural algorithms. In: Proceedings of third annual conference on evolutionary programming, vol 172, pp 131–139. https://doi.org/10.1136/ewjm.172.5.335

21. Srinivasan S, Ramakrishnan S (2012) A hybrid agent based virtual organization for studying knowledge evolution in social systems. Artif Intell Res 1:99–116
22. Srinivasan S, Ramakrishnan S (2013) A social intelligent system for multi-objective optimization of classification rules using cultural algorithms. Computing 95:327–350. https://doi.org/10.1007/s00607-012-0246-4
23. Bache K, Lichman M UCI machine learning repository. www.ics.uci.edu/mlearn/MLRepository.html
24. Oishi S, Roth DP (2009) HEXACO personality predicts counterproductive work behavior and organizational citizenship behavior in low-stakes and job applicant contexts. J Pers 43:107–109. https://doi.org/10.1016/j.jrp.2008.11.002

Credit Card Fraud Detection Using AES Technic

C. Sudha and D. Akila

Abstract With the quick update of e-business, level of trades by credit cards is growing quickly. As e-shopping changes into the maximum basic trade mode, occasions of trade weight are tied in with augmenting. We propose a new press introduction structure that makes out four stages. To revive a consumer's basic impact models, we at first apply the cardholders' chronicled trade details to design all the customers with various get-togethers to such a degree, to point that trade practices of packed structure in a comparative party are relative. We from this time forward suggest a window sliding structure to mean the trades each social affair. Next, we void a party of specific individual direct measures for each consumer subject to the totaled trades and the consumers' chronicled trades. By then, we train the method of classifiers for every party on this basis of all rules of direct. Finally, we use the classifiers set to see mutilation on the Web and if another trade is coercion, an information instrument is taken in the prominent proof present with the incredible old shaped focus to regard the issue of thought skim. The yielded consequences of our basics show up that our structure is better than various individuals; here, we are using AES algorithm to maintain the data securely.

Index Keywords Patterns · Sliding window · Machine learning

1 Introduction

The development of PDAs and Web shopping changes into a mistaking structure for especially created buys. Regardless, the Web conditions are open, Web shopping structures have too much of bugs, and punks can utilize some unpalatable help [1, 2]. All these entire outcomes in a legitimate time of credit card misdirecting a particular event [3]. Right when a criminal takes or on the other hand obviously undeniably swindles the data of the Mastercard of a consumer, some of the criminals can utilize the energized card to eat [4, 5]. Agreeing of the Nilson Report in October 2016,

C. Sudha (✉) · D. Akila
School of Computing Sciences, Vels Institute of Science & Advanced Studies (VISTAS), Chennai, India

S.-L. Peng et al. (eds.), *Intelligent Computing and Innovation on Data Science*, Lecture Notes in Networks and Systems 118,
https://doi.org/10.1007/978-981-15-3284-9_10

approximately $31 trillion and more were made in all over the world by online part structures in 2015, seeing the chance to be 7.3% than 2014. Everything thought about changes from visa perplexity rose to $21 billion out of 2015 and will maybe reach by $31 billion, and 8 charge card blackmails distinguishing proof are a basic technique to preclude distortion events which is commonly characterized into two systems: (1) irregularity revelation and (2) classifier-based acknowledgment [6, 7]. Variation from the norm ID revolves around figuring the partition among data centers. By figuring the partition between the moving toward trade and the cardholder's profile [8], an anomaly acknowledgment system can channel any moving toward trade.

This is conflicting with the consumer's profile. The next system uses some guided learning strategies to prepare which is clashing with the consumer's profile [9]. This approach consumes some guided learning system and to get ready a classifier dependent on the assumed basic trades and terrorizing ones. Controlled learning turns around segregating deception features from compulsion trades [10]. At any rate, those two have destinations. For the irregularity divulgence, it has no limitation to lay out bowing features despite the way in which it can depict consumer's trade hones. For the classifier-based confirmation, it flops to see particular standard practices from different cardholders expelling the way in which it can get fraudsters' practices. As revealed in [11], trade affinities for a man change once in a while; meanwhile, they are adequately impacted by its wage, resources, age and characters. Thusly, their course of action prompts after some time in light of the way that of consistency and new strike structures [12]. This is known as the issue of thought skim [13] that is difficult to be settled by the above particular affirmation systems. A classifier is subjected to the given ordinary exchanges and compulsion ones. The controlled learning spins around separating misdirection highlight from shakedown exchanges. At any rate those two have objectives.

2 Literature Survey

With the movement of web shopping, trade impulse is rising truly [2]. In like way, the examination of terrorizing assertion is charming and basic. A basic methodology for seeing winding is to clear the quick profiles (BPs) of customers reliant on their honest to goodness trade records and after that to check if a pushing toward trade is a squeeze or not in setting of their BPs. Markov joint models are phenomenal to address BPs of customers, which is outrageous for those customers whose trade hones are persevering tolerably. Regardless, with redesign and advancement of e-shopping, it is all more valuable for customers to eat up by procedures for the Internet, which confines the trade practices of customers. In this way, Markov chain model is blocked for the portrayal of these practices. We propose true blue diagram of BP (LGBP) which is an absolute interest-based model to address the reasonable relationship of qualities of exchange records. In light of LGBP and clients' exchange records, we can pick way-based development likelihood from and a sound delegate for another. At that point, we depict an information entropy-based planned accumulation coefficient with the veritable objective to portray the not too horrendous game plan of trade practices of a customer. Likewise, we portray a state change probability structure to get transient features of trades of a customer. Therefore, we can build up a BP for each customer and a short time allotment later uses it to verify if a pushing toward trade is a twisting or not. Our examinations over a good of fashioned illumination get together structure that our system is better than three best in class ones.

Dal Pozzolo et al. identified fakes in Mastercard trades which is perhaps remarkable contrasted with other proving grounds for computational information figurings. Truth be told, this issue includes various significant challenges, specifically: idea float (clients' propensities develop, and fraudsters change their methodologies after some time), class irregularity (veritable exchanges far dwarf cheats) and confirmation dormancy (just a little arrangement of exchanges is opportunely checked by examiners). Be that as it may, by far most of learning calculations that have been proposed for misrepresentation identification depend on suppositions that barely hold in a genuine misrepresentation discovery framework (FDS). The absence of authenticity concerns two principle perspectives: (1) the way what's more, skill with which administered data is given, and (2) the measures used to survey misrepresentation location execution.

Existing System

Here in existing system nowadays, most of the extensive systems are applying distinctive charge cards. If they have money, they are paying return mean the bank; else they are not paying. Around then, the bank people are getting disaster. In case of losing a segment of data, the hindrance money from that account can be overcome using the recognizing undeniable proposed methodology.

Problem Statement

The Credit Card Fraud Detection Problem appearing in the past card exchanges for the learning of the ones that injury up being shakedown. This model is then used to see whether another exchange is false or not. Our point here is to perceive 100% of the precarious exchanges while limiting the off-kilter double-dealing groupings.

3 Proposed System

Here to overcome this issue, first customer needs to fill those bits of data about the individual honest parts that all inspirations driving premium will share to all banks; in case we share like these, they can keep up that data and they will not allow to apply indisputable records. Another issue is if the customer lost that charge card, a bit of the customer will hack that record and they will control that money. To beat this issue, bank social event will suit each customer particular access framework. They have to use basically indistinguishable system, the customer can get the notice at any rate mail so ordinary to find if they attempted again they will be allowed for three times, after that they will be blocked.

4 Module and System Architecture

1. User's interface design
2. Data uploading
3. Key generation and file sharing
4. Clients' key requests to data owner (Fig. 1).

Advanced Encryption Standard Algorithm

Advanced encryption standard may be a bilaterally symmetrical block cipher to guard and to classify the information, and it is enforced in the package and hardware throughout the globe to inscribe the sensitive information.

Most importantly, AES will execute all its work on bytes instead of bits. Henceforth, advanced encryption standard takes care of the 128 bits of a normal text part as sixteen bytes (Fig. 2).

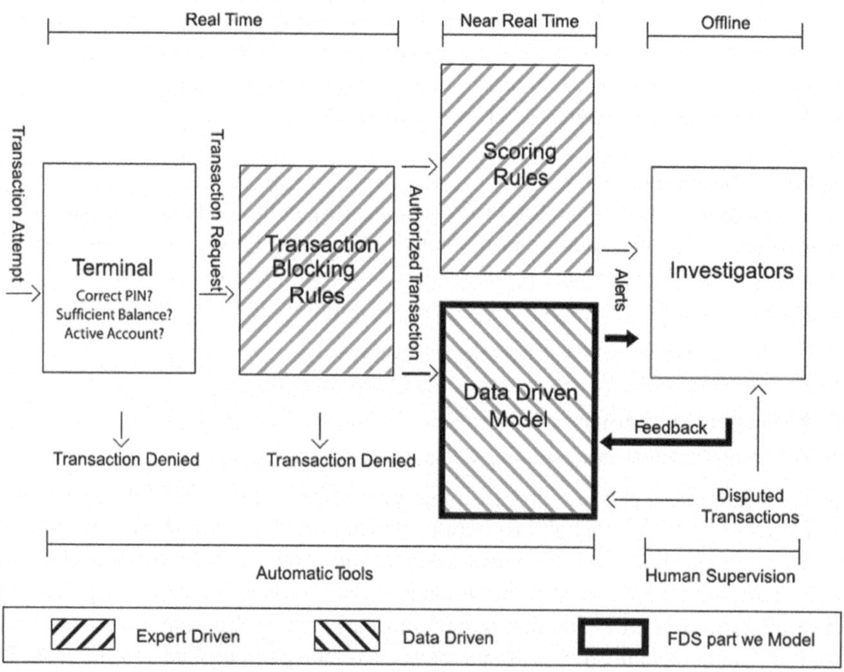

Fig. 1 System architecture

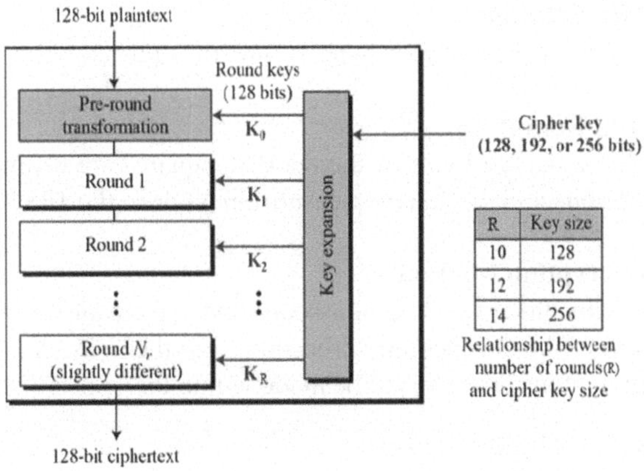

Fig. 2 Advanced encryption standard structure

Encryption Method
Here, we tend to disallow to portrayal of a run of the mill circular of AES encoding. Each circular contains four sub-forms (Fig. 3).

Byte Substitution (Sub-bytes)
The sixteen info byte units of measurement are substituted by needing up a gathering table (S-box) given stylish. The result is during the matrices of four rows and four columns.

Shift Rows
Every one of the bits of the four columns of the network is moved to one side. Any passages that 'tumble off' zone unit are re-embedded on the best possible side of line.

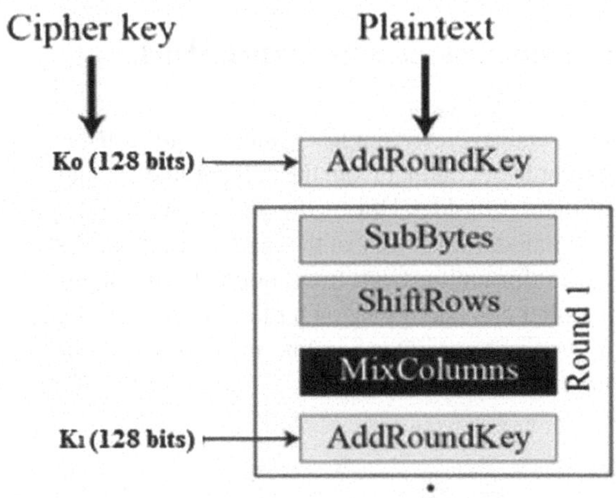

Fig. 3 First round procedure

Mix Columns
All the section of four bytes is directly adjusted utilizing an uncommon connection.

Add Round Key
These sixteen bytes of the network squares are estimated right now considered as 128 bits and the square measures in XOR to the 128 bits of the circular key.

Decryption Method
The method of coding of associate degree advanced encryption standard figure content is like the encoding procedure the other way. All the circular operation consists of the four processes to be made within the opposite directions.

- To add spherical key
- Then to do mix columns
- And then shifting the rows
- At last to substitute bytes.

Then, the sub-forms in each round square measure backward, dislike for a Feistel Cipher, the encoding and coding calculations must be severally upheld, however they're horribly firmly associated.

Result Analysis
After that, the above techniques are supported to discriminate analysis and multivariate analysis is widely used which may discover fraud by credit rate for cardholders and Mastercard dealings.

Chart
See Fig. 4.

5 Conclusions and Future Work

In our endeavor, we suggest a unique extortion location procedure. We tend to use the standards of conduct from the comparable customers to make an ongoing social profile of a cardholder. During these ideas, we tend to propose an approach to unwind the accommodating ability of the model. An input system will top off utilization of truth mark information from transactions to disentangle the idea of float downside. The classifier can change its own rating score with regard to a progression of approaching transactions. These online misrepresentation recognition approaches will be a powerful correction.

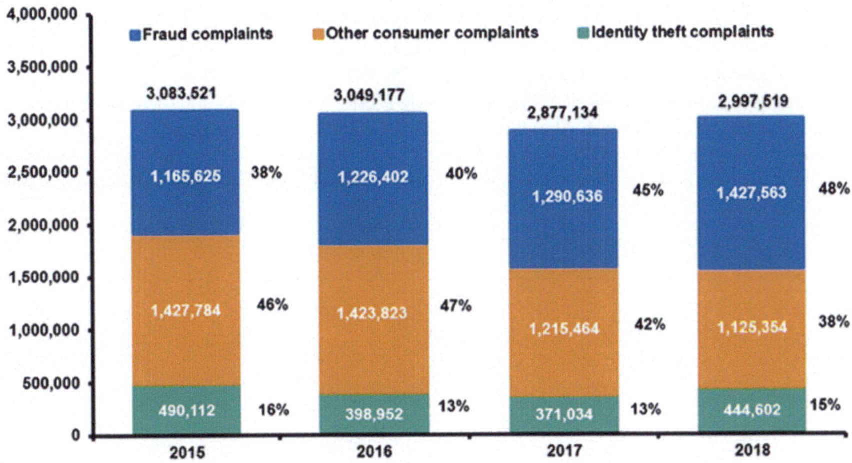

Fig. 4 Year-wise fraud detections

References

1. Nilson Report (2016) The Nilson report. https://www.nilsonreport.com/upload/contentpromo/TheNilsonReport10-17-2016.pdf. Oct 2016
2. Chen RC, Luo ST, Liang X, Lee VCS (2005) Personalized approach based on SVM and ANN for detecting credit card fraud. In: Proceedings of international conference on neural networks and brain, Beijing, China, pp 810–815
3. Shen A, Tong R, Deng Y (2007) Application of classification models on credit card fraud detection. In: Proceedings of IEEE international conference on service systems and service management, Chengdu, China, pp 1–4
4. Quah JTS, Sriganesh M (2008) Real time credit card fraud detection using computational intelligence. In: Proceedings of IEEE international conference on neural networks, Orlando, FL, USA, pp 863–868
5. Srivastava A, Kundu A, Sural S, Majumdar A (2008) Credit card fraud detection using hidden markov model. IEEE Trans Depend Secure Comput 5(1):37–48
6. Bahnsen AC, Aouada D, Stojanovic A, Ottersten B (2016) Feature engineering strategies for credit card fraud detection. Exp Syst Appl Int J 51(C):134–142
7. Vlasselaer VV, Bravo C, Caelen O et al (2016) APATE: a novel approach for automated credit card transaction fraud detection using network-based extensions. Decis Support Syst 65:38–48
8. Gurjar RN, Sharma N, Wadhwa M (2014) Finding outliers using mutual nearness based ranks detection algorithm. In Proceedings of IEEE international conference on reliability optimization and information technology (ICROIT), Faridabad, India, pp 141–144
9. Ganji VR, Mannem SNP (2012) Credit card fraud detection using anti-k nearest neighbor algorithm. Int J Comput Sci Eng 4(6):1035
10. Sudha C, Akila D (2019) Detection of AES algorithm for data security on credit card transaction. Int J Recent Technol Eng (IJRTE) 7(5C). ISSN:2277-3878
11. Masud M, Gao J, Khan L et al (2015) Classification and novel class detection in concept-drifting data streams under time constraints. IEEE Trans Knowl Data Eng 23(6):859–874
12. Malekian D, Hashemi MR (2013) An adaptive profile based fraud detection framework for handling concept drift. In: Proceedings of international conference on information security and cryptology, Yazd, Iran, pp 1–6

13. Wei Q Yang Z, Junping Z, Yong W (2003) Mining multi-label concept drifting data streams using ensemble classifiers. In: Proceedings of IEEE international conference on fuzzy systems and knowledge discovery, Tianjin, China, pp 275–279
14. Panigrahi S, Kundu A, Sural S, Majumdar AK (2009) Credit card fraud detection: a fusion approach using Dempster C Shafer theory and Bayesian learning. Inf Fusion 10(4):354–363
15. Seyedhossein L, Hashemi MR (2011) Mining information from credit card time series for timelier fraud detection. In: Proceedings of IEEE international conference on telecommunications (IST), Tehran, Iran, pp 619–624

An Improved Travel Package Framework Utilizing (COPE)

A. Ambeth Raja and J. Dhilipan

Abstract Main classification goals are to make adapted transportable platform endorsements for the sightseers founded on the concerned POI and to progress the recognition exactness and endorsement concert. Lastly, supports toward treasure correct portable correspondences based on the POI. It progresses the endorsement scheme by choosing optimum travel suites constructed on user modified POI. Assistances to determinate the tricky provided an instinctive modified journey scheduling and plans. In this work, we pronounce the evaluations among the prevailing approaches and the planned and proposed algorithm of COPE, technique through treating interruption, exactness, and amount of repetitions besides the treating period aimed at the quantity of records.

Keywords NP-Hard technique · GA models · Journey scheduling · TSP · COPE

1 Introduction

Actually most noteworthy measure of trades finished over remote telecom for instance adaptable in winning region organizations need stretch out take to discover the nearest organizations driven by customer request, notwithstanding certain everything and outcomes are insufficient. Toward beat this issue past everything are made all through the Location recognized, misusing Locate plot customer move the zone care built data dissemination, between the different groupings of region made solicitation. To convey a noticeable tourist metropolitan [1], it incorporates cautiously taking a gander at the numerous Points to visit (PIV) to choose the PIV that not at all short of what unique inclinations, deciding the request wherein they are really to move toward becoming visited, likewise.

A. Ambeth Raja (✉)
Thiruthangal Nadar College, Selaivayal, Chennai, India

J. Dhilipan (✉)
SRM Institute of Science and Technology, Ramapuram Campus, Chennai, India
e-mail: hod.mca.rmp@srmist.edu.in

S.-L. Peng et al. (eds.), *Intelligent Computing and Innovation on Data Science*, Lecture Notes in Networks and Systems 118,
https://doi.org/10.1007/978-981-15-3284-9_11

Frameworks remain extremely helpful instruments aimed at portable manufacturing since container bolster clients in arranging extended, average and small outings, proposing areas, convenience, transport, etc. For instance, a traveler remaining in the Apulia area could be keen on investing energy visiting the primary urban areas, scenes or social occasions (celebrations, parades, uncommon markets, and so on).

The examination endeavors in this area have yielded various recommender frameworks expecting to help the vacationer in settling on the best decisions. A large portion of the recommender frameworks exhibited in writing can propose single things (flights, structures, urban communities, and so forth.) based on the traveler's solicitations. Now, the traveler ought to make her/his very own schedule, picking the things that fit her/his necessities. This suggests they produce a succession of magnetisms or commercials to be stayed, separating information as indicated by the client's requirements (day of visit, cost, etc.) determined in the happening of the solicitation.

Contextual

NP-Hard: NP-Hard hunt is grounded happening the proof that dangerous splitting in order towards prevail by method for brainy and fundamental incorporate versatile memory and congenial assessment. NP-Hard is created on announcing elastic recollection constructions in aggregation by calculated limitations and objective stages as resources for misusing exploration places (Fig. 1).

Tabu Method In Journey Scheduling: It proposes strategies for the movement bundle determination to gauge and describe the effect of agenda arranging over an enormous scale travel data set. This thinks about the most extreme use with various points to visit over various arrangements of bundles. This incorporates the estimation of everyday bundle and multiday bundle determination and investigation of point to visits. So as to improve the exactness and execution, the framework breaks down the ways and its separation to fuse the combinatorial calculations.

Toward defeat advancement issue, impact bundle circulation and bundle determination are upgraded dependent on specific components. The framework performs way

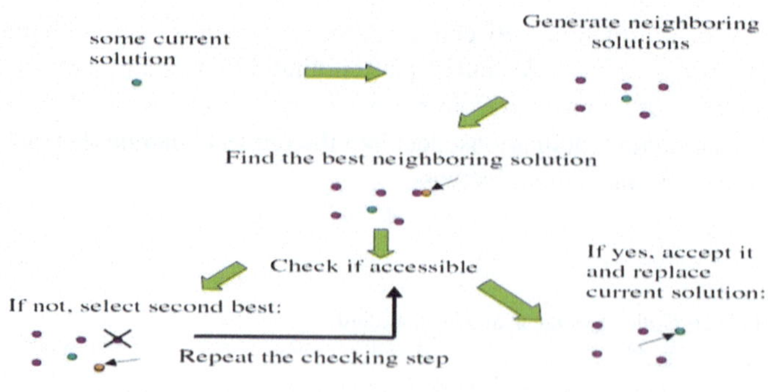

Fig. 1 Combinatorial Tabu search

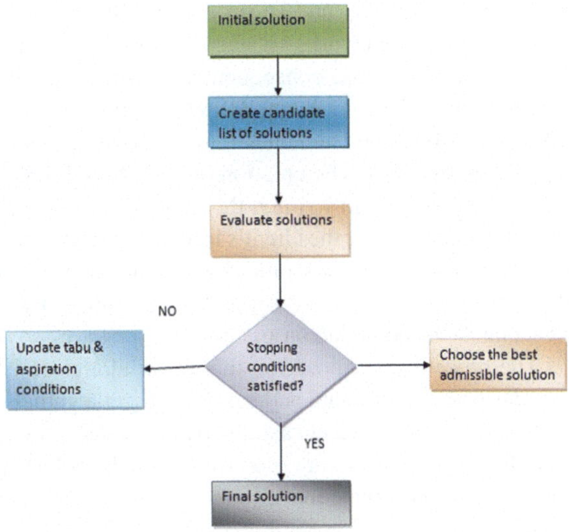

Fig. 2 Final solution with NP-Hard

examination and investigation over every point to visit and its association. All visiting points' necessity remains evaluated besides handed-off through that arrangement of circumstances besides standards.

Be that as it may, to upgrade the bundle determination with ideal arranging and dissemination, the neighborhood assessments should be ceaselessly refreshed. This will take care of by the NP-Hard problems (Fig. 2).

After the overhead second Figure demonstrations, the combinatorial problems treasure the concluding resolution through NP, wherever the clarifications determination remains specified hooked on the subsequent progression. Compassionate of journey progression removes those imprecise than inappropriate consequences of involuntary journey scheduling.

2 Related Work

Recent proposed frameworks aimed at travel industry space container stand isolated in binary primary gatherings: frameworks that propose solitary things (or travel items) and frameworks that recommend a gathering of things. They select travel goals by separating the items gathered inside a virtual list as per the client's needs and inclinations. The frameworks that recommend a gathering of things (the following gathering) bolster travelers in arranging their excursions, organizing transportation, convenience, and whatever else required, for instance of these assortments of frameworks are Expedia (www.expedia.com), DieToRecs [2, 3]. In cases, for example, this proposal procedure works like a virtual travel office: It recommends a prepackaged offer, or else it manages the client in their decision of the movement parts (flight, convenience, vehicle, and so on.) as per the client's needs and inclinations. Such recommender frameworks bolster visitors in structuring their own kind of schedules; frameworks that could consequently construct agendas starting from the traveler's solicitations still are not many. A few creators report this is frequently "because of the intricacy of the errand and questions concerning adaptability to encourage huge voyages databases" [4]. The creation of agendas requires producing an arrangement of things to be visited. In writing, the issue of creating schedules is adapted by separating information as per the client's limitations (day from visit, cost, etc.) determined in the solicitation. The Electronic Travel Planner (ETP) model [5], for instance, stores data, (for example, span, cost, and accessibility, dates and times) about movement items inside a social database.

The procedure proposed in this paper alongside its application inside a recommender framework enables the visitor to get a gathering of schedules included in a gathering of associated occasions (parade, celebration, unique market, and so forth.). The chains of occasions are produced by methods for transitive conclusion calculation of a given space-time connections.

3 Methodology

Main aspects of commitments given that researches are given as pursues: investigates utilization target highlights toward demonstrate that demonstrative impression of n number of Point to holiday has been acquired after each client [6]. That demonstration a Point to visit-based package Reclamation structure has been advances then schedules various package closeness procedures aimed at various clients. In particular, a client provided POI enables the framework to figure out which subset of a lot of target highlights approximates all the more proficiently the emotional bundle closeness of a particular client. In this segment, we talk about the correlations concerning current and suggested combinatorial technique through deference preparing interruption, exactness, and quantity of emphases besides handling time for the measure of records.

Quantity of Point to visit develops exponentially through quantity of things. In any case, this unpredictability is handled with some most recent calculations which can proficiently prune the pursuit space. Also, the issue of discovering results from guidelines, for example picking ideal outcomes from set of outputs. In general, the principle inspiration for utilizing GAs in the agenda choice is that they play out an ideal pursuit and adapt preferable to point to visit association over the covetous standard acceptance calculations regularly utilized in information mining. The utilization of COPE in agenda arranging predicts ideal bundle dependent on the given point to visit. This area talks about a few parts of GAs for agenda arranging. The primary zones of exchange incorporate individual portrayal of point to visit generally increasingly successful arrangements are chosen to have progressively off springs which are, somehow or another, identified with the first arrangements.

4 Assessment Progression

The characterization utilization's various measurements to amount the stage repossession capability along the side by the sanctioning of the techniques. For all correspondences in the data set, this acquires a level gradient of correspondences besides its price, which are calculated by all point to visit and its detachment constructed on the COPE previously signifying correspondences [7]. The concluding mark aimed at every platform is considered as the price of every compendium. Correspondence that wherever is no arranging records. The structure scholarly to improvement the associating correspondences made on the controller recognized Point to visit. In the directive to demonstrate the effectiveness of combinatorial algorithms which are Iterative Genetic Algorithm with a Tabu search and collaborative filters, this has done a series of experiments of combinatorial. There are ten sizes of the traveling salesman problem that have used. The itinerary planning for multiday employed an initial package size of 10, and a set of point to visit is 5 (Fig. 3).

4.1 Outcomes Besides Conversation

(A) **Enactment of Anticipated Classification**

In this segment quantity, the concerts of the surviving greedy before portioning the outcomes of the combinatorial optimal package evocation constructed evolutionary algorithm. Multiday journey variety accurateness calculated through associating correspondences is dispersed through the point in the table of first. After the overhead fourth diagram, demonstrations have period occupied by every phase of Combinatorial Optimal Packages Evocation. The quantity of facts quantified from 15 to 60. Aimed at every procedure, the interval deliberates and then lastly classification offers the complete period engaged by the proposed algorithm. This shows for ten POIs, the system takes 63.2 frequency interval duration (Fig. 4; Table 1).

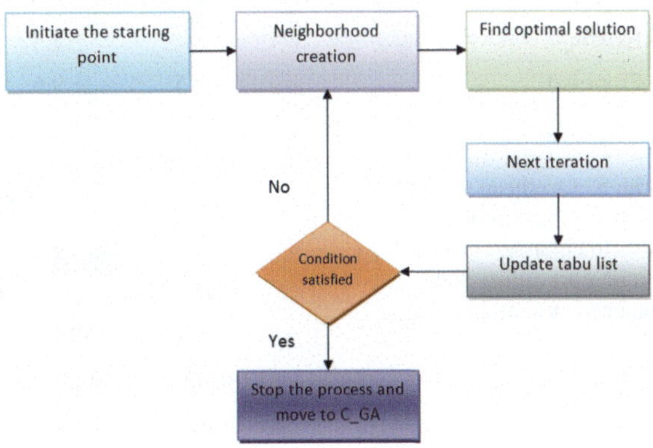

Fig. 3 Initial C_TS process

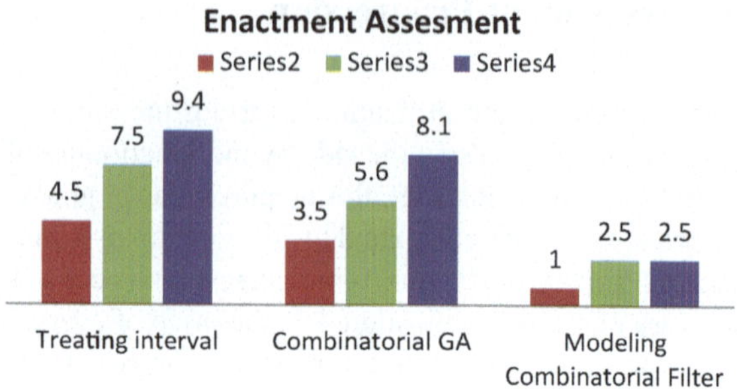

Fig. 4 Enactment assessment

Table 1 Enactment assessment

Numeral of points	Treating interval	Combinatorial GA	Modeling combinatorial filter
5	4.5	3.5	1.0
10	7.5	5.6	2.5
15	9.4	8.1	2.5

Proportional Training

Main research aspects scrutiny by succeeding diagram besides desk defines enactment assessment among prevailing and planned classifications founded on interval interruption, exactness, and point of visit consequence obtainability (Table 2).

To assess the exhibition of the intentional plans, performance period besides price is occupied as the principle of execution assessment. Deprived of damage of all-inclusive statement, it characterizes handling postponement besides grouping interruption for conveyed bunching. Preparing postponement demonstrates the execution time for way choice to create proper bundle for relating POI (Fig. 5).

Table 2 Enactment assessment (interval interruption, exactness, and point of visit outcome obtainability)

Classical name	Interval interruption	Exactness	Point to visit	Outcome obtainability
Existing	9.5	8	7.5	5.5
Proposed	7.5	8.5	9.6	9.0

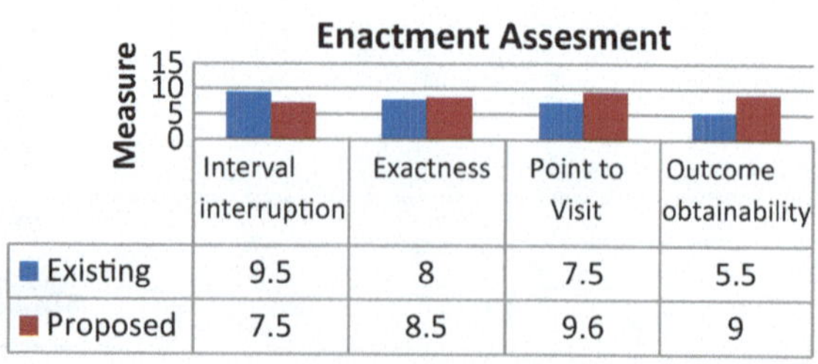

Fig. 5 Overhead diagram displays the complete enactment assessment among prevailing and planned structures

Table 3 Enactment assessment (interval interruption, exactness, and point of visit consequence obtainability)

Measures	Prevailing	Planned
Treating interruption	3604.06	2018.06
Cost	78	89
Accuracy	68.5	49.61

It additionally assessed estimating period consumed on each procedure of proposed system. Additional measure is price assessment. Price assessment includes capacity and then calculation angles, which are for most part embroil the support stockpiling and ordering expenses.

Presentation has recommended effort combinatorial utilizing traveling salesman and Genetic Structure is contrasted and current methodology covetous calculation. The assesment of interval interuption, accuracy and point of visit, user location based cost, displayed in Table 3.

Enactment assessment of anticipated COPE consuming Genetic and Tabu search with prevailing approaches is based on dispensation interruption (325 Point to Visit) (Fig. 6; Table 4).

After the sixth diagram demonstrations, the concert quantity based on the computational interruption and the suggested method of combinatorial optimum packages evocation appropriated a smaller amount period, whereas associating other approaches and the poorest interval complication are a prevailing greedy technique (Figs. 7 and 8).

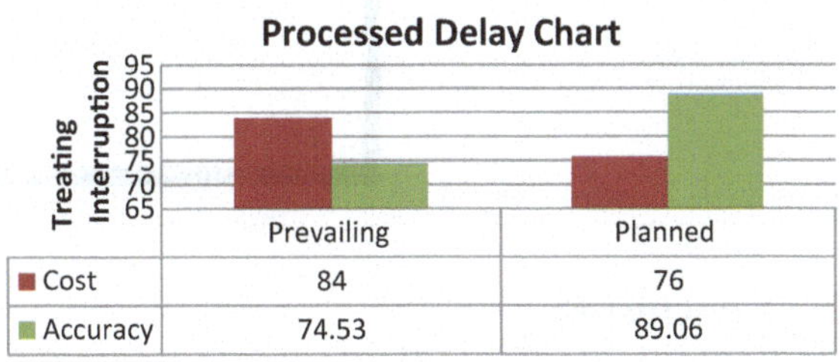

Fig. 6 Enactment quantity constructed on the computational interruption and the planned method combinatorial

Table 4 Enactment assessment of planned COPE consuming Genetic and Tabu search

Measures	Prevailing	Planned
Treating interruption (325 point to visit)	786.07	326.08
Cost	84	76
Accuracy	74.53	89.06

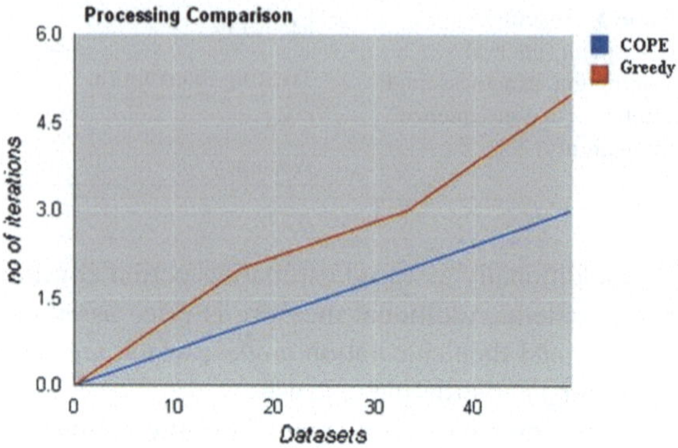

Fig. 7 Enactment assessment of planned proposed

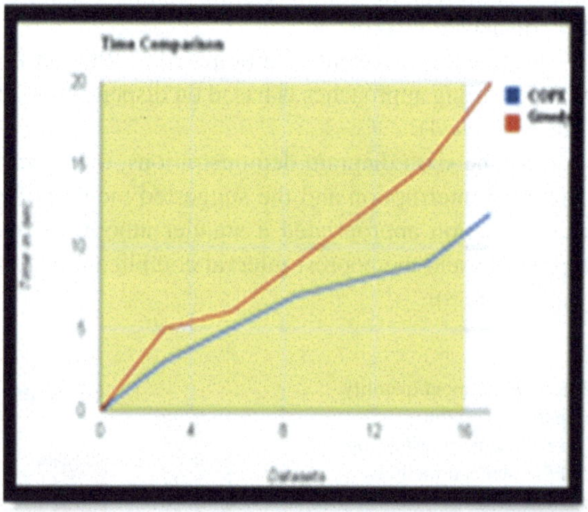

Fig. 8 Enactment assessment of recommended COPE with prevailing methods based on time

5 Conclusion

The recommended reflex journey planning and endorsement system benefits to circumvent the platform collection concern in sightseer purview. Here stand various procedural and dominion experiments essential in manipulative and executing an operative scheduler classification for modified portable platform assortment based on manipulator quantified opinion of concentration in LBS. To overwhelm the encounters trendy the development besides endorsement classification, the present effort projected a delicate classical to incredulous enormous aspects. The prescribed reflex excursion arranging and support framework advantages to dodge the stage assortment worry in tourist domain.

References

1. Adomavicius G, Tuzhilin A (2005) Toward the next generation of recommender systems: a survey of the state-of-the-art and possible extensions. IEEE Trans Knowl Data Eng 17(6):734–749
2. Ricci F, Fesenmaier DR, Mirzadeh N, Rumetshofer H, Schaumlechner E, Venturini A, Wöber KW, Zins AH (2006) DieToRecs: a case-based travel advisory system. Destination recommendation systems: behavioral foundations and applications. CABI Publisher International, Wallingford
3. Venturini A, Ricci F (2006) Applying trip@advice recommendation technology to www.visiteurope.com. In: 17th European conference on artificial intelligence. IOS Press, Amsterdam, pp 607–611
4. Dunstall S, Horn M, Kilby P, Krishnamoorthy M, Owens B, Sier D, Thiebaux S (2004) An automated itinerary planning system for holiday travel. J Inf Technol Tourism 6:1–33
5. Petrone G, Ardissono L, Goy A (2003) INTRIGUE: personalized recommendation of tourist attractions for desktop and handset devices. J Appl Artif Intell 17:687–714
6. Ricci F (2002) Travel recommender systems. J IEEE Intell Syst 17:55–57
7. Goy A, Magro D (2004) Dynamic configuration of a personalized tourist agenda. In: IADIS international conference WWW/Internet 2004. IADIS, Madrid, pp 619–626
8. Corchado JM, Pavón J, Corchado ES, Castillo LF (2004) Development of CBR-BDI agents: a tourist guide application. In: 7th European conference on case-based reasoning. Springer, Berlin, pp 547–559
9. Biuk-Aghai RP, Fong S, Si Y-W (2008) Design of a Recommender System for Mobile Tourism Multimedia Selection. In: 2nd International conference on internet multimedia services architecture and application. IEEE Press, pp 1–6
10. Biuk-Aghai RP (2004) MacauMap: next generation mobile travelling assistant. In: Proceedings of Map Asia
11. Hart PE, Nilsson NJ, Raphael B (1968) A formal basis for the heuristic determination of minimum cost paths. J IEEE Trans Syst Sci Cybern 4:100–107
12. Chen G, Wu S, Zhou J, Tung AK (2014) Automatic itinerary planning for traveling services. IEEE Trans Knowl Data Eng 26(3):514–527
13. Basu Roy S, Das G, Amer-Yahia S, Yu C (2011, April) Interactive itinerary planning. In: 2011 IEEE 27th International Conference on Data Engineering (ICDE). IEEE, pp 15–26

An Empirical Study on Neuroevolutional Algorithm Based on Machine Learning for Crop Yield Prediction

E. Kanimozhi and D. Akila

Abstract Machine learning has been come out with high performance computation power leads to create a great prospect in multi-disciplinary domain. Here, we present a novel machine learning method for predicting crop yield. The classification technique using machine learning algorithm demonstrated the performance improvement in prediction of crop yield. It depends on the factors of weather which have relationship with climate change data, soil of that area, and water irrigations. Here, we have illustrated an approach of implementing neuroevolution model based on ANN for predicting wheat crop yield. Crop yield prediction at different months is considered from June to September; the yields predictions are computed based on weather and fertilizer utilized data. A major improvement in the prediction ability is observed that yield diverge as for the season changes based on weather data. Therefore, the result of the proposed model assists in decision making in advance for planting wheat crop. The outcomes are more functional for decision making as well as in transplantation of wheat in advance with various farm activities throughout various stages of the wheat crop growing. Also, the same model can also be utilized for predicting various agricultural data such as disease prediction and weather prediction.

Keywords Neuroevolution · Prediction · Crop yield · Artificial neural network

1 Introduction

The idea of using genetic algorithm that is evolving is called "Neuroevolution." Neuroevolution algorithms are mainly utilized in analyzing network topology; neuroevolution algorithm can obtain optimal solution, and it can be used for prediction of crop yield. In India, the major part of agriculture falls in cultivation of wheat and

E. Kanimozhi (✉)
Department of Computer Science, School of Computing Sciences, Vels Institute of Science, Technology & Advanced Studies (VISTAS), Chennai, India

D. Akila
Department of Information Technology, School of Computing Sciences, Vels Institute of Science, Technology & Advanced Studies (VISTAS), Chennai, India

S.-L. Peng et al. (eds.), *Intelligent Computing and Innovation on Data Science*, Lecture Notes in Networks and Systems 118,
https://doi.org/10.1007/978-981-15-3284-9_12

wheat plays an important role in the global economy. Due to population increase, the pressure is on improvement in agri-production. Technological development in agriculture and farming has been arisen in new scientific areas that can intense in agricultural productivity. Whereas machine learning algorithm supports with some scientific approach which provides machines with the capability to learn by itself without being programmed exactly [1]. The wheat types are chosen with specific genes based on the nutrients it contain and adaptation to climate change. Machine learning algorithms help in agriculture field by analyzing the yield of the crops based on various climates, soil type, and irrigation type can help in building a predictive model that can be contributed for plantation of crops.

1.1 Role of Soil Type in Wheat Yield

Soil is natural resource that plays major role in crop yield; therefore, we considered the type of soil of that place as one of the features. The type of soil is responsible for the complex processes and indistinct mechanisms [2]. Based on the soil type, it can also have the evaporation processes, soil moisture, and temperature to recognize the mighty of ecosystems along with the impingement in agriculture [3].

1.2 Role of Irrigation in Wheat Yield

Water is also a natural resource that has an major impact on cultivation of crops, and based on irrigation type, the yield of crop differs; therefore, we have considered the type of irrigation as one of the factors for yield prediction. Water management in agriculture impacts on hydrological, climatologically, and agronomical balance (Fig. 1).

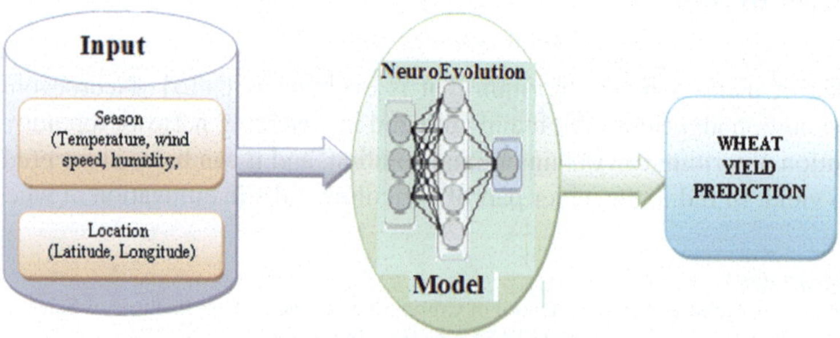

Fig. 1 Neuroevolution model for wheat yield prediction

2 Literature Review

Harvey et al. 1997 propose an controller using neural system which leads to the development of drive portable robots, cars, and rockets. This control framework advances the framework for PC development. Lucas, 2005, proposes similar methodology in developing counterfeit detection system, neuroevolution algorithm can also be applicable in developing complex practices and it can also regulate continuously [2]. Acharya et al. [1] proposes nonparametric neural network technique to predict corn yield in the location of US Midwest, and the author demonstrated that nonparametric neural network outperforms in both statistical methods as well as in fully-nonparametric neural networks while predicting the corn yield. Based on the scenarios such that climate models show large negative impacts on corn production, whereas less impacts are anticipated using statistical methods. Therefore, his approach is less gloomy in the warmest regions. Gravina et al. [4] neuroevolution technique developed based on biologically inspired for generating artificial intelligence. Many researches are conducted based on constructing effective and scalable evolutionary algorithms which leads to high level intelligence, Cognitive behaviors such as multimodal actions, communication, and lifetime learning are evolved due to scaling the neuroevolution algorithm. Intelligent robot controllers and video game controllers are all developed because of practical approaches to evolving as such as biological neural networks and the evolution of intelligence itself. Satisfying the challenges may yield better neuroevolution approach. The leading technique implemented for training neural networks is backpropagation, and it helps in calculating loss function's gradient more efficiently. This technique has demonstrated extremely effective for supervised learning, and has also produced impressive reinforcement learning results. In recent days many successful applications like image processing involved in construction of tiny neural networks using modern standards, which is composed of only few numbers of connections that are comparatively in less number than modern deep neural network (DNN) research.

3 Materials and Methods

3.1 Neuroevolution Algorithm

Neuroevolution algorithm is an artificial neural networks technique that was developed using the evolution of natural organic systems. Neuroevolution strategy is more helpful and appropriate in solving problems and it is utilized in development of evolutionary robotics: Neuroevolution can be used to explore how the knowledge changed in nature for designing artificial neural systems to achieve needed tasks. Neuroevolution has unusual strategies that lead to inconsistent spaces and fortification learning. These areas are incorporated with few provable uses that are reinforcement in learning; the most evident application is versatile, nonlinear control of physical gadgets.

3.2 *Yield Prediction Factors*

Yield prediction is mainly implemented to obtain agriculture precision; therefore, it identifies the mapping and estimation of crop yield with demand. The modern techniques have gone further than simple predictions based on the historical information, but later incorporating with computer vision technologies, it provide comprehensive information about multi-dimensional analysis of crops, and weather conditions which can help most farmers and population to improve crop yielding. Disease diagnosing in plants helps in pest and disease control by evenly spraying pesticides over the cropping area [5]. This approach needs major amounts of pesticides that lead for high financial cost. Machine learning is used as a part of the general precision in management of agricultural activities, where a Grow Chemicals are given as an input and targeted in terms of time, place, and affected plants [1] like disease, weeds also affect to crop production. The major issue in weed removing is that it is more difficult to spot and distinguish from crops. Computer vision and ML algorithms can improve the accuracy of spotting and distinguishing of weeds without affecting the crops. Similar to crop management, machine learning algorithm can provide accurate prediction for optimizing the cost-effective stock production systems.

3.3 *Factors Considered for Model Construction*

These experiments deal with prediction of wheat yields based on location and weather. The dataset consist of information on location, time, and wind speed. Season is computed using the feature such as the start date the month. The wheat yields are predicted based on those features. The model is built such that in a specific duration of months, how much wheat got yielded with that; it is predicted which month is better for wheat yield; and later, based on the production of wheat on every location which location is better suited for yield is predicted. The dataset are preprocessed by cleaning and missing or NULL/NaN values [2]. Here, weather-related data are changed with time and location. The missing value is replaced by the average value obtained from previous and subsequent days records at the target location. Features such as location (latitude and longitude) and the season (weather, temperature, humidity and month) are considered for the experiments. The final features that were used in the modeling are shown in Table 1.

Table 1 Final features

Feature	Description
Location	The geographical latitude and longitude of the location in degrees
Humidity	Average humidity of the location in g/m^3
Rainfall	Average rainfall in the location within the specific duration measured in mm
Mintemp	Minimum temperature in a 60-day period/months having minimum temperature
Maxtemp	Maximum temperature in a 60-day period/months having minimum temperature
Wind_speed	The mean value of the computed wind speed in a specific location
Mean_temp	The mean value of the differences in daily temperature
Yield	The maximum yield value for any crop at the end of the season

3.4 Experimental Setup

In the process of optimization of neural network, backpropagation plays major role and it can be replaced by an genetic algorithm, which has the capability to choose better pair, therefore, to optimize by itself. In this technique, each network tries to carry out its deliberate task which is provided with its fitness score. The next generation is obtained with better networks that are mutated and adjusted with the biases. Among various machine learning algorithms, neuroevolution algorithm has been successful in recent years and explored by better performance. It is quite complex compared with the relevant hyper-parameters number. So there was no reason to choose one over the other based on complexity [5], it is previously pointed out that pure linear models cannot be the better choice of algorithms. But regularization can assist in removing the issues regarding co-linearity, whereas introducing higher-order features helps in nonlinearity.

3.4.1 Algorithm

In neuroevolution, any mutation techniques can be applied such as add or subtract values from mutated weights.

Step 1: Input data values are taken;
Step 2: Specific features considered are computed (Location and Season);
Step 3: Neural network (neuroevolution) is constructed;
Step 4: Mutation technique is applied for the neural network;
Step 5: Random values nearest to 1 are multiplied to the biases;
Step 6: Replace previous biases value with new value;
Step 7: Negate some values for biases;
Step 8: Substitute some weight values with other weight values
Step 9: Train the Model
Step 10: Evaluate the performance using test set.

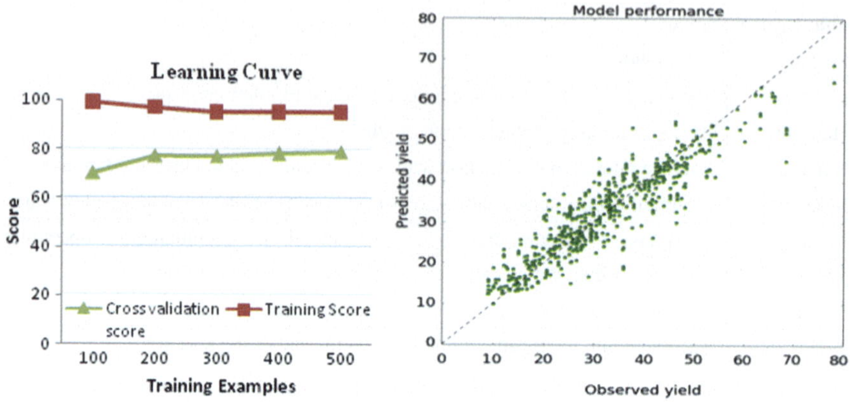

Fig. 2 Learning curve for training score and cross-validation

Various mutation techniques are employed and the results are compared [6]. The best result obtained are demonstrated as a result in Fig. 3.

4 Result and Discussion

The score curve of the proposed model demonstrates that few issues like overfitting are still happening, but in general, the performance measure and variance measure appear to be promising. Increasing the parameter in the proposed model would help in further increase in the performance score, but at the cost of increased overfitting, it likely seems to be reducing the problem of overfitting by including more training data. The proposed neuroevolution model also provides access to feature importance, and the absolute performance measure of the proposed model was depicted using a test set (80/20 split) for which compared model predictions to actual yield numbers (Figs. 2 and 3).

The proposed RMSE model yields 24% for the given testing dataset. As better yields observed with the proposed model appear to be constant under prediction, the prediction model for lesser yields seems to be well balanced.

5 Conclusion

In this study, a neuroevolution technique is proposed for predicting the wheat productivity considering various features such as location, amount of rainfall, temperature, wind speed, and humidity level [7] by considering the monthly report that gives a throughout value for the specific period. Though the prediction model for the wheat yield requires daily weather data, a disaggregation method is used for generation of

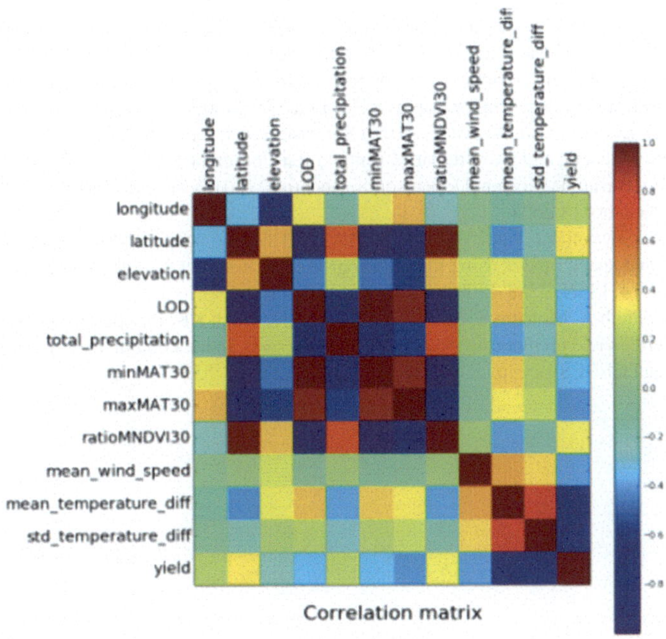

Fig. 3 Model prediction for wheat yield

such daily time series. To evaluate the wheat yield productivity at different stages of the various seasons with respect to the amount of rainfall and the temperature for the remaining period, wheat yields have been simulated. The monthly and seasonal data obtained are analyzed with wheat production, and they showed significant efficiency prediction of yields. Moreover, improvement of ability in predicting wheat productivity has been observed prior to the specific season. Due to subsequent integration of experiential weather data of successive months, uncertainty in yield prediction diminishes, as crop experiences the weather conditions that are more of observed for the predicted nature. In the present study, we used five-year data for training the proposed model. The results of the study will be useful in decision making in coastal regions of India, for choosing of suitable period for planting wheat. This study also further extends other major crops like rice, corn, sugarcane, etc., of that region to evaluate performance of these crops under expected weather conditions, which also helps in making decision for selecting the type of crop that is suitable for cultivation in monsoon season.

References

1. Acharya N, Kar SC, Kulkarni MA, Mohanty UC, Sahoo LN (2011) Multi-model ensemble schemes for predicting northeast monsoon rainfall over peninsular India. J Earth Syst Sci 120(5):795–805

2. Stanley KO, Clune J, Lehman J, Miikkulainen R (2019) Designing neural networks through neuroevolution. Nat Mac Intell 2522–5839. https://doi.org/10.1038/s42256-018-0006-z
3. Crane-Droesch A (2018) Machine learning methods for crop yield prediction and climate change impact assessment in agriculture. Environ Res Lett 13:114003
4. Gravina D, Liapis A, Yannakakis G (2016) Surprise search: beyond objectives and novelty. In: Proceedings Genetic and Evolutionary Computation Conference (GECCO), ACM, pp 677–684
5. Liakos KG, Busato P, Moshou D, Pearson S, Bochtis D (2018) Machine learning in agriculture: a review. Sensors 18:2674
6. Tejani GG, Savsani VJ, Patel VK, Savsani PV (2018) Size, shape, and topology optimization of planar and space trusses using mutation-based improved metaheuristics. J Comput Des Eng 5(2):198–214, ISSN: 2288-4300
7. Ghosh K, Singh A, Mohanty UC, Acharya N, Pal RK, Singh KK, Pasupalak S (2015) Development of a rice yield prediction system over Bhubaneswar, India: combination of extended range forecast and CERES-rice model. Meteorol Appl 22(3):525–533

Automatic Pruning of Rules Through Multi-objective Optimization—A Case Study with a Multi-objective Cultural Algorithm

Sujatha Srinivasan and S. Muruganandam

Abstract Classification algorithms create an overwhelmingly large number of rules, sometimes exceeding the number of data instances in the data set. Large number of rules hinders decision making. Therefore, there is a need for decreasing the rules before presenting it to the user. The process of removing unwanted rules is known in the data mining literature as rule pruning. In this pruning process, care should be taken to preserve the accuracy of the classifier. Mining compact classifiers with less number of rules that are also accurate and novel is a challenge. Thus, rule mining is a multi-objective optimization problem. Finding the best combination of subjective and objective metrics to present the user with compact, accurate and novel rules is the problem taken for study in the present study.

Keywords Classification · Rule mining · Rule pruning · Data analytics · Multi-objective optimization · Cultural algorithm

1 Introduction

Classification is a classical data mining process that mines data sources for patterns and presents the user with "If -Then" type rules. The rules presented to the user should be accurate and comprehensible for better decision making. Thus, the problem of rule mining requires that the number of rules presented to the user should be minimized while accuracy must be maximized. In order to present users with more comprehensible knowledge, rule pruning techniques are used that remove unwanted, redundant and more specific rules. Rule pruning should be done without compromising on the accuracy of the classifier [1]. Another challenge facing classification rule mining is that the rules produced must be interesting enough to present users with previously unseen, rare and interesting patterns. These properties of the rules form the objectives to be optimized in generating rules. The generated rules from the rule mining algorithm undergo pruning process using a variety of heuristics. The most frequently used pruning technique is rule coverage that chooses rules that cover a

S. Srinivasan (✉) · S. Muruganandam
SRM Institute for Training and Development, Chennai, Tamil Nadu, India

S.-L. Peng et al. (eds.), *Intelligent Computing and Innovation on Data Science*, Lecture Notes in Networks and Systems 118,
https://doi.org/10.1007/978-981-15-3284-9_13

maximum number of training data instances or choosing rules which satisfy a minimum threshold value of the rule evaluation metric. There are more than forty metrics as reported by Abe and Tsumoto [2, 3]. Both subjective and objective rule metrics are used for rule evaluation. The complexity of the classifier is measured using the cardinality of the classifier or the length of the rule depending upon whether the Michigan style classifier or the Pittsburg style classifier is used to represent the rule. The problem here thus reduces to that of optimizing the rules based on the metrics chosen for evaluation. Taking this as the basic idea, the present article tries to find the best combination of metrics for evaluating a classifier to mine a compact classifier with accurate and novel rules using a multi-objective cultural algorithm to mine classifiers as well as prune them on the go. The proposed methodology is an extension of the technique used in [4] to explore the possibility of different combinations of metrics for mining better rules with desired properties.

The paper is organized as follows. The second section gives the state of the art in rule pruning techniques. Section 3 gives the problem definition, the motivation behind the study and the methodology used. Section 4 explains the experiments conducted, results and discussion. Section 5 brings out a summary of the present study and future research directions.

2 Review of Literature

Pruning literally means cutting the overgrown branches of a plant or tree. Hence, removing the unwanted branches from a decision tree resulted in coining the word classifier pruning. Pruning of classifiers applies to decision trees as well as removing unwanted rules, associations or patterns from the solution space.

Dimitrijevic and Bosnjak [5] have mined association rules from web usage data using confidence and support. An efficient data structure known as frequent closed enumeration table is used in [6] to find association rule sets from previously generated rules using support and confidence values for pruning. An evolutionary approach is used in [7] for constructing associative classifiers and uses the elitist approach to prune rules. Stahl and Bramer [8] have described J-pruning techniques known as Jmax-pruning. Understandable and interpretable rules are mined from support vector machines in [9] from consistent regions in the solution space. While Manda et al. [10] have used post processing techniques using rule metrics to prune classifiers. Haralambous and Lenca [11] have used three pruning methods, whereas in [12] the authors have used time as an attribute in mining relational rules from web log data while the authors of [13] use data coverage rule pruning technique for mining business rules.

Qabajeh et al. [14, 15] propose a dynamic rule induction system that uses as frequency and rule strength to reduce the search space. The authors of [16] propose a methodology for explaining the prediction models for predicting type 2 diabetes risks in patients. The authors of [17] have proposed an improved Apriori algorithm that uses deep pruning strategies. The authors of [18] have used three standardized interestingness measures to mine and prune the rules. Pruning strategies using minimum support is used in three stages in the frequent item set mining algorithm in [19], whereas in [20] metadata in the form of metarules is used to prune the unwanted rules.

Using multi-objective evolutionary systems for rule mining is sparsely found in the literature. The proposed study is an extension of the work by [4]. The authors have used a multi-objective cultural algorithm for rule induction and have compared the performance of the algorithm using different combinations of 2, 4 and 5 metrics on three data sets. The article deals with how the proposed system is an interactive one in allowing the user to choose the different metrics of their choice. But the study lacks explanation in terms of rule pruning and further exploration in this area. Thus, an attempt has been made in this paper for pruning rules by using a good combination of rule evaluation metrics as the optimization criteria and avoids a separate post processing phase for pruning. The proposed methodology has been demonstrated using benchmark data set from UCI ML [3] repository using different combinations of accuracy and novelty metrics.

3 The Problem and Methodology

Problem definition

> Given a data set and a set of rule metrics in an evolutionary environment that produces rules and chooses dominators using Pareto optimization strategy, the challenge is finding the best combination of metrics that will enable the algorithm to mine a compact classifier with interesting rules without compromising on the accuracy of the classifier.

3.1 Methodology

An extended cultural algorithm was introduced for rule mining and is explained in detail in [4, 21, 22]. The steps in the algorithm are as follows:

1. The proposed algorithm takes a data set and creates training and test data sets.
2. The metadata is stored in the belief space consisting of the normative (NKS), domain (DKS), situation (SKS), topographical (TKS) and history knowledge source (HKS) and rule knowledge source (RKS). The normative knowledge source is initialized with the range of values of the different attributes, and SKS is initialized with the instance chosen by the user from the data set.
3. The algorithm then uses the training data to create an initial population.
4. The individuals are evaluated based on the chosen metrics, and the values are represented as the objective vectors and stored in DKS.
5. Pareto optimal optimization strategy is used for choosing better individuals.

6. The data in DKS is used to choose dominators that are stored in HKS.
7. The rule difference between pairs of rules is calculated and stored in TKS. The belief space is thus updated to add the new knowledge to the different knowledge sources.
8. The belief space data influences the agents to create new generations.
9. Steps 4 to 8 are repeated until the stopping criterion is reached. HKS contains the dominators in the chosen metrics and forms the classifier.
10. The algorithm then uses the test data to test the classifier.
11. The rules along with the metric values and the number of misclassified test data instances are presented to the user.

Objectives of the study are as follows:

- To test the influence of different types and number of rule evaluation metrics on the quality of the generated classifier
- To study the effect of introducing novelty metrics to find its influence on the classification accuracy and number of rules in the classifier
- To study the performance of the multi-objective optimization system under different number and combination of metrics in choosing dominators and generating compact classifiers.

4 Experiments, Results and Discussion

The experiments conducted for measuring the performance of the proposed methodology consisted of running sets of experiments with different number and combination of metrics. The Iris and Wisconsin breast cancer (WBC) data set from the UCI machine learning repository [3] was taken to test the algorithm. Experiments were conducted with a combination of 2, 3, 4 and 5 metrics. The hypothesis to be tested is as follows.

Hypothesis: "Use of a good combination accuracy and novelty metrics in the objective vector enables discovering a compact set of novel and accurate rules."

4.1 Results

Table 1 summarizes the results of the experiments conducted on WBC data set and Table 2 on Iris data set with various number and combinations of rule metrics in the objective vector. Figure 1 gives a comparison of using 2, 3, 4 and 5 metrics in the objective vector in returning rules, and Fig. 2 gives the comparative performance of the classifiers. The X-axis represents the combination of rule metrics represented as follows: Sup—Support, Con—Confidence, Int—Interest, Sur—Surprise and RD—Rule difference.

Table 1 Summary of results of combination of rule metrics (WBC data set)

Rule metrics used	Measures	RKS	HKS	Time (s)	Accuracy%
Sup, Con	Average	319.1	23.7	12.97	95.40
	Stdev	11.01	11.14	6.27	1.31
	Minimum	303	6	6.42	93.86
	Maximum	335	39	25.51	97.37
Int, Sur, RD	Average	317.5	16.2	13.15	94.87
	Stdev	13.79	10.53	2.15	1.26
	Minimum	297	3	7.78	92.54
	Maximum	336	35	15.46	96.49
Sup, Con, Int, Sur	Average	211	11.5	11.71	95.18
	Stdev	12.91	6.88	2.94	1.13
	Minimum	197	5	6.10	93.86
	Maximum	241	24	15.78	97.37
Sup, Con, Int, Sur, RD	Average	214.6	3.1	10.29	93.55
	Stdev	10.18	1.29	2.34	1.146
	Minimum	203	1	5.46	91.92
	Maximum	232	5	12.66	95.60

Table 2 Summary of results of combination of rule metrics (Iris data set)

Rule metrics used	Measures	RKS	HKS	Time (s)	Accuracy%
Sup, Con	Average	47.4	24.1	8.02	94.8
	Stdev	5.58	5.51	0.93	2.53
	Minimum	38	15.0	6.85	92
	Maximum	57	31.0	9.48	98
Int, Sur, RD	Average	83.1	36.9	42.42	93.6
	Stdev	3.63	18.5	3.53	3.63
	Minimum	77	4.0	34.54	88
	Maximum	88	64.0	46.93	98
Sup, Con, Int, Sur	Average	28.5	7.4	6.98	94.6
	Stdev	2.27	6.2	1.04	2.67
	Minimum	26	1.0	5.48	90
	Maximum	32	16.0	9.27	98
Sup, Con, Int, Sur, RD	Average	47.9	8.3	12.76	92
	Stdev	4.79	3.2	1.31	2.49
	Minimum	40	2.0	10.47	88
	Maximum	54	13.0	14.18	98

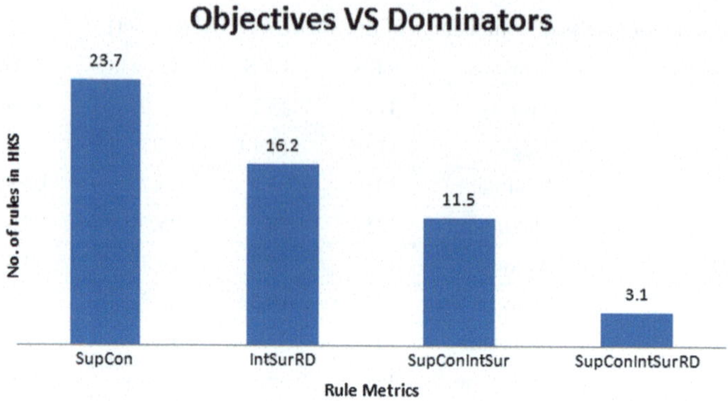

Fig. 1 Rule metrics (objectives) versus dominators (rules in HKS) for WBC data

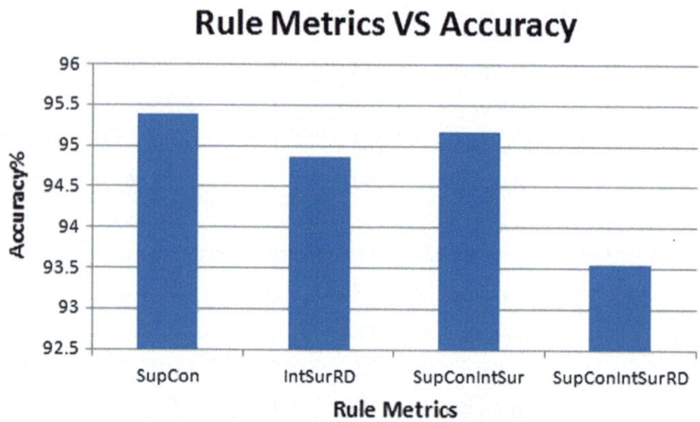

Fig. 2 Rule metrics versus accuracy (WBC data)

Table 3 gives a comparative summary of the proposed CA with other techniques in the literature. Some algorithms do not specify the time taken which are noted as Not Specified (NS), while Accuracy (Acc) is specified in some algorithms for both train and test data. Time is represented in seconds (sec).

4.2 Discussion

Tables 1 and 2 give interesting observations obtained as the result of the experiments. It can be observed that the best average accuracy of 95.40% is obtained with the combination of accuracy metrics of support and confidence. However, an accuracy of 95.18% which is much closer to the maximum obtained accuracy is observed when a combination of accuracy metrics of support and confidence along

Table 3 Comparison of the proposed algorithm with other algorithms

Reference	Pruning technique	Data set	Rule evaluation metrics	Number of rules (Average)	Acc%	Time (s)
Stahl and Bramer [8]	Jmax-pruning based on J measure	WBC	Jmeasure	24	96	NS
Zhu and Hu [9]	SVM	WBC	Volume maximization and Point coverage maximization	Train 28.4 Test 18.3	Train 94.4 Test 95.3	NS
Proposed multi-objective cultural algorithm	Multi-objective optimization using accuracy and novelty metrics	WBC	Support, Confidence, Interest, Surprise	11.5	95.18	11.61

with novelty metrics is taken for optimization. As far as compactness of the classifier is concerned, it is natural to note that the number of metrics and the number of rules in the classifier are inversely proportional. However, increasing the number of metrics compromises the accuracy of the classifier. It was also observed that when the number of metrics was increased to 6 the algorithm literally produced no rules in the historical knowledge source which forms the Pareto optimal front. Similarly, it is observed from Table 2 that the maximum number of 336 unique rules was produced when three metrics were taken, closely followed by a maximum of 335 unique rules when two metrics were taken. However, the number of unique rules drastically reduced to 241. The time taken for choosing rules was reduced as the number of metrics increased. It can also be observed from the table that taking novelty metrics alone did not produce a classifier that was both accurate and compact. Finally from the table, it can be observed that a combination of accuracy and novelty metrics produces a compact set of 11.5 rules without compromising on the accuracy of the classifier. From Figs. 1 and 2, it can be observed that the dominators in the classifier are inversely proportional to the number of metrics taken in the objective vector for optimization. This gives us the insight that novelty metrics can be included in the objective vector for optimization of the classifier without affecting the accuracy of the classifier and in returning a novel and compact rule set.

Table 3 gives a comparative summary of the performance of the proposed algorithm with other recent algorithms that have also used the WBC data set. There are only a few studies which try to investigate deeply in the area of rule pruning with multi-objective optimization. Therefore, the proposed algorithm was compared with other algorithms using Michigan style rules which use WBC data set for comparison purposes. From Table 3, it can be observed that Stahl and Bramer [8] have reported an accuracy of 96% on Wisconsin data set. However, as observed from their result, the number of rules generated by their method is 28 as compared to 11.5 rules by the proposed cultural algorithm also with an average accuracy of 95.18. Again Zhu and Hu [9] have used volume maximization and point coverage maximization as rule evaluation criteria and have obtained an average number of 28.4 and 18.3 rules with an accuracy of 94.4% and 95.3% on train and test data sets, respectively.

Finally, it could be concluded that the combination of accuracy metrics of support and confidence and novelty metrics of interest and surprise produced the best set of rules without affecting the accuracy of the classifier. This has also been observed by [23] who have proposed a combination of five complimentary criteria chosen from correlation studies that enables a complete way of evaluating rules. The chosen criteria proposed by them include Support, Confidence, Jmeasure, Interest and Surprise. Thus, the hypothesis that "A *good combination of objectives in the fitness vector improves compactness and accuracy of the rules*" has been proved.

5 Conclusion and Future Research Directions

The study proposed a multi-objective cultural algorithm for rule mining which uses the Pareto optimal optimization strategy for rule pruning. The algorithm was tested on benchmark sets and the results reported. The hypothesis that a good combination of rule metrics in the objective vector will produce a compact and novel set of classifier that is also accurate in classifying unknown data instances was tested. It was observed that a combination of accuracy and novelty metrics produced better results. The study takes Support, Confidence, Interest, Surprise and Rule Difference as metrics. However, there are more than forty rule evaluation indices. Therefore, studying other combination of metrics on different datasets are further considerations for future research.

References

1. Zaïane OR, Antonie M-L (2005) On pruning and tuning rules for associative classifiers. In: Knowledge-Based Intelligent Information Engineering Systems, vol 3683, pp 966–973. https://doi.org/10.1007/11553939
2. Abe H, Tsumoto S (2008) Analyzing behavior of objective rule evaluation indices based on pearson product-moment correlation coefficient. In: Foundations of intelligent systems, pp 84–89
3. Bache K, Lichman M UCI Machine Learning Repository. http://www.ics.uci.edu/~mlearn/MLRepository.html
4. Srinivasan S, Ramakrishnan S (2012) Cultural algorithm toolkit for multi-objective rule mining. Int J Comput Sci Appl 2:9–23
5. Dimitrijevic M, Bosnjak Z (2010) Discovering interesting association rules in the web log usage data. Interdiscip J Inf Knowl Manag 5:191
6. Wu C-M, Huang Y-F (2011) Generalized association rule mining using an efficient data structure. Expert Syst Appl 38:7277–7290. https://doi.org/10.1016/j.eswa.2010.12.023
7. Ibrahim SPS, Christopher JJ (2011) An evolutionary approach for ruleset selection in a class based associative classifier 50: 422–429

8. Stahl F, Bramer M (2012) Jmax-pruning: a facility for the information theoretic pruning of modular classification rules. Knowl -Based Syst 29:12–19. https://doi.org/10.1016/j.knosys.2011.06.016
9. Zhu P, Hu Q (2013) Rule extraction from support vector machines based on consistent region covering reduction. Knowl -Based Syst 42:1–8. https://doi.org/10.1016/j.knosys.2012.12.003
10. Manda P, McCarthy F, Bridges SM (2013) Interestingness measures and strategies for mining multi-ontology multi-level association rules from gene ontology annotations for the discovery of new GO relationships. J Biomed Inform 46:849–856. https://doi.org/10.1016/j.jbi.2013.06.012
11. Haralambous Y, Lenca P (2014) Text classification using association rules, dependency Pruning and Hyperonymization 16
12. Khairudin NM, Mustapha A, Ahmad MH (2014) Effect of temporal relationships in associative rule mining for web log data. Sci J 2014:1–19. https://doi.org/10.1155/2014/813983
13. Kliegr T, Kuchař J, Sottara D, Vojíř S (2014) Learning business rules with association rule classifiers. In: Bikakis A, Fodor P, Roman D (eds) Rules on the web. From theory to applications. In: Proceedings on 8th international symposium, RuleML 2014, Co-located with the 21st European conference on Artificial Intelligence, ECAI 2014, Prague, Czech Republic, Springer International Publishing, Cham, pp 236–250, 18–20 August 2014
14. Qabajehb I, Chiclana F, Thabtah F (2015) A classification rules mining method based on dynamic rules' frequency. In: Proceedings of IEEE/ACS International Conference on Computer Systems and Applications, AICCSA 2016 July
15. Qabajeh I, Thabtah F, Chiclana F (2015) A dynamic rule-induction method for classification in data mining. J Manag Anal 2:233–253. https://doi.org/10.1080/23270012.2015.1090889
16. Luo G (2016) Automatically explaining machine learning prediction results: a demonstration on type 2 diabetes risk prediction. Heal Inf Sci Syst 4:2. https://doi.org/10.1186/s13755-016-0015-4
17. Li L, Li Q, Wu Y, Ou Y, Chen D (2018) Mining association rules based on deep pruning strategies. Wirel Pers Commun 102:2157–2181. https://doi.org/10.1007/s11277-017-5169-0
18. Shaikh MR, McNicholas PD, Antonie ML, Murphy TB (2018) Standardizing interestingness measures for association rules. Stat Anal Data Min 11:282–295. https://doi.org/10.1002/sam.11394
19. Han X, Liu X, Chen J, Lai G, Gao H, Li J (2019) Efficiently mining frequent itemsets on massive data. IEEE Access 7:31409–31421. https://doi.org/10.1109/ACCESS.2019.2902602
20. Djenouri Y, Lin JCW, Djenouri D, Belhadi A, Fournier-Viger P (2019) GBSO-RSS: GPU-based BSO for rules space summarization. In: Advances in intelligent systems and computing, pp 123–129
21. Srinivasan S, Ramakrishnan S (2012) Nugget discovery with a multi-objective cultural algorithm. Comput Sci Eng An Int J 2:11–25
22. Srinivasan S, Ramakrishnan S (2013) A social intelligent system for multi-objective optimization of classification rules using cultural algorithms. Computing 95:327–350. https://doi.org/10.1007/s00607-012-0246-4
23. Khabzaoui M, Dhaenens C, Talbi E-G (2008) Combining evolutionary algorithms and exact approaches for multi-objective knowledge discovery. RAIRO Oper Res 42:415–431. https://doi.org/10.1051/ro:2008004

A Machine Learning Approach in Medical Image Analysis for Brain Tumor Detection

K. Aswani, D. Menaka, and M. K. Manoj

Abstract Brain tumor is a serious medical condition if not detected early will reduce the life span of humans. Magnetic resonance imaging (MRI) is a common method nowadays for abnormality detection and classification. But the manual detection is less accurate also a large amount data to be processed. Thus, manual detection leads to error in tumor segmentation. The data obtained from MRI images also inherent to noise produced by the MRI machine parts. The detection and removal of this noise play a vital part in the detection accuracy of tumor. So, here, we propose a Total Variation (TV) homomorphic filter to reduce the noise and enhance the edges of MRI data. SVM classifier is employed for learning and classification. The method produces better results than conventional methods like median filtering, anisotropic filtering, etc.

Keywords MRI · Total Variation · Machine learning · SVM

1 Introduction

Brain tumor is found to be two types low grade gliomas (LGG) and high grade gliomas (HGG). In this, HGG is very serious if not treated will reduce the life time of an affected person considerably. Based on the type and location of tumor the symptoms may vary. Losing weight, fever, and fatigues are some of the symptoms of tumor.

The majority of research in developed countries shows that the number of people having brain tumors was killed due to the fact of inaccurate detection. Among cancer, brain tumor is the most rapidly developing cancer and, therefore, its correct and early

K. Aswani (✉)
Research Scholor, Noorul Islam Center for Higher Education, Kanyakumari, India

D. Menaka
Department of AE, Noorul Islam Center for Higher Education, Kanyakumari, India

M. K. Manoj
Department of ECE, MEA Engineering College, Malappuram, India
e-mail: manojmk@meaec.edu.in

S.-L. Peng et al. (eds.), *Intelligent Computing and Innovation on Data Science*, Lecture Notes in Networks and Systems 118,
https://doi.org/10.1007/978-981-15-3284-9_14

identification will be a challenge. Brain tumor is a major type of cancer. The cause of brain tumor is abnormal growth of cells in the brain. The growth and death of cells in brain are balanced. If that balance alters, brain tumor occurs. These tumors may be broadly classified into two, namely malignant (cancerous) or benign (non-cancerous). Gliomas are a common and malignant tumor, which may lead to short life span in their highest grade. Based on the type and location of tumor, the symptoms may vary.

Gliomas are cancerous tumors which can be further classified into high grade gliomas (HGG) and low grade gliomas (LGG) [1, 2]. In these, HGG is deadly compared to LGG. HGG can reduce the life span of a person to less than a year even if it is detected [3]. The early detection of cancer will certainly improve the life of an oncology patient, which is a major step of treatment. There are a lot of techniques to detect cancer but most of them detect the cancer in advanced stage, so the chance of recovery of the patient will be less. Due to overlapped structure of cancer cells, the early detection of tumor is challenging. If the affected person is in an advanced stage, Doctors suggest surgery, chemotherapy, or Radiotherapy as treatments to cure the disease.

By using magnetic resonance imaging (MRI), early detection of tumor is possible. But the large amount of MRI data to be processed is an overhead. Thus, manual detection will be challenging. Image processing is used to segment the tumor parts in MRI images accurately. For early detection of tumor, we use biopsy, expert's opinion, etc., human prediction of tumor is less accurate and the existing biopsy test may take one or two weeks to produce the result. Thus, automatic detection of tumor using image processing techniques is getting popularity.

The major benefit of using image processing technique is the time for detection will comparatively much lesser than manual detection. The location of the tumor can be detected by using image processing on MRI data. But the noise produced in the MRI data leads to false identification of tumors [4]. So to reduce the noise present in MRI data, novel methods have to be used. Anisotropic filters are a good example for noise removal. But here, we took Total Variation (TV) as a method which is better than anisotropic filtering [5]. For detecting and classifying the features in an MRI data, many techniques are used like support vector machine (SVM), nearest neighbor (K-NN) [6], neural networks (NN) [7], deep learning-based convolutional neural network (CNN) [8, 9], etc. Before applying classification techniques, pre-processing techniques should be applied.

2 Literature Survey

Natarajan and Krishnan [10] suggest a simple thresholding-based tumor identification. First, the image is preprocessed to remove noise. Median filtering is employed for noise reduction and edge sharpening is also employed. Simple thresholding is used to segment the image. Morphological operations are also done to fine-tune the

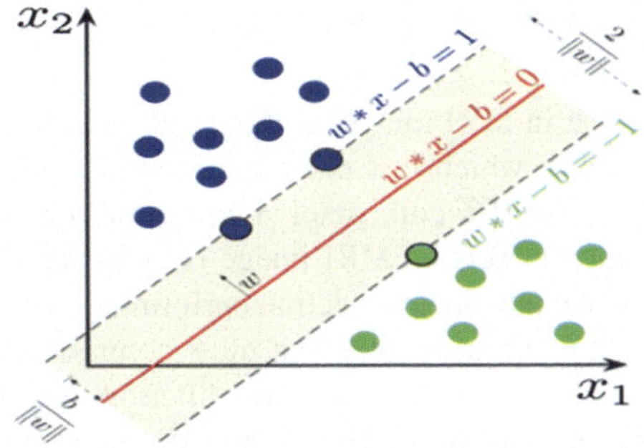

Fig. 1 A typical linear SVM classifier

result. Tumor is identified using image subtraction method. The method fails due to the luminance invariance present in the MRI data.

Sahoo et al. [11] use support vector machine (SVM) for tumor classification. SVM is used either as a classifier or it is used for regression. In classification, it creates a hyper plane to separate the features obtained from an image. Figure 1 shows the SVM hyper plane for classification. SVM is also be used for regression. SVM algorithms analyze and recognize the special patterns present in the data. When the data is given, the hyper plane is used to separate the features into two, thus it works as binary classifier [12].

The major disadvantages of SVM are the optimal features are not easily identifiable when there is nonlinearly separable data is present and the method is likely to give poor performance if the number of features is much less than the number of samples.

Arriaga-Gomez et al. [13] proposed k-NN as one of the major distance-based algorithms where given k as a positive integer and a sample feature vector (sample template), the k training features with the smallest distance to the sample is selected. The sample is identified as the most repeated among the selected k feature vector.

Ramteke and Monali et al. [14] proposed automatic classification for medical images. Abnormality detection in medical images is performed to classify whether tumor is detected or not. The major advantages of k-NN are simpler to implement and understand. The result will not be accurate always as it determines its class assignment by either getting a majority vote for them or averaging the class numbers of nearest k points.

Pereira et al. [15] use convolutional neural networks (CNN) to produce some great results. Convolving an image with kernels to obtain lower level to higher level features is the main application of CNN. In paper [8], single-layered CNN is used so that the features used for classification is less when compared to deep networks.

Deep learning methods use highly complex algorithm to extract the features automatically from the data provided. The importance is given to developing the architecture of the network rather than finding special features in the data. Thus, here, we propose machine learning-based SVM for classification and Total Variation (TV) filtering for noise removal. The rest of the paper is organized as system model in the next section and result and discussion in fourth and conclusion.

3 System Model

Noise in MRI images is due to the malfunction of the equipment and the environment in which it is placed. The MRI equipment noise is caused by field strength, RF pulse, RF coil, voxel volume, and receiver bandwidth. The most common noise which affects an MRI image is Gaussian or salt and pepper. Both of these noises reduce the tumor detection performance of SVM classifier. Several de-noising techniques are available. The most common methods are median filtering, histogram equalization, and anisotropic filtering. But for de-noising and to enhance MRI data, we propose Total Variation (TV) filtering. Method like median filtering or linear smoothing reduces the effect of noise but affect import details such as edges. But TV filtering is based on the fact that images affected by sharp or spurious noise will have high variation. So the absolute gradient of the image will be high on these regions. In order to reduce the affect noise, the gradient is to be minimized. This reduces the noise but also keeps edge information than ordinary noise removal algorithms.

The Total Variation de-noising is expressed as

$$y = x + n; \quad y, x, n \in R^n \tag{1}$$

where x is the original image and y is the image corrupted by noise n. To estimate x from y is by minimizing the objective function

$$E(x) = \min_x \|y - x\|_2^2 + \lambda \|\nabla x\| \tag{2}$$

where λ is called the regularization parameter which controls the smoothing operation performed on the image and ∇ is the gradient operator.

The entire process in the proposed method is shown in Fig. 2.

3.1 *Tumor Image*

A publically available dataset is used for training and testing. The database contains 789 tumor images of four different categories. The images are in ordinary PNG format and can be easily processed.

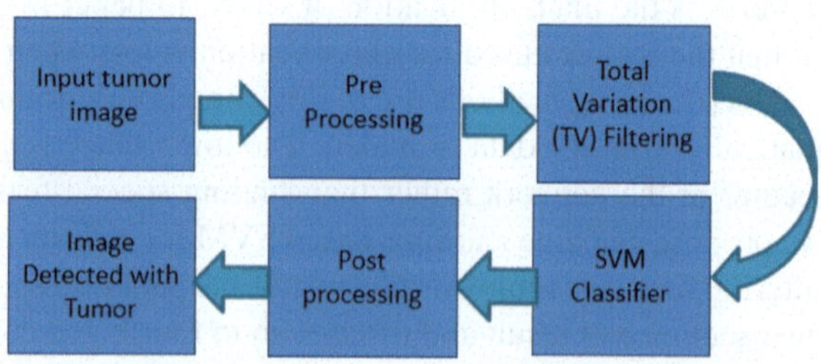

Fig. 2 Block diagram of proposed method

3.2 Preprocessing

Medical data is difficult to obtain. Also the data obtained will be taken by MRI machine using different conditions. So illumination variation is inherent. Preprocessing of the images reduce the illumination variation. Here, normalization of the image is performed. The pre-processing operation is shown below

$$I(x, y) = \frac{I(x, y) - \mu(x, y)}{\sigma(x, y)} \tag{3}$$

where μ is the mean value of the image and σ is the variance. The process converts the image to zero mean and unit variance.

3.3 Total Variation

Total Variation filtering is performed on the normalized image using Eq. 2. It is based on the principle that images having spurious noise have high total variation. So the gradient of the image is high. Reducing the gradient also minimize the total variation, hence noise.

The process is iterative. We did 100 iterations per image. Noises like Gaussian and salt and pepper are successfully reduced. Here, λ is the regularization parameter controlling the amount of de-noising, smaller value implies more aggressive de-noising and hence smoothed results. Some examples on tumor images have shown below. Fig. 3 is the noisy MRI image.

Figure 4 clearly shows a smaller value of $\lambda = 0.01$ with 100 iterations reducing the Gaussian noise considerably.

Fig. 3 Noisy image

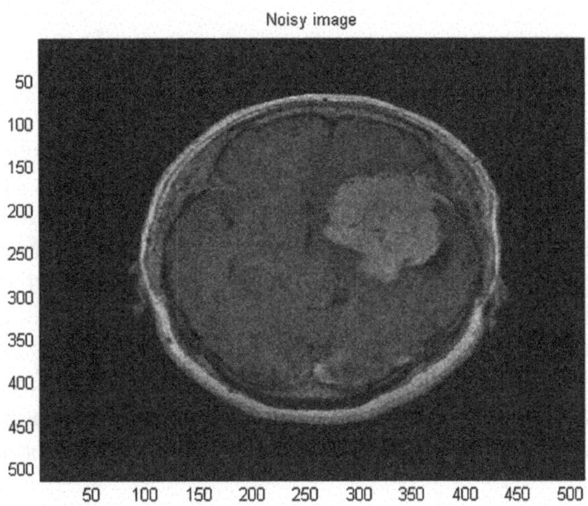

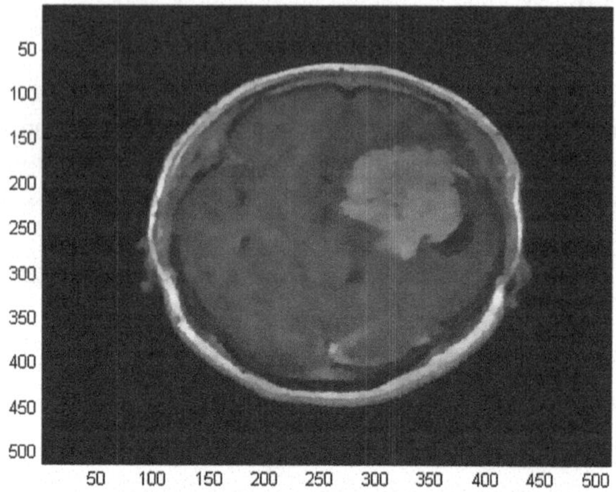

Fig. 4 Total variation de-noising with $\lambda = 0.01$

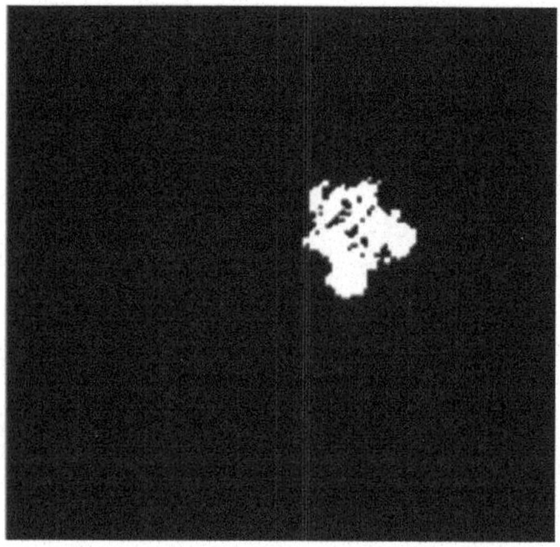

Fig. 5 Tumor area detected using TV de-noising, and SVM classifier

3.4 SVM Classifier

Machine learning is an application of artificial intelligence which learns from the observed data rather programed for finding a result. Machine learning makes use of patterns in the data for learning and classification. Machine learning can be classified into supervised learning and unsupervised learning. In supervised learning, a previously learned data and their labels are used for the future events prediction. But in unsupervised learning, there is no previous data or labels. Unsupervised learning is mainly used for regression analysis.

SVM comes under supervised learning models. Here, we used a binary SVM classifier which makes use of pixel intensity for classification. A threshold value is used for labeling the pixels either as tumor pixels or non-tumor pixels. The dataset contains 789 tumor images and training of SVM classifier is performed on 600 images. To get more images, data augmentation can also be performed.

The learned model is applied on the remaining images and got good results. Along with total variation de-noising, SVM performs better than common methods like median filtering and machine learning. The training is done with fivefold cross-validation for better accuracy. Some of the results obtained are shown below.

The above results in Fig. 6 show the need for removing the noise present in the MRI data before going for tumor detection. Figure 6 leads to inaccurate results in the presence of noise. Figure 5 shows better tumor segmentation with TV de-noising. Some morphological operations are performed if required after the classification. Figure 7 shows the classifier result for TV de-noising.

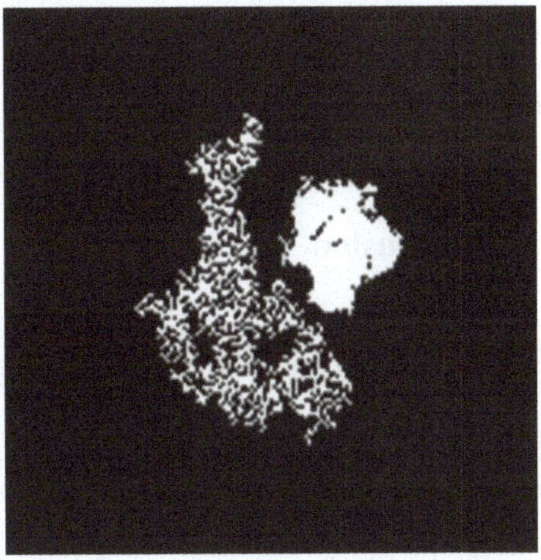

Fig. 6 Tumor area detected in the presence of noise and SVM classifier

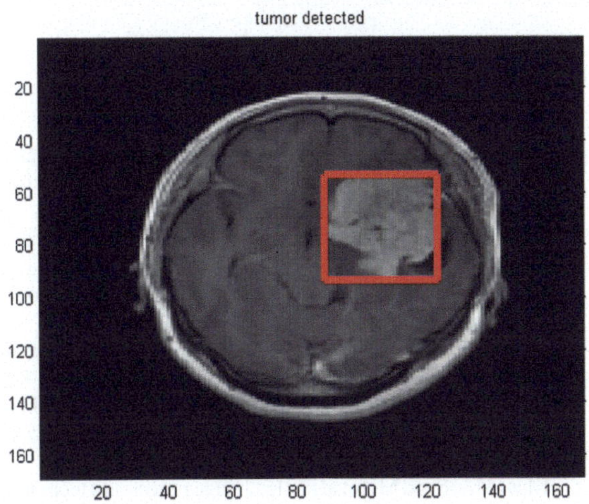

Fig. 7 SVM classification

Table 1 Comparison of SVM with de-noising and without de-noising along with median filtering and anisotropic filtering

Method	Sensitivity	Specificity	Accuracy
SVM with noise	0.85	0.75	0.75
SVM + TV-proposed	0.90	0.98	0.92
SVM + Median Filter	0.75	0.78	0.78
SVM + Anisotropic	0.90	0.95	0.90

4 Experimental Setup and Discussion

The proposed method is validated on a publically available database. The dataset contains around 3000 tumor images along with the ground truth data. The proposed method is validated on the first dataset contains 766 images with tumor. Table 1 shows the average value of evaluation parameters for 769 images. Six hundred images are used for training the classifier and 169 images are used for testing. The result obtained is compared with SVM classifier.

5 Conclusion

The factors causing MRI noise is mainly due to the hardware. So it can not be avoided. Noisy data can lead to false tumor detection and leads to inaccurate results. The only method is to reduce the noise before going for tumor classification. Since SVM is a pixel-based binary classifier, each pixel is important. The proposed method outperforms the commonly available noise reduction methods like median filtering, anisotropic diffusion, etc. The obtained result clearly shows the advantage of the proposed method.

References

1. Bauer S (1892) A survey of MRI-based medical image analysis for brain tumor studies. Phys Med Biol 58(13): R97: Clerk Maxwell J (1892) A treatise on electricity and magnetism, 3rd edn., vol 2. Oxford, Clarendon, pp 68–73
2. Louis DN (2007) The 2007 who classification of tumors of the central nervous system. Acta neuropathologica 114(2):97–109
3. Van Meir EG (2010) Exciting new advances in neyrooncology: the avenue to a cure for malignant glioma. CA Cancer J Clin 60(3):166–193
4. Kamavisdar P, Saluja S, Agrawal S (2013) A survey on image classification approaches and techniques. Int J Adv Res Comput Commun Eng 2(1):1005–1009
5. Parveen, Singh A (2015) Detection of brain tumor in MRI images, using combination of fuzzy c-means and SVM. In: 2015 2nd International Conference on Signal Processing and Integrated Networks (SPIN)
6. Khan MA, Syed MN et al (2015) Image processing techniques for automatic detection of tumor in human brain using SVM. Int J Adv Res Comput Commun Eng 4(4)

7. Kamavisdar P, Saluja S, Agrawal S (2013) A survey on image classification approaches and techniques. Int J Adv Res Comput Commun Eng 2:1005–1009
8. Elleuch M, Tagougui N, Kherallah M (2015) Arabic handwritten characters recognition using deep belief neural networks. In: 12th International Multi-Conference on Systems, Signals & Devices (SSD), IEEE pp 1–5
9. Das D, Chakrabarty A (2016) Emotion recognition from face dataset using deep neural nets. In: 2016 International Symposium on Innovations in Intelligent Systems and Applications (INISTA), IEEE
10. Natarajan P, Krishnan N, Kenkre NS, Nancy S, Singh BP (2012) Tumor detection using threshold operation in MRI brain images. In: 2012 IEEE International Conference on Computational Intelligence & Computing Research (ICCIC), pp 1–4, 18–20 Dec 2012
11. Sahoo L (2016) Alternate machine validation of early brain tumor detection. In: Information Communication and Embedded Systems (ICICES)
12. Dhaware C, Wanjale KH (2013) Survey on image classification methods in image processing. Res Comput Commun Eng 2(1):1005–1009
13. Arriaga-Gomez MF et al (2014) A comparative survey on supervised classifiers for face recognition. In: 2014 International Carnahan Conference on Security Technology (ICCST), IEEE
14. Ramteke RJ, Monali YK (2012) Automatic medical image classification and abnormality detection using k-nearest neighbour. Int J Adv Comput Res 2(4):190–196
15. Pereira S et al (2016) Brain tumor segmentation using convolutional neural networks in MRI images. IEEE Trans Med Imag 35(5):1240–1251

A Review of Recent Trends: Text Mining of Taxonomy Using WordNet 3.1 for the Solution and Problems of Ambiguity in Social Media

Ali Muttaleb Hasan, Taha Hussein Rassem, Noorhuzaimi Mohd Noor, and Ahmed Muttaleb Hasan

Abstract Text processing has been playing a great role in information retrieval to solve the problem of ambiguity in natural language processing, e.g., internet search, data mining, and social media. In semantic similarity, it will be used to analyze the relationships between Word-Pairs on social media. Organizing a huge number of unstructured text documents into a small number of concepts of word sense disambiguation is essential so that the lexical source could incorporate the features for capturing more semantic evidence. Text mining involves the pre-processing of documents collections, text categorization and classification, and extracting information and terms from golden standard data sets. This work proposed the lexical sourced from the semantic representation. The paper contained an evaluation of the advanced measures, which include shortest path, depth, and information content measures. In this paper, we used the same set of measures as previous studies, but different methods such as taxonomy on social media by semantic similarities, such as Synonymy (https://github.com/alimuttaleb/Ali-Muttaleb/blob/master/Synonym.txt), Non-taxonomy, Hypernym, and Glosses. This paper has focused to address the synonymy and ambiguity by incorporating the knowledge in the lexical resources. Thus, each word in a document is linked to its corresponding concept in the lexical resources. To build the semantic representation, these approaches can be classified into two main approaches: knowledge-based and statistical approaches. The knowledge-based approaches depend on structured information that is normally available in forms of dictionaries, thesaurus, lexicons, WordNet 3.1, and ontologies.

A. M. Hasan · T. H. Rassem · N. M. Noor (✉) · A. M. Hasan
Faculty of Computing (Fkom), University Malaysia Pahang, DarulMakmur, Gambang, 26300 Kuantan, Pahang, Malaysia
e-mail: nhuzaimi@ump.edu.my

A. M. Hasan
e-mail: alimatlab65@yahoo.com

T. H. Rassem
e-mail: tahahussein@ump.edu.my

A. M. Hasan
e-mail: ahmed.matlab11@gmail.com

S.-L. Peng et al. (eds.), *Intelligent Computing and Innovation on Data Science*, Lecture Notes in Networks and Systems 118,
https://doi.org/10.1007/978-981-15-3284-9_15

The statistical approaches are based on finding the semantic relations among words using the frequencies of words in a given corpus.

Keywords Semantic similarity · Shortest path measures · Depth measures · Information content measures · WordNet 3.1 · Text mining · Knowledge-based approach

1 Introduction

The starting of the internet coincided with that of social media networks like Facebook, Tweeter, and LinkedIn in 2003. Following the emergence of YouTube in 2005, just a few dozen Exabyte of information existed on the Web at the time. However, nowadays, great volume of information is churned out weekly on social media. As such, the advantage of social media is to offer people new and improved contents including sharing information that would enable them to capture and share their own stories, opinions, and ideas on everyday life issues in time with nearly millions of people who are connected to the World Wide Web in a cost-efficient way. In addition, text mining plays an important part in natural language processing (NLP).The work of measures used to determine similarity of semantic is to solve the issue of artificial intelligence in the analysis of the text of semantic representation in order to find the ambiguity between words. This has received wide usage in natural language processing [1], information retrieval [2], word sense disambiguation to reduce the similarity between words [3, 4], information extraction [5], and many others. In contemporary times, there is an enticing great concern for the measures based on WordNet. They display their talent and make this application more intelligent [6–9]. Several parameters have been proposed to determine the similarity of semantic. Accordingly, these measures are grouped into three classes such as depth, shortest path, and information content parameters, in line with the new semantic of relations of Synonymy, Non-taxonomy,[1] Hypernym,[2] and Glosses.[3] This paper examines the characteristics, advantages, performance, and shortcomings of different methods, and finally proposes areas for subsequent research.

The paper is subsequently structured in this order: First, Sect. 2 discusses text mining in social media while Sect. 3 presents literature review and problem identification. In Sect. 4, the WordNet 3.1 is presented while there is a discussion on semantic similarity measure based on text mining of taxonomy using WordNet 3.1 in Sect. 5. While comparison and evaluation of the analysis of the advanced measure are provided in Sect. 6, a description of the characteristics, performance, advantage, and shortcoming, a conclusion as well as more research is done in Sect. 7.

[1] https://github.com/alimuttaleb/Ali-Muttaleb/blob/master/Non-taxonomy.

[2] https://github.com/alimuttaleb/Ali-Muttaleb/blob/master/Hypernym.

[3] https://github.com/alimuttaleb/Ali-Muttaleb/blob/master/Glosses.

2 Text Mining on Social Media

The knowledge-based text mining on data mining is a method that uses the lexical source to represent the semantics of the textual documents as the drive over the notions from the lexical source used, rather than over the words derived from the joint information, since the lexical source has a huge number of concepts which is normally more than the number of words in the collocations of documents. However, the knowledge-based text mining method leads to high-dimensional semantic representation. To reduce this high-dimensional semantic representation, the feature-based method is proposed in semantic features to incorporate the features for capturing more semantic evidence from lexical source in order to locate the ambiguity between words of the concepts. Although few methods [10] have been recommended to address the great dimensionality issue in the semantic representation, they depend on external tools and ignore the structural features of the semantic network itself. To achieve correspondence with the notions in the knowledge base, the central thought in this research behind knowledge-based methods is that the documents are evaluated to map individual term in the documents. In this mapping, synonyms words in distinct tokens with related meaning in a certain event are documents plotted to specific concepts. However, the experimental results of previous studies are inconclusive. Based on the previous studies, we have created the same setting but different method as we added for methods "Synonymy, Non-taxonomy, Hypernym, and Glosses." It is worth noting that some studies [11, 12] demonstrated that exploiting lexical sources is helpful for document clustering, while others [13, 14] showed that the conventional text clustering technique without lexical sources outperforms knowledge-based techniques. This is because of the absence of uniformity in comparing the previous experimental results in terms of the exploited knowledge sources. In this paper, we examined the features based while placing attention purely on the semantic representation of the concepts. Nevertheless, this method has made a tacit supposition that all characteristics in the semantics related representation of specific concepts have the same weight of ambiguity. For instance, in the semantic of the *car*, this method deals with the features (*vehicle, convertible, accelerator, train, and cable car*) as they have the same semantic associations with the concept of the *car*.

3 Literature Review and Problem Identification

After reviewing previous literature of the present study, recent works about using the lexical sources for representing the textual documents in text mining task are considered elementary and essential phase in the current research. It is an important phase, which can help formulate the research hypotheses to study the cause and effect relationships for each experiment in the research. In this phase of the study, framework of the study is recognized based on the studies conducted on the previous

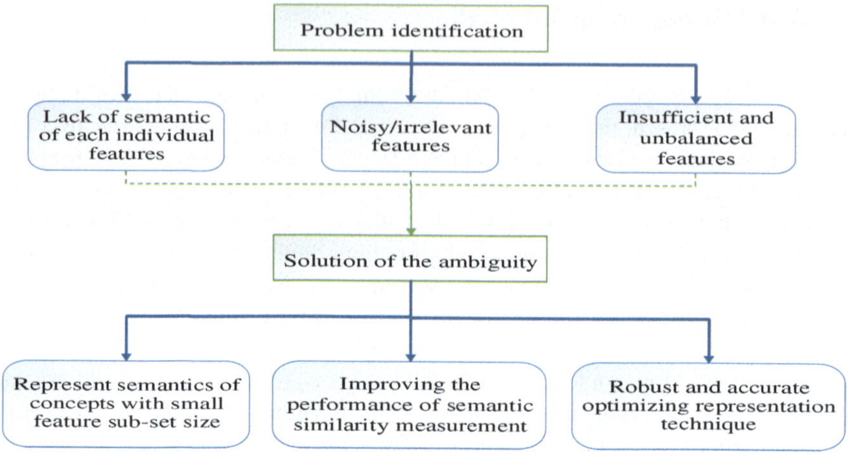

Fig. 1 Direction of the research highlight based on features

research works that have been done for semantic representation and semantic similarity measures. The figure shows three parts of the solution to the problems based on the previous studies as they are subjective and did not make it. The complications identified are "Lack of semantic of individual features," "Noisy/irrelevant features," and "Insufficient and unbalanced features." As revealed in Fig. 1, the direction of this study is emphasized based on the evaluation of the features of essential features that incorporate the problem in this research to locate the ambiguity with their intended outcomes.

In this paper, previous studies have been reviewed comprehensively on semantic representation approaches including the distributional-based and knowledge-based features, the semantic similarity measurement, and knower-based on text mining. Based on the literature review and problem identification, there are inputs, activates, and deliverables, which are as follow:

Inputs of Journals

1. Online journals, science direct, books, published and unpublished papers
2. Periodical journals
   - Knowledge-Based Systems of the Information System
   - Expert Systems with Applications Elsevier
   - The Journal of Artificial Intelligence Research
   - Engineering Applications of Artificial Intelligence in Natural Language Processing
   - Natural Language Engineering of Cambridge University
   - IEEE Transactions on Knowledge and Data Engineering
3. Other documents for publishing

Activates of papers

- Literature review

  1. Identify the scope of research
  2. Review associated papers
  3. Identify the factors related to the knowledge-based semantic representation
  4. Identify the limitations of the feature-based methods
  5. Exploring the optimizing techniques for feature selection tasks
  6. Identify the issues in knowledge-based text mining

Deliverables by previous studies

1. Knowledge in the types of semantic representations: distributional-based and knowledge-based methods
2. Identifying limitations, challenges, and issues of previous studies
3. Defining common factors for semantic representation, semantic similarity measurement, and knowledge-based text mining
4. Insight for building and developing methods for overcoming the limitations in previous works.

In the reviewing studies, the recent works about using lexical source for representing the text on documents to extract the concepts in text mining task focus on three topics. In the first topic, there is focus on the semantic information to provide the efficient semantic measure by introducing a new approach for semantic representation [10, 15], proposing a new measure on [16, 17] and combining the semantic relations in the semantic representation by using lexical source also, to improve this performance on the semantic measure by using comparative evaluation on approach measures. The second topic focuses on using knowledge approach to measure the similarity between words [18], or between lexical sourced from the knowledge approach [19]. The third topic on the literature review focuses on how to handle the semantic knowledge acquisition issues [20, 21] in a bid to improve the quantity and quality of lexical resources to make the semantic more evident and effective. Based on the above agreement on the problem identified in this research, as our figure revealed, and the literature review, we have been able to identify three main directions. In the first direction of this paper, the lack of semantic similarity of individual feature and the semantic of a concept is represented based on the electronic sources of the natural language processing. These sources have been classified as part of the approach in order to extract each approach from the semantic of representation by taxonomy of the words/concepts to get many features. This very big feature will create semantic problem on how to solve it, by representing semantics of concepts with small feature sub-set size. The second direction of our study has been to observe the feature-based method, which focuses purely on semantic representation of the concepts. It is assumed that all features in the semantics related representation of the specific notion have the same significance that will make it very noisy/irrelevant. For instance, *vehicle, convertible, accelerator, train, and cable car* have the same semantic associations with the Concept *Car*. Solving this problem will improve the performance of measuring similarity of semantic by three parameters; "shortest path measures," "depth measures," and "information content measures." The third direction of our problem is to analyze the text mining of a sub-field of artificial intelligence as which aimed to extract the information from the large corpus as collocation of document. Extracting the information from the data is a complex issue, which requires deep analysis of the text mining to insufficient and unbalanced features. Because of the complexity of the natural languages, this issue is more complex in the textual data.Thus, in our paper, this idea has been used to compute the semantic analysis for representing the semantics of concepts in the knowledge source of languages as a graph network. However, the graph-based method has measured the similarity of semantic that exists between concepts. The greater part of this work is built on the graph-based method to analyze the words and focus on how the structural elements can be exploited to improve the total performance of the measures used. Nevertheless, the improvement of the semantic representation on the features based is missed in the previous works, which makes their work inconclusive. To compare words between these approaches, previous studies have been introduced. Even though the rate of fraud occurrence is minimal, but still it is required to invent a technique for detecting the fraudulent cases before the transaction has been completed. The central aim of this paper is to propose the knowledge-based semantic representation of text mining by using Synonymy, Non-taxonomy, Hypernym, and Glosses, in the knowledge-based approach to select the concepts by using it to solve issues (synonymy and ambiguity issues) existing in social media pertaining to the text mining tasks. In the knowledge-based semantic representation, the method can be classified into two types of semantic representation to analyze the words from the concepts. The types are as follow graph-based and feature-based method. This graph represents the words to analyze by graphing map. Figure 2 shows the classification of our knowledge-based methods of semantic representation methods.

The graph-based or network-based methods depend on the semantic representation graph of the knowledge lexical source of semantic taxonomy by using weighting to select the best features from the concepts of words or semantic ontology so as to present the meaning in the concepts. The idea behind this method comes from the

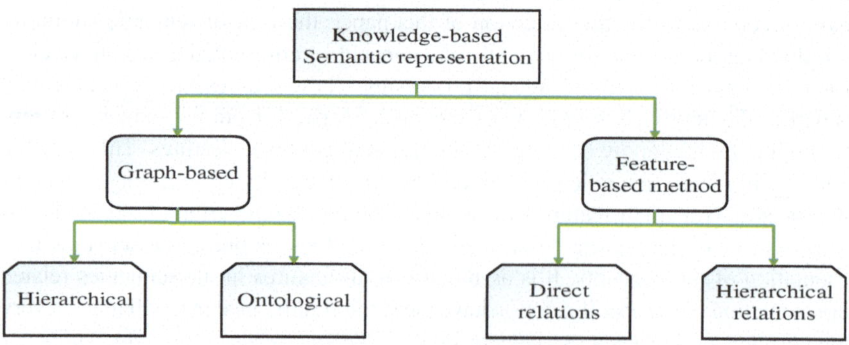

Fig. 2 Graph of knowledge semantic representation between words

perceptive sense as the human brain depends on linking the concepts to form the semantic of some given concepts. When the human brain received a concept, it first brings other-related concepts and then links between these concepts to understand the degree of the direct relationships between the words of the concepts in order to receive the concepts and other concepts. In knowledge-based semantic representation, the graph-based will have two types of semantic relations, one of which is hierarchical. It merges the feature methods and structure-based to assess the relationship between two words/concepts. Therefore, the hierarchical-based process will be adopted to signify the knowledge in a certain lexical source as a taxonomy using the "Is-A" relationships, synonyms, holonym, and meronym, and each notion will semantically be represented by a certain characteristic recursively derived from our taxonomy by using the weighting under the measures. In the second type of this graph is ontological, the ontology in the relations will help prevent random arrangement of the words of the concepts in order to choose the right concepts. In the feature-based method of the concepts, two types are identified. One is direct relations. This direct relation represents the semantics of the concepts as a set of mixed attributes from the semantic relations in the knowledge sources.

4 WordNet 3.1 Based on Semantic Representation

The relations of WordNet[4] 3.1 version used for meaning relation: These connections will be linked with words based on hierarchical structure, which serves as the instrument for computational linguistics as well as NLP. It is claimed that the language of semantic representation is mostly denoted by nouns or noun phrases so that some researchers are using or focusing on noun in the semantic relation calculated by the statistics. There are four relation methods used in this paper: "Synonymy, Non-taxonomy, Hypernym, and Glosses." The following relations are used in relation domain hypernym/hyponym (is-a), part-holonym/part-meronym (part of), member-holonym/member-meronym (member-of) and substance-holonym/substance-meronym (substance-of). For instance, tomato is a vegetable (is-a) and the head is part of the body (part of). The relation between concepts in WordNet 3.1 based on the taxonomy shown in Fig. 3 is a frame of "Is-a Relation" in WordNet 3.1. The concepts are more specific. For instance, the vehicle is more conceptual than bicycle. In addition, conveyance is more conceptual than the vehicle. The object is much more abstract in concepts.

[4] https://wordnet.princeton.edu.

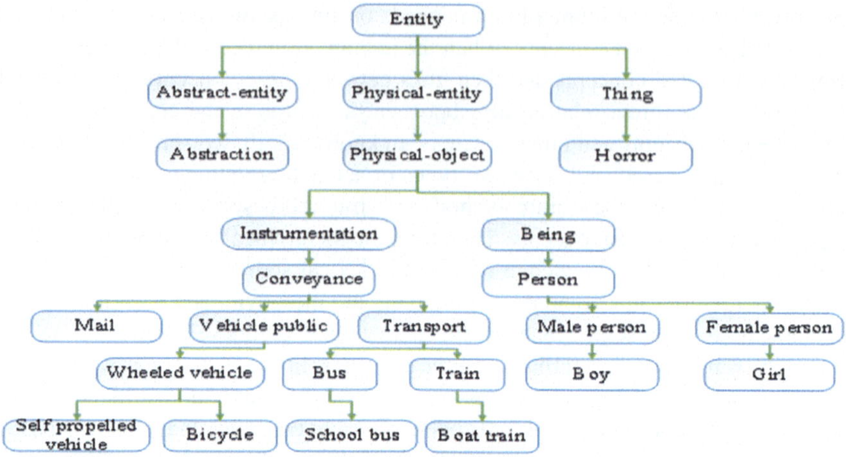

Fig. 3 Frame of "Is-a" and relationships between words and concept in the WordNet 3.1

4.1 Semantic

The semantic is that field which focuses on the investigation of the meaning of concepts and how natural language processing is normally intercepted. The psychology of semantic representation is defined as the topic which studies the technique of using the mental lexicon to interpret the meaning of words in natural languages.

4.2 Semantic Relation

The semantic relation is a link between two words/concepts to reveal the relevance of their semantics. The semantic knowledge that will be used all through the phase of this paper is extracted from English WordNet. The WordNet is a system, which comprises of the lexical files that can be converted into a database in order to describe the relations between synonyms. Two main types of relations are signified by pointers: lexical and semantic relation. While the lexical relations exist between semantically related word forms, the semantic relations exist between word meanings. These relations comprise hypernym/hyponym (superordinate/subordinate), attribute, antonym, domain-topic, domain-region, domain-usage, hyponym/hypernym, instance-hypernym, instance-hyponym, member-meronym, member-holonym, member-of-domain-topic, member-of-domain-region, member-of-domain-usage, pertainym, part-meronym, part-holonym, related-form, substance-meronym, substance-holonym, similar to, verb group also see, attribute, domain of synset (*region*, *topic*, *and usage*), member of this domain (*region*, *topic*, *usage*) and meronym/holonymy. On the other hand, the evaluation of these relations dataset is covered by the gold standard of yardsticks, which have been used for evaluating the semantic similarity of text mining tasks [22, 23].

4.3 Semantic Taxonomy

This is the relations used to represent the knowledge-based semantic in the lexical source as it is centered on the weighting of the features. The semantic is a network among concepts in the lexicon in which the nodes representing the concepts by edges representing the hypernyms and hyponyms relations. The hypernyms and hyponyms relations are those nominal and verbal synset as the inheritance taxonomy. The part of semantic taxonomy of WordNet called structure-based method is used to signify the knowledge in a certain lexical basis as a taxonomy using "Is-a" relationships [23]. However, each of the concepts is represented semantically by a number of characteristics reclusively gotten from a taxonomy.

4.4 Semantic Ontology on Conceptual Graphs

A conceptual graph can be described as a mutually aligned chart in which cases of notions are presented as an oblong while that of ideal relationships are shown as a spheroid. Aligned edges connect the tips and represent the orientation and existence of relationship. A relationship may have several boundaries; in such cases, the boundaries are given numbers. An instance of semantic ontology on conceptual pictorial depiction is known as display forum (DF) of a sentence: "a Rabbit is on a Table." The basic conceptual graph in the depictive illustration, which uses text denotation linear form (LF) in this group of words, is entered as [Rabbit]-(On)-[Table]. LF and DF are meant as illustration and performance structure for a human. There exist a formal language CG interchange form (CGIF) defined too. Using the CGIF model, this group of words would be written thus, [Rabbit: **x*] [Table: **y*] (on ?*x* ?*y*). **x* is an adjustable definition and ?*x* is referring to the defined adjustable. Applying structural alternative, the same group of words may be expressed using the same model as (On [Rabbit] [Table]). The variation among the three models is identified as with direct variation between Conceptual Graphs Interchange Format (CGIF) and Knowledge Interchange Format (KIF) Using KIF model, this sentence may be represented as (exists ((?*x* Rabbit) (?*y* Table)) (On ?*x* ?*y*)). All the formats have similar semantics in the predicate logic: $\exists\, x, y$: Rabbit(x) $\wedge$Table (x) $\wedge$ on(x, y). Conceptual graphs possess the same expressive capability as predicate logic. Figure 4 shows the conceptual graphical representation of relation between words. For instance of *Rabbit* and *Table* are $\exists(x, y) = (NP(Det(the))) \wedge (N(\text{“}Rabbit\text{”}) \wedge (VP(V(sleep))) \wedge (NP(Det(onthe) \wedge (N(\text{“}Table\text{”})\neg(NP(Det(and)) \wedge (VP(V(leave)) \wedge (NP(N(\text{“}hair\text{”})).$

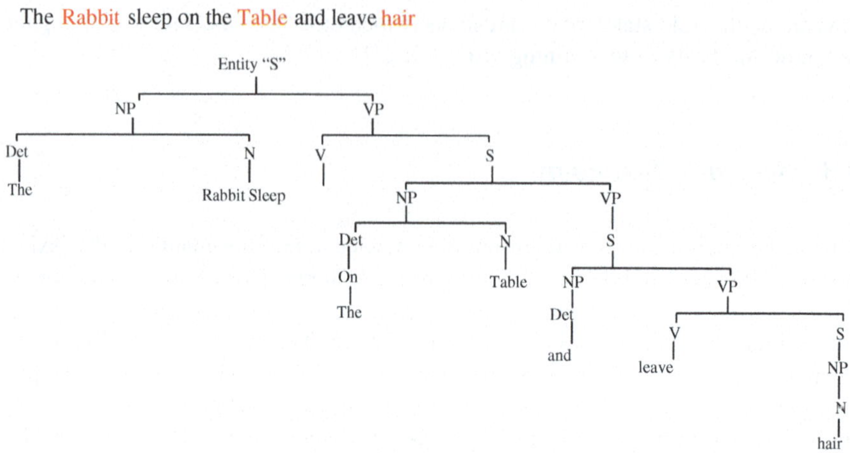

Fig. 4 Conceptual graph of the semantic representation in ontology

5 A Semantic Similarity Measure Based on Text Mining of Taxonomy Using WordNet 3.1

In this paper, three measures are introduced based on the previous study, which compares with them in the comparison and their evaluation. In this work, the statistical of the hypothesis in the significance of the feature-based method is different from the proposed method and human judgment is the three measures, which are taxonomy relations, Non-taxonomy relations, and Glosses. Each one has the hypothesis, and the scores of the similarity and the computed similarity values are unrelated. The measures will follow each one as proposed and evaluated:

5.1 Shortest Path Measures

The central thought of these measures' short path says the relationship between two ideas corresponds to the length of the path connecting the ideas and their place in the taxonomy.

5.2 Depth Measures

The depth measures are based on topological parameters to a constant weighting of assumption. Most of the promising topological parameters could be exploited from the features on the taxonomy to find the concepts. Although the units in the semantic representation are taxonomically extracted, the topological metrics (e.g., depth) are ignored when computing the semantic similarity.

5.3 *Information Content Measures*

In the information content measure, we referred to the semantic measure to use the words of concepts to organize the features to quantify the information content (IC) to a particular concept and make use of the (IC) to identify the semantic relationship between two concepts. In this work, we proposed the information content measure for weighting the semantic representation in order to make the weighting assumption in a feature-based method constant by using topological parameter to select the feature and taxonomy to find the weighting of features. To get the semantic evidence by using to extract the features and weight it on social media, we incorporated the features and captured more semantic features for weighting each feature. In previous work, they focused on the semantic measure in computing the features by weighting elements in the semantic representation of the vector.

6 Comparison and Evaluation of Semantic Similarity Measures Between Our Judgments

In this table, the comparison of semantic similarity, as we compare our work with them to find the weakness and strength and then have to apply on our new method on feature-based method, to find the problems and the ambiguity in the social media. The knowledge-based text mining can be built the synsets of words using data mining methods as which mentioned in this paper, the data mining based on methods can be implemented with the standard classification, clustering, or machine learning algorithm. However, internet/intranet semantic representation and corporate information are stored with unstructured text documents like Facebook, YouTube, Tweeter, and LinkedIn. To extract knowledge from text, some complicated text mining and learning algorithms are required (Table 1).

To determine the difference between the semantic similarity measures, which have different performance measures with distinct strengths and weaknesses, we have to use a distance measure that is suitable for comparing vectors of different lengths. The shortest path considers the path length connecting the concept and the position of the concepts, as the related words to a given word are the length of the shortest path from the root to the concept *c*. To use the nexus or edge as a factor to refer to the relationship between concepts of nodes, almost all of the methods are simpler as the local of the density of pairs cannot be as displayed. Depth: the depth of a node in the taxonomical hierarchy refers to the distance between the specific node and the root of the taxonomical hierarchy. Information content (IC): It is based on the measure

Table 1 Statistical of the significance of feature-based method different between the "proposed method" and human judgment in the taxonomy of relations of different semantic similarity (SS) between measures

Category	Scale of weighting	Measure	Feature	Advantages	Disadvantages
Shortest path	The part of the path length in the notions and the position of the notions in the taxonomy	Rada [24]	Semantic taxonomy and depth between concepts	Simple/easy to implement the competitive results on measuring semantic relatedness	Two pairs that have the same lengths of the shortest path will possess the same semantic relationship
Depth	LCS, length, depth	Wu and Palmer [25]	The part of the length of semantic of subsumes path to the root of the entity from the LCS, length, and depth find the linear of the functioning of the shortest path and depth of LCS, length, and depth	Simple	Two pairs that have the same LCS and the same lengths of the shortest path will possess the same semantic relationship
		Li and McLean [26]		Simple	Two of these pairs with the same lengths of the shortest path will possess the same relationship
		Leacock and Chodorow [27]	Length, depth of semantic taxonomy	Simple	Two pairs of words from the concepts have equal lengths of the shortest path will possess the same relationship

(continued)

Table 1 (continued)

Category	Scale of weighting	Measure	Feature	Advantages	Disadvantages
Information content	The common the information content that refer two concepts share, the more related the concepts for measuring the semantic similarity between words via concepts	Sebti and Barfroush [28]	Information content metric and edge counting based on tuning function	Simple	Two pairs with LCS will have the same similarity in semantic of taxonomy
		Sánchez and Batet [29]	Leaves of concept subsume, and amount of leaves in the semantic taxonomy	Take the IC compared with concepts and consider the words between concepts	To make a pair with the same IC for C_1 and C_2 as they possess the same relationship
		Meng and Zhou [16]	The hyponyms of the concept *c*, depth of concept, number of nodes, depth of semantic taxonomy	By taking the concepts of features to reckon the relationship between words	Computationally expensive and high dimensionality to unfiltered features of many redundant concepts in the representation
		Taieb and Aouicha [30]	Descendants, depth of concept, hypernyms, ancestors	Make a measure and combine the semantic relations to enhance the operation of measuring the semantic relationship	High-quality results in wide experimental
		Aouicha and Taieb [31]	Depth, ancestors, hypernyms, descendants, and hyponyms	Same with make a measure and combining the semantic relations to enhance the operation of measuring the semantic relationship	Highly relevant features with low redundancy High feature-to-sample ratio High dimensionality

as a given concept and then uses the IC to examine the semantic similarity between two concepts. These measures have been classified in the previous researches as the information content-based measures.

7 Conclusion

This paper explored recent trends to discover the semantic similarity in advanced measures. The similarity measures are based on feature-based method to capture more the biggest number that which relevance between words in the vectors from the documents based on social media to solve the problems and ambiguity in social media by using WordNet 3.1 based on shortest path measures, depth measures, and information content measures, to proposed these lexical sourced from the semantic features (taxonomical relation, meronym relation, gloss (definition), and Is-A relation) to extraction technique to retrieved the documents based on user query to find the semantic relations that which existing between concepts in the social media. We also analyzed the scale of weighting, features, advantages, and disadvantages of different measures. However, we presented the commonly by using IC metric in information-based content measure. Lastly, we discussed how to measure the performance of the similarity measure on social media. As a different measure will reveal the different performance indifferent application, in the specific application a measure will contain all other aspects system of another factor. Moreover, the WordNet is common sense ontology. There exists a host of other domains of taxonomically oriented ontologies. How to efficiently solve the varied problem especially in social media and apply the measure in other ontology is required in further study. We are currently conducting further research and studies working on the potential application of our method on the English language for documents on social media using WordNet 3.1 versions. We are currently conducting further research and studies working on the potential application of our method on the English language using WordNet 3.1 versions. The future research involves knowledge-based approach of semantic representation of WordNet 3.1 for documents using dictionaries to improve these concepts and analyzing the words based on semantic similarity. The work is underway and results will be available soon.

Acknowledgements This work is supported by the University Malaysia Pahang (UMP) via Research Grant UMP (RDU1803141, PGRS190398).

References

1. Al Ajeeli AT (2016) An intelligent framework for natural language stems processing. Glob J Comput Sci Technol

2. Zhang H et al (2014) Online social network profile linkage. In: Asia information retrieval symposium. Springer, Heidelberg
3. Chen X, Liu Z, Sun M (2014) A unified model for word sense representation and disambiguation. In: Proceedings of the 2014 conference on Empirical Methods in Natural Language Processing (EMNLP)
4. Lopez-Arevalo I et al (2017) Improving selection of synsets from WordNet for domain-specific word sense disambiguation. Comput Speech Lang 41:128–145
5. Beliga S, Meštrović A, Martinčić-Ipšić S (2015) An overview of graph-based keyword extraction methods and approaches. J Inf Organ Sci 39(1):1–20
6. Al-Tashi Q, Hasan AM (2019) Word sense disambiguation: a review. Southern Connecticut State University, Hilton C. Buley Library, 1, 2, pp 20–458
7. Hasan AM, Rassem TH, Karimah M (2018) Pattern-matching based for Arabic question answering: a challenge perspective. Adv Sci Lett 24(10):7655–7661
8. Hasan AM, Rassem TH, Noorhuzaimi M (2018) Combined support vector machine and pattern matching for Arabic Islamic Hadith question classification system. In: International conference of reliable information and communication technology. Springer, Heidelberg
9. Hasan AM, Zakaria LQ (2016) Question classification using support vector machine and pattern matching. J Theor Appl Inf Technol 87(2)
10. Taieb MA, Ben Aouicha M, Ben Hamadou A (2013) Computing semantic relatedness using Wikipedia features. Knowl-Based Syst 50:260–278
11. Saif A, Ab Aziz MJ, Omar N (2016) Reducing explicit semantic representation vectors using latent dirichlet allocation. Knowl-Based Syst 100:145–159
12. Wei T et al (2015) A semantic approach for text clustering using WordNet and lexical chains. Expert Syst Appl 42(4):2264–2275
13. AlAgha I, Nafee R (2016) Investigating the efficiency of WordNet as background knowledge for document clustering. J Eng Res Technol 2(2)
14. Vijaymeena M, Kavitha K (2016) A survey on similarity measures in text mining. Mach Learn Appl Int J 3:19–28
15. Hassan S, Mihalcea R (2011) Semantic relatedness using salient semantic analysis. In: Proceedings of AAAI 2011 (25th AAAI Conference on Artificial Intelligence). Association for the Advancement of Artificial Intelligence, San Francisco
16. Meng L, Gu J, Zhou Z (2012) A new model of information content based on concept's topology for measuring semantic similarity in WordNet. Int J Grid Distrib Comput 5(3):81–94
17. Rassem TH et al (2017) Restoring the missing features of the corrupted speech using linear interpolation methods. In: AIP conference proceedings. AIP Publishing
18. Li P et al (2017) A graph-based semantic relatedness assessment method combining wikipedia features. Eng Appl Artif Intell 65:268–281
19. Horsmann T, Zesch T (2016) LTL-UDE $@ $ EmpiriST 2015: Tokenization and PoS tagging of social media text. In: Proceedings of the 10th web as Corpus workshop
20. Matuschek M, Gurevych I (2013) Dijkstra-wsa: a graph-based approach to word sense alignment. Trans Assoc Comput Linguist 1:151–164
21. Pilehvar MT, Navigli R (2015) From senses to texts: an all-in-one graph-based approach for measuring semantic similarity. Artif Intelll 228:95–128
22. Hasan AM, Rassem TH, Noorhuzaimi M, Hasan AM (2019) A semantic taxonomy for weighting assumptions to reduce feature selection from social media and forum posts. In: International conference of reliable information and communication technology. Springer, Heidelberg, pp 154–291
23. Hasan AM et al (2020) A proposed method using the semantic similarity of WordNet 3.1 to handle the ambiguity to apply in social media text. In: Information science and applications. Springer, Heidelberg, pp 471–483
24. Rada R et al (1989) Development and application of a metric on semantic nets. IEEE Trans Syst Man Cybern 19(1):17–30
25. Wu Z, Palmer M (1994) Verbs semantics and lexical selection. In: Proceedings of the 32nd annual meeting on Association for Computational Linguistics

26. Li Y, Bandar ZA, McLean D (2003) An approach for measuring semantic similarity between words using multiple information sources. IEEE Trans Knowl Data Eng 15(4):871–882
27. Leacock C, Chodorow M (1998) Combining local context and WordNet similarity for word sense identification. WordNet: an electronic lexical database, vol 49, issue 2, pp 265–283
28. Sebti A, Barfroush AA (2008) A new word sense similarity measure in WordNet. In: International multiconference on computer science and information technology, IMCSIT 2008, IEEE
29. Batet M et al (2014) An information theoretic approach to improve semantic similarity assessments across multiple ontologies. Inf Sci 283:197–210
30. Taieb MAH, Aouicha MB, Hamadou AB (2014) A new semantic relatedness measurement using WordNet features. Knowl Inf Syst 41(2):467–497
31. Aouicha MB, Taieb MAH, Hamadou AB (2016) Taxonomy-based information content and wordnet-wiktionary-wikipedia glosses for semantic relatedness. Appl Intell 45(2):475–511

Identification of Appropriate Filters for Preprocessing Palm Print Images

S. Kavitha and P. Sripriya

Abstract Efficient preprocessing of image is necessary in all classification of biometrics to obtain an enhanced image. In this research study, CASIA database consisting of numbered palm print is used. Original left and right palm print images are subjected to various stages of filtering and pre-processed image is obtained. The PSNR values are computed and results are tabulated for the filters applied and the graph is plotted. The comparison between tabulated values indicated that preprocessed image has the higher values of PSNR and an improvement in the image quality is observed. The appropriate filter based on PSNR values is identified to proceed further with segmentation and feature extraction process.

Keywords Preprocessing · Palm print images · Filters · PSNR

1 Introduction

Preprocessing helps to improve the quality of image by changing the orientation, removal of noise, texture, and brightness. The principle of preprocessing is applied in all the image-related applications like intelligent transportation, moving object tracking, and defense surveillance. Preprocessing refers to the correction of image with respect to brightness by considering the position of pixels that exists in original image. In preprocessing approach, difference between the pixel values of image is greater than the threshold is assumed as criteria for filtering.

The traditional belief about human palm print lines is that they help to predict the future. Fortune prediction using palm print lines is familiar even today in countries

S. Kavitha (✉)
Research Scholar, Department of Computer Science, Vels Institute of Science, Technology & Advanced Studies, Chennai, India
e-mail: kavitha.phdcs@gmail.com

P. Sripriya
Professor, Department of Computer Applications, Vels Institute of Science, Technology & Advanced Studies, Chennai, India
e-mail: sripriya.phd@gmail.com

S.-L. Peng et al. (eds.), *Intelligent Computing and Innovation on Data Science*, Lecture Notes in Networks and Systems 118,
https://doi.org/10.1007/978-981-15-3284-9_16

like India and China. In Palmistry, predictions are done with the lines running on a palm such as fate line, health line, and by palm length, size, and shape defined by their position and thickness. The same principle is applied technically and scientifically by subjecting to various research studies which progressively indicated the biometric feature of each line in the palm is distinct and hence it can be utilized to identify humans. The human palm is unique because of the running palm lines, geometric patterns, ridges, datum points, minutiae feature, etc., when fused together can give an accurate biometric results. The areas of palm are larger and wider when compared to finger print recognition and more precise details can be extracted and are cost effective.

To capture the palm print area, digital scanners and CCD-based scanners are available. The scanners used for palm print have its own advantages and disadvantages and hence online palm print datasets are mostly used. Images for palm can be collected by properly aligning the scanner or infrared camera. Choosing a good quality palm print images for study to achieve better results is an important criteria. Nowadays, the cameras in certain mobile phones like Xiami Redmi, Vivo, etc., also provide good quality images and can be used for preprocessing. The left and right palm images should be taken separately and numbering should be done. CASIA dataset is used in this paper as it has proper numbered left and right images of palm and hence preprocessing is implemented with the existing dataset. It consists of 7200 palm print images. PSNR values calculation is done using MATLAB software. The values obtained for right palm print images are tabulated and graph is generated.

The research paper is divided into following sections: Section 1 highlights on the introduction about preprocessing and palm print study, Sect. 2 discuss about review on literature survey and Sect. 3 details on the various filters available and the methods used in this paper, Sect. 4 describes about the experimental methods and discussion, and Sect. 5 deals with the conclusion and the future enhancement.

2 A Review on Literature Survey

Zhong et al. [1] conducted a survey on the progress of palm print recognition research over a decade. Michalak and Okarma [2] improved the preprocessing methods and binarized the image using entropy filtering method for alphanumeric character recognition. Giełczyk et al. [3] proposed a light weight method for palm print recognition. Caledron et al. [4] worked with digital X-ray images for biomedical applications and analyzed the importance of preprocessing in a convolutional network model. George et al. [5] used CASIA palm print database that consists of 5502 images and applied preprocessing to reduce the impact of noise and detected the geometrical position for ROI feature extraction. Bajracharya et al. [6] in his research study provided an improved method for digital images by compressing preprocessing and non preprocessing images and calculated PSNR values and improved the compression ratio of images. Sharma et al. [7] used Hadoop denoising interface and calculated PSNR and MSE to achieve better qualitative output results. Manju et al. [8] using the lossless image compression in CCSDS algorithm had done the PSNR and MSE value calculations, concluded that higher PSNR value provides better transmission and reception of images. George et al. [9] discussed the better filters for preprocessing and methods for mass segmentation based on the parameters of MSE and PSNR values. Aswin Kumar et al. [10] preprocessed IRIS images for human recognition using various filters and tabulated the PSNR and MSE values. Victor and Ghalib [11] detected and classified skin cancer cells with hybrid segmentation approach based on PSNR values from the different filters used. Sowmiya et al. [12] captured the satellite images by applying sensors and then preprocessed it to eliminate the atmospheric and geometric distortions present in the image. Pinki and Mehra [13] estimated the quality of image by applying different distortions using the PSNR values. Rajkumar and Malathi [14] analyzed the quality of image for real-time satellite images with image quality metrics such as PSNR and MSE. Ali, Mouad et al. [15] used palm print lines and measured PSNR by applying edge detection algorithms for various online datasets.

Based on the literature review, it is observed that palm print image quality with existing dataset need to be identified by preprocessing image using filters and also applying any parameters such as mean, standard deviation, PSNR, and MSE to substantiate the resultant values.

The objective of this research work is to apply filters to every stage of preprocessing and to derive PSNR values and analyze the results.

3 Discussion on the Filters Used

Filtering is the basic step in any image preprocessing. The choice of filter for any particular application is entirely dependent on the nature of image and the application for which it is used. Image filters can be of type linear or nonlinear. In linear filters, output pixel values are linear combination of input given, whereas in nonlinear filters, output is purely based on the type of application it is used.

3.1 Median Filter

The noise from an image is totally eliminated or reduced efficiently, image edges preserved with this nonlinear filter. It functions by calculating the median of the surrounding pixels by sorting them in ascending order and replaces the middle pixel value. The neighboring pixels are patterned in window and slides from one pixel to another.

3.2 *Smoothening Filters*

These filters help to achieve better and fine details of the image. There are many types of smoothening linear filters, namely low pass, high pass, Kalman, Butterworth, Wiener, and Gaussian Filters.

3.2.1 Gaussian Filter

It is a type of linear filter used mostly to reduce noise keeping edges relatively sharp, to reduce contrast and blur the images. The performance of the filter is faster when compared to other filters where as sharpening filters remove completely image blurring and are used to highlight finer details of the image, that is it works by emphasizing the edges. It is more popular and used in finding the edges in an image and spatial analysis. Gaussian filters are the approximation of Gaussian function and it modifies the output signal with convolution with minimum possible group delay. Salt-and-pepper noise when added to the image Gaussian filter removes noise better compared to other smoothening filters. It also reduces the high frequency details from an image and hence it is also called as low pass filter.

3.3 *Sharpening Filters*

There are wide range of filters for image sharpening like Laplacian filter, unsharp mask filter, high boost filter, Gradient mask filter etc., for image sharpening. Laplacian filter used highlights discontinuities in gray-level images and eliminates or less emphasizes the regions with least changing gray levels. It is a derivative filter and the image is improved by ameliorating the background features.

4 Experimental Analysis

A brief outline of the proposed preprocessing work is presented as a flow diagram as shown in Fig. 1.

The proposed system uses palm print dataset and median filter is applied to expel the image noise and smoothening is done with Gaussian filter and sharpening using Laplacian filter is carried out and the preprocessed image is obtained. The left and right palm print PSNR values are determined and the graph is produced.

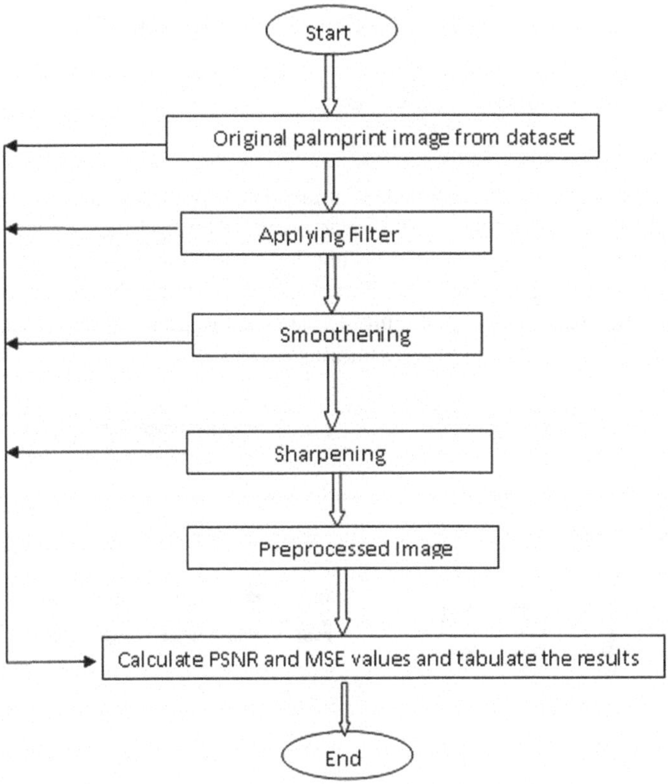

Fig. 1 Proposed work flow diagram

4.1 PSNR

Peak signal-to-noise ratio is interpreted as ratio of quality measure between the source and regenerated image. The equation for **PSNR** is described in mathematical form as follows:

$$\mathrm{PSNR} = 20\log_{10}\left(\frac{\mathrm{MAX_f}}{\sqrt{\mathrm{MSE}}}\right) \quad (1)$$

4.2 MSE

Mean Square error is the aggregate of the squared error values between the regenerated and the source image. It is employed to determine the mean square difference between the expected outcome and predicted results. Lesser the value of MSE indicates the error is lower and the equation is represented below:

$$\text{MSE} = \frac{1}{mn}\sum_{0}^{m-1}\sum_{0}^{n-1}\|f(i,j) - g(i,j)\|^2 \tag{2}$$

where m and n represent the row and columns in an input images.

The analysis of experimental results is presented below as follows. Figure 2 represents sharpening and smoothening filters applied for left and right palm print images. Figure 3 represents the tabulated PSNR values of left and right palm print images before and after preprocessing. Figure 4 displays the graphical representation of PSNR values of left and right palm print images.

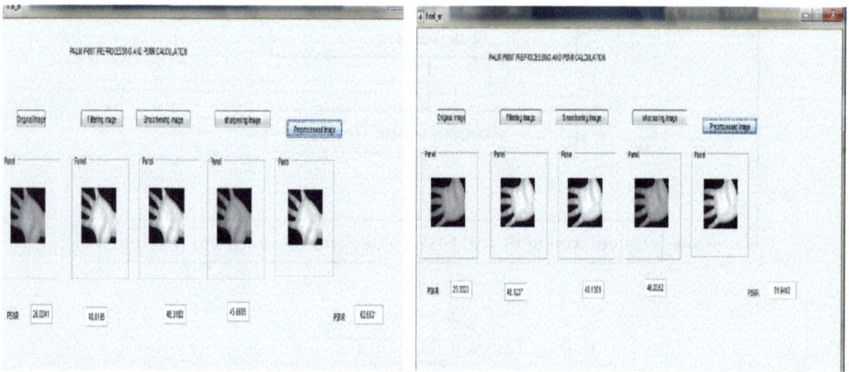

Fig. 2 Filters applied to left and right palm print images

	CALCULATION OF PSNR				
	Original Image	Median Filter	Smoothening	Sharpening	Preprocessed Image
left1	26.0995	47.8928	47.9301	46.0308	61.7963
left2	26.0341	48.0195	48.0183	45.8685	62.6831
left3	25.9187	48.3865	48.3961	45.8813	62.0481
left4	25.9582	48.3012	48.3052	45.7505	61.728
left5	25.9693	48.9701	49.0174	46.0398	61.3022
left6	25.927	45.8096	45.8522	46.0238	61.3517
left7	25.8274	45.2887	45.3142	46.003	61.3046
left8	25.839	47.6279	47.6312	46.0832	61.9177

	CALCULATION OF PSNR				
	Original Image	Median Filter	Smoother	Sharpenir	Preprocessed Image
right1	25.8323	48.1237	48.1361	46.0352	61.9402
right2	25.6791	45.5863	45.6058	45.9203	62.18
right3	25.725	45.46	45.4842	45.9663	61.7884
right4	25.7531	42.4982	42.512	45.6016	61.2999
right5	25.8415	48.3257	48.3639	46.1843	61.8103
right6	25.7347	48.5304	48.5834	45.8778	61.0969
right7	25.7565	48.7845	48.8083	45.9717	62.2949
right8	25.6435	48.2283	48.2649	46.1844	61.5976

Fig. 3 Tabulated PSNR values of left and right palm print images

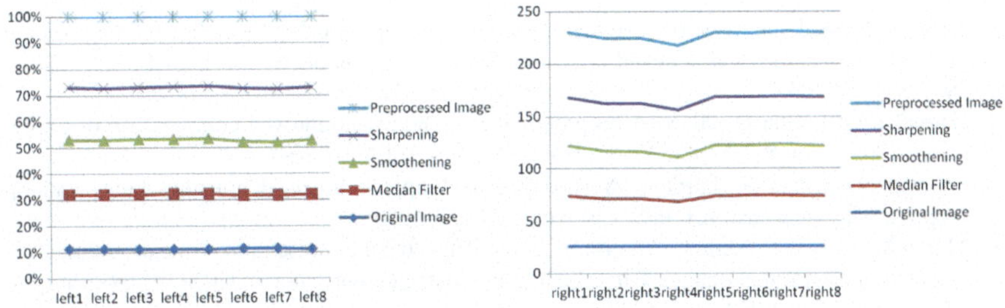

Fig. 4 Graphical PSNR of left and right palm images with the tabulated values

5 Conclusion

Preprocessing plays a very important role for palm print recognition. Preprocessing step helps to proceed further for efficient segmentation and feature extraction. Objective of applying filters in the palm print images is to eliminate the noise and to further smoothen and sharpen the image to preserve the edges. In this paper, the output of median, Gaussian, Laplacian with respect to quality parameters PSNR is discussed. Based on the output values obtained for right and left palm images using Gaussian filter imply the high value of PSNR even when compared with the original images from the dataset used in this paper. From the experimental results and observation, the usage of Gaussian filter will be appropriate when compared with other filters taken for study and is concluded based on higher PSNR values.

Additional parameters like MSE and SD can also be used with different datasets available to enhance the work in the future. Determination of PSNR for the existing dataset can be done by combining features of iris, finger, palm, hand, and so on to provide better results.

Data Availability
The CASIA palm print dataset is used in this article for experimental purpose.

References

1. Zhong D, Du X, Zhong K (2019) Decade progress of palmprint recognition: a brief survey. Neurocomputing. https://doi.org/10.1016/j.neucom.2018.03.081
2. Michalak H, Okarma K (2019) Improvement of image binarization methods using image preprocessing with local entropy filtering for alphanumerical character recognition purposes. Entropy 21:562
3. Giełczyk A, Choraś M, Kozik R (2019) Lightweight verification schema for image-based palmprint biometric systems. Mobile Inf Syst 9
4. Calderon S et al (2018) Assessing the impact of the deceived non local means filter as a preprocessing stage in a convolutional neural network based approach for age estimation using digital hand X-ray images. In: 2018 25th IEEE International Conference on Image Processing (ICIP), Athens, pp 1752–1756

5. George LE, Shakour AA (2018) ROI extraction for palmprint using local thresholding, region growing and geometrical centroid criterion. Técnica Vitivinícola Sci Technol J 33(3):63–69. ISSN: 0254-0223
6. Bajracharya B, Hua D (2018) A preprocessing method for improved compression of digital images. J Comput Sci Appl 6:32–37. https://doi.org/10.12691/jcsa-6-1-4
7. Sharma N, Bagga S, Girdhar A (2018) Novel approach for denoising using Hadoop preprocessing interface. ICCIDS, Elseiver, pp 1327–1350
8. Manju M, Abarna P, Akila U, Yamini S (2018) Peak signal to noise ratio & mean square error calculation for various images using the lossless image compression in CCSDS algorithm. Int J Pure Appl Math 119:14471–14477
9. George MJ, Sankar SP (2017) Efficient preprocessing filters and mass segmentation techniques for mammogram images. In: IEEE International Conference on Circuits and Systems (ICCS), Thiruvananthapuram, pp 408–413
10. Aswin Kumar S, Paneer Selvam S (2017) Preprocessing of IRIS image using High Boost Median (HBM) for human personal identification. IJCSMC 6(2):142–151
11. Victor A, Ghalib MR (2017) A hybrid segmentation approach for detection and classification of skin cancer. BioMed Res 28(16)
12. Sowmya DR, Deepa Shenoy P, Venugopal KR (2017) Remote sensing satellite image processing techniques for image classification: a comprehensive survey. Int J Comput Appl 161(11)
13. Pinki RM, Mehra R (2016) Estimating image quality under different distortions. IJECS 05(07):17291–17296
14. Rajkumar S, Malathi G (2017) A comparative analysis on image quality assessment for real time satellite images. IJST 9(34)
15. Ali M, Yannawar P, Gaikwad A (2016) Study of edge detection methods based on palmprint lines. https://doi.org/10.1109/iceeot.2016.7754900

Feature Extraction of Metastasis and Acrometastasis Diseases Using the SVM Classifier

A. Vidhyalakshmi and C. Priya

Abstract Metastasis and acrometastasis diseases were presented as the tumor in the tongue and hands of the human body. These two diseases may cause cancer to the lungs part of the human body. Metastases have the base in parts of the lung, breast or kidney. In this paper, the lung cancer symptoms can be diagnosed based on the metastases to the tongue images and acrometastases to the hands' images. The proposed effort composed of feature extraction of tongue and hand images as input by using the Wiener filter processes. A support vector machine in supervised learning models with learning steps can explore the data which can be used for classification and regression scrutiny. The classifier techniques sustain to classify the metastases and acrometastases data. Finally, the classification and regression process can increase accuracy to predict the metastasis and acrometastasis diseases for lungs cancer. The proposed work is implemented using the MATLAB software and classifier increases an accuracy of feature extraction.

Keywords Lungs cancer · Metastases to tongue · Acrometastasis to hand · Support vector machine · Symptom

1 Introduction

Lung plays the majority recurrent location of metastases which start from the head and neck cell with the carcinomas wax that is being true for the oral tongue cancers. The type of solitary metastasis over the parts of squamous cell carcinoma may be reported as the legitimate option w2–4x [1]. Commonly, hands may have the acrometastasis involvement. The primary report discussed for elderly women, with carcinoma of the

A. Vidhyalakshmi (✉)
Department of Computer Science, Vels Institute of Science, Technology and Advanced Studies (VISTAS), Chennai, India
e-mail: vidhar07@gmail.com

C. Priya
Department of Information Technology, Vels Institute of Science, Technology and Advanced Studies (VISTAS), Chennai, Tamil Nadu, India

S.-L. Peng et al. (eds.), *Intelligent Computing and Innovation on Data Science*, Lecture Notes in Networks and Systems 118,
https://doi.org/10.1007/978-981-15-3284-9_17

breast with the multiple metacarpal bone. The greatest opportunity for the metastases grows throughout the body [2]. A type of classifier known as the support vector machines (SVM) has the deposit of associated supervised.

Learning methods need for the classifications. The SVM built the separate hyperplane in the space with the maximum boundary between the two datasets. The condition for support vector machines, which indicates the data points of p dimensional vector known to support p-1 dimensional hyperplane termed as the linear classifier [3]. The SVM is one kind of linear classifier, which enables the optimized hyperplane, which maximizes the boundaries between patterns. These features make the SVM as a powerful tool for the task of pattern recognition. The SVM has a definite expression for the data analysis [4, 5]. A cluster of SVMs has the necessary kernel functions in use. The fivefold terms of cross-validation are agreed in the SVM for the training dataset.

The study includes the CV accurateness required for all the datasets with the smallest CV error. A machine learning (ML) can make the learning model with the history data to foretell the upcoming data [6]. Next section discusses the literature survey, Sect. 3 discusses feature extraction of the metastasis and acrometastatic, Sect. 4 discusses the result and discussion. Finally, this work discusses the conclusion of the proposed work.

2 Related Work

Han and Jiang [7] proposed a sparse-coding kernel approaches can overcome the SVM needed for fitting the diagnosis of diseases. The traditional ad hoc tuning approach is based on the parametric conquers for the overfitting problem of diagnosing the accuracy. Based on the knowledge, the first rigorous method proposed to overcome the SVM overfitting. Finally, the author proposed a novel biomarker discovery algorithm such as the Gene-Switch-Marker (GSM), which captures the meaningful biomarkers for the advantage of SVM over a fitting on single genes.

Sawada et al. [8] proposed a lung by the renal cell, breast and colon cancers, which will be the widespread primary tumors required for the acrometastasis. The system for the acrometastasis fully elucidated, which can accept the bone metastasis that may include the acral regions needed for the hematogenous. There may be a difficulty in the distributed cancer cells that can adapt, proliferate the metastases in side-line bones. The lesser amount of red marrow content may compare with the axial bone. The worse temperature based on the location becomes closes to the surface of skin. Zhang et al. [9] proposed the research, which develops a diagnostic method for analysis using the standard tongue image by the support vector machine. The tongue body coating may be alienated using the partition by merger and the chrominance threshold method. The color extraction and the texture feature of the tongue images are given as the input variables, method for the investigtion of diabetes the SVM were trained.

The input variables needed for the influences are the combination of method in analysis. Eccles and Welch [10] proposed a cancer death reason for the enlargement of metastases; the majority in morbidity and mortality results shows the prevention of disseminated disease. The direct treatments desired for the metastasis become the cells which are ready for the primary tumor.

Liu et al. [11] anticipated the approaches for the account of imbalanced costs needed for the breast cancer judgment in an intelligent way. This information provides the reasonable results for the previous work and more conventional model for the classification. The evaluation performances for the proposed approaches using the Wisconsin Breast Cancer and the Wisconsin Diagnostic Breast Cancer (WDBC). The breast cancer dataset composed from the University of California at Irvine. The machine line depository considered necessary for the studies.

3 Feature Extraction of Metastasis and Acrometastatic

3.1 Metastasis

The secondary malignant growth based on the distance is obtained from the sites of primary cancer. The metastasis is a type of cancer present in the mouth cavity, which particularly spread over the tongue rarely in nature. The reviews conducted for the 6881 cancers results show the 0.2% occurs as the lingual metastasis. The lingual metastasis can be possible for the explanation in the form of inhospitable nature for the site. The inhabit colonization obtained the process of mechanical, chemical and the variation in the thermal basis. The idea behind the recognition of skeletal muscle secretes the several factors in the anticancer activity. Cancer can commonly occur with the metastasis to the tongue which includes the 28% of kidney and lung, 11% of the skin melanoma and 9% of the breast cancer [12].

3.2 Acrometastatic

The metastasis in the digits form can occur infrequently, with a non-specific clinical presentation. The report contains the patient details of the persistent swelling and ulceration involves the right thumbnail. The information provided not fully diagnosis cancer but they have the delay in the eventual diagnosis of cell like the lung cancer representing the acrometastasis. The treatment needs with the amputation for more measures of palliative. The person should have the acrometastasis for the differential diagnosis that relates to the patients for non-healing digital injuries or ulcers. The patients may have limited treatment options with very poor prediction.

Figure 1 shows the proposed method.

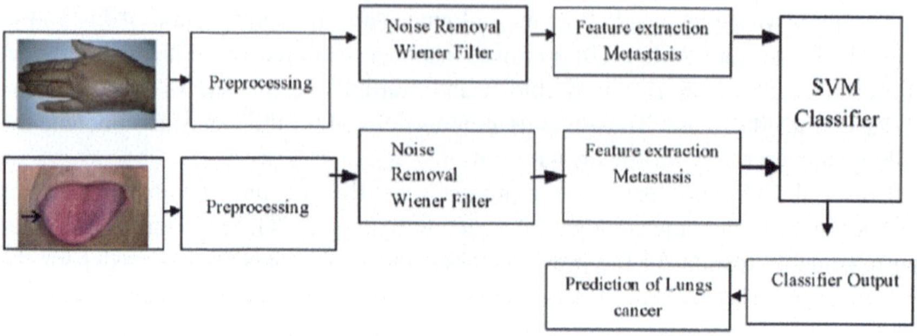

Fig. 1 Feature extraction for metastasis and acrometastatic

3.3 *Input I*

The feature extraction, which involves the reduction of quantity of resources needed for the description of data in the bulky set. The performance analysis for the complex majority data has the problems that identify a large number of variables involved in the process. The analysis of variable in the large number basically

- Read image
- Resize image
- Remove noise (Denoise)
- Segmentation
- Morphology (smoothing edges).

3.4 *Preprocessing*

The preprocessing techniques required for the color, gray level or binary level documentation of images may contain the text and the graphics. The character recognition of system with the most applications uses gray or binary images. The color images and the document described with the preprocessing of the input image may contain the binary images only. There are several steps available for the processes; the first step will be the image enhancement techniques which are capable of removing the image noise, adjust the contrast in the image. The second step includes the thresholding facilities that can remove any scenes, watermark and noise present in the image. The third steps include the page segmentation in separating the graphics presented in the text, the fourth character with the segmentation processes. The next step involves the morphological dealing out to the characters present with the thresholding and further morphological processing. This process enhances the preprocessing techniques' parts for the character pixels from them. Techniques above follow the presentation of a few character recognition techniques in which some applications with few of these are taken as the character recognition system for some applications.

The feature extraction may be a general step followed in the construction of combining these variables that may be around these problems which describes the sufficient data accuracy. Many machine learning practices may believe the property of optimized feature extraction that may be the key role for the effective applications. An excision of growth present in the anterior tongue revealed a 2.6 cm tumor may be composed of the sheets in additional with the clear cells placed around the oval nuclei. An enhancement of the CT scans for the portion of abdomen with guided FNAC was suggested. The chest portions for the part-time routine metastatic workup with the discrete nodular opacities for the lungs are the secondary deposits.

The Gaussian blur known as the Gaussian smoothing may result in the blurring an image with the Gaussian functions. The visual effect of these blurred techniques represents the smooth image. The Gaussian smoothing also uses the preprocessing stage for the version algorithm, which can enhance the image structures for the different scales. The additional use of the threshold compared with the suitable filters can remove the local noise for the smoothing of rank filters.

3.5 *Removal of Noise*

The Wiener filter uses the filter toward the desired target random process which uses the linear time-invariant filtering of noise known as stationary and the noise signal. The Wiener filter can reduce the error connecting random process and the preferred process. The Wiener filter has solutions for cases: one non-causal filter requiring a quantity of both data, next to where filter is desired with an infinite amount data, and finite impulse response case where input data used in the result which is not fed back to the filter as in the IIR case.

Non-causal solution

$$G(s) = \frac{S_{x,s}(s)}{S_x(s)} e^{\alpha s}. \quad (1)$$

where S is the spectral densities. Provided that $g(t)$ is optimal, then minimum mean-square error equation reduces to

$$E\left(e^2\right) = R_s(0) - \int_{-\infty}^{\infty} g(\tau) R_{x,s}(\tau + \alpha) \mathrm{d}\tau. \quad (2)$$

And the solution $g(\tau)$ is the inverse two-sided Laplace transform of $G(s)$.

3.6 SVM Classifier Algorithm

The intention of the support vector machine is to locate a hyperplane in *N*-dimensional space (*N*—the number of features). Our work is to get a plane that has the highest margin in which the utmost distance in data points of both classes is the maximizing distance that provides some reinforcement. Support vectors are record points that are closer to hyperplane and persuade the position and orientation of the hyperplaner.

Result Analysis

The extractions of metastasis and acrometastatic on tongue and hand images are taken from the databases. The proposed work is implemented using the MATLAB software and the simulation results were shown in Figs. 2, 3, 4, 5, 6 and 7.

Here, the feature extraction is done using the basic image processing techniques such as the input image, filtering, contrast stretched on image, edge detection taken as the feature extraction for both the tongue and hand images. The measurement values are shown in Table 1.

Table 1 shows the result of metastasis and acrometastatic details in the spread of hand and tongue. The volume, thinnest tumor thickness, thickness, area, perimeter, accuracy, sensitivity and specificity measured using the MATLAB software. The tabulation gives the measured values for the input image. These values are taken as the feature extraction of the given image.

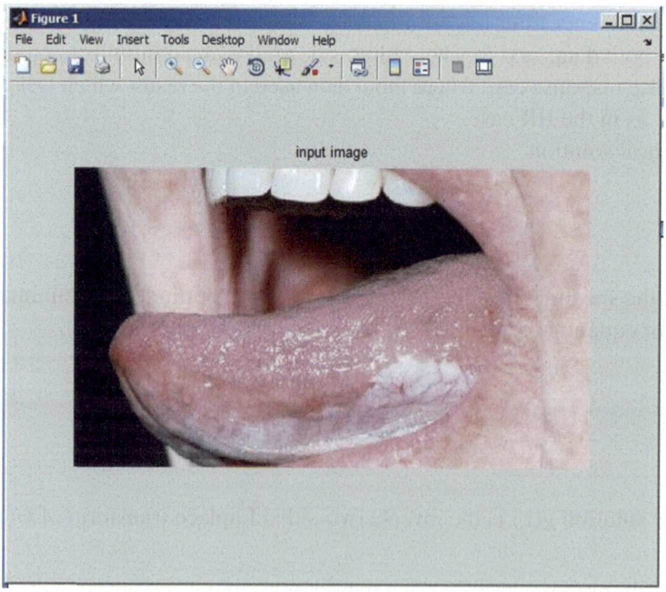

Fig. 2 Metastasis input tongue image

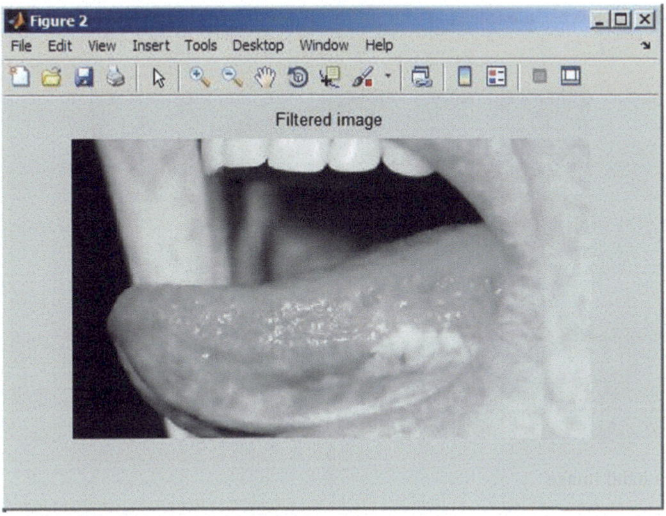

Fig. 3 Filtered image

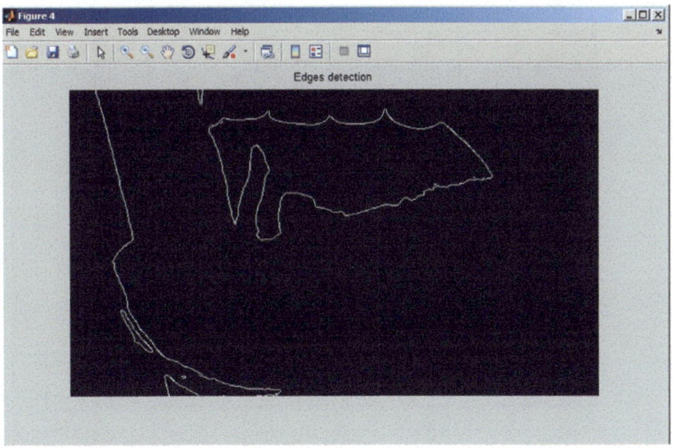

Fig. 4 Feature extraction details of the tongue image

4 Conclusion

This paper describes the feature extraction techniques for the two diseases such as the metastasis and acrometastatic for the tongue and hand images. The fundamental image processing techniques such as filtering, contrast stretching, edge detection process for the feature selection. These features are trained by means of support vector machine classifiers headed to obtain the measurements like volume, thinnest

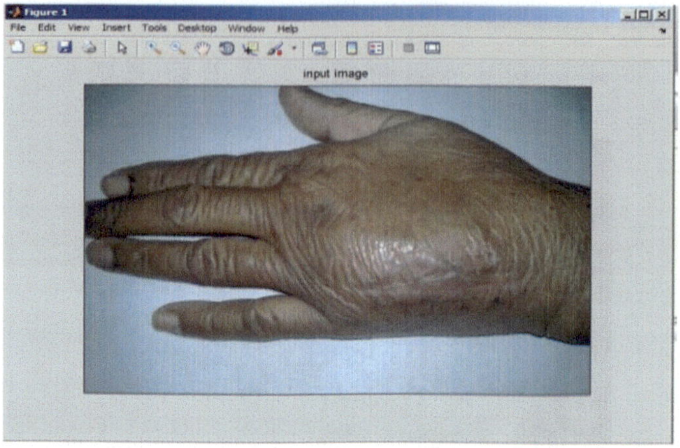

Fig. 5 Input hand image

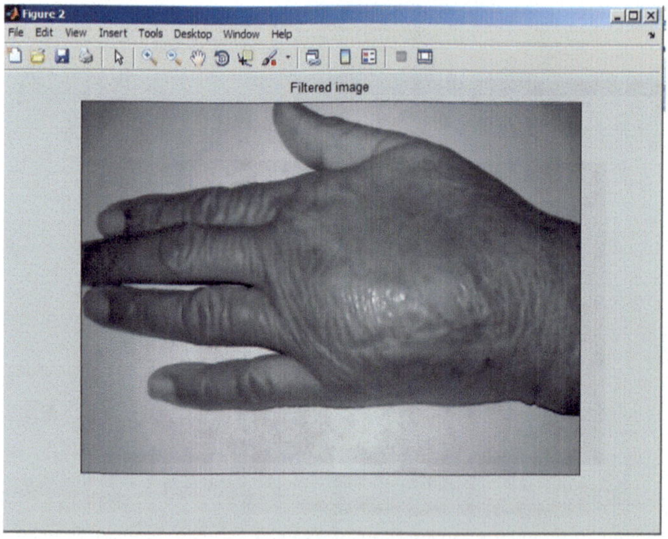

Fig. 6 Contrast stretched image

tumor thickness, thickness, area, perimeter, accuracy, sensitivity and specificity for the input image.

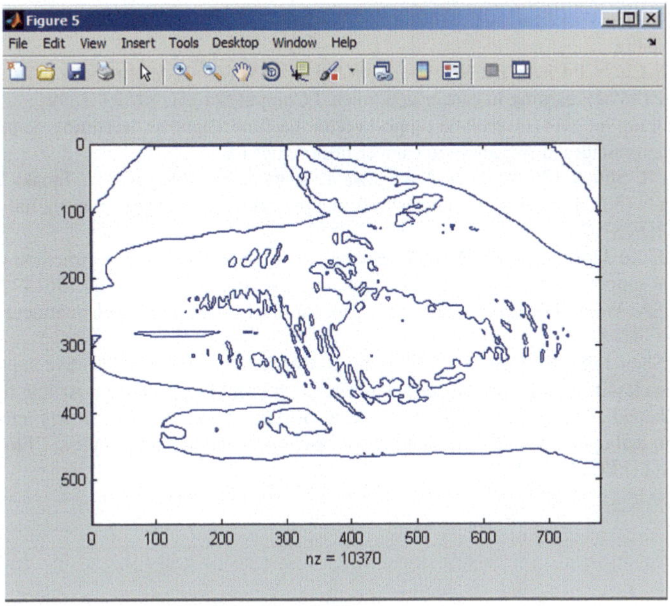

Fig. 7 Edge detection details to feature extraction of hand

Table 1 Metastasis and acrometastatic

S. No.	List of parameter	Metastasis in tongue image	Acrometastatic in hands
1.	Volume	1.109376e−01	3.409414e−01
2.	Thinnest tumor thickness	152	213
3.	Thickness	32	0
4.	Area	513	12
5.	Perimeter	1.061429e+03	2.156000e+01
6.	Accuracy	5.714286e−01	5.714286e−01
7.	Sensitivity	2.571429e−01	2.571429e−01
8.	Specificity	8.857143e−01	8.857143e−01

References

1. Muñoz-Mahamud E, Combalia A, Carreño A, Arandes JM (2016) Five cases of acrometastasis to the hand from a carcinoma and review of the literature. 36(1):12–16
2. Flynn CJ, Danjoux C, Wong J, Christakis M, Rubenstein J, Yee A, Yip D, Chow E (2008) Two cases of acrometastasis to the hands and review of the literature. Acrometastasis Hands Curr Oncol 15(5):51–58
3. Bharathi A, Natarajan AM (2011) Cancer classification using support vector machines and relevance vector machine based on analysis of variance features. J Comput Sci 7(9):1393–1399
4. Kavitha M, Lavanya G, Janani J, Balaji J (2018) Enhanced SVM classifier for breast cancer diagnosis. Int J Eng Technol Manage Res 5(3)

5. https://www.researchgate.net/deref/http%3A%2F%2Fhttps://doi.org/10.1109/OCEANS.2003.178498
6. Huang S, Cai N, Pacheco PP, Narandes S, Wang Y, Xu W (2011) Applications of support vector machine (SVM) learning in cancer genomics. J Comput Sci 15(1):1393–1399
7. Han H, Jiang X (2014) Overcome support vector machine diagnosis overfitting. Supplementary Issue: Computational Advances in Cancer Informatics (A)
8. Sawada R, Shinoda Y, Niimi A, Nakagawa T, Ikegami M, Kobayashi H, Tanaka S, Homma Y, Haga N (2017) Multiple acrometastases in a patient with renal pelvic urothelial cancer. 29, Article ID 7830207
9. Zhang J, Xu J, Hu X, Chen Q, Tu L, Huang J, Cui J (2017) Diagnostic method of diabetes based on support vector machine and tongue images, Hindawi, Article ID 7961494
10. Eccles SA, Welch DR (2007) Metastasis: recent discoveries and novel treatment strategies. Lancet 1742–1757
11. Liu N, Shen J, Xu M, Gan D, Qi ES, Gao B (2018) Improved cost-sensitive support vector machine classifier for breast cancer diagnosis. Math Probl Eng, Hindawi, Article ID 3875082
12. Li W, Zhang L, Huang Y (2010) Multiple distant metastasis of tongue squamous cell carcinoma after surgical operation and radiotherapy—a case report and literature review. Chin Ger J Clin Oncol 9(11):P669–P673

Knowledge Genesis and Dissemination: Impact on Performance in Information Technology Services

S. Karthikeyan

Abstract Knowledge either exists in abundance or needs to be generated in a tech savvy environment that we live in. Genesis per se would not suffice. Knowledge needs to be disseminated to concerned stakeholders to achieve set goals and to foster performance and productivity. The knowledge creation process espoused by Nonaka and Takeuchi has been hailed by many albeit with the criticism that certain things were abstract and may not suit non-Japanese scenarios. However, knowledge can either be tacit or explicit and the transformations keep spiraling. This paper examines the components of the SECI model and their impact on knowledge genesis and dissemination and subsequent impact on employee behavior and employee performance in information technology services. Structural equation modeling has been applied to assess the causal relations. The effect of certain demographics like cadre and educational level are also examined. The primary data stems from survey of 829 IT professionals serving in IT firms at Chennai, South India.

Keywords Knowledge · Behavior · Performance

1 Models Pertaining to Knowledge Management

Several models [1–6] pertaining to knowledge evolution and management have being doing the rounds. The popular ones are:

(i) SECI: This model traced the creation of knowledge and classified the process as a spiral.
(ii) Three Worlds: The domains comprise physical objects or materials, the produce of the mind, and psychological processes and events.
(iii) Capability Maturity: Processes in an enterprise are managed at five levels and consistency gives rise to repeatable phase.

S. Karthikeyan (✉)
Accenture, Chennai, India

S.-L. Peng et al. (eds.), *Intelligent Computing and Innovation on Data Science*, Lecture Notes in Networks and Systems 118,
https://doi.org/10.1007/978-981-15-3284-9_18

(iv) Pyramid to Wisdom: Relationships can be functional or structural depending upon the relations between four entities (wisdom, knowledge, information, and data).
(v) Business Intelligence: Management information systems could be applied in myriad ways to foster better decision-making.
(vi) Knowledge Life Cycle: This process is continuous and dynamic and comprises production, integration, and innovation.
(vii) Johari Window: Involves attributes like how information is received or disseminated and how these influence perceptions.
(viii) Knowledge Management Method: Knowledge and eventually experiences can be relished through efficient capture, sharing, exploring, and comprehension.
(ix) Bridging Epistemologies: Knowledge is neither stagnant nor predictable in an organization. It is consumed and depends on the nature and the context.
(x) Six Knows Knowledge: Knowledge can be processed and consumed and understood better when the following dimensions are applied: when, where, who, how, why, and what.

Popularity of SECI Model: Finley and Sathe [7] as well as Martin and Root [8] dwelt on Nonaka's SECI model and its applicability. Nonaka was credited with having discerned explicit from implicit knowledge in an organizational setting. The extension of this model for the construction industry was examined. The manager is the core of the spectrum and individuals go about the conversion process, and hence the significance of knowledge conversion is paramount.

Natek and Zwilling [9] averred that the management of knowledge stems from myriad disciplines; however, its core constituents are information technology and communication with individuals. The genesis and processing of knowledge comprise phases like discovery, capture, sharing, and application. Genesis of knowledge comes from various platforms, digital or otherwise (e.g., communities of practice) and has been collectively coined as 'socialization.' 'Externalization' propels the transformation to explicit processing (e.g., decision support systems). 'Combination' helps in collating knowledge and disseminates it for better availability (e.g., data mining). 'Internalization' helps in applicability by further transformation to tacit knowledge (e.g., usage of wikis). These processes may occur rapidly multiple times and the spiral continues.

Mustapha [10] highlighted the applicability of the SECI model. The domains comprised armed forces, regional application, collaborative learning, virtual learning, product development, and development of personal knowledge and computational platforms.

Kassem et al. [11] researched the cyberspace leaning environment to ascertain how as to how the SECI model could be applied. Greater the capacity of the online learning environment better are the chances of fostering the creation of knowledge. Knowledge needs to be shared and utilized and further knowledge needs to be triggered. Persons involved with similar learning tasks could be encouraged to collaborate as a group. This would facilitate faster comprehension and could trigger more ideas and thoughts. Eventually, deeper understanding would occur and new knowledge or unexplored knowledge could evolve. Synergy and productivity go hand in hand.

Kaur [12] stressed on the fact that the SECI model may not be viewed as a circular process; rather it is to be perceived as a spiral. The philosophical tag to the SECI model hampers a serious empirical research. The four components of the model stem from Japanese thought. Hence, cultural fit in other zones is debatable.

Rice and Rice [13] studied the significant constituents of the SECI model and described it as a self-transcending phenomenon. Knowledge dissemination through sharing happens when there is a need to go beyond knowledge boundaries and more clarity is required in ambiguous assignments. Organizations thrive and are remembered for their capabilities as well as knowledge capital. Knowledge transfer transcends physical entities and projects are now multi-organizational. There is so much data waiting out there to be mined and warehoused. The basic necessity is to identify the type of knowledge required for projects as well as organizational operations. Other things follow from there until knowledge is applied for sustenance and growth.

Hitherto Research on Performance and Behavior: Performance: Martin [14] investigated Government entities in Zambia with special interest on how information technology could be effectively used. Author reiterated the need for performance management system and highlighted their impact on business processes. He recommended that systems need to be put in place to capture reports and this should be implemented on a real-time basis. The access to performance data should be enabled through mobile devices.

Liviu [15] highlighted that all problems do not warrant the same IT solution. The complexity of the situation warrants an appropriate approach, whereby the proper use of information technology can produce better products, customer service, and eventually productivity.

Barbosa et al. [16] focused on information technology governance in Brazil with emphasis on the performance measurement in financial enterprises. The study highlighted that IT governance was still in a nascent phase and there was plenty of scope for its betterment and applicability. They suggested that the study could be extrapolated to other industries and services across the globe.

Zehir et al. [17] stated that linkages between information technology and productive growth have been mixed. The research model proposed by them comprised information technology at the time of decision-making, perception, usage and investment levels in information technology as well as technology and futuristic orientation. The study was undertaken in Turkey, a country in a developing stage. Among all the variables studies, the stress was on information technology investments. It was espoused that better investments would better the probability of good outcomes.

Behavior: Individual behavior (both negative and positive) as well as organizational behavior and such influence on performance has been studied by many. Some pertained to behavior modifications [18], outcomes of positive and negative behavior [19], role of leaders and their behavior [20], organizational behavior [21], and negative behaviors [22].

Research Gaps and Current Focus: Research on the knowledge creation process and its components have been executed but there needs to be more investigation on how knowledge genesis and dissemination is linked to employee behavior. Behavior may be influenced by many factors but a lot depends upon knowledge possessed, how it is communicated and how it is perceived. Behavior, in turn, affects performance and what organizations desire is discipled and committed behavior to foster growth. The current research endeavors to address these issues.

2 Need for the Research

Knowledge keeps evolving and has no pre-defined route. Knowledge may emanate from within the organization or from outside (stakeholders, media, groups). The essence of knowledge can be truly felt only when it is available in a desired form and for its relevancy. Knowledge and further processing could result in positive of negative behavior (thoughts, action and emotions). There is a need to ascertain whether knowledge in the organization is channeled for productive behavior and whether there is a productive impact on the employees' performance by such behavior. Knowledge, when manipulated or misused, could prove catastrophic leading to downfall in various aspects. Information technology is an apt domain to investigate the knowledge creation process and its consequents.

3 Objectives

The objectives of the current research:

(a) To dwell into the knowledge evolution process by comprehending information technology service employees' perception about the components of the process.
(b) To analyze the causal elements involving knowledge evolution process, employee behavior, and employee performance.
(c) To ascertain the effect of educational level and cadre on employee behavior and performance.

4 Methods and Design

Causal research enables better comprehension of how relationships function as provides a more holistic view. Employees serving in information technology services surveyed. Information technology is one of the significant services in an economy that predominantly depends on the services sector. IT services involve organizations that have multiple teams and the proper coordination and functioning of these teams depend on the knowledge expertise as well as team behavior and performance. 829 information technology service employees stemming from different cadre were

Table 1 Variables

Variables	Initial items	After CFA
Socialization	7	5
Externalization	6	6
Combination	6	6
Internalization	6	5
Behavior	3	3
Performance	4	4

requested to voluntarily participate in the study that involved purposive sampling (stage one) and random sampling (stage two, among cadre). Table 1 describes the variables [23, 24].

5 Analysis and Outcome

The sample comprised 22 (2.7%) employees serving in junior cadre, 427 (50.8%) in middle, and 386 (46.6%) in senior cadre. 130 (15.37) were drawing a salary of less than INR 30,000; 181 (21.8%) between INR 30,000 to INR 50,000 and 518 (62.5%) more than INR 50,000. 6 (0.7%) had diploma or certificate; 351 (42.3%) had a bachelors' degree and 472 (56.9) had masters' degree or above. 126 (15.2%) were serving in the IT domain for up to five years; 179 (21.6%) between six and ten years; and 524 (63.2%) for more than ten years. 620 (74.8%) were male employees while 209 (25.2%) were female employees.

H_1: Good fit is enjoyed by employee performance model (Table 2).

KGD gets augmented by 1 unit when SCN gets an accruement of 1 unit. KGD gets augmented by 1 unit when ESN gets an accruement of 0.730 units. KGD gets augmented by 1 unit when CMN gets an accruement of 0.760 units. KGD gets augmented by 1 unit when ISN gets an accruement of 0.865 units. EBH gets augmented by 1 unit when KCD gets an accruement of 0.288 units. C EPM gets augmented by 1 unit when EBH gets an accruement of 0.329 units. Employee Performance Model Fit

Table 2 Performance model paths and coefficients

Structural model path	Unstandardized coefficient	*p*
Socialization ← KGD	1.00	
Externalization ← KGD	.730	***
Combination ← KGD	.760	***
Internalization ← KGD	.865	***
Behavior ← KGD	.288	***
Performance ← Behavior	.329	***

KGD—Knowledge genesis and dissemination; *** $p < 0.001$

Table 3 Effect of educational level on employee behavior and performance

	Sun of squares	df	Mean square	F	p
Employee behavior					
Between education groups	14.526	2	13.726	1.984	0.138
Within education groups	5715.358	826	6.919		
Total	5742.779	828			
Employee performance					
Between education groups	19.496	2	9.748	1.963	0.141
Within education groups	4101.749	826	4.966		
Total	5121.245	828			

Outcome: Educational level has no effect on employee behavior and performance as p value is not significant

Table 4 Effect of educational level on employee behavior and performance

	Sun of squares	df	Mean square	F	p
Employee behavior					
Between cadre groups	5.426	2	2.713	0.391	0.677
Within cadre groups	5737.353	826	6.946		
Total	5742.779	828			
Employee performance					
Between cadre groups	5.240	2	2.620	0.526	0.591
Within cadre groups	4116.005	826	4.983		
Total	4121.245	828			

Outcome: Cadre has no effect on employee behavior and performance as p value is not significant

Indices were: CMIN/df(3.047); GFI(0.997), AGFI(0.978), NFI(0.996), CFI(0.998), RMSEA(0.057).

H_2: Educational level influences employee behavior and performance (Table 3).

H_3: Cadre influences employee behavior and performance (Table 4).

6 Conclusion

The four components had a positive relation with knowledge genesis and dissemination. Knowledge genesis and dissemination had a positive yet negligible impact on behavior of the employee. The relation between employee performance and employee behavior was also found to be mild, yet positive. Performance necessitates many skills in multiple domains to propel ahead but knowledge is indeed a critical element. Behavior too has many influences like parenting, reference groups, educational institutions, and role models. However, greater the knowledge an individual as well as organization possesses, greater would be the empowerment and commitment. Industries and services function using open-source platforms. Knowledge goes wasted unless it is facilitated in a usable and applicable form.

7 Future Scope

More research is needed about how knowledge can be optimally utilized, especially in today's age of artificial intelligence and machine language. All innovations and creativity stem from knowledge and wisdom, and more research should be undertaken, not in general, but specific to industries and services.

References

1. Mohajan H (2017) The impact of knowledge management models for the development of organizations. J Environ Treat Tech 5–1:12–33
2. Spangler SC, Skovira RJ, Kohun FG (2015) Key factors in a successful knowledge management model. Online J Appl Knowl Manage 3(1):51–60
3. Greve L (2015) Knowledge sharing is knowledge creation: an action research study of metaphors for knowledge. J Organ Knowl Commun 2(1):66–80
4. Bratianu C, Orzea I (2010) Organizational knowledge creation. Manag Mark 5(3):41–62
5. Pei NS (2008) Enhancing knowledge creation in organizations. Commun IBIMA 3:1–6
6. Web 1. Retrieved from https://warwick.ac.uk/fac/soc/wbs/conf/olkc/archive/oklc3/id151.pdf. 24 July 2019
7. Finley D, Sathe V (2013) Nonaka's SECI framework: case study evidence and an extension. Kindai Manag Rev 1:59–68
8. Martin L, Root D (2009) Knowledge creation in construction: the SECI model. In: Dainty A (ed) Proceeding of the 25th annual ARCOM conference, Association of Researchers in Construction Management, Nottingham, UK, pp 749–758, 7–9 Sept 2009
9. Natek S, Zwilling M (2016) Knowledge management systems support SECI model of knowledge creating process. In: International conference managing innovation and diversity in knowledge society through turbulent time, Timisoara, Romania, 25–27 May 2016
10. Mustapha S (2016) Towards building monolithic computational platform for SECI model. Int J Intel Serv 6:29–41
11. Kassem S, Hammami S, Alhousary T (2015) Applying SECI model to encourage knowledge creation in learning environment. Ind J Econ Res 12(4):1601–1611
12. Kaur H (2015) Knowledge creation and the SECI model. Int J Bus Manag 2(1):833–839
13. Rice J, Rice B (2005) The applicability of the SECI model to multi-organisational endeavours: an integrated view. Int J Org Beh 9(8):671–682
14. Martin M (2017) Effective use of information technology for performance management in Zambian government institutions. World Sci News 61(1):1–55
15. Liviu B (2015) Information technology and the company performance in the sector of services. Academica Brancust Publisher, pp 127–133
16. Barbosa SCB, Rodello IA, Padua SIDD (2014) Performance measurement of information technology governance in Brazilian financial institutions. J Inf Syst Tech Manag 11(2):397–414
17. Zehir C, Muceldidi B, Akyuz B, Celep A (2010) The impact of information technology investments of firm performance in national and multinational companies. J Glob Strateg Manag 4(1):143–154

18. Obiageli OL, Uzochukwu OC, Leo O, Angela AAI (2016) Behaviour modification and employee performance in selected paint manufacturing companies in Anambra state. IOSR J Bus Manag 18(9):44–53
19. Kattara HS, Weheba D, Ahmed O (2015) The impact of employees' behaviour on customers' service quality perceptions and overall satisfaction. Afr J Hosp Tour Leisure 4(2):1–14
20. Mubarak E (2014) Leadership behaviors and its effects on employees' happiness. Int J Sci Eng Res 5(10):622–624
21. Renuka P, Frederick H (2014) Organisational behaviour and its role in management of business. Glob J Fin Manag 6(6):563–568
22. Burnes B, Pope R (2007) Negative behaviours in the workplace: a study of two primary care trusts in the NHS. Int J Pub Sec Manag 20(4):285–303
23. F Al Mulhim A (2017) The effect of knowledge creation process on organizational performance: evidence from Saudi banking sector. Int J Manag Sci Bus Res 6(1):11–22
24. Akhavan P, Ghojavand S, Abdali R (2012) Knowledge sharing and its impact on knowledge creation. J Inf Knowl Manag 1–17

Examining the Acceptance of Innovations in Learning Technologies in Higher Education—A Malaysian Perspective

Dinesh Rajassekharan, Ali Ameen, and Divya Midhunchakkravarthy

Abstract Higher education institutions in Malaysia are increasingly moving towards being part of the regional higher education hub through significant progressions to stimulate learning through digital learning technologies. The interpretation of opinions gathered from similar studies provides us with many instances of technology adoption practices which is deemed necessary to overcome the challenges faced by institutions. To provide justification to the merits of learning technologies, it is critical to inspect significant elements which will be helpful in giving an appropriate framework model from within and outside learning institutions. An exploratory study is conducted here on existing work on the provisions and usage of computer learning tools in various higher education institutions (HEI) in Malaysia. The purpose of this study is to explore the framework model as a way to assist teachers improve teaching performance using computing tools so as to support progressive learning in institutions. The study findings could firstly provide pragmatic data on the technology innovations usages in HEI from a Malaysian perspective. Secondly, the study outcome may perhaps guide the HEI authority to establish practical boundaries on current and emerging ICT implementations. Thirdly, this may assist in developing strategies in line with the National Educational Policy on the use of technology-assisted learning for twenty-first century higher education. The study concludes that the use of digital technology tools can be considered as a resourceful alternative to conventional learning and teaching providing suitable recommendations for learning orientation and extend opportunities for future empirical research.

Keywords Digital learning technologies · Higher education · Learning objects · eLearning · Social media

D. Rajassekharan (✉) · A. Ameen · D. Midhunchakkravarthy
Faculty of Computer and Multimedia, Lincoln University College, 47301 Petaling Jaya, Selangor, Malaysia

A. Ameen
e-mail: abdulbaqi@lincoln.edu.my

D. Midhunchakkravarthy
e-mail: divya@lincoln.edu.my

S.-L. Peng et al. (eds.), *Intelligent Computing and Innovation on Data Science*, Lecture Notes in Networks and Systems 118,
https://doi.org/10.1007/978-981-15-3284-9_19

1 Introduction

The instructive mediums in the twenty-first century have turned out to be progressively reliant on advanced technological innovations to give a tangible and productive learning setting. The fast rate at which new innovations are incorporated and changed suggests that advanced education system modelling must keep pace with progressions in learning, abilities, requests and necessities for end users so that the colleges in the locale can prepare their pupils with the befitting information, abilities and aptitudes to be viable [8]. The data utilizations in teaching are intently connected with the development of the Internet and Web technologies. In non-developed economies in the Far East Asian region, there still exists a rift in technology usages in the learning environment. The use of information in education is related to the evolution of the Web and Internet technologies. The education system, especially in Far East Asia, points to the existence of a digital divide contrasting those in advanced nations.

The fundamental indicators in the education sector across six South-East regional countries are shown in Fig. 1. As per the World Bank ordering, Singapore and Brunei are named top-level economy; Malaysia and Thailand designated higher middle-level economy, whereas Indonesia and the Philippines are given lower middle-level economy across the Association of Southeast Asian Nations member countries.

The indicators above provide a benchmark on the relationship between the environmental variables and ICT that could encourage schools in developing strategies to improve education. This paper will try to identify the additional features and forms of ICT integration practices in the curriculum in the day-to-day operations and the choices made by administrators in the implementation and execution of ICT usages in educational institutions in Malaysia. Further, we will also identify the challenges that create reluctance in educators to integrate ICT and find ways around these barriers to promote a conducive learning environment.

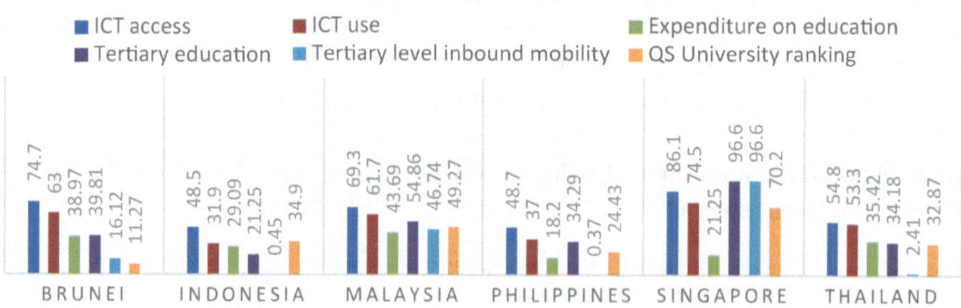

Fig. 1 Educational indicators score across ASEAN economies. *Source* Global Innovation Index, 2018

2 Information Technology and Education

The Malaysian education ministry defined strategies for ICT instruction to make sure ICT for all learners be utilized as an empowering agent to decrease the technology rift amongst the schools, emphasize the job and capacity of ICT in training as an instructing and learning instrument and increment the profitability, productivity and adequacy of the administration framework [4]. However, many tertiary institutions are still incapable to give unbiased right to use teaching tools because of rigid societal credence.

The Malaysian government has invested substantially to equip modern ICT facilities in schools to gauge the efficiency and capacity of technology use in pedagogy, the access and availability of amenities and assessment of teachers and students in Malaysian schools [9]. However, their results show that the bulk of instructors affirmed that ICT can facilitate school management although ICT policy was absent at school level and that they were unaware of the national policy on ICT. A similar study done by Lau and Sim [19] demonstrates that numerous educators do not utilize ICT in their instructions as they are not completely ICT proficient and like to utilize class-based directives for advancing participation and replication in learning. The work by Singh and Muniandi [32] suggests that pre-service training and programs in ICT for instructors empower them to have sufficient subject information, a collection of various training practices allowing self-learning to occur with a readiness of deep learning.

Educational curriculum that demands high level skillset is traditionally carried out using the conventional teaching process. With regards to Malaysia, the institutional strategy planners are facilitating information knowhow and giving advantages to the dynamic imbuement of computerized innovations in higher learning.

3 Learning Objects Facilitating Education

The inception of innovations in technologies and its assimilation in the pedagogy and learning is making good progress for IT usages although there seems to be a lack in viability and efficacy on student learning results. In academic institutions, the medium of instruction is moving away from conventional study hall strategies to innovation encouraged training. The TPACK framework (Fig. 2) as suggested by Koehler and Mishra [18] shows the interrelationship between the components within the classroom context.

McGreal [25] suggests a functional description of learning objects as entities that facilitate learning in metadata enabled online education. Object like printed book, articles or paper report that can possibly advance learning and teaching [26]; any element, advanced, non-digital, that could be recurrently utilized or emphasized in innovation education (IEEE LTSC). The above descriptions leave us to ponder that

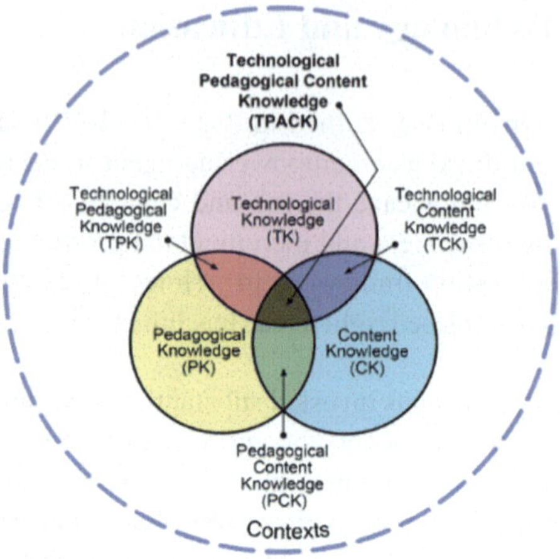

Fig. 2 The TPACK framework and its Knowledge components. *Source* Koehler and Mishra (2009)

the education item(s) can occur in an electronic or non-digital format which may be utilized to help in learning.

4 Effects of Teaching with Technologies

Educational technologies can be suggested as equipment or potential programming devices that are utilized in some helpful manner to help and support instruction and learning. As suggested by Masrom [23], both apparent simplicity and practicality of technology are distinct features influencing the user's approach towards social objectives and tangible use of technology. Similarly, work done by Lye [21] finds that although implementing ICT in education increases commitment, group work and ease in learning, the low rate of ICT implementations causes challenges in developing teaching materials and also to the instructors' awareness, techniques and skills.

The present cohort of innovation clients is consolidating more prominent coordinated effort, socialization, student-centred pedagogy, practical values through technologies like e-Portfolios, Web journals, user tagging and bookmarking, virtual reproductions and portable registering [3]. Since most instructors and students have rudimentary understanding and abilities in ICT, it has been suggested that the usage of e-Portfolio in education to identify the readiness of vocational education in Malaysia can be achieved provided the students are given more exposure to the ICT use in learning so as to evaluate their level of knowledge and performance [29].

5 Related Work

A number of interrelated studies in Malaysia have shown that educational institutions can improve the nature of instruction and education by utilizing innovative tools to produce graduates with top skills in a globally competitive employment environment. Investigation done by Letchumanan and Tarmizi [20] on the utilization of digital books as educational objects amongst students at UPM using technology acceptance model find that apparent worth and positive attitudes of students are some significant aspects contributing to the students' intention to use e-book but is limited to perceived enjoyment, cost and facilities to predict intent to use digital book.

Observations done by Hussin et al. [11] on administrators, educators and students in two universities in Malaysia (UTM, UKM) find that the student group is ready and dexterous with technologies and communicating activities by means of mobile phone although the administrators and educators need to be ready to provide infrastructure support and innovative pedagogical techniques in using mobile phones in m-learning mode. However, a study by Ismail et al. [13] finds that although many male gender teachers agreed to the benefits and willingness of portable computing device as assisting tool for education at institutions, they were sceptical towards m-learning and did not agree or were not ready to embrace mobile phone technologies for their teaching activities. With the intention of supplementing the study of computer games in education on students' motivation and engagement in tertiary education, Ibrahim and Jaafar [12] propose the creation of an educational Internet-based computer game focussing on the ease of use, multimodal approach for enjoyment and finding solutions in the learning curriculum.

Similar studies were conducted by Shuib et al. [31] at USM on the growth of portable learning technologies to help English language learning by means of cell phones could provide a customized learning setting that accommodates person's learning process and the portability needs. In order to boost learning process, utilizing innovation in electronic learning entities can enable eLearning to provide equilibrium to the needs of learners and establishment.

6 Paradigms of eLearning Initiatives

The progression in the educational delivery from offline PC centred to electronic and blended learning has personified the requirement to develop strategies for training and pedagogy. The work done by Jamian et al. [14] exploring the conduciveness and enrichment of eLearning in mixed learning setting in UPM finds that hidden affordance occurring in learning management system and slideshow presentation programs need to be transferred to perceived affordance by training and education to good ICT practices.

The examination by Hsbollah et al. [10] on the attributes of lecturers in UUM on decision making adoption in education line finds that lecturers should not only improve participation in using eLearning, be proactive towards innovation usages but should also be provided incentives by the institution for the time and energy they spend using the technological innovations. The survey done by Poon et al. [27] at eight universities in Malaysia on innovation acceptance found that the primary factors that affected the adequacy of Web-based learning method were learners' particular values and the quality of the instructors' technical skills but not disregarding the policy maker's guidelines and exertions towards recognizing trust and confidence of Malaysians towards eLearning.

The observation by Karim and Hashim [16] on the implementation of eLearning at UPSI found issues/challenges relating to courseware design and development team, system maintenance and safety, lack of technical skills and management support, deficiency in training and coaching of teachers, learners and non-teaching workforce, etc. The authors conclude that some guidelines provided by the Ministry of Education to aid schools and HEIs to realize eLearning effectively and productively could support educators and learners improve their practical and perspective abilities in ICT to progress into real consumers of innovations.

7 The Singularity of Learning Management System

The Internet supplements locally accessible data, improves and quickens information streams, and can be utilized to convey ingenious training models to remote locations [15]. Learning management systems (LMS) are used by specific professions through Internet-based platforms offering various functionalities to events subject to the intricacies of the programming [24]. The complicated choreography of activities within the system could cause apprehension and misperception to the inexperienced client. Then again, Vrazalic et al. [34] attests that online human communication can mitigate specialized issue and time-based obstacles to improve inspiration and commitment in the learning procedure. Learning institutions should take opportunities of freely available open source LMS systems in building prototype and testing the model according to their customized needs.

In an attempt to provide insights on the impact of LMS usages amongst various university students, Adzharuddin and Ling [1] find that many generally use the academia learning management system and express positive views about LMS in helping them enhance their learning process as long as the universities provide training and guidance for students and lecturers and provide on-call team to solve any problems in LMS usages. However, the availability of just the essential technical innovations does not ensure the ideal execution of LMS by the employees and administrators of the organization. The irregularities of the pupils' utilization of innovation in schools and their homes may deliver indecisive result in learning. Study done by Kaur and Sidhu [17] in a local Malaysian university on user learning autonomy and views in using LMS propose that all learners who expect to set out on Internet learning should be facilitated, so they are provided with the correct learning devices, for example, having the capacity, information and aptitudes to design, arrange, observe and assess their own learning before setting out on a Web-based learning experience.

8 Social Networking Tools for Learning

Societal networking tools in Web 2.0 technologies are being progressively infused as a technology support tool in education instead of a tool used for personal communication. Many studies have highlighted the importance of communication technologies in education in developed countries and the contradictory empirical evidence on the benefits. Survey conducted by Hafiz et al. [5] to gather the utilization of Web 2.0 applications by Malaysian pupils for both formal and casual kinds of learning is observed to be encouraging even though some level of unfamiliarity and reluctance was observed for specific learning devices. Similarly, comparative examinations demonstrate that learners perceive and revere learning advantages of utilizing social media in tertiary education [7], have optimistic perception on interpersonal interaction and its impact in their lives as learners although students commonly use societal media apps for casual knowledge gathering or which is focused on the unpremeditated parts of learning [6].

The effect of online networking to improve academic productivity is shown by Al-Rahmi et al. [2] that socializing apps encourages cooperative learning and commitment which improves the intellect of learners and scholars. On the other hand, the exploration of Manca and Ranieri [22] on the evolving usages of Facebook to discover its teaching prospects discovers that learners are not able to accept familiar and casual tools like Facebook as exclusive devices for learning purposes. Nevertheless, Rasiah [28] finds proof that the use of Facebook outside class to supplement direct instructional methods could benefit the current generation learners as instructional medium to assist them in group work project collaborations. Although a preliminary survey study by See Yin Lim et al. [30] stated that the students mainly utilize social media technologies (SMTs) for socializing, the information collected from their surveys showed that although learners and trainers had started to assess and support the use of media technologies as a means for engagement as well as for teaching and learning, limited knowledge was known of its usage and the outcome within educational settings.

9 Conclusions

This exploration of studies across various HIE in Malaysia on the use and implementation of digital learning technologies has addressed various theoretical and practical issues aimed at supporting progressive education in institutions. Empirical studies from many authors find that ICT policy was absent at school level or they are yet to build up a procedure for fair access to learning tools to benefit from the observed effectiveness of technology. Education is subjective to the student's personal characteristic and the quality of the instructors' technical skills. Most of the facilitators lack sufficient training and are not fully ICT literate to assist, prepare and guide the students in utilizing eLearning.

The change in outlook from the conventional education to computerized learning systems is gaining headway in establishments of learning the world over. The advantages of actualizing advanced learning innovations are many considering the organization necessities, practical and managerial sustenance. Yet, a large number of the establishments in Malaysia do not have or have not effectively commissioned ICT. The disparities in the community of those who have and those who do not have tangible reach to technology innovations, abilities and aptitudes and that accomplishing such access for all would take care of specific issues in the economy and society [33]. There is a convincing worldwide concurrence for an adjustment in the learning delivery methods, from traditional techniques to ones that exploit the advances in computerized innovation. Besides, the more current improvements in eLearning appear to give progressively advantageous, completely intelligent learning sensation, providing extra appeal for present and new generations.

References

1. Adzharuddin NA, Ling LH (2013) Learning management system (LMS) among university students: does it work. Int J E-Educ, E-Bus, E-Manag E-Learn 3(3):248–252
2. Al-Rahmi W, Othman M, Yusuf L (2015) The role of social media for collaborative learning to improve academic performance of students and researchers in Malaysian higher education. Int Rev Res Open Distrib Learn 16(4)
3. Buzzetto-More N (2008) Student perceptions of various eLearning components. Interdiscip J ELearning Learn Objects 4(1):113–135
4. Chan FM (2002) ICT in Malaysian schools: policy and strategies. In a workshop on the promotion of ICT in education to narrow the digital divide, pp 15–22
5. Hafiz Zakaria M, Watson J, Edwards SL (2010) Investigating the use of Web 2.0 technology by Malaysian students. Multicult Educ & Technol J 4(1):17–29
6. Hamat A, Embi MA, Hassan HA (2012) The use of social networking sites among Malaysian university students. Int Educ Stud 5(3):56–66
7. Hamid S, Waycott J, Kurnia S, Chang S (2015) Understanding students' perceptions of the benefits of online social networking use for teaching and learning. Internet High Educ 26:1–9
8. Hong KS, Songan P (2011) ICT in the changing landscape of higher education in Southeast Asia. Australas J Educ Technol 27(8)
9. Hoque KE, Razak AZA, Zohora MF (2012) ICT utilization among school teachers and principals in Malaysia. Int J Acad Res Proglve Educ Dev 1(4):17–34
10. Hsbollah HM, Idris KM (2009) ELearning adoption: the role of relative advantages, trialability and academic specialisation. Campus-Wide Inf Syst 26(1):54–70
11. Hussin S, Manap MR, Amir Z, Krish P (2012) Mobile learning readiness among Malaysian students at higher learning institutes. Asian Soc Sci 8(12):276
12. Ibrahim R, Jaafar A (2009) Educational games (EG) design framework: combination of game design, pedagogy and content modeling. In: 2009 international conference on electrical engineering and informatics vol 1. IEEE, pp 293–298
13. Ismail I, Azizan SN, Azman N (2013) Mobile phone as pedagogical tools: are teachers ready? Int Educ Stud 6(3):36–47

14. Jamian M, Ab Jalil H, Krauss SE (2012) Malaysian public university learning environments: assessing conduciveness through ICT affordances. Procedia-Soc Behav Sci 35:154–161
15. Kamba MA (2009) Problems, challenges and benefits of implementing eLearning in Nigerian Universities: an empirical study. Int J Emerg Technol Learn 4(1)
16. Karim MRA, Hashim Y (2004) The experience of the eLearning implementation at the Universiti Pendidikan Sultan Idris, Malaysia. Malays Online J Instr Technol (MOJIT) 1(1):50–59
17. Kaur R, Sidhu G (2010) Learner autonomy via asynchronous online interactions: a Malaysian perspective. Int J Educ Dev Using ICT 6(3):88–100
18. Koehler M, Mishra P (2009) What is technological pedagogical content knowledge (TPACK)? Contemp Issues Technol Teach Educ 9(1):60–70
19. Lau BT, Sim CH (2008) Exploring the extent of ICT adoption among secondary school teachers in Malaysia. Int J Comput ICT Res 2(2):19–36
20. Letchumanan M, Tarmizi R (2011) Assessing the intention to use e-book among engineering undergraduates in Universiti Putra Malaysia. Malaysia. Library Hi Tech 29(3):512–528
21. Lye LT (2013) Opportunities and challenges faced by private higher education institution using the TPACK model in Malaysia. Procedia-Soc Behav Sci 91:294–305
22. Manca S, Ranieri M (2013) Is it a tool suitable for learning? A critical review of the literature on Facebook as a technology-enhanced learning environment. J Comput Assist Learn 29(6):487–504
23. Masrom M (2007) Technology acceptance model and e-learning. Technology 21(24):81
24. Mattheos N, Stefanovic N, Apse P, Attstrom R, Buchanan J, Brown P … Walmsley AD (2008) Potential of information technology in dental education. Eur J Dent Educ 12(s1):85–92
25. McGreal R (2004) Learning objects: a practical definition. Int J Instr Technol Distance Learn 1(9):21–32
26. Naidu S (2006) E-learning: a guidebook of principles. Procedures and Practices, 2nd Revised Edition, CEMCA
27. Poon WC, Lock-Teng Low K, Gun-Fie Yong D (2004) A study of web-based learning (WBL) environment in Malaysia. Int J Educ Manag 18(6):374–385
28. Rasiah RRV (2014) Transformative higher education teaching and learning: using social media in a team-based learning environment. Procedia-Soc Behav Sci 123:369–379
29. Ruhizan MY, Bekri M, Faizal AN (2014) Vocational education readiness in Malaysia on the use of e-portfolios. J Tech Educ Train 6(1)
30. See Yin Lim J, Agostinho S, Harper B, Chicharo J (2014) The engagement of social media technologies by undergraduate informatics students for academic purpose in Malaysia. J Inf, Commun Ethics Soc 12(3):177–194
31. Shuib M, Abdullah A, Azizan SN, Gunasegaran T (2015) Designing an intelligent mobile learning tool for grammar learning (i-MoL). Int J Interact Mob Technol 9(1)
32. Singh TKR, Muniandi K (2012) Factors affecting school administrators' choices in adopting ICT tools in schools—the case of Malaysian schools. Int Educ Stud 5(4):21–30
33. Van Dijk JA (2006) Digital divide research, achievements and shortcomings. Poetics 34(4):221–235
34. Vrazalic L, MacGregor R, Behl D, Fitzgerald J (2009) eLearning barriers in the United Arab Emirates: preliminary results from an empirical investigation. IBIMA Bus Rev 4(1):1–7

Web Content Classification Techniques Based on Fuzzy Ontology

T. Sreenivasulu, R. Jayakarthik, and R. Shobarani

Abstract Web content classification utilizes for converting the organization document into predefined classes with the help of machine learning algorithms. This is mostly applied in the industries which utilize the unstructured text format information largely. Web content classification is often used to filter email, classify Web content and manage Web browser results. The word collections represent the documents in the traditional Web content classification. The terms are obtained from their finer context which presents in a document or sentence. Detailed semantic classification of the Web content is discussed here. This study examines past and past achievements in the semantic Web content classification. This approach is based on SVM, a fuzzy ontology. In addition, this study shows the advantages of semantic Web content classification algorithms compared to tradition.

Keywords Semantic Web content · Classification technique · SVM · Fuzzy ontology

1 Introduction

Today, the data are approachable to everyone because of the World Wide Web (WWW). Huge amount of data are exposed in currently by the Internet users. It is enhanced for the intelligent software agents for improving the capability of filtering, finding and sorting of available content. Web mining is termed as data mining and is of three kinds: (a) Web usage mining (b) Web structure mining and (c) Web content or site content mining. 'Web usage mining' also termed as the application of data mining. Web content classification considers as the supervised learning task for

T. Sreenivasulu (✉)
VELS Institute of Science, Technology and Advanced Studies (VISTAS), Chennai 600117, India

R. Jayakarthik
Department of Computer Science, VELS Institute of Science, Technology and Advanced Studies (VISTAS), Chennai 600117, India

R. Shobarani
Department of CSE, Dr. MGR Educational and Research Institute, Chennai 600095, India

S.-L. Peng et al. (eds.), *Intelligent Computing and Innovation on Data Science*, Lecture Notes in Networks and Systems 118,
https://doi.org/10.1007/978-981-15-3284-9_20

choosing the content of the Web to one or more of post defined classifications. The above-mentioned classes can be linked, for instance, the presence of spam or to the topic of the content or pornographic content or function of the content on the page. The key goal of classifying Web content is to automatically categorize different types of academic Web content. This is achieved by first grouping Web content into categories that are in line with the three missions of Higher Education Institutions (HEIs) and then into functions of its Website and finally by automating the classification system with managed machine learning techniques.

Many studies have applied machine learning or statistical classifications like support vector machines, Bayesian classifier, neural networks and K-nearest neighbour classifiers to resolve the difficulty of classification. The above-mentioned problem is eliminated by the utilizing of the meta-heuristic approach. Supervised machine learning techniques helped to carry out the text classification process. In order to categorize new and unlabelled instances, a classification algorithm is trained over a quantity of labelled documents so that it captures the most distinctive category patterns.

2 Related Work

Kadhim AI conducts analysis for supervising the automated text classification by the utilization of machine learning techniques. The text classification efficiency was enhanced by specific rules of appropriate weights designed with the help of term weighting methods [1].

Sharma S et al. have proposed a semantic approach for Web service categorization. Hybrid approach independent of service models is applied in this work. Proposed approach greatly helps the registry administrator and users to register and receive service betterment. Support vector machine (SVM) and K-nearest neighbour classifiers are utilized for categorizing the various classifications. OWL-X dataset is provided with estimation and separation by implementation of the recommended approach to discover and reuse the existing services [2].

Pradhan VM et al. by working sentiment analysis algorithms from rapid growing areas like opinion mining provided survey result. The feedback of the specific product was obtained with the help of e-commerce sites which provide more help to the individuals as well as organization to buy a specific product and to predict sentiments. The pre-processed analysis provides an object on which opinion is presented by considering opinion words. The polarity of re-exam is obtained by the various opinion mining techniques. In this work, various algorithms sentiments analysis have been analysed because of its arisen of challenge and application [3].

Khader AT and Abualigah LM conduct a work by the help of unsupervised text feature selection technique method that uses an algorithm for optimizing a swarm of particles using a genetic operator for character selection problems. Effectiveness of the obtained subsets is evaluated by using the k-means clustering. This paper feature suggests a hybrid of particle swarm optimization algorithm (PSO) with genetic operators. Also, for performance improvement of the clustering algorithm, a fresh subset of the more descriptive feature is generated by the hybrid algorithm (H-FSPSOTC). The proposed algorithm is weighed against the other algorithms for comparison in the literature for publishing purpose [4].

Fdez-Glez J et al. had proposed a dynamic model to integrate simple categorization techniques of Internet spam. In this context, current reexamine introduced a new framework for Internet spam filter which use WSF2, along with variety of categorization instructions and schemes. They designed an experiment that includes publicly available corpus and various simple well-known classifiers and group approaches to calculate the performance of the dynamic model. The WSF2 read the results correctly, availability of each category and achieved better performance compared to other replacements [5].

3 Overview of the Re-examine Method

Below is a brief overview of some of the data mining techniques.

Decision tree (DT): In machine learning and classifications, decision trees are one of the most preferred algorithms. The decision trees are mainly used for partition of instance space into subspace. The root nodes of incoming edges have no incoming edges, and therefore, it is known as the directive trees. There is just one incoming edge for other nodes of the decision tree, while the internal node has one outgoing edge. The remaining nodes of decision trees are termed as leaf nodes. The choosing of the internal node instead of root node conducted by the using of multiple criteria.

C4.5 or J48 is known as the gain ratio which used to induce the decision tree. The information gain is used to modify the C4.5 in which bias effect is reduced towards multi-valued attributes. J48 is mostly used because of its better properties in the handling of complex values and accuracy measures than Gini and information gain index. The effect of bias was reduced in J48 for allowing the consistency and breadth of values. The highest gain ratio attribute is considered as splitting attribute in decision tree construction [6].

Adeniyi et al. [7] had adopted the use of regression tree and classification. Entropy value and Gini indexes are applied as the splitting indexes when the construction of the decision tree. The decision tree method has the restriction that the training tuples ought to dwell in memory, in this way, on account of exceptionally enormous information, decision tree develop, in this manner, gets wasteful due to swapping of the training tuples all through the primary and cache memories. The result of this research work shows scalable approaches like KNN method, fit for dealing with enormous training data and memory [7].

Decision tree classifier is a tree that consists of nodes tagged by; the branches that proceed by them are tagged by checks on the weight which has the term in the test document, and also, the terminals are tagged by classifications. Divide and conquer strategies are utilized by DT for enhancing the development of tree classifiers and

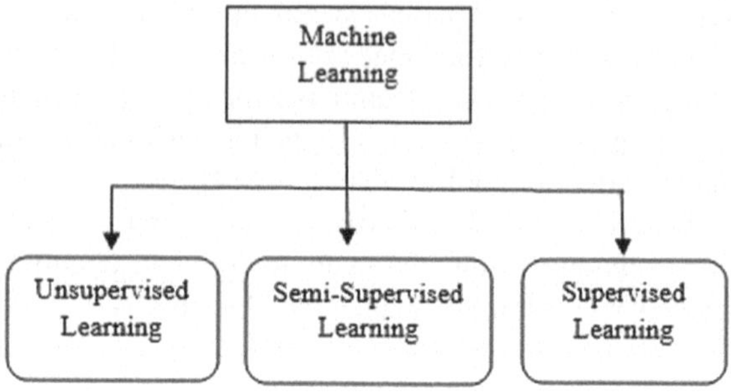

Fig. 1 Machine learning-based opinion classification techniques

also recursively partition. The developing of DT was very simple and interpretation also very easy which are the main advantages. In addition, without any retraining of sample data, DT has the capacity to "grow" in adaptive manner [8].

4 Proposed Methods

The aim of the machine learning techniques is to achieve a good prediction by achieving low bias and low variance. This section explains how to convert text messages to vectors and use training methods to tailored Web content sentiment analysis.

4.1 Classification Techniques

Classification is the method of assigning a class name to a group of unclassified cases. There are three types of classifications:

- Unsupervised Classification
- Semi-supervised Classification
- Supervised Classification (Fig. 1) [11].

4.2 Unsupervised Classification

The labels of every class will not be known for an unsupervised type of classification. After classification, group the records by natural similarity and assign the records to class labels. The clustering is also called unsupervised classification.

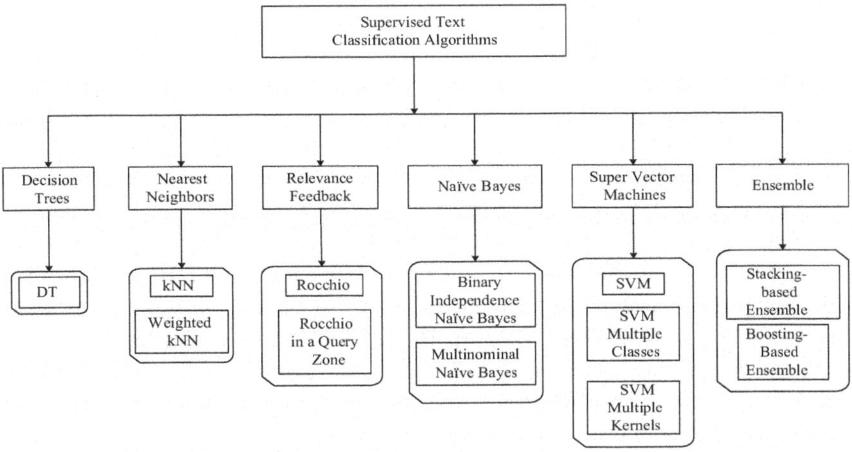

Fig. 2 Supervised learning classification algorithm

4.3 Semi-supervised Classification

During the training process of basic classifiers, they accept only labelled data. As they require the endeavors of experienced human annotators marked examples, be that as it may, much of the time troublesome, costly, or tedious to acquire. It needs high domain knowledge for analysing the comments by labelling the scenarios. There are not many approaches to utilize them, meanwhile, it is easy to collect the unlabelled data because it is abundant and mostly available in Internet. The problem of classifying such unlabelled data has been solved by semi-supervised learning technique. It classifies the unlabelled data based on the minimal amount of data already tagged.

4.4 Supervised Classification

The labels of every class will be known from the initial stage on supervised type of classification. Here, on the training data, several records with various attributes are predefined with their class labels. It creates the learning model by examining training dataset for in this technique. On testing data, the class labels are assigned by using the created model (Fig. 2).

4.5 Support Vector Machine (SVM)

The support vector machine (SVM) is a type of machine learning technique works with any type of learning. SVM is mainly utilized for the problem of classification and regression. The SVM is also considered as the binary linear classifier [6]. Use kernel equations to compile data in multidimensional space to create a hyperplane that separates data patterns from one view to another. The main function is the ability to convert data that cannot be linearly separated from one domain to another. Virtual machines are linearly divided into new domains. It also includes different types of kernel equations includes Gaussian, quadratic, linear and so on. Once it is separated into two, the better hyperplane is defined to separate the data into several instances. During the categorization process, the SVM only uses binary data. The main aim of the SVM is to identify the boundary of the hyperplane to complete the classification of the performance analysis. A well-known and best standard library called LIBSVM is developed by Ali F. et al. in 2016 [9]. The LIBSVM is a combined software, and it helps on supporting various tasks like regression, classification and distribution. LIBSVM classifies several classes, also achieves and does two things. The SVM method contains several steps which are given as follows:

- Based on the SVM package, the format of data will be transformed.
- Perform a scaling operation for scaling the data.
- Examine the RBF kernel.
- Perform cross-validation for identifying the best parameter and train the complete training set.
- Test.

After pre-processing, the listed steps are performed for training SVM. The proposed system finds an appropriate support vector for the classification of these two classes. The classification function then determines the best result for returning the hyperplane with the maximum margin. After a positive and negative evaluation of the review, the polarity of each function is estimated by using FDO.

The learning process of linear hyperplane is used by support vector machine (SVM) from the training set for differentiating the negative and positive examples of dark and darkness. Hyperplanes are known as support vectors and are located at the point of super space, which maximizes the distance between positive and negative numbers. There are two factors are in linear classification: perpendicular to the hyperplane and vertical weight (including training with the participation of the components). The offset of the hyperplane from the source is decided by a bias $\dot{b}$. The classification of untagged example $\dot{x}$is given as positive if $f(\ddot{x} = \ddot{W}\ddot{X} + \dot{b} \geq 0)$, else, it is categorized as negative.

The main benefits of 'support vector machine' (SVM) as text classifier are that it can easily deal with various feature space even it is exponential or substantially, it does not need any trained samples for representing it on its transformed space and estimate the similarity more efficiently. The SVM does not need any aggressive feature selection process because of having capability on handling high dimensional and redundant features [8]. For example, the efficiency of SVM classifier is shown by analysing the C4.5 in the domain of DW and comparing it with various classifiers. Most of the approaches based on SVM proved that the SVM has the ability to provide more exact results on classifications.

In the problem of text classification, the most accepting results are demonstrated by SVM-based powerful learning techniques. It reduces the generalization error more than the local error on training data, and meanwhile, it is based on the theory of Structured Risk Maximization. The SVM-based learning method developed by Abdelbadie B et al. [10] for text categorization, the efficiency of SVM is proved by using a linear kernel while having higher dimensional feature vector. They utilized two dataset for experimentation, and the second dataset earned better performance than the first one.

5 Discussion

There exist numerous noticeable problems in the Web content classification process. The data collection process has found that the previous steps of the pre-processing, feature extraction and finally the text categorization have been affected.

Most of the recent studies utilized support vector machine to find out an objectionable or unwanted content, and further, fuzzy logic is employed to categorize the relevant category of Web page such as adult, medical and normal. The specific Web page type is recognized through feature and pre-processing extraction process. During the pre-processing stage, the Web page elements are transformed into a format of HTML file. The Web content classification system extracts this information and detects the different forms of words through morphological analysis using a lexicon. To split a compound text into chunks, tokenization process is employed. In order to compare the tokenization outcomes with the dictionary of supporting lexicon and words, they are kept in array form.

The next phase is feature extraction, and then, the following the recovery of each word (ordinary, restorative and obscene words), the exactness rate is low. In this way, the SVM classifier is utilized to discover significant features with important words while evacuating disconnected ones. A unique function is connected to recognize significant features. If a content value is greater than 0, that states that the content is related to pornography or medicine; otherwise, the content is filtered out.

Since FPv and FNv ratings are the key to adult content detection, renowned methods were used to evaluate classification system. We have selected 4646 Web pages of three different types randomly. The system examined the information in these Web pages and categorized them into three types: normal, medical and adult content. The proposed system categorized Web pages have 1787 as normal, 2251 as pornographic and 608 as medicine. For each request to Web page, the indicator value was evaluated for Web page classification by making use of lexicon dictionary along with certain set of features. As and when the indicator value was evaluated and confirmed, the access to adult content was blocked.

To evaluate the discussion of the proposed fuzzy ontology, of broadened metrics while the precision, recall, accuracy, and function types are characterized additionally used to break down by n-gram, KNN, and SVM classifiers.

Table 1 Precision, recall, accuracy and function measure for the fuzzy ontology with SVM, n-gram, KNN and SVM techniques

Classification method	Precision (%)	Recall (%)	Accuracy (%)	Function measure (%)
Fuzzy ontology with SVM	98	94	97	96
n-gram	85	80	94	82
KNN	84	94	95	89
SVM	97	81	94	87

$$\text{Precision (P)} = \text{TPv}/((\text{TPv} + \text{FPv})) \times 100\% \tag{1}$$

$$\text{Recall (R)} = \text{TPv}/((\text{TPv} + \text{FNv})) \times 100\% \tag{2}$$

$$\text{Accuracy} = ((\text{TPv} + \text{TNv}))/((\text{TPv} + \text{FPv} + \text{FNv} + \text{TNv})) \tag{3}$$

$$\text{Function Measure} = 2 * (P * R)/((P + R)) \tag{4}$$

In adult content classification, abbreviations denoted as TNv—true negative, FNv—false negative, FPv—false positive and TPv—true positive. Table 1 refers real-time analysis results of the n-gram, fuzzy ontology, SVM and KNN related to the detection of adult content. It is evident that the n-gram method accomplishes 94% accuracy which is good. Nevertheless, values of precision—85%, recall—80% and function measure is 82%. Alternatively, KNN mechanism achieves good results with accuracy up to 95% and recall up to 94%. KNN method, nevertheless, yields function measure—89% and precision—84%, whereas SVM yields good results with precision—97% and accuracy—95%. However, function measure—87% and recall measure—81% only.

In distinction, the metrics are best with proposed fuzzy ontology with SVM. The relative results highpoint the dominance of the proposed fuzzy ontology with SVM system compared to other three techniques—SVM, KNN and n-gram in detecting and classifying adult content.

6 Conclusion

Executing a semantic 'Web content classification algorithm' involves numerous challenges for researchers: In Web content classification, text document should try to prefer composite semantics in the form of natural language data by utilizing feature vector or structure. Because of complex computations, most of the Web content classification methods are unsuitable to many applications. In such scenarios, there should be a considerable decrease in the number of features, which may in turn affect the performance of Web content classification. Due to dynamic processing, the data that was considered in the previous instant as non-relevant may now become relevant and vice versa. Hence, it is a serious issue when considering context-based feature extraction process.

It is required that the researches must apply various text mining approaches in order to filter and pre-process data in corpus-based and learning-based systems. However, it is still a challenging task to identify the use of corpus-based or learning-based method as it depends on the availability of dataset, knowledge bases, size of dataset and the nature of issue under examination. In upcoming days, Web content mining tools can be used as intelligent agents for extracting the reporting relevant study results and textual materials to the users without any need of explicit request.

For successfully achieving content classification systems, it is critical to extract semantic connections from such unstructured form. Various inter-related factors contribute to the identification of semantic content classification method. Every approach has certain advantages over the other; however, in the meantime, certain restrictions result in sufferance.

References

1. Kadhim AI (2019) Survey on supervised machine learning techniques for automatic text classification. Artif Intell Rev 1–20 (2019)
2. Sharma S, Lather JS, Dave M (2016) Semantic approach for Web service classification using machine learning and measures of semantic relatedness. Serv Oriented Comput Appl 3(10):221–231 (2016)
3. Pradhan VM, Vala J, Balani P (2016) A survey on sentiment analysis algorithms for opinion mining. Int J Comput Appl 9(133):7–11
4. Abualigah LM, Khader AT (2017) Unsupervised text feature selection technique based on hybrid particle swarm optimization algorithm with genetic operators for the text clustering. J Supercomput 11(73):4773–4795
5. Fdez-Glez J, Ruano-Ordas D, Méndez JR, Fdez-Riverola F, Laza R, Pavón R (2015) A dynamic model for integrating simple web spam classification techniques. Expert Syst Appl **21**(42):7969–7978 (2015)
6. Qamar U, Niza R, Bashir S, Khan FH (2016) A majority vote based classifier ensemble for web service classification. Bus Inf Syst Eng 58(4):249–259
7. Adeniyi DA, Wei Z, Yongquan Y (2016) Automated web usage data mining and recommendation system using K-Nearest Neighbor (KNN) classification method. Appl Comput Inform 12(1):90–108
8. Sabbah T, Selamat A (2014) Modified frequency-based term weighting scheme for accurate dark web content classification. In: Information retrieval, pp 184–196
9. Ali F, Kwak K-S, Kim Y-G (2016) Opinion mining based on fuzzy domain ontology and support vector machine: a proposal to automate online review classification. Appl Soft Comput 47:235–250
10. Abdelbadie B, Mohammed B (2014) A clique based web page classification corrective approach. In: Web intelligence (WI) and intelligent agent technologies (IAT), vol 2, pp 467–473
11. Hemmatian F, Sohrabi MK (2017) A survey on classification techniques for opinion mining and sentiment analysis. Artif Intell Rev 52:1495–1545. https://doi.org/10.1007/s10462-017-9599-6

Field Programmable Gate Array (FPGA)-Based Fast and Low-Pass Finite Impulse Response (FIR) Filter

R. Raja Sudharsan and J. Deny

Abstract The finite impulse response (FIR) filters are one among the digital filters which are widely proposed in field programmable gate array implementations. This paper presents the design of 4-tap and 8-bit fast low-pass FIR filter design under FPGA background using hardware description language (HDL). This design leads many applications like biomedical signals, pattern recognition, image processing and communications fields. The main attention of this FIR filter is focused towards the noise and performance constraints. In light of FPGA to accomplish FIR filter, not just considered the fixed capacity DSP-explicit chip constant, yet in addition the DSP processor adaptability. The blend FPGA and DSP innovation can further improve integration, increment work speed and framework abilities.

Keywords Field programmable gate array (FPGA) · Finite impulse response (FIR) · Hardware description language (HDL)

1 Introduction

Signal processing plays a crucial role in modern electronic systems, and it is one of the emerging areas especially in biomedical, mobile applications, etc. In DSP applications, digital filters like infinite impulse response (IIR) and finite impulse response (FIR) play a significant role in design. Partition is required when impedance of commotion or other signal pollutes the actual signal, and rebuilding is useful when a signal is distorted by certain methods [1]. Each of these assignments requires exact recurrence particulars to evacuate noise or other distortions which can be accomplished with higher order finite impulse response (FIR) filter.

R. Raja Sudharsan (✉) · J. Deny
Department of Electronics and Communication Engineering, School of Electronics and Electrical Technology, Kalasalingam Academy of Research and Education, Krishnan Koil, Srivilliputtur (via), Virudhunagar (Dt), Tamil Nadu, India
e-mail: rajasudharsan@klu.ac.in

J. Deny
e-mail: j.deny@klu.ac.in

S.-L. Peng et al. (eds.), *Intelligent Computing and Innovation on Data Science*, Lecture Notes in Networks and Systems 118,
https://doi.org/10.1007/978-981-15-3284-9_21

FIR filters are generally utilized because of the intense various design algorithms that already exist and their stability [2]. The non-recursive implementation of the FIR filter is a straightforward way to accomplish the linear phase. Equi-ripple filter, least mean square method, windowing technique for planning of low-pass FIR filter exist. A windowing technique is utilized, where Kaiser, rectangular, Hamming, Hanning and Blackman capacities are composed. In general, FIR filter performs signal pre-conditioning, low-pass filtering, anti-aliasing, decimation, band-select filtering and interpolation. The main advantage of using FIR filter is that it is faster in speed and consumes low power [3].

With the broad utilization of the rapid digital signal processing (DSP) innovation, PLD has additionally expanded speed and incorporation, which gives another approach to digital signal processing. By and by, the signal processing must be auspicious and exact or else it will lose its significance. Along these lines, the continuous digital signal processing, speed and dependability become a significant marker of sign preparing [4]. In addition to eliminating noise, FIR filter provides better digital signal output using Filed Programmable Gate Array (FPGA) [5]. FPGA is one of the common PLDs; its consistent logical block arrangement is standard, and the association asset is rich; at the same time, its engineering of LUT is relevant to actualize continuous, rapid and dependable FIR channel. Therefor, FIR channel module dependent on FPGA chip configuration has incredible point of interest.

One of the key advances of EDA is to plan computerized equipment framework with equipment portrayal language (HDL). At present, Verilog HDL is the most broadly utilized equipment depiction language. Verilog HDL is one of the most generally used hardware description languages. Its hardware portrayal capacity is extremely amazing from the rationale entryway level, circuit level to framework level, and different levels can be depicted and displayed, for example a counter, a capacity framework, a chip, turning into the business standard equipment depiction language. In particular, the plan of the Verilog HDL language has nothing to do with the particular equipment, consequently lessening the trouble of planning the equipment circuit, permitting free structure and adaptable portrayal. Therefore, under the support of EDA technology, reconfiguring the inner equipment structure and working method of the FPGA chip is more time-saving, cost-saving, good adaptability and great transplant capacity. By utilizing Verilog HDL as a depiction strategy, a technique for actualizing a FIR filter with a FPGA is contemplated, which results in higher execution, lower scale and lower cost.

2 Design Idea and Implementation

The main idea is to design a fast, low-pass, 4-tap and 8-bit finite impulse response (FIR) filter. In the present, electronic circuit configuration involves a generally modest quantity of assets, and assets in this moderately quick with the quickest speed are the structure heading [6]. The structure of the filter, based on the number of bits obtained from the FPGA calculations, increases the speed by reducing the delay time [7, 8].

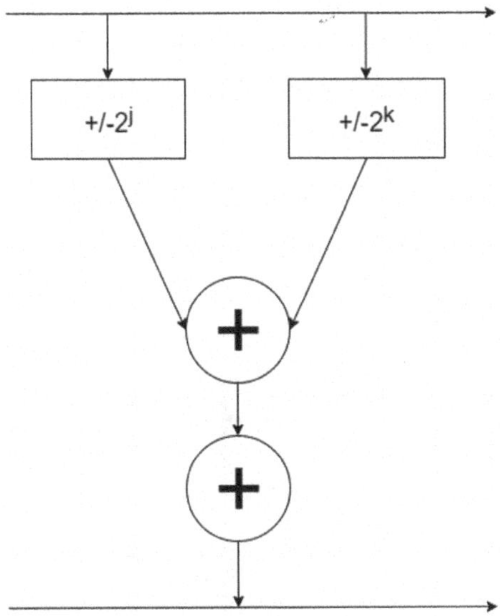

Fig. 1 Finite impulse response (FIR) arithmetic unit

The FPGA gadgets utilize the Spartan-3 group of EP2S60F1020I4 gadgets, which highlight very superior and thickness and are improved for absolute gadget control [9]. Altera one of a kind excess innovation incredibly improves the throughput and lessens segment costs. Also, simultaneously shows signs of improvement signal integrity.

Figure 1 depicts the arithmetic unit of FIR filter, in which the multipliers can be replaced by shifters. Usage of frameworks with multiplications might be streamlined by utilizing just a predetermined number of powers of two terms, so just few numbers of shifting and adding operations are required.

These improvements are, be that as it may, accomplished to the detriment of a weakening in the frequency response attributes, the degree of which relies upon the quantity of power of two terms utilized in approximating every coefficient term, the design of the filter, what is more, the advancement procedure used to determine the discrete space coefficient term [10]. The combinations of many numbers of taps with more numbers of input bits are depicted as follows.

Figure 2 shows the four taps finite impulse response filter architecture design. Here, the windowing technique is used for building the low-pass FIR filters [9, 11]. This technique is convenient, robust and can easily be integrated with frequency sampling. In theory, the window technique [12] of impulse response $y(n)$ and windowing function $v(n)$ is given in Eq. (1).

$$X_v(n) = y(n).\,v(n) \tag{1}$$

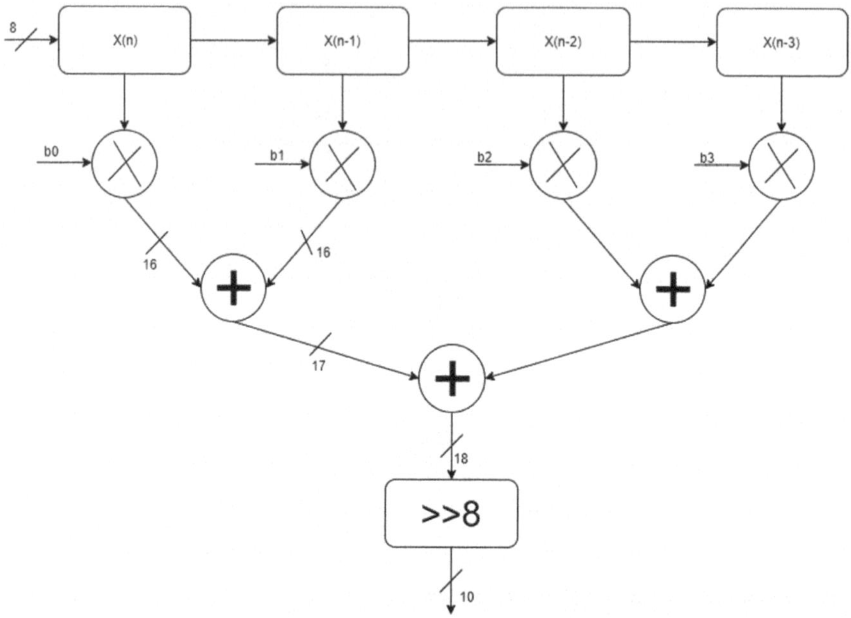

Fig. 2 4-tap and 8-bit input finite impulse response (FIR) filter architecture

This architecture design can be built with flip-flops, multipliers and adders in the field programmable gate array (FPGA) [13, 14]. These flip-flops will be acting as registers. The 8-bit value is given as the input to the registers. The output from each flip-flop (registers) is multiplied with output coefficient of registers $y(n)$. The output from each multiplier is summed with the adders, and as a result, it produces 10-bit value.

3 Simulation Results

A finite impulse response filter tap as shown in Fig. 4 can be actualized in two cluster sections of Xilinx XC3100-arrangement FPGAs. On account of the high level of spatial and temporal region, most signal routing deferrals are not basic, as they are with commonplace elite FPGA structures [15]. Each one of the bit cuts for the tap requires two combinational logic blocks (CLBs) in the array for execution. The broad nearby directing ability of run of the field programmable gate arrays can be utilized for most of the signals within and between taps.

Compose and summon code for fast and low-pass FIR filter utilizing the VHDL language and after that compose the top-level module code to interface with all submodules. As shown in Fig. 2, the FIR filter with flip-flops, adders and multipliers is implemented in Quartus-II Altera ModelSim 10.1b. Figure 3 shows the RTL view

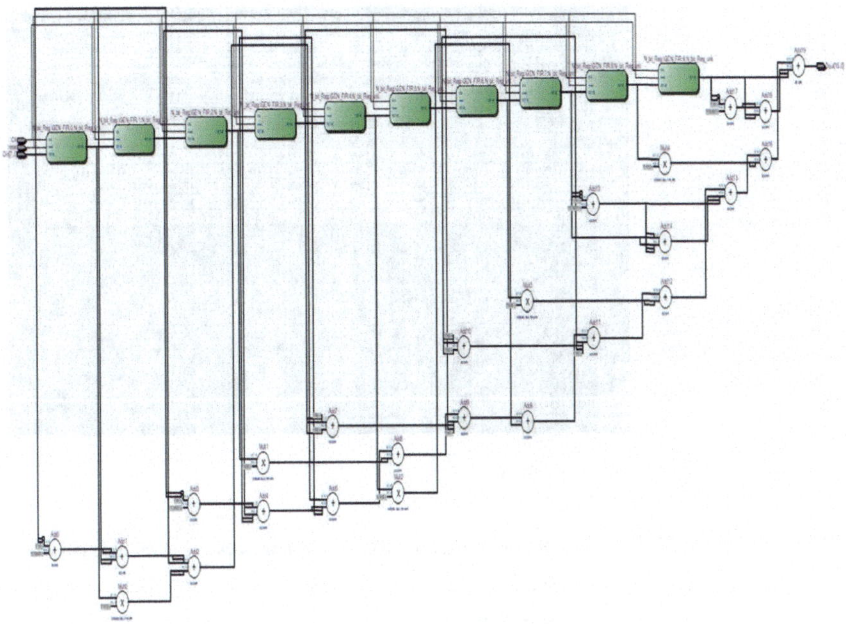

Fig. 3 RTL viewer of 4-tap 8-bit FIR filter

of 4-tap, 8-bit, fast and low-pass filter using windowing technique. And also, it clearly explains the flow of registers with 8 bits as input bits and 10 bits being the output bit. In the wake of finishing the VHDL code for all modules and top-level modules, will in any case need the top-level modules inspiration test document, which gives you input motivating forces with the goal that you can mimic it. At last, the VHDL testbench planning is completed.

The waveform of FIR filter using Quartus-II Altera ModelSim 10.1b for corresponding RTL viewer in Fig. 3 is depicted in Fig. 4. The waveform shows the 8-bit input with clock in positive edge, and reset is fixed as '1', resulting the output of 10 bits. The corresponding coefficients, registers, multipliers and adders are also shown in Fig. 4.

These simulations are compared with the active filter [1, 16] and prove this filter gives the efficient results compared to the active filter which is depicted in Table 1. This table represents some parameters such as the number of slices, flip-flops, LUTs, IOBs, GCLKs. Figure 5 shows the available devices versus used devices. This represents the total number of available devices and total numbers used from available devices which denote the device utilization. The delay time and frequency of the finite impulse response (FIR) filter is 13.42 ns and 74.5156 MHz, respectively.

Table 1 depicts the device utilization of FIR filter obtained from FPGA background, which illustrates finite impulse response filters. Figure 5 shows the number of available devices from the used devices which are synthesis of FIR filter.

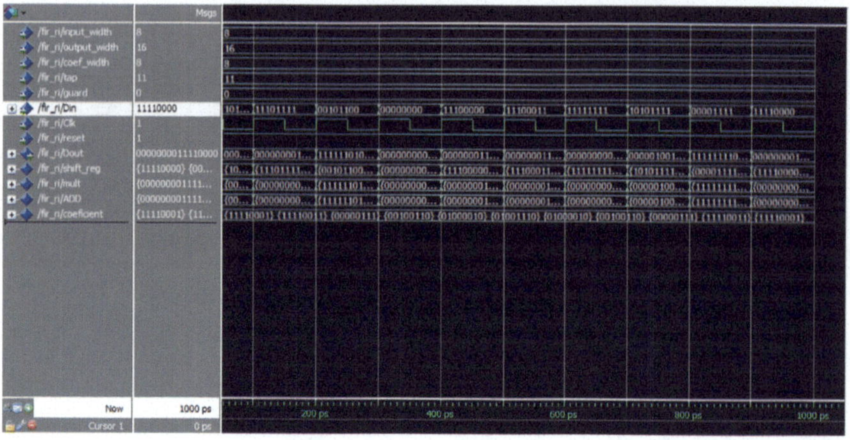

Fig. 4 Simulated waveform of FIR filter

Table 1 Synthesis of FIR filter

Device utilization (#)	Used devices	Available devices
Slices	25	5887
Flip-flops	30	11,778
LUTs	25	11,775
IOBs	18	371
GCLKs	2	25

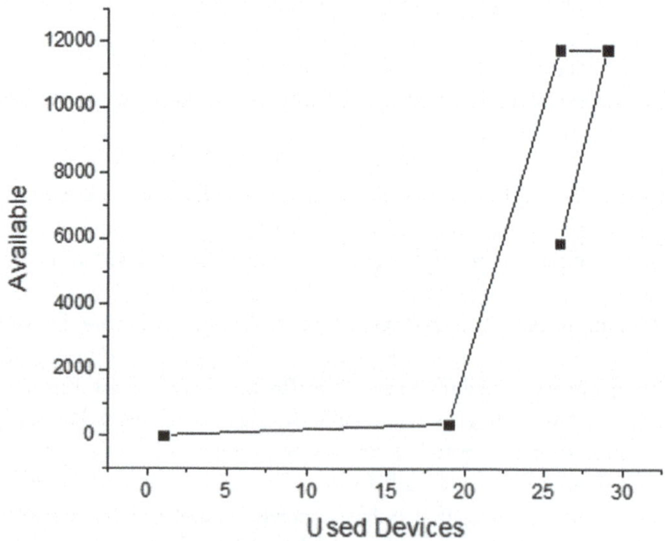

Fig. 5 Used devices versus available devices

4 Conclusion

A fast, low-pass 8-bit finite impulse response (FIR) filter is implemented in Quartus-II Altera ModelSim 10.1b using hardware description language (HDL). VHDL is used for programming the FIR filter with Xilinx (xc3s700a-5 fg-484) for FPGA implementation. The result shows the efficient fast and low-pass filter in which there is a less delay during operation and execution. This type of FIR filter can be used for biomedical signal applications. In near future, this type of filter can be better option for reducing the noise and distortion with high speed for low frequency signals. In future, this filter can be designed for various orders (fourth order, fifth order and so on) for biomedical applications for filtering purposes.

References

1. Bhalke S, Manjula BM, Sharma C (2014) FPGA implementation of efficient FIR filter with quantized fixed point coefficients. In: 2013 international conference on emerging trends in communication, control, signal processing and computing applications (C2SPCA). IEEE Xplore, pp 1–6. https://doi.org/10.1109/c2spca.2013.6749406
2. Firat Kula Y, Ayhan T, Altun M: Approximate implementation of FIR filters on FPGA In: 2018 26th signal processing and communications applications conference (SIU). IEEE Xplore, pp 1–4. https://doi.org/10.1109/siu.2018.8404611
3. Anand V, Kaur A (2018) Implementation of energy efficient FIR Gaussian filter on FPGA. In: 2017 4th international conference on signal processing, computing and control (ISPCC). IEEE Xplore, pp 431–435. https://doi.org/10.1109/ispcc.2017.8269717
4. Meyer-Baese U (2007) Digital signal processing with field programmable gate arrays, 3rd edn. Springer-Verlag Berlin Heidelberg, Springer (India) Pvt. Ltd.
5. Trimale MB, Chilveri PG FIR filter implementation using MCM Technique. In: 2017 international conference on circuits, controls, and communications (CCUBE). IEEE Xplore, pp 231–217. https://doi.org/10.1109/ccube.2017.8394140
6. Khan S, Jaffery ZA (2016) Lower order FIR filter implementation on FPGA using parallel Distribute Arithmetic. In: 2015 annual IEEE India conference (INDICON). IEEE Xplore, pp 1–5. https://doi.org/10.1109/indicon.2015.7443314
7. Jiang X, Bao Y (2010) FIR filter design based on FPGA. In: International conference on computer application and system modeling (ICCASM 2010). https://doi.org/10.1109/iccasm.2010.5622482
8. Rengaprakash S, Vignesh M, Syed Anwar N, Pragadheesh M, Senthilkumar E, Sandhya M, Manikandan J (2018) FPGA implementation of fast running FIR filters. In: 2017 International conference on wireless communications, signal processing and networking (WiSPNET). IEEE Xplore, pp 1282–1286. https://doi.org/10.1109/wispnet.2017.8299970
9. Das R, Guha A, Bhattacharya A (2017) FPGA based higher order FIR filter using XILINX system generator. IEEE Xplore, pp 111–115
10. Evans JB (1994) Efficient FIR filter architectures suitable for FPGA implementation. IEEE Trans Circuits Syst-II: Analog Digit Signal Process 41(7). https://doi.org/10.1109/82.298385
11. Xie J, He J, Tan G (2010) FPGA realization of FIR filters for high-speed and medium-speed by using modified distributed arithmetic architectures. Microelectron J 41:365–370 (Elsevier)
12. Jinding G, Yubao H, Long S (2011) Design and FPGA implementation of linear FIR low-pass filter based on kaiser window function. In: 2011 fourth international conference on intelligent computation technology and automation. IEEE Xplore, pp 496–498. https://doi.org/10.1109/icicta.2011.408

13. Nguyen MS, Kim J, Kim I, Choi K (2010) Design and implementation of flash ADC and DBNS FIR filter. In: 2009 international SoC design conference (ISOCC). IEEE Xplore, pp 325–328. https://doi.org/10.1109/socdc.2009.5423784
14. Cui Y, Huang J, Wu L, Cui X, Yu D (2010) An optimized design for a decimation filter and implementation for Sigma-Delta ADC. In: 2009 IEEE international conference of electron devices and solid-state circuits (EDSSC). IEEE Xplore, pp 338–341. https://doi.org/10.1109/edssc.2009.5394251
15. Lin K, Zhao K, Chui E, Krone A, Nohrden J (1996) Digital filters for high performance audio delta-sigma analog-to-digital and digital-to-analog conversions. In: Proceedings of third international conference on signal processing (ICSP'96). IEEE Xplore, pp 59–63. https://doi.org/10.1109/icsigp.1996.566972
16. Firat Kula Y, Ayhan T, Altun M (2018) Approximate implementation of FIR filters on FPGA. In: 2018 26th signal processing and communications applications conference (SIU). IEEE Xplore, pp 1–4. https://doi.org/10.1109/siu.2018.8404611

Block Rearrangements and TSVs for a Standard Cell 3D IC Placement

J. Deny and R. Raja Sudharsan

Abstract The block rearrangements for vertically stacked integrated circuits (ICs) are done moving the blocks as per the cost function which in return will reduce the overall wire length and allocate white spaces for a standard cell benchmark circuit. This white space allocation will be the most adaptable way for routability when the multiple layers in 3D IC are designed. In this paper, the 3D IC has three layers which are vertically stacked, and through-silicon vias (TSVs) are placed in between the layers for interconnection between the block and also between the layers. Moreover, the thermal level of a TSV is computed using COMSOL Multiphysics. As a result, the wire length between the layers is optimized to 8% using JAVA background, and thermal level is computed as 10%. The input is taken as IBM-PLACE benchmark circuits.

Keywords Block rearrangements · Through-silicon via (TSV) · Thermal level · IBM-PLACE benchmarks

1 Introduction

As the advancements in technology and design standards, for example emerging of vertically stacked integrated circuits which are otherwise termed as 3D ICs, and the structural design of future VLSI circuits are turning out to be more troublesome, new objectives and constraints should be considered. For instance, with increasing the transistor numbers and power densities, the subsequent higher temperatures and gradients are prompting to execution and unwavering quality concerns. In three-dimensional integration circuits, through-silicon via (TSV) is experimented to be the

J. Deny (✉) · R. Raja Sudharsan
Department of Electronics and Communication Engineering School of Electronics and Electrical Technology, Kalasalingam Academy of Research and Education, Krishnan Koil, Srivilliputtur (via), Virudhunagar, Tamil Nadu, India
e-mail: j.deny@klu.ac.in

R. Raja Sudharsan
e-mail: rajasudharsan@klu.ac.in

S.-L. Peng et al. (eds.), *Intelligent Computing and Innovation on Data Science*, Lecture Notes in Networks and Systems 118,
https://doi.org/10.1007/978-981-15-3284-9_22

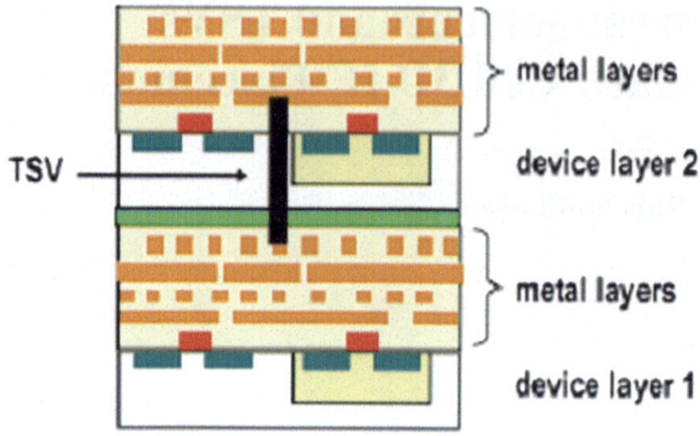

Fig. 1 Interconnect layers of 3D ICs

most prominent technology for connecting the vertical layers. However, through-silicon vias (TSVs) are one of the most important blocks of 3D IC because they consume smaller footprint or chip area and which leads to good routing in physical design [1]. These TSVs will provide optimized wire length and increase density level of integrated circuits compared with the other alternatives such as wire bonding, micro bump. Patterns demonstrate that these thermal issues will be much more articulated in the creating innovation of 3D ICs. Likewise, rise in transistor numbers makes effective calculations a need in electronic design automation. Constraints on the quantity of interlayer vias that can be manufactured in 3D ICs cause limitations on the wire length change that can be acquired with three-dimensional innovations. To address these issues in and around placement process, the research is focused towards the trade-off between interlayer vias and wire length in 3D ICs and net weighting to reduce the total area [2, 3] (Fig. 1).

The primary benefit of vertically stacked integrated circuits is the optimization of wire length that can be accomplished by utilizing vertical interconnects between various layers. However, the end goal to make vertical associations and interlayer vias must be made through device layers, and more noteworthy are thicknesses. Subsequently, these vias are limited in number due to their area consumption. They contend with transistors for area, are considerably bigger in size than consistent vias and require a specific measure of range for resilience in wafer arrangement [4, 5]. With limitations on interlayer by means of numbers, 3D placement techniques must be produced, so that the wire length can be limited without utilizing excessively numerous interlayer vias. Comparing to other computational techniques, the half-perimeter wire length (HPWL) is the best technique because it finds the optimized connecting blocks which cover all the pins of the net to be connected, and also it always underestimates the wire length for congested nets. So, this technique is used to compute the total number of wire length between the blocks and also between the layers of three-dimensional integrated circuits (3D ICs) [6]. The IBM-PLACE benchmark circuits are made as inputs to the problem.

2 Problem Description

The block rearrangements are done by considering the entire rectangles (blocks) in a tree structure. Assign the notations for each rectangle (blocks) for computation such as i_{jH} represents the rectangle j on top of rectangle i and i_{jV} represents the rectangle j on left of rectangle i. Then, move the blocks according to the cost function which is represented in Eq. (1). By doing this block rearrangements, the total wire length can be optimized based on the cost function using HPWL technique.

$$\text{Cost} = W_1 * \text{area} + W_2 * \text{Len} \tag{1}$$

Here, area represents the total area of the blocks (rectangle) enclosing the given basic blocks, Len represents the total interconnect length, and this W_1 and W_2 are the user-specified parameters. For two-terminal wire length nets, we can use Manhattan distance for estimating the number of wire length between the two terminals. Let's be the end coordinates (f_1, g_1) and (f_2, g_2), then the wire length.

These suppositions are needed to be made ready before computing the wire length. It incorporates treating through-silicon via as a block/cell and after that ascertains the wire length layer by layer in half-perimeter model [7]. The through-silicon vias are utilized for interconnection of two layers in vertically stacked IC design.

$$L_t = \Sigma l(\text{En}) \tag{2}$$

The L_t represents the total wire length in vertically stacked integrated circuits and En represents the wire length in the n layers. The ultimate aim is to optimize the wire length and to define the actual white spaces between the cells/blocks. Once the white spaces are allocated, then the TSVs can be positioned with appropriate shape and size. These TSVs are modelled using COMSOL Multiphysics tool. The HPWL was generally used that is sensibly exact and effectively computed. The bounding box of a net with p pins is the smallest block that encases the pin locations. The wire length is accessed as a large portion of the bouncing box. For two and three pin nets (70–80% of all nets in most present-day designs), this is precisely the same as the rectilinear Steiner least tree (RSMT) cost [8]. Thermal issues are anticipated to be more suggested in three-dimensional integrated circuits (ICs) [8, 9] for two different motives. First, regardless of the energy dissipation per transistor being lesser in three-dimensional integrated circuits, the higher packing densities will necessarily cause better energy densities in 3D ICs [10]. Second, vertically stacked ICs have extra heat resistance alongside heat-conducting paths to the heat sink which results in increase in temperature. This analysis of thermal in TSVs can be done by using COMSOL Multiphysics software.

3 Experimental Procedure

The block rearrangements are done in 3D ICs to reduce the wire length in between the layers or dies and between the blocks/cells which are defined as the objective function mentioned in Eq. (3).

$$\text{Min}\,(L\,(o,p,r) + \lambda \sum b\,(B_b\,(o,p,r) - A_b\,(o,p,r))^2) \tag{3}$$

The major constraint of this objective function is that $B_b < A_b$ where b represents bins. $L\ (o, p, r)$ represents the length of the wire length along e, f and g directions. Bb (o, p, r) represents sum of areas of the movable cells in bins b along o, p and r directions. $A_b\ (o, p, r)$ represents maximum area of the movable cell in bins b along o, p and r directions. The even spreading of cells in all respective layers is named as overflow ratio. This ratio plays a vital role in vertically stacked integrated circuits, and it can be computed by using Eq. (4).

$$\text{Overflow ratio} = \sum (\max(B_b(o,p,r) - A_b(o,p,r)))/\sum \text{total area} \tag{4}$$

The bins can be stated as each cell of a rectangular grid. This is calculated by using Eq. (5)

$$\text{No. of bins} = \surd(N/L) \tag{5}$$

where N represents the total number of cells in ICs and L represents the number of stacked layers in ICs. The algorithm for the block rearrangements and also to optimize the total wire length is as depicted as follows (Table 1).

Table 1 Algorithmic flow of 3D IC placement

Standard cell 3D IC placement algorithm for block rearrangement (Optimization of wire length)
Step: 1 Start the process with the positioning of cells using random placement
Step: 2 Compute the total numbers of bins
Step: 3 Compute the base potential of each grid block
Step: 4 Initialize value of λ,
Step: 5 While overflow ratio >0.01 do
Step: 6 Compute objective function
Step: 7 Estimate overflow ratio
Step: 8 If then overflow ratio <0.1
Step: 9 Legalize and save best result

4 Results and Discussions

This research speaks, the total net counts (wire length) has been reduced by rearranging the blocks in 3D IC. The experiment was undergone in the background of JAVA platform with a RAM of 2 GB. The input of this is taken from standard cell IBM-PLACE benchmark circuits (Fig. 2).

TSVs can be designed or modelled with the help of COMSOL Multiphysics software in order for the placement of TSVs between the layers. The thermal level analysis of TSVs using COMSOL Multiphysics software. This TSV has a diameter of 25 nm which is made up of copper core of 15 nm, and SiO_2 shell has 5 nm, and other materials (air and silicon) has 5 nm. The materials used in the TSV are as shown in Table 2 with their respective properties in simulation (Fig. 3, 4, 5 and 6).

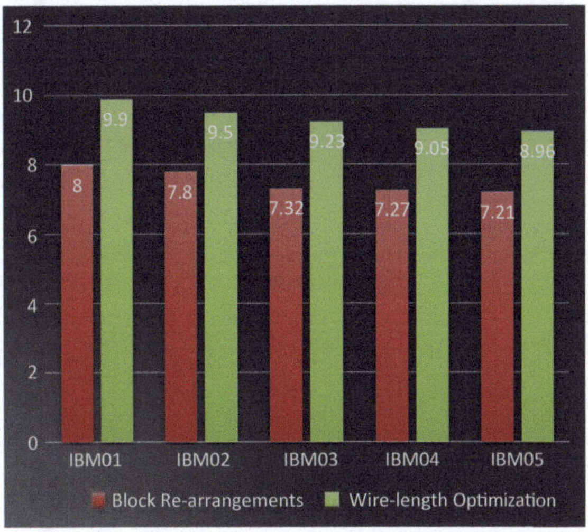

Fig. 2 Block rearrangements versus wire length optimization for different IBM-PLACE benchmark circuits

Table 2 Materials and their properties for modelling TSV

Materials used	Thermal conductivity (W/mK)	Congestion level (kg/m^3)	Heat capacity (J/Kg K)	Dielectric constant
Air	0.048	0.625	1112	1
Si	166	2328	711	12.3
SiO_2	1.33	2242	711	3.11
Cu	402	8710	389	–

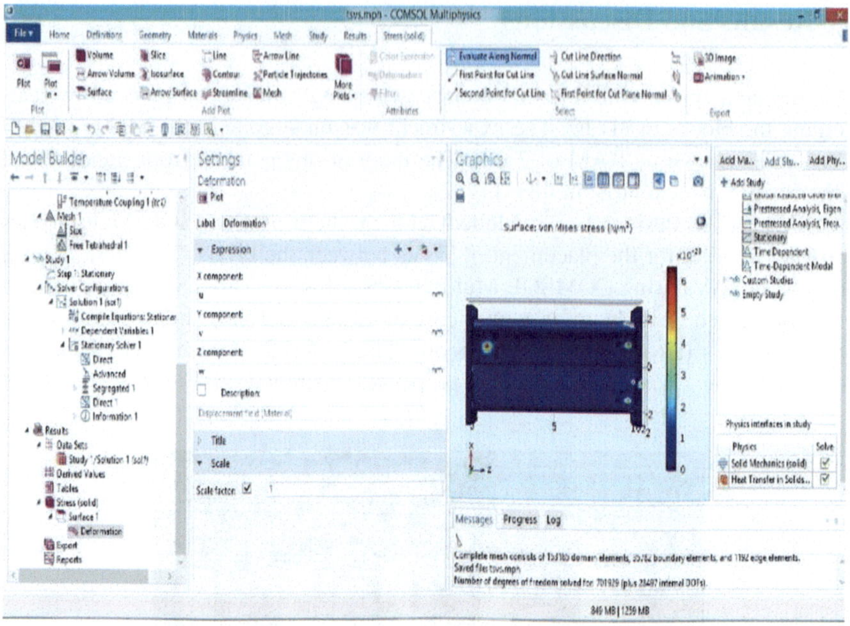

Fig. 3 Side views of thermal TSVs

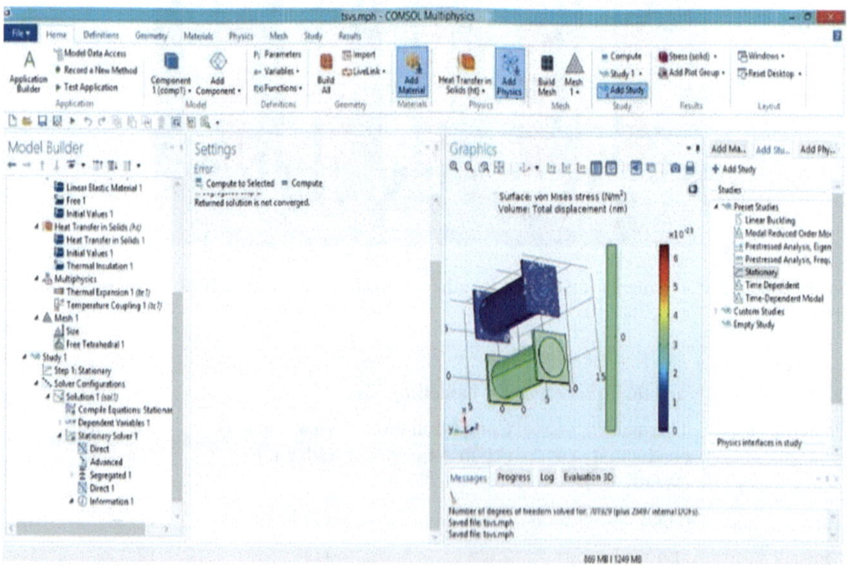

Fig. 4 Cross-sectional view of thermal TSVs

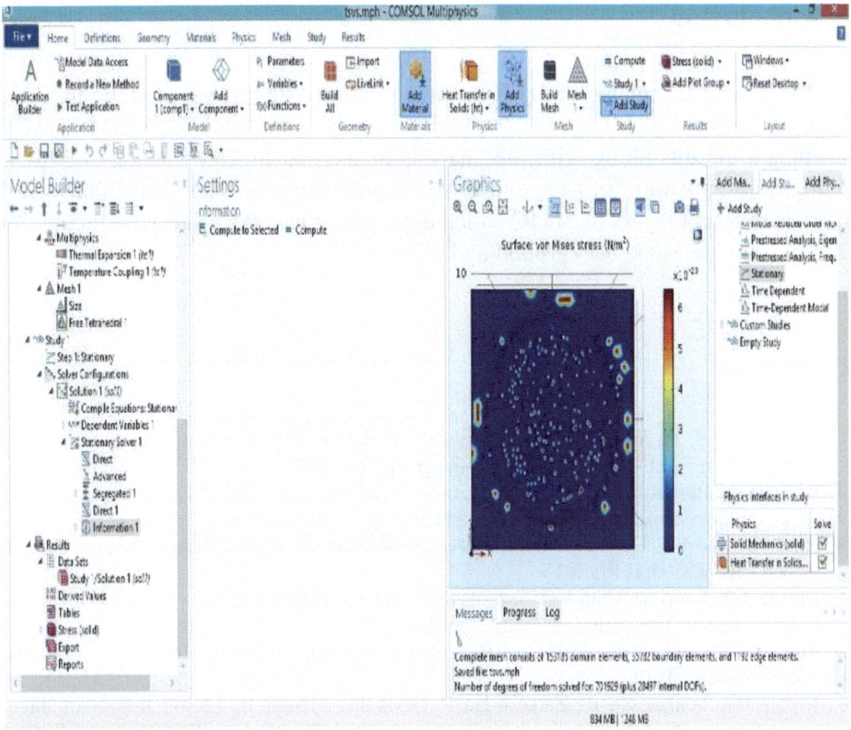

Fig. 5 Top view of thermal TSVs

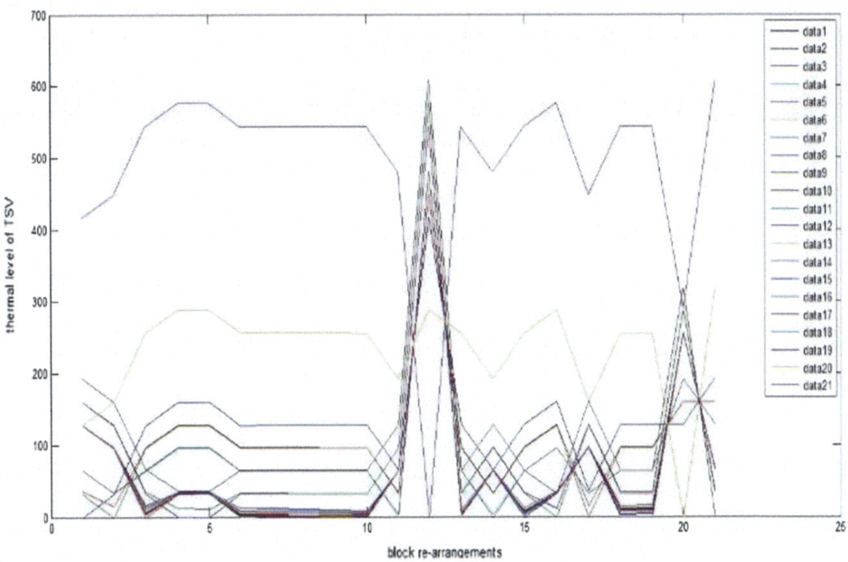

Fig. 6 Block rearrangements versus thermal level in TSV for various materials

5 Conclusion

The block rearrangements of 3D IC with the optimization of wire length are done in the background of JAVA platform with the inputs of standard cell IBM-PLACE benchmark circuits along with the analysis of thermal in TSVs using COMSOL Multiphysics software. In near future, this experiment may lead to the analysis of stress of different layers of 3D IC which are interconnected by through-silicon vias (TSVs) using the same background and inputs.

References

1. Xu Q, Chen S, Xu X, Yu B (2017) Clustered fault tolerance TSV planning for 3D integrated circuits. Clust Fault Toler TSV Plan 3D Integr Circuits 36(8)
2. Rahman A, Reif R (2001) Thermal analysis of three-dimensional (3-D) integrated circuits (ICs). In: Proceedings of the interconnect technology conference, pp 157–159
3. Patel SB, Ghosh T, Dutta A, Singh S (2013) Stress analysis in 3D IC having thermal through silicon vias (TSVs). IEEE 2013
4. Chiang T-Y, Souri SJ, Chui CO, Saraswat KC (2001) Thermal analysis of heterogeneous 3D ICs with various integration scenarios. Int Electron Devices Meet Tech Dig 681–684
5. Mak W-K Rethinking the Wirelength Benefit of 3D Integration Member, IEEE, and Chris Chu, Senior Member, IEEE
6. Joyner JW, Venkatesan R, Zarkesh-Ha P, Davis JA, Meindl JD (2001) Impact of three-dimensional architectures on interconnects in gigascale integration. IEEE Trans VLSI Syst 9(6):922–928
7. Radeep Krishna R, Siva Kumar P, Raja Sudharsan R (2017) Optimization of wire-length and block re-arrangements for a modern IC placement using evolutionary techniques. IEEE xplore
8. Sheeba MR, Gracia Nirmala Rani D (2017) Placement of TSVs in three dimensional integrated circuits (3D IC). Int J Pure Appl Math 117(16):179–184
9. Pawanekar S, Trivedi G (2015) TSV aware standard cell placement for 3D ICs. IEEE
10. Banerjee K, Souri SJ, Kapur P, Saraswat KC (2001) 3-D ICs: a novel chip design for improving deep sub micrometer interconnect performance and systems-on-chip integration. Proc IEEE 89(5):602–633

Artificial Intelligence-Based Load Balancing in Cloud Computing Environment: A Study

Janmaya Kumar Mishra

Abstract The discussion is based in the vicinity of load balance technique by using the artificial intelligence for cloud computing system. Cloud load balancing is a series of actions for distributing workloads to underutilized VMs for computing and sharing the resources in a more effective way for a cloud computing environment. The research is still finding for a more robust technique to distribute the workloads among the servers in this environment. The acceptable way of artificial neural network (ANN) model along with back propagation technique has been studied for an efficient proposed system. Objective of this article is for evaluation of load balancing algorithms in view of the proficiency of each virtual machine or VM, each of the requested task, and its interdependency on multiple jobs.

Keywords Artificial neural network · Back propagation · Load balancing · Artificial intelligence · Cloud computing

1 Introduction

There is one key feature in the study of cloud environment is Load balancing, which is having a significant accomplish on the framework effectiveness for the work allocated to a time period. It is the way to distribute workload among system assets to enhance system performance [1]. Load among data centres can be considered with CPU restrictions, memory capabilities, and network. Load balancer must balance the workloads by following some steps for overcoming the situation with data centre overload or unburden [2]. The steps are as follows:

- Receive incoming service requests
- Load size calculation of the incoming requests then build a request queue
- Periodically load status calculation in server pool with help of the server monitor daemon

J. K. Mishra (✉)
Department of Insights & Data, Capgemini Technology Services India Limited, Hyderabad, India
e-mail: janmaya@gmail.com

S.-L. Peng et al. (eds.), *Intelligent Computing and Innovation on Data Science*, Lecture Notes in Networks and Systems 118,
https://doi.org/10.1007/978-981-15-3284-9_23

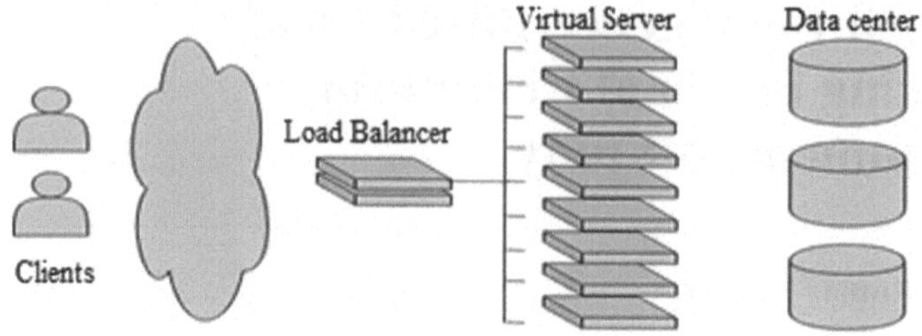

Fig. 1 Load balancing in cloud environment

- Use load balancing algorithm towards the assortment of appropriate server resources.

Load balancing strategies for clouds:

There are mainly two categories in load balancing algorithms: static and dynamic.

Static algorithm: All nodes and the properties of nodes are recognized in advance and the algorithm works which is based on the prior knowledge, it is easy to implement because it does not require current system status information. Example: round robin algorithm (which selects the first node on a random basis then assigns jobs for all other nodes with round robin fashion).

Dynamic algorithm: It works as per the dynamic changes in state of nodes. This algorithm is complex in nature for the implementation; however, it distributes the load in an effective way. Example: Biased Random Sampling Algorithm: It is one of the dynamic approaches, where Migration time as well as Response time is less so improve the overall performance [3].

There are numerous goals in load balancing to maintain or balance the resources or assets [4, 5] (Fig. 1).

Typically, there are three types of techniques for load balancing:

- Centralized
- Decentralized
- Hierarchical.

Centralized load balancing: The major factor for decision making on the algorithm is central load balancer, used as static or dynamic based on global information and all the allocation with decisions are handled by one of the single nodes. This method works smoothly up to a several thousands of processors. The limitation is the single point failure as a central node, which is very challenging for recovery [6, 7].

Decentralized load balancing: There are a number of load balancers to make the decisions. In this framework, each one of the processors shares the workload information to other nearby processors by consuming less memory. The load balancing judgement can be assembled once after collecting all the workload statistics from others (the neighbour) so it may cause poor decisions [8].

Hierarchical load balancing: The load balancers are allocated with a tree structure format. All the requests are accepted through root node and afterwards allocated to

Table 1 Comparison: workload balance architectures [2]

Type of architecture	Workload information status	Advantage	Disadvantage
Centralized	Single node enables with the global workload information	Works well even if in small-scale networks	Non-scalable, single point failure, and no longer fault isolation or fault tolerant
Decentralized	Information sharing with one another nodes as multiple nodes	Less node failure potency, more reliable, fault isolation or fault tolerant	Load balancing is poor for large machines, ageing of load information
Hierarchical	Information sharing between nodes and child nodes	Provides both advantage for centralized as well as decentralized	Less fault tolerant and complex to implement

the load balancer and different load balancer can start processing with different algorithm. It combines both centralized and decentralized by providing benefits of both. The limitation is the implementation architecture for its complexity and additional overhead for dealing among load balancers [9].

The cloud deployment strategies are categorized specifically into three types of environments: Public, private and hybrid [10].

Public cloud: In case of public cloud environment, same hardware, same storage as well as same network devices can share with cloud tenants or multiple organizations.

Private cloud: This environment comprises the resources for computing, which is used only by one organization.

Hybrid cloud: In hybrid cloud environment, it integrates with on-premises infrastructure, or else private cloud environment along with public cloud environment. Data can move between private and public clouds.

Nowadays, there are a number of market leading cloud providers are expanding their data centres due to increase of demand in industry. The cloud data centres are different in multiple ways from the traditional network-based system. Virtualization is one of the best features in the environment of cloud computing for maximizing the use of shared resources in an effective manner. There are multiple users avail cloud resources by dynamic reallocation on demand (Table 1).

2 Artificial Neural Network

ANN incorporates with the group of nodes or artificial neurons, interconnected with each other. ANN methodology is developed by looking towards the large network structure of neurons inside human brain. The internal adjustable parameters of ANN mentioned as weights that refer to the actual strength of a connection between two nodes (neurons). By modifying the weights, ANN can learn an arbitrary vector mapping from the space of input to output. There are three layers in ANN (Fig. 2), input, hidden and output layers.

The representation mentioned below (Fig. 3) is the view of a single hidden unit, taking inputs from multiple input units and introduces with three specific functions.

- *Transfer Potential*: It accumulates the inputs and the weights.
- *Activation Function*: It applies to a transfer function which is non-linear or "activation" over the transfer potential.
- *Threshold Function*: It depends upon the activation function, either to "activate" or not to "activate" the neuron.

The transfer potential, which is a summation function, implies with a sum input inner dot products to weights of the connection.

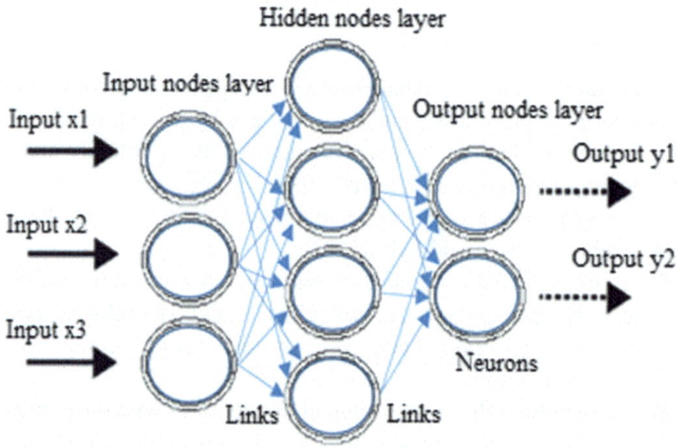

Fig. 2 ANN structure

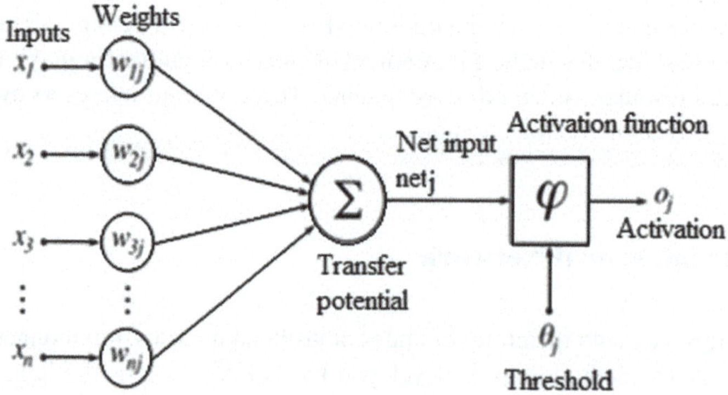

Fig. 3 Activation function of ANN [11]

$$\sum_{i=1}^{n} x_i * w_i \tag{1}$$

Mostly, the activation function used is a logistic sigmoid as follows:

$$f(\theta) = \frac{1}{1 + e^{\theta}} \tag{2}$$

Here, theta is "logit" which is equal to the function (transfer potential) as follows:

$$\theta = \sum_{i=1}^{n} x_i * w_i$$

Artificial intelligence technologies implementation for the cloud computing is changing the topography of the corporate world. Artificial intelligence has the potentiality for further streamline, the substantial capabilities of cloud computing. It helps machines to identify patterns, to make real-time decisions by exploring and learning from historical data, which leads to process automation and eliminate the human errors. It enables machines to take a decision, think and learn like a human being. So, the neural level of approach indicates biological concepts for the machines to recognize the patterns because of the artificial neural networks invention as it is setting-up the recreation of physiology or the study of organism's knowledge in human brain [12]. The process of recognition has been accomplished once after the proper establishment of the network by using the target vectors and the stable equilibrium points are stored in the network for these target vectors [13].

The combination of both cloud computing and artificial intelligence presents a very distinctive circumstance in cloud and artificial intelligence professions for exploring the endless achievability for future.

3 Artificial Intelligence for Cloud and Business World

As per the current scenario on cloud-artificial technology, it can be classified into two categories:

- Cloud Machine Learning Platforms: Recently, the machine learning technologies for cloud environments in AWS, Google and Azure cloud platforms use the specific technology which is responsible towards machine learning model creation.
- Artificial Intelligence Cloud Services: The technologies that support artificial intelligence platform for business like Google cloud vision as well as Microsoft cognitive services, and IBM Watson or natural language application programming interfaces empower abstract complex artificial intelligence capabilities via simple application programming interface calls. By using this, it can incorporate artificial

intelligence capabilities without investing a high amount for artificial intelligence infra.

There is a massive investment for implementing artificial intelligence in the cloud platforms from last few years. The tech-giants like Amazon, Google and Microsoft are leading the environment by integrating artificial intelligence capabilities in Platform as a Service (PaaS) solutions.

Artificial intelligence needs support towards brand new programming paradigm as well as requires a new computing infrastructure. So, artificial intelligence capabilities can be incorporated. Businesses are increasingly implementing artificial intelligence for interaction with end-user through chatbots by enhancing their online versions. One of such ESDS chatbot service is based on natural language processing system for using into multiple sectors including banking. The conversation with chatbots is making more real as these are enhanced with ANN.

One of such neural network tools called neural-network-pattern-recognition tool or NPR Tool, leads to solve the "*two layer*" and "*feed forward*" type of pattern recognition network with the sigmoid output neurons. Patternet network is the feed forward network, can be trained for classification of inputs is basically based on the specified target classes. Again, the target data should comprise of vectors with all the 0 values apart for a 1 inside element *i*, where *i* is the class which must entitle in the case of pattern recognition networks. The input data can be categorized into three sets for the purpose of training: Training, Validation and Testing. Here, Training data is useful for the adjustment of the weight and biases. Validation data is useful to take a decision for stopping the training process, testing-data is useful for the evaluation of performance over the trained-network [14]. There is a two-way relationship between ANN and cloud computing means ANN can be used in cloud computing environmental studies and cloud computing feature can be used in ANN-based projects (Fig. 4).

Business proprietors need artificial intelligence technology for improvising the strategies to operate and help stay them ahead from their contestants. The combination of cloud and artificial intelligence is shaping up for becoming a disruptive force over the multiple industry verticals. Artificial intelligence and cloud computing relation not only forms one of the new strategies of think process over the other existing methodologies, however, also provides a different degree of accessibility towards artificial intelligence technology [15].

4 Load Balancing Challenges

There is the list of metrics, can be improved to provide a better way for cloud computing load balancing [16, 17].

Response time: Total time required for a certain load balancing method to acknowledge on a request in distributed system.

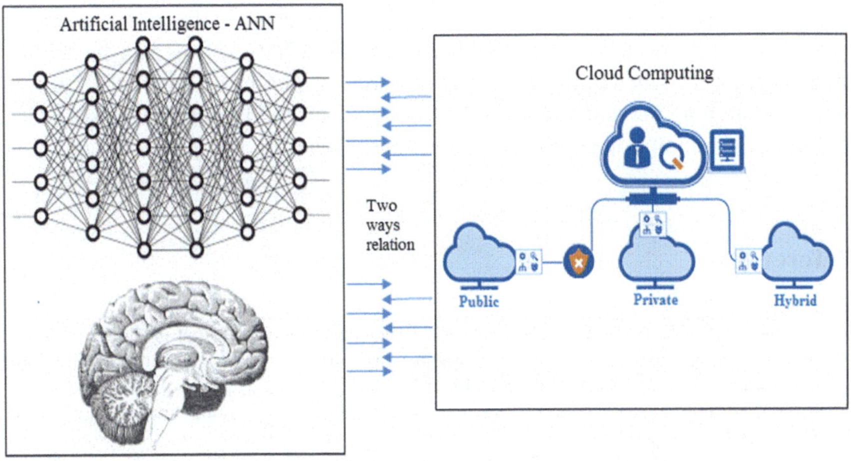

Fig. 4 ANN cloud relation

Resource Utilization: The parameter provides statistics about what extant the resource can be manipulated. Optimum resource can be considered for efficient load balancing activity.

Migration time: Overall time required for a process to transfer between system nodes for execution. Less time taken by the system implies better performance.

Throughput: The total executed task counts in a certain period. It is needed for the high throughput towards system performance improvement.

Fault tolerant: The ability to provide or perform load balancing through a specific algorithm by avoidance of node failure. There should have a proper fault tolerance approach for each load balancing algorithm.

Associated Overhead: It represents the overhead at the time of implementation of algorithm of load balancing. There should be a minimum overhead for the load balancing method to work properly. It is an action or activity of tasks composition, inter-processor, and similarly the other one is inter-process communication.

Performance: The overall system effectiveness and system performance, which can be increased by the improvement of all the parameters.

Scalability: The effectiveness of an algorithm to manage the workload with any fixed count of machines and processors. This can be improved by applying an enhanced strategy for getting a better system performance.

5 Conclusion and Future Scope

One major challenge in cloud computing paradigm is to build the high-throughput architecture of load balancer. The implementation of load balancing with a most optimized algorithm can enhance the efficiency throughput. There is a huge scope to enhance the load balancing and algorithm performance can be enhanced by adjusting with the multiple parameters. Artificial neural network (ANN) forecasts the demand and hence, it keeps the ability to allocate resources as per the demand. ANN always keeps maintaining the active servers based on the current demand, in consequence, low energy consumption compared to other approaches in over-provisioning.

References

1. Mesbahi MR, Hashemi M, Rahmani AM (2016) Performance evaluation and analysis of load balancing algorithms in cloud computing environments. In: 2016 second international conference web research (ICWR), pp 145–151
2. Rajat D, Kumar S (2017) Cloud computing based load balancing architecture: a study. IJCSC 8(2):112–116
3. Aditya A, Chatterjee U, Gupta S (2015) A comparative study of different static and dynamic load balancing algorithm in cloud computing with special emphasis on time factor. IJCET 5:3
4. Garg A, Patidar K, Sexana GK, Jain M (2016) A literature review of various load balancing techniques in cloud computing environment. Int J Enhanc Res Manag Comput Appl 5(2):11–14
5. Desai T, Prajapati J (2013) A survey of various load balancing techniques and challenges in cloud computing. Int J Sci Technol Res 2(11):158–161
6. Vig A, Kushwah RS, Kushwah SS (2015) An efficient distributed approach for load balancing in cloud computing. 2015 international conference on presented at the computational intelligence and communication networks (CICN), Jabalpur, India, pp 751–755
7. Phillips JC, Zheng G, Kumar S, Kalé LV (2002) NAMD: biomolecular simulation on thousands of processors. In: Supercomputing, ACM/IEEE 2002 conference, pp 1–18
8. Yang J, Ling L, Liu H (2016) A hierarchical load balancing strategy considering communication delay overhead for large distributed computing systems. Math Probl Eng 2016:1–9
9. Khiyaita A, El Bakkali H, Zbakh M, El Kettani D (2012) Load balancing cloud computing: State of art. In: 2012 national days of network security and systems (JNS2), pp 106–109
10. What are public, private and hybrid clouds? https://azure.microsoft.com/en-in/overview/what-are-private-public-hybrid-clouds/
11. Mathematical foundation for activation functions in artificial neural networks. https://medium.com/autonomous-agents/mathematical-foundation-for-activation-functions-in-artificial-neural-networks-a51c9dd7c089/
12. Mishra JK, Alam K (2014) Computer transcription of handwritten english pitman's shorthand. Int J Softw & Hardw Res Eng 2(3). ISSN No: 2347-4890
13. Mishra JK, Alam K (2014) A neural network based method for recognition of handwritten english pitman's shorthand. Int J Comput Appl (0975–8887) 102(6)
14. Sahoo AK, Ravulakollu KK (2014) Indian sign language recognition using skin colour detection. Int J Appl Eng Res 9(20):7347–7360. ISSN 0973-4562
15. The future belongs to Cloud and Artificial Intelligence. www.esds.co.in/blog/future-belongs-cloud-artificial-intelligence By ESDS | June 26, 2018
16. Foster I Zhao Y, Raicu I, Lu S (2008) Cloud computing and grid computing 360-degree compared. In: proceeding Grid computing Environments Workshop, pp 99–106 (2008)
17. Buyya R, Ranjan R, Calheiros RN (2010) InterCloud: utilityoriented federation of cloud computing environments for scaling of application services. In: Proceeding 10th international conference on algorithms and architectures for parallel processing (ICA3PP), Busan, South Korea

An Investigation Study on Secured Data Storage and Access Control in Cloud Environment

P. Calista Bebe and D. Akila

Abstract Cloud computing is surroundings for imparting the information and resources which might be brought as the service to an end users over Internet on call for. Cloud allowed the users to get way into their stored information from any environmental places at any time. Cloud comprised of the key problems like safety, data confidentiality, network dependency and centralization. When storing the client sensitive information into cloud data storage, security plays an essential part. Providing the security to sensitive information is a key issue in cloud computing. In existing works, numerous methods were introduced for securely storing data into the cloud. But, the security level was not improved, and data accessing time was not reduced. Our research work concentrated on the cryptographic and data structure techniques for solving the existing problems during cloud storage and data access.

Keywords Cloud computing · Security · Data integrity · Geographical location · Data access · Cryptographic techniques

1 Introduction

Cloud computing is a type of figuring where the mutual assets and IT-related capacities are provided as the supplier to external customers by the utilization of Internet methodologies. Cloud computing is primarily based on data sharing and computing resources than the usage of local servers to manage there quests. Cloud computing allowed the users to receive benefit without any requirement for deep knowledge or expertise.

P. Calista Bebe (✉)
Department of Computer Science, School of Computing Sciences, Vels Institute of Science Technology & Advanced Studies (VISTAS), Chennai, India

D. Akila
Department of Information Technology, School of Computing Sciences, Vels Institute of Science Technology & Advanced Studies (VISTAS), Chennai, India

S.-L. Peng et al. (eds.), *Intelligent Computing and Innovation on Data Science*, Lecture Notes in Networks and Systems 118,
https://doi.org/10.1007/978-981-15-3284-9_24

2 Literature Survey

A Dynamic Proof of retrievability method was designed in [1] for public auditability and for communication efficient restoration since information corruption [2]. A secure disintegration protocol (SDP) was introduced in [3] for protection of privacy on-site in cloud. Probabilistic analysis was carried out for finding the intrusion tolerance abilities. But, the key management method was not introduced for secure data integration in cloud. The designed method failed to utilize cryptography method effectively. A new large DAC-MACS scheme (NEDAC-MACS) was introduced in [4] to guarantee the secure attribute revocation. However, the attack detection rate was not increased by NEDAC-MACS scheme.

A new cloud storage encryption scheme was introduced in [5] to convince false client secrets and to improve the client privacy level. But, the encryption time was not minimized using new cloud storage encryption scheme. The multi-tenant networked cloud infrastructure architecture was introduced in [6] for securing the hosted services. The designed architecture was based on trusted virtual domains with security policies of tenant domains and security policies of virtual machines. But, the access control was not carried out in enhanced manner using multi-tenant networked cloud architecture. A new security assessment methodology was designed in [7] for examining the safety of critical services in cloud. But, the security level was not enhanced using security assessment methodology [10]. Broker-based structure was introduced in [8] [9]. However, the information confidentiality rate was not more suitable for using broker-based framework.

3 Secured Cloud Data Storage and Access Control Techniques in Cloud Computing

The main objective of designed scheme was to assure the redistributed information honesty and information accessibility in distributed storage. The key aim was to guarantee that cloud server stores the data in secured manner. The cloud storage systems were not used by peoples when his data changed randomly by CSP or different entities with no approval.

NEDAC-MACS assured safe quality revocation, information confidentiality and protection besides the stationary corruption of authorities. NEDAC-MACS enhanced security devoid of reducing the effectiveness.

The media cloud structure was introduced and used as manual in procedure of accumulation safety features or new media clouds. The designed structure was partitioned into three security limitations among every layer organizing the subsequent system safety characteristic on border to attain dissimilar levels of local safety protection.

4 Comparison of Techniques in Cloud Environment and Suggestions

4.1 Space Complexity

Space complexity is given by,

$$SC = \text{Total memory} - \text{unused memory space in cloud server} \quad (1)$$

From (1), the space complexity is calculated. When the space complexity is lesser, the approach is stated as greater efficient.

Table 1 describes the space complexity with respect to range of cloud user requests ranging from 10 to 100. Space complexity comparison takes place on existing dynamic Proof of retrievability scheme, NEDAC-MACS and media cloud framework. The graphical analysis of space complexity is described in Fig. 1.

High coding granularity was achieved through encoding at information square dimension than part enormous information record. Information gets parceled into little information squares and encodes each datum square independently. With coding approach, a redesign inside information square influenced the current information square and connected images without refreshing huge information document. This in turn helps to reduce the space complexity. Therefore, the space complexity of dynamic Proof of retrievability scheme is 21% lesser than NEDAC-MACS scheme and 38% lesser than media cloud framework.

Table 1 Space complexity

Number of cloud user request (number)	Space complexity (MB)		
	Dynamic Proof of retrievability scheme	NEDAC-MACS scheme	Media cloud framework
10	25	33	42
20	28	36	45
30	31	40	49
40	29	38	47
50	27	35	44
60	25	31	42
70	22	28	38
80	26	32	43
90	30	36	46
100	34	41	50

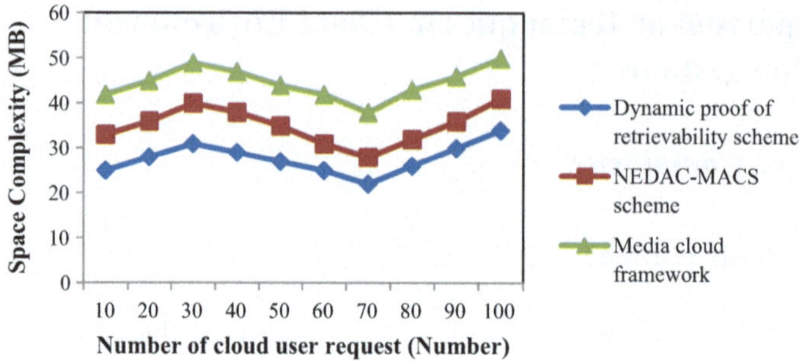

Fig. 1 Measure of space complexity

4.2 Security Level

Security level is defined as the ratio of number of cloud user data can be correctly accessed by authorized cloud users to the total number of cloud user data. It is calculated in terms of percentage (%). The formula can be

$$SL = \frac{\text{Number of cloud user data correctly accessed by authorized cloud users}}{\text{Total number of cloud user data}} \quad (2)$$

From (2), the security level is calculated.

Table 2 illustrates the security level with the esteem number of cloud user data ranging from 10 to 100. The graphical analysis of security level is illustrated in Fig. 2.

Figure 2 explains the security level comparison for different number of cloud user data. From figure, it is observed that the security level using NEDAC-MACS

Table 2 Security level

Number of cloud user data (number)	Security level (%)		
	Dynamic Proof of retrievability scheme	NEDAC-MACS scheme	Media cloud framework
10	76	85	70
20	79	87	73
30	82	89	76
40	80	86	74
50	85	90	79
60	87	93	82
70	84	89	80
80	81	87	77
90	83	91	79
100	87	94	82

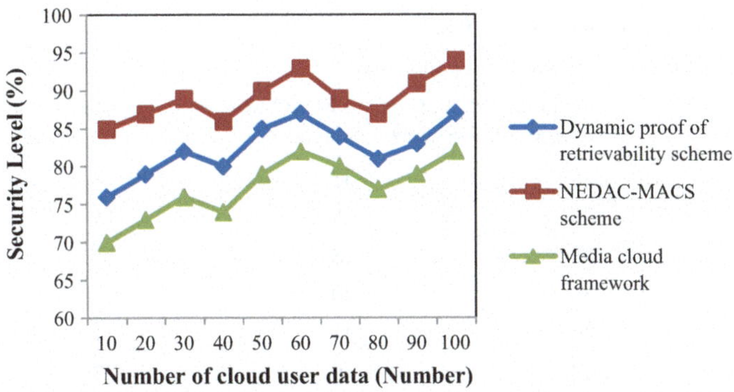

Fig. 2 Measure of security level

is higher when compared media cloud framework. NEDAC-MACS enhanced safety without reducing the performance. Therefore, the security level of NEDAC-MACS scheme is 8% higher than Dynamic Proof of retrievability scheme and 16% higher than media cloud framework.

4.3 Data Retrieval Time

The data retrieval time can be calculated in phrases of milliseconds (ms). It is given by,

$$\text{Data Retrieval Time} = \text{Ending Time} - \text{Starting time of data access} \quad (3)$$

When the data accessing time is lesser, the approach is believed as greater efficient (Table 3).

Data retrieval time comparison takes place on existing Dynamic Proof of retrievability scheme, NEDAC-MACS and media cloud framework. The graphical representation of data accessing time is explained in Fig. 3.

Figure 3 illustrates the data retrieval time comparison for different number of cloud user requests. From figure, it is clear that the data accessing time using media cloud framework is lesser when compared to NEDAC-MACS. This is because the designed framework uses three protection limitations with every layer arranging equivalent system protection events on border to attain different local security protection level. By this way, the data accessing time gets reduced. As a result, the data retrieval time consumption of media cloud framework is 38% lesser than Dynamic Proof of retrievability method and 45% lesser than NEDAC-MACS scheme.

Table 3 Tabulation for data retrieval time

Range of cloud user request (number)	Data accessing time (ms)		
	Dynamic Proof of retrievability scheme	NEDAC-MACS scheme	Media cloud framework
10	25	33	13
20	28	35	15
30	32	38	19
40	36	40	21
50	39	43	24
60	42	46	27
70	45	49	30
80	41	45	28
90	44	49	31
100	47	52	33

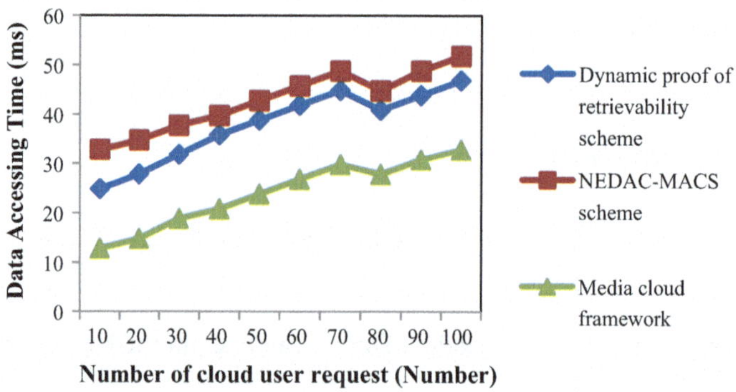

Fig. 3 Measure of data retrieval time

5 Discussion on Limitation of Secured Cloud Data Storage and Access Control Techniques in Cloud Computing

The data confidentiality rate was not improved using Dynamic Proof of retrievability scheme. NEDAC-MACS addressed two vulnerabilities though the nonrevoked users disclosed obtained key update keys to the revoked user. But, the attack detection rate was not enhanced using NEDAC-MACS scheme.

A security media cloud framework was introduced for preserving the multimedia data and services. The existing media cloud structure comprised of three protection limitation in media cloud to guarantee cloud protection. Sec-ABAC access manage protocol guaranteed the access manage of cloud resources. The operation of structure on Amazon Web Services failed to examine exact performance of Sec-ABAC protocol. The space complexity was not reduced using security media cloud framework.

6 Conclusion

A comparison of different existing secured data storage and access control techniques for improving the security is studied in cloud computing. This survey paper also discussed the methodologies and different methods to store the data in efficient manner. From the study, it is clear that the existing techniques failed to improve the data confidentiality rate. In addition, the attack detection rate was not improved using NEDAC-MACS scheme. The huge range of experiments on current techniques calculates the relative overall performance of many secured data storage and access control techniques with its restrictions. The future research can be carried out using cryptographic and data structure techniques for performing the secured data storage and access control in cloud computing.

References

1. Ren Z, Wang L, Wang Q, Xu M (2018) Dynamic proofs of retrievability for coded cloud storage systems. IEEE Trans Serv Comput 11(4):685–698
2. Du M, Wang Q, He M, Weng J (2018) Privacy-preserving indexing and query processing for secure dynamic cloud storage. IEEE Trans Inf Forensics Secur 13(9):2320–2332
3. Rawal BS, Vijayakumar V, Manogaran G, Varatharajan R, Chilamkurti N (2018) Secure disintegration protocol for privacy preserving cloud storage. Wirel Pers Commun 103(2):1161–1177
4. Wu X, Jiang R, Bhargava B (2017) On the security of data access control for multi-authority cloud storage systems. IEEE Trans Serv Comput 10(2):258–272
5. Chi P-W, Lei C-L (2018) Audit-Free Cloud storage via deniable attribute-based encryption. IEEE Trans Cloud Comput 6(2):414–427
6. Varadharajan V, Tupakula U (2018) Securing services in networked cloud infrastructures. IEEE Trans Cloud Comput 6(4):1149–1163
7. Hudic A, Smith P, Weippl ER (2017) Security assurance assessment methodology for hybrid clouds. Comput Secur 70:723–743 (Elsevier)
8. Halabi T, Bellaiche M (2018) A broker-based framework for standardization and management of cloud security-SLAs. Comput Secur 75:59–71 (Elsevier)
9. Li H, Yang C, Liu J (2018) A novel security media cloud framework. Comput. Electr. Eng. 1–11 (Elsevier)
10. Sudha C, Akila D (2019) Detection of AES algorithm for data security on credit card transaction. Int J Recent Technol Eng (IJRTE) 7(5C):283–287

A Comparative Study and Analysis of Classification Methodologies in Data Mining for Energy Resources

M. Anita Priscilla Mary, M. S. Josephine, and V. Jeyabalaraja

Abstract Retrieval of right sequence in extensive quantity of records in irrelevant, unreported and concealed data by applying the process of data mining methods. Classification is a procedure used for building classification models for a set of input data. This study is about to compare and analyze the various classification algorithms for energy resources using Weka tool. In this study, it uses five different data mining methods, namely iterative classifier optimizer, Bayes net, classifier via regression, LMT and JRip. The diverse attainment of the algorithm is found by the assess of variables like true conclusive, false conclusive, exactness, reminiscence and ratio. Meticulousness of algorithm is examined using the values of correctly classified occurrences and incorrectly classified occurrences.

Keywords Attainment · Exactness · Occurrence

1 Introduction

The key feature which affects the demand of power is the generation of energy resources. We need to inspect the diverse power generations which will help in future demand. It can be obtained using the methods of data mining. Inorder to discover the hidden patterns in database, classification is one the method used in data mining. In classification, data objects are classified as predefined several classes. This characteristic feature helps in the performance of the classifier. Classifier contains

M. Anita Priscilla Mary (✉) · M. S. Josephine
Department of Computer Applications, Dr. M.G.R. Educational and Research Institute, Chennai, Tamil Nadu, India
e-mail: anitapriscilla80@gmail.com

M. S. Josephine
e-mail: josebr@yahoo.co.in

V. Jeyabalaraja
Department of Computer Science & Engineering, Velammal Engineering College, Chennai, Tamil Nadu, India
e-mail: jayabalaraja@gmail.com

S.-L. Peng et al. (eds.), *Intelligent Computing and Innovation on Data Science*,
Lecture Notes in Networks and Systems 118,
https://doi.org/10.1007/978-981-15-3284-9_25

repugnant percept; if the percept is individualistic of each other, then every record is covered by at most one rule. The dataset used for the comparative study is 12 years of statistical data used in power generation, and it contains six attributes to define a class. Each test is suspected to belong to a predetermined class, as determined by the class label impute. The set of trial used for model building in training sets. The recognized label of the given sample is correlated with the classified outcome of the model; high standard of results is attained in most of the instances. The accuracy rate is calculated by the correctly classified occurrences using the model for the given of test data.

2 Literaturer Survey

Asumedu et al. specified that sustainable energy has a direct relationship with sustainable development through its impact on human development and economic productivity [1].

Estenban and Leary described four ways of obtaining energy resources from ocean areas, namely from breeze, whirlpool, stream, thermal and shoals [1].

Cardinale has investigated on solar plant for hot water manufacture using lifespan savings method (LCS) by taking into account of three long-established fuels such as gas oil, LPG and electricity to estimate financial feasibility [1].

Adaptation of sunlight-based techniques. Forecasting model was developed to portend the requirement on sunlight water warming systems and their saving methods in domestic sectors.

Sfetsos has done a similar work on the forecasting methods to mean the hourly wind speed implementing the methods of time-sequential examination, conventional fixed models, feed presumptuous and sporadic neurological set of connections [1].

3 Tools and Techniques

3.1 Dataset

A data is approximately identical to a two-dimensional spreadsheet or database table [2]. A classifier model is a complex bounding into one dataset characteristic of a class attribute [3]. Energy resources are the principal sources of power generation. Energy resources are of two types: one is conventional and nonconventional; based on this category, the attributes are selected for the datasets. This study uses 12 years of statistical data. The resources are calculated in million units, which are used for analysis to find which algorithm is the best one out of the five. The class uses different input attributes, namely Hydro ($X1$), Wind ($X2$), Thermal ($X3$), Gas ($X4$), Private Purchase ($X5$) and Sale of Power/Swap return ($X6$), in order to determine the output variable (Y) The classification algorithms are compared for the given sets of data with different parameters to find which is the best one out of five.

3.2 *Tools*

3.2.1 Weka Software

Weka is a software which is built on the foundation of Java. This software also provides a Java appetizer for the use in applications and to connect databases. Weka was developed in the University of Waikato. It is primarily used for data mining functions such as data prehandling, grouping, reversion, visual representation and characteristic selection. Weka classifies every attribute as numeric, so we need to manually transform them to nominal if needed.

3.2.2 Categories of Classification Algorithms

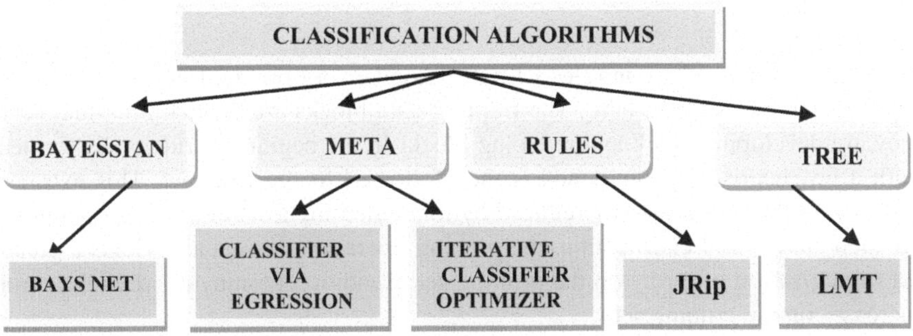

3.3 *Techniques*

3.3.1 Bayes Net

Bayesian networks are perfect for fetching an event that has taken place and predicting any one of the possible known causes in the contribution factors. In Bayesian network, model in which sequences of variables are named dynamic Bayesian network. It can be represented to solve decision problems. Bayes network uses various learning research algorithms and quality efforts [4, 5].

3.3.2 Classification Via Regression

Classification is applied using the regression techniques. Class is compounded by two parts. Regression is a procedure applied to construct for each of the class value. It is used to prepare the algorithm for supervised learning [6, 7]. More or fewer cases may be provided, but this gives accomplishments a chance to specify a preferred batch size. This method is a bit ponderous, but the result works quite well as a separator.

3.3.3 Iterative Classifier Optimizer

This algorithm has the qualities of a nexus of efferent neuron. Hence, it can be measured to a structure that is a part of the nervous system. In this network, the version is passed once this procedure is renewed until all the records are presented. This algorithm is coinciding to the features of artificial neurological interconnections. This is used to implement to a meticulous function by adopting the concepts of neurological interconnections. The basics of the data originate from picking the entail eights haphazardly.

3.3.4 JRip

It is stationed in association rules with decreased error pruning. Decision tree algorithm technique is found here. The training data blocks are divided into dual sets, they are developing block and reducing block. In the beginning, rule set is formed with a few heuristic methods and is then followed for growing set. This extreme order set is then continuously reduced by implementing a set of pruning operators. At each platform of clarification, the pruning operant is chosen and yields a greatest minimization of errors on the pruning set. Rendition terminates when the error increases on the pruning set.

3.3.5 LMT

Logistic model tree (LMT) is a classifying miniature with a connected managed learning principle that connects logistic regression and decision tree training algorithm. The basic LMT enthronement algorithm uses cross-checking to find a number of LogitBoost iterations that does not overfit the training data.

4 Experimental Studies

4.1 Comparison of Classification Algorithms

Using *F*-measure: The input data given in million units for various resources. We need to analyze the different specifications [6].
TP: Positive Rate
FP: Positive Rate
Precision (P) = Number of relevant documents retrieved/Total number of documents retrieved
Recall (R) = Number of relevant documents retrieved/Total number of relevant documents
***F*-measure** = 2 * Recall * Precision/Precision + Recall (Table 1).

In Fig. 1, *x*-axis specifies the types of algorithm and *y*-axis specifies the values of different parameters such as TP, FP, P, R and *F*-measure.

Table 1 *F*-measure of various algorithms

Sl. No.	Parameter (algorithms)	TP	FP	P	R	*F*-measure
1	Iterative classifier optimizer	0.958	0.008	0.959	0.958	0.958
2	Bayes net	0.958	0.008	0.959	0.958	0.958
3	Classifier via regression	0.958	0.008	0.959	0.958	0.93
4	LMT	0.917	0.017	0.921	0.917	0.918
5	JRip	0.875	0.025	0.877	0.875	0.873

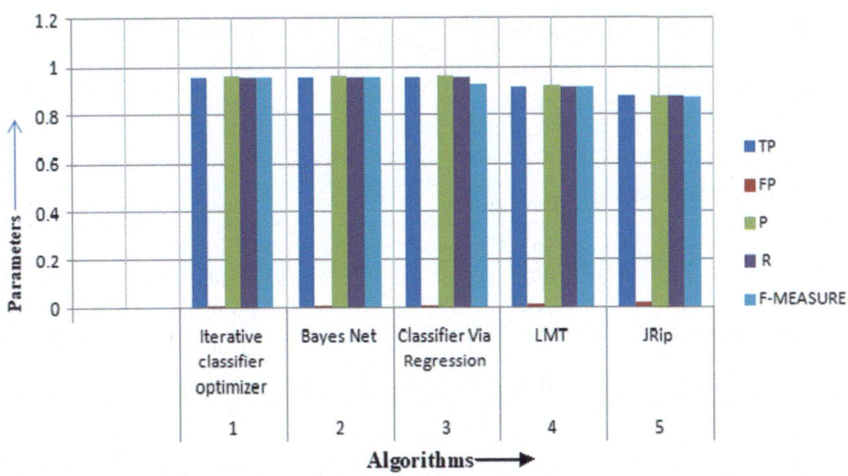

Fig. 1 Comparison of algorithms using various parameters (TP, FP, P, R) and *F*-measure

4.2 Comparison of Classification Algorithms Using Kappa Statistic and Errors

Iterative classifier optimizer, Bayes net, classification Via regression, LMT and JRip algorithm are used for the energy resources. The dataset is calculated in million units; after applying the algorithm, the following result is obtained in the form of a table.

Using the parameters:

KS—Kappa statistic
MAE—Mean absolute error
RMSE—Root Mean Squared Error
RAE—Relative Absolute Error
RRSE—Root Relative Squared Error (Tables 2, 3, 4, 5 and 6).

Table 2 Iterative classifier optimizer

Parameters	Values
KS	0.95
MAE	0.0232
RMSE	0.1186
RAE	8.34%
RRSE	31.74%

Table 3 Bayes net

Parameters	Values
KS	0.95
MAE	0.1522
RMSE	0.2228
RAE	54.66%
RRSE	59.64%

Table 4 Classification via regression

Parameters	Values
KS	0.9167
MAE	0.114
RMSE	0.1854
RAE	40.95%
RRSE	49.64%

Table 5 LMT

Parameters	Values
KS	0.9
MAE	0.0277
RMSE	0.1661
RAE	9.95%
RRSE	9.95%

Table 6 JRip

Parameters	Values
KS	0.85
MAE	0.0461
RMSE	0.2024
RAE	16.54%
RRSE	54.20%

4.2.1 Analysis of Algorithms Using Kappa Statistic

Energy resources are calculated in million units with different attributes such as Hydro ($X1$), Wind ($X2$), Thermal ($X3$), Gas ($X4$), Private Purchase ($X5$) and Sale of Power/Swap ($X6$) return which are used to find the kappa statistics using these algorithms. The kappa statistic is the same for iterative classifier optimizer and Bayes net (Table 7).

In the Fig. 2 x-axis specifies the types of algorithm and y-axis specifies the assesses of Kappa Statistic where the values lie between 0 to 1.

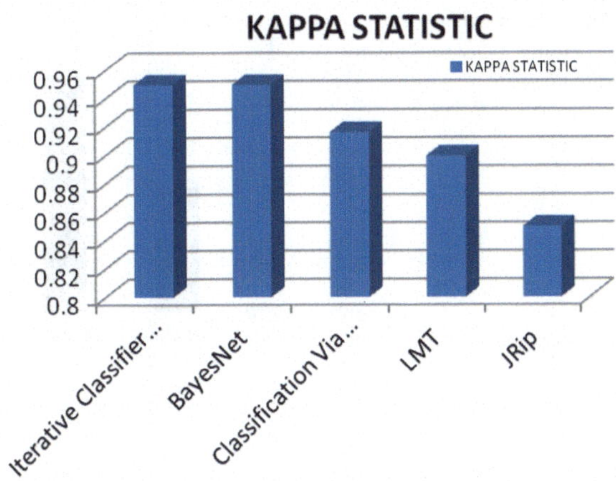

Fig. 2 Analysis of algorithms using kappa statistic

Table 7 Analysis of algorithm using kappa statistic

Algorithm	Kappa statistic
Iterative classifier optimizer	0.95
Bayes net	0.95
Classification via regression	0.9167
LMT	0.9
JRip	0.85

Table 8 Using root mean squared error

Algorithm	RMSE
Iterative classifier optimizer	0.1186
Bayes net	0.2228
Classification via regression	0.1854
LMT	0.1661
JRip	0.2024

4.2.2 Comparative Study on Root Mean Squared Error

The kappa statistic is same in iterative classifier optimizer and Bayes net. RMSE values are calculated for data in million units. To find which one is best out of the two algorithms (Table 8).

In Fig. 3, *x*-axis specifies the types of algorithm and *y*-axis specifies the values of RMSE.

The error value is less in the iterative classifier optimizer. Hence, it is the best algorithm out of the five algorithms for classification

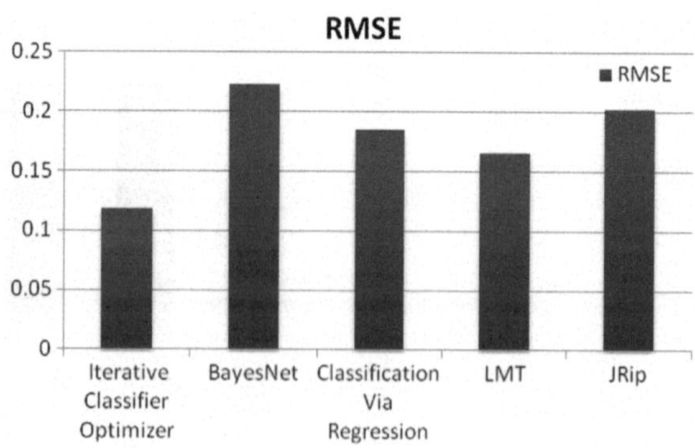

Fig. 3 Comparison of algorithms using root mean squared error

5 Results and Discussion

5.1 Comparison of Energy Resources Using Iterative Classifier Optimizer

Iterative classifier optimizer algorithm is used to find the values of Matthew's correlation coefficient (MCC) in knowledge engineering. It is used to estimate the character of two classes in categorization. It holds two relations, one is the accurate constructive values and other is the fake constructive values. In correlation coefficient, MCC is betwixt of the detected and estimated categorization. It is paired; hence, it returns the values between −1 and +1. A coefficient of +1 refines the estimation, and −1 illustrates the total disagreement among estimation and inspection. The MCC is calculated by applying the procedure

$$\mathbf{MCC} = \frac{\mathbf{TP{+}TN{-}FP{+}FN}}{\sqrt{\mathbf{(TP{+}FP)(TP{+}FN)(TN{+}FP)(TN{+}FN)}}}$$

If each of the four sums within the divisor is zero, the denominator is a discretional set to at least one. This ends up in a Matthew's correlation constant to zero. MCC can be computed by applying the formula

$$\mathbf{MCC} = \sqrt{\mathbf{PPV * TPR * TNR}} * \sqrt{\mathbf{NPV - FDR * FNR * FPR * FOR}}$$

PPV Positive Predictive Values
TPR True Positive Rate
TNR True Negative Rate
NPV Negative Predictive Values
FDR False Discovery Rate
FNR False Negative Rate
FPR False Positive Rate
FOR False Omission Rate (Table 9).

In Fig. 4, x-axis specifies the various energy resources and y-axis specifies the values of MCC.

Table 9 Comparison of energy resources using iterative classifier optimizer

Class	MCC
Hydro	0.942
Thermal	0.956
Gas	0.971
Wind	1
Private purchase	0.985
Sales of power/swap return	1

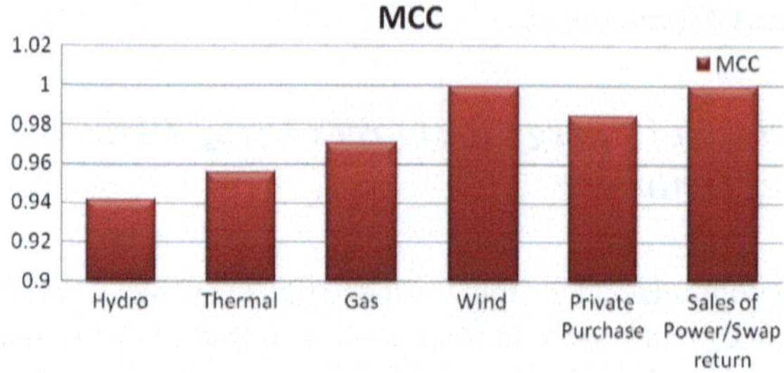

Fig. 4 Comparison of energy resources using iterative optimizer algorithm using the values of MCC

6 Conclusion

Classification is a method used to solve the traditional problems, extensively learned by statisticians and machine learning researchers. The operation of a model may depend on other features besides the training algorithm such as class allocation, cost of miscategorization, size of training and test sets. Each training example can progressively increase/decrease depending on the possibility that a hypothesis is right. Given data is chosen by chance into partitions of dual separate sets. It is then divided as training block for model construction and test block for accuracy estimation. Grouping and forecasting are the two methods used for the data analysis. It can be applied to retrieve the models that outline the significant data classes to forecast future data patterns. The MCC parameter helps to predict which attribute will help for the future generation of power resources to meet over the demand.

References

1. https://www.researchgate.net/publication/309479657_Is_there_a_causal_effect_between_agricultural_production_and_carbon_dioxide_emissions_in_Ghana/citation/download
2. Han J, Kamber M (2006) Data mining: concepts and techniques. University of Illinois, Urbana Champaign
3. Gorunescu F (2011) Data mining: concepts, models, and techniques. Springer
4. Han J, Kamber M (2001) Data mining: concepts and techniques. Morgan-Kaufman series of data management systems. Academic Press, San Diego
5. Vikram K, Upadhayaya N (2011) Data mining tools and techniques: a review. Comput Eng Intell Syst 2(8):31–39
6. Kishore1 GDK, Dr. Babu Reddy2 M (2017) Comparative analysis between classification algorithms and data sets (1: N & N: 1) through WEKA. ISSN (Online) 2456–3293
7. Mitchell TM (2009) Machine learning. McGraw-Hill. ISBN 0-07-042807–7
8. https://doi.org/10.1016/j.apenergy.2011.06.011
9. https://www.researchgate.net/publication/222516933_Economic_optimization_of_low-flow_solar_domestic_hot_water_plants/citation/download
10. https://www.researchgate.net/publication/222206878_A_comparison_of_various_forecasting_techniques_applied_to_mean_hourly_wind_speed_time_series/citation/download

A Survey on Feature Fatigue Analysis Using Machine Learning Approaches for Online Products

Midhunchakkaravarthy, Divya Midhunchakkravarthy, D. Balaganesh, B. Vivekanandam, and Albert Devaraj

Abstract On recent days, the leading manufacturers are very keen in online business processing; the products are brought into heavy competitive online market. To be successful in the online market, the manufacturers launch the new products with many features to compete with other manufacturers' products. The online customers are also interested in choosing the products with more attractive features, but the customers face problems in the features of the product by using it. Here, the term "feature fatigue" refers to the concept of customers' dissatisfaction. In the real market, the products with maximum attractive features are preferred by the customers in the initial stage. Later, the customers realize that some of the features are inconsistent which results in dissatisfaction on the product. Moreover, this dissatisfactory of the customers on the product majorly affects the goodwill and growth of the manufacturer. Many researchers have contributed several techniques to overcome the concept of feature fatigue. In this paper, the most significant feature fatigue analysis methods are discussed as a review to elevate the feature fatigue.

The original version of this chapter was revised: The authors "Midhunchakkaravarthy" and "B. Vivekanandam" names are corrected. The correction to this chapter is available at https://doi.org/10.1007/978-981-15-3284-9_86

Midhunchakkaravarthy (✉) · D. Midhunchakkravarthy · D. Balaganesh · B. Vivekanandam · A. Devaraj
Faculty of Computer Science and Multimedia, Lincoln University College, Kuala lumpur, Malaysia
e-mail: midhun.research@gmail.com

D. Midhunchakkravarthy
e-mail: divya.phd.research@gmail.com

D. Balaganesh
e-mail: balaganesh@lincoln.edu.my

B. Vivekanandam
e-mail: vivekresearch2014@gmail.com

A. Devaraj
e-mail: albert@lincoln.edu.my

S.-L. Peng et al. (eds.), *Intelligent Computing and Innovation on Data Science*, Lecture Notes in Networks and Systems 118,
https://doi.org/10.1007/978-981-15-3284-9_26

Keywords Feature fatigue · FF analysis · Usability evaluation · Online product · Customer equity

1 Introduction

In the current scenario, the manufacturers are interested in providing products with increasing features to the customers intending to increase the attractiveness and capacity of the product [1, 2]. At the moment of purchase, customers are drawn to products having the maximum number of features. But when customers use the product, they feel dissatisfied due to the features' complexity. This dissatisfactory leads to bring down the god-will of the manufacturer through negative word of mouth (WOM) [3, 4].

A classic instance is the BMW 745 model, which was brought out with 700 features on the dashboard alone. This brought a good opening to the car in the market initially, but later after using it most of the users are dissatisfied due to the multi-function displays feature and iDrive system feature [1, 5, 6]. A survey made in USA specified that after using the products with more features, 56% of the customers felt dissatisfied on the features of the product [1]. A survey in UK reflected 63% of dissatisfactory on features of smartphone, where there were no issues with the hardware and software of the smartphone [6]. Hence, it is more important to a manufacturer to decide the features to be added on a new product while development should be attractive and also easy to use.

2 Feature Fatigue

Feature fatigue (FF) is the conception of customers' dissatisfaction of the product features based on the usability. Customers choose the products with more features while buying and that may lead to dissatisfaction of the customers if they face difficulty in using those features [7]. So, in the development of a new product, the process of adding the features is more difficult as a double-edged sword. In one side, addition of more attractive features improves the product usability and results good in the market. On the other side, the same process of adding more attractive features to the product will make the customers to face complexity in using the product and results in failure due to FF [8]. FF spreads the negative word of mouth (WOM) due to the customers' dissatisfaction on the product, and this spoils the market of the products and the growth of the manufacturer as well [9]. And due to the impact created by FF, the customers will move on to other manufacturers' products which affect the growth of the organization [2].

In the current competitive globalized business world, it is very important for all the organizations to keep hold on the potential customers, who will continue in repurchasing the products whenever a new product is launched by the manufacturer.

So, it is also more important to avoid FF of the product, else it will tend the potential customers to move on other brand in the market [6]. Hence, even though the addition of more attractive features to the products brings good result in earlier stage, later FF may result in the failure of the product in the market and spoil the reputation of the organization which would decrease long-term goals by decreasing the customer equity (CE) of the manufacturer [5]. It is very important for the manufacturers to make decision in selecting the optimal combination of features to be added in developing a new product. This will result in the development of a new product with enough attractive and optimal features and reliable to use, and this removes FF.

3 Various Feature Fatigue Analysis Models

3.1 FF Analysis Based on Norton–Bass Model [10]

Norton–Bass model is one of the efficient techniques to perform feature fatigue analysis. This model discovers a set of feature combinations for the construction of new products through analyzing the successful features of previous products. The model is combined with product's usability, capability, and WOM properties to calculate the necessary features to be added based on costumer's expectation. The model helps to estimate the product usability from customer view. Through these analyses, the CE is pictured out in the initial stage of new product designing, considering end users' perception model. These aspects contribute to the manufacturer in making decision to add the essential features into the product in order to increase the CE and decrease FF.

The structural design feature fatigue analysis using on NB model is provided in Fig. 1. The model contains two phases. Phase 1 is customer purchase analysis and Phase 2 is feature combinations' optimization. Phase 1 shows WOM impacts on a customer's purchase behavior, and an evaluation of the customer's purchase conduct based on the verity in the feature combinations is suggested for a quantitative consumer transition model. In Phase 2 to maximize the CE, a GA is employed to find the best feature combination, in which the expected CE calculated in the Phase 1 is utilized by the GA as essential function. As a result, from GA, a chromosome will represent a feature combination which will be again passed into Phase 1 as input. The flow will be continued as an iterative process till the finest feature combination is found.

WOM is an important insight into buying customers. Positive WOM will increase the buy-out of customers, since the negative WOM will "remove" many prospective customers, which will decrease customer buying in exchange [2].

In Norton–Bass model, two limitations are implemented to operate with WOM impacts when buying customers. The individual client first began purchasing the item and those who were distressed by the WOM in a single phase. Secondly, WOM

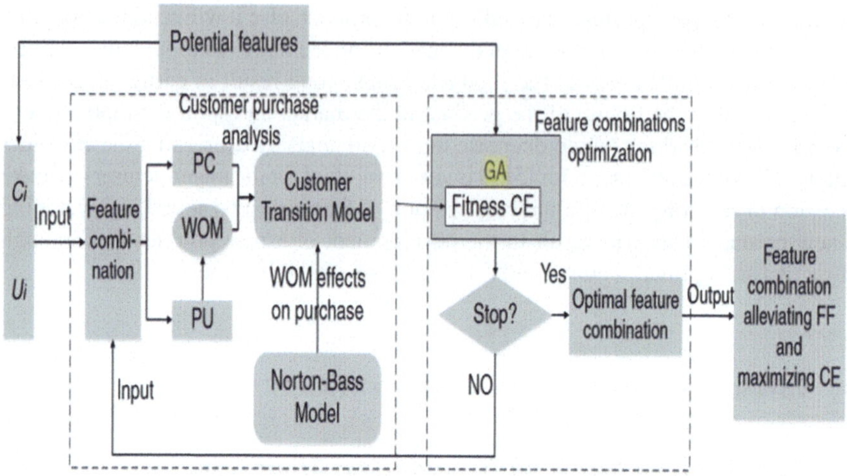

Fig. 1 Structural design feature fatigue analyses using Norton–Bass model

is limited for those clients who previously bought the product and used the item [5]. WOM impacts for client buying are shown on the model of Norton–Bass (Fig. 2).

The customers of a product are classified into four groups stated with the relationship on the manufacturer:

- Perspectives: prospective customers
- First timers: new customers
- Repetitive: retainable customers
- Customers' loss: Not retained and never be acquired.

The positive effect from the WOM is that the perspectives are converted into first timers, and also they retain as repetitive, whereas the negative WOM results in the loss of customers. The impacts from the WOM on CE with Norton–Bass parameters reflect customer behavior.

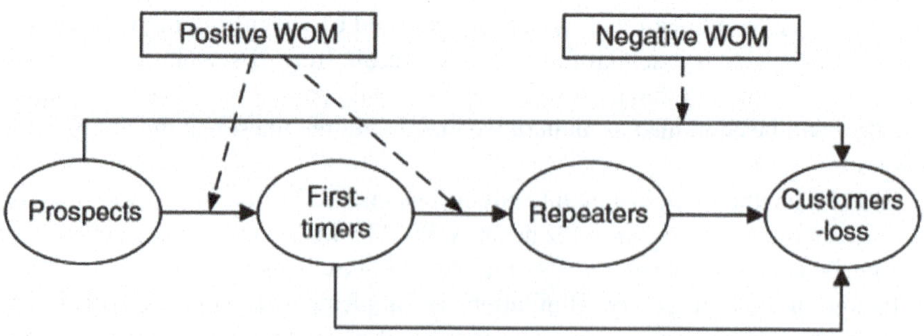

Fig. 2 WOM impacts Norton–Bass model

3.2 FF Analysis Based on SIR Model [11]

In consideration for previous studies, an epidemic susceptible–infected–recovered technique with GA is implemented to help manufacturers to identify an advanced combination of functionalities to remove FF.

Since WOM is comparable to an epidemic illness for client purchases and the spread of WOM, the SIR model is tested to demonstrate WOM behavioral impacts. In order to analyze the buying conduct of customers with distinct function combinations, a client transition model that combines WOM impacts is developed.

The FF analysis architectural design based on the GA models is provided in Fig. 3. It comprises Phase 1: purchase analysis of customers and Phase 2: optimization of feature combinations. The SIR model is used in the first phase to list the impacts of WOM on client buying conduct. The identified WOM effects are analyzed through a range of feature combinations using a customer transition model. In the second phase, a GA is used to discover the best mix of features to boost CE.

The effects of WOM on buying customers are comparable to the SIR model (Fig. 4).

Customers are divided into three groups of people with their perceptions of the goods, depending on their behavior. These individual groups are typed as vulnerable as prospects, customers and defectors-S, infected-I, and retrieved-R. Prospects (S) are received from clients (I) in the early stages. A number of these beneficial opportunities are affected by WOM (I), as the epidemiology infection phase. Several clients (I) are also translated into defectors (R) comparable to the epidemiological recovery method. In order to assess the CE, the parameter identified is then transferred into the client model.

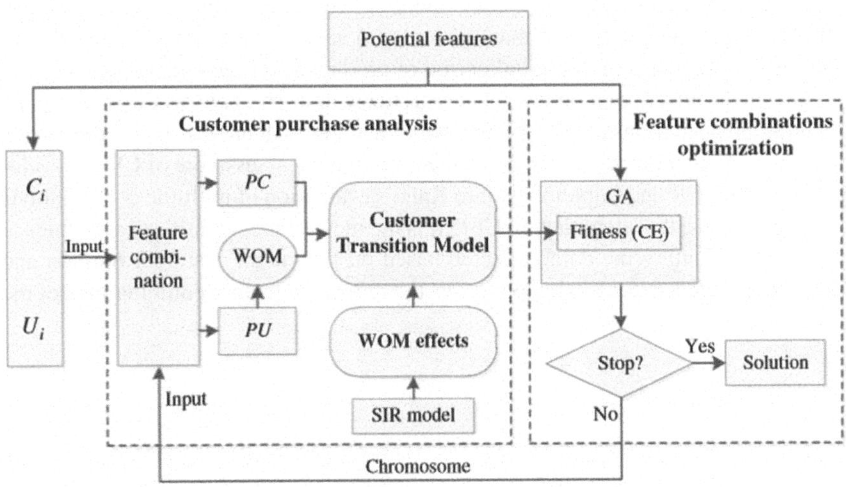

Fig. 3 SIR model-based FF analysis with GA

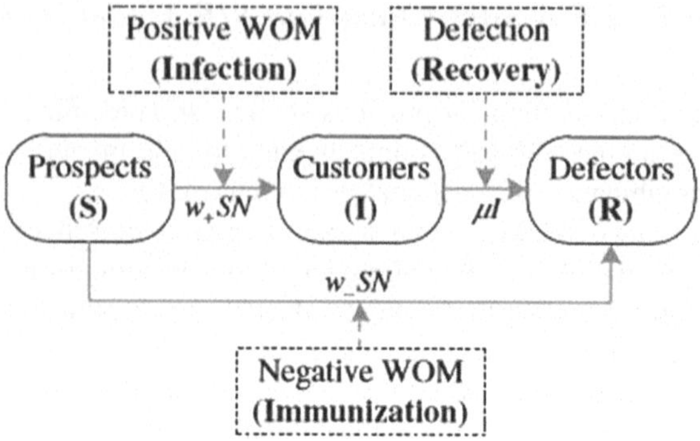

Fig. 4 SIR model of WOM effects

3.3 FF Analysis Using Continuous Fuzzy Kano's Model (CFKM) [11]

The continuous fuzzy Kano's model (CFKM) considers the indistinctness and hesitations of customer based on their requirements to ease unwanted features of the product. The effective way to deduct and remove FF is to evaluate the customer requirements (CRs) on a new product development which helps to find out the limitations in product capability and usability.

Kano's model is effective and implemented for CRs analysis in most research works [11]. The traditional Kano's model is an effective technique [12] that removes the indistinctness and hesitations in CRs [13]. The CFKM is implemented to analyze the indistinctness and hesitations in CRs of real-world circumstances which ease the unwanted features of the product.

The CFKM incorporates the fuzzy methods with Kano's model and formulate a new method which helps the manufacturers to analyze CRs precisely in difficult situations, particularly in the initial period of product development [14].

The CFKM advances a new Kano's with fuzzy assessment to work on the indistinctness and hesitations in CRs. In which the customer will provide their opinions on the feature of the product, the data collected through the assessment of CFKM evaluates individual CR and implements into Kano's evaluation of multiple combinations of features. Thus, the advantage of CFKM is the method that identifies an "Influence Value" by evaluating the data collected based on individual customer opinion and provides into the Kano's model to remove the unwanted feature combinations of the product [15].

3.4 FF Analysis Using Non-dominated Sorting Genetic Algorithm II (NSGA—II) [16]

NSGA-II is considered as an efficient technique among the multi-objective evolutionary algorithms and implemented in any different research areas [17, 18]. By implementing NSGA-II in multiple combinations of features, an optimal combination of features can be obtained through strong analysis. The methodology of NSGA-II algorithm compares all the feature combination set with each other to find the dominated feature combination. The non-dominated feature combinations are brought into the front and removed temporarily, and then the process of comparing the feature combinations with each other continues to find and remove the remaining non-dominated feature combination temporarily in an iterative process. At the end, no dominated feature combinations that are temporarily removed are ranked with its level of non-dominations.

By sorting the non-dominated feature combination in an order, the crowding distance is analyzed by fixing the first and last feature combinations as infinite value object function F and all other ranked feature combination in the sorted list are evaluated as follows:

$$p(x).q = s(x).q + \frac{(x+1).F - (x-1).F}{F_{\max} - F_{\min}} \quad (1)$$

where $x = 1, 2, n$, and n is the combination number at top, $p(x)$. q in objective function F is the distance of the individual X, $p(x + 1)$. F is the object function value F of $(x + 1)$ F_i and $p(x + 1)$. F is the $(x - 1)$ value, $F_{\max}$ is the highest value of F in FI, and $F_{\min}$ is the lowest value. The non-dominated feature combination sorted list and crowding distance help in selecting the best possible feature mix in NSGA-II in all decision-making countries.

3.5 FF Analysis Using Usability Evaluation [19]

As the FF is obtained based on its capability through usability, it can be removed by analyzing its capability. FF is evaluated based on the differences between product capability and product usability. Product usability is all about the efficiency and the effectiveness and the customers' satisfaction as well [20, 21]. In this model, the capacity of the feature of product is analyzed based on five score points as in Table 2. The score points are collected from the customers based on their impact upon usability of the product. Based on the score points, FF analysis is performed by balancing the capability and the usability of product [22, 23].

In the usability evaluation technique to analyze FF, the following equation is implemented.

Table 1 Usability measures [8]

Scores	Description
9	Very poor impression
7	Poor impression
5	Normal impression
3	Good impression
1	Very Good impression

Table 2 Capability measures [8]

Scores	Description
9	Highly impressed
7	Decidedly impressed
5	Impressed
3	Slightly impressed
1	Not impressed

$$\mathrm{FFD} = U - C \tag{2}$$

$$U = \frac{\mathrm{FU} - \mathrm{FU}_{\mathrm{min}}}{\mathrm{FU}_{\mathrm{max}}} \tag{3}$$

$$C = \frac{\mathrm{FC} - \mathrm{FC}_{\mathrm{min}}}{\mathrm{FC}_{\mathrm{max}}} \tag{4}$$

where U is the score points of usability (Table 1) and C is the score point of capability (Table 2) given by the customers based on the impact they have on the products. FU is the obtained value of usability according to Table 1. FC is the obtained value of capability according to Table 2.

Thus, if the FFD is greater than zero, then the feature is declared as FF feature. This helps the manufacturer to identify the feature with FF. Upon the FF analysis using usability evaluation, the methodology lists out the following results through evaluating Tables 1 and 2.

i. Features with high points in capability and low points in usability
ii. Features with high attractive points and low usability points
iii. Features with low attractive points but good in usability
iv. Features with low attractive and poor usability

From the above-obtained analysis, the manufacturers are benefitted to identify the features which should be removed from the product. This method also helps the manufacturer to concentrate on other levels of feature combinations on the product.

4 Conclusion

In this paper, various FF analysis methods are discussed to remove FF. Addition of more features to a product will increase its sales initially, but the overloaded features result in FF which brings down the growth of the brand and leads to the loss of potential customers which automatically reduces the CE. To remove FF, the manufacturer should design the product with optimal feature combination and easy to use as well. As a conclusion, upon the study made for this paper, all the FF analysis methods are performed on the feedback data collected from the existing customers based on their impact on product features, which does not include the data of publics' impact. The view of non-customers about the product and features is also very important to the manufacturer in designing a new product with optimal features. To overcome this limitation, the FF analysis methods should also include the sales flow data of the product by analyzing why the customers choose the product or why the customers does not choose the product, what are the features are more attractive, or what are the unwanted features of the product. As the impact of the market varies periodically, geographically, and socially, it would be more effective if the FF analysis methods also evaluate the sales flow data in removing FF.

References

1. Thompson DV, Hamilton RW, Rust RT (2005) Feature fatigue: when product capabilities become too much of a good thing. J Mark Res 42:431–442. https://doi.org/10.1509/jmkr.2005.42.4.431
2. Midhunchakkaravarthy J, Brunda SS (2016) An enhanced web mining approach for product usability evaluation in feature fatigue analysis using LDA model and association rule mining with fruit fly algorithm. Indian J Sci Technol 9. https://doi.org/10.17485/ijst/2016/v9i8/84592
3. Li M, Wang L (2011) Feature fatigue analysis in product development using Bayesian networks. Expert Syst Appl 38:10631–10637. https://doi.org/10.1016/j.eswa.2011.02.126
4. Midhunchakkaravarthy J, Brunda SS (2019) A novel approach for feature fatigue analysis using HMM stemming and adaptive invasive weed optimisation with hybrid firework optimisation method. Int J Comput Aided Eng Technol 11:411. https://doi.org/10.1504/IJCAET.2019.100442
5. Midhunchakkaravarthy J, Selva Brunda S (2017) Feature fatigue analysis of product usability using Hybrid ant colony optimization with artificial bee colony approach. J Supercomput 1–18 (2017). https://doi.org/10.1007/s11227-017-2178-4
6. Li M, Wang L, Wu M (2013) A multi-objective genetic algorithm approach for solving feature addition problem in feature fatigue analysis. J Intell Manuf 24:1197–1211. https://doi.org/10.1007/s10845-012-0652-7
7. Midhunchakkaravarthy D (2018) Product Usability and capability evaluation using modified BAT-ARM algorithm to alleviate feature fatigue. In: 2018 IEEE conference -international conference on recent innovations in electrical, electronics & communication engineering (ICRIEECE). IEEE
8. Midhunchakkaravarthy D, Bhattacharyya D, Kim T (2018) Evaluation of product usability using improved FP-growth frequent itemset algorithm and DSLC–FOA algorithm for alleviating feature fatigue. Int J Adv Sci Technol 117:163–180

9. Midhunchakkaravarthy J, Selvabrunda S (2014) A survey on various mining types, different text mining approaches and applications. Int J Recent Sci Res 5:665–668
10. Chai J, Wang L, Shi Q, Wu M (2015) Alleviating feature fatigue of multi-generation products. Ind Manag Data Syst 115:1435–1456. https://doi.org/10.1108/IMDS-03-2015-0104
11. Wu M, Wang L, Li M, Long H (2015) An approach based on the SIR epidemic model and a genetic algorithm for optimizing product feature combinations in feature fatigue analysis. J Intell Manuf 26:199–209. https://doi.org/10.1007/s10845-013-0773-7
12. Divya S, Padmavathi G (2016) Malicious Traffic detection and containment based on connection attempt failures using kernelized ELM with automated worm containment algorithm. Indian J Sci Technol 9. https://doi.org/10.17485/ijst/2016/v9i41/86922
13. Divya S, Padmavathi G (2014) Computer Network worms propagation and its defence mechanisms: a survey. In: Proceedings of international conference on advances in communication, network, and computing, CNC. Elsevier Ltd., pp 643–658
14. Divya S, Padmavathi DG (2014) Internet Worm detection based on traffic behavior monitoring with improved C4. 5. In: Proceedings of international conference on cryptography and security. ASDF, pp 48–56
15. Doppala BP, Midhunchakkaravarthy D, Bhattacharyya D (2019) Early stage detection of cardiomegaly: an extensive review. Int J Adv Sci Technol 125:13–24 (2019). https://doi.org/10.33832/ijast.2019.125.02
16. Li M, Wang L, Wu M (2014) An integrated methodology for robustness analysis in feature fatigue problem. Int J Prod Res 52:5985–5996. https://doi.org/10.1080/00207543.2014.895443
17. Bhattacharjee S, Chakkaravarthy M, Midhun Chakkaravarthy D (2018) GPU-based integrated security system for minimizing data loss in big data transmission. In: Balas V, Sharma NCA (eds) Data management, analytics and innovation. Advances in intelligent systems and computing. Springer, Singapore, pp. 421–435. https://doi.org/10.1007/978-981-13-1274-8_32
18. Bhattacharjee S, Chakkaravarthy M, Midhun Chakkaravarthy D, Rahim LBA (2019) An integrated technique to ensure confidentiality and integrity in data transmission through the strongest and authentic hotspot selection mechanism. In: Balas V, Sharma NCA (eds) Data management, analytics and innovation. Adv Intell Syst Comput. https://doi.org/10.1007/978-981-13-9364- 8
19. Wu M, Wang L, Li M, Long H (2014) An approach of product usability evaluation based on Web mining in feature fatigue analysis. Comput Ind Eng 75:230–238. https://doi.org/10.1016/j.cie.2014.07.001
20. Vardhan KA, Rao NT, Raj SNM, Sudeepthi GD, Bhattacharyya D, Kim T (2019) Health advisory system using IoT technology. Int J Recent Technol Eng 7:183–187 (2019)
21. Myint YY, Mithunchakravarthy D, Raju V, Bhaumik A (2019) Impact of budget participation on job performance in myanmar private commercial banks with mediation effect of budget goal commitment. Int J Innov Technol Explor Eng 8:579–587
22. Myint YY, Mithunchakravarthy D, Raju V, Bhaumik A (2019) Budget participation and employees' motivation in myanmar private commercial banks. Int J Innov Technol Explor Eng 8:573–578
23. Thirupathi Rao N, Debnath Bhattacharyya M, Kim T-H (2019) Steady state analysis of M/G/1 and M/Er/1 line models with MATLAB environment in cloud computing applications. J Eng Appl Sci 14:2016–2021. https://doi.org/10.3923/jeasci.2019.2016.2021

A Hybrid of Scheduling and Probabilistic Approach to Decrease the Effect of Idle Listening in WSN

Ruchi Kulshrestha, Prakash Ramani, and Akhilesh Kumar Sharma

Abstract Wireless sensor network (WSN) has now become a newfangled area of research. WSN is an interconnection of sensor nodes which are installed in a region for a particular task. Sensors sense their surrounding physical attributes and transmit the acquired information to the dedicated node or base station. There are some challenges associated with WSN: Energy efficiency, security, coverage, connectivity, responsiveness, etc. WSN has large number of application areas. Energy efficiency of the network is a major concern for some applications because battery replacement is not easy. Clustering is used to enhance energy efficiency of the network. When sensor nodes transmit information, some amount of energy dissipated. But when nodes are not in transmitting state, still some amount of energy is consumed in listening to data. In this paper, we proposed a hybrid of probabilistic and Sleep–wakeup clusterings to decrease the energy consumption due to idle listening. The proposed method adds the energy efficiency by using adaptive probabilistic approach with a sleep–wakeup schedule. Results are simulated in MATLAB. The analysis of results reveals that the duration of first-node dead and last-node dead while considering round number is increased in comparison of LEACH, which ultimately shows increased energy efficiency of the system.

Keywords WSN · Energy efficient · Clustering · Sleep–wakeup · Idle listening

R. Kulshrestha (✉) · P. Ramani
Department of Computer Science and Engineering, Manipal University Jaipur, Jaipur, India

P. Ramani
e-mail: prakash.ramani@jaipur.manipal.edu

A. K. Sharma
Department of Information Technology, Manipal University Jaipur, Jaipur, India

S.-L. Peng et al. (eds.), *Intelligent Computing and Innovation on Data Science*,
Lecture Notes in Networks and Systems 118,
https://doi.org/10.1007/978-981-15-3284-9_27

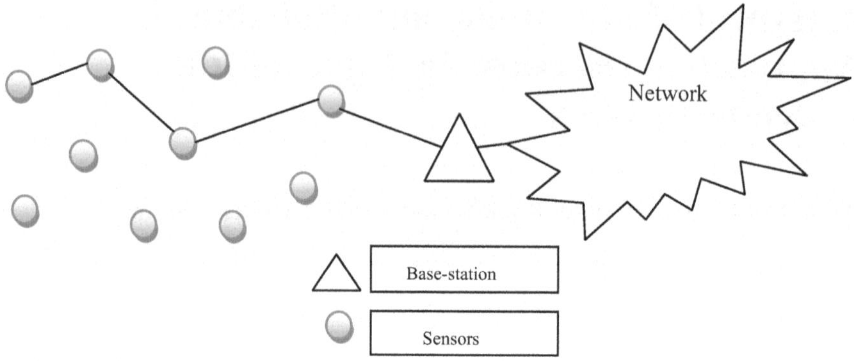

Fig. 1 WSN architecture

1 Introduction

1.1 Brief View of WSN

WSN is an interconnected network of sensor nodes. Sensors are light-weighted physical devices that are deployed in a specific area to gather information about some physical attributes. Sensors sense the change in physical parameters of the external environment and send this information to a dedicated device. This information is further being used in the analysis to take further actions (Fig. 1).

1.2 WSN Challenges

WSN has some challenges:

(1) Energy efficiency
(2) Data security
(3) Time synchronization
(4) System responsiveness
(5) Sensors deployment.

1.3 Clustering

Clustering is used to make an energy-efficient network. In clustering, groups of sensors are made. Each group or cluster has a head node called cluster-head (CH) [1]. Sensed information is destined to the CH from all members (sensor nodes) of a cluster. Now, CH becomes responsible to aggregate collected data and transmit it to the base station (Fig. 2).

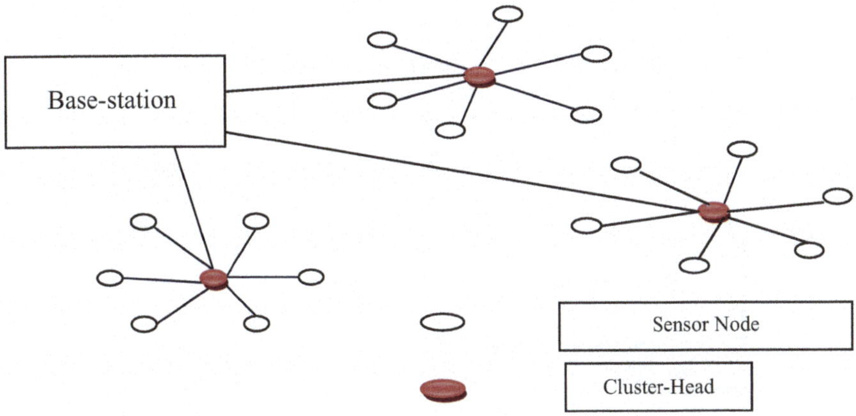

Fig. 2 Clustering in WSN

1.4 Energy Efficiency

Wireless sensor network has so many application areas like healthcare, forest fire detection, underwater surveillance, habitat monitoring, etc. All applications have their specific constraints like healthcare application requires data security; dense forest applications require energy efficiency as battery replacement is not possible [2]. To design a system as energy-efficient, energy consumption should be decreased.

1.5 Idle Listening

Idle listening is the energy consumption problem during an idle state of sensor nodes. When sensors transmit collected information to the destined base station or other intermediates, some energy has been consumed. But during no transmission of data or sitting in idle state, nodes dissipate some amount of energy. This energy consumption is called idle listening energy consumption. This consumption can be avoided in the process of prototyping an energy adequate clustering protocol. Sleep–wakeup approach is suitable to minimize energy drainage due to idle listening.

1.6 Low Energy Adaptive Clustering Hierarchy (LEACH)

LEACH is a standard clustering protocol stemmed from probabilistic approach for WSN. It comprises following cycle: Setup and steady [3]. Cluster-head selection for a cluster is based on probabilistic approach. When a node is selected as CH, it advertises itself as a cluster-head through broadcasting its id [4]. Both TDMA and CDMA are used in clustering and data transmission (Fig. 3).

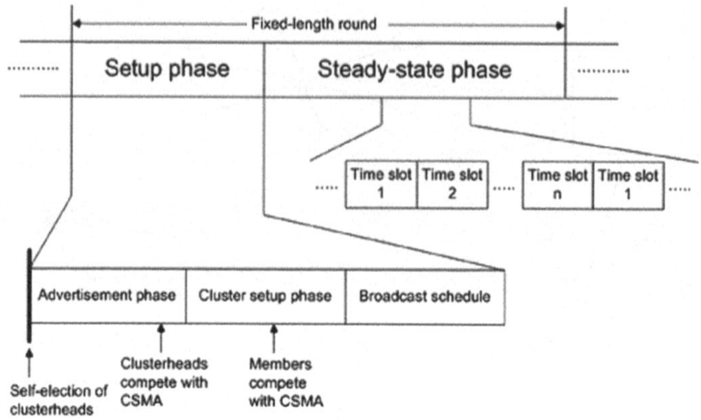

Fig. 3 LEACH phases [5]

2 Related Work

When sensor nodes are not transmitting any data, still dissipate some energy, this is called idle listening. To save energy due to this passive listening, sleep–wakeup approach can be used. Sasikala and Chandrasekar [6] urged a method grounded on residual energies. In this method, nodes are kept inactive or sleep mode. Some nodes are dedicated to select cluster heads. Clusters remain in an active state while its surviving energy is more significant than a predefined threshold value. Nazir et al. [7] adopted an approach of sleep–wakeup in WSN. It considers three parameters: Transmission distance between a node and destined base station, location significance of node, and the existing environment where an event occurs. According to these three parameters, a sleep–wakeup approach is applied. Ye and Zhang [8] proposed a technique in which the operation manner of each node is decided by its own that whether it will sleep, listen, or transmitting. The allotted time axis is divided into some slots. Location and surrounding of each node decide it's appropriate time slot. Sivakumar et al. [9] proposed a method of sleep–wakeup in which CH can create a path to the BS. The request message for data is initiated by BS and is sent to the destined cluster-head. Now, this cluster-head transmit a message to the required awaken nodes and sends sleep message to other nodes. Yang et al. [10] proposed a protocol named TCH-MAC. It is based on hybrid TDMA/CSMA. It uses sleep–wakeup to reduce energy consumption. To manage data traffic of a network, an adaptive approach of TDMA is used. Also, CSMA is used for energy efficiency. Shah et al. [11] proposed protocol EESAA. It is based on the pairing of nodes considering the distance from the base station. Sleep–wakeup state transition is applied to decrease the effect of idle listening.

3 Proposed Protocol

3.1 Sleep–Wakeup Approach

In this paper, we proposed a scheduled probabilistic clustering protocol. Clustering is done using a probabilistic approach. When sensor nodes are not in the transmitting phase, still dissipate some amount of energy. This energy consumption can be reduced by applying sleep–wakeup approach at sensor nodes. In sleep–wakeup approach, sensor nodes remain inactive (sleep mode) until they have no data to transmit/receive. And, thus, energy drainage due to idle listening is reduced.

3.2 Proposed Algorithm

Step 1: Random deployment of sensors in the sensing field.
Step 2: For i=1 to R (No. of rounds) Repeat step 3 to 6
Step 3: Cluster-head selection using probabilistic approach considering residual energy.
Step 4: For i=1 to n (Total no. of nodes)
Step 5: If Node[i]!=transmit data since opt_time
    Node[i].State=sleep;
    n=n+1;
Step 6: If Node[i].receive Data=true;
  if (Node[i].State=sleep)
   Node[i].State=Wakeup;
Step 7. End.

3.3 Method

See Fig. 4.

4 Results

Simulations have been performed in MATLAB to obtain results. Table 1 shows considered parameters for simulation. The first simulation is done for 100 nodes and 5000 number of rounds. Table 2 shows the comparison between LEACH and proposed protocol. Comparison parameters are FND and LND at the round number of algorithm. Latest FND and LND present an increased lifetime of the network.

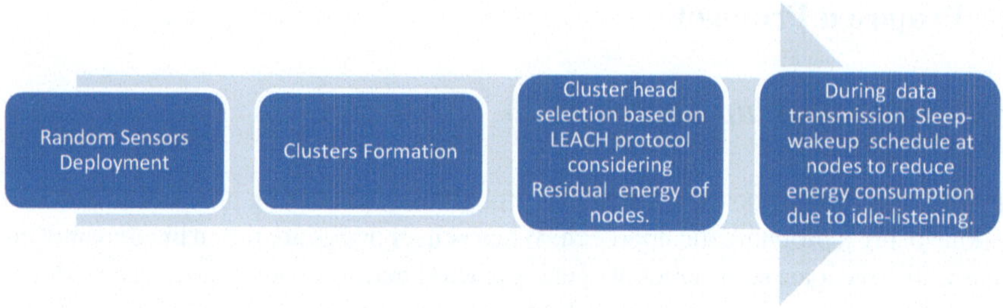

Fig. 4 Sequential phases of the proposed protocol

Table 1 Parameters for simulation (MATLAB) with respective values

Parameters for simulation	Value
Total rounds	5000
EDA (Energy consumption for data aggregation)	0.5 J
E0 (Initial energy of nodes)	0.5 J
Efs (Free space dissipated energy)	10 pJ
Emp (Multipath transmission dissipated energy)	0.0013 pJ

Table 2 Comparative results of LEACH and proposed (for 100 nodes)

	LEACH	Proposed
Round number of FND (First node become dead)	995	1203
Round number of LND (Last node become dead)	3260	4516

Figure 5 shows the lifetime comparison of LEACH and proposed protocol. The figure clearly shows that proposed protocol has the latest FND and LND so a lifetime of network is increased, which results in the energy-efficient network (Fig. 5).

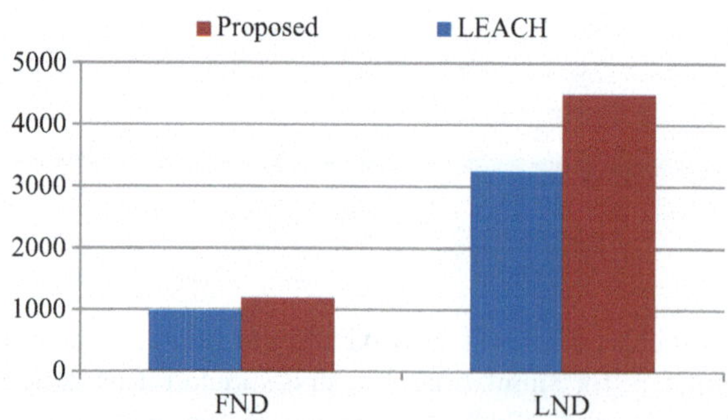

Fig. 5 Lifetime of network comparison between LEACH and proposed (for 100 nodes)

Table 3 Comparative results of LEACH and proposed (for 200 nodes)

	LEACH	Proposed
Round number of FND (First node become dead)	998	1198
Round number of LND (Last node become dead)	3303	4545

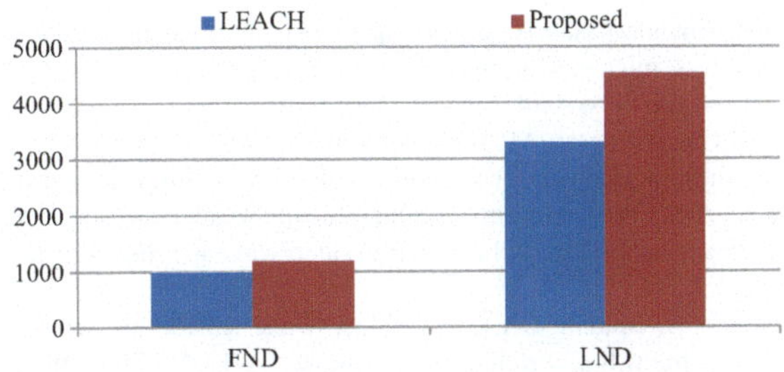

Fig. 6 Lifetime of network comparison between LEACH and proposed (for 200 nodes)

Again simulation is done for 200 nodes and 5000 number of rounds for simulation parameters given in Table 1. Comparison between LEACH and proposed protocol is presented in Table 3. Comparison parameters are FND and LND at the round number of algorithm. Result analysis from Table 3 and Fig. 6 shows that for 200 nodes also, the proposed protocol gives better results in terms of energy efficiency.

5 Conclusion

Wireless sensor networks have an immense number of applications. Numerous applications among them are energy constraint. There are some existing techniques to achieve energy efficiency in the system. Clustering is an apt method to make a network as energy efficient. There are some excellent clustering protocols existed, LEACH among them, giving good results in terms of energy efficiency. Existing protocols perform well, but still, there are some challenges to be considered. Energy consumption due to idle listening is one of them. Sensor nodes dissipate some amount of energy in an idle state for listening data apart from transmitting data. This energy consumption can be reduced by keeping nodes in sleep mode when they aren't transmitting data and make them awaken when they have to transmit data. With the help of this research, an algorithm has been designed which is hybrid of LEACH-based probabilistic clustering and sleep–wakeup schedule to achieve energy efficiency in the network. Results show that the proposed approach gives effective results.

References

1. Qing L, Zhu Q, Wang M (2006) Design of a distributed energy-efficient clustering algorithm for heterogeneous wireless sensor networks. Comput Commun 29(12):2230–2237
2. Chen W, Li W, Shou H, Yuan B (2006) A QoS-based adaptive clustering algorithm for wireless sensor networks. In: 2006 International conference on mechatronics and automation. IEEE, pp 1947–1952
3. Heinzelman WR, Chandrakasan A, Balakrishnan H (2000) Energy-efficient communication protocol for wireless microsensor networks. In: Proceedings of the 33rd annual Hawaii international conference on system sciences, 10pp. IEEE
4. Heinzelman WB, Chandrakasan AP, Balakrishnan H (2002) An application-specific protocol architecture for wireless microsensor networks. IEEE Trans Wirel Commun 1(4):660–670
5. Karl H, Willig A (2007) Protocols and architectures for wireless sensor networks. Wiley
6. Sasikala V, Chandrasekar C (2013) Cluster based sleep/wakeup scheduling technique for WSN. Int J Comput Appl 72(8)
7. Nazir B, Hasbullah H, Madani SA (2011) Sleep/wake scheduling scheme for minimizing end-to-end delay in multi-hop wireless sensor networks. EURASIP J Wirel Commun Netw 2011(1):92
8. Ye D, Zhang M (2017) A self-adaptive sleep/wake-up scheduling approach for wireless sensor networks. IEEE Trans Cybern 48(3):979–992
9. Senthil M, Sivakumar P, Indhumathi T (2014) Sleep & wakeup technique based clustering protocol-performance evaluation in wireless sensor network. J Theor Appl Inf Technol 68(3)
10. Yang X, Wang L, Xie J, Zhang Z (2018) Energy efficiency TDMA/CSMA hybrid protocol with power control for WSN. Wirel Commun Mobile Comput
11. Shah T, Javaid N, Qureshi T N. (2012) Energy efficient sleep awake aware (EESAA) intelligent sensor network routing protocol. In: 2012 15th International multitopic conference (INMIC). IEEE, pp 317–322

Prediction of Consumer's Future Demand in Web Page Personalization System

V. Raju, N. Srinivasan, and S. Muruganandam

Abstract Furtherance in Information Technology has progressed usage of Web services to extensive and exhaustive. From the gigantic volume of data, predicting optimal Web page is the cumbersome process. This paper presents an innovative methodology of minifying Web page recommendation systems by employing hybrid Levenberg–Marquardt firefly neural network algorithm along with improved fuzzy c-means clustering. Hybrid Levenberg–Marquardt firefly neural network algorithm is used to categorize the prospective and non-prospective consumer data of the Web log where improved fuzzy c-means clustering clusters the prospective data. Before a user begins the path of navigating through Internet for fetching relevant Web page, the cluster recommends the most relevant Web pages by analyzing the interests of similar users. A comparative analysis of proposed system with prevailing fuzzy k-means clustering technique exhibits that the performance of projected system is better.

Keywords Classification · Prediction · Artificial neural network · Web page personalization · Improved k-means clustering · Levenberg–marquardt firefly

1 Introduction

Information generated from various sources is accumulated progressively to make the Internet, a gigantic repository of data. Mining relevant information from the Internet becomes the hot topic.

V. Raju (✉)
Department of Science and Humanities, Sathyabama Institute of Science and Technology, Chennai, India

N. Srinivasan
Department of Computer Applications, B. S. Abdur Rahman CRESCENT Institute of Science and Technology, Chennai, India

S. Muruganandam
Department of Computer Science, SRM Institute for Training and Development, Chennai, India

S.-L. Peng et al. (eds.), *Intelligent Computing and Innovation on Data Science*, Lecture Notes in Networks and Systems 118,
https://doi.org/10.1007/978-981-15-3284-9_28

The prime process in data mining is to identify, investigate, and interpret to mine relevant information [1]. There are many current researches on optimizing Web mining with single criteria to search the desired content in World Wide Web. Content searching under multi-criteria approach using dominance principle mines qualitative and quantitative content from Web [2]. E-commerce is the modernized technology of doing efficient business transactions. Personalization of e-commerce system makes the entire system as an intelligent system with the help of radial basis neural network [3]. Through e-commerce transactions, huge amount of stored information of the consumers may lead to difficulty in seeking the guidance from the repository [4]. In improving the service prominence and consumer's satisfaction, e-commerce recommendation system enforces e-commerce platform with progressive suggestion toward the heterogeneous desires and the objectives of consumers [5].

Because of mammoth amount of data hosted on the Web, segregating data according to their access frequency is the demanding task [6]. Rest on their entity disposition, comforts and necessities, Web personalization system has materialized to compact through the difficulty of mining suitable requirement and to mine the personalized knowledge to consumer [7]. From the elucidation, we institute an attainable Web page recommendation practice comprises underneath limitations. The significant restriction is that it uses unadventurous collaborative filtering process. Therefore, accuracy and the exposure of personalized recommended penalties are not appropriate to please consumer [8]. Furthermore, amplification on Web page suggestion in regard to personalization and contextualization is measured as essential characteristic to gather disposition of diverse customers [9]. Some methods of Web recommendation are exactly derived from learning Web logs and give the suggestion subsequently to the consumers to optimize the search result by re-ranking process which result in diminishing the investigation time of desired Web pages [10]. Web page personalization is a noteworthy process in making Web systems to an intelligent system that incorporates various procedures and prediction algorithms such as Page Prioritize algorithm, Markov Analysis Models and Hyperlink-Induced Topic Search (HITS) for fetching quality of data on the Internet [11].

Page Prioritize algorithm uses recursive scheme to measure the page ranking by knowing the Web page linking relationships. It gives lesser rank to new page though it has very good contents than existing old page having higher rank [12]. HITS uses the link structure of the Web to rank the page appropriate for a specific searching keyword [13]. The notable difficulty of HITS is that neighborhood graphs must be assembled in "On-Fly" procedure.

2 Problem Definition

Some fundamental restrictions such as inadequate access and flexibility exist in the existing Web page recommendation systems. Many of the Web page recommendation systems do not consider the rarely indexed and lately included pages in the existing content. It is a strenuous task of spotting relevant pages according to the interest of the consumer and sequencing them in the prioritized order. These limitations stimulated us to accomplish the study and propose advancements on Web page suggestion systems.

This article is constructed with four sections of promoting the proposed methodology of Web personalizing system. 1. Literature review, 2. Illustration of proposed methodology, 3. Comparative analysis and result, and 4. Conclusion.

3 Related Works

Page weight in the user clusters and user's average evaluation on pages were taken into consideration for online page recommendation by Lian [14]. Jalali et al. [15] have suggested a model with the application of Web Server Log Mining to predict the navigational path for online forecasting.

Among the Web recommendation models such as association rule mining, Markov analysis, and collaborative filtering method, mutual filtering method mines the exact Web pages. It is proved by Suguna and Sharmila [16] in their recommended model.

Romil et al. [17] have suggested a model for data categorization with the application of naive Bayesian classification. Jafari et al. [18] proposed a complete preface to Web usage mining (WUM) to enhance the functionality of Web consumer navigational model. An analysis was conducted by Waykule and Gupta [19] with the procedures like weighted association rule mining and sequential pattern mining. They have recognized that the pages which were lately appended are rarely browsed by the customers.

Saleh et al. [20] have suggested intelligent adaptive vertical recommendation with neuro-fuzzy model and k-nearest neighbors (KNN) model to promote commerce consumer to categorize in upgrading Web customized system.

Raju et al. [21] have proposed a Web personalized system to predict consumer's future request with the application of hybrid Levenberg–Marquardt firefly neural network (LMFF-ANN) and fuzzy k-means clustering (FKCM) algorithms.

Muruganandam et al. [22] have suggested the methodology of personalizing of e-learning system which mines the preferred Web pages using the log history of the previous learners of the system.

4 Proposed Methodology

Endorsing opt Web page is a noteworthy process in all intellect Web system. Its fundamental functionalities are

(i) Discover knowledge (K) from Web usage data of Web log file
(ii) Depict the Web page suggestion system from the knowledge (K).

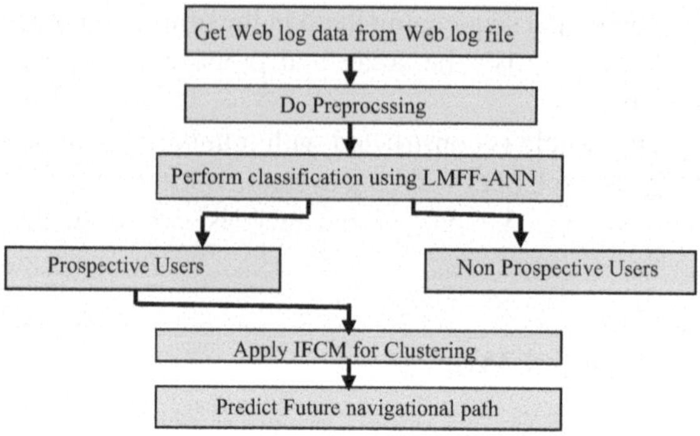

Fig. 1 Projected future demand-based Web recommendation

The proposed system uses clustering and classification procedures to cluster mutual consumers and to classify prospective and non-prospective consumers.

The proposed methodology performs the following steps:

1. Web log data is identified and preprocesses the related Web log file.
2. Apply firefly-based artificial neural network (FANN) to classify the consumers into prospective and non-prospective. In the network, Levenberg–Marquardt firefly (LM + Firefly) algorithm is to be applied.
3. Apply Fuzzy c-means clustering algorithm to cluster data of prospective users.
4. Predict the upcoming demand (prediction of future request) from equivalent consumers and is compared with existing k-means clustering algorithm.

The suggested methodology is systemized with JAVA and Web databases (Fig. 1).

5 Result and Discussion

The proposed system is built with JAVA platform with CloudSim with the applications of LMFF-ANN and improved fuzzy c-means clustering algorithms. Performances of existing and suggested are measured in term of time and accuracy.

Clustering accuracy is estimated with different number of iterations and is tabulated in Table 1. This table exhibits the dimension values of our projected analysis. Figure 2 depicts the graphical representation of cluster accuracy in the proposed

Table 1 Accuracy of clustering in proposed model

Number of iterations	Accuracy of cluster
10	71.24
15	72.48
20	73.23
25	76.53

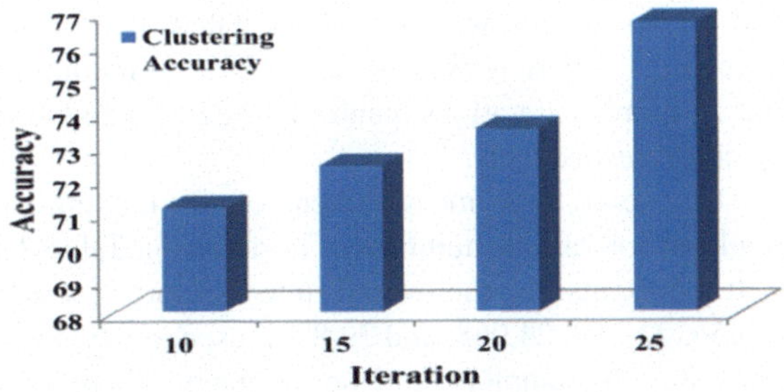

Fig. 2 Chart—Accuracy of clustering accuracy of proposed model

Table 2 Clustering time measures for proposed Web recommendation System

Number of iterations	Clustering time
10	9658
15	11,256
20	12,369
25	14,569

methodology for various numbers of iterations using. In 10 iterations, clustering accuracy of 71.24% is reached, whereas clustering accuracy increases to 72.48, 73.23, and 76.53 (in %) for iterations 15, 20 and 25, respectively.

Table 2 shows the data of estimated clustering time of proposed methodology for various numbers of iterations. The graphical representation of tabulated data is depicted in Fig. 3. Executing 10 iterations takes 9658 ms, while 15, 20 and 25 iterations take 11,256, 12,369, and 14,569 ms, respectively.

Table 3 displays the data of estimated accuracy of proposed model of future request prediction for various numbers of iterations and is tabulated in Table 3.

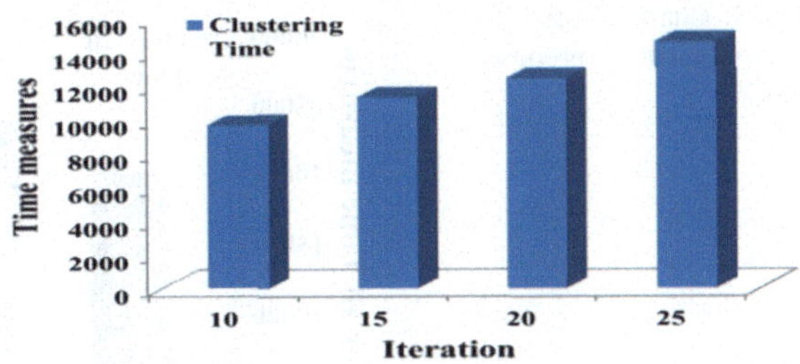

Fig. 3 Chart—Clustering time measurement of proposed model

Table 3 Accuracy measurement in proposed model

Number of iterations	Accuracy
10	78.23
15	79.27
20	80.38
25	82.26

The graphical representation of tabulated data of accuracy is depicted in Fig. 4. Executing 10 iterations achieves 78.23% of recommended accuracy, while 15, 20, and 25 number iterations result in 79.27%, 80.38%, and 82.26% of recommended accuracy, respectively.

The overall time measurement of the recommended model is estimated and tabulated for various numbers of iterations in Table 4.

It represents that the overall time measures for 10, 15, 20, and 25 iterations are 18,456, 22,369, 24,968, and 29,874, respectively. Figure 5 shows the graphical representation of complete time measurements for proposed methodology in executing various numbers of iterations.

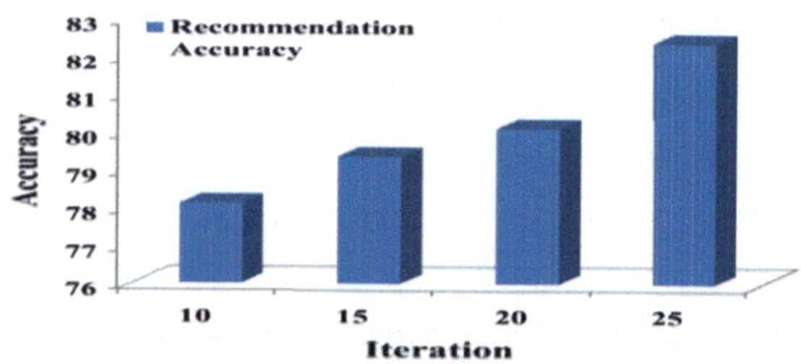

Fig. 4 Chart—Accuracy measurement in proposed model

Table 4 Overall time measures for our proposed research

Number of iterations	Overall time (in ms)
10	18,456
15	22,369
20	24,968
25	29,874

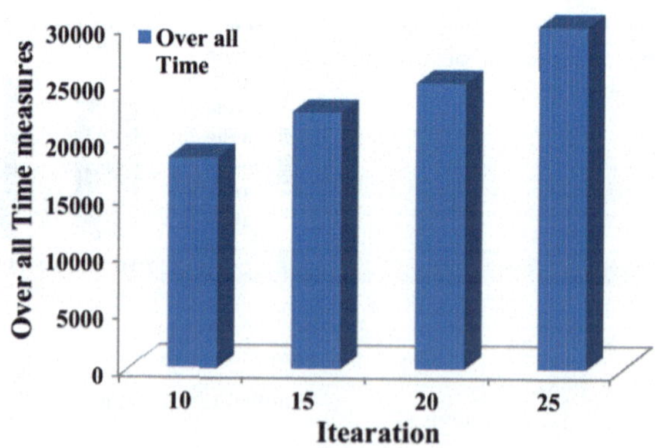

Fig. 5 Chart—Time measurement for proposed model

Table 5 Comparison of time measurement for proposed and existing models

Number of iterations	Calculated time of existing model (in ms)	Calculated time of proposed model (in ms)
10	18,988	18,457
15	23,567	22,368
20	25,125	24,969
25	29,985	29,873

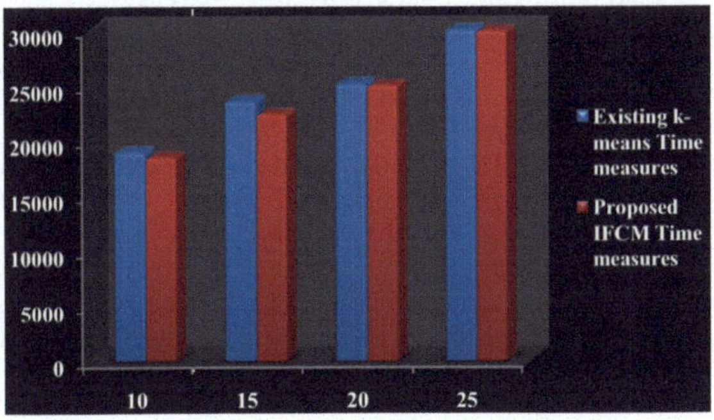

Fig. 6 Comparison chart of time measurement for proposed and existing models

5.1 Comparative Analysis

Data regarding to overall time of existing and proposed are estimated and analyzed. The existing k-means clustering takes 18,988 ms for 10 assessments. In contrast, the projected model with IFCM takes only 18,457 ms for the same number of assessments. In the subsequent 15 assessments, time requirement for k-means is 23,567 ms and is higher than proposed model which takes only 22,368 ms. The outcome of the evaluation states that proposed projected function is superior to the existing k-means analysis. Comparison data are tabulated in Table 5 and depicted in Fig. 6. The scenario is same for all other iterations as well.

6 Conclusion

The proposed model predicts the future request (navigational path) with the help of IFCM algorithm to mine the relevant pages. Data regarding to clustering time, utilization of memory, and performance of existing and proposed are analyzed. The result shows that 70% of clustering accuracy is achieved which proves that the suggested recommender system is more efficient than existing model with k-means technique.

References

1. Sunena KK (2016) Web usage mining-current trends and future challenges. In: International conference on electrical, electronics, and optimization techniques (ICEEOT), IEEE, India, pp 1409–1414
2. do Couto ABG, Gomes LFAM (2016) Multi-criteria web mining with DRSA. In: Information technology and quantitative management (ITQM), Korea, vol 91, pp 131–140
3. Pushpa CN, Patil A, Thriveni J, Venugopal KR, Patnaik LM (2013) Web page recommendations using radial basis neural network technique. In: 2013 IEEE 8th international conference on industrial and information systems, ICIIS, Sri Lanka, pp 18–20
4. Bhavsar MR, Mrs. Chavan PM (2014) Web page recommendation using web mining. Int J Eng Res Appl 4(7 Version 2):201–206
5. Lin S, Wenzhen X (2015) E-commerce recommendation system based on web mining technology design and implementation. In: International conference on intelligent transportation, big data and smart city, Halong Bay, Vietnam, pp 347–350
6. Moawad IF, Talha H, Hosny E, Hashim M (2012) Agent-based web search personalization approach using dynamic user profile. Egyptian Inf J 13:191–198
7. Hawalah A, Fasli M (2015) Dynamic user profiles for web personalisation. Expert Syst Appl 42(5):2547–2569
8. Ying Z, Zhou Z*, Han F, Zhu G (2013) Research on personalized web page recommendation algorithm based on user context and collaborative filtering. In: 4th IEEE international conference on software engineering and service science (ICSESS), pp 220–224
9. Sneha YS, Dr Mahadevan G, Madhura Prakash M (2011) An online recommendation system based on web usage mining and semantic web using LCS Algorithm. In: AMCEC, Bangalore. IEEE, India, pp 223–226
10. Bhushan R, Nath R (2013) Recommendation of optimized web pages to users using web log mining techniques. In: IEEE 3rd international conference on advance computing conference (IACC), pp 1030–1033
11. Gohil PBJ, Patel K (2015) A study of various web page recommendation algorithms. Int J Eng Comput Sci 4(3):10608–10610. ISSN 2319-7242
12. Yang B, Chen H, Zhao X, Naka M, Huang J (2015) On characterizing and computing the diversity of hyperlinks for anti-spamming page ranking. Knowl-Based Syst 77:56–67 (Texas)
13. He H, Li Z, Yao C, Zhang W (2016) Sentiment classification technology based on markov logic networks. New Rev Hypermedia Multimedia 22(3):243–256
14. Lian R (2011) The construction of personalized web page recommendation system in e-commerce. In: 2011 International conference on computer science and service system (CSSS), pp 2681–2690
15. Jalali M, Mustapha Sulaiman MN, Mamat A (2010) WebPUM: a web-based recommendation system to predict user future movements. IEEE J Expert Syst Appl 37:6201–6212
16. Suguna R, Sharmila D (2013) An efficient web recommendation system using collaborative filtering and pattern discovery algorithms. Int J Comput Appl 70(3):37–44
17. Romil E, Patel V, Singh DK (2013) Pattern classification based on web usage mining using neural network technique. Int J Comput Appl 71(21):13–17
18. Jafari M, Soleymani Sabzchi F, Irani AJ (2014) Applying web usage mining techniques to design effective web recommendation systems: a case study. Int J Comput Sci 3(8):78–90
19. Waykule V, Prof. Gupta SS (2014) Review of web recommendation system and its techniques: future road map. Int J Comput Sci Inf Technol 5(1):547–551
20. Saleh AI, El Desouky AI, Ali SH (2015) Promoting the performance of vertical recommendation systems by applying new classification techniques. IEEE Int J Knowl Based Syst 75:192–223

21. Raju V, Srinivasan N (2018) Prediction of user future request utilizing the combination of both ANN and FCM in web page recommendation. J Intell Syst
22. Muruganandam S, Srinivasan N (2017) Personalized e-learning system using learner profile ontology and sequential pattern mining-based recommendation. Int J Bus Intell Data Mining 12(1):78–93. E-ISSN 1743-8195

Analysis on SLA in Virtual Machine Migration Algorithms of Cloud Computing

T. Lavanya Suja and B. Booba

Abstract Cloud computing is a transforming field, which has grown into multi-dimensions because of contributions from academia and industry in research and development. The cloud services provide the flexibility to achieve the operational excellence of the modern applications in all domains. Moving all legacy applications into cloud utilize the advantage of advance features in cloud infrastructure like security, reliability, and scalability. As storage and computation are not done in physical machines, they are termed as virtual machines (VM). Virtual machine migration is inevitable in all the services, so there are various VM migration algorithms in the market. All algorithms need load balancing component and adhere to the service-level agreement (SLA) signed between the cloud provider and cloud consumer. SLA's importance for the provider and consumer in terms of profit and benefits, respectively, is analyzed in detail.

Keywords Cloud computing · VM · Migration algorithms · Load balancing · SLA · Uptime · Downtime

1 Introduction

A cloud refers to a distinct information technology (IT) environment that is designed for the purpose of remotely provisioning scalable and measured IT resources. Cloud computing is the delivery of computing services—servers, storage, databases, networking, software, analytics, intelligence, etc., over the Internet to offer faster innovation, flexible resources, and economies of scale in a pay per use basis [1].

Nearly, every business is using some form of cloud computing or storage service. Cloud computing has transformed so many businesses throughout the past decade with its scalability, versatility, and reliability [2].

In cloud business, the provider and the consumer are bonded by the service-level agreement (SLA), an agreement between them on the list of services called Quality of Services (QoS). As the consumer pays on the basis of his usage, the quality standards

T. Lavanya Suja (✉) · B. Booba
Department of CSE, Vel's University (VISTAS), Chennai, Tamil Nadu, India

S.-L. Peng et al. (eds.), *Intelligent Computing and Innovation on Data Science*,
Lecture Notes in Networks and Systems 118,
https://doi.org/10.1007/978-981-15-3284-9_29

are followed to the core. Any violation in QoS will cost the provider so the factors listed in it play a key role in the business lifetime.

Cloud providers invest a lot of time and energy in devising an algorithm for their own requirements, so there are numerous algorithms in the market. In the previous work [3], an elaborate survey on VM migration algorithms was done. The next step in the work is to analyze the monetary benefits and losses for the cloud consumer and cloud provider so as to throw light on the importance of SLA.

The rest of the paper is organized like this. Section 2 talks about the related work notably done for developing a VM migration algorithm and inclusion of factors in SLA and its importance. Section 3 registers the implications from the wide literature review, then the findings are projected as graphs in Sect. 4, and finally, Sect. 5 gives the conclusion and the future work of this study.

2 Related Work

A pre-emptive scheduling algorithm [4] which improves the efficiency and makes the job done before their deadline is met. It is claimed a better option than traditional scheduling and other similar approach algorithms.

The CloudSim tool kit provides a platform to test the working of algorithms in cloud environment and analyze the performance metrics. Without plunging into the cloud directly, we are able to simulate the results of the performance of the IaaS algorithms and also the applications in SaaS. This paves way for improvement in our development and also a cost reduction factor in the SDLC [5].

Load balancing is an important concept in VM migration algorithms. There are various approaches which are classified under static and dynamic. The metrics include throughput, fault tolerance, response time, migration time, and scalability. Many algorithms for load balancing [6] also have inspired by nature like ant colony and bees foraging behavior just like migration algorithms. Here, we find a similarity in the metrics and inspiration for algorithms between migration and load balancing algorithms.

An improved Round Robin [7] was proposed to improve the overall task completion time and number of VM migrations both in space-shared and time-shared scenario. The other parameters measured are idle time of tasks, number of million instructions re-executed, and number of delayed tasks. In all the parameters, the improved Round Robin algorithm performed well which showed dynamic load balancing and knowing the length of the task is very essential. Hence, those parameters are considered.

Chien et al. propose an efficient virtual migration algorithm [8] with a smaller number of migrations. The authors also bring the advantage of better SLA compliant during the VM migration. Apart from the fact they propose, the experimental results show that the total migration has increased, time to select VM needed to migrate has increased, and so is the percent time of SLA violations.

SLA's importance and need for detection and cost of violation are discussed in this paper [9]. The authors throw light on formulation of SLA, burdens laid on consumer to detect and report SLA violations so as to claim the benefits like money refund within the stipulated time. They also appreciate cloud providers like Verizon for automatic credit done to the consumer account in case of SLA violation. Azure is stringent claimed by the authors as it asks the consumers to report SLA violation within 5 days compared to 30 days by other cloud providers.

Saravanan et al. portray the trade-off between the minimization of migration time and consumption of energy in artificial bee colony algorithm [10]. Hence, they propose SALMonADA model which includes a system for monitoring the SLA violations and report immediately. They say that it relieves the burden of the cloud consumer from monitoring and reporting to the cloud provider and thereby reaping the benefits of payment credit. This stresses the importance of SLA and its monitoring both by the cloud provider and consumer.

SLA monitoring has got very much importance as it costs much for the cloud provider and the cloud consumer. As cloud providers lay this responsibility of reporting SLA violation within 30 days of the next payment cycle, it becomes mandatory for the cloud consumer to be vigilant on this. Accuracy and fast detection are the main characteristics of this SLA monitoring. Apart from this, inclusion of SLA factors plays a major role. In [11], the author emphasizes on the frequency of times and the interval between each checking is done.

Analyzing the guarantee terms and identifying the test requirements done in a proactive and a reactive way [12] bring out a Testing suite for SLA violations. This may serve as a benchmark for measuring the performance and helps the cloud provider in avoiding the SLA penalty amount. On the other hand, it helps the business up and running for the cloud user thereby their service is uninterrupted.

The work [13] has reviewed five different SLA-based architectures and brought out their merits and demerits. CloudWatch is taken as an example to analyze the SLA parameters and studied for their usage in finding out the SLA violations.

In an attempt on a comprehensive study on architecture of SLA [14], the authors have given a detailed explanation on its lifecycle, pricing, and parameter. There are roughly 12 parameters to be considered for SLA in IaaS which is the maximum number of parameters compared to SaaS and PaaS.

They have also compared the performance metrics of five different providers and have brought in light to the penalty cost incurred by the cloud providers because of SLA violations. As less than 30 min of downtime cost a 10% of credits to a maximum of 100% thereafter. This clearly shows that the credits given by the cloud provider are certainly a loss in his profit and good will.

Hussain et al provide a detailed review of SLA requirements and the SLA violation penalty paid by the cloud providers. The situation of a small and medium cloud provider is mentioned of importance as their resources should be thoroughly utilized. They propose optimized personalized viable SLA [15] to generate a viable SLA and predict the violation before happening thereby avoiding the violation and penalty.

GIPL's SLA document clearly gives the percentage of uptime as 99.721% and a decrease in 0.5% of it cost a penalty of 1% of its quarterly payment [16]. The overall penalty cap is 15% of quarterly payment and it increases by 5% for every SLA violation. When it reaches the 20% of quarterly payment, the consumer has the right to terminate the contract between them. Such is the situation and importance of SLA terms laid down in the agreement. This gives a clear picture of money gained as profit by SLA compliance and loss incurred because of SLA violation.

In a detailed literature survey [17], the findings clearly say that there are five different groups of cloud computing service composition and the most attention needed research objectives are algorithm improvement and user requirement satisfaction which call for new and improved algorithms which includes SLA parameters.

The authors [18] propose that the cloud application needs some time, for example, 5 min to get steady and thereafter the net utility of the provider becomes constant. They have proposed a model CASVid and an algorithm for monitoring SLA violation at application level. They were successful in doing so for a single application and have to extend it for heterogenous applications too as cloud provider has more than one application for service.

A detailed taxonomy of load balancing algorithms is presented in [19] by categorizing into two, viz. static and dynamic type. The authors agree that load balancing is a NP complete problem and is of much importance as it saves in terms of cost and time. Among the approaches discussed, the simulation results show that minimum compilation time (MCT) gives an optimal solution. This finding gives a clue that compilation time should be given more importance than execution time.

A conventional approach of dynamic load balancing [20] is claimed to be 30% more average research utilization rate and 225% less makespan than first come first serve and shortest job first algorithms. In order to achieve this, more migration of tasks from one VM to another VM takes place which specifies VM migration not only improves performance but also inevitable. This gives a clear indication that VM migration plays a major role in any algorithm and approach.

In this soft-computing-based stochastic hill climbing approach [21], the authors claim it to be a better load balancing algorithm than FCFS and RR algorithms. The overall response time is 30% less than the existing approaches. Cloud Analyst is the simulation tool used to measure the performance metrics.

This [21] is a centralized approach in contrary to the many decentralized approaches [22] many in the market. While it has advantages like minimum time to take a decision, there are disadvantages too like crashing of the central node shuts down the service itself.

3 Implications

Going through the literature review reveals the fact that factors included in SLA play a crucial role in deciding the profit of the cloud provider and increase the business of the cloud user. Every minute of uptime is counted and every hour of downtime is monitored. After including the factors in SLA, it should be strictly adhered and violation of it costs money for the cloud provider and loses the goodwill of the cloud user. Data from various sources were compiled in the form of graph for clear picture and better understanding to bring out the loss of profit in dollars. The loss of cloud provider turns into a bonus income for cloud user as it is paid in percentage of quarterly payment (QP).

4 Findings

Average downtime and uptime of major cloud providers like AWS, Google Cloud, Microsoft, and IBM in hours/year are projected in the below graph [23, 24]. The graphs have been plotted on data acquired from 2007 to 2013 (Figs. 1 and 2).

4.1 Cost Benefits in Terms of Repayment for the Cloud Consumer

When a factor in SLA like uptime percentage is not met then the cloud provider pays the cloud consumer 0.5–5% of quarterly payment (QP) [16] and is shown in the graph below. Here, the SLA includes 99% of uptime, so the minimum loss of 0.5% starts when uptime goes in the range of 98.5–98.99% and thereafter it increases by 0.5% (Fig. 3).

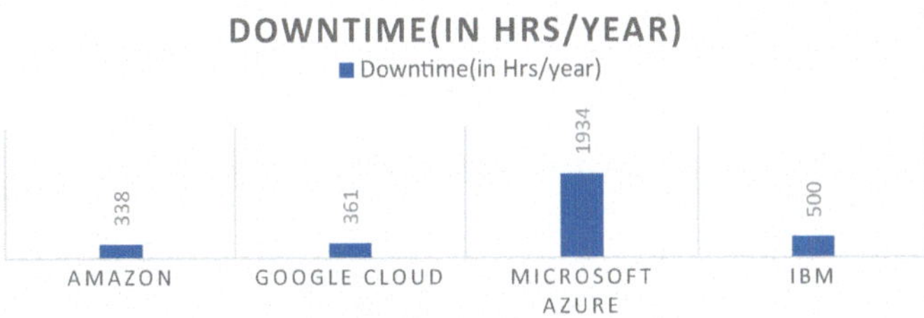

Fig. 1 Average downtime in hours/year of four major cloud providers

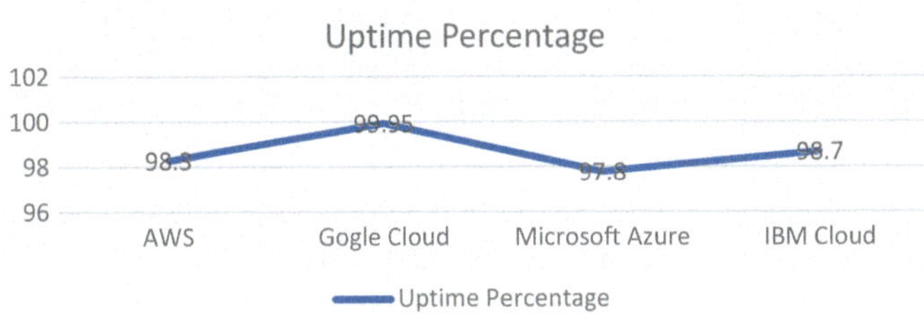

Fig. 2 Average uptime percentage of four major cloud providers

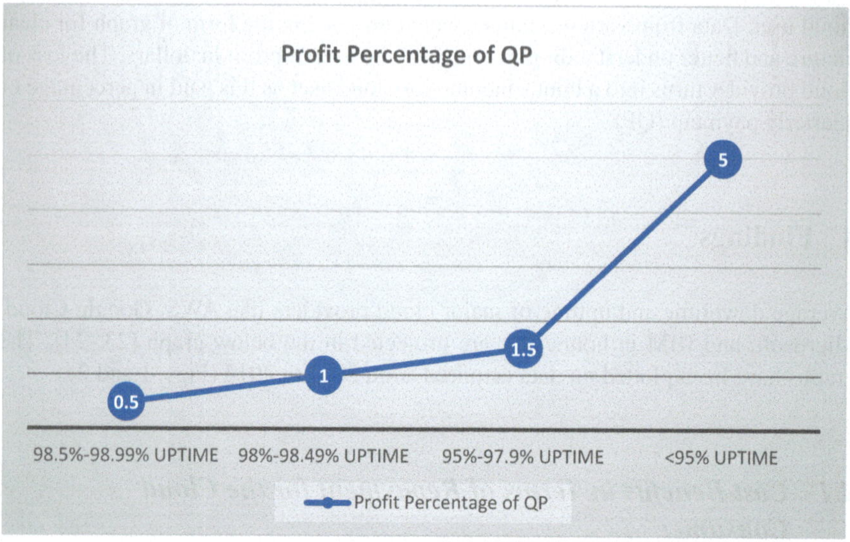

Fig. 3 Profit percentage of a cloud consumer

4.2 *Incurred Loss for the Cloud Provider*

The below graph plots the lost money in million USD for the major four cloud providers in the duration years 2007–2013 due to unavailability of service and SLA breach [24–26] (Fig. 4).

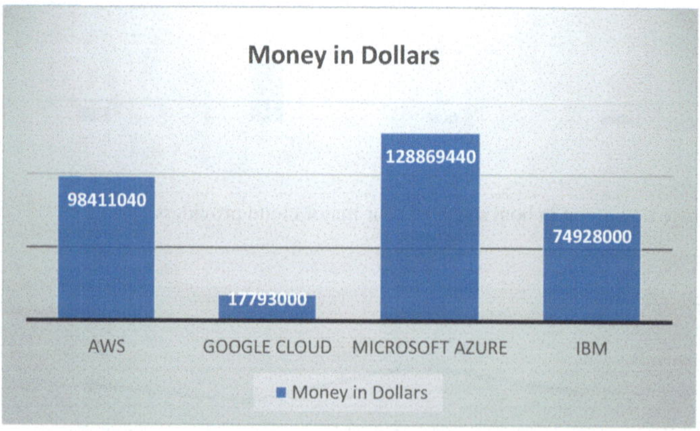

Fig. 4 Average loss incurred by cloud provider for SLA breach

5 Conclusion and Future Work

Having analyzed the key components of VM migration algorithm reveals the fact that load balancing, energy efficiency, and complying to SLA are the most important. Though SLA is followed to 95–98% on an average, the rest of the 5–2% breached time incurs a lot of money loss for the cloud providers. On the other hand, even though the cloud consumers get a monetary benefit by the 5–2% SLA breach, their business and goodwill get affected is a great concern especially for cloud start-ups. In the next work, it is planned to propose a detailed review on performance and profit of cloud start-up companies in terms of SLA.

References

1. azure.microsoft.com/en-au/overview/what-is-cloud-computing. Date of access: 10/07/2019
2. techgenix.com/future-of-cloud-computing. Date of access: 10/07/2019
3. Lavanya Suja T, Booba B (2019) A study on virtual machine migration algorithms in cloud computing. Int J Emerg Technol Innovative Res (JETIR) 6(3):337–340
4. Santhosh R, Ravichandran T (2013) Pre-emptive scheduling of on-line real time services with task migration for cloud computing. In: 2013 international conference on pattern recognition, informatics and mobile engineering, IEEE, pp 271–276
5. Calheiros RN, Ranjan R, Beloglazov A, De Rose CA, Buyya R (2011) CloudSim: a toolkit for modeling and simulation of cloud computing environments and evaluation of resource provisioning algorithms. Softw Pract Experience 41(1):23–50
6. Kaur R, Luthra P (2012) Load balancing in cloud computing. In: Proceedings of international conference on recent trends in information, telecommunication and computing, ITC
7. Devi DC, Uthariaraj VR (2016) Load balancing in cloud computing environment using improved weighted round robin algorithm for nonpreemptive dependent tasks. Sci World J
8. Chien NK, Dong VSG, Son NH, Loc HD (2016) An efficient virtual machine migration algorithm based on minimization of migration in cloud computing. In: Vinh P, Barolli L (eds) Nature of computation and communication. ICTCC 2016. Lecture Notes of the Institute for Computer Sciences, Social Informatics and Telecommunications Engineering, vol 168. Springer, Cham
9. Juneja R, Sharma D (2014) Service level agreement comparison in cloud computing. Int J Eng Manage Res (IJEMR) 4(3):126–131
10. Saravanan S, Venkatachalam V, Malligai ST (2015) Optimization of SLA violation in cloud computing using artificial bee colony. Int J Adv Eng 1(3):410–414
11. Jamail NSM, Mogos G (2019) Cost-benefit evaluation of service level agreement in cloud environment—dynamic monitoring interval tool. Int J Eng Res Technol 12(1):107–112. ISSN 0974-3154
12. Palacios M, García-Fanjul J, Tuya J, Spanoudakis G (2012) Identifying test requirements by analyzing SLA guarantee terms. In: 2012 IEEE 19th international conference on web services, IEEE, pp 351–358
13. Absa S, Benedict S (2016) A survey on SLA based cloud architectures. J Convergence Inf Technol (JCIT) 11(1)
14. Aljoumah E, Al-Mousawi F, Ahmad I, Al-Shammri M, Al-Jady Z (2015) SLA in cloud computing architectures: a comprehensive study. Int J Grid Distrib Comput 8(5):7–32
15. Hussain W, Hussain FK, Hussain OK, Damiani E, Chang E (2017) Formulating and managing viable SLAs in cloud computing from a small to medium service provider's viewpoint: a state-of-the-art review. Inf Syst 71:240–259
16. Guj Info Private Limited Service Level Agreement(SLA) & Penalties Document

17. Jula A, Sundararajan E, Othman Z (2014) Cloud computing service composition: a systematic literature review. Expert Syst Appl 41(8):3809–3824
18. Emeakaroha VC, Ferreto TC, Netto MA, Brandic I, De Rose CA (2012) Casvid: application level monitoring for SLA violation detection in clouds. In: 2012 IEEE 36th annual computer software and applications conference, IEEE, pp 499–508
19. Mishra SK, Sahoo B, Parida PP (2018) Load balancing in cloud computing: a big picture. J King Saud Univ-Comput Inf Sci
20. Kumar M, Sharma SC (2017) Dynamic load balancing algorithm for balancing the workload among virtual machine in cloud computing. Proc Comput Sci 115:322–329
21. Mondal B, Dasgupta K, Dutta P (2012) Load balancing in cloud computing using stochastic hill climbing-a soft computing approach. Proc Technol 4:783–789
22. Wang X, Liu X, Fan L, Jia X (2013) A decentralized virtual machine migration approach of data centers for cloud computing. Math Probl Eng
23. http://iwgcr.org/wp-content/uploads/2014/03/downtime-statistics-current-1.3.pdf. Date of Access: 10/07/2019
24. https://www.uctoday.com/unified-communications/how-much-would-a-cloud-outage-cost-your-business/. Date of Access: 10/07/2019
25. https://www.sciencedirect.com/topics/computer-science/cloud-service-consumer. Date of Access: 10/07/2019
26. https://www.wired.com/insights/2011/12/service-level-agreements-in-the-cloud-who-cares/. Date of Access: 10/07/2019

Effective Mining of High Utility Itemsets with Automated Minimum Utility Thresholds

J. Wisely Joe, Mithil Ghinaiya, and S. P. Syed Ibrahim

Abstract Utility mining is the recent data science task in the ground of data mining. Utility mining observes profit and quantity of each distinct item present in the transactions, thus it results in productive patterns with high importance in transactional databases. There are many algorithms sketched to trace the entire set of highly productive utility itemsets using user-defined single minimum utility threshold. An efficient framework called high utility itemset mining with automated minimum utility thresholds (HUIM-AMU) is put forward in this research paper. This algorithm uses a condensed tree arrangement called utility pattern tree to store the transactions and a constant value indicating the amount of most profitable itemsets. With very large count of transactions in a database, it is very difficult to identify the importance or productivity of every item. Without the knowledge of items, threshold setting may degrade the effectiveness of the process. In our proposed work, the difficulty in the setting of minimum threshold and the time spent on the analysis of database to set threshold are reduced by automating the threshold setting process. The results clearly indicate that the HUIM-AMU generates only profitable and compact itemsets.

Keywords High utility itemset mining · Setting minimum utility threshold · Multiple · Automatic minimum utility thresholds

1 Introduction

This pattern mining research in data mining has been begun in early 1990s. They were trying to yield frequent patterns present in databases [1]. The very first of that kind was Apriori algorithm, drafted to locate frequent itemsets in the form of rules from transactional databases. A transactional database is a collection of sales transactions done with stores at different timings by the customers. The batch of

J. Wisely Joe (✉) · M. Ghinaiya · S. P. Syed Ibrahim
School Computing Sciences and Engineering, VIT University Chennai Campus, Chennai, India
e-mail: wiselyjoe.j2013@vit.ac.in

S. P. Syed Ibrahim
e-mail: syedibrahim.sp@vit.ac.in

S.-L. Peng et al. (eds.), *Intelligent Computing and Innovation on Data Science*,
Lecture Notes in Networks and Systems 118,
https://doi.org/10.1007/978-981-15-3284-9_30

items purchased together by the customers often is considered as frequent itemsets. The extracted patterns are very much used in decision making to increase the profit. The algorithms related to frequent itemset mining and their drawbacks [1, 2] have been studied. To overcome the issue of not considering the importance of items present, utility pattern mining has been risen. In utility mining, instead of frequency, we consider the profit of items and units purchased. Profit is an integer variable assigned to every item in the database based on its profitability. Utility mining is used to trace patterns producing more profit or the patterns drive to a successful business decision. A high utility itemset is a group of purchased elements with actual utility larger than the user-specified minimum utility threshold [3, 4]. This value is set by the user after analyzing the data manually. The algorithm proposed by Liu uses the weighted transaction utilization to extract productive itemsets [5]. There are many important algorithms [6–8] already explored for the betterment of the mining process with and without generating candidate itemsets. This algorithm uses a list-like structure called utility lists. Lin et al. [9] proposed an algorithm to trace HUIs with a different structure called HUP-tree which integrates the transaction-weighted-utilization model and the FP-tree proposed in [9]. In HUP-tree approach, every distinct element in the database has a unique minimum utility threshold. Setting same threshold for all elements is more difficult without knowing the significance of elements. So there is a need for an efficient framework to trace high utility itemsets with different thresholds at every level of process. The threshold can be incremented at levels with some calculated measure of existing information. Lin et al. [10] and Gan et al. [11] proposed the algorithms HUI-MMU and HIMU with many threshold values. Our approach is completely divergent from the existing algorithms in the literature. We propose a novel technique to calculate the threshold value at every item level and the experiment results show that this algorithm gives more productive patterns than the existing approaches when applied on authentic and artificial datasets.

2 Related Literature

2.1 *Association Mining: With One or Many Support Thresholds*

The first association mining algorithm was designed by Agrawal et al. [11] and it dealt the consistent problem in this field. The association between the items is obtained in two basic steps. Initially, the combination of items that occurs frequently in customer transactions is chosen. Later, using confidence, measure the correspondence between the frequent items is listed. There are many more prime algorithms proposed by various researchers to extract recurrent itemsets efficiently. The most desired algorithms are Apriori algorithm, FP-Growth algorithm, and Eclat algorithm proposed by Zaki et al. Liu et al. discussed a new problem called rare profitable item problem faced by frequent itemset mining algorithms. The reason behind this problem was that the single minimum threshold usage for the entire process. To overcome the issues, multiple minimum support thresholds were proposed in some papers with some modifications and novelty. Eminent works done in this topic are: MS Apriori, CFP-Growth, CFP-Growth++, and FP-ME. All the above-mentioned algorithms are extended versions of primary algorithms mining frequent itemsets like Apriori and FP-Growth. Here, the support threshold is set by the user at every item level. But these approaches can not be enforced to the utility mining issues directly as they deal with profit and quantity of involved items in their mining process.

2.2 *High Utility Itemset Mining: Using One or Many Utility Thresholds*

The trending investigated area in the scope of association mining is HUI mining for last ten years. Liu et al. were the first to introduce this matter [5] in the year 2005 and dealt with the limitations of the existing algorithms by not considering the characteristics of the item purchased by the user. The two-phase algorithm explained how the candidate generation is done in level-wise approach. The algorithm computes the utilities of the itemsets after mining them based on their transaction-weighted utility in the first phase. With user set minimum utility threshold, the non-HUIs are filtered. In the same research area, some more approaches have been proposed which use level-wise mining. UMining algorithm is a level-wise mining algorithm and works well with moderate sized and sparse databases not with large or dense databases. UMining_H algorithm was also given by the same author Yao and works better than UMining. IHUP, an tree structure-based algorithm was designed by Ahmed et al. [3], to prune low utility itemsets from the database. The performance of this algorithm is not to the level because of the huge number of mined HUIs. All the above-discussed algorithms have the same minimum utility threshold for all the levels of pruning. The recent algorithms are quicker than the traditional algorithms on the standard datasets.

Multiple minimum threshold concept in utility-based mining was an initiative by Lin et al. HUI-MMU and HUI-MMU-TID [12] framework were the improved algorithms proposed by his team for effective HUI mining. In stead of TWU-property, a modern ranked downward closure property was introduced. It can work well on items arranged in the order of minimum utility threshold values. Some algorithms got improved efficiency because of the data structures used. Lin et al. proposed an efficient HUI-MMU-TE algorithm which improves the productivity of utility mining as it follows vertical data representation. The authors proved that the above algorithms with level-wise multiple minimum utility thresholds generate very accurate, productive, and quality high utility itemsets when compared with existing single and several minimum threshold utility algorithms. To enhance the efficiency of existing algorithms, an approach is suggested in this paper (HUIM-AMU) which automatically initializes the base utility threshold to a positive integer based on the utility of 1-itemsets and increases the threshold at every needed places during it's execution. The algorithm in Sect. 3 shows how the threshold has been initially set and incremented.

2.3 Definitions

1. Total transaction utility(TTU) of any row is the summation of item utility of the items in transaction.
2. Transaction-weighted utilization (TWU) of an itemset is the summation of total transaction utility having that itemset. If TWU (itemset) is not lesser than the threshold, then the taken itemset is considered as a high transaction-weighted utilization itemset (HTWUI).
3. If the itemset is not an HTWUI, then none of it's superset can be.
4. An item's minimum item utility in the given set of transactions is minimum of all its utilities.
5. An itemset's minimum item utility is the result of items's minimum item utility X its support count in the set of transactions.
6. Item's maximum item utility in dataset is maximum of all its utilities.
7. An itemset's maximum item utility is the result of item's maximum item utility X its support count in the set of transactions.

3 Proposed Work

3.1 Improved Approach of Utility Mining with Automated Minimum Utility Threshold

The new efficient algorithm named high utility itemset mining with automated minimum utility threshold algorithm (HUIM-AMU) discovers only productive itemsets. The strategy used to boost up the threshold is based on the utility of itemsets. It improves the productivity of the algorithm by increasing the profit, reducing execution time, and memory consumption.

3.2 The General Approach

HUIM-AMU takes the set of transactions and profit of individual items as input and gives back the most profitable high utility itemsets. It reduces the difficulty in setting minimum utility threshold. The data structure used to maintain the customer transactions in the database is UP-Tree. This proposed algorithm HUIM-AMU is the extension of HUI-MMU algorithm proposed in [4, 12]. This algorithm works in three phases. (1) transaction representation as UP-Tree, (2) automatic update of threshold, (3) identifying high utility itemsets which satisfy the threshold.

3.2.1 Construction of UP-Tree

The primary database has to be read twice to construct the tree structure. Transaction utility of every entry in the dataset and TWU of items present in the dataset are calculated. Based on the calculated value, every transaction is sorted in descending order. Compute the actual utilization of every 1-itemset. We tried many strategies to set threshold automatically. If minimum of all 1-itemset utilization is set as initial minimum utility threshold, number of candidate generation will be more. If maximum 1-itemset utilization is set as initial threshold, very less number of candidates are generated and we may lose productive high utility itemsets. In practice, we tried different values as threshold in trial and error method based on the requirement. Here, we calculated average utility of singe items and used that measure to set initial minimum utility threshold. The elements which are not satisfying the initial minimum utility threshold are removed from transactions and not included in further processing. After this pruning, the reorganized transactions are read and processed in order and inserted into the UP-Tree [7, 8]. UP-Tree construction procedure is clearly explained in the paper [7].

3.2.2 Generating HUIs and Automated Threshold

The procedure used to mine the productive itemset is a variation of UP-Growth explained in the paper [7]. Conditional pattern bases are constructed from the UP-Tree and stored in the memory. Every node in UP-Tree is connected to the other node in its path as links. So construction of pattern base is simpler by following the links maintained by linked lists. For every candidate from conditional pattern base minimum itemset utility (MIU), maximum itemset utility (MAU) and transaction-weighted utilization (TWU) are calculated. If the MAU and TWU exceed the present threshold, then the candidate will be considered as an HUI. When the minimum utility threshold becomes lesser than MIU, threshold will be raised automatically by finding the average utility of itemsets in that level. From that point, pruning is done with this new threshold. The chosen high utility itemsets are maintained separately in a list and used for productive decision-making process.

3.3 Algorithm: HUIM-AMU Algorithm

```
Input   :  UP Tree T,header table HT , initial
threshold  m_ut_th
Output : L1,Productive  high utility itemsets(PHUIs)
for each entry Y  in header table HT  do
  Construct Y's CPB
  Construct search space tree(SST) for the
   items from leaf item Y to top in its CPB with Y as
   root
  for every new item i inserted into SST
    Calculate MIU,MAU, TWU of i
          Output i  to list
              if MIU(i) ≥ m_ut_th
                 Calculate AVG(TWU) of itemsets in
                  current level of SST
                    Raise  m_ut_th  to AVG(TWU)
              end if
    end if
  end for
  Output itemsets  in L1
end for
```

4 Experimental Results

To determine our algorithm's performance, experiments have been extensively done on many distinct databases. Leading algorithms to extract HUIs with multiple user-defined utility thresholds and our algorithm with automated minimum utility threshold are compared. We have reported and discussed the results. Three dense databases are used in our experimental process and the results are given in the graph. The databases accidents, chess, and mushroom are real datasets and they are customized for our requirement. The running time of our algorithm for different minimum utility thresholds are recorded and shown in the graph. The output of previous and new approach is relatively same. The lesser minimum threshold makes the higher count of high utility itemsets. We can prove this variation in count of high utility itemsets for different thresholds very clearly in the graphs given. From the graph, we can see that the outcome is not varying much with the running algorithms for multiple dense datasets. But the effort in terms of analysis, knowledge, and time for finalizing the threshold is very much reduced. For proposed and existing approaches, the curves are relatively same, which shows that the automatic minimum utility thresholds set by our proposed system makes relatively same quantity of HUIs as that of the pioneering algorithm HUI miner.

5 Conclusion

In the previous methods, the user has to fix the minimum utility threshold by trying different choices. Because process of threshold setting is difficult with zero knowledge of the data stored in transactional databases. In the approach, we proposed an algorithm which initializes and increments the minimum utility threshold automatically. As we are using average of utility of itemsets present in search space tree at that level, we always set a minimum utility threshold to a mean value. This algorithm improves the productivity of the high utility mining process drastically for the dense input datasets than the sparse. We can also note from the above results that this tested results show that this approach's performance is healthy when all the transactions are having number of items in a same range. The HUIM-AMU algorithm is very convincing than most of the utility mining algorithms which mine high utility itemsets on every real and synthetic datasets.

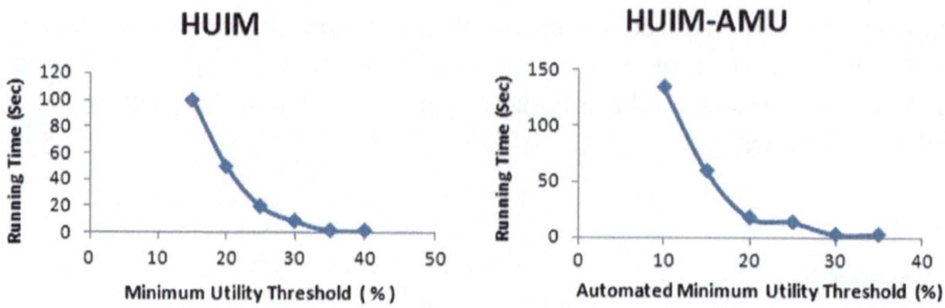

(a) Accidents

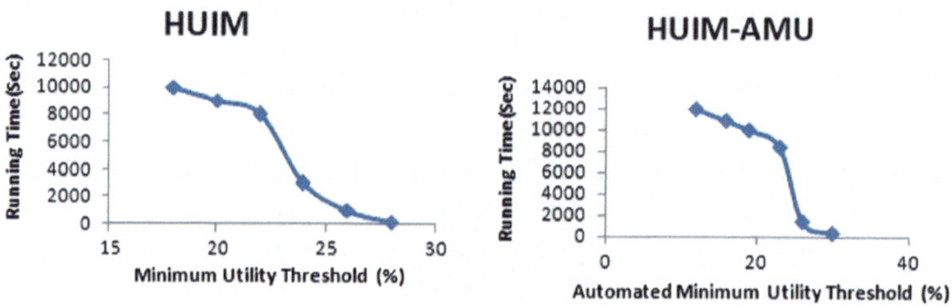

(b) Chess

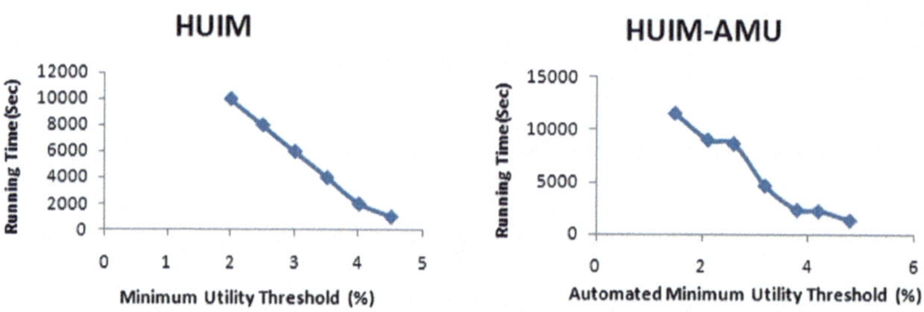

(c) Mushroom

References

1. Han J, Pei J, Yin Y (2000) Mining frequent patterns without candidate generation. ACM Sigmod Rec 29(2):1–12
2. Hu Y-H, Chen Y-L (2006) Mining association rules with multiple minimum supports: a new mining algorithm and a support tuning mechanism. DSS42 (1):1–24
3. Ahmed CF, Tanbeer SK, Jeong B-S, Lee Y-K (2011) HUC-Prune: an efficient candidate pruning technique to mine high utility patterns. Appl Intell 34(2):181–198
4. Kiran RU, Reddy PK (2011) Novel techniques to reduce search space in multiple minimum supports-based frequent pattern mining algorithms. In: Proceedings of the 14th international conference on extending database technology, pp 11–20

5. Liu Y, Liao WK, Choudhary AN (2005) A two-phase algorithm for fast discovery of high utility itemsets. In: Proceedings of the 9th Pacific-Asia conference on knowledge discovery and data mining. Springer, Heidelberg, pp 689–695
6. Liu M, Qu J (2012) Mining high utility itemsets without Candidate generation. In: Conference on information and knowledge management. Association for Computing Machinery, pp 55– 64
7. Tseng VS, Wu C-W, Shie B-E, Yu PS (2010) UP-Growth: an efficient algorithm for high utility itemset mining. In: Proceedings of the 16th ACM SIGKDD ICKDDM, pp 253–262
8. Tseng VS, Shie B-E, Wu C-W, Yu PS (2012) Efficient algorithms for mining high utility itemsets from transactional databases. IEEE Trans Know Data Engg 25(8):1772–1786
9. Lin C-W, Hong T-P, Lu W-H (2011) An effective tree structure for mining high utility itemsets. Expert Syst Appl 38(6):7419–7424
10. Liu B, Hsu W, Ma Y (1999) Mining association rules with multiple minimum supports. In: Proceedings of the 5th ACM SIGKDD ICKDDM, pp 337–341
11. Gan W, Lin JC-W, Fournier-Viger P, Chao H-C, Zhan J (2017) Mining of frequent patterns with multiple minimum supports. Eng Appl Artif Intell 60:83–96
12. Lin JCW, Gan W, Fournier-Viger P, Hong T-P (2015) Mining high-utility itemsets with multiple minimum utility thresholds. In: Proceedings of the eighth international conference on computer science & software engineering, pp 9–17

Implementation of Fuzzy Clustering Algorithms to Analyze Students Performance Using R-Tool

T. Thilagaraj and N. Sengottaiyan

Abstract The special techniques like clustering and classification exist in data mining to handle any number of datasets that are available in the education field. The main use of data mining is to take out valuable information to create new knowledge in the field of education. The detection of low performers, improving the pass percentage and employment opportunities, is the main goal of every educational institution. In data mining, the well-known technique is to deal with disjoint and noisy data is clustering. This technique used for distance calculation between similar group objects and the different cluster centers is also found. In this paper, the implementation of fuzzy models like Fuzzy C-Means (FCM), Fuzzy Possibilistic C-Means (FPCM), Modified Fuzzy Possibilistic C-Means (MFPCM) and Fuzzy Possibilistic Product Partition C-Means (FPPPCM) clustering algorithms is used to measure the student's levels and low performers identification through its size.

Keywords Data mining · Fuzzy clustering · FCM · FPCM · FPPPCM · MFPCM

1 Introduction

The Gathering of information from a vast storage area have been taken out to make predictions is the main purpose of the data mining [1]. To find the best accuracy in the prediction process will make by using different classifiers available in data mining [2]. The common approach to finding the center for each cluster is made in clustering techniques which may take place different levels of performance [3]. The k-means and fuzzy are the clustering techniques that may execute easily to extract the data from the different educational repository [4]. Providing an opportunity for academia to concentrate the low performers for improving their level in placement will create growth for institutions [5]. The process of clustering students into different groups

T. Thilagaraj (✉)
Department of Computer Applications, Kongu Arts and Science College, Erode, Tamil Nadu 638107, India

N. Sengottaiyan
Sri Shanmugha College of Engineering and Technology, Sankari, Tamil Nadu 637304, India

S.-L. Peng et al. (eds.), *Intelligent Computing and Innovation on Data Science*,
Lecture Notes in Networks and Systems 118,
https://doi.org/10.1007/978-981-15-3284-9_31

and displays various perceptions will help predictors to make needy decisions [6]. The different factors are required from the student's performance details to understand their skill sets and learning aspects [7]. The fuzzy clustering models will produce better results when one object comes under more than one group and it results in a good performance on different variety of data [8]. In fuzzy clustering, the boundaries may not sharp in many situations [9].

2 Methodology

2.1 Fuzzy C-Means Algorithm

The Fuzzy C-Means clustering algorithm (FCM) will allocate the membership by measuring its distance to all data points from the cluster center. The least-square value will be general among all groups and it is minimized [10]. The object closeness and cluster center are measured in a membership degree to analyze the fuzzy center [11]. This FCM algorithm partitions the given data frame in k partitions; also, it works iteratively to obtain the best solution. The FCM algorithm objective function follows in Eq. (1).

$$J_{\text{fcm}}(P, Q, R) = \sum_{i=1}^{n} r_{ij}^{m} d^2\left(\vec{p}_v, \vec{q}_j\right) \tag{1}$$

Here, P represents the data set, Q implies cluster centers, R represents membership degrees and m is to mention fuzziness in the clustering.

$$1 \leq m \leq \infty$$

The normal value of m is 2. The higher value of m will show more fuzziness and if it is low that implies hard clusters. If the value of m is 1, then the result of FCM and k-means is the same and it is called a hard algorithm. The FCM algorithm will verify the following conditions.

$$r_{ij} = [0, 1];\ 1 \leq i \leq n;\ 1 \leq j \leq k$$

$$0 \leq \sum_{i=1}^{n} r_{ij} \leq n;\ 1 \leq j \leq k$$

$$\sum_{j=1}^{k} r_{ij} = 1;\ 1 \leq i \leq n$$

Table 1 shows the academic, interpersonal and add-on marks of 20 students who have opted for placement. The special course add-on for the student is considered here. Figure 1 shows the level of cluster 3 as a low performer, cluster 4 as a

Table 1 Pre-assessment marks of 20 students before placement training

Stu. Id	Academic	Interpersonal	Add-on
1	37	35	46
2	45	70	31
3	41	49	54
4	55	67	78
5	64	85	17
6	78	75	74
7	40	37	45
8	78	45	56
9	67	41	58
10	40	45	54
11	31	85	85
12	78	86	52
13	47	88	37
14	34	29	45
15	37	32	41
16	45	93	24
17	41	75	26
18	65	77	65
19	52	66	62
20	17	45	42

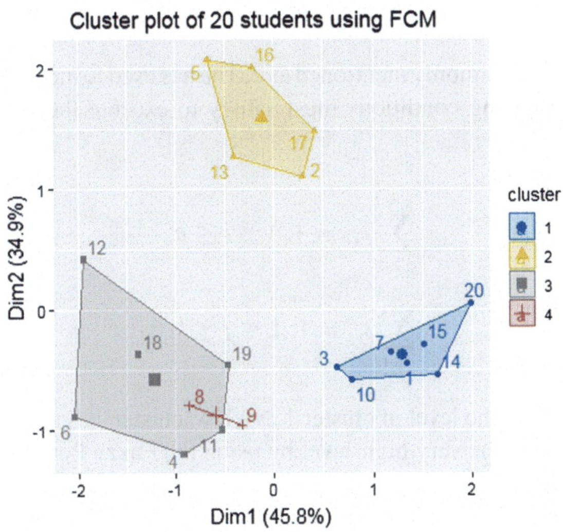

Fig. 1 Result of the four clusters using the FCM algorithm

medium performer, cluster 1 as a high-level performer and cluster 2 as very high-level performers using the Fuzzy C-Means clustering algorithm.

2.2 *Fuzzy Possibilistic C-Means Algorithm*

The combination of the FCM algorithm and the Possibilistic C-Means (PCM) clustering algorithm will be formed the Fuzzy Possibilistic C-Means clustering algorithm (FPCM). This also named a mixed c-means clustering algorithm [12]. The objective function of the FPCM clustering algorithm is as follows.

$$J_{\text{fpcm}}(P, Q, R, S) = \sum_{i=1}^{n} \left(r_{ij}^{m} + s_{ij}^{n}\right) d^2\left(\vec{p}_i, \vec{q}_j\right) \tag{2}$$

In the above equation,

$P = \{\vec{p}_1, \vec{p}_2, \ldots \vec{p}_n\}$ represents the data set,
$Q = \{\vec{q}_1, \vec{q}_2, \ldots \vec{q}_n\}$ is the prototype of cluster matrix,
$R = \{r_{ij}\}$ is the matrix for fuzzy partition of P,
$S = \{s_{ij}\}$ is the matrix for possibilistic partition of P,
$d^2(\vec{p}_i, \vec{q}_j)$ is the distance of squared Euclidean between $\overrightarrow{p_j}$ and $\overrightarrow{q_i}$.

The fuzziness is mentioned using m and it checks the below condition.

$$1 \le m \le \infty$$

The value of m is commonly mentioned as 2. The n is used to mention the typicality exponent. The following conditions must satisfy to execute the FPCM clustering algorithm.

$$\sum_{j=1}^{k} r_{ij} = 1;\ 1 \le i \le n$$

$$\sum_{i=1}^{n} s_{ij} = 1;\ 1 \le j \le k$$

Figure 2 represents the level of cluster 1 as a low, cluster 4 as a medium, cluster 2 as a high and cluster 3 as very high-level by using the Fuzzy Possibilistic C-Means clustering algorithm.

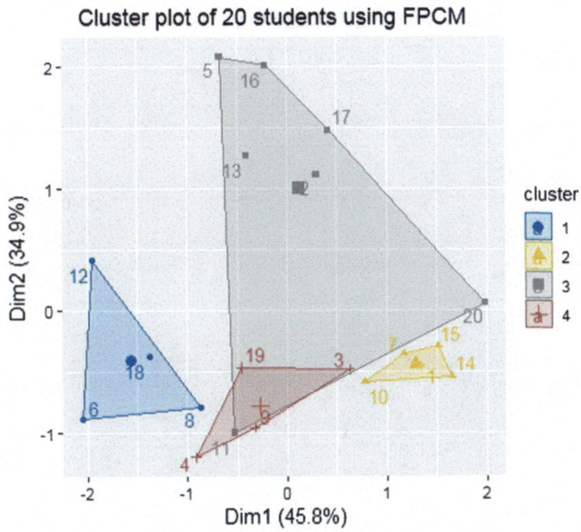

Fig. 2 Result of the four clusters using the FPCM algorithm

2.3 *Modified Fuzzy Possibilistic C-Means Algorithm*

The Modified Fuzzy Possibilistic C-Means clustering (MFPCM) algorithm is used to calculate the weight of all clusters to incorporate the parameters of weight it uses every data object. The levels of cluster centers and descriptive membership degrees are found to help for better classification while dealing with unstructured data. The objective function of MFPCM is as follows (3).

$$J_{\text{MFPCM}}(P, Q, R, S) = \sum_{i=1}^{n} r_{ij}^{m} t_{ij}^{m} d^{2m}(\overrightarrow{p_i}, \overrightarrow{q_j}) + s_{ij}^{m} t_{ij}^{n} d^{2n}(\overrightarrow{p_i}, \overrightarrow{q_j}) \tag{3}$$

The weight is calculated by using Eq. (4).

$$t_{ij}^{m} = \exp\left[-\frac{d^2(\overrightarrow{p_i}, \overrightarrow{q_j})}{\sum_{i=1}^{n}\left(d^2(\overrightarrow{p_i}, \overrightarrow{q_j})\frac{k}{n}\right)}\right] \tag{4}$$

Figure 3 represents the level of cluster 1 as a low, cluster 4 as a medium, cluster 3 as a high and cluster 2 as very high-level by using the Modified Fuzzy Possibilistic C-Means clustering algorithm.

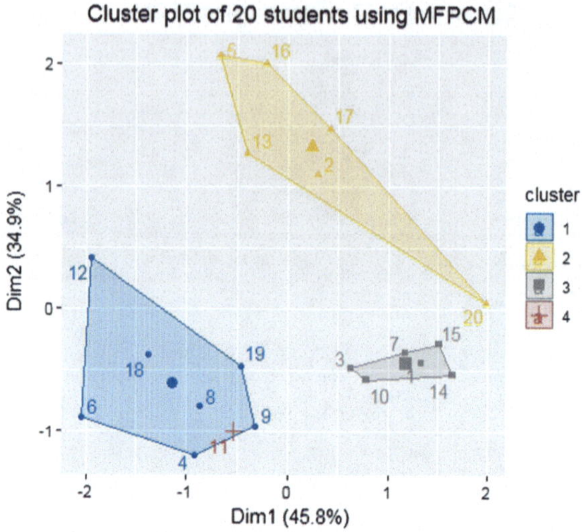

Fig. 3 Result of the four clusters using the MFPCM algorithm

2.4 *Fuzzy Possibilistic Product Partition C-Means Algorithm*

The Fuzzy Possibilistic Product Partition C-Means clustering algorithm will eliminate outlier effects and the multiplicative way is used. The main objectives of the algorithm will show in (5).

$$J_{\text{FPPPCM}}(P, Q, R, S) = \sum_{j=1}^{k} \sum_{l=1}^{n} r_{ij}^{m} \left[s_{ij}^{n} d^{2}(\vec{p_l}, \vec{q_j}) + \Omega_j \left(1 - s_{ij}\right)^{n} \right] \quad (5)$$

The fuzzifier m will specify clustering fuzziness $1 \leq m \leq \infty$. The usual assignment value chosen is 2. The typicality exponent n is to specify the amount of typicality in clustering. $1 \leq n \leq \infty$ Here, also 2 is the default value. The possibilistic penalty is used to control the cluster's variance.

Figure 4 represents the level of cluster 2 as low, cluster 4 as medium, cluster 1 as high and cluster 3 as very high by using Fuzzy Possibilistic Product Partition C-Means clustering algorithm.

Table 2 represents the sizes of FCM, FPCM, MFPCM and FPPPCM with various levels. While comparing the result of the low performer's cluster size is high in MFPCM.

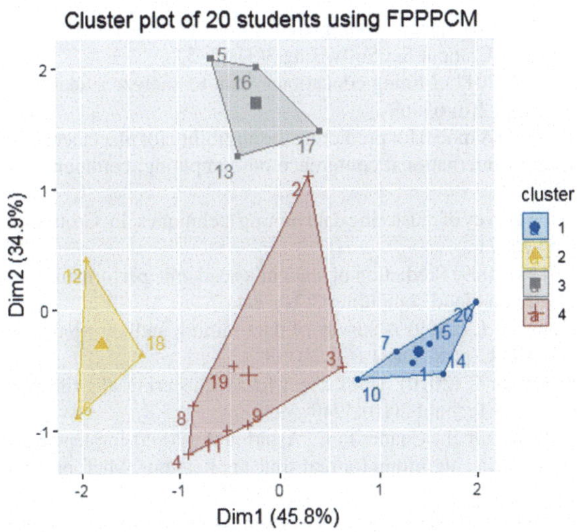

Fig. 4 Result of the four clusters using the FPPPCM algorithm

Table 2 Comparison of fuzzy clustering algorithms using its cluster sizes

Description	Low	Medium	High	Very high
FCM	6	2	7	5
FPCM	4	4	5	7
MFPCM	7	1	6	6
FPPPCM	3	7	6	4

3 Conclusion

Nowadays, the biggest challenge for every educational institution is to focus on low-level performers to improve them on placement factors. Here, the FCM, FPCM, MFPCM and FPPPCM clustering algorithms are implemented to analyze the different factors. The Modified Fuzzy Possibilistic C-Means clustering algorithm will find a high range of low performers while comparing with other models and this will help the academia to provide timely training for low performers to get proper placement in the future.

References

1. Varghese BM, Unnikrishnan A, Sciencist G, Kochi N, Kochi C (2010) Clustering student data to characterize performance patterns. Int J Adv Comput Sci Appl 2:138–140
2. Wu X et al (2008) Top 10 algorithms in data mining. Knowl Inf Syst 14(1):1–37

3. Joshi A, Kaur R (2013) A review: comparative study of various clustering techniques in data mining. Int J Adv Res Comput Sci Softw Eng 3(3):55–57
4. Baradwaj BK, Pal S (2011) Mining educational data to analyze students' performance. Int J Adv Comput Sci Appl 2(6):63–69
5. Gera M, Goel S (2015) A model for predicting the eligibility for placement of students using data mining technique. In International conference on computing, communication & automation, IEEE, pp 114–117
6. Berkhin P (2006) A survey of clustering data mining techniques. In: Grouping multidimensional data. Springer, pp 25–71
7. Saxena PS, Govil M (2009) Prediction of student's academic performance using clustering. In: National conference on cloud computing & big data
8. Goebel M, Gruenwald L (1999) A survey of data mining and knowledge discovery software tools. ACM SIGKDD Explor Newsl 1(1):20–33
9. Vanisri D, Loganathan C (2010) An efficient fuzzy clustering algorithm based on modified k-means. Int J Eng Sci Technol 2(10):5949–5958
10. Lazaro J, Arias J, Martín JL, Cuadrado C, Astarloa A (2005) Implementation of a modified Fuzzy C-means clustering algorithm for real-time applications. Microprocess Microsyst 29(8–9):375–380
11. Izakian H, Abraham A (2011) Fuzzy C-means and fuzzy swarm for fuzzy clustering problem. Expert Syst Appl 38(3):1835–1838
12. Rubio E, Castillo O, Melin P (2016) Interval type-2 fuzzy possibilistic c-means clustering algorithm. In: Recent developments and new direction in soft-computing foundations and applications. Springer, pp 185–194

Comparative Analysis of Various Algorithms in ARM

J. Sumithra Devi and M. Ramakrishnan

Abstract Data mining is a process in which useful information is discovered from large volumes of data using various tasks such as classification, clustering, association rules. Frequent items are the sets of items or structures which occur in a transaction. It gives the information about how frequently the specific item appears in a transaction. Though there are many mining tasks, one of the finest methods is association rule mining which finds the correlation, frequent patterns and rules from a various large amount of dataset. Association rule mining uses various scalable and efficient algorithms which predicts the rules to find the occurrence of an element in the dataset. This paper compares various association rule mining algorithms based on the data support and speed.

Keywords Data mining · Association rule mining · FP growth · Apriori algorithm · Fast distributed mining · Elcat

1 Introduction

Data has become one of the important phenomena which revolves around each and every single organization. These data are organized and analyzed to retrieve useful knowledge. Mining is the process which discovers more useful knowledge and brings out the hidden details in the data by analyzing the database under different methods. This process is carried out under two categories of databases, namely centralized and distributed databases. Centralized databases are the database where the data has been stored as one database which is accessible by everyone. A distributed database is those databases where the data is stored separately and then they are combined indifferently to acquire knowledge hidden in it. A distributed database

J. Sumithra Devi (✉)
Research Scholar, Department of Computer Science, Bharathiar University, Coimbatore, Tamil Nadu, India

M. Ramakrishnan
Chairperson, School of Information Technology, Madurai Kamaraj University, Madurai, Tamil Nadu, India

S.-L. Peng et al. (eds.), *Intelligent Computing and Innovation on Data Science*, Lecture Notes in Networks and Systems 118,
https://doi.org/10.1007/978-981-15-3284-9_32

is further divided into horizontally distributed database and vertically distributed database. Many methods are used to mine the data. Though there are many methods are available, the fast emerging digitized system creates a massive database with a very large amount of undiscovered data hidden in it. One of the most effective methods of mining is association rule mining finding the frequent patterns and rule generation. This method helps us to understand the associations among the data and also models with specific types of data associated with each other. Since there are many volumes of data, mining has to be done carefully because there are possibilities of leakage of confidential information.

For instance, in the field of biomedical, very crucial or sensitive information is represented in patterns associating with the nature of its occurrences, genetical configuration with severe ailment. These kinds of data are very sensitive personal knowledge of the individual and leakage of prevented from the intruders. Beyond preserving the privacy of confidential data, it is also equally important to hide our sensitive knowledge from other access. In business, provisions are made to access the information from distributed databases but their individual strategies and important data are always hidden from unauthorized accessing [1].

1.1 Association Rule Mining

Association rule mining is one of the technique to find interesting associations between the items in the database. In 1993, a technique is introduced to identify the symmetry in huge volumes of databases. The data has different accessible levels of hierarchy. These data can be accessed according to the assigned authorization level persons. But there are possibly of identifying sensitive data from nonsensitive data known as 'interference problem.' Hence, privacy has to be maintained without affecting the creditability of the database [2].

It also generates a set of rules with a specific threshold. These rules are extracted from the transaction item set among the large volumes of data. Since there are many rules generated, security must be high in sharp boundary, and hence, various algorithms are deployed to maintain security and to hide sensitive data without any leakage of valuable information.

The association rules are classified depending on different criteria including the abstraction level employed in the set of rules, usage of dimensions, pattern types, different values handled by the rules and its extensions. The efficiency of the algorithms used for mining associative rules increases with the decrease in the number of passes made in the database, sampling the database, the addition of constraints on structure patterns by parallelization [3].

1.2 Problem Study

Mining frequent item sets have become one of the areas of consideration as it has very broadly applied in generation of associations' rules, finding correlations, forming the patterns in graph. Frequent pattern is one of the tasks in data mining and patterns in sequence are used accordingly to produce better outcome. Most importantly, the algorithm should be more efficient for mining frequent item sets are required for mining association rules as well as for many other data mining tasks. The frequent pattern generation is one of the major challenges in where more number of patterns are generated in large volumes of data. As the threshold decreases exponentially, there is an increase in the generation of item sets. Therefore, one of the main tasks is to pruning of unimportant patterns has to be performed effectively.

The main aim is to optimize the process of finding efficient, scalable and should be able to detect important patterns used in various forms.

1.3 Apriori Algorithm

Among the best mining algorithm, Apriori algorithm plays a vital role in mining large volumes of data using ARM. Multiple passes are made over the database which employs breadth-first search to explore new item sets. Apriori represents candidate key generation approach. The frequency of the item set is based on the number of occurrences in the transactions. Apriori algorithm can be used with data structure like FP tree reducing memory to a large extent with the help of parallel algorithm [4]. Pruning leaves less item sets with easy implementation and minimal memory consumption. But it scans the database repeatedly to know the availability of a particular item in the database allowing only one minimum support threshold at a time [5]. To avoid repeated scanning, a new method with improved Apriori algorithm scans the database only one time so that the identifier set for each item can be obtained. So that pruning is done only with the limited item set. Then the candidate items support is counted using the TID set, hence reducing the number of candidate items which in turn minimized the time greatly [6].

To improve the efficiency of Apriori algorithm, the following methods are used.

(1) Hash-based item set counting
(2) Partitioning
(3) Transaction reduction
(4) Sampling
(5) Dynamic item set counting.

1.4 FP Growth

Another important efficient data mining technique is frequent pattern growth algorithm (FP growth) which retrieves the hidden knowledge from huge databases and can be compressed using a prefix tree structure. Divide and conquer strategy is applied to get the frequent item set. The database is first compressed into FP tree with items and their associations which in turn is divided into smaller conditional FP trees and then frequent item sets are derived finally. Repeatedly, the databases are scanned for frequent item sets [7]. The association rule mining algorithms designed for single-core machines do not match when dealing with huge volumes of databases which increase the computational cost. If the tasks are joined in multicore machines, then the database can be more utilized in parallel. The work-stealing algorithm can be used where preassigned tasks to the current core are not essentially required so that the undone tasks are stolen from the busy core so that all the core will be busy with the tasks while nothing remains idle. Hence, it results in increased utilization of resource. The tree structures are merged so that the operation will be faster with less time consumption [8].

1.5 Elcat

Equivalence class clustering popularly known as Elcat is a bottom-up traversal for generating frequent items by the intersection of all distinct atoms pairs TID lists. It also checks the cardinality of the outcome TID lists. Item set of the current level is repeated frequently by a called recursive procedure. All the frequent item sets are enumerated by repeating the process. The performance decreases with the exponential increase along multiple numbers of transactions with limited pruning in Apriori or FP growth algorithm. Elcat algorithm improves its performance by storing the descending order of support efficiently and generation of candidate is kept ascending order of support to reduced redundancy [10]. Some of the applications in which Elcat algorithm used are map-reduce framework and in the implementation of Java. In online shopping, this algorithm plays an important role in finding out the customers' details buying few products frequently [11]. Elcat reduces the memory consumption during the process. Though similar to Apriori algorithm, it generates large set of rule without generating the candidate set [12].

1.6 Fast Distributed Data Mining

Fast distributed mining is a distributed unsecured form of the Apriori algorithm. Frequent item sets which are both global and local generate minimum set of candidates during each iteration which results in reduced count of messages exchanged. After the candidate sets generation, the global and local reduction techniques were applied to remove some candidate sets at each site. Then in order to find whether an item is frequent, the support counts and the number of sites are considered [9]. FDM has the following stages namely

(1) Initialization where all the items are present already and it has to be processed and calculated.
(2) Candidate sets are generated at each iteration
(3) Pruning is done locally.
(4) The candidate item sets are unified
(5) Local support is computed.

2 Performance Analysis of Apriori Algorithm

In Apriori algorithm, array-based data structure is used and follows Apriori property with join and prune method. Since huge amount of candidate sets were generated in this algorithm, it requires less memory space and involves multiple scans for the generation of candidate set with high execution time, less accuracy for sparse, dense datasets and best for closed item set.

3 Performance Analysis of FP Growth Algorithm

In FP growth, tree-based data structure is used and follows intersection of transaction TID list for the generation of candidate sets. Since compact data structure is involved in this algorithm, it requires less memory space and involves twice scan for the generation tree. It has less execution time and high accuracy. It is well suited for large, medium datasets and best for large item set.

4 Performance Analysis of ECLAT Algorithm

In Elcat, array data structure is used and follows conditional frequent pattern tree property with minimum support. It requires less memory space (when item set count is small) and involves continuous scan to update the database with less execution time, more accuracy (compared to Apriori and well suited for dense, medium datasets and best for large free item set.

Table 1 Datasets used in comparison

FileName	Division	Dist/Rand	Records	I/P columns
Supermarket dataset	5	Yes	48,842	15
Census dataset	0	No	48,842	14
Mushroom dataset d90.n8124.c2.num	5	Yes	8124	23

5 Result and Discussion

On account of the survey done on performance of above algorithms, we conclude that the Apriori algorithm performs less when compared to the other two algorithms. Since the performance of Apriori is low, we compare FP growth and Elcat algorithm under result and discussion using different dataset applications.

6 Dataset

For the evaluation of performance of FP growth and Elcat algorithm, the dataset was retrieved from the repository of UCI, machine learning database. The table (Table 1) portraits the dataset characteristics selected for performance evaluation.

7 Performance Comparison

Weka tool is a tool which deals with the collection of machine learning algorithms, written in Java is used to generate items sets in association rule mining. In this study, it is used in FP growth and Elcat algorithm separately and the execution time is calculated. The execution time of both algorithms is calculated from the dataset with same support values and with different threshold values ranges from 30 to 70%.

The Table 2 displays information about the execution time for FP growth and Eclat algorithms with different support value (threshold) for adult dataset (Fig. 1).

Table 2 Performance of (execution time) adult dataset for various threshold values

Support (threshold value)	Execution time (seconds)	
	FP growth	Elcats
30	0.57	0.53
40	0.49	0.47
50	0.5	0.49
60	0.48	0.44
70	0.45	0.4

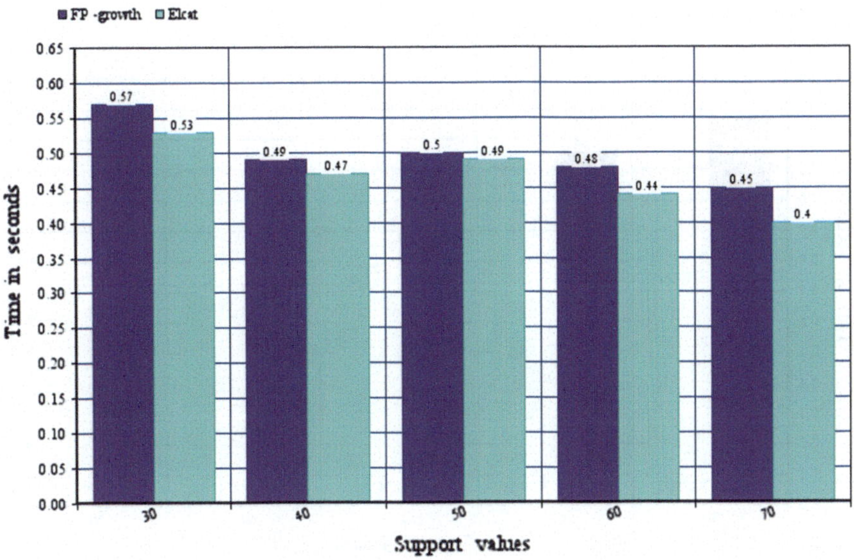

Fig. 1 Execution time (in seconds) for the FP growth algorithm and Eclat algorithm for adult dataset. The execution time decreased when the threshold value (support) values are increased. Finally, we observe that Eclat algorithm produce the best performance for supermarket dataset

Table 3 Performance of (execution time) census dataset for various threshold values

Support (threshold value)	Execution time (seconds)	
	FP growth	Elcats
30	1.27	0.85
40	1.16	0.70
50	0.89	0.75
060	0.73	0.69
70	0.66	0.64

The Table 3 displays the execution time (in seconds) for FP growth and Eclat algorithms with different support value (threshold) for census dataset (Fig. 2).

The Table 4 shows the execution time for FP growth and Eclat algorithms with different support value (threshold) for Mushroom dataset (Fig. 3).

8 Conclusion

The performance of the association rule mining techniques helps in selecting the appropriate algorithm to find the minimum item sets. In which Apriori algorithm, counting method iterates through all the transaction each time resulting in huge

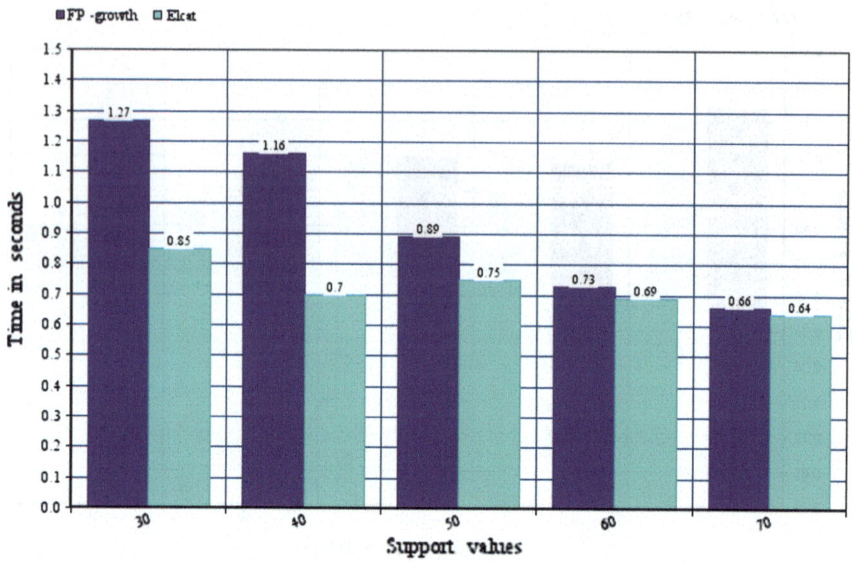

Fig. 2 Execution time (in seconds) for the FP growth algorithm and Eclat algorithm for adult dataset. The execution time decreased when the threshold value (support) values are increased. Finally, we observe that Eclat algorithm produces the best performance for census dataset

Table 4 Performance of (execution time) Mushroom dataset for various threshold values

Support (threshold value)	Execution time (seconds)	
	FP growth	Elcats
30	0.13	0.11
40	0.12	0.11
50	0.09	0.09
60	0.08	0.07
70	0.08	0.06

memory consumption and processing speed is slow. The FP growth algorithm inserts sorted items by frequency into a pattern tree. It is more scalable with less runtime and memory usage. Elcat algorithms are best suited only for medium and dense databases. Fast distributed mining algorithms generate fewer candidate item sets even on heterogeneous database and give the results according to the datasets used in the database. Further, the work can be extended by implementing privacy preservation in ARM using cryptographic techniques in homogeneous database.

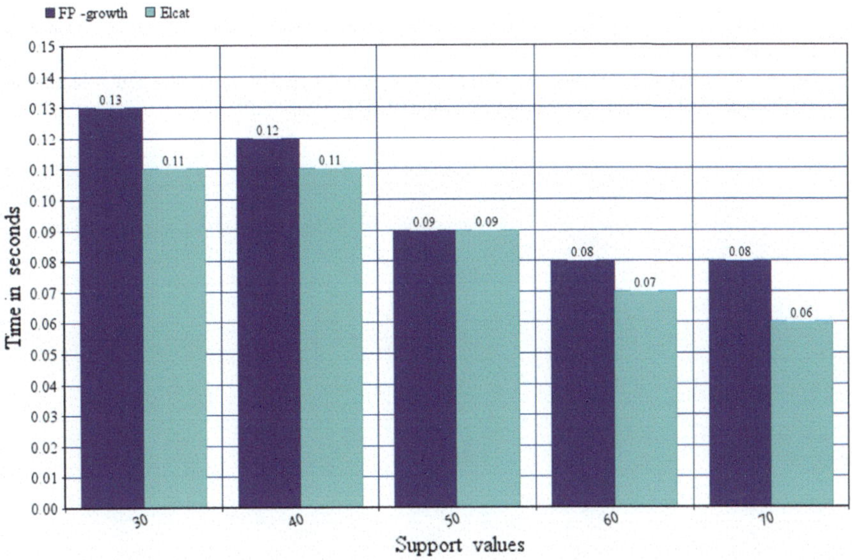

Fig. 3 Execution time (in seconds) for the FP growth algorithm and Eclat algorithm for Mushroom dataset. The execution time decreases with the increase in the support values (threshold values). So, it is observed that both Elcat and FP growth algorithms give better performance for Mushroom dataset

References

1. Abul O, Bonchi F, Giannotti F (2010) Hiding sequential and spatiotemporal patterns. IEEE Trans Knowl Data Eng 22(12)
2. Gayathiri P, Poorna B (2017) Association rule hiding for privacy preserving data mining: a survey on algorithmic classifications. Int J Appl Eng Res 12(23):13917–13926. ISSN 0973-4562
3. Kotsiantis S, Kanellopoulos D (2006) Association rules mining: a recent overview. GESTS Int Trans Comput Sci Eng 32(1)
4. Bhandari A, Gupta A, Das D (2014) Improvised apriori algorithm using frequent pattern tree for real time applications in data mining. In: International Conference on Information and Communication Technologies (ICICT 2014)
5. Kaur G (2014) Association rule mining: a survey. Int J Comput Sci Inf Technol 5(2):2320–2324
6. Yuan X (2017) An improved Apriori algorithm for mining association rules. AIP conference proceedings 1820, 080005
7. Narvekara M, Syed SF (2015) An optimized algorithm for association rule mining using FP tree. In: International Conference on Advanced Computing Technologies and Applications (ICACTA2015)
8. Gadia K, Bhowmick K (2015) Parallel text mining in multicore systems using FP-Tree algorithm. In: International Conference on Advanced Computing Technologies and Applications (ICACTA2015)
9. Győrödi C (2003) A comparative study of distributed algorithms in mining association rules. In: International Symposium on System Theory, 2003
10. Yu X, Wong H (2014) Improvement of eclat algorithm based on support in frequent itemset mining. Int J Comput 9(9)

11. Kaur M, Grag U (2014) ECLAT algorithm for frequent itemsets generation. Int J Comput Syst 1(3). ISSN 2394-1065
12. Omana J, Monika S, Deepika B (2017) Survey on efficiency of association rule mining techniques. Int J Comput Sci Mobile Comput (IJCSMC) 6(4)

Implementation of Statistical Data Analytics in Data Science Life Cycle

S. Gomathi, R. P. Ragavi, and S. Monika

Abstract The paper focussed on showing how the data science life cycle can be implemented with the real time data. Rain fall data is used in this research to show how to apply data science life cycle. The dataset consists of multivariate data were visualization has become easier and effective. The data description is shown in the table for better understanding. The various visualization like bar graph, tree map, line graph are shown by using tableau software.

Keywords Statistics · Data analytics · Tableau · Data · Rainfall · Predictive analytics · Diagnostic analytics

1 Introduction

Statistical analytics is the method of generating statistics from stored data and analyzing the result. Data analytics is the method of extracting info from data. It includes multiple stages as well as preparing the data for processing, applying models, establishing a data set, identifying key results and creating reports [2] (Fig. 1).

The various types of data analytics is analyzed using tableau. Tableau is a quickest rising information visual image tool. It helps in shortening data into the clear format. Data analytics is very fast in Tableau and the visualizations formed are in the method of dashboards and worksheets. The data that's formed using Tableau may be understood by expert at any level in an organization. It allows a non-technical operator to create a custom-made dashboard.

The top features of Tableau are

(1) Real time analysis
(2) Collaboration of data
(3) Data Blending.

S. Gomathi (✉) · R. P. Ragavi · S. Monika
Department of Computer Science (PG), PSGR Krishnammal College for Women, Coimbatore 641004, India
e-mail: mailtogomathisrinivasan@gmail.com

S.-L. Peng et al. (eds.), *Intelligent Computing and Innovation on Data Science*, Lecture Notes in Networks and Systems 118,
https://doi.org/10.1007/978-981-15-3284-9_33

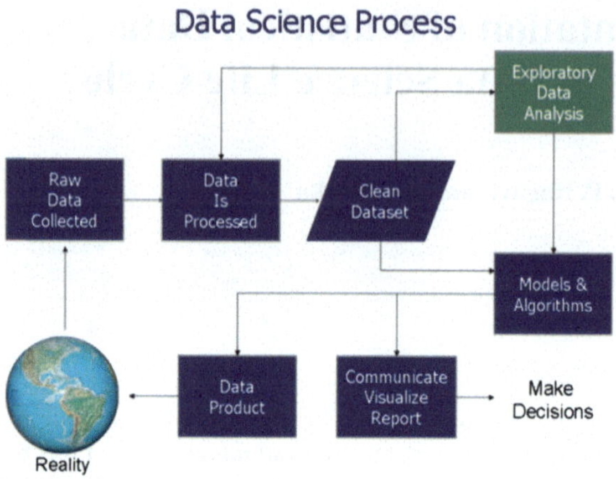

Fig. 1 Data science process

Tableau software doesn't require any technical or any kind of programming skills to operate [3].

2 Life Cycle of Data Analytics

Prescriptive analytics can process new data to improve the accuracy of estimates and offer better decisions [4]. Diagnostic analytics is used to determine the reason for its occurrence. It examines data to answer the question "Why did it happen". Predictive analytics is a region of statistics that deals with the extracting info from data and to predict the future. Its statistical techniques include artificial intelligence, data mining, machine learning, data modeling and deep learning algorithms. Predictive analytics can be useful to any type of unknowns, it may be in the present, past or future [5]. Descriptive analytics is a initial step of data processing. It summarizes raw data and convert it into a form that can be easily understood by humans. It makes a outline of historic data to yield a helpful info and it may prepare the data for future analysis [6] (Fig. 2).

3 Results and Discussion

The rainfall data records has been obtained from government data [1] where the dataset of 32 districts were available, from this we have analyzed 19 districts using tableau (Table 1).

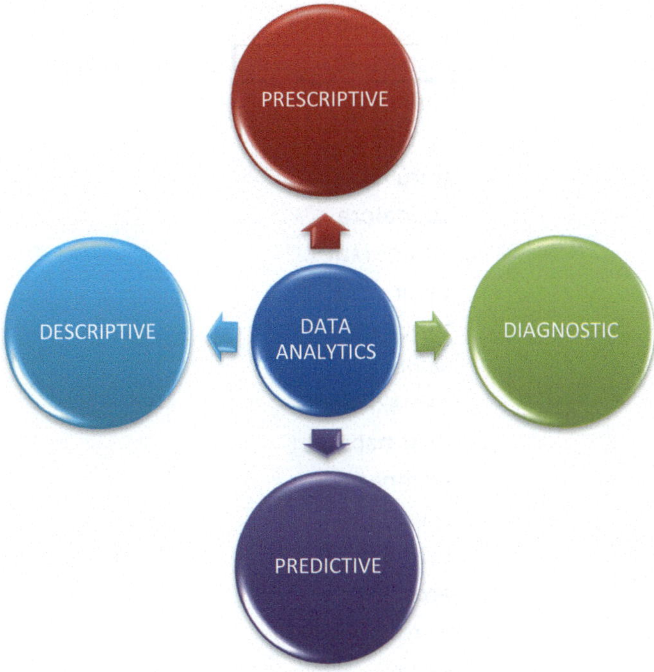

Fig. 2 Types of data analytics

Table 1 Rain fall data description [1]

Attribute	Data type	Description
District	Categorical	Name of the district
Actual hot weather season	Numerical	Hot climate
Actual winter season	Numerical	Winter climate
Actual southwest monsoon	Numerical	Monsoon climate
Population	Numerical	Male, female

3.1 Prescriptive Analysis

Prescriptive analytics is helpful in finding the best course of action in a given condition.

Step-1: Import the data from the excel to tableau (Fig. 3).

Step-2: (Fig. 4).

Fig. 3 Actual hot weather dataset [1]

A	B
district	actual hot weather season
chennai	1.8
kancheepuram	143
thiruvallur	15.8
cuddalore	23.4
villupuram	49.8
velur	106.8
thiruvannamalai	57.5
salem	167.9
namakkal	141.7
dharmapuri	240.9
krishnagiri	285.2
thiruppur	106.2
coimbatore	170.5
erode	258.4
thirucharappalli	107.8
karur	127.8
perambalur	122.2
pudukkottai	54.4
thanjore	60.4

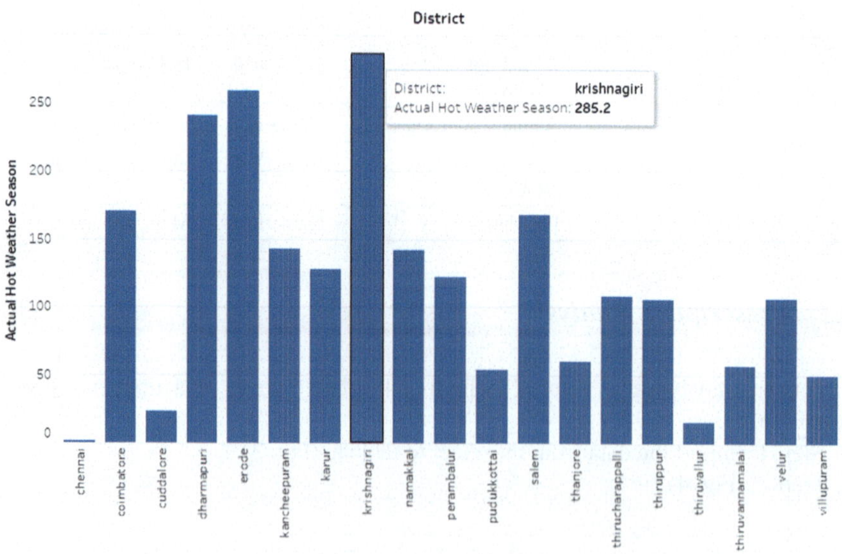

Fig. 4 Prescriptive analytics shows the bar graph of actual hot weather seasons of various districts

district	population rate	actual hot weather season
chennai	46,46,732	1.8
kancheepuram	39,98,252	143
thiruvallur	37,28,104	15.8
cuddalore	26,05,914	23.4
villupuram	34,58,873	49.8
velur	39,36,331	106.8
thiruvannamalai	24,64,875	57.5
salem	34,82,056	167.9
namakkal	17,21,179	141.7
dharmapuri	15,06,843	240.9
krishnagiri	18,79,809	285.2
thiruppur	24,71,222	106.2
coimbatore	34,58,045	170.5
erode	22,51,744	258.4
thirucharappalli	27,22,290	107.8
karur	10,64,493	127.8
perambalur	5,65,223	122.2
pudukkottai	16,18,345	54.4
thanjore	24,05,890	60.4

Fig. 5 Population and actual hot weather dataset [1]

3.2 *Diagnostic Analysis*

This is used for discovery or to determine the reason for its occurrence.

Step-1: Import the data from excel to tableau (Fig. 5).

Step-2: (Fig. 6).

Figure 6 shows that, the population rate and actual hot weather season is analyzed in comparison as due to the highest population rate in Chennai the rainfall is very low in actual hot weather season.

3.3 *Predictive Analysis*

Predictive analytics is the process of extracting info from current dataset in order to decide patterns and predict future outcomes and trends.

Step-1: Import data from excel to tableau (Fig. 7).

Step-2: (Fig. 8).

Figure 8 predicted that the district cuddalore and thanjore has the highest rainfall during the actual winter season. The rainfall in cuddalore during winter season is expected in future.

Fig. 6 Diagnostic analytics shows the bar graph of population rate and actual hot weather seasons

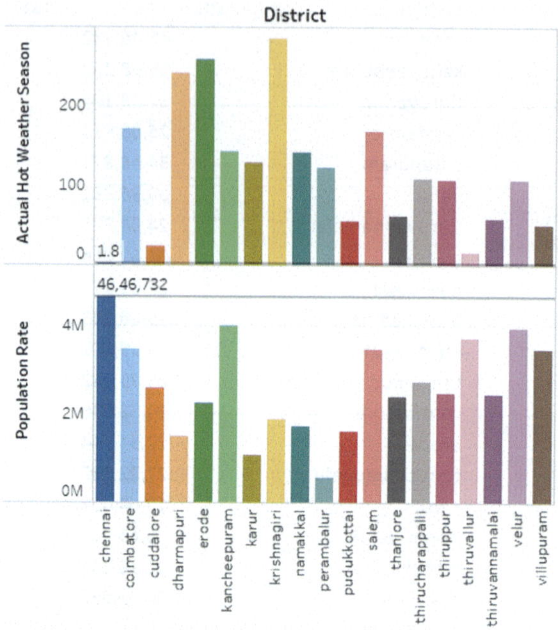

Fig. 7 Actual winter season dataset [1]

district	actual winter season (jan'17 and feb'17)
chennai	4.5
kancheepuram	16.5
thiruvallur	6
cuddalore	114.3
villupuram	40.2
velur	23.1
thiruvannamalai	47
salem	12.2
namakkal	6.3
dharmapuri	8.4
krishnagiri	5.7
thiruppur	7.6
coimbatore	10.4
erode	10.8
thirucharappalli	24.7
karur	17.2
perambalur	24.3
pudukkottai	55.2
thanjore	103.5

Fig. 8 Predictive analytics shows the tree map of actual winter season for various districts

3.4 Descriptive Analysis

Descriptive analysis describes the basic features from the collection of data. Example mean, median, mode.

Step-1: (Fig. 9).

Step-2: (Fig. 10).

The attribute district and actual south west monsoon season is visualized from Fig. 10. Through that the line graph has been generated. In actual south west monsoon season the average rainfall of these districts is 299.3.

4 Conclusion

In this paper the statistical data analytics in the data science life cycle have been discussed. Tableau is used to answer the questions like which is the best rainfall area, why the rainfall will be low in a particular district, what might happen in future and the average rainfall of the districts were shown graphically.

A	B
district	actual south west monsoo
chennai	495.9
kancheepuram	482.1
thiruvallur	406.2
cuddalore	346
villupuram	329
velur	420
thiruvannamalai	419
salem	346
namakkal	239.8
dharmapuri	269.2
krishnagiri	326.8
thiruppur	81.1
coimbatore	222.6
erode	174.4
thirucharappalli	211
karur	125
perambalur	270.9
pudukkottai	233.8
thanjore	288.6

Fig. 9 Actual southwest monsoon data [1]

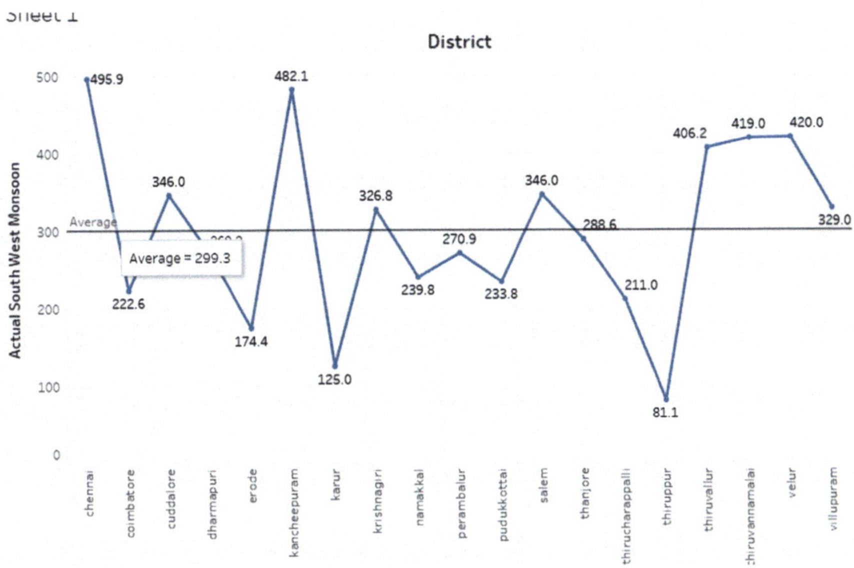

Fig. 10 Descriptive analytics shows the line graph of actual south west monsoon season

References

1. www.data.gov.in
2. Tsai C-W et al (2015) Big data analytics: a survey. J Big Data 2(1):21
3. Nair L, Shetty S, Shetty S (2016) Interactive visual analytics on Big Data: Tableau vs D3. js. J e-Learning Knowl Soc 12(4)
4. Lepenioti K et al (2018) Prescriptive analytics: a survey of approaches and methods. In: International conference on business information systems. Springer, Cham
5. Schoenherr T, Speier-Pero C (2015) Data science, predictive analytics, and big data in supply chain management: current state and future potential. J Bus Logistics 36(1):120–132
6. Raja B et al (2019) Market behavior analysis using descriptive approach. Available at SSRN 3330017

Performing Hierarchical Clustering on Huge Volumes of Data Using Enhanced Mapreduce Technique

K. Maheswari and M. Ramakrishnan

Abstract Among the various methods of clustering, hierarchical clustering is advantageous in many aspects. The implication of hierarchical clustering on large volumes of data is difficult as these data are normally unstructured, heterogeneous, in huge volumes, contains various types of noise and volatile. The Mapreduce framework is used to analyze huge volumes of data under parallel and distributed fashion. The efficiency of the algorithm can be improved by two optimization techniques viz. co-occurrence based feature selection and batch updating are used. Hence this paper presents a hierarchical clustering method using enhanced version of mapreduce framework for huge volumes of data. The research is conducted on web access log file containing 512 GB of data. The outcome of the results conducted by the algorithm show that the proposed method outperforms traditional clustering methods in terms of execution time and number of clusters formed.

Keywords Hierarchical clustering · Data mining · Mapreduce · Hadoop · K-means

1 Introduction

Clustering plays an important role in the data mining domain where a particular sets of objects are grouped based on their features and accumulate them corresponding to their homogeneity [1]. After partitioning the set of data into groups, it assigns labels to each group. Clustering is more preferable than classification because it is adaptable to changes and provides useful features that distinguish different groups. Clustering is an unsupervised classification that has no predefined classes [2].

Clustering methods are broadly classified as partition, hierarchical, density-based, grid-based, model-based and constraint-based [3]. Clustering technique is having wide applications such as market-basket analysis, pattern recognition, data analysis, image processing, etc. [4]. It is also having applications in specific domains like

K. Maheswari (✉)
Department of Computer Science, Bharathiyar University, Coimbatore, Tamil Nadu, India

M. Ramakrishnan (✉)
School of Information Technology, Madurai Kamaraj University, Madurai, Tamil Nadu, India

S.-L. Peng et al. (eds.), *Intelligent Computing and Innovation on Data Science*,
Lecture Notes in Networks and Systems 118,
https://doi.org/10.1007/978-981-15-3284-9_34

biology in gene categorization, animal taxonomies and insight structure inherent populations can be analyzed [4]. Apart from this, discovering information from web, credit card fraud detection, etc. are other important usage of clustering [5]. Due to the usage of new technologies, huge amount of data are generated which are heterogeneous in nature. Termed as big data, it can become a new pillar for emerging economies as well as for scientific research [6]. Many real life applications such as health, bio-medical, marketing, transportation, finance, banking, business, government services, social media, etc. often generate huge volume of data. Big data are collected from multiple sources, in multiple tables with different views and perceptions [7]. This complexity increases when there is an increase in volume of data. Big data are dynamic in nature and having time impact.

Big data are unstructured. Ordinary database management tools are not capable enough to handle big data [8]. Generally there is a growing demand that either can increase the capacity and performance of existing tools to analyze big data or new solutions to fully analyze and extract useful information from big data. With the data growing in exponential phase, normal data mining tools become inefficient in extracting knowledge from data and hence special techniques are the required to solve the current problems [9].

Big data Clustering can be performed in many ways including single machine clustering and multiple machine clustering. Mapreduce is a mechanism that partitions the given task into smaller sub tasks and results are consolidated [10]. It would bean good idea to use this mapreduce technique for clustering big data. Hence this paper presents a novel big data clustering technique using mapreduce technique. The organization of the paper is given as follows. It starts by giving brief introduction about clustering and big data in Sect. 1. A lucid literature survey defining the related works done by various researchers have been presented in Sect. 2. Basic modalities of mapreduce algorithm and big data analysis are thoroughly analyzed in Sect. 3. Section 4 elaborately defines the proposed methods. The investigational results and following discussion are presented in Sect. 5. Conclusion remarks is discussed at the end of the paper in Sect. 6.

2 Related Work

Cai et al [11] presented a method for clustering big data using multi-viewed k-means algorithm. It is a robust, large-scale multi-view clustering algorithm which integrates heterogeneous large scale data. Optimization algorithm is used which iteratively solves non-smooth objective problem with proved convergence. The procedure is experimented with six types of standard data sets and the comparison is done based on the performance with traditional clustering techniques. Experimental results show higher performance. Rehioui et al. [12] proposed a modification of traditional DENCLUE algorithm called DENCLUE-IM that is a unique method big data clustering. DENCLUE allows grouping of complex big data. However, it lacks in performance and speed. The important step in DENCLUE algorithm is Hill Climbing which consumes so much of computation time. DENCLUE-IM avoids this step thereby increasing response time. The quality of clustering is also good and better than DENCLUE-SA and DENCLUE-GA.

Tsapanos et al. [13] proposed fastest kernel matrix calculation for big data clustering. If the number of sample increases, the entire kernel matrix cannot be stored in a single computer's memory. The proposed kernel matrix k-means algorithm operates using a little part of kernel matrix. Moreover, experimentation is done on exceptional-lyquickcurrent hardware and BLAS library which allows quick computing of kernel matrix. This method is tested for clustering YouTube Faces data set which contains approximately 62,116 data samples and the performance results are satisfactory.

Santi et al. [14] proposed a method for clustering of data with diversed is similarity. The data required for forming cluster is always $n \times n$ dissimilarity matrix. But many applications generate data with more instances of dissimilar matrix. Aggregation is not the perfect solution as it sometimes overwrites the actualnature of data. This method handles heterogeneous data and identifies group of individual clustering objects. The method introduces variable neighbourhood search heuristic algorithm to provide solutions. Investigational outcome show that this model is well-organized and well suited for recovering heterogeneous data.

Fahad and Alam [15] presented a customized k-means algorithm for clustering of big data. Since the structure of big data is not static viz. unstructured, semi-structured and structured, traditional k-means clustering algorithm cannot be used. Their method first finds initial centroid and a range is created between clusters that will change and may not change. This reduces the unwanted computation and increases the working of k-means algorithm significantly. The method is compare with traditional methods. It is observed that this method is efficient, performs better clustering and less time is required.

3 Background

3.1 Mapreduce

Mapreduce is a parallel distributed programming interface that can develop enormous amounts of data in a parallel manner. It adopts the functional programming concepts of mapping and reducing [16]. The user supplies two pieces of code to process the data. i.e. a map code and a reduce code. Mapreduce algorithm works in three different phases viz. map phase, shuffle phase and reduce phase [17]. The advantage of using mapreduce is that data can be easily scaled over multiple computing nodes.

Let F_1 is a huge file containing input data with a simple list of pairs of type (keys and values). The user supplies Key_1, Value_1 to the mapreduce algorithm. The output after performing map function is keys_2, values_2

$$\text{Map}(\text{keys}_1, \text{values}_1) = (\text{keys}_2, \text{values}_2)$$

In phase where map method is used, every record in the input data set is called. The phase shuffle accepts keys_2, values_2 from the map phase and combines them together so that all of the pairs form the clusters. The output of shuffle phase is given as

$$\text{Shuffle}(\text{key}_2, \text{value}_2) = (\text{key}_2, \text{list}\langle\text{value}_2\rangle)$$

The reduce phase is called for every key_2 value, which is the output of shuffle phase. This phase is executed simultaneously. The output of these executions is collected in a huge output file. This phase summarizes the complete data set.

3.2 Big Data Analysis

Big data is playing an important role in our life and in our societies. It is a huge collection of data with high velocity and variety [18]. Big data mining refers to the process of extracting purposeful information from big data. Big data analysis works in multiple, distinct phases from data acquisition to extracting knowledge interpretation [19]. Each phase is having challenges. The general hazards of big data mining are heterogeneity and incompleteness of data, scalability and complexity of data, response time of the process, maintaining security and privacy during analysis, usage of proper programming framework, and so on [20].

4 Proposed Method

In the proposed method, we adopt the hierarchy clustering scheme because the output is a hierarchy of clusters that are more informative. Moreover, hierarchy clustering will never need specifying list of clusters in the beginning as well as they are deterministic in nature. Agglomerative clustering, which is a type of hierarchical clustering, generates small clusters that are helpful for discovery and very informative.

As stated in the introduction, we use mapreduce framework for clustering. Map reduce provides good scalability. But it does not address efficiency and optimization issues during the analysis process. And huge memory is required to manage large dimensions of feature vectors to avoid paging penalty. Hence two optimization techniques are proposed that can address the efficiency issues.

The occurence base attribute range technique is used in the preprocessing of data. This characteristic records the co-occurrence frequency of the feature vectors and simplifies the computation. The efficiency of hierarchical clustering is improved by, unwanted data need to be removed thereby reducing the dimension of user based keyword matrix. This removes the noisy keywords repetition. A list is created with keywords fetched from the title and metadata. From this, semantically related keywords are grouped. Using co-occurrence based feature selection, dimension of feature vectors are reduced and keywords with most interestingness are selected. The attention degree of two keywords is calculated and keyword with higher attention degree value is selected. This keyword represents both occurrence value and co-occurrence frequency of pair of keywords. Minimized user-keyword matrix is reused in a large number of cluster iterations.

To calculate the attention degree of keywords, a count is made on a single keyword, its pairs and availability in metadata. It is represented as

$$R_1 : \langle \text{user; keyword } i\text{; count } i \rangle$$
$$R_2 : \langle \text{user m; keyword i; keyword } j\text{; count } i, j \rangle$$

where R_1 represents that keyword i appear count I number of period accessed by the user and R_2 represents the pair of keywords in keyword i; keyword j appears count ij number of times in the pages accessed by the user. In metadata, number of times both keywords have appeared is represented in R_3.

$$R_3 : \langle \text{urls; keyword } i, \text{ keyword } j\text{; count } i, j \rangle$$

The proposed method generates huge amount of intermediate data that cannot be loaded in the computer memory for efficient processing. Hence, only few dimensions are selected which are vital for clustering. Co-occurrence feature selection is a unified framework that can be used both in supervised and unsupervised learning. It calculates the attention degree of the dimensions by estimating feature consistency forming a matrix derived from similarity matrix. It uses radical_base functions to calculate the similarity among two dimensions. It is represented by

$$S_{i,j} = \mathrm{e}^{-\frac{\|x_i - x_j\|^2}{2\sigma^2}}$$

where x_i and x_j are two samples and 'σ' represents the standard deviation among them. A graph is constructed suing S and from this graph, adjacency matrix is tabulated. The attention degree matrix $\bar{D}$ is computed from adjancency matrix.

$$\overline{D_{\iota,j}} = \sum_{j=1}^{n} W_{i,j}$$

Using $\bar{D}$ and adjacency matrix, Laplacian matrix L is computed as

$$L = \bar{D} - W$$

And normalized Laplacian matrix, represented as L' is given as

$$L' = \bar{D}^{-\frac{1}{2}}$$

The attention degree of each dimension is calculated using $\varphi_1, \varphi_2, \varphi_3$ where

$$\varphi_1(f_i) = \hat{f}_i^T \mathrm{L}\, \hat{f}_i$$

$$\varphi_2(f_i) = \frac{\hat{f}_i^T \mathrm{L}\, \hat{f}_i}{1 - \hat{f}_i^T \in 0}$$

$$\varphi_3(f_i) = \sum_{j=1}^{k-1} (2 - \gamma_j)\alpha_j^2$$

Clustering is performed with many numbers of iterations and in each iteration, manipulating the comparison of each pair of user, eliminating similar users, merging the users and updating of characteristic vectors in keyword-matrix file are performed. Simultaneous execution of hierarchical clustering can result in poor performance as huge I/O cost is incurred. Hence, the distribution of load over many different groups can minimize the cost. To overcome this, batch updating is used which improves the efficiency of clustering algorithm on big data. Batch processing combines a number of repetitions of clustering into one that updates keyword matrix and change similarity values. Combining many iterations will reduce the response time of the clustering process. It selects the highest N pairs of user groups, performs updation of in-memory batch and updates the customer keyword matrix file. In order to perform batch processing, two special data structures are used. The C-queue contains pairs of users with highest similarity index. It is a sorted queue and each element in the queue denotes pairs containing user groups. The B-queue contains all the C-queue elements and they are also sorted in order. To ensure the genuinely of the process of clustering, each element in the C-queue is in process in three different options. Inserting an element in the B-queue, eliminating the element from B-queue and stopping the updation of the keyword matrix file.

4.1 Mapreduce Phase

The application is composed of functions that perform filtering, sorting and also procedures to reduce the dimensionality of data. Mapreduce marshals distributed servers, runs multiple tasks in hierarchical order, manages all communication between various tasks, avoids redundancy and provides fault tolerance. Mapreduce deploys a huge number of computers, collectively referred as clusters, forms a distributed architecture.

The map function computes the distance entry for each row clusters. The map function takes tuple entry and tuple id pair as input, provides row id and entry distance tuple as output. The reduce function computes the sum of distances of individual row elements and finds total distance of the row. This total distance is for all possible row clusters is assigned with minimum distance. Choosing the pairs of user groups is done in the first phase. Processing elements in C-queue and filling them in the B-queue is done in the second phase. Updating the user keyword file is done in the final phase. It also involves modifying the similarity value file.

5 Results and Discussion

The effectiveness of the proposed method is proved by conducting two tests viz. scalability and cluster quality tests. First test is evaluated by noticing the execution time by increasing the number of objects and clusters. Second test performance depends on calculating NRMSE. Root Mean Square Error (RMSE) is calculated first by taking square root of Euclidean distance between objects and global mean. From this, NRMSE is calculated by dividing individual RMSE with global RMSE. In order to perform scalability test by increasing the total count of objects, two datasets containing 1 million and 16 million pixels are used. For simplicity, we calculate the clusters close to 10. Since it is difficult to define the production of number of clusters, we run the method multiple times with an order of *K* between 10 and 100 objects. During every iteration, the number of clusters produced on each level is recorded. For the 1 million objects dataset a *K*-Tree with order 15 was used in all cases and hence the execution time is constant (Tables 1 and 2).

NRMSE results for 1 million records are as follows (Tables 3 and 4).

Table 1 Execution time—Scenario I

Number of clusters	Execution time (s)		
	k-means	Parallel k-means	Our method
10	3.85	3.47	3.261
112	25.69	23.80	22.37
1180	260.57	247.69	232.82
11,236	2482	2395	2251.3
1,13,784	23,774	22,486	21,136.84

Table 2 NRMSE results—Scenario I

Number of clusters	NRMSE		
	k-means	Parallel k-means	Our method
10	0.4182	0.4432	0.4786
112	0.1823	0.1932	0.2086
1180	0.1012	0.1072	0.1157
11,236	0.0626	0.0663	0.0716
1,13,784	0.0368	0.0390	0.0421

Table 3 Execution time—Scenario II

Number of clusters	Execution time (s)		
	k-means	Parallel k-means	Our method
10	59.67	53.78	50.54
104	398.19	368.9	346.73
1012	4038.83	3839.19	3608.71
10,228	38,471	37,122.5	34,895.15
1,00,688	368,497	348,533	327,621.02

Table 4 NRMSE results—Scenario II

Number of clusters	NRMSE		
	k-means	Parallel k-means	Our Method
10	0.4182	0.4432	0.46875
112	0.1823	0.1932	0.20431
1180	0.1012	0.1072	0.1133
11,236	0.0626	0.0663	0.0730
1,13,784	0.0368	0.0390	0.0429

NRMSE results for 16 million records are as follows

For our next experiment, we have taken a large data set containing 1,16,846 keywords. By using a feature selection process, number of keywords considered for clustering is reduced to 55,849 thereby achieving around 48% reduction in keywords. Since most of the feature vectors are sparse, there is a reduction in memory requirement and efficiently similarity values are calculated.

Feature selection process brings changes in attention degree values of keywords. To test the efficiency of feature selection process, top keywords of the list are selected. In our experiment, it is quite clear that number of keywords selected decreases as the user groups. For our experiment, we have chosen $N = 100$, where 'N' denotes the number of top keywords selected. The evaluation of efficient hierarchical clustering algorithm, it is compared with updation of batch with different size of N. The below table shows the details of number of iterations, execution time for different values of N.

From the Table 5, it is perceived that the number of iterations is almost same regardless of the value of N. The difference in number of iterations between $N = 10$ and $N = 1000$ is just 30 which is very less. This shows that the proposed method is efficient.

Batch updating is having huge impact of the performance of any clustering technique. The Table 6 shows the performance of the method proposed using the same data set with different N values under batch processing.

From the above table, it is viewed that as much as 57% of the overall execution time is improved. Moreover, usage of batch updation also increases the performance

Table 5 Results on keywords

N	No. of iterations	Execution time
10	1566	25,615
50	1573	22,313
100	1609	26,846
250	1611	28,303
500	1578	43,982
1000	1596	41,687

Table 6 Execution time (with batch updation)

N	No. of iterations	Execution time
10	901	14,601
50	905	12,718
100	926	15,302
250	928	16,133
500	907	25,070
1000	918	23,762

of the proposed method by reducing number of I/O operations and communication costs.

6 Conclusion

This paper provides a novel hierarchical clustering algorithm using mapreduce technique. It is implemented in a distributed fashion that groups the internet users based on their web logs. The data are preprocessed and the important attributes are selected for clustering process. Experimental results show that this technique is capable and improves the mapreduce performance. Batch updating technique used in our approach merges user groups and updation is done in one iteration. This reduces the overall I/O and communication cost.

References

1. Jain AK (2010) Data clustering: 50 years beyond K-means. Pattern Recogn Lett 31(8):651–666
2. Hastie T, Tibshirani R, Friedman J (2009) Unsupervised learning. The elements of statistical learning. Springer, New York, pp 485–585
3. Jensi R, WiselinJiji G (2014) A survey on optimization approaches to text document clustering. arXiv preprint arXiv:1401.2229

4. Han J, Jian P, Kamber M (2011) Data mining: concepts and techniques. Elsevier
5. Srivastava A et al (2008) Credit card fraud detection using hidden Markov model. IEEE Trans Dependable Secure Comput 5(1):37–48
6. Van Dijck J (2014) Datafication, dataism and dataveillance: big data between scientific paradigm and ideology. Surveill Soc 12(2):197
7. Bizer C, Heath T, Berners-Lee T (2009) Linked data-the story so far. Semantic services, interoperability and web applications: emerging concepts 205–227
8. Katal A, Wazid M, Goudar RH (2013) Big data: issues, challenges, tools and good practices. In: 2013 Sixth international conference on contemporary computing (IC3). IEEE
9. Ferrara E et al (2014) Web data extraction, applications and techniques: a survey. Knowl Based Syst 70:301–323
10. Kittur A et al (2011) Crowdforge: crowdsourcing complex work. In: Proceedings of the 24th annual ACM symposium on user interface software and technology. ACM
11. Cai X, Nie F, Huang H (2013) Multi-view K-means clustering on big data. In: Rossi F (ed) Proceedings of the twenty-third international joint conference on Artificial Intelligence (IJCAI'13), AAAI Press, pp 2598-2604
12. Rehioui H et al (2016) DENCLUE-IM: a new approach for big data clustering. Procedia Comput Sci 83:560–567
13. Tsapanos N et al (2015) Fast Kernel matrix computation for big data clustering. Procedia Comput Sci 51:2445–2452
14. Santi É, Aloise D, Blanchard SJ (2016) A model for clustering data from heterogeneous dissimilarities. Eur J Oper Res 253(3):659–672
15. Fahad SKA, Alam MM (2016) A modified K-means algorithm for big data clustering. Int J Sci Eng Comput Technol 6(4):129
16. Bu Y et al (2010) HaLoop: efficient iterative data processing on large clusters. Proc VLDB Endowment 3(1–2):285–296
17. Verma A, Cherkasova L, Campbell RH (2011) ARIA: automatic resource inference and allocation for mapreduce environments. In: Proceedings of the 8th ACM international conference on autonomic computing. ACM
18. Sagiroglu S, Sinanc D (2013) Big data: a review. 2013 International conference on Collaboration Technologies and Systems (CTS). IEEE
19. Jagadish HV, Gehrke J, Labrinidis A, Papakonstantinou Y, Patel JM, Ramakrishnan R, Shahabi C (2014) Big data and its technical challenges. Commun ACM 57(7):86–94
20. Hashem IAT et al (2015) The rise of "big data" on cloud computing: review and open research issues. Inf Syst 47:98–115

Real Time Virtual Networks Monitoring Based on Service Level Agreement Requirements

Mohammed Errais, Mohamed Al-Sarem, Rachdi Mohamed, and Muaadh Mukred

Abstract Network virtualization is the ideal solution for the ossification internet phenomena. However, the multitude of actors involved poses significant challenges to the virtual network monitoring. For this purpose, we propose in this work a new approach for monitoring the services based on SLA established during the supply operation. An approach that aims to ensure an acceptable level of performance during all phases of the development and operation of virtual networks.

Keywords Network virtualization · Network administration · eTOM frameworks

1 Introduction

Network virtualization [1–3] is an emerging concept that aims to bring the benefits of system virtualization to networks. The concept takes its interest in deployment of several heterogeneous networks on existing physical media. Thus, the ossification of the Internet will become old history.

Setting up virtual networks requires migrating the current business model of the telecommunications industry to a new model [2, 4–6]. The latter divides the business of the traditional operator between several actors, the most important of which are

M. Errais (✉)
Research and Computer Innovation Laboratory, Hassan II University, Casablanca, Morocco
e-mail: mahammed_errais@yahoo.fr

M. Al-Sarem
Department of Information System, Taibah University, Medina, Saudi Arabia
e-mail: mohsarem@gmail.com

R. Mohamed
Faculty of Science Ben M'sik, Hassan II University, Casablanca, Morocco
e-mail: rachdi.simo@gmail.com

M. Mukred
Faculty of Information Science and Technology, Universiti Kebangsaan Malaysia, 43600 Bangi, Malaysia

S.-L. Peng et al. (eds.), *Intelligent Computing and Innovation on Data Science*, Lecture Notes in Networks and Systems 118,
https://doi.org/10.1007/978-981-15-3284-9_35

the infrastructure provider (InP), the virtual network provider (VnP) and the service provider.

The new business model poses several challenges [6], including the dynamic provisioning of resources, especially the selection and administration of virtual nodes. Several research works have focused on communication between actors for the supply of resources [6, 7]. However, the administration of the nodes taking in account various actors' requirements defined in the Service Level Agreement (SLA) [8] remains a quite treated area.

The idea of this work is to develop an autonomous system for virtual node monitoring in real time. Monitoring is carried out on the basis of the SLA requirements which are defined during the creation of these nodes for the two main actors: the InP and the VnP. The proposed approach, in this work, is essentially based on the sequence of the proposed business processes within the framework of the eTOM [9] by TmFORUM [10]. Accordingly, this will unify the exchanges between different actors and raise the challenges of a multitude of actors in the virtualization of the network.

The migration of the architecture of the Internet towards the virtualization of the networks faces several challenges [11] which can be summarized as follows:

- *Resource provisioning*: When creating virtual nodes, the VnP should look for the best InP offers. Before starting the process of creating and configuring these nodes, a set of operations for integration of several actors as well as the selection of the best offer are required. Several works have tackled these operations and have led to acceptable solutions [6, 7].
- *The instances of the business model*: the business model has been the subject of several works that have given rise to several model [5–7]. The latter differ in terms of the number of stakeholders, as well as the role of each stakeholder.
- *Control and monitoring*: methods for virtual nodes: to support the implementation of a virtual network, it is important to create the necessary devices to monitor virtual networks via the various deployed nodes. In this sense, few studies have focused on this aspect.

In this work, we will focus on the monitoring of virtual networks based on SLA defined between InP and VnP when creating nodes. This will allow to set up an autonomous system being able to detect any type of violation in real time. Consequently, ensuring a permanent monitoring as well as in real time.

This document is organized as follows. At the beginning we will present the business model of the network virtualization and the organization of the business processes of the eTOM framework. Before presenting our system for monitoring and experimentation for testing and validation.

2 Background and Related Works

2.1 Network Virtualization

The virtualization becomes a necessity to ensure the survival of the architecture of the Internet before the phenomenon of ossification. It offers significant benefits such as minimizing deployment costs and the ease of integration of new technologies. However, the implementation of such an architecture requires the migration to a new business model in the telecommunications industry. Surveying the literatures, there are several models for deploying virtual networks [1–3, 6].

These models differ according to the number of speakers and the ease of the usual operations, in particular the discovery and the selection of the resources, during the dynamic creation of the virtual networks. Thus, the respected model in this work consists of the following actors [6]:

- *Infrastructure Provider (InP)*: This entity is responsible for the deployment and administration of the physical infrastructure as well as it represents an autonomous authority that must ensure the proper functioning of the resources and tools required for the virtualization of network components.
- *Virtual Network Provider (VnP)*: which is The responsible for deploying virtual networks on physical devices. It provides basic protocols and tools for network operation. For this, the VnP must negotiate the use of resources with one or more VnP.
- *Service Provider (SP)*: it is responsible for providing value-added services for end-users.
- *Final user* who plays the role of service consumer in the current model.

The multitude of stakeholders poses several challenges to the deployment of network virtualization. The biggest challenge is to establish transparent communication between the different actors to establish the usual operation, including the supply of resources, quality of service assurance and billing.

2.2 eTOM Process

The eTOM framework is a grouping of business processes established by TmFORUM. The framework aims to provide an unified model for modeling all the usual operations in the telecommunications industry.

The business division of the business processes simplifies the organization of the usual operations. The sequence of processes leads to an operation involving different actors needs. Thus, the different actors can communicate in a transparent and efficient way independently of the internal business organization.

The eTOM framework offers generic business processes with an abstract description of actions. Thus, the use of these processes requires the nominal definition of the actions as well as the content and structure of the exchanged messages.

3 Real-Time Virtual Networks Monitoring

3.1 Work Description

Real-time monitoring is essentially based on the SLA verification operation [9]. That is, in regular intervals, the system checks the nodes' indicators based on the requirements defined on the SLA. the verification steps are as follows:

- *Collection of indicators from resources*: The first phase consists of collecting pre-defined performance indicators directly from the resources.
- *KPIs Mapping*: Mapping is a key operation which consists, in extracting from, the performance indicators and the appropriate quality indicators.
- *Requirements verification*: In this step, it is necessary to proceed to the verification of the quality of the performance indicators according to the established thresholds pre-defined in the SLA.

For implementation of the monitoring system, we have followed the following steps:

(i) **eTOM Adaptation of the Scenario According to the eTOM Framework**:

The second step is to sequentially establish the eTOM business processes. The objective of this step is to define the communication interfaces between the different actors, as well as the tasks of each process. Grouping business processes results in the execution of an operation by one or more actors. Accordingly, we define the processes and the sequence of the messages as described in Sect. 3.2.

(ii) **The Technical Choices**:

The last step is to define the technical choices for implementation of the monitoring system. In order to meet the need for communication between different actors and components of the system, we opted for the SOA architecture [12]. The business processes will be exposed as web services. The system components will be deployed as independent EJB modules [13], see Sect. 3.3.

3.2 *Modeling Processes According to ETOM Business Processes*

In order to optimize the exchanges, the processes have been grouped into two levels:

- *The resource level* which contains the business processes needed to collect and process performance indicators. This level includes two processes, the process known as "Resource data collection & Proceeding" and "distribute information & management". The former process is responsible for collecting the KPIs via the different logical nodes deployed in the InP concerned. Whilst, the latter is responsible for processing and structuring the performance indicators.
- *Insurance level*, which includes the business processes that are responsible for mapping the quality indicators, loading the VnP SLA and checking the requirements. In addition, this level is act as supervision layer that is responsible for establishing audit reports and detecting anomalies.

Figure 1 illustrates the sequence of processes during the verification operation. The operation can be initiated by the VnP or periodically by the system. In the first case, the "Customer Interface Management" process receives the VnP request before loading its profile via the "Retention & Loyalty" process. The produced report is then that includes the VnP identity and the SLA of the virtual network established during the creation of the virtual network. Then, this report is forwarded to the Service Quality Management (QSM) process. The QSM latter identifies the identity of the

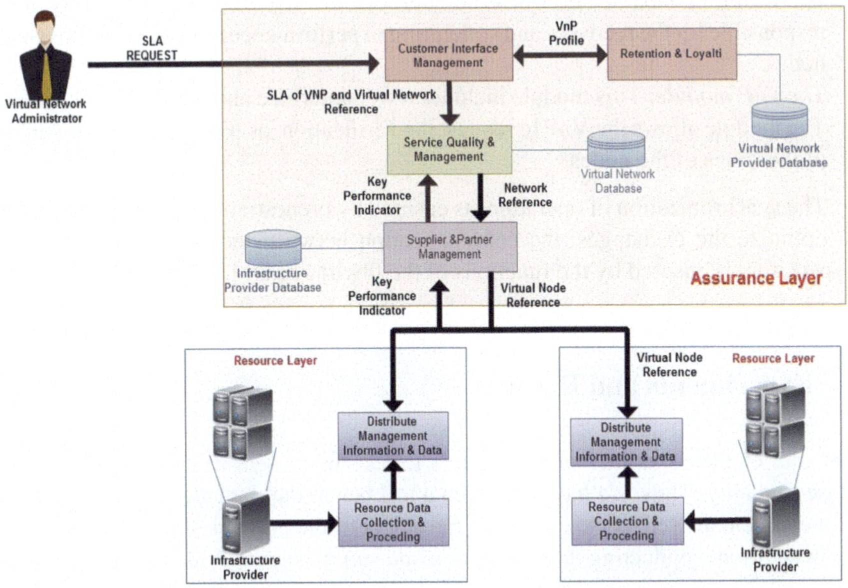

Fig. 1 Modeling the verification operation according to eTOM business processes

virtual nodes included in the network concerned by the verification. In the following, the SQM launches the operation of collecting the performance indicators.

The KPI collection operation is initiated by the "Supplier & Partner management" process. The latter identifies the InP that deploy deploys the different virtual nodes involved in the verification. The collection request, then, is sent to the different distribute information & data (DID) process. The DID process collects the indicators directly from the resources. These indicators are then structured and transmitted to the SPM.

3.3 System Architecture of the Autonomous System

The supervision system must be able to ensure transparent communication between the different actors of the business model of network virtualization. For this end, the system consists of several EJB modules. The system consists of the following modules:

- *The Assurance module*: the EJB module that includes the business processes is responsible for indicator mapping, loading the VnP SLA and SLA verification. It also includes three databases: (i) SLADB which contains the different VnP associated with the systems; (ii) DNDB which groups the information of the virtual nodes deployed as well as the SLA defined during the creation; and finally (iii) the InPDB which includes all InP related data associated with the system.
- *The Resource module*: EJB module deployed directly on the InP. For this, it is responsible for the recovery and structuring of performance indicators via physical nodes.
- *The VnP module*: This module includes a web interface and a web service client. The module allows the VnP to request the verification, as well as the visualization of the states of the nodes.

The synchronization of operations is ensured by orchestration processes. In order to optimize the exchanges, the communication between the collection processes hypervisors is ensured by the functions of the libvirt library [14].

4 Experiments and Results

The aim of the experimentation phase is to test the system in experimental cases close to reality. Thus, we have deployed a test bench that includes all the elements of the system as well as the entities of the actors of the network virtualization.

In addition, conducting the verification and supervision to evaluate the execution time and the success rate of the verification. The test bench consists of:

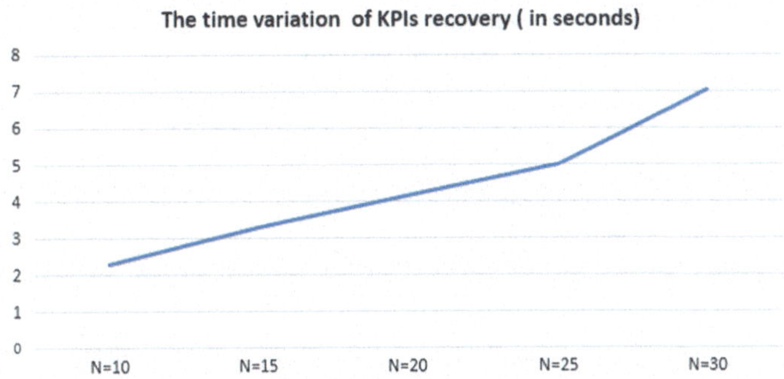

Fig. 2 Variation in collection time indicator according to the number of virtual nodes

- *Administration server*: The administration server contains the assurance module and two VnP modules. Each module is deployed in an independent virtual machine.
- *Infrastructure Provider Server*: This server contains several Resource modules of the system architecture. Each module is deployed as a virtual machine.
- *Hypervisors*: Hypervisors are the virtualization layer. They comprise the virtual nodes deployed under the authority of the VnP in the infrastructure provider. Each hypervisor is under the responsibility of one of the InP that is deployed in the provider infrastructure server. The used virtualization tool is KVM.

Before beginning the experiment, each VnP creates several virtual nodes, in order to constitute several virtual networks. The nodes are created on multiple InP for the same virtual network.

The evaluation of the system is based essentially on two criteria: (i) the collection time indicator which is based on the number of virtual nodes on the physical nodes, and (ii) the verification time according to the number of InP.

Figure 2 illustrates the variation in the indicator of collection time from resources. This is the most difficult step of the audit, as it requires several exchanges on one side between the business processes and on the other side between the virtualization entities.

The collection time is significantly affected by the number of virtual nodes deployed in the physical nodes. This is explained by the resources needed to deploy these nodes and thus the degradation of the response time during the recovery of the indicators.

However, using the libvirt API has reduced the load on physical nodes by avoiding deploying physical agents to compute these indicators. Accordingly, this made it possible to reduce the time needed to collect the indicators.

Figure 3 illustrates the variation in the verification time according to the number of partner provider infrastructures. In this experiment, we used a single hypervisor from each provider infrastructure. The time shown is the overall verification time of the SLA check for the virtual network. For this purpose, the verification time is acceptable ($\approx$8 s for 16 InP). This is explained by the technological choices of the

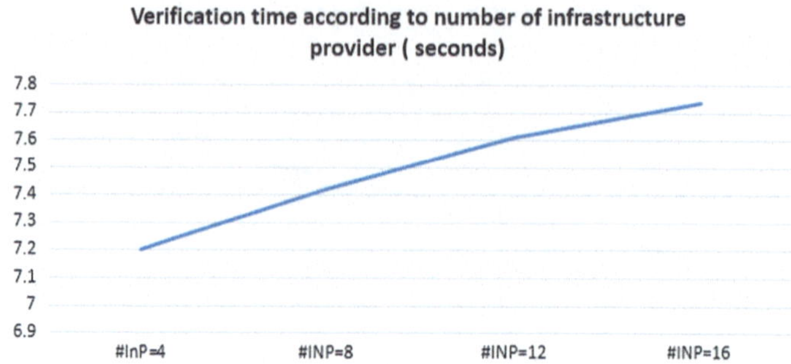

Fig. 3 Evolution of the verification time according to the number of infrastructure provider

system that allow the exchange of messages in a simple and transparent way between the various actors via the SOAP protocol.

5 Conclusion

The deployment of virtual networks in the telecommunications industry faces several challenges, including the provision of resources, the assurance and control of nodes and virtual networks. For this purpose, in this work, we proposed a solution for controlling nodes based on the eTOM business processes.

The solution has made it possible to deploy an autonomous system capable in real time of detecting any anomaly in virtual networks. Degradation detection is based on the thresholds defined in the SLA between the provider of the virtual networks and the infrastructure provider. Experimentation of the proposed solution allowed to validate the approach in experimental cases close to reality. However, the verification is not sufficient to complete the operation of the insurance. For this purpose, the correction of anomalies is an operation that is necessary.

References

1. Ferguson, B. (2012). The official VCP5 certification guide. VMware Press
2. Chowdhury NMMK, Boutaba R (2009) Network virtualization: state of the art and research challenges. IEEE Commun Mag 47(7):20–26
3. Amarasinghe H, Belbekkouche A, Karmouch A (2012) Aggregation based discovery for virtual network environments. In: Proceedings of the IEEE International Conference on Communications (ICC 2012). IEEE Press, Ottawa, ON, pp 1276–1280
4. Elliott C (2008) Geni–global environment for network innovations. In: 33rd IEEE conference on local computer networks, p 8 (2008)
5. Xu Y, Han Y, Niu W, Li Y, Lin T, Ci S (2012) A reference model for virtual resource description and discovery in virtual networks. In: Proceedings of ICCSA. Springer, Brazil, pp 297–310

6. Rabah S, El Barachi M, Kara N, Dssouli R, Paquet J (2015) A service oriented broker-based approach for dynamic resource discovery in virtual networks. J Cloud Comput Adv Syst Appl 4:3. https://doi.org/10.1186/s13677-015-0029-5
7. Mohammed E, Abderrahim S (2017, May). Implementation of new broker based on the eTOM framework for dynamic supply of resources during the composition of virtual networks. In 2017 International Symposium on Networks, Computers and Communications (ISNCC) (pp. 1-6). IEEE
8. Mohammed E, Mostafa B, Ranc D (2015). Autonomous system for network monitoring and service correction, IN IMS Architecture. Int J Comput Sci Appl 12(1)
9. Business Process Framework (eTOM), Enhanced Telecom Operation management, GB921, version 7.2
10. The TM FORUM. http://www.tmforum.org/
11. Tutschku K, Zinner T, Nakao A, Tran-Gia P (2009) Network virtualization: implementation steps towards the future Internet. In: Proceedings of the workshop on overlay and network virtualization at KiVS
12. Errais M, Mostapha B, Brahim R, Daniel R (2012, December). Cost optimization of monitoring and supervision of IP Multimedia Subsystem networks. In 2012 Next Generation Networks and Services (NGNS) (pp. 74–80). IEEE
13. Panda D, Rahman R, Lane D (2007). EJB 3 in Action (Vol. 15). Manning Publications Company
14. Takemura C, Crawford LS (2010). The book of Xen: a practical guide for the system administrator. No Starch Press

Wireless Sensor Network-Based Hybrid Intrusion Detection System on Feature Extraction Deep Learning and Reinforcement Learning Techniques

K. C. Krishnachalitha and C. Priya

Abstract A Wireless Sensor Network (WSN) is one of the most huge parts of the field of correspondence innovation. A Wireless Sensor Network is one kind of remote framework that consolidates endless coursing, self-composed, minute, low controlled contraptions named sensor center points called motes. This innovation has numerous application zones like therapeutic, ecological, transportation, military, amusement, country guard, emergency the board and furthermore keen spaces. Security is one of the most vital aspects concerned with WSN. Intrusion detection (ID) is one of the main issues while concerning about security. This paper is concerned with the comparative study on the existing hybrid intrusion detection method with their advantages in addition to disadvantages. The article also proposes a hybrid interruption recognition system on the basis of feature extraction, deep learning and reinforcement learning techniques which reduces human dependency and takes most decisions automatically. The proposed system integrates anomaly based as well as signature mechanisms for detecting attacks.

Key Terms Deep learning · Feature extraction · Hybrid intrusion · Reinforcement learning

1 Introduction

Wireless Sensor Networks (WSNs) have turned out to be a standout among the most indispensable and provocative territory of research. WSN is another technology which is ending up more across the board and helpful in numerous zones like military applications, environmental observing, home application, wellbeing or

K. C. Krishnachalitha (✉)
School of Computing Sciences, Vels Institute of Science, Technology and Advanced Studies (VISTAS), Chennai, India

C. Priya
Department of Information Technology, School of Computing Science, Vels Institute of Science, Technology and Advanced Studied (VISTAS), Chennai, India

S.-L. Peng et al. (eds.), *Intelligent Computing and Innovation on Data Science*, Lecture Notes in Networks and Systems 118,
https://doi.org/10.1007/978-981-15-3284-9_36

medicinal application, modern checking, basic quality observing and so forth. Substantially, more rivalries happen in the field of WSN. WSN takes a shot at the idea of sensor hubs which are sorted out over a wide range with the assistance of vitality hotspots for powerful working. While discussing WSN, the premier thing to be well-thought-of is the security. Intrusion detection has a critical influence on the matter of security. There happen two guideline classifications of intrusion detection system (IDS). Anomaly based IDS identifies PC interferences and maltreatment by watching structure activity and requests it into conventional or unusual lead. Signature constructed IDS distinguishes interruptions situated in light of particular arrangements. It gains from the past assaults and the information is put away in the database for future references. Hybrid IDS is a crease of anomaly and signature-based IDS for a solid interruption location.

2 Related Works

The authors Indira et al. [1] had arranged a cluster-based hybrid intrusion discovery framework with a Hybrid Energy-Efficient Distributed Clustering (HEED) convention. The principle aim behind this framework was to structure a hybrid IDS that downsizes utilization of energy, along these lines expanding the system timeframe. A cluster node is chosen to gather data from all elective gadget hubs; along these lines, it is solely communicating with the base station rather than all other exchange hubs working together with the base station which clear approach to expand the system timeframe. The blend of anomaly detection method and a collection of signature rules are utilized to distinguish resentful outbreaks. The fundamental preferred standpoint of this framework is that it recognizes the assaults in a conservative way. Besides the energy expended, identification is relatively less. SVM calculation has been joined in anomaly identification for grouping WSN. The primary disadvantage of this framework is that however it goes for vitality productive hybrid IDS, the framework is powerless for obscure assaults or in other words the framework absolutely relies upon the effectiveness of the calculations consolidated.

The authors Jinhuia et al. [2] had projected a hybrid IDS strategy dependent on energy trust in WSN. The primary expectation behind this system is to anticipate the energy utilization and to build the relationship figuring of energy utilization to assess the security conditions of nodes. The current model usefully pinpoints a more disguised hybrid DoS assault. The principle favorable position of this model is that it can decrease the energy devoured. It combines both anomaly and signature centered detection methods. The standard drawback is that if the energy reins the nodes and produces the vitality data, the hub escapes from viewing.

3 Proposed Model

The paper proposes a hybrid intrusion detection framework which fuses both anomaly and signature-based IDS. The proposed model chips away at the premise of feature extraction, deep learning and reinforcement learning. In addition, the current frameworks rely upon the productivity of the calculation utilized; the new framework is proposed to work by lessening the human reliance and consequently pointing the framework to take choices naturally.

4 Techniques Used

4.1 Feature Extraction

Feature extraction begins from a shrouded strategy of assessed information and makes chose attributes (highlights) needed to enlighten and non-plenitude, enabling the ensuing learning and speculation steps, and every so often inciting better human clarifications. Feature extraction is an abatement process, where a basic game plan of rough factors is lessened to increasingly reasonable gatherings (feature) for planning, while still correctly and absolutely portraying the main informational record mandatory for the corresponding author.

4.2 Deep Learning

Deep learning is a bit of a gradually expansive congregation of AI strategies contingent on learning data depiction, as opposed to task explicit calculations. Learning can be coordinated, semi-controlled or unaided. It utilizes a course of various layers of nonlinear dealing with units for feature extraction and change. Each powerful layer utilizes the yield from the past layer as information.

4.3 Reinforcement Learning

Reinforcement learning (RL) alludes to a sort of machine learning technique in which the operator gets a postponed reward in whenever venture to assess its past activity. RL setup consists of an agent and an environment (Fig. 1).

Activity (An) is all the conceivable moves that the operator can take. State (S) is the present circumstance returned by the environment. Reward (R) is a quick return sends once again from the environment to assess the last activity.

Fig. 1 Illustration of reinforcement learning

5 Algorithms Used

5.1 SVM Algorithm

SVM calculation for directly distinguishable double sets. The objective is to structure a hyperplane that characterizes all preparation vectors in two classes, i.e., a1, a2. The best choice will be the hyperplane that withdraws the most extraordinary edge from the two classes. The SVM technique is suited to characterize the high-measurement information in IDS. Amid the preparation stage, which happens disconnected at a framework with bottomless assets, information is gathered from the physical, medium access control. At that point, the gathered preparing information is pre-handled utilizing an information decrease process, which goes for diminishing their size with the end goal to be prepared by SVM. Characterization hyperplane of preparing information may be partitioned by straight grouping plane or not by means of mapping the preparation information vector to higher dimensional space with some capacity and exchanging the issue to a direct arrangement issue in that space. After the mapping technique, SVM discovers a straight isolating hyperplane with the most extreme edge in the space.

$$w \cdot +b = 0[1]$$

where w is an ordinary vector and the parameter b is balanced. The preparation tests on the hyperplane are called support vectors, since they bolster the ideal order hyperplane. So our concern can be figured as

$$\min \emptyset(w) = 12||w||2 = 12(w \cdot w)\min \emptyset(w) = 12||w||2 = 12(w \cdot w)\,[1]$$

5.2 Self-taught Learning

In self-taught learning and unverified element learning, we will give our computations a ton of unlabeled data with which to take in an average component depiction of the information. On the off chance that we are endeavoring to fathom a particular arrangement undertaking, at that point we take this scholarly component portrayal and whatever (maybe little measure of) marked information we have for that order assignment, and apply managed learning on that named information to tackle the grouping errand. These thoughts presumably have the most intense impacts on issues where we have a considerable measure of unlabeled information and a littler measure of marked information. Be that as it may, they commonly give great outcomes regardless of whether we have just named information (in which case we more often than not play out the element learning step utilizing the marked information, however disregarding the names). An autoencoder can be utilized to take in highlights from unlabeled information.

Solidly, assume we have an unlabeled preparing set $\{x_u^{(1)}, x_u^{(2)}, \ldots x_u^{(m)}\}$ with mu unlabeled models. (The subscript "u" remains for "unlabeled.") We would then be able to prepare an inadequate autoencoder on this information (maybe with suitable brightening or other pre-handling): presently, assume we encompass a marked preparing set $\{(x_l(1), y(1)), ..(x_l(\mathrm{ml}), y(\mathrm{ml}))\}$ of models. (The subscript "l" means "marked.") We would now be able to locate a superior portrayal for the data sources. Specifically, instead of speaking to the main preparing model as xl(1), we can feed xl(1) as the contribution to our autoencoder and acquire the relating vector of initiations al(1). To speak to this model, we can either simply supplant the first element vector with al(1). On the other hand, we can connect the two element vectors together, getting a portrayal (xl(1), al(1)). The concatenated representation becomes $\{(x1^{(1)}, a1^{(1)}), y^{(1)}), \ldots (x1^{(m1)}, a1^{(m1)}), y^{(m1)})\}$. At long last, we can prepare a managed learning calculation, for example, a SVM, logistic regression and so forth to acquire a capacity that makes expectations on the y esteems (Fig 2).

5.3 Deep Q Network (DQN)

The estimation consolidates Q-Learning with significant neural frameworks to allow RL to work for incredible, high-dimensional conditions, like PC amusements or apply self-governance. We train the system dependent on the Q-learning refresh condition. DQN uses a neural network to evaluate the Q-esteem work. The contribution for the system is the current, while the yield is the relating Q-esteem for every one of the activities (Fig. 3).

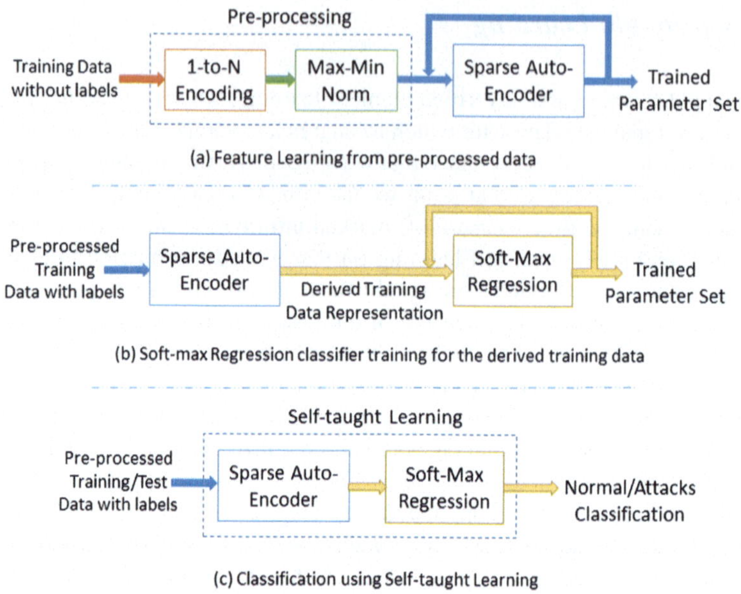

Fig. 2 Working of self-taught learning [3]

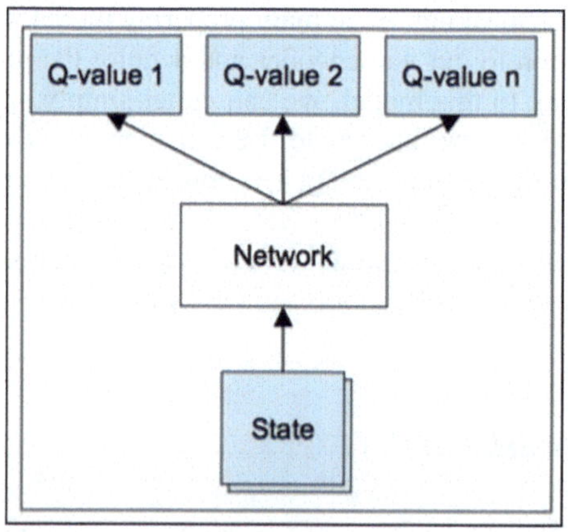

Fig. 3 Illustration of DQN [4]

6 Conclusion

The key test of advancing interruption identification framework in WSN is to distinguish assaults with high precision and fulfilled the required imperatives and difficulties, to draw out the lifetime of the whole system. These points could be accomplished from a few different ways. Right off the bat giving careful consideration to recognition methods utilized for assaults identification is portrayed by effectiveness and capacity and also recreating identification instrument with an appropriated way, to lessening the correspondence overhead. This paper has proposed a hybrid IDS that combines SVM, self-taught learning and DQN for hybrid intrusion detection in order to minimize the dependency of human skills. It not only helps to detect the intrusion efficiently but also helps to act to the unknown situations. Notwithstanding, numerous issues are as yet open and need additionally look into endeavors, for example, progressive bunching designs, utilizing machine learning in asset administration issue of remote sensor systems, building up a classifier that is prepared well with system designs, choosing and preprocessing a suitable dataset. Likewise, considering brilliant methodologies, for example, packing the info dataset, narrowing the size of traits set and streamlining the strategy of investigation and choice could gain heaps of ground for IDS to fulfill the prerequisite requirement of WSN without losing the security and unwavering quality.

References

1. Indira K, UshaNandini D, Sivasangari A (2018) An efficient hybrid intrusion detection system for wireless sensor networks. Int J Pure Appl Math 119(7):539–556
2. Jinhuia X, Yang T, Feiyue Y, Leina P, Juan X, Yao H (2018) Intrusion detection system for hybrid DoS Attacks, ScienceDirect. Procedia Comput Sci 131:1188–1195 (8th International Congress of Information and Communication Technology (ICICT-2018))
3. Priya C et al (2011) Monitoring system using smart phones. Int J Comput Eng Technol (3). ISSN: 0976-6375, 1:2011
4. https://ufldl.stanford.edu/wiki/index.php/Self-Taught_Learning
5. El Mourabit Y, Bouirden A, Toumanari A, El Moussaid N (2015) Intrusion detection techniques in wireless sensor network using data mining algorithms: comparative evaluation based on attacks detection. Int J Adv Comput Sci Appl (IJACSA) 6(9)
6. Niyaz Q, Sun W, Javaid AY, Alam M (2015) A deep learning approach for network intrusion detection system. BICT 2015, December 03-05, New York, United States. https://doi.org/10.4108/eai.3-12-2015.2262516

Green Technology to Assess and Measure Energy Efficiency of Data Center in Cloud Computing

C. Priya, G. Suseendran, D. Akila, and B. Vivekanandam

Abstract Our survey audit uncovers a couple of imperatives capability structures for server farms which join a green IT plan with express exercises and procedure is incited decline the effect on condition and less CO_2 floods. The current accessible structures have a few upsides and downsides that is the motivation behind why there is an earnest requirement for a coordinated foundation for choosing and embracing energy efficiency system for data centers. The required proficiency structure is vital for criteria should in like manner consider the casual association applications as a key related factor in raising imperatives usage, just as talent in data centers for better vitality proficiency. Furthermore, the featured significance of the recognizable proof of proficient and viable vitality effectiveness estimation of measurement can be utilized and confirmation of the estimation of data centers productivity and their execution joined with complete and observationally energy efficiency (EE) framework.

Key words Energy efficiency · Green cloud · Datacenter · Cloud computing

1 Introduction

Distributed computing is a promising locale in appropriated figuring. The essential piece of distributed computing is server farm, server ranches essentialness use

The original version of this chapter was revised: The author's name has been corrected from "V. Vivekanandam" to "B. Vivekanandam". The correction to this chapter is available at https://doi.org/10.1007/978-981-15-3284-9_86

C. Priya (✉) · G. Suseendran
Department of Information Technology, School of Computing Sciences, Vels Institute of Science, Technology and Advanced Studies (VISTAS), Chennai, Tamil Nadu, India

D. Akila
Department of Information Technology, School of Computing Sciences, Vels Institute of Science, Technology and Advanced Studies (VISTAS), Pallavaram, Chennai, Tamil Nadu, India

B. Vivekanandam
Faculty of Computer Science and Multimedia, Lincoln University College, Kota Bharu, Malaysia

S.-L. Peng et al. (eds.), *Intelligent Computing and Innovation on Data Science*, Lecture Notes in Networks and Systems 118,
https://doi.org/10.1007/978-981-15-3284-9_37

cost and biological effect are energetic test to distributed computing. In like manner, the making use of e-business requires an expansion in the measure of server farms. Regardless, the blend of an earth-wide temperature help and conflicting condition make the expense of vitality a basic test for the sensibility of e-business [1, 2]. They broadened the point of confinement of united document for taking care of, empowering, limit, the managers, checking, systems association and strategy of information.

With the snappy augmentation in the cutoff and size of server farms, there is an industrious addition in light of a legitimate concern for essentialness usage and the yearly report initiate that conveyed registering came to $41 bn in 2013 and the pay of cloud in 2014 was $151 bn [3, 4]. Firstly, in data center vitality effectiveness is enhanced, secondly, utilize clean vitality supply. The diverse methods to take care of vitality productive issue by limiting the effect of distributed computing on the earth by cloud computing. These procedures manage vitality proficiency utilization is similar to virtualization, equipment base, base of operating system and data centers [5]. Some new highlights emerge like time astute and energy performance. In any case, the worries ought to be to swap issue between Energy Efficiency and execution [6].

2 Literature Review

In our writing survey depends on past investigations of explored data center innovation and energy efficiency on cloud computing. Sabbaghi et al. explored past looks into and presented energy efficiency system on data innovation that empowered green supply chain management [7]. The theoretical scientific categorization of data innovation for maintainability are proposed. They additionally recognized the connection amid information flow in green supply chain management, governance of IT plus green foundation segments [8]. PriyaC [9] proposed the system to support amplifying asset usage by utilizing dynamic and inactive vitality utilization by completed minimization of time. This mechanism enables the power utilization of spare servers elects out of gear condition. The record QoS of cloud datacenter is located in this mechanism.

Rajkumar Buyya et al. [10], is proposed a innovative device in three dissimilar methods: (a) engineering standards in cloud management for energy-efficient; (b) a novel software technology of clouds in energy-efficient management; and (c) resource allotment of energy-efficient policies and scheduling algorithms considering QoS and gadgets control use attributes.

Beloglazov has developed a noteworthy system that ropes the energetic cementing of VMs subject to flexible limits [11].

Nguyen Quang Hung [12] proposed one of a kind server determination strategy, and four algorithms taking care of the let scheduling problem. This methodology lessens 7.25% and 7.45% vitality utilization than the current ravenous mapping calculation.

Uddin et al. acquainted a special system with enhance the execution and energy efficiency of data centers. They built up an order mechanism for data center segments relying upon various asset pools [13] and distinctive parameters like vitality utilization, asset usage, remaining task at hand and so on. The structure includes the centrality of executing green estimations to check the capability of server ranch the extent that imperativeness.

Sharma et al. [13] built up another two stages of mechanism: firstly, right off the bat, they built up an investigation of various virtual machine(VM) [14] load balancing algorithms, secondly, presented another load balancing algorithm VM that has been created as well as executed within virtual machine environment of cloud computing toward the accomplish well retort time and price.

S. Kontogiannis built up a special mechanism described adaptive workload balancing algorithm (AWLB) in support of cloud-based data center in web frameworks which manages operators hooked on two measurements the web servers plus web data center. AWLB algorithm additionally underpins convention determination for flagging purposes among web switch and data center hubs and furthermore uses different conventions, for example, SNMP along with ICMP for its adjusting procedure. Execution improvements are appeared from the trial. The outline of Literature review on cloud computing vitality productivity structures and systems as shown in Table 1 [15].

MueenUddin, clarified the arranged naiveté IT framework conjointly encourages IT business ventures explicitly information focus exchange to pursue a virtualized green IT structure, to abstain from squandering Brobdingnagian amount of vitality and simultaneously back the ozone-depleting substance discharges that eventually lessens an unnatural weather change impacts. It comprises of five stages to be pursued to appropriately implement virtualization at various layers and levels. Afterwards, utilize unpracticed measurements to live the productivity of information focus regarding vitality intensity and ozone-depleting substance emissions.

3 Requirement for Energy Effectiveness for Data Centers in Cloud Computing

Decreasing transmissions of carbon dioxide (CO_2) and vitality utilization in server farms speak to open difficulties in server ranches. We examine uncovers the crushing essential for formed hugeness ability system for server ranches which joins a green IT structure with unequivocal exercises and technique that actuated insignificant effect on condition and less CO_2 outpourings. The required imperativeness efficiency framework ought to comparatively consider the easygoing system applications as a fundamental related factor in raising centrality use, likewise as high potential for essentialness viability.

Table 1 Techniques in data centers energy efficiency

No.	Author	Approach	Strengths	Limitation
1	Sabbaghi	Calculated scientific classification of data innovation	Supply management	Hub on infrastructure only
2	Zhiming Wang	Resource utilization is increased	Put into account QoS	Amount of time is taken for job performance as Sleep-in- and Waking up- ready
3	Priya	Scheduling the resource allocation	QoS	No indication in parameter
4	Beloglazov	Adaptive utilization	Service level agreements (SLA) is obtained	Veto parameter shows the energy efficiency level
5	Sharma	LBA	High quality to decrease vitality, valuing and time	A large amount of calculation need more opportunity to take choice
6	Uddin	VM	Increase the utilization ratio	Skyscraping utilization leads to introduce CO_2
7	Kontogiannis	WBA	Poise the remaining task at hand in multidimensional assets	Augment the web traffic

4 Energy Efficiency of Datacenter to Measure and Assess by Green Technology

Universally, the data center in energy consumption [16] is consistently on the expansion [16]. The vitality tasks cost will keep on multiplying each five years somewhere in the range of 2005 and 2025. This expansion prompted higher outflow of CO_2 that considers adversely a worldwide temperature alteration and natural well being [17].

Estimating vitality utilization of server farms has turned into a significant worry of all datacenters partners to meet end-client understanding [18]. Energy effectiveness measures an apparatus used to gauge vitality proficiency in server farms [19]. The imperative test during data centers [20, 21] industry is the restriction of powerful standard vitality productivity measurements, which bolsters enhancing vitality effectiveness. For a compelling vitality productivity appraisal on segments, data center, evaluate viable measurements and quantify data center energy effectiveness [22]. If the measurements are either powerful nor to survey measurements in planned objectives, a scope of basic utilized cases toward decides the estimations and its viability regarding revealing, targets, instruction, examination and choice help.

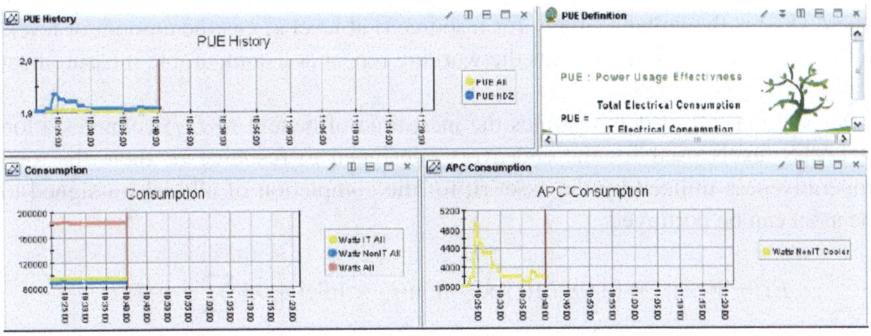

Fig. 1 Market value green data center

Writing survey on normal metrics in energy efficiency are at present being used in data centers uncovers that not a bit of these measurements are reunion the earlier referenced criterion. In this manner, the examination is not just presenting a relative audit of the most widely recognized utilized measurements and their highlights (criteria) yet, in addition, endeavoring to prescribe better measurement to be utilized in the evaluation of data centers energy efficiency.

In most recent couple of years, administrators have received PUE measurements since the data center is extent of vitality effectiveness for the mechanized and electrical establishment. The technique of assessment has displayed a focus and for all intents and purposes indistinguishable extent of execution, which has engaged server farms directors to make liberal updates. In any case, until now no consensus about IT or software energy efficiency. Figure 1 depicts the separation among the addresses in arranged goals of centrality ability estimations [23].

Energy-efficient optimization

In this area, we will, in general, propose the vitality effective improvement model upheld the dynamic voltage and recurrence scaling (DVFS) [13] that the electrical marvel intensity of a given asset hub relies upon the voltage offer and asset recurrence. Dynamic power utilization is done by the hub capacitance brought about by charging and releasing; its fundamental articulations are regularly sketched out as pursues:

$$P = \gamma \times v2 \times f \tag{1}$$

where $\gamma = A \times C$, A is the amount of switches per clock cycle, C is the load capacitance, v is the stock voltage and f is the repeat of the asset hub.

Expect that si addresses the voltage supply class of advantage ri, and si has k DVS level; by then the inventory voltage and repeat relationship network of si can be depicted:

$$Vi = [(v1(i), f1(i));\ (v2(i), f2(i));\ \ldots;\ vk(i), fk(i))]^{\mathrm{T}} \tag{2}$$

where vk(i) is the voltage supply for resource ri at level *k*, *k* is the amount of levels in the class si, and $f\,k\ (I)$ implies the working repeat at a comparable measurement $k, 0 \leq fk(i) \leq 1$.

Expect that idle *I* demonstrates the inert time of benefit ri, *L*(*j*) connotes a ton of DVFS levels used for the endeavors consigned to resource ri; then, the total imperativeness utilized by the asset ri; for the completion of all tasks assigned to the asset can be portrayed;

$$Ei = \gamma \times f \times \{+vmin(i) \times f\min(i) \times \text{Idlei} + \lambda\} \sum j \in T(i),$$
$$k \in L(j)([(vk(i))j]2 \times CT(i,j)) \tag{3}$$

where *v*min(i) and *f*min(i) address the voltage and repeat when resources ri progress to rest mode in the dormant time, independently, and λ is the load factor of benefits ri.

5 Conclusion

The central commitment of this paper is our writing survey on current vitality productivity system. The examination uncovers that there are starting at now a couple of vitality productivity structures for server farms which join a green IT structuring with unequivocal exercises and approach that will actuate rot the effect on condition and the decreasing of CO_2 radiations. The present open structures have two or three of intrigue and obstacles (Table 1) that is the explanation there is an edgy essential for a combined essentialness capability system for server homesteads and passed on handling. The system ought to think about a typical and made arrangement out of criteria. The decision and adoption of such framework should be according with the data center and its surrounding environment.

The subsequent responsibility was the composing review on imperativeness viability estimations that are at present utilized for the assessment of essentialness efficiency in server ranches (portrayed in Fig. 2). This bit of our assessment built up a nearby assessment of the most normally utilized estimations and their highlights (criteria), other than we embraced the use of FVER instead of PUE as a common estimation for the appraisal of server ranches vitality capacity which depended upon certain necessary criteria including its usage and programming applications in server. Our future work will focus on the enhancement and accurate endorsement of an incorporated energy efficiency framework.

Fig. 2 PEU Vs VER

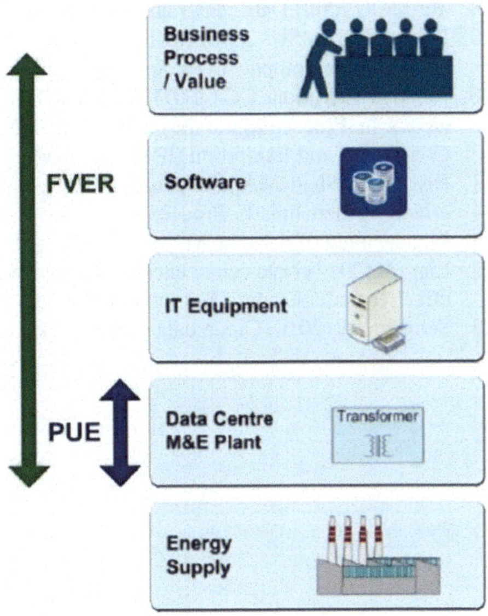

References

1. Mell P, Grance T (2009) The NIST definition of cloud computing
2. Priya C et al (2011) The next generation of cloud computing on information technology. Int J Comput Sci Inf Technol 2(5)
3. Uddin M (2012) Framework for energy efficient data centers using virtualization
4. IDC (2013) Press Release
5. Smith JW (2011) Green cloud a literature review of energy-aware computing
6. Sabbaghi A (2012) Green information technology and sustainability (2012)
7. Woods E Data center electricity consumption 2005–2010: The Good and Bad News (2011)
8. Sabbaghi A (2012) Green information technology and sustainability: a conceptual taxonomy. 13(2): 26–32
9. Priya C (2017) TaaS: trust management model for cloud-based on QoS. J Adv Res Dyn Control Syst 9(18):1336–1345
10. Buyya R et al (2010) Energy-efficient management of data center resources for cloud computing: a vision, architectural elements, and open challenges
11. Beloglazov A (2010) A survey on power management solutions for individual systems and cloud
12. Nguyen Quang Hung, Performance constraint and power-aware allocation for user requests in virtual computing,2011
13. Uddin M (2012) Green Information Technology (IT) framework for energy efficient data centers using virtualization
14. Sharma M (2012) Performance evaluation of adaptive virtual machine load balancing algorithm
15. Kontogiannis S (2011) A probing algorithm with adaptive workload load balancing capabilities for heterogeneous clusters. J Comput 3(7): A probing algorithm with Adaptive workload load balancing capabilities for heterogeneous clusters. J Comput 3(7) July 2011
16. Lacity MC, Khan SA, Willcocks LP (2009) A review of the IT outsourcing literature: insights for practice. J Strateg Inf Syst 18:130–146
17. Sisó L, Fornós R, Napolitano A, Salom J (2012) Energy- and heat-aware metrics for computing modules

18. Teresa T (2008) Data center energy forecast. Silicon valley leadership group, San Jose, CA
19. Wang L, Khan SU (2013) Review of performance metrics for green data centers: a taxonomy study. J Supercomput, 1–18. Springer
20. Belady CL, Malone CG (2007) Metrics and an infrastructure model to evaluate data center efficiency. In: Proceedings of the Pacific Rim/ASME International Electronic Packaging Technical Conference and Exhibition (IPACK), ASME
21. Rivoire S, Shah MA, Ranganathan P, Kozyrakis C, (2007) JouleSort: a balanced energy-efficiency benchmark. Proceedings of the 2007 ACMSIGMOD international conference on Management of data, ACM, pp 365–376
22. Liam N (2013) Data center energy efficiency metrics existing and proposed metrics to provide effective understanding and reporting of data center energy
23. Schwartz A (2010) Green data center market to more than triple over next five years

Reliable and Consistent Data Collection Framework for IoT Sensor Networks

K. Kavitha and G. Suseendran

Abstract In IoT sensor networks, in the course of statistics accumulation, the statistics severance and stowage overhead may perhaps be augmented. Likewise, the dependability and steadiness of radar information need to be mentioned. Therefore in this research, Reliable and Consistent Data Collection Framework for IoT sensor networks is aimed. In this agenda, a group of applicant nodules are nominated depending upon the vitality suitability feature and bumper place accessibility. As soon as the information is detected at period interim t, it will be transferred to the nominated applicant nodule depending on the complete discrepancy rate. If the package inaccuracy amount at the sink nodule is greater than the brink rate, then the foundation will choose to direct the simulated statistics to a nominated group of applicant nodules. Through replication outcomes, we demonstrate that the suggested method verifies both the steadiness and idleness of statistics, thus resolving the trade-off. It also decreases the quantity of simulated statistics.

Keywords IoT · Sensor · Reliable · Data · Framework

1 Introduction

The Internet of Things (IoT) mentions the fast-developing system of devices/equipment. IoT enables Internet connection to a varied series of strategies and equipment that use the entrenched mechanism to interconnect and interrelate with the exterior atmosphere. IoT has been accepted by numerous request parts, for instance, distant detecting, actual congestion observing, climate observing, army observation,

K. Kavitha (✉)
Research Scholar, Department of Computer Science, School of Computing Sciences, VELS Institute of Science Technology & Advanced Studies (VISTAS), Chennai, India

Assistant Professor, Chellammal Women's College, Chennai, India

G. Suseendran
Assistant Professor, Department of Information Technology, School of Computing Sciences, VELS Institute of Science Technology & Advanced Studies (VISTAS), Chennai, India

S.-L. Peng et al. (eds.), *Intelligent Computing and Innovation on Data Science*,
Lecture Notes in Networks and Systems 118,
https://doi.org/10.1007/978-981-15-3284-9_38

health care, etc. [1]. The motive of IoT is to generate a dispersed system of intellectual radar nodules that is able to calculate several factors to handle the urban further proficiently. By the fast growth of radio frequency identification (RFID), publics and substances in the bodily creation are armed with all types of radars and wireless devices to validate their individuality and position [2]. At present, wireless sensor network (WSN) has turned out to be a permitter technology for the IoT solicitations that spread the somatic extent of the observing ability. In IoT-centred WSN, the elementary problems are apprehensive with the tool to decrease the dynamic ingestion of nodules that will consequence in extending the lifespan of the nodules [3, 4]. Many electrical expedients comprise radars which produce numerous statistics. These statistics can be utilised to afford beneficial data in actual. These collected radar statistics can be utilised for data mining [5].

1.1 Problem Identification

In general, the challenges need to be addressed in data collection and delivery in IoT sensor network are: buffer management and storage, consistency of sensor data, sensor heterogeneity, processing bottlenecks and transmission overheads and reducing data redundancy. But maintaining the data consistency and reducing the redundancy have a trade-off, that is while trying to increase data consistency, redundancy will be increased and vice versa.

In data distribution agenda [6], the identified statistics is spread and simulated through the system to lessen the possibility of statistics harm. This can be completed by allocating the statistics to the designated adjacents which have advanced obtainable retention and dynamism ranks. On the other hand, this presents enormous stowage overhead, as every single package has to be simulated numerous times.

In CG-E2S2 [7], the ideal snooze period of IoT devices and accumulation period in gateway (GW) are together resolved depending on the kind of the IoT device. The IoT GW categorises statistics along with IoT manoeuvres and agrees whether to combine statistics or to transfer to the solicitation server. Conversely, this method does not confirm the dependability and steadiness of radar statistics.

2 Reliable and Consistent Data Collection Framework (RCDC)

2.1 Overview

This work aims to design a Reliable and Consistent Data Collection Framework for IoT-based WSN. In this framework, a set of candidate nodes are selected based on the energy eligibility factor and buffer space availability. Once the data is sensed at time interval t, it will be transmitted to the selected candidate node based on the absolute differential value. If the packet error rate at the sink node is more than the threshold value, then the source will decide to send the replicated data to a selected set of candidate nodes.

2.2 Derivation of Various Metrics

2.2.1 Energy Eligibility Factor

The energy eligibility factor (EF_{Ri}) is defined as:

$$\mathrm{EEF}_{\mathrm{Ri}} = \frac{E_{\mathrm{res},i}}{E_{\mathrm{init},i}} \tag{1}$$

where $E_{\mathrm{res},i}$ is the available battery capacity of node N_i (in Joules) and $E_{\mathrm{init},i}$ is the initial battery capacity of N_i (in Joules)

2.2.2 Absolute Differential Value

The absolute differential value (D) is defined as the relative difference between the present and previously obtained measurements, which is by

$$\mathrm{ADF} = \begin{cases} If\ A = \left|Q_t - Q_{(t-1)}\right| > \alpha, & 1 \\ \text{Otherwise}, & 0 \end{cases} \tag{2}$$

where

α	threshold value
Q_t and $Q_{t\text{-}1}$	current and previous sensor measuring value.

2.2.3 Packet Error Rate

The following equation illustrates the packet error rate (PER)

$$\mathrm{PER}(\rho)\frac{1}{1 + (w_n\rho)^{v_n}} \quad \forall\rho \geq 0 \tag{3}$$

where v_n and w_n are parameters that depend on AMC mode (adaptive modulation and coding) and packet size, ρ—signa- to-noise plus interference ratio (SNIR).

2.3 Candidate Node Selection

Let N_i be the sensor node, $i = 1,2\ldots.N$. Let (Ne_i_L) be the one-hop neighbours list of node $N_{i.}$ Let EEF_{th} and BA_{th} are the threshold values of EEF and BA, respectively. The COLLECT message contains the fields node id, seq no, EEF_{Ri} and BA.

The steps involved in the selection of candidate nodes are.

Algorithm: Candidate Node selection

1. Each N_i of the network broadcast a COLLECT message to its one-hop neighbours

$$\mathrm{Ni} \xrightarrow{COLLECT} \mathrm{One-Hop\ Neighbos}$$

2. Each node estimates its EEF and BA using (1) and replies back to N_i
3. Based on the reply to the COLLECT message, N_i maintains Ne_i_L.
4. For each $N_j \in Ne_i\_L$
5. If $EEF_{Ri}(N_j) > EEF_{th}$ and $BA(N_j) > BA_{th}$, then
6. Select N_j as the candidate node CN_j
7. Else
8. If $EEF_{Ri}(N_j) > EEF_{th}$ or $BA(N_j) > BA_{th,}$ then
9. Select N_j as the candidate node CN_j
10. End if
11. End For

In this algorithm, from the one-hop neighbours of node N_i, the nodes having higher EEF and BA are selected. If no such node is present, then the nodes with either higher EEF or BA are selected.

2.4 Data Collection

Let DF_{min} and DF_{max} be the minimum and maximum threshold values of ADF. Let PER_{th} be the threshold value of PER. Let S_j and D be the source and sink node. Let K_j be the number of candidate nodes for S_j

Algorithm: Redundancy and Consistency Check

1. S_j senses the data $D_j(t)$ at time interval t
2. S_j estimates $ADF_j(t)$ using Eq. (2)

3. If ($ADF_j(t) < DF_{min}$), then
   Data is considered as redundant and will not be transmitted.
4. Else If ($DF_{min} < ADF_j(t) < DF_{max}$) then
   Data is considered as non-redundant and consistent.
5. Else If ($ADF_j(t) > DF_{max}$) then
   Data is considered as an outlier or error and will be dropped.
6. End if
7. If the data is consistent, then S_j will transmit $D_j(t)$ to its nearest CN_i.

$$S_j \xrightarrow{data} CN_i$$

8. If CN_i receives $D_j(t)$ from all S_j, then
9. CN_i aggregates $D_j(t)$ to AggD(t)
10. CN_i transmits AggD(t) to CN_{i+1} towards D

$$CN_i \xrightarrow{\sum data} D$$

11. End if
12. If D receives AggD(t) from all CN_i, then
13. D retrieves $D_j(t)$ from AggD(t)
14. D estimates $PER_j(t)$ using Eq. (3)
15. If $PER_j(t) > PER_{th}$, then
16. D broadcast an error message with PER_j to S_j.

$$D \xrightarrow{Error} S_j$$

17. End if
18. End if
19. If S_j receives error message from D, then
20. S replicates $D_j(t)$ into K_j times
21. For each CN_i, $i = 1,2\ldots K_j$
22. S transmits $D_j(t)$ to CN_i
23. CN_i forwards it to D
24. End For
25. End if

Note: The amount of replication (i.e. value of N) will be decided based on PER at the sink.

3 Experimental Results

3.1 Experimental Settings

The simulation of RCDCF is conducted in NS-2, and it is compared with content-centric networking (CCN) [1]. The performance is measured with respect to packet delivery ratio (PDR), packet drop, average residual energy (RE) and data accuracy. The number of IoT sensors is varied from 20 to 100 with a central coordinator. The size of the topology is fixed as 50 m × 50 m. The IEEE 802.15.4 MAC protocol is used. There are four CBR and exponential traffic flows with transmission rate of 50 Kbps.

3.2 Experimental Results

This section presents the results for varying the nodes from 20 to 100.

The graph showing the results of PDR is shown in Fig. 1. The figure depicts that the PDR of RCDCF ranges from 0.82 to 0.75 and PDR of CCN ranges from 0.73 to 0.61. Ultimately, the PDR of RCDCF is 15% high when compared to CCN.

The graph showing the results of residual energy is shown in Fig. 2. The figure depicts that the residual energy of RCDCF ranges from 8.2 to 7.2 J and residual energy of CCN ranges from 7.6 to 6.7. Ultimately, the residual energy of RCDCF is 5% high when compared to CCN.

Fig. 1 PDR for varying the nodes

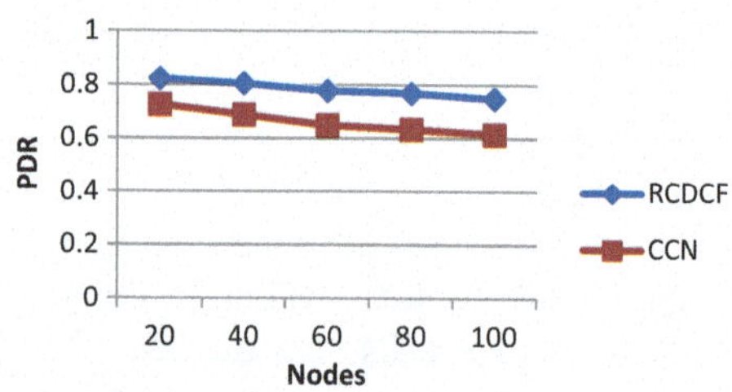

Fig. 2 Residual energy for varying nodes

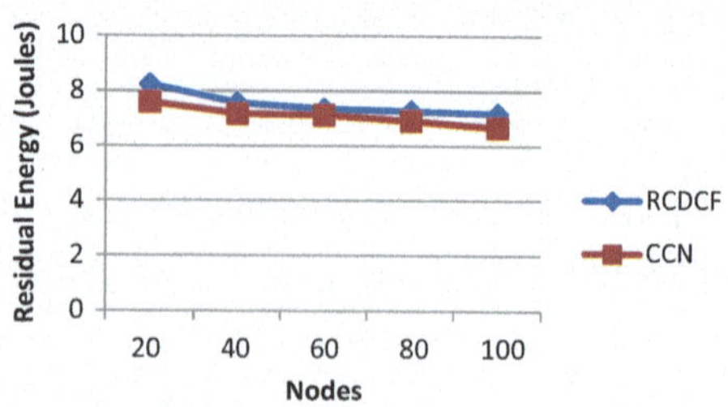

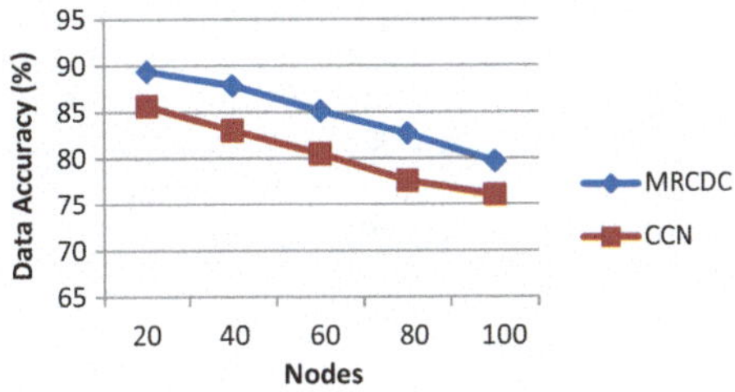

Fig. 3 Data accuracy for varying nodes

The graph showing the results of data accuracy is shown in Fig. 3. The figure depicts that the overhead of RCDCF ranges from 89 to 79% and the accuracy of CCN ranges from 85 to 76%. Ultimately, the overhead of RCDCF is 5% high when compared to CCN.

4 Conclusion

An RCDC Framework has been designed for IoT sensor networks. In this framework, a set of candidate nodes are selected based on the energy eligibility factor and buffer space availability. By simulation results, we have shown that the proposed technique checks both the consistency and redundancy of data thereby solving the trade-off. It also reduces the amount of replicated data. Since the candidate nodes are selected based on energy and buffer space values, the reliability of data is ensured at the cost of reduced energy consumption.

References

1. Bosunia MR, Hasan K, Nasir NA, Kwon S, Jeong S-H (2016) Efficient data delivery based on content-centric networking for Internet of Things applications. Int J Distrib Sens Netw 12(8)
2. Zhang Q, Huang T, Zhu Y, Qiu M (2013) A case study of sensor data collection and analysis in smart city: provenance in smart food supply chain. Int J Distrib Sens Netw Vol 2013, Article ID 382132, p 12
3. Alduais NA, Abdullah J, Jamil A, Audah L (2016) An efficient data collection and dissemination for IOT based WSN. In: IEEE 7th annual information technology, Electronics and Mobile Communication Conference (IEMCON), Canada
4. Plageras AP, Psannis KE, Stergiou C, Wang H, Gupta BB (2018) Efficient IoT-based sensor BIG data collection-processing and analysis in smart buildings. Future Gener Comput Syst 82:349–357
5. Song J, Kim K, Lee M (2017) Implementation of an IoT sensor data collection and analysis library. Int J Comput Syst Eng 11(12):1324–1328

6. Amrutha S, Mohanraj T, Chakrapani Ramapriya N, Sujatha M, Ezhilarasie R, Umamakeswari A (2016) Data dissemination framework for IoT based Applications. Indian J Sci Technol 9(48). https://doi.org/10.17485/ijst/2016/v9i48/108022
7. Ko H, Lee J, Pack S (2017) CG-E2S2: Consistency-guaranteed and energy-efficient sleep scheduling algorithm with data aggregation for IoT. Future Gener Comput Syst. https://doi.org/10.1016/j.future.2017.08.040

A Cloud of Things (CoT) Approach for Monitoring Product Purchase and Price Hike

Muhammad Jafar Sadeq, S. Rayhan Kabir, Rafita Haque, Jannatul Ferdaws, Md. Akhtaruzzaman, Rokeya Forhat, and Shaikh Muhammad Allayear

Abstract Price hike is common and one of the major issues all over the world. The prices of daily necessities are increasing day by day. Many shops, restaurants and transport systems are charging extra price from the customers over the products' expected or maximum retail price. Moreover, in special occasions, such as festivals, the sellers take high price from customers. Furthermore, unauthorized VAT or other taxes are charged on products on which such taxes are exempted by the government. This study proposes a Cloud of Things (CoT)-based model for monitoring the products' price during transactions, where a cloud server maintains historical pricing data and maximum retail prices. Two algorithms run on the server based on live invoice data from sellers and customer feedback to detect unfair price hike.

M. J. Sadeq (✉) · S. R. Kabir · R. Haque · Md. Akhtaruzzaman
Department of Computer Science and Engineering, Asian University of Bangladesh, Dhaka, Bangladesh
e-mail: jafar@aub.edu.bd

S. R. Kabir
e-mail: rayhan923@aub.edu.bd; rayhan561@diu.edu.bd

R. Haque
e-mail: rafitahaque@aub.edu.bd

Md. Akhtaruzzaman
e-mail: azaman01@aub.edu.bd

S. R. Kabir · J. Ferdaws · R. Forhat · S. M. Allayear
Department of Software Engineering, Daffodil International University, Dhaka, Bangladesh
e-mail: jannatul458@diu.edu.bd

R. Forhat
e-mail: rokeya35-1028@diu.edu.bd

S. M. Allayear
e-mail: drallayear.swe@diu.edu.bd

S. M. Allayear
Department of Multimedia and Creative Technology, Daffodil International University, Dhaka, Bangladesh

S.-L. Peng et al. (eds.), *Intelligent Computing and Innovation on Data Science*, Lecture Notes in Networks and Systems 118,
https://doi.org/10.1007/978-981-15-3284-9_39

Keywords IoT · Cloud of Things (CoT) · Price hike · Smart government

1 Introduction

Raising product price or price hike is a problem all over the world. Moreover, monitoring the product or food price is a vital topic of research for reducing the impact of price hike [1]. Changes in sustenance costs have critical effects on the poor. In Vietnam, higher prices of main products such as rice may impact on the poverty level [2]. Rising food price has a greater effect on lower-income countries and the poorer families are the most adversely affected [3]. A study found that 23% were in a state of hunger due to rising food prices in Addis Ababa, Ethiopia [4]. The middle-income families of Sylhet division of Bangladesh are fighting against price hikes for buying daily necessary products [5]. VAT fraud is another problem in regular life [6]. Sometimes supermarkets and restaurant authorities take unlegislated VAT from customers [7, 8]. In this paper, we propose the use of information technology for detecting unfair price hikes using a smart Cloud of Things (CoT)-based system [9].

At present, 'Smart Governance' is a domain of study that describes the use of IT for the interaction between a government and its citizens [10]. A recent study has disclosed an IoT-based Smart Government approach for adopting technologies in the public sectors [11]. A digital platform has been proposed where different government services are delivered to the poor [12]. Besides, social innovation has should be the main aspect of long-term solutions that involve a smart nation [13]. Moreover, nowadays, cloud storage and IoT technology have become a strategic direction for many e-government sectors [14]. An experiment has introduced the concept of IoT technology for monitoring agricultural products' safety [15]. The National Institute of Food and Agriculture (an agency of the United States Department of Agriculture) describes the importance of cloud server across various governmental organizations and food industries [16]. In this vein, the development of a cloud server-based model is described in this paper where a smart application can monitor price hike.

In previous research, a conceptual blockchain and IoT-based model have been demonstrated for monitoring corruption [17], where a cloud server maintains the transaction history between buyers and sellers. Based on that model, a price hike monitoring system (PHMS) is proposed in this paper, containing two algorithms: (1) price difference identification (PDI) algorithm and (2) price hike identification (PHI) algorithm. Mid-Level networking happens between customers' and sellers' devices for uploading purchase information into the cloud server. Three types of datasets have been used: (1) supermarket data, (2) restaurant data and (3) transport ticket data. These datasets have been collected from MerinaSoft, a company that provides inventory management systems, restaurant management systems and travel agency management systems among different sellers and has stored the data in a cloud server. The proposed algorithms are applied to this data to detect price hikes.

2 Proposed Model for PHMS

In the proposed model, for identifying the price hike, the customer and seller are connected to a cloud server that monitors the product prices at the moment of transaction. A mid-level network between mobile or any other computing system used by the seller uploads invoice price information into the cloud. The cloud server informs the customer of any problems in the invoice and the customer can also give feedback. Any issues detected by the algorithms are alerted to a monitoring team. The steps of the proposed model are demonstrated are as follows:

1. The cloud server contains the customer and seller information such as ID, name and mobile number. Customers and sellers are connected to the cloud through a smart application. The cloud server also contains dataset that has different product price information.
2. The seller's system has preset information of the seller's product prices. Some sellers may also input their selling price at the moment of transaction, including details such as product, quantity, price and VAT. In addition, the seller inputs the customer ID. All this information is communicated to the cloud server.
3. The cloud stores the recent price of all products and VAT information. Price difference identification (PDI) (Algorithm 1) compares the sales price and VAT with the historical price and authorized VAT. The PDI algorithm attempts to detect extra price or unauthorized VAT.
4. After the PDI algorithm execution, the customer gets a message or notification from the server.
5. The customer can give positive or negative feedback.
6. Once the transaction is completed by confirmation from both buyer and seller, price hike identification (PHI) algorithm attempts to detect the price hike by acquiring extra price, VAT from PDI algorithm and customer feedback.
7. Finally, the price hike monitoring team is alerted to any suspected price hike.

3 System Architecture of PHMS

This section gives details of the architecture, showing the data flow diagram (DFD) and class diagram of PHMS. Figures 1 and 2 show the DFD, class diagram and architecture for PHMS.

The DFD shows that the PDI and PHI algorithms are two functions that operate in the cloud server. The invoice information is fed to these functions by the seller. The PDI algorithm compares the seller's information to the existing information in the product and VAT databases. After processing the information, the PDI algorithm gives an output notification to customer through notification database. The PHI algorithm uses invoice information from the seller, the customer feedback and information from the PDI to determine whether price hike occurred. The output of

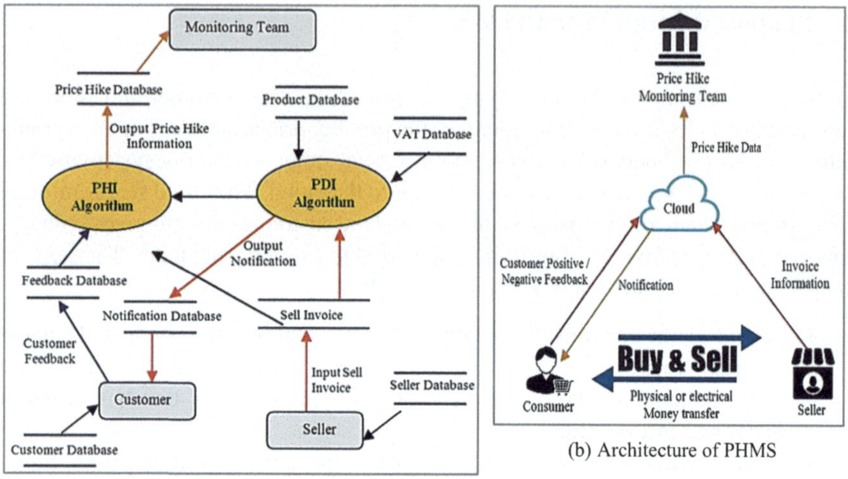

(a) Data Flow Diagram (DFD) of PHMS

(b) Architecture of PHMS

Fig. 1 Data flow diagram and architecture of PHMS

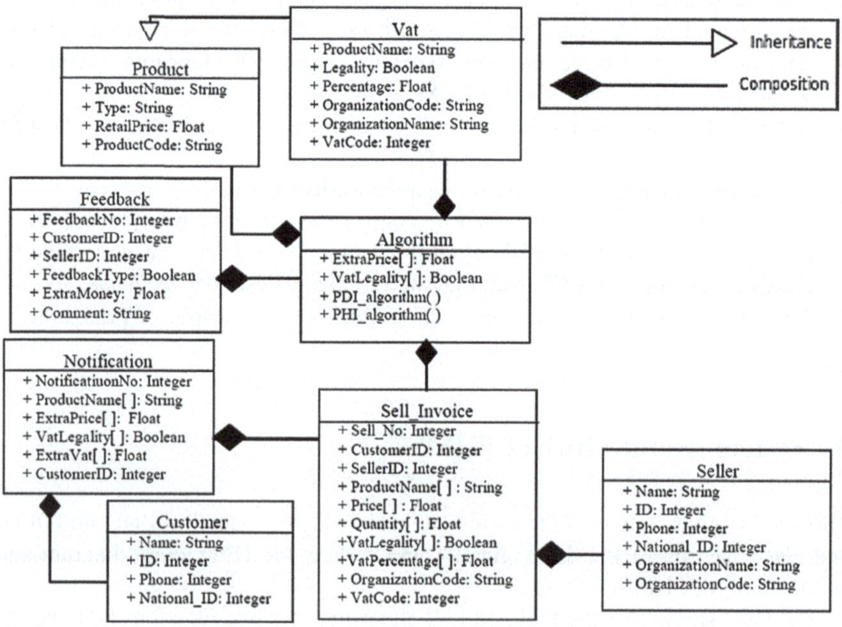

Fig. 2 Class diagram of PHMS

the PHI algorithm function goes to the monitoring team for viewing the price hike. At the proposed class diagram, the *Algorithm* class contains *PDI_algorithm()* and *PHI_algorithm()* functions (see Algorithms 1 and 2). The algorithm class has composition relationship with five classes: *Sell_Invoice, Product, Vat, Notification* and *Feedback*. Besides, for acquiring VAT data, *Vat* class has inherited the *Product* class. The *Sell_Invoice* class gets information from the *Seller*, *Customer* and *Notification* classes.

4 PDI and PHI Algorithms

4.1 Price Difference Identification (PDI) Algorithm

The PDI algorithm's purpose is to identify the extra price and VAT of different products. Additionally, it determines VAT legality in the seller invoice. It then informs the customer and the PHI algorithm of its results. The structure and formation of PDI algorithm are given below:

- Three variables (*ExtraPrice, VatLegality* and *ExtraVat*) are declared.
- Three objects (sell, *product* and *vat*) are created for calling *Sell_Invoice, Product* and *Vat* classes (see Fig. 2). *CustomerID* and *SellerID* variables are declared and set from *Sell_Invoice* class by using the *sell* object.
- For each product in the invoice, price, VAT legality and VAT amount are compared between the invoice and the *Product* and *Vat* classes. The feedback for each product is sent to the customer.

```
Algorithm 1: PDI Algorithm
PDI_algorithm( ) :
   float ExtraPrice[ ], ExtraVat[ ];
   boolean VatLegality[ ];
   Sell_Invoice  sell  =  new  Sell_Invoice( );
   Product product  =  new  Product( );
   Vat vat  =  new  Vat( ); /*See Fig.2 Class Diagram */
   int CustomerID = sell.CustomerID;
   int SellerID = sell.SellerID;
   for (i = 0; i<sell.ProductName.length; i++)
     if(sell.Price[i] > product.RetailPrice) then,
         ExtraPrice[i] = sell.Price[i] - product.RetailPrice;
     else ExtraPrice[i] = 0;
     if(sell.VatLegality[i] != Vat.Legality) then,
        VatLegality[i] = false;
     else VatLegality[i] = true;
     if(sell.VatPercentage[i] > Vat.Percentage) then,
        ExtraVat[i] = sell.VatPercentage[i]-Vat.Percentage;
     else ExtraVat[i]  = 0;
   end for
   Notification Database←CustomerID,SellerID,ExtraPrice[],
    VatLegality[],ExtraVat[];
```

4.2 *Price Hike Identification (PHI) Algorithm*

Identification of price hike is important and the PDI algorithm's purpose is to identify the price hike from product purchase. Customer feedback along with the extra price and extra VAT are used to recognize price hikes. The PHI algorithm is given below:

Algorithm 2: PHI Algorithm

```
PHI_algorithm( ) :
    boolean PriceHike,FeedbackType, VatLegality[];
    float ExtraPrice[], ExtraVat[], ExtraMoney;
    String Comment;
   ExtraPrice[],VatLegality[],ExtraVat[] ← Notification Database;
   Feedback feedback  =  new  Feedback ( );
   for (i= 0; i <ExtraPrice.length; i++)
       if(ExtraPrice[i] > 0) then, PriceHike= true ;
        else if(VatLegality[i]=false) then, PriceHike=true;
       else if(ExtraVat[i]>0) then, PriceHike = true;
   end for
   if(FeedbackType == false) then,
       Comment = feedback.comment;
       PriceHike = true;
   else if (ExtraMoney > 0) then,
       ExtraMoney = feedback.ExtraMoney;
       PriceHike= true;
   if(PriceHike==true)then, Database ← ExtraPrice[],
               VatLegality[], ExtraVat[],Comment,ExtraMoney;
   else print ("Price Hike not found") ;
```

Descriptively, the PHI algorithm is as follows:

- Seven variables (*FeedbackType*, *PriceHike, ExtraPrice, VatLegality, ExtraVat, Comment* and *ExtraMoney*) are declared for identifying price differentiation. The values for the variables are acquired from the *Notification Database*, where the PDI algorithm had sent the data and the customer has sent feedback.
- For each product, the price hike is checked using the product's base price and the VAT.
- The feedback of the customer is taken into account regarding whether the seller has made any extra demands that are not found in the product price, and also the customer's own perception of whether any increase in price is justified.
- If the PHI algorithm finds any price hike it sends all the data of the transaction to the *Price Hike Database* so that the monitoring team can review.

5 Results and Findings

The dataset used for testing PHMS does not include customer feedback, so some test cases of customer feedback were added. With regards to measuring price hike, a baseline price was required, for which the average historical price was used from the data, which was taken to be the maximum retail price (MRP). Based on this, results of some test cases are shown in Tables 1, 2 and 3. Table 1 shows a sample dataset of purchased products in a supermarket, where the seller takes extra price from sugar and extra VAT from milk. Firstly, the PDI algorithm identifies the Extra Price, VAT Legality and Extra VAT that are shaded in the table. Then, PHI algorithm identifies

Table 1 Test case 1: identification of price hike from shop data

Test case 1: Price hike from shop data								
Sell invoice				Retail price	VAT		Customer feedback	
Serial	Product name	Price	Vat.%	Price	VAT legality	Vat.%	Positive/negative feedback	Unauthorized extra money
1.	Sugar	70	4%	60	True	4%	True	N/A
2.	Tea	40	4%	40	True	4%	False	N/A
3.	Milk	250	6%	250	True	4%	False	N/A
PDI algorithm				PHI algorithm (*PriceHike* is **Boolean**type; True means identify price hike)				
Serial.	Extra price	VAT legality	Extra VAT	Extra price	VAT legality	Extra VAT	Positive/negative feedback	Unauthorized extra money
1. (Cont.)	10	True	0%	True			True	
2. (Cont.)	0	True	0%					
3. (Cont.)	0	True	2%		True	True		

Table 2 Test case 2: identification of price hike from restaurant data

Test case 2: Price hike at restaurant								
Sell invoice				Retail price	VAT		Customer feedback	
Serial	Product name	Price	Vat.%	Price	VAT legality	Vat.%	Positive/negative feedback	Unauthorized extra money
1.	Burger	350	2%	N/A	False	N/A	True (positive)	N/A
2.	Pepsi	240	2%	N/A	False	N/A		N/A
PDI algorithm				PHI algorithm (*PriceHike* is **Boolean**type; True means identify price hike)				
Serial.	Extra price	VAT legality	Extra VAT	Extra price	VAT legality	Extra VAT	Positive/negative feedback	Unauthorized extra money
1. (Cont.)	0	False	2%		True	True		
2. (Cont.)	0	False	2%		True	True		

the price hikes that are also shaded. Note that due to positive customer feedback, the extra price of sugar is appended with a special note that positive customer feedback was received.

Table 2 shows purchase data of a restaurant, where the sellers have not been authorized to charge VAT but are doing so. Thus, the PHMS detects a price hike due to unauthorized VAT. Table 3 shows price hike detected in transport data due to the customer feedback of extra money being taken.

Table 3 Test case 3: identification of price hike from transport tickets

Test case 3: price hike at transport ticket								
Sell invoice				Retail price	VAT		Customer feedback	
Serial	Product Name	Price	Vat.%	Price	VAT legality	Vat.%	Positive/negative feedback	Unauthorized extra money
1.	Dhaka to Natore ticket	500	N/A	500	N/A	N/A	False (negative)	200
PDI algorithm				PHI algorithm (*PriceHike* is **Boolean**type; True means identify price hike)				
Serial.	Extra price	VAT legality	Extra VAT	Extra price	VAT legality	Extra VAT	Positive/negative feedback	Unauthorized extra money
1. (Cont.)	0	False	0				True	True

6 Conclusion and Future Works

Raising product and food cost or price hike is an issue throughout the world. Identifying whether prices have been raised unfairly would be extremely helpful to ameliorate increasing poverty in many places. This paper has proposed a model for a price hike monitoring system that uses a Cloud of Things (CoT)-based system to identify the price hike in product bases prices and VAT. It is hoped that this model will contribute to poverty alleviation and sustainable development. In future work, the maximum retail price MRP should be designed as a combination of a flat government/producer price for some products and a statistical price based on historical data for other products. The effect of customer feedback may be varied, allowing margins of price to be dynamically adjusted based on store, product, historical data, etc. Additionally, detailed feedback may be analyzed through natural language processing and machine learning to handle more complex real-world cases.

References

1. Kane GQ, Piot-Lepetit I, MabahTene GL, Ambagna JJ (2019) Managing the impact of food price rise in Sub-Saharan Africa. In: Farazmand A (eds) Global Encyclopedia of Public Administration, Public Policy, and Governance
2. Ivanic M, Martin W (2008) Implications of higher global good prices for poverty in low-income countries. Agric Econ 39(1):405–416
3. Rosemary G, Laura C et al (2013) The effect of rising food prices on food consumption: systematic review with meta-regression. BMJ 346:f3703
4. Birhane T, Shiferaw S, Hagos S, Mohindra KS (2014) Urban food insecurity in the context of high food prices: a community based cross sectional study in Addis Ababa Ethiopia. BMC Public Health 14:680

5. Latif MA, Hanif M (2016) Impact of price hike on the standard of living of middle income people: a study on sylhet city Bangladesh. Manag Stud Econ Syst 2(4):279–286
6. Report VAT fraud. GOV.UK. https://www.gov.uk/report-vat-fraud
7. Restaurant Owners Want 20 percent VAT on Gross Profit. bdnews24.com (2006)
8. VAT Hike on Superstores 'A Discriminatory Barrier to Expansion'. (2018) Dhaka Tribune
9. Farahzadi A, Shams P, Rezazadeh J, Farahbakhsh R (2018) Middleware technologies for Cloud of things- a survey. Digital Commun Netw 4(3):2389–2406
10. Pereira GV, Parycek P, Falco E, Kleinhans R (2018) Smart governance in the context of smart cities: a literature review. Inf Polity 23(2):143–162
11. Kankanhalli A, Charalabidis Y, Mellouli S (2019) IoT and AI for smart government: a research agenda. Gov Inf Q 36(2):304–309
12. Mukhopadhyay S, Bouwman H, Jaiswal MP (2019) An open platform centric approach for scalable government service delivery to the poor: the aadhaar case. Gov Inf Q 36(3):437–448
13. Kar AK, Ilavarasan V, Gupta MP et al (2019) Moving beyond smart cities: digital nations for social innovation & sustainability. Inf Syst Frontiers 21(3):495–501
14. Clohessy T, Acton T, Morgan L (2014) Smart City as a Service (SCaaS): a future roadmap for E-Government smart city cloud computing initiatives. In: IEEE/ACM 7th International Conference on Utility and Cloud Computing, pp 836–842
15. Ping H, Wang J, Ma Z, Du Y (2018) Mini-review of application of IoT technology in monitoring agricultural products quality and safety. Int J Agric Biol Eng 11(5):35–45
16. Onyegbula F, Dawson M, Stevens J (2011) Understanding the need and importance of the cloud computing environment within the national institute of food and agriculture, an agency of the united states department of agriculture. J Inf Syst Technol Plan 4(8):17–42
17. Akhtaruzzaman M, Kabir SR et al (2018) A combined model of Blockchain, price intelligence and IoT for reducing the corruption and poverty. In: 5th international conference on poverty and sustainable development

Hyperspectral Image Classification by Means of Suprepixel Representation with KNN

D. Akila, Amiya Bhaumik, Srinath Doss, and Ali Ameen

Abstract In real-world application, especially in remote sensing based on image processing hyperspectral imaging (HSI) shows promising results. Superpixel-based image segmentation is the powerful tool in hyperspectral image processing. Series of neighboring pixels composes superpixel which may belong to different classes but can be regarded as homogenous region. Extraction of more representative feature is considered to be most important thing in hyperspectral imaging. Training and testing samples that are more representative are found by proposing a new method for selecting two *k* values for representing optimal superpixels. This paper starts with superpixel shifting as first step and followed by KNN classifier. Which is performed by pixels with minimal spectral features in HSI are clustered together in the same superpixel. Followed by spatial-spectral feature is extraction by a domain transformation from spectral to spatial. For each superpixel, training and test samples are selected to eliminate classification within the same class. An average distance between test and training samples are used for determining class label. Finally, by the results from most common hyperspectral images Indian pines, Salinas, Pavia show that this method shows a better classification performance.

Keywords Hyperspectral image classification · Superpixel segmentation · K-nearest neighbor classification (KNN)

D. Akila (✉)
Department of Information Technology, Vels Institute of Science, Technology and Advanced Studies (VISTAS), Chennai, India
e-mail: akiindia@yahoo.com

A. Bhaumik
Faculty of Business and Accounting, Lincoln University College, Kota Bharu, Malaysia
e-mail: amiya@lincoln.edu.my

S. Doss
Faculty of Computing, Botho University, Gaborone, Botswana
e-mail: srinath.doss@bothouniversity.ac.bw

A. Ameen
Faculty of Computer Science and Multimedia, Lincoln University College, Kota Bharu, Malaysia
e-mail: abdulbaqi@lincoln.edu.my

S.-L. Peng et al. (eds.), *Intelligent Computing and Innovation on Data Science*,
Lecture Notes in Networks and Systems 118,
https://doi.org/10.1007/978-981-15-3284-9_40

1 Introduction

In the field of remote sensing, there is an endless development of spectral resolution which has been achieved by spontaneous research in hyperspectral imaging. For earth observation and space exploration, hyperspectral images play a major role. Number of spectral bands is extended to several hundreds. With their reflectance the material present in the pixel detected. The collected information are recorded by the hyperspectral sensors by series of images. The scene of observation's reflected solar radiation provides the spatial distribution. Narrow and contiguous spectral bands image acquisition are made possible only through rapid development in hyperspectral imaging and remote sensor technology. Reliable and rich spectral information can be given by HSI and has been used widely in remote sensing field. Extra computing burden and Hughes phenomena are felt over high dimensionality of hyperspectral imaging. Chaos theory and Manifold learning are said to be some classical method for nonlinear HIS dimension reduction used for feature extraction in HSI. Most of these methods perform classification of HIS images only based on spectral information which may not give satisfactory results.

Nowadays, in feature extraction process, some spatial-spectral feature-based classification methods have been carried out that uses contextual information into account for classification process. For HSI classification process, one of a popular method to be used is KNN classifier. A label is assigned simply to the test sample that occurs most frequently in its KNN so it is said to be nonparametric classifier. The inter-sample relationship cannot be accurately described in the input by a solo Euclidean distance subspace which has been implied by Manifold [1]. The following major steps are followed in the proposed method for KNN representation of superpixels (KNNRS). On the first three principle components of hyperspectral, an entropy rate superpixel (ERS) segmentation technique is executed initially. Secondly for extracting the spectral-spatial features, a filtering method based on edge preserving has been adopted effectively to remove the texture and noise removal from the image. For different superpixels, different numbers of training and test samples can be selected by using KNN method. Existence of pixels that belong to dissimilar classes may occur in one superpixel region is the main problem addressed which is main objective of the paper. With minimal distance, the superpixel region is assigned with label by calculating the average distance among the selecting testing and training samples.

2 Literature Survey

Benediktsson et al. [2] described that for high-spatial resolutions urban areas, hyperspectral data classification has been concerned. Mathematical morphology has been used for preprocessing of hyperspectral images. For separation of bright and dark structures in the proposed approach, morphological opening and closing functions have been used. For original image starting with single, by using repetitive use of opening and closing structure elements that have high dimensions a morphological profile was created. Wang et al. [1] designate that for terrestrial laser scanning point clouds classification in a precise and efficient manner of a significant necessity is to mine shape features valuably. It is still a challenging task to analyze noisy and unstable TLS point and get robust and discriminative features. For classification of TLS point clouds for the cluttered and urban scenes, hierarchical and multi-scale outline were presented by the authors. Multi-scale and hierarchical point clusters (MHPC) were discussed by the authors.

Lefevre, Aptoula, Courty, [3] conveys by means of manifold learning and morphological features a new method of spectral classification of hyperspectral images was presented by the authors. The investigational indications were earlierly on the attention of class-wise orderings were followed and the issue was mentioned. A superior performance model has been demonstrated by the authors by comparing extended morphological profiles Pavia dataset to the proposed method.

3 Related Work

A. *Feature extraction using recursive filtering (RF)*

The domain transform recursive filter (DTRF) by preserving sharp edges and boundaries eliminates noise and texture in an image and it is a real-time EPF. Transformation of a signal to domain transformed signal

$$Z_m = I_0 + \sum_{m=1}^{n}(1 + \frac{r_s}{r_r}|I_m - I_{m-1})$$

I_m = mth input signal; Z_m = mth domain transformed signal; r_s = spatial parameter of the filter; r_r = range parameter of the filter; pixels that lie on identical flank of sturdy edge and those lie on different sides of strong edge are nearby and far coordinates, respectively, in the transformation domain. Then comes the processing of transformed signal by RF as follows

$$K_m = \left(1 - f^l\right)I_m + f^l K_{m-1}$$

where f^l = feedback coefficient; l = measure of distance between two neighbor samples; K_m = transform domain by which the signals sharp edges are preserved the propagation chain will be stopped as increase in l results in making f^l zero. Output from the *mth* signal is filtered by K_m.

$$f = \exp\left(\frac{-\sqrt{2}}{r_x}\right)\epsilon[0, 1]$$

On each dimension, a one-dimensional filtering should be performed on every two-dimensional image. By performing one-dimension filtering on images with three iterations shows satisfactory results [4].

For processing images 1D field transformation RF is agreed for three iterations.

B. *Superpixel image segmentation*

In computer vision based on graph theory the superpixel segmentation algorithms have been widely used [5]. Based on different spatial structures the size and shape of each superpixel can be adaptively changed. The number of superpixels can be given by the following equation

$$I = \bigcup_{i=0}^{N} Y_i, \, and \, Y_i \cap Y_j = \emptyset, \, (i \neq j)$$

Y_i = ith superpixel; with ERS segmentation algorithm the superpixels are generated. Initially, the graph $G = (V, E)$ was constructed where V and E represent the corresponding to pixels of image by which adjacent pixels pairwise similarities can be measured. The original graph is divided into N connected subgraphs by selecting subset of edges $M \subseteq B$. A balancing term $B(N)$ and an entropy rate term $E(N)$ were introduced to obtain homogenous and compact superpixels into the superpixel segmentation [6].

$$E(N) = -\sum_m \mathrm{sd}_m \sum_n p_{m,n}(N) \log\left(p_{m,n}(N)\right)$$

$$B(N) = E(C_N) - CC_N = -\sum_i PC_N(i) \log(PC_N(i)) - CC_N$$

where $p_{m,n}$ = transition probabilities; sd_m = stationary distribution; CC_N = the number of connected components in the graph;C_N = cluster membership distribution; The superpixel segmentations detached function is presented as follows:

$$\max N\{E(N) + \mathrm{sd}B(N)\} \text{ subject to } M \subseteq B$$

Entropy rate term and balancing contributes the weight controlling by sd > *0*. To solve the optimization problem efficiently, a greedy algorithm has been used.

4 Proposed System

In this paper, a classification method based on KNNRS with spectral-spatial has been introduced. This method consists of four parts

- Hyperspectral image partitioning into superpixels;

- Feature extraction of HSI by domain transform RF;
- For each superpixel, the training and testing samples that are most representative are selected by using the KNN;
- Based on decision function decision function the class labels of superpixels are obtained.

Tessellation of an image into "superpixels" has become a basic thing for many kinds of object recognition, segmentation, etc. In advancement to rectangular patch, the patches are aligned better with edge intensities [7]. The superpixel partitioning problem is formulated with optimized graph cuts and minimized energy framework. Regular superpixels are explicitly encouraged by our energy function [8] (Fig. 1).

Superpixel segmentation was initially introduced by Ren and Malik [9] in which a perceptually meaningful connection of pixels that similar in color or other feature in a group. Many algorithmic approaches are proposed in subsequent year [10, 11].

KNN-based superpixels representation: Pixels with similar structural information are clustered together in the same superpixel which is referred as superpixel segmentation method [13, 14]. Training samples that are more representative for each class is given by selection of k_I by which within class variations are effectively overcome.

Detailed description is given by the following equation.

Training samples $X^i = X_1^i, X_2^i, X_3^i \ldots\ldots..X_j^i$

$$X_j^i = \text{belongs to } jth \text{ class}$$

J = number of training samples.

$$Y_n = y_{n,1}, y_{n,2}, \ldots\ldots y_{n}, k_n$$

k_n = pixels count in nth superpixel

Due to spectral mixing samples belonging to similar session shows spectral variations even due to environmental factors such as cloud and shadow [5]. And the Euclidean distance $E\left(y_{n,i}, X_j^i\right)$ and mean $E_{\text{mean}}\left(y_{n,i}, X^i\right)$ are given by

$$E\left(y_{n,i}, X_j^i\right) = \left\| y_{n,i} - X_j^i \right\|_2^2$$

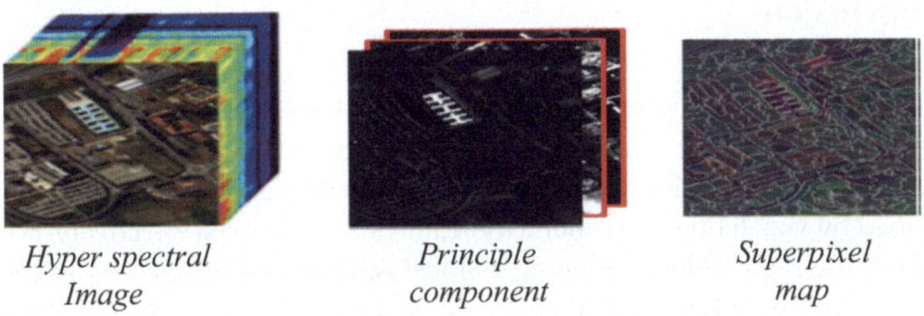

Fig. 1 Input HSI image its principle component and superpixel map

$$E_{\text{mean}}\left(y_{n,i}, X^i\right)\frac{\sum \hat{E}\left(y_{n,i}, X^i\right)}{k_1}$$

Test samples that are most representative in each superpixel is adopted by k_2 selection rule instead of using all the pixels in each superpixel. Distance for discrimination is given by the following equation [12, 16].

$$d\left(y_{n,i}, X^i\right)\frac{\sum \hat{E}\left(Y_n, X^i\right)}{k_2}$$

By K selection method pixels with same distance are in one superpixel region [15]. The resulting distance for the test superpixel Y_n used for discrimination is described as

$$d\left(Y_n, X^i\right) = d\left(y_{n,i}, X^i\right)$$

Labeling based on distance: in this step, the superpixel **Y***n* that gives the minimal distance is assigned with the label as follows:

$$\text{Class}(Y_n) = \arg \min_{i=1,2,....C} d\left(y_{n,i}, X^i\right)$$

5 Result

For this experiment Salinas *C* and Indian pine hyperspectral images have been chosen for classification purpose. Airborne visible/infrared imaging spectrometer sensor was used for acquisition of images. Salinas image has 224 spectral bands with 217 × 512 as pixel size with 3.7 m resolution. The image was acquired in Salinas valley, California at October 8, 1998. Indian pines image has 220 spectral bands with 125 × 125 pixels. Acquired at November 1992 (Figs. 2 and 3: Tables 1, 2 and 3).

6 Conclusion

The work for HSI transformation a KNN-based representation of superpixels (KNNRS) which is a novel method has been proposed. First spectral and special features are extracted by RF by which the complete spatial information in superpixels are used by which Edge and boundary features are enhanced effectively. For each superpixel, the representative training and test samples are selected by KNN algorithm. In the proposed method on real hyperspectral database, higher classification accuracy is obtained in limited number of training samples. The main limitation of

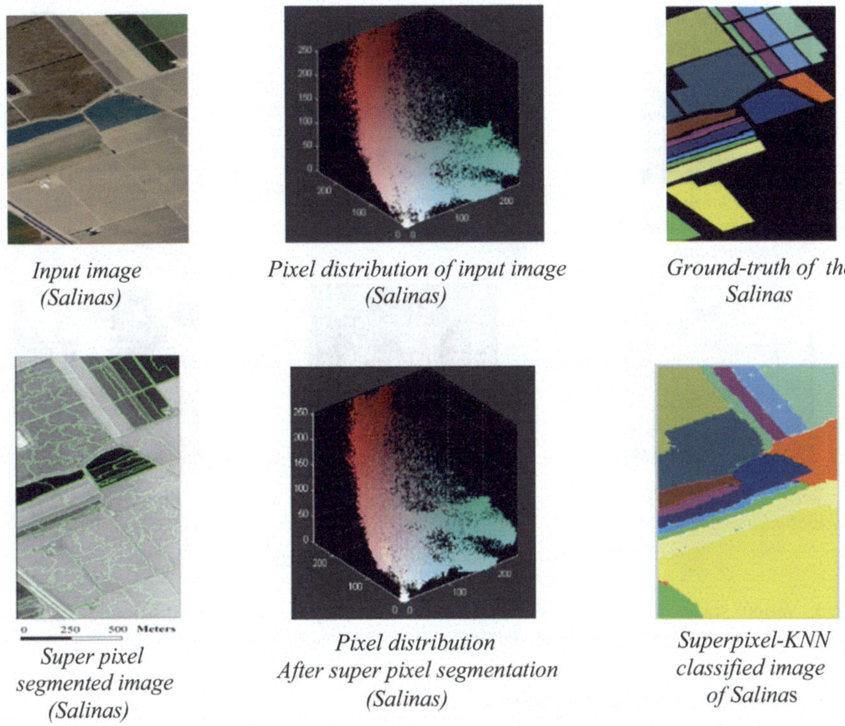

Fig. 2 Various stages of hyperspecral image classification—input image, pixel distribution, ground truth image, superpixel segmentation, and finally classification of Salinas image

this method is the performance of segmentation decides the performance of classification. Classification may not be efficient when pixels of different class lie in same pixel region. Another challenging problem is the computational burden of KNN operation. Still the system shows a better performance compares to state of art methods.

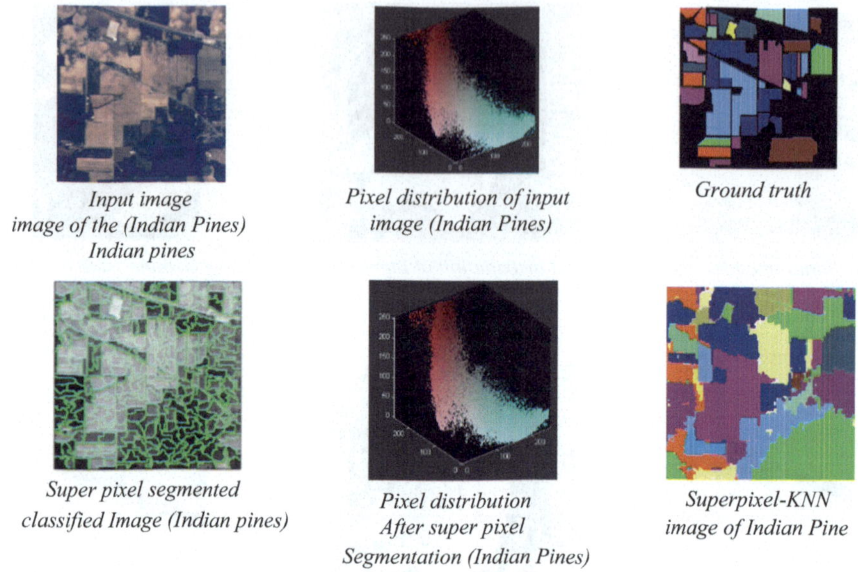

Fig. 3 Various stages of hyperspecral image classification—input image, pixel distribution, ground truth image, superpixel segmentation, and finally classification of Indian pines image

Table 1 Details of Hyperspectral data used

Class	Indian pines	Salinas	Pavia University
	16	16	9
Band	200	204	103
Size	145 × 145	512 × 217	610 × 340
Sensor	AVIRIS	AVIRIS	ROSIS
Resolution (m)	20	3.7	1.3
Sample	10,249	54,129	42,776

Table 2 Train and test samples in Indian pines image

Class	Name	Training		Test	
		2%	0.2%	98%	99.8%
1	Weeds_1	40	4	1969	2005
2	Weeds_2	76	8	3650	3718
3	Fallow	38	4	1938	1972
4	Fallow_P	26	3	1368	1391
5	Fallow_S	52	5	2626	2673
6	Stubble	79	8	3880	3951
7	Celery	70	7	3509	3572

(continued)

Table 2 (continued)

Class	Name	Training		Test	
		2%	0.2%	98%	99.8%
8	Grapes	225	21	11,046	11,250
9	Soil	124	11	6079	6192
10	Corn	21	3	1047	3275
11	Lettuce_4wk	21	3	1047	1065
12	Lettuce_5wk	38	4	1889	1923
13	Lettuce_6wk	18	2	898	914
14	Lettuce_7wk	20	2	1050	1068
15	Vinyard_U	140	13	7128	7255
16	Vinyard_T	36	4	1771	1803
Total		1024	102	53,105	54,027

Table 3 Train and test samples in Indian pines image

Class	Name	Training		Test	
		10%	1%	90%	99%
1	Alfalfa	10	3	36	43
2	Corn_N	143	14	1285	1414
3	Corn_M	83	8	747	822
4	Corn	34	3	203	234
5	Grass_M	48	6	435	477
6	Grass_T	23	7	707	723
7	Grass_p	2	2	26	26
8	Hay_W	28	5	450	473
9	Oats	2	2	18	18
10	Soyabean_N	150	10	822	962
11	Soyabean_M	246	24	2209	2431
12	Soyabean_C	60	6	533	587
13	Wheat	21	2	184	203
14	Woods	127	13	1138	1252
15	Building	35	4	351	382
16	Stones	10	3	83	90
Total		1022	112	9227	10,137

References

1. Wang Z, Zhang L, Fang T, Mathiopoulos PT, Tong X, Qu H, Xiao Z, Li F, Chen D (2015) A multiscaleand hierarchical feature extraction method for terrestrial laser scanning point cloud classification. IEEE Trans Geosci Remote Sens 53(5):2409–2425
2. Benediktsson JA, Palmason JA, Sveinsson JR (2005) Classification of hyperspectral data from urban areas based on extended morphological profiles. IEEE Trans Geosci Remote Sens 43(3):480–491
3. McKeown DM Jr, Cochran SD, Ford SJ, McGlone JC, Shufelt JA, Yocum DA (1999) Fusion of HYDICE hyperspectral data with panchromatic imagery for cartographic feature extraction. IEEE Trans Geosci Remote Sens 37(3):1261–1277
4. Gastal ESL, Oliveira MM (2011) Domain transform for edge aware image and video processing. ACM Trans Graph vol 30, Art. no 69
5. Jiao L, Liang M, Chen H, Yang S, Liu H, Cao X (2017) Deep fully convolutional network-based spatial distribution prediction for hyperspectral image classification. IEEE Trans Geosci Remote Sens 55(10):5585–5599
6. Ma X, Geng J, Wang H (2015) Hyperspectral image classification via contextual deep learning. EURASIP J Image Video Process 2015:20
7. Maierhofer G, Heydecker D, Aviles-Rivero AI, Alsaleh SM, Schonlieb C (2018) Peekaboo-where are the objects? structure adjusting su- ¨ perpixels. In: IEEE International Conference on Image Processing (ICIP)
8. Lu T, Li S, Fang L, Jia X, Benediktsson JA (2017) From subpixel to superpixel: A novel fusion framework for hyperspectral image classification. IEEE Trans Geosci Remote Sens 55(8):4398–4411
9. Ren X, Malik J (2002) A probabilistic multi-scale model for contour completion based on image statistics. In: ECCV'02, vol 1, pp 312–327
10. Achanta R, Shaji A, Smith K, Lucchi A, Fua P, Susstrunk S (2012) SLIC superpixels compared to state-of-the-art superpixel methods. IEEE Trans Pattern Anal Mach Intell 34(11):2274–2282
11. Liu YJ, Yu C, Yu M, He Y (2016) Manifold slic: a fast method to compute content-sensitive superpixels. Proc IEEE Conf Comput Vis Pattern Recog pp 651–659
12. Alzubi JA, Alzubi OA, Suseendran G, Akila D (2019) A novel Chaotic map encryption methodology for image cryptography and secret Communication with steganography. Int J Recent Technol Eng 8(1C2):1122–1128
13. Veganzones MA, Grana M (2012) Lattice auto-associative memories induced supervised ordering defining a multivariate morphology on hyperspectral data. In: 2012 4th Workshop on Hyperspectral Image and Signal Processing: Evolution in Remote Sensing (WHISPERS), pp 1–4, 4–7 June 2012
14. Stutz D, Hermans A, Leibe B (2018) Superpixels: an evaluation of the state-of-the-art. Comput Vis Image Underst
15. Guo X, Huang X, Zhang L, Zhang L, Plaza A, Benediktsson JA (2016) Support tensor machines for classification of hyperspectral remote sensing imagery. IEEE Trans Geosci Remote Sens 54(6):3248–3264
16. Ma X, Wang H, Wang J (2016) Semisupervised classification for hyperspectral image based on multi-decision labeling and deep feature learning. ISPRS J Photogramm Remote Sens 120:99–107

Prediction of Bottom-Hole Pressure Differential During Tripping Operations Using Artificial Neural Networks (ANN)

Shwetank Krishna, Syahrir Ridha, and Pandian Vasant

Abstract Tripping in or out drill string/casing with a certain speed from the wellbore will result in downhole pressure surges. These surges could result in well integrity or well control problems which can be avoided if pressure imbalances are predicted before this operation engaged. To predict these pressure imbalances, number of analytical models have been developed but require time-consuming cumbersome numerical analysis. In this paper, an intelligent model (ANN) is developed which can predict the surge pressure under varying rheological and geometrical parameters. ANN is developed with six neurons in input layer representing six input parameters (pipe velocity, PV, YP, diameter of hole, outer diameter of pipe and mud weight) and one neuron in output layer which represents surge pressure. Now, to find the most optimum neural network structure (number of hidden layer and neurons), total 108 ANN configuration is trained and tested. Performance analysis on these configurations indicates network structure with two hidden layers including ten and 16 neurons in first and second layer, respectively, as the most optimum. Since the selected model is complex, another trained model with one hidden layer containing 14 nodes can be considered due to its satisfactory prediction result. The trained intelligent model can be utilized when tripping operation is carried out in low-pressure margin wells where repetitive calculation of surge/swab pressure is required.

Keywords Surge/swab pressure · ANN · Prediction · Sensitivity analysis · Statistical analysis

---

S. Krishna (✉) · S. Ridha
Petroleum Engineering Department, Universiti Teknologi PETRONAS, Seri Iskandar, Perak, Malaysia
e-mail: shwetank_17007790@utp.edu.my

S. Ridha
e-mail: syahrir.ridha@utp.edu.my

P. Vasant
Fundamental & Applied Science Department, Universiti Teknologi PETRONAS, Seri Iskandar, Perak, Malaysia

S.-L. Peng et al. (eds.), *Intelligent Computing and Innovation on Data Science*, Lecture Notes in Networks and Systems 118,
https://doi.org/10.1007/978-981-15-3284-9_41

1 Introduction

Drill string/casing moves up or down longitudinally while drilling for various reason like changing the borehole assembly, cleaning the wellbore, bit removal, well stability etc. The drill string is assembled or dismantled by adding or removing the tubulars from the assembly which will move up or down axially in the hole. This centroidal action of drilling hydraulics in the wellbore is stated as tripping operations.

Tripping operations are considered as one of the costly aspects while drilling due to its role in increasing non-productive time, additionally, it could also result in various unforeseen well integrity or well control problems. Moving pipe in or out of the borehole will result in surge or swab pressure, respectively. In case of surging, equivalent circulation density (ECD) will be increased, and if it is higher than the fracture gradient, crack could also be incurred on the wellbore wall which subsequently leads to loss of fluid in the formation. Specifically, while moving casing down, excessive surge pressure is encountered which results in lost circulation. During swabbing, if induced pressure is higher than the collapse gradient, it will result in caving which can block the space around the bit and affects the well integrity [1, 2]. Furthermore, excessive tripping-out velocity can induce the flow of fluid inside the wellbore, particularly in deep wells where formation pore and fracture pressure difference are comparatively small. All these potential issues can be avoided if these pressure imbalances are studied properly and monitored regularly.

The effect of axial movement on bottom-hole pressure is studied since the drilling discovery. First research on this effect is done by Cannon [3]. Since then, number of researches has been done in this area. Hussain and Sharif [4] showed a decline in bottom-hole pressure differential during tripping-out operation with increasing eccentricity. For different well geometry, an independent work is carried out to predict the pressure imbalance due to longitudinal movement of pipe inside the wellbore [5, 6]. Also, different models are developed for predicting surge and swab pressure for different rheology and well geometry [7–11]. Most of these models lack simplicity and require complex input parameters and extensive numerical analysis to determine the result. The primary objective of this work is to develop an intelligent model (ANN) for forecasting surge and swab pressure with less time-consuming numerical analysis. To accomplish this, published system of equations is utilized to determine pressure differential numerically under varying rheological and well geometry, which are easier in understanding and application. Then, to reduce the cumbersome procedure of calculating pressure imbalance whenever rheology or well geometry changes, an intelligent model (ANN) is developed using data generated which predicts the surge pressure at different input parameters. Second objective of this study is identifying the effect of various rheological parameters and tripping speed on wellbore differential pressure.

2 Methodology

For the development of ANN model, four rudimentary steps must be followed. First, generation of data by utilizing the numerical analysis for training the model, second step is to perform sensitivity analysis to identify the prominent input parameters, third step is to train the ANN configuration, and fourth step is to validate and check the evaluation of the model. Each step is explained in detail below.

2.1 Generation of Data

For the generation of data, the numerical calculation method based on hydraulic behavior of power-law fluid in wellbore is used which is presented by Hussain [12]. These equations utilize the rheological properties and well configuration in combination with the displaced fluid flow rate due to inner pipe velocity to calculate surge/swab pressure. The calculation scheme includes

The average flow velocity (u_a) in the annulus is calculated by

$$u_{\mathrm{a}} = \frac{24.5\,q}{d_{\mathrm{h}}^2 - d_{\mathrm{od}}^2} \tag{1}$$

where 'q' is the flow rate of displaced fluid, 'd_h' is the diameter of hole, and 'd_{od}' is the outer diameter of drill string or casing.

Flow rate (q) in the annulus due to displaced fluid is determined by multiplying the pipe area and pipe tripping velocity (u_p), represented as

$$q = \pi d_{\mathrm{od}} u_{\mathrm{p}} \tag{2}$$

The flow critical velocity (u_c) around the drill string/casing is calculated by,

$$u_{\mathrm{c}} = \left(\frac{3.878K10^4}{\rho}\right)^{\frac{1}{(2-n)}} \times \left(\frac{24(2n+1)}{(d_{\mathrm{h}} - d_{\mathrm{od}})3n}\right)^{\left(\frac{n}{1-n}\right)} \tag{3}$$

where 'n' is power-law index, 'ρ' is mud density, and K is fluid consistency index.

$$n = 3.32\log\left(\frac{2\mathrm{PV} + \mathrm{YP}}{\mathrm{PV} + \mathrm{YP}}\right) \tag{4}$$

$$K = \frac{\mathrm{PV} + \mathrm{YP}}{(511)^n} \tag{5}$$

where PV and YP represent fluids plastic viscosity and yield point, respectively.

In case, $u_a < u_c$, in annulus, it shows laminar flow regime, and to calculate loss in pressure due to this flow regime, use Eq. (6),

Table 1 Input data

Parameter	Value or range	Unit
Mud weight, ρ	8–12	ppg
Plastic viscosity, PV	10–50	cP
Yield point, YP	10–200	$lb/100ft^2$
Diameter of hole, d_h	17–7.875	Inches
Outer diameter of pipe, d_{od}	13.37–6.13	Inches
Tripping velocity, u_p	6–60	ft/min

$$\Delta p_l = \left(\frac{K\,h}{300(d_h - d_{od})}\right) \times \left(\frac{24\,u_{a(1+2n)}}{(d_h - d_{od})3n}\right)^n \tag{6}$$

In case, $u_a > u_c$, in annulus, it shows turbulent flow regime, and to calculate loss in pressure due to this flow regime, use Eq. (7),

$$\Delta p_t = \frac{8.91 \times 10^{-5}\rho^{0.8} q^{1.8} \mu^{0.2} L}{(d_h - d_{od})^3 (d_h + d_{od})^{1.8}} \tag{7}$$

Now, to calculate the surge and swab pressure, the following equations are used

$$\text{Surge pressure, } P_{su} = 0.052\rho\,h + \Delta p \tag{8}$$

$$\text{Swab pressure, } P_{sw} = 0.052\rho\,h - \Delta p \tag{9}$$

Above given system of equation and the range of parameters given in Table 1 are used to generate the dataset for training, testing and validation of ANN model. Before the development of intelligent model, a sensitivity analysis is done for better understanding the effects of pressure imbalance and to identify the primary parameter affecting these pressure differences.

2.2 *Sensitivity Analysis*

Figure 1 represents the results of sensitivity analysis. In this analysis, tripping velocity, plastic viscosity, yield point, mud weight and diameter ratio are considered because other data is directly or indirectly depending upon these functions. According to all these results, it could be inferred that with increase in tripping velocity, surge pressure also increases. Figure 1a depicts the effect of PV on the bottom-hole pressure differential with increasing tripping velocity. It depicts that with increase in plastic viscosity, surge pressure is also increased at 15% rate at high pipe velocity whereas at low pipe velocity around 1%. Figure 1b, c shows the effect YP and density of mud on surge pressure, respectively. Both the figures showed the increase in surge

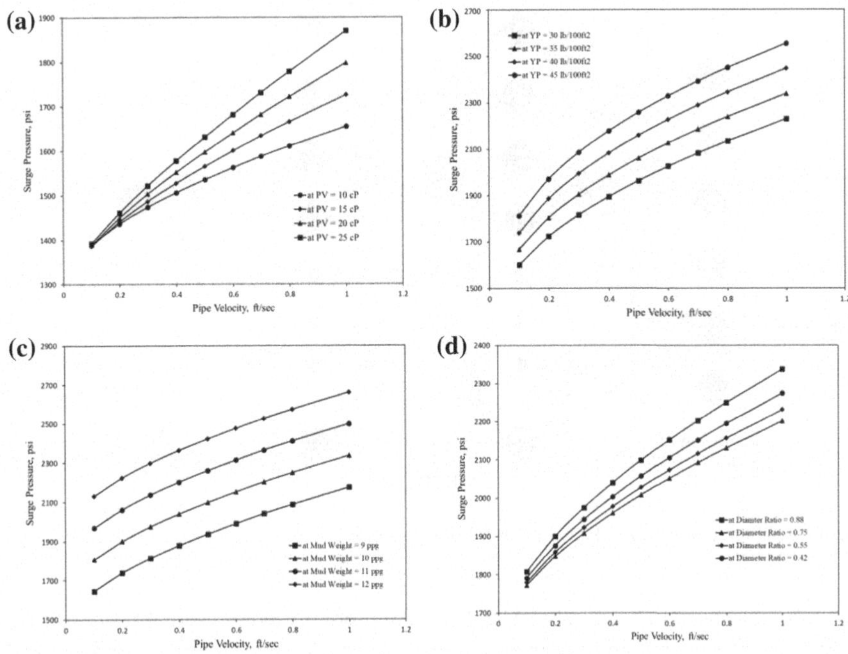

Fig. 1 Sensitivity analysis of different parameters on surge pressure; **a** effect of plastic viscosity (PV) on surge pressure; **b** effect of yield point (YP) on surge pressure; **c** effect of mud weight on surge pressure and **d** effect of diameter ratio on surge pressure

pressure with increasing YP/mud weight. Figure 1d depicts the effect of diameter ratio $\left(\frac{d_{\text{od}}}{d_{\text{h}}}\right)$ on surge pressure, and it clearly showed the increase in surge pressure with increasing diameter ratio because large diameter ratio represents low annular space. Lower annular clearance will result in high shearing effect between the fluid and the wellbore, which results in high surge pressure.

2.3 Training of ANN

In this stage, feed-forward network structure is used, shown in Fig. 2. In the input layer, six nodes are present which represents six input parameters (PV, YP, d_{h}, d_{od}, u_{p} and mud weight), whereas in output layer, only one neuron representing surge pressure. MATLAB educational package is used to develop the intelligent model. In neural network, backpropagation algorithm was used for network building and training. For the development of backpropagation network, total number of hidden layers with nodes had to be defined. Addition to this, several other parameters like training function, transfer function, adaption learning function and number of learning

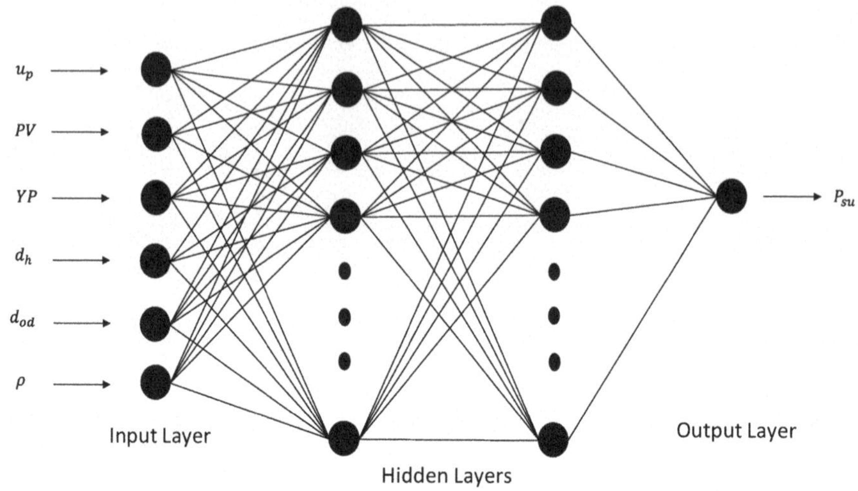

Fig. 2 Schematic diagram of feed-forward ANN structure

iterations also need to be selected. Due to lack of literature available in this topic, trial-n-error process is applied to select the most appropriate combination of above given parameter. The best combination was log-sigmoid transfer function, scaled conjugate gradient backpropagation (training function), gradient descent with momentum weight, bias learning function and 100,000 number of iterations. The structure of the neural network changes according to the complexity of the sample data. In this work, 1–3 hidden layer was varied with each layer consists of neuron in the range of 2–16. Total of 108 networks with variations of hidden layer and number of neurons were developed and trained.

2.4 Performance of ANN Networks

To check the performance of the different networks, statistical analysis was carried out to calculate mean absolute error (MAE), root mean square error (RMSE), relative absolute error (RAE), root relative square error (RRSE) and coefficient of determination (R^2). If the value of RAE and RRSE is closer to zero, performance of the model will be excellent, whereas for R^2, value needs to be closer to 1.

Equations (10–14) of all performance measuring tools are mentioned below:

$$\text{MAE} = \frac{1}{P}\sum_{1=1}^{P}(\text{PD} - \text{DD}) \tag{10}$$

$$\text{RMSE} = \sqrt{\frac{1}{P}\sum_{1=1}^{P}(\text{PD} - \text{DD})^2} \tag{11}$$

$$\text{RAE} = \frac{\sum_{1=1}^{P}|\text{PD} - \text{DD}|}{\sum_{1=1}^{P}\left|\text{DD} - \overline{\text{DD}}\right|} \tag{12}$$

$$\text{RRSE} = \sqrt{\frac{\sum_{1=1}^{P}(\text{PD} - \text{DD})^2}{\sum_{1=1}^{P}\left(\text{DD} - \overline{\text{DD}}\right)^2}} \tag{13}$$

$$R^2 = 1 - \frac{\sum_{1=1}^{P}(\text{DD} - \text{PD})^2}{\sum_{1=1}^{P}\left(\text{DD} - \overline{\text{DD}}\right)^2} \tag{14}$$

where P is the number of samples, PD represents the predicted data given by the developed model, and DD represents the desired data calculated by using the above-given system of equations according to flowchart given in Fig. 1.

3 Results and Discussions

Dataset of total 1513 scenario was generated, out of which 454 scenarios were set aside for testing the model once it was trained. The training dataset consisted of 957 scenarios which were calculated by utilizing the data provided in Table 1. Using these dataset, ANN models are trained, and its performance is analyzed using the test dataset. The network configurations provide the most reduced result for all error measuring equation mentioned in last section, and optimized result of the R^2 is selected as the most favorable network structure. This procedure is repeated for 108 network structures, and its performance data is presented in Table 2. Due to page constraint, all the data is not presented in Table 2, instead only 24 best network structures are presented. The most optimum result given by an artificial neural network structure includes two hidden layers with ten neurons in first layer and 16 neurons in second layer. The root absolute error (RAE) for this configuration is 5.76% with the root relative square error (RRSE) of 6.16% and the coefficient of determination (R^2) of 0.987 (Fig. 3). Although, this network is most optimum, it is very complex and will pose a difficulty to present a large set of weights and bias coefficient in linear equation. Considering this issue, another trained model with one hidden layer containing 14 nodes can be considered due to its satisfactory prediction result that shows relative absolute error (RAE) is 5.6% with relative root squared error (RRSE) of 7.82% and the coefficient of determination (R^2) of 0.985.

Table 2 Summary of ANN configuration and its performance data

Hidden layers	Neurons	MAE	RMSE	RAE	RRSE	R2		
	first layer	second layer	third layer					
1	2	–	–	2863.78	3673.67	14.07	15.17	0.952381
	4	–	–	1614.87	2432.68	7.93	10.04	0.976342
	6	–	–	1524.924	2351.48	7.496	9.71	0.975156
	8	–	–	1496.913	2226.19	7.358	9.19	0.978517
	10	–	–	1562.44	2303.53	7.68	9.51	0.975156
	12	–	–	1240.04	1979.54	6.09	8.17	0.983072
	14	–	–	1140.28	1892.98	5.6	7.82	0.985056
	16	–	–	1529.56	2255.41	7.51	9.31	0.973182
2	10	16	–	1172.73	1491.95	5.76	6.16	0.987837
	12	2	–	1646.36	2400.67	8.09	9.91	0.97733
	8	10	–	1915.6	2780.4	9.41	11.48	0.976539
	12	14	–	1822.63	2660	8.95	10.98	0.974761
	12	10	–	1833.68	2693.77	9.01	11.12	0.974564
	16	16	–	1709.44	2511.46	8.4	10.37	0.974366
	8	12	–	1775.84	2590.75	8.72	10.7	0.97338
	10	12	–	1745.62	2586.64	8.58	10.68	0.973182
3	6	8	2	1473.086	2218.87	7.24	9.16	0.977132
	14	14	14	1843.01	2603.27	9.05	10.75	0.974169
	8	8	14	1880.34	2685.52	9.24	11.09	0.97338
	12	6	8	1846.7	2763.23	9.07	11.41	0.971604
	16	16	16	1758.76	2595.63	8.64	10.72	0.968846
	6	2	10	1990.949	2981.03	9.78	12.31	0.966092

3.1 Validation of the Trained Model

The most optimum model with two hidden layers with ten neurons in first layer and 16 neurons in second layer is authenticated with a dataset of 506 scenarios (random untrained data). The best model predicted the surge pressure with the relative absolute error (RAE) of 8.44%, root relative square error (RRSE) of 10.32% and the coefficient of determination (R^2) of 0.974. The same dataset is also used to validate the simple structure of ANN with one hidden layer including 14 neurons. This ANN structure predicted the surge pressure with the relative absolute error (RAE) of 14.07%, root relative square error (RRSE) of 15.17% and the coefficient of determination (R^2) of 0.952.

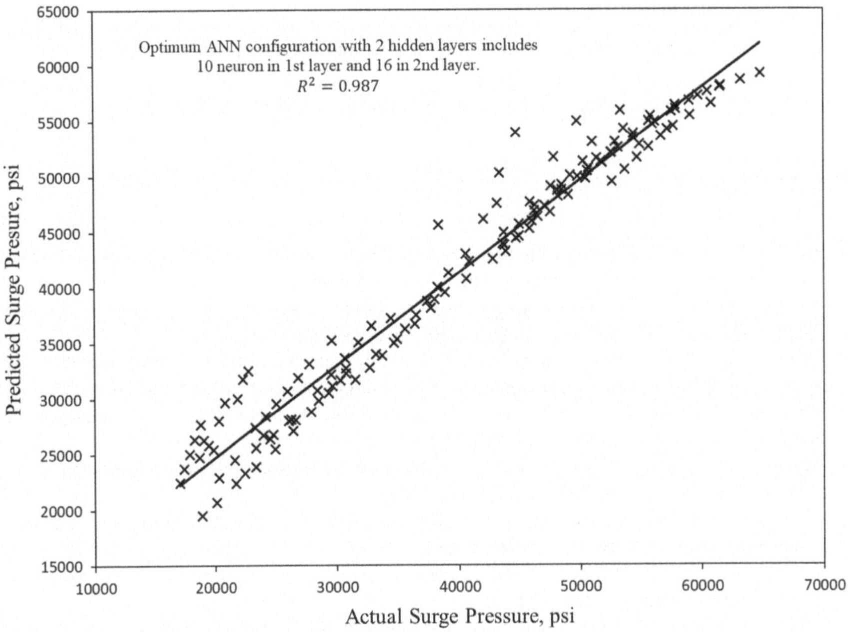

Fig. 3 Predicted versus actual data of surge pressure for the optimal ANN configuration

4 Conclusions

An artificial neural network (ANN) is developed for predicting surge/swab pressure for power-law fluid while conducting tripping operations. This model has the ability to predict surge pressure under a wide range of varying rheological and wellbore geometrical parameters. After performing trial-n-error method, two network structures were selected as the best predictive ANN configuration. First one has complex configuration with two hidden layers includes ten and 16 neurons in first and second layer, respectively, and the other one with simpler network structure with one hidden layer including 14 neurons. Complex ANN configuration showed coefficient of determination of (R^2) of 0.987 (0.974 for validation data), and simpler ANN configuration showed coefficient of determination of (R^2) of 0.985 (0.952 for validation data). Both models are predicting the surge pressure satisfactorily without using the cumbersome time-consuming numerical analysis. This intelligent model may give vital advantages when tripping operation is carried out in low-pressure margin wells where repetitive calculation of surge/swab pressure is required. Also, this model can be extremely useful during real-time tripping operations because it will help well engineers to set the appropriate pipe tripping speed and rheological parameters so that surge/swab pressure due to centroidal action of drilling hydraulics pipe axial movement will be in between the range of drilled formation fracture and pore pressure.

Acknowledgements The author acknowledges the support of Universiti Teknologi PETRONAS (UTP) for providing the financial support from YUTP project number: 0153AA-E27.

References

1. Goins WG Jr et al (1951) Down-the-hole pressure surges and their effect on loss of circulation. In: Drilling and production practice. American Petroleum Institute, New York, p 8
2. Cardwell WT Jr (1953) Pressure changes in drilling wells caused by pipe movement. In: Drilling and production practice. American Petroleum Institute, New York, p 16
3. Cannon GE (1934) Changes in hydrostatic pressure due to withdrawing drill pipe from the hole. In: Drilling and production practice. American Petroleum Institute, New York, p 7
4. Hussain QE, Sharif MAR (1997) Viscoplastic fluid flow in irregular eccentric annuli due to axial motion of the inner pipe. Can J Chem Eng 75(6):1038–1045
5. Haige W, Xisheng L (1996) Study on steady surge pressure for yield-pseudoplastic fluid in a concentric annulus. Appl Math Mech 17(1):15–23
6. Filip P, David J (2003) Axial Couette-Poiseuille flow of power-law viscoplastic fluids in concentric annuli. J Petrol Sci Eng 40(3):111–119
7. Crespo F, Ahmed R (2013) A simplified surge and swab pressure model for yield power law fluids. J Petrol Sci Eng 101:12–20
8. Ettehadi A, Altun G (2018) Functional and practical analytical pressure surges model through herschel bulkley fluids. J Petrol Sci Eng 171:748–759
9. Ahmed RM et al (2010) The effect of drillstring rotation on equivalent circulation density: modeling and analysis of field measurements. In: SPE annual technical conference and exhibition. Society of Petroleum Engineers, Florence, Italy, p 11
10. Schuh FJ (1964) Computer makes surge-pressure calculations useful. Oil Gas J 31(62):96–104
11. Burkhardt JA (1961) Wellbore pressure surges produced by pipe movement. J Petrol Technol 13(06):595–605
12. Hussain R (2001) Well engineering and constructions. Entac Consulting

Exploration on Revenue Using Pioneering Technology in Infrastructure Facilities of Luxury Hotels

H. M. Moyeenudin, R. Anandan, and Shaik Javed Parvez

Abstract The existing innovations in technology developed a framework towards the development of luxury hotels with various feasible facilities, and this will permit the promising and creative ideas to implement during the development process of a hotel as well as it is useful in preparing the innovative infrastructures for hotel properties. This study demonstrates that the technology in hotel foundations is continually working up according to the guest desires of hotel concerning towards the comfort with new and advanced technology that is increasing day by day, and secondly, it focuses on the recent tools used to maintain the room occupancy in hotel. The infrastructure of the hotel gives permits for complete synchronization with recent technology along with the tools like cloud-based property management system (CBPMS) for guest occupancy in coordination with the revenue management system (RMS) and energy management system (EMS) of hotel.

Keywords Hotel infrastructure · Cloud technology · PMS · Wi-fi connectivity

1 Introduction

The urge of luxury hotels is to satisfy the guest who visits the property, regardless with the expense on implementing the latest technology in hotel [1], and the hotel industry is developing in such manner to adopt and explore the hypothetical and functional purposes of infrastructure with technology provisioning for new luxury hotels with an edge on earth for the best guest experience with a huge interests in recent technology which is not only to meet the needs of hotel guest also it supports in maintaining and increasing the revenue of the hotel simultaneously, in order to

H. M. Moyeenudin (✉)
School of Hotel and Catering Management, Vels Institute of Science, Technology and Advanced Studies, Pallavaram, Chennai, India
e-mail: moyeenudin@velsuniv.org

R. Anandan · S. J. Parvez
Department of Computer Science and Engineering, Vels Institute of Science, Technology and Advanced Studies, Pallavaram, Chennai, India

S.-L. Peng et al. (eds.), *Intelligent Computing and Innovation on Data Science*,
Lecture Notes in Networks and Systems 118,
https://doi.org/10.1007/978-981-15-3284-9_42

transfer, share and to receive business information for registration and reservation the cloud-based property management system (CBPMS) is used in connection with a hotel Website [2]. The facilities provided to guest are focused to meet their needs with maximum comfort, the Wi-fi is used for ordering their food from room as well as to access the Internet at any point of time, similarly, the RFID and near-field communication (NFC) technology assist in giving guest security and accessibility [3], the guest room of luxury hotel may have energy and movement sensor which assist the hotel guest towards comfort by giving a balanced in-room condition and climate as per usage by identifying the distance and presence in room, and there are voice recognizing sensors that actuate voice for regulating the functions of electronic gadgets, for example, drape, lightings, room temperatures. The room temperatures are measured through temperature sensor that is fixed in guest rooms to guarantee the experience of guest at hotel be able to remain by means of agreeable condition [4]. The guest room doors can be locked and opened with an Android phone application makes the hotel guest to go as without key or card for accessing entry in room, the sensor is available to screen guest health and comfort condition at the time of exercise, the structure of hotel property has outside temperature sensor to measure the outer temperature conditions in order to adjust the power consumption, and sensor is used to detect the daylight and alter room shades, screens with an intellectual lighting framework [5]. In any case, the benefits of shrewdness remain the key for the development of guest satisfaction in luxury hotel, and thus, many hotel properties are involved in reengineering their facilities to meet the guest expectations [6].

2 Materials and Methods

This research is carried out through collecting qualitative and quantitative data from star categorized hotels. A total of 50 questionnaires are used to obtain these data from three-star, four-star and five-star hotels accordingly, but only 40 questionnaires were completed and used for analysis. This study was carried out from Jan 2019 to July 2019 at hotels, respectively. The Statistical Package for Social Sciences (SPSS 24 version) is used to investigate the relationship between quantitative variables using Pearson correlation and graph board parallel line graph to complete this study.

3 Results and Discussion

3.1 Figures

Chart 1 shows that the luxury hotels using PMS with revenue management have higher room occupancy percentage almost close to 85–90%, whereas the hotels without automated PMS can able to make lesser than 80–65% of room occupancy as

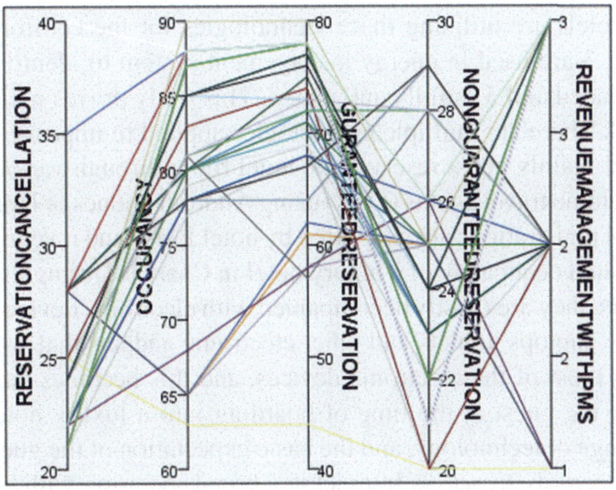

Chart 1 Room occupancy with PMS

well as the PMS with revenue management system has also reduced the reservation cancellation to 20–22% and increased the guaranteed reservation from 70 to 80%, and thus, this indicates a stronger relationship between automated revenue management of PMS in room occupancy [7].

The Chart 2 demonstrates the influence of technology in infrastructural facilities of hotel with energy management of property management system, RFID, temperature control, smartphone locking, electronic door locking system and Wi-fi Technology is used as a variables to recognize the availability in luxury hotels. The variables are measured according to the usage in hotel, whether the facility is available or not, the value 1 indicates that the facility is used and 2 as not available. The majority

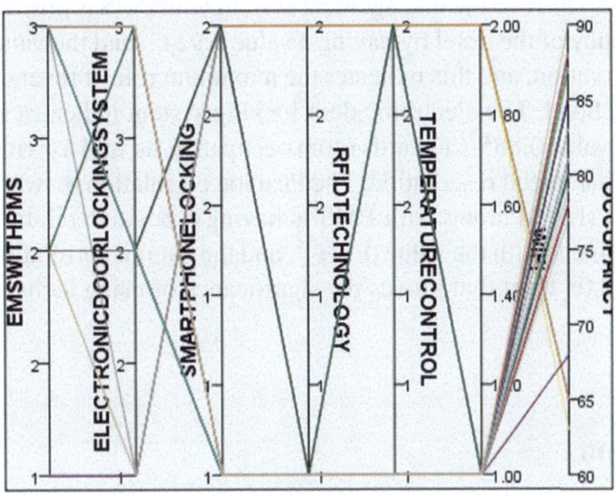

Chart 2 Technology used in infrastructural facilities

of the luxury hotels are utilizing these technologies for the comfort of the guest. The values 1, 2, 3 are used in energy management system to identify 1 as manual, 2 as semi-automated and 3 as fully automated. This study proves majority of luxury hotels use EMS. There are multiple technologies applied to improve the experience of guest at hotel mainly guest reserves the hotel room through various sources like hotel Website, online travel agents (OTA) using Android phones or laptops [8]. These reservations and registrations are supported by hotel PMS, and revenue management is positively related occupancy of a luxury hotel in Chart 1. During the time of guest check into hotels, they are mostly accompanied with electronic devices, for example, Android phones, laptops, tablets and other electronic gadgets that work with Wi-fi, thus it supports most of the electronic devices, and this becomes an unconditional requirement for the guest at the time of boarding into a luxury hotel [9]. Chart 2 indicates the usage of technology, and the basic expectation of the guest at hotel is to get a Wi-fi connectivity to access Internet in a broad spectrum with high bandwidth [10].

3.2 Correlations

In Table 1, the correlation between hotel occupancy with reservation cancellation shows significant value $r = 0.131$, the value 0.045 indicates that there is a strong significance with guaranteed reservation, $p = 0.000$ shows the probability and reservation cancellation, and most importantly, the significant value $r = 0.096$ is shown with reservation cancellation and the revenue management with cloud-based property management system (CBPMS). The Pearson correlation value 0.892^{**} indicates a positive association between room occupancy and guaranteed reservation and the non-guaranteed reservation value −0.243 with occupancy indicates that it is negatively associated. The revenue management system with PMS is positively associated with the occupancy of the hotel by having a value 0.924^{**} and the value 0.844^{**} with guaranteed reservation, and this indicates the maximum role of revenue management in occupancy of hotel. The electronic door locking system indicates a positive association with the value 0.885^{**} towards room occupancy, as well as with value 0.806^{**} in regard with guaranteed reservation. The Pearson correlation between energy management system (EMS) along with PMS is having a positive relationship between revenue management with the value 0.711^{**}, and the overall correlation is significant at the 0.01 and 0.05 level that proves the significance of using technology in luxury hotels.

4 Conclusion

There are numerous advancements with basic improvements should be actualized by entrepreneurs of luxury hotels to rebuild according to meet the expectation of hotel

Table 1 Correlation between hotel occupancy and technology

Correlations

		Occupancy	GR	NGR	RM with PMS	EDLS	EMS with PMS
Occupancy	Pearson correlation	1	0.892**	−0.243	0.924**	0.885**	0.815**
	Sig. (two-tailed)		0	0.131	0	0	0
	N	40	40	40	40	40	40
Guaranteed reservation	Pearson correlation	0.892**	1	−0.319*	0.844**	0.806**	0.757**
	Sig. (two-tailed)	0		0.045	0	0	0
	N	40	40	40	40	40	40
Non-guaranteed reservation	Pearson correlation	−0.243	−0.319*	1	−0.267	−0.275	−0.211
	Sig. (two-tailed)	0.131	0.045		0.096	0.086	0.192
	N	40	40	40	40	40	40
Revenue management with PMS	Pearson correlation	0.924**	0.844**	−0.267	1	0.820**	0.711**
	Sig. (two-tailed)	0	0	0.096		0	0
	N	40	40	40	40	40	40
Electronic door locking system	Pearson correlation	0.885**	0.806**	−0.275	0.820**	1	0.707**
	Sig. (two-tailed)	0	0	0.086	0		0
	N	40	40	40	40	40	40
EMS with PMS	Pearson correlation	0.815**	0.757**	−0.211	0.711**	0.707**	1
	Sig. (two-tailed)	0	0	0.192	0	0	
	N	40	40	40	40	40	40

**Correlation is significant at the 0.01 level (two-tailed)
*Correlation is significant at the 0.05 level (two-tailed)

guest and to improve their hotel business in the cutting edge of innovative world [11]. The synchronization of technology with various facilities of hotel property is carried out in hotel industry in most of which the upgrades are engaged with the CBPMS with RMS [12]. This study shows significant value of 0.096 on reservation cancellation with RMS towards guest occupancy. In addition, RMS with EMS indicates a positive relationship with a value of 0.711**. The future luxury hotels may also adopt technologies to meet the guest expectation in order to improve their revenue by implementing the cloud-based property management system in connection with an automated RMS and EMS to manage the hotel capacity. The guest comfort is obtained by implementing sensors to maintain room temperatures, voice control and wearable body zone sensors for health may increase their satisfaction, the facilities like NFC are used in security and protection, and the Wi-fi provides convenience to the guest. Most of the luxury hotels use CBPMS which supports hotel in managing their property with high revenue through EMS and RMS.

References

1. Neuhofer B, Buhalis D, Ladkin A (2015) Smart technologies for personalized experiences: a case study in the hospitality domain. Int J Networked Bus 25(3):243–254
2. Bilgihan A, Okumus F, "Khal" Nusair K, Kwun DJW (2011) Information technology applications and competitive advantage in hotel companies. J Hospitality Tourism Technol 2(2):139–153
3. Davis MM, Spohrer JC, Maglio PP (2011) How technology is changing the design and delivery of service. Oper Manage Res 4(1–2):1–5
4. Jung S, Kim J, Farrish J (2014) In-room technology trends and their implications for enhancing guest experiences and revenue. J Hospitality Tourism Technol 5(3):210–228
5. Kapiki S (2012) Current and future trends in tourism and hospitality. The case of Greece. Int J Econ Pract Theor 2
6. Leung D, Fong LHN, Law R (2013) Assessing the visibility of hotels on smartphones: a case study of hotels in Hong Kong. https://doi.org/10.1007/978-3-319-03973-2_61
7. Kimes SE (2011) The future of hotel revenue management. J Revenue Pricing Manage 10(1):62–72
8. Erdem M, Jiang L (2016) An overview of hotel revenue management research and emerging key patterns in the third millennium. J Hospitality Tourism Technol 7:300–312
9. Buhalisa D, Leun R (2018) Smart hospitality—interconnectivity and interoperability towards an ecosystem. Int J Hospitality Manage 71:41–50
10. Kansakar P, Munir A, Shabani N (2018) Technology in hospitality industry: prospects and challenges
11. Sherri S (2015) Growing trends in the tourism and hospitality industry. J Tourism Hospitality 4(3):1000162
12. Law R, Jogaratnam G (2005) A study of hotel information technology applications. Int J Contemp Hospitality Manage 17:170–180. https://doi.org/10.1108/09596110510582369

A Big Data Analytics-Based Design for Viable Evolution of Retail Sector

Neha Malhotra, Dheeraj Malhotra, and O. P. Rishi

Abstract Retailing is one of the world's prominent and most diversified commercial activities, which has considerably transformed business strategies for earning more profit. Today, the retailing definition is a synonym to attractive and appropriately managed merchandise stores with incredible comfort and ambience rather than randomly stacked traditional stores. Also, the modern customer is focused towards quality/brands and expects for services delivered to them by different vendors at the ease of home with a single click. As a result, customers prefer to shop from various online shopping Websites rather than physically moving to a retail store, which in turn leads to the downfall in the sales of retailers which has become a significant threat to them. Therefore, this paper highlights this current problem faced by retailers and suggests some corrective measures, which retailers should deploy. Consequently, the retailers are required to practice corrective measures towards meeting all customers' expectations by providing their necessary goods under the same roof. Besides, the retailers should provide several benefits and lucrative offers like discounts, cash-backs, buy one get one, free home delivery, combo purchase and other tailor-made offers to attract customers via targeted marketing by meeting their specific needs and hence to overcome diversion of their customers towards E-Commerce Websites.

Keywords Big data · Association rule mining · Cloud computing · Hadoop · MapReduce · MR–Apriori algorithm

1 Introduction

Retailing is one of the world's leading and most diverse businesses. In today's world, retailing has changed various aspects of doing a successful business. Retailing is the field of purchasing merchandises in bulk capacities either from a wholesaler or directly from a manufacturer and further selling these merchandises and associated

N. Malhotra · D. Malhotra (✉)
VSIT, Vivekananda Institute of Professional Studies, GGSIPU, Delhi, India

O. P. Rishi
University of Kota, Rajasthan, India

S.-L. Peng et al. (eds.), *Intelligent Computing and Innovation on Data Science*, Lecture Notes in Networks and Systems 118,
https://doi.org/10.1007/978-981-15-3284-9_43

services to the customers for satisfying their personal needs. In other words, retailing is the set of business activities that complement value to the products for the benefit of customers. A retailer may be an individual/organisation with expertise in the sale of a particular category of goods and facilities to their consumers for their personal use, besides it, retailer act as an interface between manufacturer and customer. However, there exist numerous types of retail stores which include supermarket stores, chemist stores, toys stores, cosmetics stores, etc., and there exists neck-to-neck battle among all kinds of the retailers which intensifies the emerging competition among all. This competition may be in the form of same-store types regarding price discounts and other types of offers to attract customers [1–7, 20].

Therefore, the primary objective for today's retailer is to protect their present customer base yet not get influenced by aggressive competition. Moreover, the modern generation of consumers is quite educated, more demanding and focused towards value for money relationships. As a result, retailing has become a highly competitive business to satisfy the dynamic demands of modern customers. Therefore, this paper combines an approach of integrating big data with scalable data mining algorithms to assist retailers, where is deployed over Google Cloud multi-node cluster using Hadoop and MapReduce platform [9–18].

2 Big Data Analytics and Market Trends

Presently, enterprises are moving to the cloud environment at unbelievable speed and propagating a rule like '*you are not going to be in the business if you are not on the cloud*'. Further, American research and advisory firm, *Gartner*, investigated and anticipated the market of big data with the retail industry. According to the Gartner report, the retailers need advanced analytical competencies to face the challenges possessed by E-Commerce boom [8]. Gartner reviewed around 500 organisations in addition to IT firms where it was apparent that organisations are rapidly moving to big data marketing. Few predictions are as follows:

- *Retail big data analytics* is estimated to generate new magnitudes and additional gain approximate US$ 60 billion by 2020 [23].
- *The market of Hadoop with big data analytics* which was approximated $8.48 billion in 2015 and is forecasted to touch $99.31 billion near 2022 emerging at a compound annual growth rate (CAGR) of 42.1% from 2015 to 2022 [19].
- By 2019, maximum of the top 200 most significant organisations will deploy smart and efficient applications as well as explore the entire arena of big data analytics and its associated tools to enhance their suggestions for customers which lead to improvement in customer purchase experience [21].

3 Research Methodology

3.1 System Design

The efficiency of association mining algorithm is usually measured by the time taken to find frequent itemsets. One of the conventional approaches is to use a pre-determined number of resources for transaction processing irrespective of current input load. This research work proposes Apriori–MapReduce (AMR)-based approach which can work in the *intelligent* and *advanced* mode for generation of association rules. As depicted in Fig. 1, initially unprocessed (unstructured) datasets will be deployed over Google Cloud multi-node cluster environment, which will be given to the processing module, where all datasets will be arranged in the form of the input queue. Engine manager module based on the instantaneous load will increase or decrease the processing engines to generate candidate sets and frequent itemsets. These itemsets will be generated by MapReduce (MR)-based Apriori algorithm. Later, the results will be generated in the output queue, and association rules are generated.

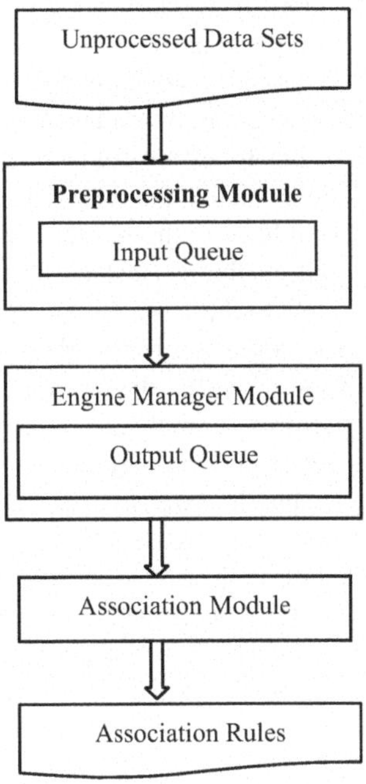

Fig. 1 System design

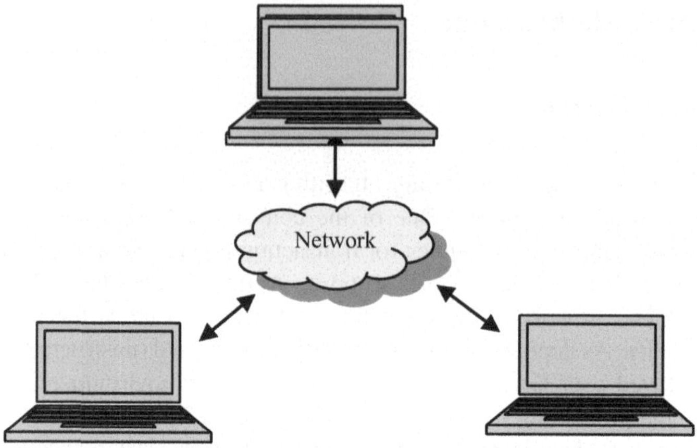

Fig. 2 Multi-node cloud architecture

3.2 Implementation

For the implementation of big data, this research deploys *Google Cloud* environment where a multi-node cluster is set up where more than one machine is deployed using virtual machine instances over Google Cloud. In Fig. 2, one machine will act as a master node (or NameNode), whereas two other machines are referred as slave nodes (or DataNode). All these machines are connected through network, data will be distributed to master and slaves, later, the data will be processed, and with the help of JobTracker, it will be sent back to the master node. Finally, the master node will display the results. Once all machines (instances) are ready, there is a need to install Hadoop and Java separately to these machines, and later, all machines will be connected to each other to make a cluster. In the Hadoop environment, the master node is referred to as NameNode, whereas slaves are commonly known as DataNodes, wherein MapReduce platform, NameNode is recognised as JobTracker and DataNodes are referred to as TaskTrackers. Further, in a multi-cluster environment, when NameNode is instructed to handle extensive datasets, then NameNode is going to distribute these datasets among various DataNodes (slaves), and finally, results from all machines are collected through JobTracker [22, 24, 25].

Figure 3 depicts the interface of the engine where scalable MR–Apriori is deployed at the background, and using this interface, the instantaneous load can be managed by using appropriate replication factor, i.e. processing engine, and finally, association rules are generated.

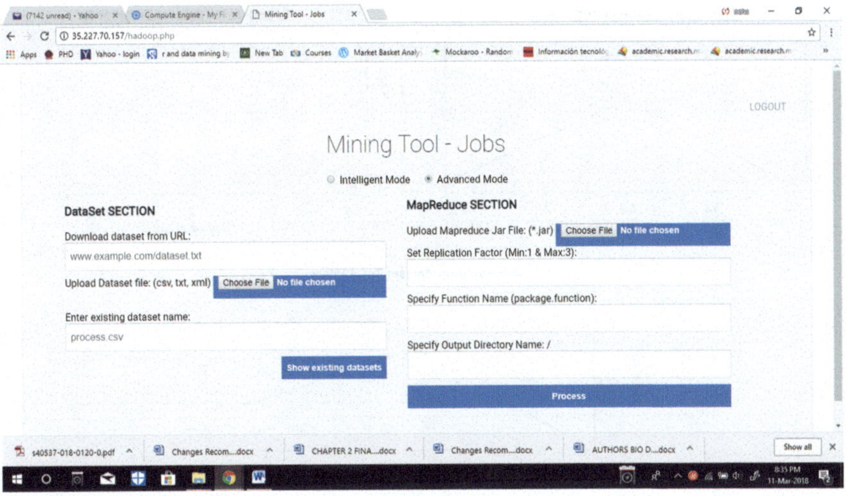

Fig. 3 Practical interface of the proposed IRM tool

4 Experimental Evaluation

The performance of the proposed methodology is evaluated based on the speedup evaluation metric. This factor is essential in association rule mining and will help to justify the need for the proposed MR–Apriori algorithm implemented on HDFS MapReduce platform. The procured association rules can be applied to boost retail business, customer retention, product recommendation. However, due to the limited volume of real-time datasets and access constraint from an online Website, synthetic retail transactional datasets were later on generated using IBM synthetic dataset generator tool as well as by writing and implementing a Java code on Eclipse version 4.3.0 of different sizes.

During the interpretation of speedup for a single node with replication factor 1, the value of speedup is 1, but then, for replication factor 2 and 3, speedup values (Fig. 4) are more than 1. Therefore, with more number of nodes, multi-node cluster speedsup well and processes dataset faster.

5 Conclusion

Combination of cloud computing and scalable data mining algorithms not only overcomes the bottleneck of original system but also can rationally use resources and improve the efficiency of data processing and analysis. Therefore, in this research work, MR–Apriori algorithm for mining association rules from massive retail transactional databases has been proposed. The retrieved association rules will boost the

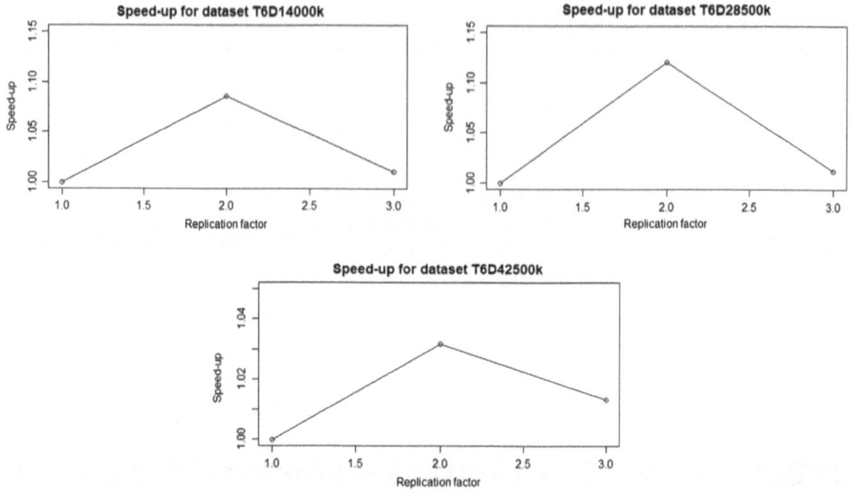

Fig. 4 Speedup for dataset T6D14000k, T6D28500k and T6D42500k

sales in supermarkets and will assist the retailers to immune their businesses from modern E-Commerce boom by avoiding the diversion of customers from their outlets to E-Commerce Websites.

References

1. Agrawal R, Imieliński T, Swami A (1993) Mining association rules between sets of items in large databases. In: ACM Sigmod Record. ACM, pp 207–216
2. Agrawal R, Shafer JC (1996) Parallel mining of association rules. IEEE Trans Knowl Data Eng 8(6):962–969
3. Davey I (2014) Technologies: consumers, big data, and online tracking in the retail industry: a case study of Walmart. Center for Media Justice
4. Farzanyar Z, Cercone N (2013) Accelerating frequent itemsets mining on the cloud: a MapReduce-based approach. In: 2013 IEEE 13th international conference on data mining workshops (ICDMW), IEEE, pp 592–598
5. Gatzioura A, Sànchez-Marrè M (2015) A case-based recommendation approach for market basket data. IEEE Intell Syst 30(1):20–27
6. Gupta D, Singh SK, Malhotra D, Verma N (2017) EPRT-an ingenious approach for E-Commerce website ranking. Int J Comput Intell Res 13(6):1471–1482
7. Li L, Zhang M (2011) The strategy of mining association rule based on cloud computing. In: 2011 international conference on business computing and global informatization, IEEE, pp 475–478
8. Malhotra D, Rishi OP (2016) IMSS-E: an intelligent approach to the design of adaptive metasearch system for E-commerce website ranking. In: Proceedings of the international conference on advances in information communication technology & computing. ACM, p 3
9. Malhotra D, Verma N (2013) An ingenious pattern matching approach to ameliorate web page rank. Int J Comput Appl 65(24):33–39

10. Malhotra D (2014) Intelligent web mining to ameliorate web page rank using back-propagation neural network. In: 5th international conference on confluence the next generation information technology summit (Confluence), IEEE, pp 77–81
11. Malhotra D, Rishi OP (2016) IMSS-E: an intelligent approach to design of adaptive meta search system for E-Commerce website ranking. In: Proceedings of the international conference on advances in information communication technology & computing, ACM. https://doi.org/10.1145/2979779.2979782
12. Malhotra D, Malhotra M, Rishi OP (2017) An innovative approach of web page ranking using hadoop- and map reduce-based cloud framework. In: Proceedings of advances in intelligent systems and computing, vol 654, CSI-2015. Springer, Heidelberg, pp 421–427
13. Malhotra D, Rishi OP (2017) IMSS: a novel approach to design of adaptive search system using second generation big data analytics. In: Proceedings of international conference on communication and networks. Springer, Heidelberg, pp 189–196
14. Malhotra D, Verma N, Rishi OP, Singh J (2017) Intelligent big data analytics: adaptive E-Commerce website ranking using Apriori Hadoop–BDAS-based cloud framework. Maximizing business performance and efficiency through intelligent systems, IGI Global, pp 50–72
15. Malhotra D, Rishi OP (2018a) An intelligent approach to design of E-Commerce metasearch and ranking system using next-generation big data analytics. J King Saud Univ-Comput Inf Sci (Elsevier). https://doi.org/10.1016/j.jksuci.2018.02.015
16. Malhotra D, Rishi OP (2018b) IMSS-P: an intelligent approach to design & development of personalized meta search & page ranking system. J King Saud Univ-Comput Inf Sci (Elsevier). https://doi.org/10.1016/j.jksuci.2018.11.013
17. Malhotra D, Rishi OP (2019) A comprehensive review from hyperlink to intelligent technologies based personalized search systems. J Manage Analytics 1–25 (Taylor & Francis)
18. Saabith AS, Sundararajan E, Bakar AA (2016) Parallel implementation of Apriori algorithms on the Hadoop-Mapreduce platform-an evaluation of literature. J Theor Appl Inf Technol 85(3):321
19. Sethi S, Malhotra D, Verma N (2016) Data mining: current applications & trends. Int J Innov Eng Technol 6(4):586–589
20. Svetina M, Zupančič J (2005) How to increase sales in retail with market basket analysis. Syst Integr 418–428
21. Tian L, Li L, Wang X (2012) Study of identifying cross-selling for online retailers in E-commerce. In: 2012 fourth international conference on computational and information sciences, IEEE, pp 417–420
22. Verma N, Singh J (2017) A comprehensive review from sequential association computing to Hadoop-MapReduce parallel computing in a retail scenario. J Manage Analytics 4(4):359–392
23. Verma N, Singh J (2015) Improved web mining for e-commerce website restructuring. In: 2015 IEEE international conference on computational intelligence & communication technology (CICT), IEEE, pp 155–160
24. Verma N, Singh J (2017) An intelligent approach to big data analytics for sustainable retail environment using Apriori-MapReduce framework. Ind Manage Data Syst 117(7):1503–1520
25. Verma N, Malhotra D, Malhotra M, Singh J (2015) E-commerce website ranking using semantic web mining and neural computing. Proc Comput Sci 45:42–51

Blockchain with Bigdata Analytics

D. R. Krithika and K. Rohini

Abstract A list of records continuously growing is known as Blockchain, linked and using cryptography. Fingerprint of data or cryptographic hash of previous block presents in every block. Blockchain is very secure and fastest transactions customized by any extent. Blockchain (BC) improves Perfect treatment and diagnosis. Bigdata analytics is a large volume of transactional data. In this paper we discussed about blockchain and bigdata analytics.

Keywords BC-Blockchain · Hash · Bigdata analytics

1 Introduction

Stuart Haber and W. Scott Stometta in 1991 initiated blockchain. Blockchain is decentralized ledger and it does not have transaction cost, It is very difficult to alter or change once the data is stored and digital currencies distributed not able copy so the data is very secure. Every block contains list of records, so the process of adding new block is mining. In this paper Part II. Previous work, Part III. Concepts of Blockchain, Part IV. Algorithms in Blockchain, V. Bigdata Analytics, Part VI. Different Analytics, Part VII. Algorithms in Bigdata, Part VIII. Bigdata in Industries, Part IX. Tools, X. Conclusion these divisions we discussed.

Block is a record data insides like string of words then it will have value which is called previous Hash, it will get in second and then will have a value which is own hash, hash is fingerprint of the block. First block is called genesis block will have value of data and no previous Hash and then this block will have own Hash. Second block have data, previous Hash and own Hash, this second block previous Hash is exactly identical to first block own Hash. Third block have data, previous Hash and own Hash. Third block previous Hash is exact identical in second block own Hash. So cryptographically linked together all the blocks. Figure 1 explains how blocks are linked. Analyzing large volume of data is bigdata analytics. The process of unstructured raw data to useful for organizations data form core of bigdata analytics.

D. R. Krithika (✉) · K. Rohini
Department of Computer Science, VISTAS, Chennai, Tamil Nadu, India

S.-L. Peng et al. (eds.), *Intelligent Computing and Innovation on Data Science*,
Lecture Notes in Networks and Systems 118,
https://doi.org/10.1007/978-981-15-3284-9_44

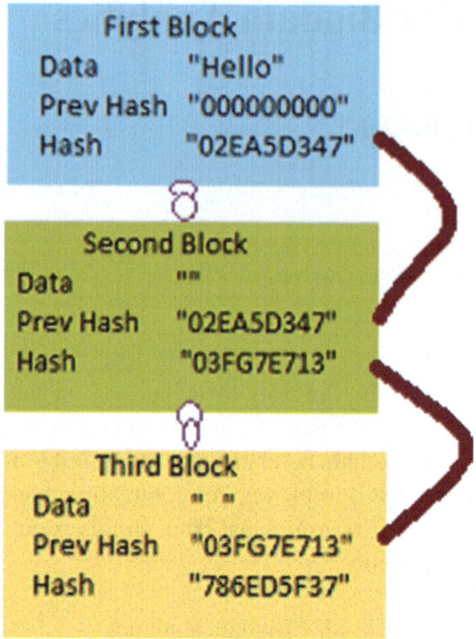

Fig. 1 Blockchain process

2 Literature Review

This paper discussed about the many organizations big dataset transactions and maintenance difficulties like banks, hospitals. So they proposed the decentralized solution for bigdata exchange, the solution aims ecosystem all participators cooperate to exchange the data peer-to-peer way [1]. The paper explains about railway department enhancing their operation in digitalization [2]. The paper discusses peer-to-peer distributed storage in blockchain and various industries popular in decentralized ecosystem. They proposed new storage in blockchain is mystiko. It is high transaction throughput, high availability, high scalability and make bigdata more secure [3]. This paper proposed hierarchy of intelligent agent and use of cryptographic keypairs, not restricted to using SHA256 algorithm and other secure hash algorithm used SHA-3 subset [4]. This paper discussed big advantage of blockchain decentralization, sharing scientific workflow data handled on chain and the approach is effective and efficient [5]. This paper said what are all the problem we are facing at the time of uploading huge amount of data to public distributed network. In this paper advantages and disadvantages discussed [6]. The paper investigates blockchain technology applications and addressed key challenges faced by domain of social business. Using semi-formal modeling approach and micro finance use case to investigated blockchain technology for SB (Social business). Another thing is this technology will provide in terms of auditability, transparency, decentralization and privacy [7]. This paper introduced lightweight manner two protocols achieve authenticity, privacy, data integrity for IOT devices. Variety of use cases applied their protocols. Identity management, sensor network data, supply chain, privacy management in BRICS-SSP and software updates in BRICS-BB used [8]. This paper discussed using consensus algorithm in blockchain to store logs, basically consensus algorithm is voting-based protocol. This paper explores implementation and architecture of log system and also verifies the feasibility [9]. This paper summarized smart grid with blockchain and decentralized transaction used in latest technology [10]. The paper discussed value of smart contract. Blockchain and smart contract disrupt well established services [11].

3 Concepts of Blockchain

- Mining
- Distributed P2P Network
- Immutable Ledger:
- Hash Cryptography
- Consensus Protocol

Mining part the Miners validate every new transaction and then save the global ledger. Miners solve the mathematical problem using cryptographic hash algorithm. Distributed peer-to-peer is computer systems connected through internet without this server files directly shared between systems. Each system acts the server and clients. Immutable Ledger is the ledger records cannot be changed is called immutable ledger. If i want to make money transaction $500 no one alter or change it. If I made mistake fix the problem. Then only make another transaction and correct the error. Actually this is good practice of accounting. Entry made into the ledger never change or removed. Hash Cryptography part the finger print is identify the person same principle is used to identify the digital documents, such a fingerprint is called SHA256 (Secure Hashing Algorithm 256 is number of bits). SHA256 Algorithm works very secure and fast move documents. It cannot go hash to documents, so one way process. We can run exactly same documents Hash will same so deterministic and Fast computation. We can take same two document change one bit of data in one document hash will be totally differ the two documents this is Avalanche effect. Consensus Protocol is the distributed processes to achieve single data value, involving the network multiple unreliable nodes.

4 Algorithm in Blockchain

4.1 SHA 256 Algorithm

It produces the output 256 bit long. SHA256 algorithm produces same arbitrary data given same inputs. Mining is finding the Nonce. Everyone run the hash and confirm it is valid or not. Because impossible to predict what the nonce will be, so proof that miner worked and get valid hash (proof-of-work). Its deterministic, one way function, fast computation, avalanche effect and must withstand collision. We can take same two text document, first one document is HAI THIS IS KRITHIKA Fig. 2 shows hash will be "26c3016f05fa282ec5ac47bdce17be81914f329aa493e8b7b510ba351dcd232b" like this.

Second document is HAI THIS IS KRITHIKA1 just 1 only added second document hash is "1bb87b327540a7821d08571e82999719565cc789e31663bc586d3952230de681" completely changed. So the difference between first and second document is 1 only but the total hash will be different for document 1 and document 2. This process is called avalanche effect (Fig. 3).

Fig. 2 Creating Hash in first document

Fig. 3 Creating Hash in second document

5 Bigdata Analytics

Bigdata is any amount of structured, semi-structured, unstructured data. If the volume of data is large difficult to process traditional technique. Data sources are enterprise, humans. Roles of Bigdata are Data scientist and Data engineer. Analytics is a process of breaking problems into simpler part and inferences based on data to take decisions. Analytics is not a tool it's a way of acting or thinking. Characteristics of Bigdata are: Volume—volume is size of the data, Velocity—velocity is change of speed, Variety—variety is data sources in different forms. Structured data is RDBMS, Oracle, MySQL, Access, Excel etc. Semi Structured Data is Email, Twitter, and Facebook etc. Unstructured Data is photo, video, music etc. Veracity—veracity is data uncertainty, and value is bigdata value attained by leveraging. Validity—validity is correctness of data.

6 Different Analytics

- Descriptive analytics
- Predictive analytics
- Prescriptive analytics

Descriptive analytics is useful for understanding already available information and help to make decisions in present. Predictive analytics is identifying patterns using statistics to make inference the future. Predictive analytics retain and grow profitable customers and unknown future events it's used to make prediction. Predictive model is to predict current data to what will be the next. Prescriptive analytics not only says what is going on and most importantly what to do about it.

7 Algorithms in Bigdata

- K-Means Clustering Algorithm.
- K-Nearest Neighbors.
- Linear Regression.
- Logistic Regression.
- Classification and Regression Trees.

K-means clustering algorithm is simple to use and very fast, It is a unsupervised learning algorithm. K-nearest neighbors algorithm very expensive depending on the size and scope of K-Nearest neighbor. Linear Regression is widely used, because easy to visualize the data and quantitative measures between two sets of data. Logistic Regression is similar to linear but problem in quantitative forecasting. It is clearly defined yes or no method. Classification tree is large and complex. The variant of these two trees is called random forests.

8 Bigdata in Industries

Financial services applications help credit card companies to secure transaction. Retail the way of buying and selling is fast. The technology used to find customer most likely certain products, this is best way to approach customers. Manufacturing Data is hugely important role plays in manufacturing. Manufacturers collecting the valuable data and monitoring efficiency of machines. Healthcare in Big Data improves quality of life and avoid preventable deaths. We used bigdata techniques to monitor heartbeats and breathing patterns.

9 Tools

Tableau, Knime, Zoho, Aquadata studio, NodeXL, Azure Hdinsight, Skytree, Talend, Splice machine, Spark, Apache samoa these tools used in Data analytics. A tableau is data visualization software, its very easy to use drag and drop, the raw data within a seconds we will result of data. People understand easily just import and get structured data, charts, maps. Zoho is also data visualization software anyone can use, many industries using this software simple drag and drop interface. Choosing various charts, pivot table etc.

10 Conclusion

We discussed blockchain and bigdata useful information from many papers in literature review. Concepts of blockchain explains mining, immutable ledger, distributed peer-2-peer network, hash cryptography and consensus protocol. So the data will be more secure, speed, efficiency and transparency for using these technologies. This paper explores the survey on Blockchain and bigdata analytics.

References

1. Chen J et al (2017) Bootstrapping a blockchain based ecosystem for big data exchange. In: 2017 IEEE international congress on big data, https://doi.org/10.1109/bigdatacongress.2017.67. Accession Number: 17188656

2. Naser F (2018) The potential use of blockchain technology in railway applications an introduction of a mobility and speech recognition. In: 2018 IEEE international conference on big data. https://doi.org/10.1109/bigdata.2018.8622234. Accession Number: 18412247
3. Maurakirinathan P et al (2018) Mystiko—blockchain meets big data. In: 2018 IEEE international conference on big data. https://doi.org/10.1109/bigdata.2018.8622341. Accession Number: 18412133
4. Wright C et al (2017) sustainable blockchain-enabled services: smart contracts. In: 2017 IEEE international conference on big data. https://doi.org/10.1109/bigdata.2017.8258452. Accession Number: 17505055
5. Wang J et al (2018) Blockchain based provenance sharing of scientific workflows. In: 2018 IEEE international conference on big data. Accession Number: 18412546. https://doi.org/10.1109/bigdata.2018.8622237
6. Preece JD, Easton JM (2018) Towards encrypting industrial data on public distributed networks. In: 2018 IEEE international conference on big data. Accession Number: 18412431. https://doi.org/10.1109/bigdata.2018.8622246
7. Vatrapu R et al (2018) Converging blockchain and social business for socio-economic development. In: 2018 IEEE international conference on big data. https://doi.org/10.1109/bigdata.2018.8622238. Accession Number: 18412388
8. Kim THJ et al (2018) BRICS: blockchain-based resilient information control system. In: 2018 IEEE international conference on big data. Accession Number: 18412082. https://doi.org/10.1109/bigdata.2018.8621993
9. Huang J et al (2018) Blockchain based log system. In: 2018 IEEE international conference on big data. Accession Number: 18412540. https://doi.org/10.1109/bigdata.2018.8622204
10. Kotsiuba I et al (2018) Blockchain evolution: from bitcoin to forensic in smart grids. In: 2018 IEEE international conference on big data. Accession Number: 18412177. https://doi.org/10.1109/bigdata.2018.8622232
11. Gilcrest J et al (2018) Smart contracts: legal considerations. In: 2018 IEEE international conference on big data. Accession Number: 18412500. https://doi.org/10.1109/bigdata.2018.8622584

Blockchains Technology Analysis: Applications, Current Trends and Future Directions—An Overview

Aisha Zahid Junejo, Mehak Maqbool Memon, Mohammed Ali Junejo, Shahnawaz Talpur, and Raheel Maqbool Memon

Abstract One of the most promising technical trends that have taken the world by storm is the blockchain technology which is the underlying mechanism of the bitcoin. It is a distributed, decentralized and innovative technology that is predicted to bring a huge change in the world. This technology is analogous to the Internet with built-in robustness. Although it is currently one of the most researched areas in the domain of information technology, it has a set of its own limitations. This paper discusses the blockchain technology from scratch, starting off with the general introduction followed by presentation of a comprehensive insight into blockchain applications, the current research trends, limitations and potential future directions. The paper concludes that blockchain networks are likely to change the outlook, perspectives and utilization techniques of the information technology industry in the future by decentralizing the World Wide Web and making the traditional business defunct. Moreover, the amalgamation of blockchain technology with IoT and data science is going to open new horizons of research and development in tech world.

Keywords Blockchain technology · IoT · Literature survey · Health care · Education · Cryptocurrency · Algorithm audit · Federated learning

1 Introduction

Blockchain is the underlying technology for a widely known cryptocurrency–bitcoin [1]. This emerging technology has eliminated the need of third parties, such as banks, for authorization and record keeping of various transactions which makes the history of transaction available to all the nodes (blocks) in the chain, ensuring

A. Z. Junejo (✉) · M. M. Memon
Universiti Teknologi Petronas, Seri Iskandar, Malaysia
e-mail: aisha.junejo@students.muet.edu.pk

A. Z. Junejo · M. M. Memon · S. Talpur · R. M. Memon
Mehran University of Engineering and Technology, Jamshoro, Pakistan

M. A. Junejo
University of Sindh, Jamshoro, Pakistan

S.-L. Peng et al. (eds.), *Intelligent Computing and Innovation on Data Science*,
Lecture Notes in Networks and Systems 118,
https://doi.org/10.1007/978-981-15-3284-9_45

transparency of data to each individual user [2]. The first widely accepted application of blockchain technology is the bitcoin, which was introduced by S. Nakamoto in 2008 [3]. This cryptocurrency offered a platform to the users for transferring tokens over a Peer-2 Peer distributed network. All these P2P transactions were secured and anonymous [4]. Blockchains were introduced with the motive of substituting the trust (by third party) with the proof (transparency). This purpose is achieved by using cryptographic hash functions. Each time a new block needs to be created, the hash functions are validated by the network first, succeeding in the creation of the block [3]. According to IBM, blockchain ensures data integrity and validity through a consensus known as smart contract, in the terminology of blockchain technology, which ensures that each transaction that is invoked must be validated by all relevant parties. Once this happens, the transaction takes place. This makes the system more efficient. Currently, the blockchain technology is being used in various applications such as patient record keeping, student record keeping, cryptocurrency, 3D printing, supply chain management, neuroscience, IoT and many more. Blockchain unit is grouping of all bitcoin exchanges executed in the past. Fundamentally, it is a decentralized database which keeps up a persistently developing proof information structure blocks which holds individual exchanges [5].

The working mechanism of a blockchain is shown in Fig. 1.

i. ***Triggering transaction***: Node 1 has to transfer some data to Node 2; after initialization of transaction represented by Node 1, the information is broadcasted to all the parties in the network.
ii. ***Validation and verification***: Blockchain module decides if the transaction is to be verified or not. Verification and validation are done on the basis of algorithm chosen by all parties of the network.
iii. ***Creation of new block***: Block is created on the basis of validation and verification.
iv. ***Addition of block to the chain***: Finally, new block is added, and information is shared over the network.

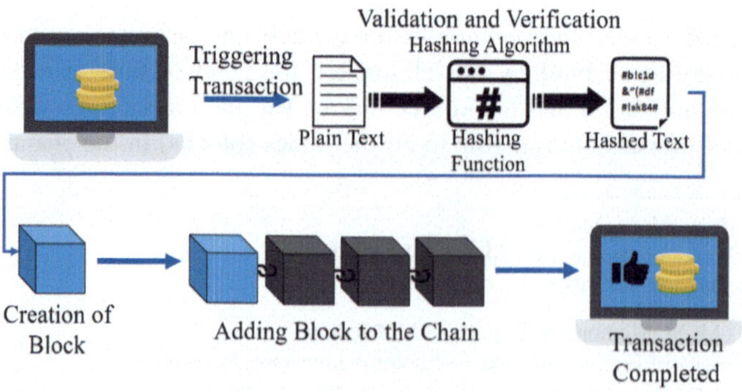

Fig. 1 Working mechanism of blockchains

2 Review of Literature

The concept behind blockchain was first introduced by S. Haber and W. S Stornetta in 1991 [6]. They presented the concept as network of blocks secured cryptographically. Few years later, Nick Szabo took this idea further and started working on decentralized digital currency, BitGold. Ten years later, S. Nakomoto finally brought the idea into real-time implementation by introducing bitcoin [7]. After it was successfully implemented in financial sector, researchers begin to study about the ways of implementing this concept in various other sectors. A brief survey of the literature is presented in this section demonstrating the acceptance of this horizontal innovation in various fields and portraying its benefits in different domains. In [8], researchers have proposed a novel blockchain architecture based on satellite chains to execute consensus protocols privately in parallel to boost up the entire scalability of entire blockchain module based on independent interconnected submodules of single blockchain system. The proposed prototype also encounters "handoff" regulations to manage entire network. Additionally, it supports heterogeneous consensus using satellite chains. The proposed module can be used to enhance and revolutionize industrial applications. Eyal and Gencer in [9] proposed next-generation bitcoin (Bitcoin-NG) to provide better scalability in the network, based on Byzantine fault-tolerant blockchain protocol.

In accordance to it, some innovative metrics to quantify the safety and effectiveness of bitcoin-based blockchain protocols are presented. In study [2], a decentralized personal data management module is proposed to ensure privacy and security of personal data. Classic blockchain protocol is turned into programmed access control manager to provide third party tolerance. Bitcoin transactions used for the proposed module are related to information transfer such as storing, querying and sharing data rather than financial transactions. Blockchain and off-blockchain storage are combined for the construction of personal data management platform concerning privacy issues. Also, legal and regulatory decisions made in the network are simplified for sensitive data sharing by implementing decentralized platform in turn. In study presented in [10], a hybrid blockchain mechanism is proposed which includes novel consensus method by means of credibility scores making the system robust enough to prevent attacks by monopolizing resources. Moreover, IBM in [11] built a system on Bluemix which is one of the best examples of the initial blockchain applications in IoT. Further, Samaniego and Deters [12] deployed blockchain, as a service for IoT applications. In [13], a higher education credit (HEC) platform based on blockchain is proposed. Another education blockchain technology based on learning outcome is proposed in [14]. Research in [15] encounters constraints of electronic health records by using blockchain module to provide innovative systems for patients to get their details of health care. Similar application in [16] describes blockchain clinical trials which help in reduction of data fraud by hiding the identity of people involved in trials, management of micro-processes and documentation involved.

Table 1 Latest trends in blockchain

Beyond cryptocurrency	IBM, Amazon, Walmart and other huge companies are experimenting with the technology. Food safety, patient record keeping, shipping and voting are someone of its other implemented applications
Algorithm auditing	Depending on the quality of data, AI learning algorithms have the tendency to amplify bias and discrimination in the results instead of benefiting. This makes it compulsory for the algorithms to be audited. Blockchains, due to their transparent record keeping and immutable nature, can hold point-to-point information decisions which can make the process of algorithm audit relatively simpler
Smart contracts	Blockchain technology has inherent flaws which are acting as obstacles in its mainstream adoption. To overcome them, the smart contract algorithms need to be improvised, and hence, this is one of the latest trends of research in blockchain technology
Federated learning	Machine learning usually occurs on a centralized system where the data and the training code reside on a single machine. This can generate trust issues between the party proposing and implementing a training model and the party providing the data to be trained. To eliminate this trust issue, a concept of federated learning comes into place. Federated learning is a kind of machine learning which is distributed and does not provide direct access to the data
Blockchain with IoT	IoT is a technology to gather data online from the devices connected in a network [17], while the blockchain records the data and makes it accessible to all the participating nodes. Combining the two concepts theoretically leaves us with a secure, verifiable and a permanent data recording method that is produced by devices in IoT network [18]

3 Latest Trends in Blockchain Technology

To understand the current position of blockchain research, we examined the current research topics of the technology and surveyed various Web blogs, research publications and Internet resources. In the end, we contemplated a few topics of the current interest in this area. Hence, some of the current trends, in terms of application areas, of this emerging technology are discussed in Table 1.

4 Blockchain Analysis

The first and most widely accepted implementation of blockchain technologies is the cryptocurrencies. Throughout years, a number of different currencies have been brought into real-time use, though, not all of them are still functional. Table 2 shows a summary of a few cryptocurrencies over the years.

Despite having a huge implementation in cryptocurrencies, the emerging technology of blockchains is not limited to them. It has only been about a decade since the technology was first used, but over the years, people have accepted the concept

Table 2 Summary of cryptocurrencies from 2008 to 2018

Cryptocurrency	Year	Author	Hash Function
Bitcoin	2008	S. Nakomoto	SHA-256
Litecoin	2011	C. Lee	Scrypt
Peercoin	2012	[online resource]	SHA-256d
Primecoin	2013	D. Schwartz, N. Youngs, A. Britto	Cunningham chain
Ripple	2014	S. King	EC digital signature
Ethereum	2014	G. Wood	Ethash
Permacoin	2014	A. Miller et al.	Floating digital signature
Blackcoin	2014	P. Vasin	Scrypt
Auroracoin	2014	D. Carway	Scrypt
Darkcoin	2014	E. Duffield and K. Hagan	X11
Namecoin	2015	H. Kalodner et al.	SHA-256d
Lisk	2016	Max Kordek, Oliver Beddows	PoS
Zcash	2016	Danny Yang, Jack G Zooko Wilcox	SHA-256
EOS.IO	2017	Daniel Larimer, Brenden Blumer	SHA-256
Bitcoin private	2018	Jacob Brutman et al.	Equihash

and are continuously working on it to spread its benefits into other sectors as well including finance departments, educational sectors, healthcare industry, online shopping, etc. The graph in Fig. 2 shows how the technology has emerged all through the years since it was first introduced. The data for the graph has been taken from Google repository of research articles/papers. It can clearly be seen from the graph that the technology has been hugely accepted and the number of publications made by people in the field of blockchains has drastically increased over the time. In fact,

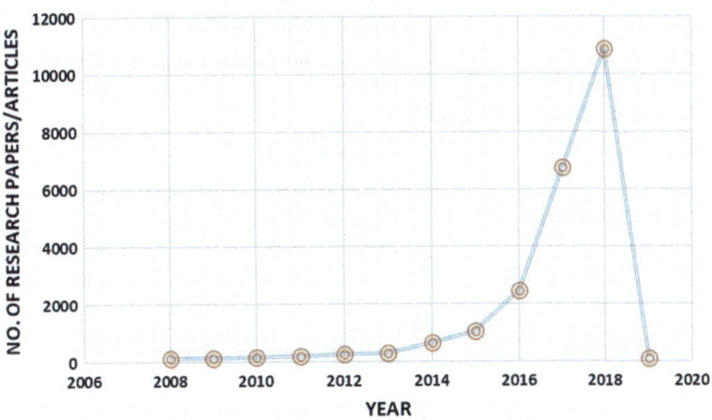

Fig. 2 Evolution of blockchain technology

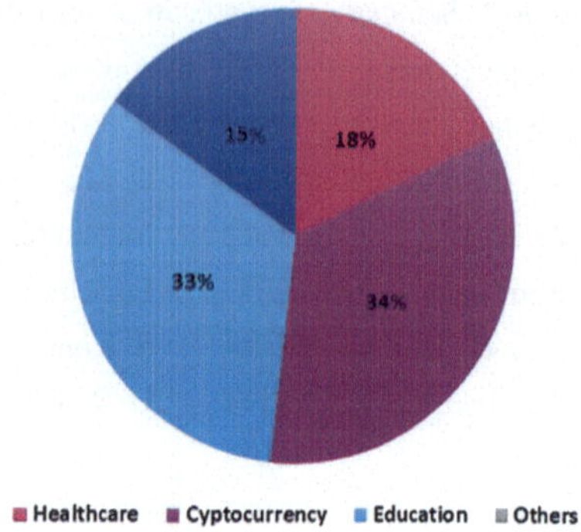

Fig. 3 Research publications in various fields

several articles for the upcoming year's (2019) journals and conferences have been prepublished. This shows that the technology, regardless of the implementation area, is an active research field.

The pie chart, shown in Fig. 3, shows the number of publications in various fields. The data is obtained on the basis of no. of records generated in search results through Google scholar. The chart depicts that the most researched topic in blockchain industry is cryptocurrency, followed by, education with almost equal importance, followed by healthcare industry and others. These other industries may include advertisement, media and marketing industries, big data and data analytics industry, IoT, etc.

Although the researches are working on it, but before implementing a blockchain network, it is very important to know about the pros and cons that it might bring with itself. Hence, a SWOT analysis was carried out which is given in Table 3.

Table 3 SWOT analysis of blockchain technology

Strengths	**Weakness**
Network decentralization Data transparency Robust architecture Autonomous network Speedy data access Availability of data on distributed ledger	Lack of scalability Poor user experience Interoperability issues Storage capacity (for huge amounts of data in future)
Opportunities	**Threats**
Cost efficiency Elimination of third-party trust Reduced risk of fraud	Hesitant adoption of technology Prone to security attacks Reduced employment in various sectors

5 Limitations

Although the blockchain technology has taken the world by storm and has proved to be a breakthrough technology, there still exists certain limitations and potential risks regarding the adoption of this horizontal innovation, like any other radical innovation.

Immutable and transparent nature of blockchains makes data privacy, a critical issue. Immutability restricts user data modification, and transparency makes the data visible to all, which is not desirable. Further, a long debate regarding the block size in blockchain has been taking place. Increasing the size of the block will allow room for more users, whereas small blocks keep the blockchain more decentralized and reduces participation costs. Another limitation includes behavioral changes. It takes a lot of time, successful experiments and trust to shift from one platform to another. People these days are so used to inclusion of third parties in day-to-day transactions that they need time to get used to the idea of safe and secure blockchains for various applications.

Another limitation arises upon evaluating the performance of blockchains in the current systems. Since there are no central systems involved, the block in the network is itself responsible for storing the data, authorization, consensus and signature verification which makes the system a bit slower. This may result in lack of data storing capacity in the future when the data to be stored in each block increases. Moreover, the blockchain applications with greater number of blocks require greater data throughput, which is one of the major concerns in creation of blockchain networks.

6 Conclusion and Future Directions

The fundamental purpose of introducing blockchains was to introduce data security, privacy, autonomy, anonymity and transparency to all the users. Another advantage that the technology has to offer is its non-hackable nature. The transactions of the blockchain network cannot be reversed, and no single server carries the entire data. This technology has also been proven to be more time efficient as compared to the third-party hosted systems. However, all the facilities induce several technical challenges and limitations in implementation that need to be addressed including the problem of data scalability, data storage capacity, interoperability, etc. Fortunately, with each passing year, more R&D specialists are working for enhancement of this technology by addressing various issues and providing viable solutions to them.

Future research can be carried out in solving the block size issue by making the size dynamically changing with respect to the application. Moreover, researchers may try to make user anonymity stronger. Blockchain can be also incorporated with big data analytics for secure data management, and transactions made on blockchain can also be used for data analytics. One of the more traditional yet compelling use cases would be the sharing of files in a more secret manner than permitted by the current P2P BitTorrent-based systems. Blockchains are predicted to be deployed in almost 20% of IoT networks in the near future, i.e., a year or two from now.

Acknowledgements The authors would like to extend their gratitude to University of Technology Petronas (UTP) for providing necessary support throughout the research work.

References

1. Tama BA et al (2017) A critical review of blockchain and its current applications. In: International conference on electrical engineering and computer science (ICECOS), IEEE, pp 109–113. 978-1-4799-7675-1/17/$31.00 ©2017
2. Zyskind G, Nathan O, Pentland A (2015) Decentralizing privacy: using blockchain to protect personal data. In: 2015 IEEE security and privacy workshops, San Jose, CA, pp 180–184. https://doi.org/10.1109/spw.2015.27
3. Florea BC (2018) Blockchain and Internet of Things data provider for smart applications. In: 7th Mediterranean conference on embedded computing (Meco), 11–14 June 2018, Budva, Montenegro
4. Nakamoto S (2009) Bitcoin: a peer-to-peer electronic cash (online). https://bitcoin.org/bitcoin.pdf. Retrieved 9 Sept 2018
5. Roe D (2018) 7 trends driving blockchain forward, June 2018 (online). Retrieved September 2018
6. Haber S, Stornetta W (1991) How to time-stamp a digital document. J Cryptol 3:99–112
7. World Crypto Index (2018) Blockchain (online). Retrieved September 2018
8. Li W, Sforzin A, Fedorov S, Karame GO (2017) Towards scalable and private industrial blockchains. In: Proceedings of the ACM workshop on blockchain, cryptocurrencies and contracts, BCC '17. ACM, New York, pp 9–14
9. Eyal I, Gencer AE, Sirer EG, Renesse RV (2016) Bitcoin-NG: a scalable blockchain protocol. In: 13th Usenix conference on networked systems design and implementation (NSDI'16), Berkeley, CA, USA, pp 45–59
10. Watanabe H, Fujimura S, Nakadaira A, Miyazaki Y, Akutsu A, Kishigami J (2016) Blockchain contract: securing a blockchain applied to smart contracts. In: 2016 IEEE international conference on consumer electronics (ICCE), Las Vegas, NV, pp 467–468. https://doi.org/10.1109/icce.2016.7430693
11. Panikkar BS, Nair S, Brody P, Pureswaran V (2014) ADEPT: An IoT Practitioner Perspective. IBM
12. Samaniego M, Deters R (2016) Blockchain as a service for IoT. In: 2016 IEEE international conference on Internet of Things (iThings) and IEEE Green Computing and Communications (GreenCom) and IEEE Cyber, Physical and Social Computing (CPSCom) and IEEE Smart Data (SmartData), Chengdu, pp 433–436. https://doi.org/10.1109/ithings-greencom-cpscom-smartdata.2016.102
13. Turkanović M, Hölbl M, Košič K, Heričko M, Kamišalić A (2018) EduCTX: a blockchain-based higher education credit platform. IEEE ACCESS, date of publication 5 Jan 2018, Digital Object Identifier. https://doi.org/10.1109/access.2018.2789929
14. Duan B, Zhong Y, Liu D (2017) Education application of blockchain technology: learning outcome and meta-diploma. In: IEEE 23rd international conference on parallel and distributed systems
15. A case study for blockchain in Healthcare: "MedRec" prototype for electronic health records and medical research data
16. http://www.ssidecisions.com/blog/how-blockchain-can-benefit-healthcare

17. Babar S, Mahalle P, Stango A, Prasad N, Prasad R (2010) Proposed security model and threat taxonomy for the Internet of Things (IoT). In: Meghanathan N, Boumerdassi S, Chaki N, Nagamalai D (eds) Recent trends in network security and applications, CNSA 2010. Communications in computer and information science, vol 89. Springer, Heidelberg
18. Marr B (2018) Blockchain and the Internet of Things: 4 important benefits of combining these two mega trends, 28 Jan 2018, Forbes (online). Retrieved September 2018

A Study on Seismic Big Data Handling at Seismic Exploration Industry

Shiladitya Bhattacharjee, Lukman Bin Ab. Rahim, Ade Wahyu Ramadhani, Midhunchakkaravarthy, and Divya Midhunchakkravarthy

Abstract Cumulative size as well as a changeable pattern of composed geographical large data boons issues in storage, handling, unfolding, studying, anticipating and proving the eminence of input data files. These issues become big challenges, especially in the oil and gas industries. At the same time, seismic exploration is to cultivate an image of the subsurface geology. The geophysical exploration in overall and seismic acquisition in specific is challenged vastly in terms of the tough logistics and intricate subsurface geology. Hence, this research proposes a unified technique to figure out time complexity in large seismic data dispensation with parallel processing, smart indexing and reducing latency time. Furthermore, this research uses a combined platform of Hadoop and Hive where MapReduce analyzes the data and HDFS stores it after processing. The result shows its high time efficiencies by offering high throughputs, I/O rates as well as low latencies.

Keywords Composed geographical large data · Seismic exploration · A combined platform of Hadoop and Hive · HDFS and MapReduce · Time complexity · High throughputs as well as I/O rates and low latencies

---

The original version of this chapter was revised: The author's name has been corrected from "Midhunchakkravarthy" to "Midhunchakkaravarthy". The correction to this chapter is available at https://doi.org/10.1007/978-981-15-3284-9_86

S. Bhattacharjee (✉) · L. B. Ab. Rahim · A. W. Ramadhani
High Performance Cloud Computing Center (HPC3), Universiti Teknologi PETRONAS, Seri Iskandar, Perak Darul Ridzuan, Malaysia
e-mail: shiladitya.b@utp.edu.my

L. B. Ab. Rahim
e-mail: lukmanrahim@utp.edu.my

A. W. Ramadhani
e-mail: ade.wahyu@utp.edu.my

Midhunchakkaravarthy · D. Midhunchakkravarthy
Computer Science and Multimedia Department, Wisma Lincoln, Petaling Jaya, Selangor Darul Ehsan, Malaysia

S.-L. Peng et al. (eds.), *Intelligent Computing and Innovation on Data Science*, Lecture Notes in Networks and Systems 118,
https://doi.org/10.1007/978-981-15-3284-9_46

1 Introduction

The oil and gas diligences are normally data-oriented business. These industries rely on information technology (IT) to enhance the speed of finding oil, intensify oil production and cut down the risks related to health, safety, as well as an environment that arises with the failure of equipment or operator dispute. Handling huge volumes of data for reservoir modeling or simulation is not new. What has different is the obtainability of new technology based on service appliances comprehends big data and analytics [1, 2]. The prospective for Big Data and analytics myths in retrieving formerly unused data, allowing the usage of geological data with seismic exploration through corrections and providing employees with institutional expertise for searching [3–5].

The production of large digital data in oil and gas industries, the information extraction from such accumulated big data and their maintenance at the same time are not easy tasks to perform. The data volume raises consequently at the same time. The diverse data setups such as DLIS, LIS, SEGx, SEGy, XLS, CSV and others are also present to explore different features using different big data analytics [6, 7]. The big size of the input data itself is a huge challenge to process with in the time. Besides, the huge assimilated various noises also make huge disruptions during such large data processing. The current literature suggests a few approaches for data collection such as collection of real-time data during drilling as well as data accumulation using different sensors [8].

This research particularly focuses on developing an integrated system that can address all these issues in a combined platform. The particular objective of this research is:

1. Reduction of processing time for data integration or accession to or from the databases during executions with an advanced parallel processing system.
2. Reduction of database traverse time to retrieve required information from the huge amount of data sets with a smart indexing system.
3. Reduction of latency time during the processing of huge sets of data in seismic exploitation.

The other sections of this article are organized as: A detail background study has been included in Sect. 2 to explore the remaining research space by analyzing the strengths and weaknesses of distinct related research work, Sect. 3 describes the construction of proposed integrated technique to fill the current research gap, Sect. 4 comprises the details of data preparation and experiment setup, Sect. 5 encompasses the result analyses to validate the current research objectives and finally, Sect. 6 concludes this research and decides the future direction.

2 Background Study

These days, oil and gas industries are facing the biggest challenges such as high abstraction cost and stormy state of international policies. These issues enhance the issues in exploration and drilling for new reserves. The collection of data and extracting valuable information are also quite challenging for managing as well as storing them. Various business organizations such as oil and gas industries have completely developed a data controlled approach for overpowering disputes and disentangling the outstanding trials [9, 10]. Several distinct types of research have been conducted to address various issues of oil and gas industries. Hence, this section discusses several important and related researches according to their tenacity as well as shortcomings for scrutinizing the present gap in these diligences.

Article [11, 12] comprises a technique that integrates cloud computing database with Internet of- Things (IoT). Management of an increasing amount of data with the variability of data types and data basics for meeting solicitation-specific presentation basics is one of the large trials in this research. Hence, in this research, the appropriateness of distinct databases for loading and retrieving IoT data in the cloud has been tested. This work has plentiful confinement; for instance, it wants to increase the current standards databases with more compound types of IoT. It examines the performances of such databases periodically. Nevertheless, this investigation needs further study to examine the strong outfit of the schema-free data example as well as the notable connection of data through distinct SQL tables.

A secluded cloud data center assistance has been designed in the article [13–15], where HBase and Cassandra are combined with an adaptable GUI. The proposed technique has been verified with the graphical image data as input during the experiment. In this work, the binary carriages decorum skill, acquired from Apache Thrift, is applied to indorse graphical user interface. It uses a command-line interface that supports to execute cross-platform processes of data comparison, read/write operations and subordinate indexing. This investigation combines Java with Thrift and it acquires a reformist skill of using graphical user interface for secondary indexing with NoSQL database. But, the time competence of the planned method may obstruct due to such type of constraint as Java has some its restraint such as Java can process only 64 MB of data in distinct repetition.

The performances of HBase and MySQL for distinct cases of random read and write operations have been compared in the article [16–18]. As per the author, HBase is a vulnerable foundation surrogate for the customary database supervision. It is extremely mountable, error-lenient, consistent, NoSQL, a dispersed database that functions over a bunch of vendible computers to manage huge data bulks. Normally, HBase offers better performance than MySQL for handling the distributed file system. However, this system does not offer any indexing system which causes lower data retrieval throughput. As a consequence, processing of large scale complex seismic data is highly time-consuming and offers comparatively higher latency time than other databases like Hive.

Hence, the entire discussion in this section shows that there are still various issues in large complex data processing. The uses of existing SQL and NoSQL databases with distinct approaches cannot solve the distinct complexed issues of big compound data. Hence, the decision making by processing the large complex data with these existing techniques is quite difficult for various organizations such as oil and gas industries. Hence, there are furthermore research scopes to resolve these issues in an integrated way for such industries.

3 Construction of Proposed Integrated Technique

The advanced indexing system of Hive further helps to retrieve the data very efficiently. A clear structure of the proposed integrated has been portrayed in the following Fig. 1.

Each of the parts is further discussed with more details at the following.

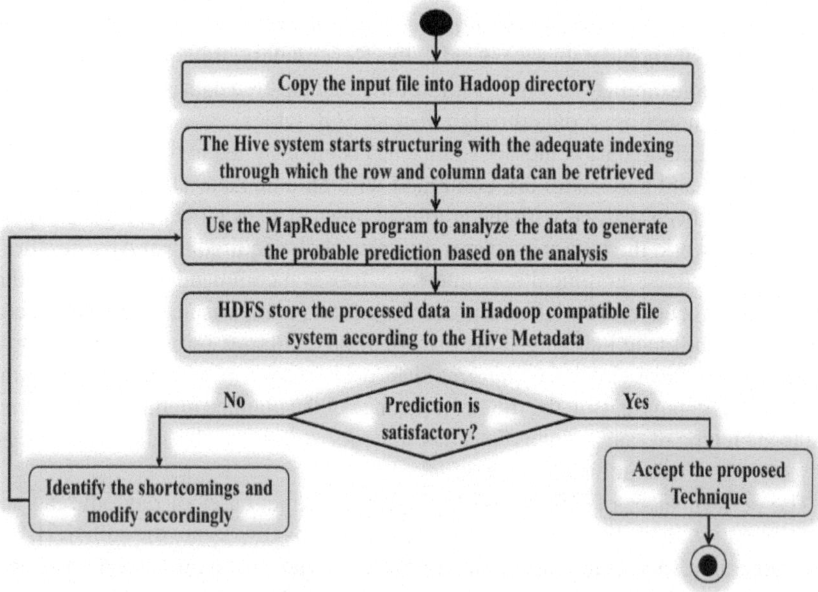

Fig. 1 Integrated technique to process data with predictive analysis and a smart indexing

3.1 Conversion of Unstructured or Semi-structured to Structured Data

The exploration of information from the complex large file such as seismic big file can be possible with the required in the Hadoop platform [1]. The MapReduce program distributes the resources in various dispersed processing units with the help of a programming model called the Hadoop Distributed File System (HDFS) [6, 13]. The HiveQL is similar to SQL coverts data into a structured format where the combination of Hive interprets queries into a sequence of MapReduce tasks. However, the existing system cannot convert the data with the use of HiveQL [17].

3.2 Smart Indexing System for Retrieving Data with Minimum Search

It does not comprise any keys such as the normal relational databases; however, it can speed up the search operations by forming the indexing data columns [17]. In this integrated system, this indexing process is further customized with the plug-in of Java code which helps to boost the distinct feature of it that it can fulfill various needs of big data processing. However, this indexing system requires additional disk space to prolong and enhances processing costs to create indexes [19, 20].

3.3 MapReduce Program for Predictive Analysis

Predicting is the process of deciding by analyzing any data set with the help of machine learning as well as deep learning. This prediction process can be many types such as failure prediction of the power network, fault detection in card-based transaction mechanism, decision making about the market by targeting the customers and so on. However, this research article does not concern about any kind of prediction; hence, how the proposed integrated technique can be applied in predicting is the extended part of this research.

3.4 Distribution of Processed Data According to Hive Meta-Data

In this framework, MapReduce programming is strictly united for processing complex data. Normally, HDFS breaks the complex large data into several blocks and distributes them in distinct nodes of a particular cluster. Thus, it enhances the parallel processing speed by the simultaneous allocation of input resources in an efficient way

[1, 6]. Thus, HDFS helps to enhance its fault-tolerant capacity. As the consequences of it, if the data node crashes, the replicated data can be found elsewhere in the cluster and the lost data can be recovered.

4 Experiment Setup and Data Preparation

The attainments in different aspects of the projected integrated technique have been tested in a Linux environment. Hadoop-2.7.4, Hive-2.1.0 and JDK-8.1 have been used as software for executing the proposed integrated technique in the Hadoop platform. The hardware has been configured with an 8-core processor, 32 GB RAM and 2 TB of the internal hard disk. The downloaded files are preprocessed and concatenated into big SEG-Y file (Max 1.2 TB) using concatenation operation.

5 Result Analysis

The execution efficiencies of the planned technique are evaluated with the following parameters.

5.1 Throughput (TP)

It is calculated to measure the competence of time to process a certain amount of data by any database. Throughput offered by any database can be calculated with the following Eq. (1) as

$$\mathrm{TP} = \frac{\text{Input File size}}{\text{processing time}} \tag{1}$$

Throughputs produced during the processing of large complexed large files by our projected technique and few former databases have been plotted in Fig. 2.

From Fig. 2, it can be comprehended that the planned combinatorial technique offers much higher TP in all cases in read–write. It fulfills our first target of this study.

5.2 I/O Rate

It calculates the read/write sequence of records from a key space. '*N*' numbers of reading/write concomitantly process a sequence of records. It can be calculated with

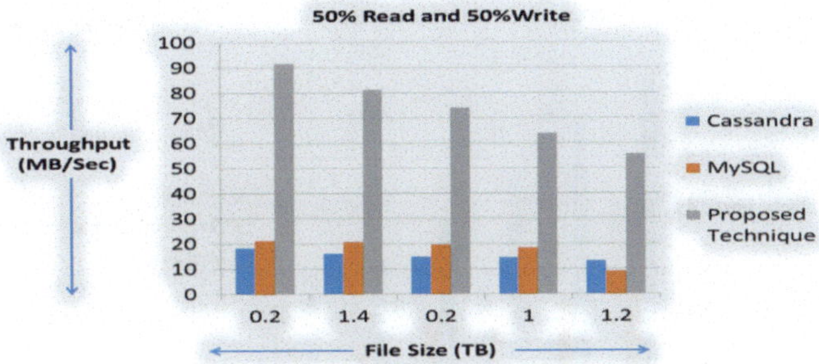

Fig. 2 Throughputs, offered by various databases

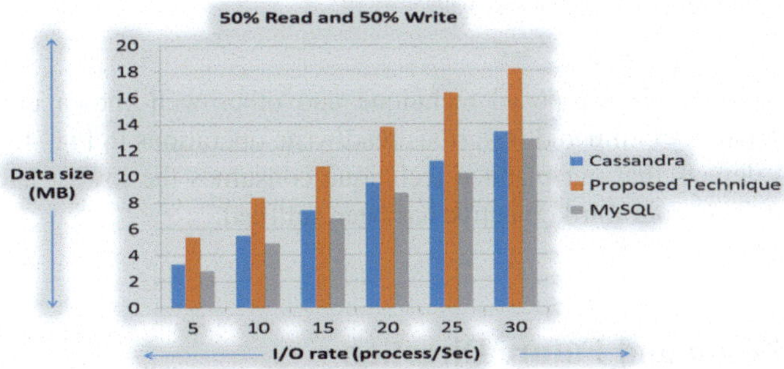

Fig. 3 I/O rates, offered by various databases

Eq. (2) as follows,

$$\text{I/O Rate} = \frac{\text{Records executed per process}}{\text{Execution time}} \quad (2)$$

The I/O rates offered via our proposed technique and former databases in terms of reading or writing are plotted in Fig. 3.

The contrast in Fig. 3 shows that our planned technique produces much greater I/O rates on each occasion. Hence, Figs. 2 and 3 fulfill the second objective.

5.3 *Latency*

Latency is the delay from input into a process to produce the desired outcome. Latency can be defined by Eq. (3) as

$$\text{Latency} = \frac{\text{Execution time}}{\text{Records executed per process}} \quad (3)$$

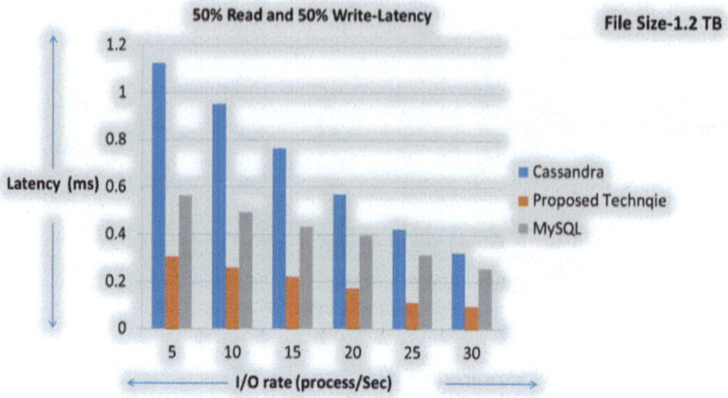

Fig. 4 Latency times, offered by various databases

The capacity of the proposed technique and other well-known databases to produce low latency compared in various read/write operations in Fig. 4.

Figure 4 depicts that our planned technique consumes the lowest latency time than the rest. Thus, our third objective has been fulfilled.

6 Conclusion and Future Direction

The proposed integrated technique combines Hadoop and Hive for processing and storing seismic big data concurrently along with the high execution speed as well as low latency time. The result shows that the projected united technique produces much-improved performances in reading, writing and retrieving data to or from the databases much faster than the prevailing individual or combinatorial techniques with the help of advanced parallel processing and smart indexing systems. The indexing system of the proposed integrated technique makes the data retrieval process faster. However, it can be faster with the further improvement of the proposed integrated technique shortly. Apart from that, the prediction with the machine learning and deep learning mechanism to make a different decision can be added to the enhanced performance of the planned combinatorial technique.

References

1. Zhonghua M (2017) Seismic data attribute extraction based on Hadoop platform. In: 2017 IEEE 2nd international conference on cloud computing and big data analysis, pp 180–184
2. Joshi P, Thapliyal R, Chittambakkam AA, Ghosh R, Bhowmick S, Khan SN (2018) Big data analytics for micro-seismic monitoring. Offshore Technol Conf Asia
3. Soupios P, Akca I, Mpogiatzis P, Basokur AT, Papazachos C (2011) Applications of hybrid genetic algorithms in seismic tomography. J Appl Geophys 75(3):479–489

4. Wu Q, Zhu Z, Yan X (2017) Research on the parameter inversion problem of prestack seismic data based on improved differential evolution algorithm. Cluster Comput 20(4):2881–2890
5. Feblowitz J (2012) The big deal about big data in upstream oil and gas. IDC Energy Insights
6. Soofi EP (2014) Drilling for new business value: how innovative oil and gas companies are using big data to out maneuver the competition. A Microsoft White Paper
7. Subhalakshmipriya C, Tamilarasi R (2014) Next generation tools for oil and gas companies-cloud computing. Int J Comput Sci Inf Technol 5:8214–8220
8. Shastri LN, Heinz D (2011) Data warehousing and mining technologies for adaptability in turbulent resources business environments. Int J Bus Intell Data Min 6(2):113–153
9. Nimmagadda SL, Dreher H (2012) On new emerging concepts of Petroleum Digital Ecosystem (PDE). Wiley Interdisc Rev: Data Min Knowl Discover 2(6):457–475
10. Skretting K, Engan K (2011) Image compression using learned dictionaries by RLS-DLA and compared with K-SVD. In: Acoustics speech and signal processing, pp 1517–1520
11. Ruxandra B, Eleonora MM, Mugurel I, Nicolae T (2012) Practical application and evaluation of no-SQL databases in cloud computing. In: Systems Conference (SysCon), pp 1–6
12. Phan TAM, Jukka KN, Mario DF (2014) Cloud databases for Internet-of-Things data, Internet of Things (iThings). In: 2014 IEEE international conference on, and Green Computing and Communications (GreenCom), pp 117–124
13. Chang BR, Tsai HF, Guo CL, Chen CY (2015) Remote cloud data center backup using HBase and Cassandra with user-friendly GUI. In: IEEE International Conference on Consumer Electronics-Taiwan, pp 420–421
14. Ramon L (2014) Integration and virtualization of relational SQL and NoSQL systems including MySQL and MongoDB. In: Computational Science and Computational Intelligence (CSCI), pp 285–290
15. Diego S, Dan H, Eleni S (2015) From relations to multi-dimensional maps: towards an SQL-to-HBase transformation methodology. In: 2015 IEEE 8th international conference on cloud computing, pp 81–89
16. Bogdan GT, Cristian B (2011) A comparison between several NoSQL databases with comments and notes. In: 2011 RoEduNet international conference 10th edition: networking in education and research, pp 1–5
17. Mehul NV (2011) Hadoop-HBase for large-scale data. Comput Sci Netw Technol 601–605
18. Timothy GA, Vamsi P, Dhruba B, Mark C (2013) LinkBench: a database benchmark based on the Facebook social graph. In: Proceedings of the 2013 ACM SIGMOD International Conference on Management of Data (SIGMOD'13), pp 1185–1196
19. Haijie D, Yuehui J, Yidong C, Tan Y (2012) Distributed storage of network measurement data on HBase. In: 2nd international conference on cloud computing and intelligence systems, pp 716–720
20. Kulshrestha S, Sachdeva S (2014) Performance comparison for data storage-DB4o and Mysql databases. In: Seventh International Conference on Contemporary Computing (IC3), pp 166–170

The Usage of Internet of Things in Transportation and Logistic Industry

K. Muni Sankar and B. Booba

Abstract Internet of Things (IoT) is quickly escalating technology. IoT is a network of objects or things surrounded with electronics, software, sensors, and network connectivity, which enables these objects to accumulate and exchange data. To develop a system which will automatically monitor the transportation and logistics applications and generate alerts, or take appropriate intelligent decisions using concept of IoT with AI. The Internet of Things has numerous opportunities in transportation and logistics sectors, like IoT vehicles can be monitored with respect to their movement, location, whether it is running or stopped, or at any risk, etc. All these can be monitored intelligently using the IoT systems. Vehicles are used for logistics purpose for carrying heavy loads which are packed inside the truck. During such times, it is very important to measure the indoor conditions of the truck like temperature, humidity, light conditions, etc. which can be monitored with sensors. Apart from the payment service near the tolls or any parking places can be automated with the vehicle tracking number, the driver id number, etc. IoT also helps in the guidance and navigation control systems of the vehicles (road transport, air transport, water transport). Transportation authority is highly possible with the use of IoT. Here, various vehicles can be monitored and controlled by means of a central control hub connected through the network. It also offers live and integrative services for monitoring the delivery status indicating the location using GIS mapping. IoT could facilitate in monitoring the traffic and gives the suggestions to take other lines. IoT has given a hopeful way to build powerful systems and applications by using wireless devices, e.g., Android, sensors. By integrating other technologies like big data analytics and artificial intelligence we can have endless applications on intelligence IOT for transportation and logistics industry. **physical objects + sensors and microprocessors = IoT**.

Keywords IoT · The internet of things · Big data analytics · Artificial intelligence · Transportation · Logistics · IoT applications

---

K. Muni Sankar (✉) · B. Booba
Department of Computer Science, Vel's University(VISTAS), Chennai, Tamil Nadu, India

S.-L. Peng et al. (eds.), *Intelligent Computing and Innovation on Data Science*,
Lecture Notes in Networks and Systems 118,
https://doi.org/10.1007/978-981-15-3284-9_47

1 Introduction

At present, the most top-priority vision and mission of any logistics companies is to make sure that the products can be delivered in on time, effective monitoring and managing of goods and controlling on supply chain management, product lifecycle of goods and providing quality services. The goal of any logistics company is embedded with effective tracking of goods, control on inventory management and monitoring of warehouses, automation of in house business operations, effective delivery of goods and take care of securing storage of goods and monitoring condition of goods while transporting with error-free operations. In the logistic industry, the accomplishment of Seven R's principles includes "right products need to move—in right measure—in right situation—in right time—at the right cost to the right places and to the right customers." This job is very difficult; therefore, it is inevitability to make use of new solutions to reach better goals. By establishing IoT technologies with help of intelligence connections, data analytics using artificial intelligence, IoT restructuring the operational process in logistics industry. By providing numerous features and applications, IoT with big data analytics using artificial intelligence together has innovative solutions that are broadly introduced in the field of transport and logistics sector. The applications are supply chain monitoring and management, transportation and vehicle tracking system, inventory management system, safeguard of goods while transporting and automation of operations are the major IoT applications of logistic systems. This paper discusses the IoT and its architecture for transportation and logistics sectors. It also outlines the various possible opportunities.

2 IoT Architecture for Logistics Industry

An IoT architecture is an integrated system with sensors, protocols, actuators, cloud servers and communication layers. Based on industry and basic business operations, the IoT architecture may vary from one industry to another industry. But the basic IoT architecture three layers are: **The client side (IoT devices layer)**. **IoT Gateway layer (Operations on the Server Side)** and **IoT Platform Layer (A pathway for connecting clients and operators)**. The main stages in the IoT architecture are four. They are sensors and actuators; Internet gateways and data acquisition systems; edge IT and data center and cloud storage. Fig. 1 proposed IoT architecture for logistics industry.

Stage 1: Wireless Sensors and Actuators using Network Things: In this stage, sensors and actuators are used to college data from the surrounding environment and having ability to convert into digital data for further analytics.
Stage 2: Internet Gateway and Data Acquisition: In this stage, the data captured by sensors are converted into digital form and store data for further analytics. With help of internet gateway, data can be collected from different sensors through different wireless network mechanisms like Wi-Fi, Bluetooth, Zigbee, etc.

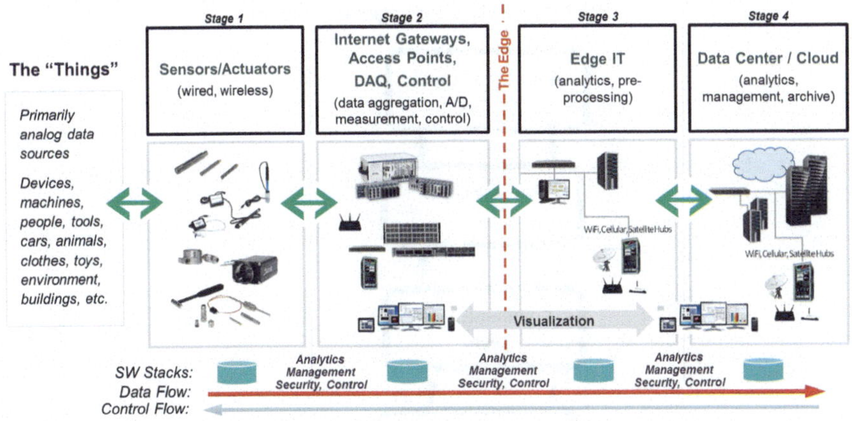

Fig. 1 Proposed IoT architecture for logistics industry

Stage 3: Edge IT Systems (for Data Analytics and Pre-processing): In this stage, the data which are collected from stage 1 and stage 2 are processed in stage 3 using edge IT environment. Here, the data refined and converted into valuable information which is used for decision support systems by using machine learning and artificial intelligence.

Stage 4: Data Centers and Cloud: In this stage, the business processed data means historic data is stored in data centers for further business analytics and business operational data means transaction data can be stored separately for business truncations using cloud-based storage.

Four Stage IOT Architecture Diagram

IoT architecture for logistics system consists of five different layers. These layers include the **infrastructure layer, service layer, communication layer, sensing layer and application layer**. Transportation is mainly used for shifting goods/cargo or living beings from one place to another place. Transportation system as a study area involves the study of so many parameters under different circumstances. All these parameters should be sensed and transfer to service layer through a proper communication channel. From the service layer, appropriate decisions were taken for controlling the system as per the requirement by using big data analytics and AI. The appropriate and sensed data are stored in the infrastructure layer, as shown in Fig. 2.

1. **Application Layer**: In this layer, requirements, components, tasks to be executed are goods, junction, terminals, service areas, people, road and vehicles.
2. **Sensing Layer**: In this layer, tasks to be executed are parking detection, compass terminals, camera, fee collection, environment monitoring, vehicle monitoring, logistics tracking, microwave detection, passenger flow detection.
3. **Communication Layer**: In this layer, tasks to be executed are 3G/4G/5G networks, **Wi-Fi, Wired Network, Optical Fiber, Public and Private Network**.

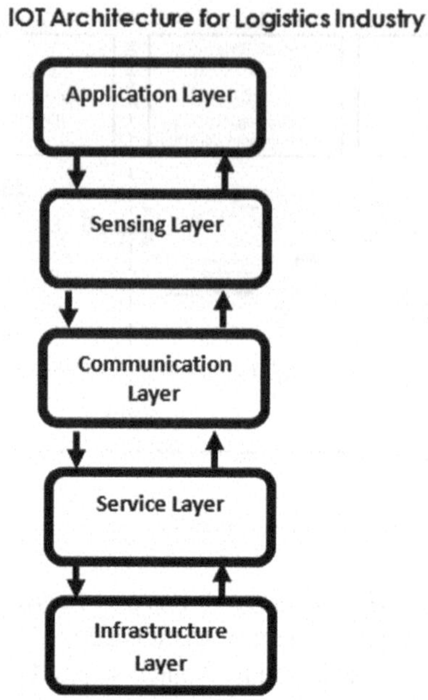

Fig. 2 Various layers in IoT architecture for logistics system

4. **Service Layer**: In this layer, tasks to be executed are logistics service platform, passenger vehicle platform, fleet vehicle service platform, highway integrated platform, intelligent traveling service platform.
5. **Infrastructure Layer**: In this layer, tasks to be executed are geographic information systems mapping service, cloud computing, cloud storage, big data.

3 IoT Opportunities for Transportation and Logistics

The Internet of Things has numerous opportunities in transportation and logistics industry. These opportunities include various applications or needs of a transportation system. Using IoT vehicles can be monitored with respect to their movement, location, whether it is on running or stopped, or at any repair, etc. All these aspects can be closely monitored intelligently using the IoT systems. In most of the cases, vehicles are used for logistics purpose or for carrying any heavy loads which are packed inside the truck. During such times, it is very important to monitor and control the indoor conditions of the truck like temperature, humidity, light conditions, etc. Apart from the payment service near the tolls or at any parking places the vehicle status can be monitor or tracking can be done automatically with the help of vehicle id number. IoT also helps in the guidance and navigation control systems of the vehicles (road transport, air transport, water transport). Transportation opeartions can be govern as per the need of the organization and it is highly recommened the usage IoT Technologies. Here, various vehicles can be monitored and controlled by means of a central intelligence connected through the network. This also helps in managing the imports and exports of materials and goods. It also offers a live and interactive service for monitoring the delivery status indicating the location using GIS mapping. IoT could help in monitoring and suggesting the possibility to avoid traffic and to take other ways to reach the destination in an optimistic way. IoT applications in transport and logistics are:

1. Logistics Applications.
2. Control and Guidance Systems.
3. Inventory Solutions.
4. Fleet Telematics and Management Solutions.
5. Security and Surveillance.
6. Commerce Applications.
7. Solutions for Supply Chain Management.
8. Passenger Entertainment.
9. Smart Vehicle Applications.
10. Navigation Tracking and GIS Mapping.
11. Tolls and Reservation Ticketing System, etc.

4 Types of Application Systems in Logistic

1. **Location Tracking Management Systems(LTMS)**:

In transportations and logistic industry, IoT can construct an intelligent smart location tracking management system which can be avail by companies to easily tracking drivers time to time activities while they are in driving, vehicle (asset) position and goods delivery status. If goods are delivered, the concern person is notified by an alert message. It assists in delivering goods, planning, compilation and screening of schedules. These activities are monitored dynamically in real time. So, Internet of Things technology improves location tracking management systems and pipelining company operations.

2. **Inventory Tracking And Warehousing Monitoring**:

Inventory tracking management and warehouse monitoring are the most important parts of the connected transport and logistic environment. The usage of small and low price sensors will permit industry simply trace inventory items, supervise current status of items, condition of the items, exact position of the items and build an intelligent warehouse system. With these technologies, workers can able to track items easily when items are needed and prevent any losses occur due to accidental damages and ensure safe storage of goods and minimize human errors in business operations.

3. **IoT Technology And Predictive Analytics**:

A predictive analytics is a center pillar and it acts as central system as backbone to help logistic industry to build effective business strategies to get better the managerial process, promote smart business insight and managing risks and many more. IoT enables electronic devices like sensor together huge amount of data and sends out them to the predictive analytics central systems for further analysis using data analytics techniques using AI. These systems can be applied for monitoring and managing business operations and identification of different faults before happing in the system and incredible leads to erroneous. The outcomes are dynamic in real time and prevent at the earlier stage of any damage.

4. **Internet Of Things And Block Chain In Supply Chain Management**:

Here, we have various challenges; for both logistic industries and their customers, they wish to have facilities to track the product (item) life cycle—from the source to till reached into the customer's hands. The RFID tag and some relevant sensors will monitor product temperature, status with the environmental factors like weather conditions, moisture, vehicle location and phases of transportation route. Here, data are captured, stored in the block chain; every item has a digital ID captures not only data about related product but also with product lifecycle.

5. **Autonomous Vehicles Or Self-Driving Vehicles**:

At present, we all witnesses the usage and implementation of self-driving vehicles or autonomous vehicles, which are used in transportation and logistics to take advantage of these technologies and integrated into business operations. IoT smart devices are accountable for collecting huge amounts of data, for decision making using data analytics and then turn them into useful smart intelligence driving routes and directions to optimize road traffic and minimize the distance to travel to deliver the goods. Using self-driving mechanism systems can minimize road accidents and cut down operating costs and optimize business operations for better improvement in industry.

5 Benefits of Integrating IoT in Transportation

By integrating IoT technology with transportation and logistics systems, many benefits are possible. These benefits include:

1. Distance can be traveled by the vehicle is optimized giving the benefits by reducing the fuel consumptions leading to the better profits in day-to-day activities.
2. Optimizing or redirecting the best possible routes during the deadly and dangerous conditions.

3. Through centrally intelligence controlled network, a service can be operated based on the demand or user request.
4. Public transportation and logistics are centrally connecting networks through a control of traffic based on the vehicle count.
5. Goods and cargo material can export, import, purchase and other shipping details can be maintained effectively.
6. Transportation and logistics revenue can be improved for the company owners.

6 Usage of Technologies in Logistics and Their Impact

See Table 1.

Table 1 Usage of technologies in logistics and their impact

The technology	The impact
Physical Internet (based on the IoT)	• Improved supply chain transparency, safety and efficiency • Improved environmental sustainability—more efficient resource planning
Data analytics	• Improvements in customer experience and operational efficiency in operations • Greater inventory visibility and management • Improved "predictive maintenance"
Cloud	• Enabling new platform-based business models and increasing efficiency
Block chain	• Enhanced supply chain security (reduction of fraud) • Reduction in bottlenecks (certification by third parties) • Reduction of errors (no more paper-based documentation) • Increased efficiency
Robotics and automation	• Reduction in human workforce and increased efficiency in delivery and warehousing (including sorting and distribution centers) • Lower costs
Autonomous vehicles	• Reduction in human workforce • Increased efficiency in delivery processes
UAVs/drones	• Increased cost efficiency (use cases: inventory, surveillance, delivery) • Workforce reduction

7 Conclusion

A detailed study is done in this paper about the IoT usage in transportation and logistics industry. IoT usage would help this sector with many opportunities and benefits. It is highly recommended to adopt the Internet of Things into transportation and logistics to make it more effective operations and profitable. At present, transportation and logistics are technologically progress and facing a lot of challenges and seek a rapid renovation and growth with upcoming inventory tracking and location management systems. IoT will revolutionize the transportation and logistics domain.

References

1. Puiu D, Bischof S, Serbanescu B, Nechifor S, Parreira J, Schreiner H (2017) A public transportation journey planner enabled by IoT data analytics. In: 2017 20th conference on innovations in clouds, Internet
2. Al-Dweik A, Muresan R, Mayhew M, Lieberman M (2017) IoT-based multifunctional scalable real-time enhanced road side unit for intelligent transportation systems. In: 2017 IEEE 30th Canadian Conference on Electrical and Computer Engineering (CCECE), Windsor, ON, pp 1–6. https://doi.org/10.1109/ccece.2017.7946618
3. Sutar SH, Koul R, Suryavanshi R (2016) Integration of Smart Phone and IOT for development of smart public transportation system. In: 2016 international conference on Internet of Things and Applications (IOTA), Pune, pp 73–78. https://doi.org/10.1109/iota.2016.7562698
4. Weis A, Strandskov M, Yelamarthi K, Aman MS, Abdelgawad A (2017) Rapid deployment of IoT enabled system for automobile fuel range and gas price location. In: 2017 IEEE international conference on Electro Information Technology (EIT), Lincoln, NE, USA, pp 452–455. https://doi.org/10.1109/eit.2017.8053404
5. Herrera-Quintero LF, Banse K, Vega-Alfonso J, Venegas-Sanchez A (2016) Smart ITS sensor for the transportation planning using the IoT and Bigdata approaches to produce ITS cloud services. In: 2016 8th Euro American Conference on Telematics and Information Systems (EATIS), Cartagena, pp 1–7. https://doi.org/10.1109/eatis.2016.7520096
6. Leal AG, Santiago A, Miyake MY, Noda MK, Pereira MJ, Avanço L (2014) Integrated environment for testing IoT and RFID technologies applied on intelligent transportation system in Brazilian scenarios. In: 2014 IEEE Brasil RFID, Sao Paulo, pp 22–24. https://doi.org/10.1109/brasilrfid.2014.712895

Evolution of Lung CT Image Dataset and Detection of Disease

C. S. Shylaja, R. Anandan, and A. Sajeev Ram

Abstract This paper presents CT scan image analysis, creation of database and evolution of content-based image retrieval technique for distinguishing lung cancer at early stages. The data are collected from the clinical environment and LIDC dataset. The features such as correlation, dissimilarity, cluster prominence and cluster shade are extracted at different orientations using GLCM features in the MATLAB environment and stored in the database as a trained phase. The testing image features are extracted and are analogized with the trained dataset and the appropriate out-turn is obtained. Minimum distance classifier is used to predict the clinical condition of the lung by matching the testing image and trained image.

Keywords Feature extraction · GLCM · CBIR · LIDC · Minimum distance

1 Introduction

Content-based image retrieval technique examines the content of the image preferably than the metadata or representations accompanied with the image [1]. In the medical field, the purpose of content-based image retrieval is to allow the radiologist to retrieve images of the same features that lead to the same judgment as to the input image. As reported by the American Cancer Society, approximately 228,150 recent cases of lung cancer will be identified in the USA in 2019 and approximately 142,670 deaths due to lung cancer will occur in 2019 [2]. This is the reason for the necessity of a lung nodule exposure system in chest CT images. Computer-aided detection

C. S. Shylaja (✉) · R. Anandan
Department of Computer Science and Engineering, Vels Institute of Science, Technology and Advanced Studies, Chennai, India

R. Anandan
e-mail: anandan.se@velsuniv.ac.in

A. Sajeev Ram
Department of Information Technology, Sri Krishna College of Engineering and Technology, Coimbatore, India

S.-L. Peng et al. (eds.), *Intelligent Computing and Innovation on Data Science*, Lecture Notes in Networks and Systems 118,
https://doi.org/10.1007/978-981-15-3284-9_48

system supports the radiologists by doing preprocessing of the images in addition to symptomatic of the most probable regions for nodules.

Finding lung nodules continues over and done with process for subduing the background patterns in the lungs which comprise blood vessels, ribs and the bronchi. The images gained after preprocessing will give improved chest image which improves prediction of regions of a nodule and classified liable to characteristics like size, contrast and shapes [3]. Computer-aided detection is used for timely identification of lung cancer. This research represents a CAD system that can significantly distinguish lung disease with a decrease in false-positive rates [4].

2 Experimental Section

Out of 400 images collected from various clinical centers and online datasets, the clinical condition of 30 different lung cancer images is collected and stored in the database. The features extraction is done for all 30 images and stored in the database and it used to discover the result of the testing image by comparing with the trained image. Table 1 shows the clinical condition of different images.

3 Feature Extraction

Feature extraction is the method of creating attributes that are used in the selection and classification process. Feature selection lessens the features used for the classification. Features are used to support in discernment as well as classification [5]. The gray-level co-occurrence matrix is a second-order analytical type and can be used in many utilization. Third, as well as higher-order textures, describes the connection between the pixels. A GLCM is a matrix where the sum of rows plus columns is same as the sum of gray levels, present in the image [6].

The GLCM matrix is the relative frequency in which pixel expanse occurs in a certain neighborhood, with '*i*' and '*j*' [7]. The matrix constituent comprises the second-order analytical possibility values for variations among gray levels at a precise displacement distance and at certain angle.

Manipulating an enormous quantity of intensity levels G indicates putting away a plenty of transitory data. Considering enormous spatial property, the GLCMs are delicate to the extent of the texture illustrations on which they are assessed [8]. Let a and b are the factors of the co-occurrence matrix, M (a, b) is the part in the co-occurrence matrix, N is the dimension of the co-occurrence matrix.

- **Correlation**:

The totaling of the correlation of a pixel and its near pixel over the entire image info out the linear dependency of gray levels on those of neighboring pixels.

Table 1 Clinical condition of lung

Sl. No.	Clinical condition
1	Adenocarcinoma
2	Lung carcinoma
3	Cystic fibrosis
4	Benign nodule
5	Malignant nodule
6	Silicosis
7	Cystadenoma of a lung
8	Splenic metastasis
9	von Hippel-Lindau (VHL) disease
10	Cancerless image
11	lymphatic vessel with carcinoma
12	Metastatic carcinoma
13	Lymphoma carcinoma
14	Invasive carcinoma
15	Small cell carcinoma
16	Non-small cell carcinoma
17	Carcinoma lung cancer
18	Mucinous adenocarcinoma lung cancer
19	Lymphoma carcinoma
20	pulmonary pseudocyst
21	Lung carcinoma
22	Cancerless image
23	Pulmonary hamartoma
24	Squamous cell carcinoma
25	Large cell carcinoma
26	Undifferentiated carcinoma
27	Undetermined carcinoma
28	Bronchoalveolar carcinoma
29	Non-cancer image
30	Cancer image

$$f = \sum_{a=0}^{N-1} \sum_{b=0}^{N-1} P_{d,\theta}(i, j) \frac{(a - \mu_x)(b - \mu_y)}{\sigma_x \sigma_y} \tag{1}$$

- **Dissimilarity**:

The dissimilarity permits principled evaluations flanked by segmentations created by dissimilar algorithms, plus segmentations on dissimilar images. The computation is asymmetric and does not represent a distance because the triangle inequality is not satisfied.

$$\sum_{a,b=0}^{N-1} P_{i,j}|i-j| \tag{2}$$

- **Cluster Prominence**:

When cluster prominence is truncated, there is a sail throughout the co-occurrence matrix nearby the mean value. This means that there is a slight difference in gray scales.

$$\text{prom} = \sum_{a=0}^{G-1}\sum_{b=o}^{G-1}\{i+j-\mu_x-\mu_y\}^4 P(i,j) \tag{3}$$

- **Cluster Shade**:

The computation of the skewness (or) deficiency of symmetry is demarcated by the feature of luster shade. When the cluster shade is in top, the image is not symmetric.

$$\text{Shade} = \sum_{a=0}^{2G-2}(i-2\mu)^3 H_s(a|\Delta x, \Delta y) \tag{4}$$

- **Homogeneity**:

A homogeneous image will result in a co-occurrence matrix with a mishmash of high and low $P[a, b]$'s.

$$C_{h=}\sum_a\sum_b \frac{P_d[a,b]}{1+|a-b|} \tag{5}$$

The GLCM shows how diverse permutations of each element brightness values happen in an image. The benefit of the co-occurrence matrix computation is the pair of pixels that are co-occurring can be correlated in many positions concerning expanse and angular spatial relationships [9]. It is described as 2D histogram of gray levels for a pair of pixels, which are separated by a static spatial connection. On the other hand, the matrix is subtle to rotation [10]. The revolution of diverse offsets describes pixel connections by changing directions, a rotation angle of 0°, 45°, 90°, 135°.

4 Minimum Distance Classification

A minimum distance classification technique is used for classifying the images. Minimum distance classification is used for classifying the images according to the closest region of interest. In minimum distance classification, the mean value for

Table 2 Difference between trained image and testing image

Sl. No	Feature extraction	0°	45°	90°	135°
1	Correlation	0.000254	0.000799	0.000056	0.000534
2	Dissimilarity	0.000303	0.000021	0.000807	0.000966
3	Cluster prominence	0.000444	0.000207	0.000557	0.000412
4	Cluster shade	0.000591	0.000241	0.000538	0.000324
5	Homogeneity	0.000508	0.000604	0.000411	0.000409

all classes of images is calculated in each band of data. The minimum distance is initialized to be the high value [11]. Minimum distance classifier allocates a pixel to the class of minimum distance. Distance measure utilizes Euclidean distance from the pixel to cluster mean [12].

$$D_{x,m} = \sqrt{x^2} - m^2 \tag{6}$$

where

x means tested pixel
M means mean value of the cluster.

Experimental Results and Discussions:

A database consisting of 30 lung images was possessed and stored in a database in the order shown in Table 1. The clinical diagnosis was determined by pathologists. Feature extraction properties are calculated for 30 images at different angles and stored in the database as shown in Table 2. The test image features are extracted and matched with the features of the trained image in the directory. When the features match with the trained image, the same clinical condition applies for test image.

Algorithm

Step 1: Create a database with images.
Step 2: Start from the first image stored in the database and proceed to Step 3 until the last image in the database.
Step 3: Extract the features of the images in four directions (0°, 45°, 90°, 135°). If the condition is not satisfied, go to Step 4.
Step 4: Select the testing image and perform Step 2.
Step 5: Terminate the process when the condition is false.
Step 6: If the testing image is matched with a trained image in the database, the result will be obtained. Otherwise, the testing image will be added in the database for further analysis.
Step 7: Display a resultant image with the details of the diagnosis.

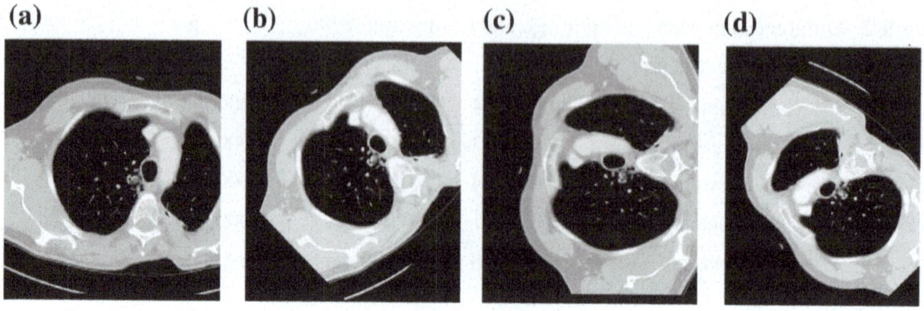

Fig. 1 **a** Shows image at 0°. 1. **b** Shows image at 45°. 1. **c** Shows image at 90°. 1. **d** Shows image at 135°

The testing image which is used for comparing is shown in Fig. 1. The sample image rotation at different angles (0°, 45°, 90°, 135°) is shown. The testing image feature is extracted using GLCM properties.

The extracted features of a sample image at different orientations are stored in a database as a training phase. The sample difference between trained image and the testing image is intimated in Table 2 and the graphical representation is shown in Fig. 2. The test image matches with the trained image benign nodule (Sl. No. 4) and minimum difference values are calculated.

It was observed that the test image has a minimum difference between correlation, dissimilarity, cluster prominence, cluster shade, homogeneity at different orientations. Table 3 shows that after extracting the testing image features, it is compared with the trained image in the database. The training image matches with the trained image, benign nodule (clinical condition 4) and the values are as follows.

The result shows that the testing image matches with the trained image benign nodule at a different orientation. The resulting image appears three times at 0°, two times at 45°, three times at 90°, three times at 135°. It shows that a minimum distance approach is best suitable for medical image diagnosis. Various studies have been conducted in much anatomical tissue abnormality by image characteristics [13].

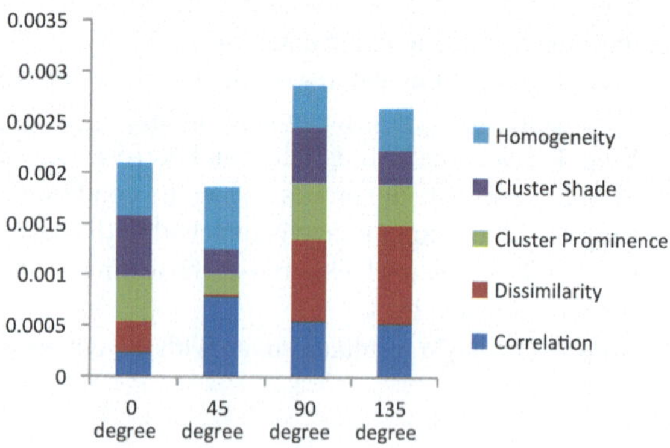

Fig. 2 Graphical representation of the difference between trained image and test image

Table 3 Results of image retrieval for different orientations

Degree	Correlation	Dissimilarity	Cluster prominence	Cluster shade	Homogeneity	Number of occurrences of image
0°	4	4	10	21	4	3
45°	16	25	4	4	3	2
90°	4	4	4	11	26	3
135°	6	4	9	4	4	3

This CAD system is used to understand image recognition by the feature extraction process and minimum distance classification technique.

5 Conclusion

A CAD system with a combination of CBIR and minimum distance approach classification is designed. The database consists of 30 lung CT images and its clinical condition is collected. The features are extracted for 30 images at different orientations that are extracted and stored in the database. The test image is compared with the trained image in the dataset and if the image matches, the lung disease is predicted. Forthcoming research is to design a 3D visual system and increase the better understanding of each disease visually.

References

1. Wadhai SA, Kawathekar SS Techniques of content based image retrieval: a review. IOSR J Comput Eng (IOSR-JCE) 75–79
2. Cancer facts and figures (2019) American cancer society. Available: https://www.cancer.org/content/dam/cancer-org/research/cancer-facts-and-statistics/annual-cancer-facts-and-figures/2019/cancer-facts-and-figures-2019.pdf. [Accessed: 02-Feb-2020]
3. Zhou W, Li H, Tian Q (2017) Recent advance in content-based image retrieval: a literature survey. arXiv preprint arXiv:1706.06064
4. Castellino RA (2005) Computer aided detection (CAD): an overview. Cancer Imaging 5(1):17
5. Chora RS (2007) Image feature extraction techniques and their applications for CBIR and biometrics systems. Int J Biol Biomed Eng 1(1):6–16
6. Mohanaiah P, Sathyanarayana P, Gurukumar L (2013) Image texture feature extraction using GLCM approach. Int J Sci Res Publ 3(5):1–5
7. Glcm architecture for image extraction 1 1 1. 3(1):75–82
8. Shijin SKP, Dharun VS (2017) Extraction of texture features using GLCM and shape features using connected regions. Int J Eng Technol 8(6):2926–2930
9. Wang J, Ren X (2014) GLCM based extraction of flame image texture features and KPCA-GLVQ recognition method for rotary kiln combustion working conditions. Int J Autom Comput 11(1):72–77

10. Rao CN et al (2013) Co-occurrence matrix and its statistical features as an approach for identification of phase transitions of mesogens. Int J Innov Res Sci Eng Technol 2(9):4531–4538
11. Jayaprakash K, Anandan R (2012) Development of pancreatic CT—scan image dataset and retrieval process for diagnosis
12. Kranz, H-G (1993) Diagnosis of partial discharge signals using neural networks and minimum distance classification. IEEE Trans Electr Insul 28(6):1016–1024
13. ŚMietański J, Tadeusiewicz R, Łuczyńska E (2010) Texture analysis in perfusion images of prostate cancer—a case study. Int J Appl Math Comput Sci 20(1):149–156

Equivalent Circuit (EC) Approximation of Miniaturized Elliptical UWB Antenna for Imaging of Wood

Tale Saeidi, Sarmad Nozad Mahmood, Shahid M. Ali, Sameer Alani, Masood Rehman, and Adam R. H. Alhawari

Abstract Antenna can be considered as an essential section of a transceiver system which acts as a filter. This filter should produce a response from the transmitter to the receiver. The equivalent circuit (EC) of an elliptical patch UWB antenna is approximately modelled with an *RLC* equivalent circuit. The EC analysis helps to understand how a UWB antenna works, and how stubs (inductors) and notches affect the reflection coefficient result of a UWB antenna. Besides, a directly optimum of an antenna based on the EM simulation is very time-consuming, particularly, when the geometry of the antenna is complex. Hence, the research and analysis of the antenna's EC model are imperative. After designing and analysis of the antenna's EC, the impedance bandwidth result of both EC and the simulated prototype of antenna are compared. It depicts an acceptable agreement in terms of the resonances and the poles in the working BW.

Keywords Equivalent circuit · UWB antennas · Stopband · Passband · Degenerated foster canonical model

T. Saeidi (✉) · S. M. Ali · M. Rehman
Electrical and Electronic Engineering Department of Universiti Teknologi PETRONAS, 32610 Bandar Seri Iskandar, Perak, Malaysia
e-mail: talecommunication@gmail.com

S. N. Mahmood
Computer Technical Engineering Department, Alkitab University, Kirkuk 36001, Iraq

S. Alani
Faculty of Information and Communication Technology, Centre for Advanced Computing Technology (C-ACT), Universiti Teknikal Malaysia Melaka, Hang Tuah Jaya, 76100 Durian Tunggal, Melaka, Malaysia

A. R. H. Alhawari
Electrical Engineering Department, College of Engineering, Najran University, Najran, Saudi Arabia

S.-L. Peng et al. (eds.), *Intelligent Computing and Innovation on Data Science*, Lecture Notes in Networks and Systems 118,
https://doi.org/10.1007/978-981-15-3284-9_49

1 Introduction

In general, antennas are considered as linear and passive elements with input impedance that can be modelled with a Foster canonical model presented in Fig. 1. Besides, this model can be degenerated to be applied to model monopole or dipole antennas [1]. One of the most critical requirements in designing the EC approximation for UWB antenna is that either the input impedance or the input admittance of the EC should be matched with the antenna's impedance or admittance. Furthermore, the antenna BW is presented as a transfer function in frequency domain, and as received pulses or signals with distortion in the time domain [2]. Many methods are presented to determine the input impedance or admittance, as shown in [3]. These broadband models were Hamid & Hamid's broadband EC (gave poor accuracy, possible EC, *RLC* components; no limitation for maximum frequency range existed) [4], Foster's canonical form [5], Long and Werner's broadband EC model [6] and Streable & Pearson's broadband EC (it showed excellent precision, achievable EC, no Darlington form, *RLC* element;no limitation for maximum frequency range) [7]. Among these models, Pearson's method presented the highest accuracy, no limitation on the frequency range, and the circuit elements were *RLC* tanks. Hamid's network had almost the same characteristics but was less accurate compared to the Pearson method. In addition to these works, many works were done to design and calculate an equivalent circuit for the microstrip and planar antennas, and filters [8–18].

After a brief introduction about the equivalent circuits that have been used to produce the EC of an UWB antenna and the matching *RLC* circuits, the design steps are started in Sect. 2. Sect. 3 shows the results and discussion. Afterwards, Sect. 4 presents the conclusion.

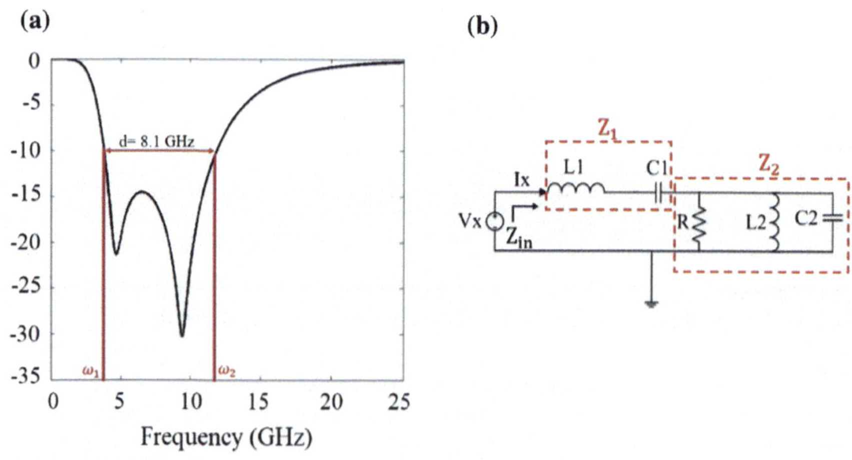

Fig. 1 Degenerated Foster canonical (DFC) model and its reflection coefficient result

2 Antenna and EC Analysis

This section demonstrates how a simple EC for the proposed UWB antenna, which required a smaller area in comparison with the other abovementioned models, is designed. According to the impedance BW results of the proposed antenna shown in Fig. 3 of [19], first, it is required to obtain the wide-band part of the BW like shown Fig. 1, then the other parts of the working BW and the other resonances are determined. A narrow-band model of the degenerated Foster canonical (DFC) model is used to define the wide-band part of the BW. Then, the other two broadband EC, namely Hamid and Pearson, are chosen and replaced with the degenerated Foster to get the wide-band part of the BW. Based on the literature, these two broadband EC showed more promising results with fewer limitations.

Figure 1 shows the degenerated Foster canonical model of UWB antenna along with its result. This model is in good agreement with the conventional elliptical patch antenna at the desired frequency. In general, UWB antenna acts as a bandpass filter (BPF); hence, in Fig. 1, the first *LC* trunk acts as a high pass filter and the parallel *RLC* as a low pass filter [1]. The antenna's input impedance (Z_{in}) (89.23 + j32.35 Ω) is assumed equal to Foster canonical model's Z_{in}, to get the value of each *RLC* components in the degenerated Foster canonical (DFC) model. Then, by applying a KVL to this matching network, the input impedance would be obtained from the following equations:

$$Z_{in} = Z_1 + Z_2 \tag{1}$$

where

$$Z_{\_1} = jX_{l2} + 1/jX_{c2}, 1/Z_2 = 1/R + 1/jX_{l1} + jX_{c1}, X_{L=}\omega l, X_c = \omega c$$

After substituting them in Z_{in}:

$$\begin{aligned} Z_{in} = &\left(j\omega_1 L_1 + 1\right)\left(R + j\omega_2 L_2 - \omega_2^2 R C_2 L_2\right) \\ &+ (jR\omega_2 L_2)(j\omega_1 C_1)/(j\omega_1 C_1)\left(R + j\omega_2 L_2 - \omega_2^2 R C_2 L_2\right) \end{aligned} \tag{2}$$

where

$$\omega_1 = 1/\sqrt{L_1 C_1}, \omega_1 = 2\pi f_1, \omega_2 = 1/\sqrt{L_2 C_2}, \omega_2 = 2\pi f_2$$

The next step is to simplify the input impedance in (2) and substitute the following associate equations in (2) (same method applied in [20]). When C_1, C_2, L_1, and L_2 are achieved, *R* needs to be found. Since the *RLC* matching network is a parallel one, *R* is achieved by using equations in (3).

$$Q = f_{r2}/BW_2, R = Q\omega_2 L_2 \tag{3}$$

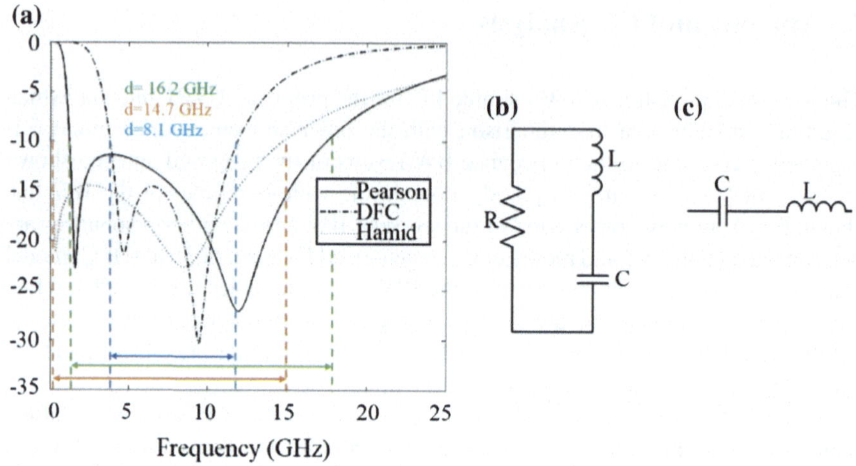

Fig. 2 Broadband models' reflection coefficient **a**, passband **b**, stopband **c**

The same Eqs. (1), (2) and their associates are applied to the Hamid and Pearson models to get the *RLC* components and to show the performance of the proposed UWB EC. Figure 2 shows the reflection coefficient results of Hamid's and Pearson's matching networks, as presented before [20]. As can be seen in Fig. 2, Pearson's network can attain a wider BW compared to both the Hamid and DFC matching networks. The reason is due to the Pearson model's advantages compared to the other models, such as having good approximation accuracy, a realizable equivalent network and having no limitations on the maximum frequency range [7].

Next, the other resonances at 1.3 and 1.8 GHz are needed to be achieved and included in the EC of the antenna presented in [19]. Therefore, for each resonance, the *LC* tank parallel with a resistor is integrated with the wideband to get the resonances at those frequencies (Fig. 3). Moreover, based on the reflection coefficient result in Fig. 3 of [19], a stopband requires around 20 GHz. A series combination of *L* and *C* is integrated into the network to achieve this (Fig. 3) [21].

Figure 3 shows how the other parts of the EC are integrated with the DFC broadband network. In Fig. 3, the resonance frequencies are $\omega_1 = 1.3$ GHz, $\omega_2 = 1.8$ GHz, $\omega_3 = 21.5$ GHz, $\omega_0 = 12.5$ GHz and $\omega_4 = 20$ GHz. To apply the other models, the DFC should be replaced with the Pearson or Hamid model. After the results are achieved from the DFC matching networks, the DFC is replaced with the Pearson and Hamid models, and their results are compared. The Pearson wide-band model is privileged from a wider BW and better matching at resonant frequencies. However, the complexity of the EC increases when it is integrated with the other parts of the network (Fig. 3).

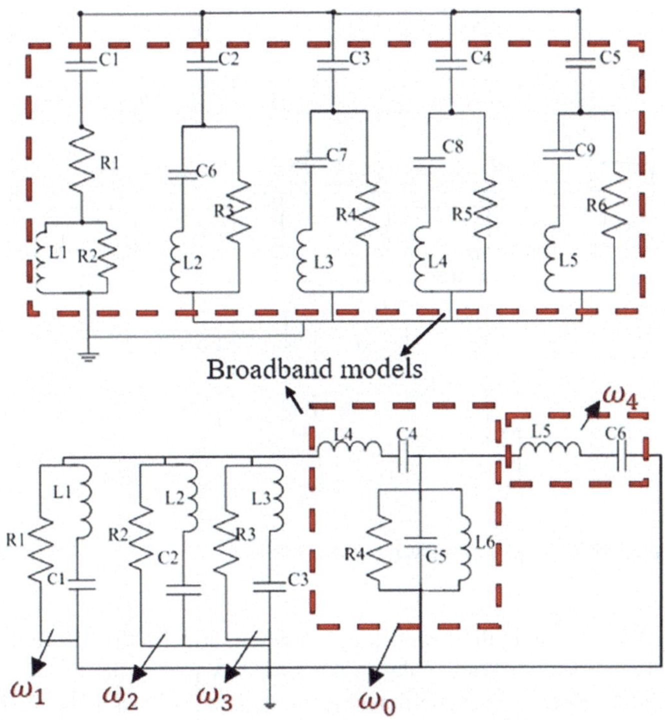

Fig. 3 EC of the proposed UWB antenna using DFC (right) and Pearson (left)

3 Result and Discussion

The UWB models used in this article, such as DFC, Hamid, and Pearson are presenting the conventional patch antenna model without any loads (stubs, slots and pins). After adding the loads on the antenna, more resonances occur inside the ultra-wide bandwidth along with the first two resonances. Thus, the improved EC shows a better agreement in reflection coefficient result and the new *RLC* tanks can represent the stubs (inductors) and the gaps (capacitor) on the patch. Besides, the printing area is reduced by optimizing the EC.

To reduce the complexity of the network and circuit footprint, the Pearson broadband model, which had the best result in terms of BW, is modified and improved. In the proposed network (Fig. 4), except keeping the complex model from the Pearson EC, two coupling capacitors (C_2, C_4) are integrated with the network between each part of the *RLC* tank. These two series capacitors decrease the total capacitance of the network and increase the total reactance. Besides, these coupling capacitors block the low-frequency signals and pass the higher frequencies.

In addition to that, the same procedures and equations presented in Sect. 2 are applied to the EC depicted in Fig. 3 to obtain the absolute values of the elements of

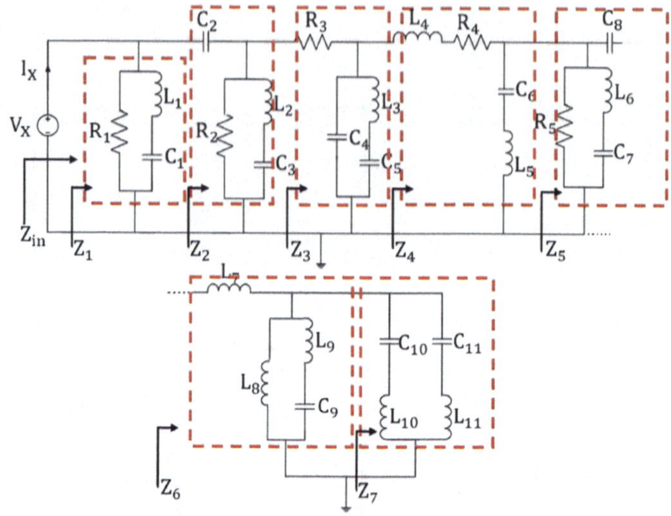

Fig. 4 Proposed EC of the proposed UWB antenna

the circuit. After attaining these values and drawing the reflection coefficient result, it is noticed that there are some poles in the wide BW part of the whole BW in the simulated antenna's reflection coefficient result that do not exist in the EC reflection coefficient result. Hence, the circuit should be improved to get those resonances.

This improvement is performed by adding the *LC* tanks presented in Fig. 2 for each stopband and passband (sidebands) that exist at the middle of the band. The following equations are exploited to calculate the *RLC* network components for the final proposed equivalent network shown in Fig. 4. Furthermore, each of the seven dotted areas in Fig. 4 has resonances that should be used in these equations, from Z_1 to Z_7 (1.3, 1.8, 4.2, 6.4 and 11.2, 16.1, 17.5 and 18.5 and 24 GHz).

$$Z_{in} = Z_1 + Z_2 + \cdots + Z_7 \tag{4}$$

where

$$\begin{aligned}
1/Z_1 &= 1/R_1 + (1/((jX_{l1} + 1/jX_{c1}))),\\
Z_2 &= 1/(jX_{c2}) + (1/R_2 + 1/(jX_{l2} + 1/(jX_{c3}))),\\
Z_3 &= R_3 + 1/(jX_{C4} + 1/(jX_{l3} + 1/(jX_{c5}))),\\
Z_4 &= jX_{l4} + R_4 + 1/(jX_{c6}) + jX_{l5},\\
Z_5 &= 1/(jX_{c8}) + 1/(1/R_5 + 1/(jX_{l6} + 1/(jX_{c7}))),\\
Z_6 &= jX_{l7} + 1/(1/(jX_{l8}) + 1/(jX_{l9} + 1/(jX_{c9}))),\\
1/Z_7 &= 1/(1/(jX_{C10}) + jX_{l10}) + 1/(jX_{l11} + 1/(jX_{C11}))
\end{aligned}$$

Figure 4 depicts the modified EC presented in Fig. 3. The first three parts, namely the first two *RLC* tanks and the coupling C_2, remain unchanged. The two lower band resonances are attainable now. A resistor (R_3) between the 2nd and 3rd *RLC* tanks is used to get the stopband-like shape (sideband) between 6.4 and 11.2 GHz, along with improving the reflection coefficient level. Then the 3rd *RLC* tank is added to resonate at 4.2 GHz. This capacitor affects the resonances at 4.2 and 11.2 GHz and degrades them. Furthermore, to acquire the resonance at 16.1 GHz and the stopband between 16.1 and 17.5 GHz, one series *LC* tank and one parallel *LC* tank are used, respectively. Next, another *RLC* tank is integrated into the circuit to resonate at 11.2 GHz, and the next series *LC* tank is attached to shift the resonance around 4.2 GHz to the desired resonance. Besides, the following two *RLC* tanks consist of two parallel *LC* tanks to resonate at 24 GHz and another *LC* tank parallel with an inductor to resonate at 17.5 GHz. The reason that the inductor (L_8) is used rather than a resistor is that the resistor degrades the reflection coefficient level of the last four resonances (16.1, 17.5, 18.5 and 24 GHz). Table 1 illustrates the final optimized values of the *RLC* components in Fig. 4.

The antenna reflection coefficient results along with its equivalent circuit (EC) are presented in Fig. 5. A good agreement between them is observed in Fig. 5 (all the passbands and stopbands are achieved). The broad BW at the middle of the frequency band performed with a slight increase by 10 MHz compared to the simulation result, both the low resonance frequencies at 1.3 and 1.8 GHz and the lower and higher ends of the UWB are obtained, as well as the stopband around 20 GHz. Furthermore, the resonance at high frequency (24 GHz) is also achieved.

Table 1 Values obtained from the final proposed circuit

Parameters	Values	Parameters	Values
R_1	250	C_{10}	0.0038
R_2	170	C_{11}	0.0038
R_3	190	L_1	24
R_4	100	L_2	88
R_5	390	L_3	7.6
C_1	0.16	L_4	3.5
C_2	0.9	L_5	8
C_3	0.1	L_6	12
C_4	0.08	L_7	6.6
C_5	0.09	L_8	45
C_6	0.01	L_9	9
C_7	0.25	L_{10}	10
C_8	0.01	L_{11}	20
C_9	0.014		

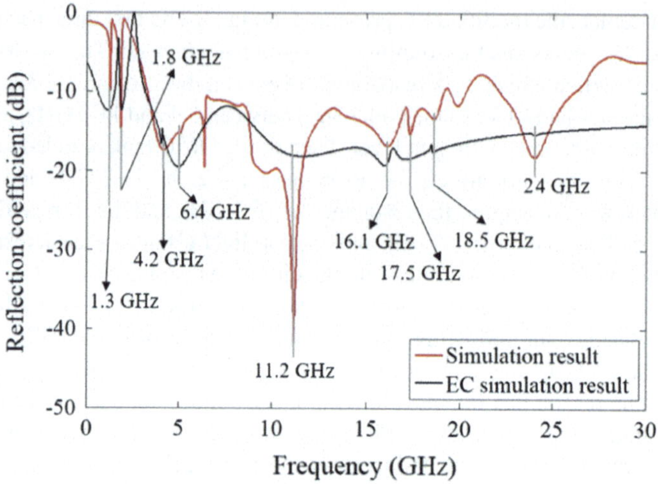

Fig. 5 Impedance BW results comparison between the proposed antenna and its representative EC

4 Conclusion

An equivalent circuit (EC) for the modified elliptical miniaturized UWB antenna published before by the author is presented in this paper. The conventional EC of the conventional UWB is designed and the results extracted from CST. Then, different models of impedance matching are utilized initially to obtain all the passbands and stopbands attained in reflection coefficient result of the antenna prototype. Finally, the simulated and the EC reflection coefficient result are compared to show the capability of the EC in presenting a model of the proposed UWB antenna.

Acknowledgements The project is supported by both Alkitab University and UTP.

References

1. Wang SBT, Niknejad AM, Brodersen RW (2006) Circuit modeling methodology for UWB omnidirectional small antennas. IEEE J Sel Areas Commun 24(4)
2. Akhoondzadeh-Asl L, Fardis M, Abolghasemi A, Dadashzadeh G (2008) Frequency and time domain characteristic of a novel notch frequency UWB antenna. Prog Electromagnet Res (PIER) 80:337–348
3. Wang Y, Liv JZ, Ran LX (2002) An equivalent circuit modeling method for ultra-wideband antennas. PIER 82:433–445
4. Hamid M, Hamid R (1997) Equivalent circuit of dipole antenna of arbitrary length. IEEE Trans Antennas Propag 45(11):1695–1696
5. Ramo S, Whinnery JR, Van Duzer T (1965) Fields and waves in communication electronics. Wiley (section 11.13)

6. Long B, Werner P, Werner D (2000) A simple broadband dipole equivalent circuit model. In: Proceedings of IEEE international symposium on antennas and propagation, vol 2, Salt Lake City, pp 1046–1049, 16–21 July 2000
7. Streable GW, Pearson LW (1981) A numerical study on realizable broad-band and equivalent admittances for dipole and loop antennas. IEEE Trans Antennas Propag 29(5):707–717
8. Woo D-J, Lee T-K, Lee JW (2013) Equivalent circuit model for a simple slot-shaped DGS microstrip line. IEEE Microw Wirel Compon Lett 23(9):447–449
9. Muroga S, Endo Y, Takamatsu M, Andoh H (2018) T-type equivalent circuit of on-chip microstrip line with magnetic film-type noise suppressor. IEEE Trans Magn 54(60
10. Zhou H-M, Zhang Q-S, Lian J, Li X-H (2016) A lumped equivalent circuit model for symmetrical T-shaped microstrip magnetoelectric tunable microwave filters. IEEE Trans Magn 52(10)
11. Park HH, Kwon JH, Ahn S (2017) A simple equivalent circuit model for shielding analysis of magnetic sheets based on microstrip line measurement. IEEE Trans Magn 53(6)
12. Ahn H-R, Nam S (2013) Compact microstrip 3-dB coupled-line ring and branch-line hybrids with new symmetric equivalent circuits. IEEE Trans Microw Theory Tech 61(3)
13. Riviere B, Jeuland H, Bolioli S (2014) New equivalent circuit model for a broadband optimization of dipole arrays. IEEE Antennas Wirel Propag Lett 13:1300–1304
14. Bojanic R, Milosevic V, Jokanovic B, Medina-Mena F, Mesa F (2014) Enhanced modelling of split-ring resonators couplings in printed circuits. IEEE Trans Microw Theory Tech 62(8)
15. Mostaani A, Mohammad Hassan Javadzadeh S (2017) Equivalent linear and non-linear circuit model of superconducting microstrip normal and enhanced T-junction structures. IET Microw Antennas Propag 11(8)
16. Cho C, Kang J-S, Choo H (2014) Improved wheeler cap method based on an equivalent high-order circuit model. IEEE Trans Antennas Propag 62(1)
17. Mandic T, Magerl M, Baric A (2018) Sequential buildup of broadband equivalent circuit model for low-cost SMA connectors. IEEE Trans Electromagn Compat 61(1):242–250
18. Sheikhi A, Alipour A, Hemesi H (2017) Design of microstrip wide stop band low pass filter with lumped equivalent circuit. Electron Lett 53(21)
19. Saeidi T, Ismail I, Alhawari ARH, Wen WP (2019) Near-field and far-field investigation of miniaturized UWB antenna for imaging of wood. AIP Adv 9(3)
20. Sattar S, Zainal Azni Zukilfli T (2017) A 2.4/5.2-GHz concurrent dual-band CMOS low noise amplifier. IEEE Access 5:2169–3536
21. Saeidi T, bin Ismail I, Saleh V, Alhawari ARH (2016) Triple band modified 90 degrees Koch Fractal H-Slot microstrip antenna. In: International Conference on Intelligent and Advanced Systems, ICIAS 2016, pp 5–9

Design of Dual-Band Wearable Crescent-Shaped Button Antenna for WLAN Applications

Shahid M. Ali, Varun Jeoti, Tale Saeidi, Sarmad Nozad Mahmood, Zuhairiah Zainal Abidin, and Masood Rehman

Abstract A new type of a circular polarized crescent-shaped button antenna is proposed for the wireless local area network applications (WLAN). The investigated design is composed of a crescent-shaped antenna. The button disc is located on the top side of a textile material. The design shows dual frequency band like a monopole in 2.5185–2.7206 GHz, and a broadside pattern in 4.4475–5.3988 GHz, so that both the off- and on-body communication can be obtained, simultaneously. Another important feature of a wearable antenna is a specific absorption rate (SAR), so that SAR is calculated, which is 0.662 W/KG for low band and 0.294 for high band, respectively. The SAR is under the limits, according to the health and safety regulations. Therefore, the proposed design is suitable for the body-centric communications.

Keywords Button antenna · WLAN applications · Dual band · On-body · Off-body · SAR

1 Introduction

Internet of Things (IoT) is an important technology and enabled by the full deployment of 5G technology, as a result, every thing will be wirelessly interconnected such as household to the daily basis devices [1]. The wearable devices are predicted to be an important part of IoT. The wearable devices are basically used to be worn by humans or animals and provided wireless connectivity to mobile phones. The wearable devices in the WBAN systems can be communicated to the central database

---

S. M. Ali (✉) · V. Jeoti · T. Saeidi · M. Rehman
Department of Electrical and Electronic Engineering, Universiti Teknologi, PETRONAS Bander Seri Iskandar, 32610 Tronoh, Perak, Malaysia
e-mail: shahid_17006402@utp.edu.my

S. Nozad Mahmood
Computer Technical Engineering Department, Alkitab Univesity, Kirkuk 36001, Iraq

Z. Zainal Abidin
Research Center for Applied Electromagnetics, Institute of Integrated Engineering, Universiti Tun Hussein Onn Malaysia, Parit Raja, Johor, Malaysia

S.-L. Peng et al. (eds.), *Intelligent Computing and Innovation on Data Science*, Lecture Notes in Networks and Systems 118,
https://doi.org/10.1007/978-981-15-3284-9_50

using sensor module, which consists of different components such as battery, micro-controller, sensor, and RF antenna. The wireless communication is rapidly increasing in the WBAN systems such as health-care, military and sports, etc. Moreover, the wearable devices got high attention from the last decade in the various operating bands such as UWB, ISM band, and WLAN for the wearable applications [2]. The antenna is playing significant roles in the circuitry system, as the whole system performance depends upon the type of transmission and receiving components like antenna [3]. However, most of the studies focused on the investigation of the design antenna and realization [4–9], whereas, some of the studies focused on the robustness such as material stability against any bending condition and its impact on the antenna performance [10–12]. Further studies emphasized the effect of the human body on wearable antenna's performance and thermal radiation, called as a specific absorption rate (SAR) [13–15]. According to these studies, the wearable antennas were investigated to be low profile, low power, flexible, robust against any bending conditions, so that textile materials and antenna planar structure were presented. The button antenna has shown clear advantages due to small patch and provides gap between the antenna and human body, thus its hard to bend in any condition. The design shape can be easily mounted on the handcuff button, jeans, shirts, and so on. Next, the rigidity of the button-shaped antenna maintained the stability of the antenna performance. Moreover, the textile conducting materials are anisotropic materials, whereas the copper-based button antenna boosts up the radiation characteristics of the wearable antennas due to high conductivity. Over the years, many studies have been published on button antennas [16–18]. In this paper, a compact dual-band wearable crescent-shaped button antenna is designed for on- and off-body communication.

This paper is organized as follows: Sect. 1 includes introduction, Sect. 2 presents the antenna design, results are presented in Sect. 3, and the conclusion in Sect. 4.

2 Antenna Structure

The design is demonstrated in Fig. 1. The radiating patch is designed on a button-type FR-4 substrate, with thickness of 1.6 mm, dielectric constant and loss tangent of 4.3 and 0.025, respectively. A large, small and hexagon conductive loops are printed on the disc top side. Besides, on the bottom side, a rectangular conductive strip with a single arm is printed. Moreover, shorting vias are used on the top side of the gapped loops to further reduce overall size. The coaxial feeding probe is connected to the feeding point on the other side. At the downside, a felt substrate material with a 3 mm thickness is used and acts as a supporting substrate. Moreover, a ShieldIt conductive material, thickness of 0.17 mm, is used as a flexible truncated ground plane at the downside of the entire design.

Moreover, a Teflon substrate is used as a shield on the bottom side of the ground, with dielectric constant and loss tangent of 0.5 mm and 0.025, respectively. The size of the ground plane and substrate are 50 mm × 50 mm^2. The air gap between textile substrate and circular button surface is 3.617 mm. On the top side of the textile

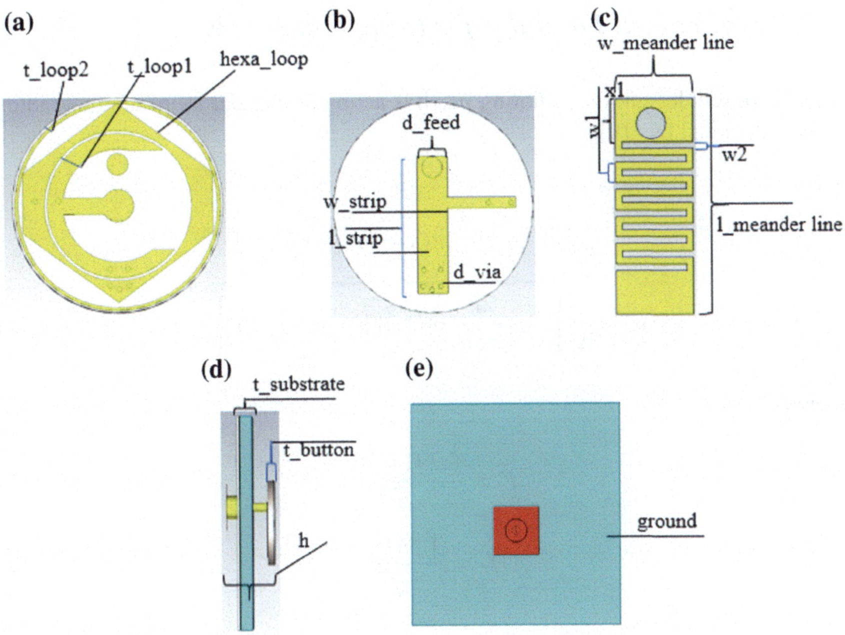

Fig. 1 Wearable design. **a** Button front side, **b** back side, **c** meandered line, **d** side view, and **e** ground plane

substrate surface, a PCB material is used, with a size of 23 mm × 16 mm^2 to design meander lines on it. Table 1 shows the important parameters of the proposed antenna.

Table 1 Antenna parameters

Parameters	Value (mm)	Parameters	Value (mm)
d_feed	1.0	t_loop2	1.04
d_via	0.25	t_Substrate	3
h	3.617	w_meander line	5.11
l_meander line	13.44	w_strip	1.84
l_stxip	11.47	wl	1.2
r_button	9.77	w2	0.4
r_loop1	5.07	x_feed	3.08
r_roop2	8.43	x_strip	4.44
t_button	1.6	x1	2.73
t_loop1	1.88	hexa_loop	9.77

2.1 Mathematical Formulas for Design Structure

In the proposed design, the radiating patch is acting as circular loops, and its radius can be calculated by [19],

$$a = \frac{F}{\sqrt{\left\{1 + \frac{2h}{\pi \varepsilon r a}[\ln(\pi a|2h) + 1.7726]\right\}}} \tag{1}$$

$$a_e = \sqrt[a]{\left\{1 + \frac{2h}{\pi \varepsilon a}[\ln(\pi a|2h) + 1.7726]\right\}} \tag{2}$$

whereas

$$F = \frac{8.791 * 10^9}{\sqrt[f]{\varepsilon r}} \tag{3}$$

$$f_r = \frac{1.841 v_o}{{}^{2\pi a_e}\sqrt{\varepsilon_r}} \tag{4}$$

v_0 speed of light (3×10^8 m/s), a = circular radius, a_e = effective circular radius, h = height.

3 Results and Discussion

The proposed antenna is intended for wearable application; the design performance can be evaluated when operating near the body model. A human model along with three different layers of size 150 mm × 150 mm × 32 mm is used for simulations. The proposed design is placed on the body model with a 15-mm gap between them, so that to mimic the condition whenever cloths or garments are present, as shown in Fig. 2. The body model is designed using CST MWS, which consists of different layers and thicknesses in [20]. Moreover, the properties of the tissues are depicted in Table 2.

Finally, the performance of the dual-band wearable crescent-shaped button antenna design which was performed using CST Microwave Studio is discussed thoroughly on the next subtopic, and it characteristic features are illustrated in Table 3.

3.1 Off-Body Communication

Figure 3 shows the reflection coefficient (S11) for off-body communication and provides an impedance bandwidth of 2.5185–2.7206 GHz for the lower band whereas

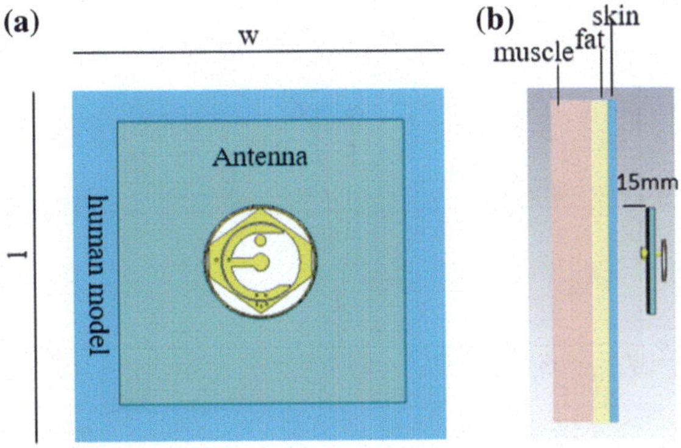

Fig. 2 Simulation model for on-body crescent-shaped button antenna. **a** On-body antenna, **b** side view

Table 2 Properties of human body tissues [20]

Tissues	Permittivity (ϵ_r)	Conductivity (S/m)	Loss tangent	Density (kg/m^3)
Skin	31.29	5.0138	0.2835	1100
Pat	5.28	0.1	0.19382	1100
Muscle	52.79	1.705	0.24191	1060
Bone	12.661	3.8591	0.25244	1850

Table 3 Summary of the performance of the investigated antenna

WLAN band	Low/high (off body)	Low/high (on body)
Bandwidth	202.6 MHz/951.3 MHz	126.5 MHz/
Directivity	2.19/1.57	8.98/5.55
Frequency points	2.5185 GHz–2.7206 4.4475–5.3988 GHz	2.4921–2.55 GHz 4.48867–5.601 GHz
Gain	1.11 dB/0.812 dB	5.76/2.86
Efficiency	77.97/83.93%	47.57/53.75
VSWR	1.120/1.183	1.26/1.034
HPBW	102.7(3 dB)	58.8/56.3 (3 dB)

for the higher band, the impedance bandwidth is 4.4475–5.3988 GHz, respectively. It is clearly shown that the performance of the off body communication doesn't affected due to large ground and shield, which provides isolation between antenna and body. However, the small shift in frequency occurred due to high dialectic constant of the human body. In the open literature review, the proposed design has achieved dual band with a wide impedance bandwidth. Table 3 shows characteristics of the proposed antenna for off-body communication.

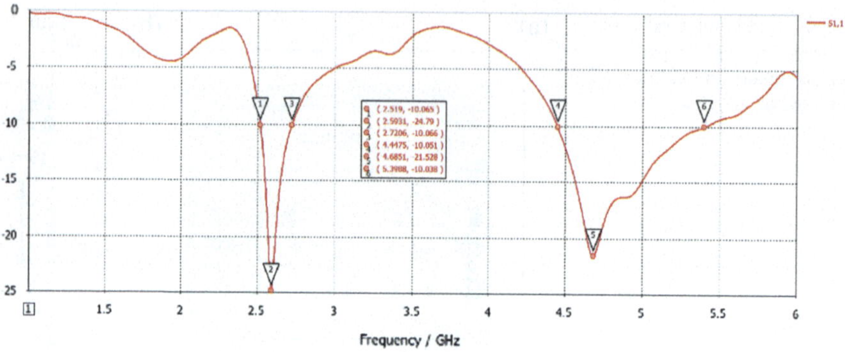

Fig. 3 Reflection of coefficient (S11) for off-body communications

3.2 *On-Body Communication*

The proposed design is tested on the body model with a size of 150 × 150 mm. It was positioned on various parts of the body model such as arm, chest, and leg (only flat condition presented here). Figure 4 shows the simulated reflection coefficients. The bandwidth is larger through the wearable button-shaped antennas such as 2.4921–2.55 GHz for lower band, respectively, whereas 4.48867–5.601 GHz for higher band, respectively. However, the −10dB impedance bandwidth is effected during on and off body communication. Therefore, the proposed investigated design is a suitable choice for WLAN applications.

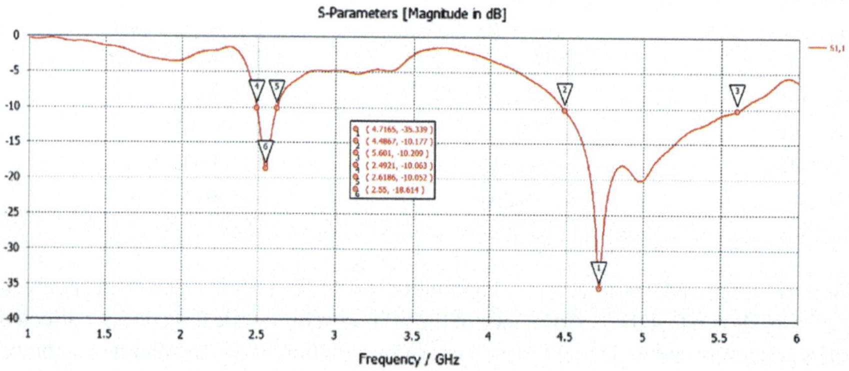

Fig. 4 Reflection of coefficient (S11) for on-body communication

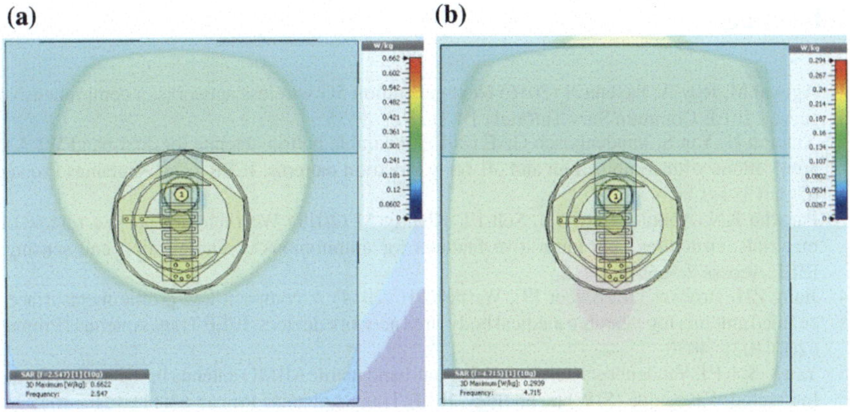

Fig. 5 Simulated SAR value of the proposed wearable crescent-shaped button antenna

3.3 Specific Absorption Rate (SAR)

The SAR is a significant feature for health and safety regulation during on-body communication. The SAR analysis can be assessed with the IEEE C95.1 standard. The investigated design was placed 15 mm away to mimic the antenna to skin distances due to garments or cloths. The design used 0.1 W(power). The simulated SAR values on 1 and 10 g of body tissues were 0.662 at 2.4921–2.55 GHz, respectively, whereas 0.294 W/kg was simulated at 4.4867–5.601, respectively, as shown in Fig. 5. Thus, the design showed the limited SAR value for on body communication.

4 Conclusion

A novel dual-band crescent-shaped button antenna is simulated for the application of the WLAN band. The design shows a compact small size as well as robust dual band during operation. The impedance bandwidth of 2.5185–2.7206 GHz and 4.4475–5.3988 GHz are obtained for off-body communication at lower and higher bands, respectively, whereas the impedance bandwidth of 2.4921–2.55 GHz and 4.48867–5.601 GHz are obtained for on-body communication at lower and higher bands, respectively, which are enough for small size button antenna. The overall radiation characteristics have shown stability during on- and off-body communication. Therefore, the proposed design has shown dual band during on- and off-body communication in lower and higher bands, respectively. Thus, the proposed design is suitable for the application of WBAN.

References

1. Agiwal M, Roy A, Saxena N (2016) Next generation 5G wireless networks: a comprehensive survey. IEEE Commun Surv Tutorials 18(3):1617–1655
2. Xiaomu H, Yan S, Vandenbosch GAE (2017) Wearable button antenna for dual-band WLAN applications with combined on and off-body radiation patterns. IEEE Trans Antennas Propag 65(3):1384–1387
3. Paracha KN, Abdul Rahim SK, Soh PJ, Khalily M (2019) Wearable antennas: a review of materials, structures, and innovative features for autonomous communication and sensing. IEEE Access 7:56694–56712
4. Jiang ZH, Brocker DE, Sieber PE, Werner DH (2014) A compact, low-profile metasurface-enabled antenna for wearable medical body-area network devices. IEEE Trans Antennas Propag 62(8):4021–4030
5. Yan S, Soh PJ, Vandenbosch GAE (2015) Dual-band textile MIMO antenna based on Substrate-Integrated Waveguide (SIW) technology. IEEE Trans Antennas Propag 63(11):4640–4647
6. Abbasi QH, Rehman MU, Yang X, Alomainy A, Qaraqe K, Serpedin E (2013) Ultrawideband band-notched flexible antenna for wearable applications. IEEE Antennas Wirel Propag Lett 12:1606–1609
7. Liu FX, Kaufmann T, Xu Z, Fumeaux C (2015) Wearable applications of quarter-wave patch and half-mode cavity antennas. IEEE Antennas Wirel Propag Lett 14:1478–1481
8. Yan S, Soh PJ, Vandenbosch GAE (2014) Wearable dual-band composite right/ left-handed waveguide textile antenna for WLAN applications. Electron Lett 50(6):424–426
9. Agneessens S, Lemey S, Vervust T, Rogier H (2015) Wearable, small, and robust: the circular quarter-mode textile antenna. IEEE Antennas Wirel Propag Lett 14:1482–1485
10. Sankaralingam S, Gupta B (2010) Determination of dielectric constant of fabric materials and their use as substrates for design and development of antennas for wearable applications. IEEE Trans Instrum Meas 59(12):3122–3130
11. Yan S, Soh PJ, Vandenbosch GAE (2014) Low-profile dual-band textile antenna with artificial magnetic conductor plane. IEEE Trans Antennas Propag 62(12):6487–6490
12. Scarpello ML, Kazani I, Hertleer C, Rogier H, Vande Ginste D (2012) Stability and efficiency of screen-printed wearable and washable antennas. IEEE Antennas Wirel Propag Lett 11:838–841
13. Yimdjo Poffelie LA, Soh PJ, Yan S, Vandenbosch AEG (2016) A high-fidelity all-textile UWB antenna with low back radiation for off-body WBAN applications. IEEE Trans Antennas Propag 64(2):757–760
14. Hu B, Gao GP, Le He L, Cong XD, Zhao JN (2016) Bending and on-arm effects on a wearable antenna for 2.45 GHz body area network. IEEE Antennas Wirel Propag Lett 15:378–381
15. Yan S, Soh PJ, Vandenbosch GAE (2015) Performance on the human body of a dual-band textile antenna loaded with metamaterials. In: 2015 9th European Conference on Antennas and Propagation, EuCAP 2015
16. Mandal B, Parui SK (2015) A miniaturized wearable button antenna for Wi-Fi and Wi-Max application using transparent acrylic sheet as substrate. Microw Opt Technol Lett 57(1):45–49
17. Mandal B, Chatterjee A, Parui SK (2014) A wearable button antenna with FSS superstrate for WLAN health care applications. In: Conference Proceeding on 2014 IEEE MTT-S International Microwave Workshop Series on RF and Wireless Technologies for Biomedical and Healthcare Applications. IMWS-Bio 2014, no 3, 1–3, 2015
18. Mandal B, Chatterjee A, Parui SK (2015) Acrylic substrate based low profile wearable button antenna with FSS layer for WLAN and Wi-Fi applications. Microw Opt Technol Lett 57(5):1033–1038
19. Sreelakshmy R, Vairavel G (2019) Novel cuff button antenna for dual-band applications. ICT Express 5(1):26–30
20. Ali U et al (2017) Design and SAR analysis of wearable antenna on various parts of human body, using conventional and artificial ground planes. J Electr Eng Technol 12(1):317–328

VEDZA: Kinect Based Virtual Shopping Assistant

Mashal Valliani, Agha Saba Asghar, and Rabeea Jaffari

Abstract Shopping trial room experiences can be time-consuming, tiring, and insecure for customers. To address these problems, this work proposes VEDZA: Kinect-based virtual shopping assistant which allows both online and traditional shopping customers to try on clothes and accessories virtually while providing realistic experiences. The aim of the system is to eliminate the hectic process of using physical trial rooms for traditional shoppers which can be insecure due to the hidden cameras installed as well as to provide real-time trial options to the online shoppers. The system framework has garnered an overall positive feedback on tests with a wide range of customers.

Keywords Virtual shopping assistant · Virtual try-on · Augmented reality · Kinect · Human–computer interaction

1 Introduction

The way toward browsing and buying things in return for cash is called shopping. There are two types of shopping: traditional and online shopping. The statistics of market line, a business organization, reveal that the apparel industry has been developed at 4.78% yearly since 2011 with an expected growth to further reach approximately US$1.65 trillion globally by 2020 [1].

With the ratio of buyers increasing, the ratio of problems while shopping has also increased. In traditional shopping, customers select the likeable products and proceed to the fitting rooms to try them out which is quite a hectic and time-consuming

M. Valliani (✉) · A. S. Asghar · R. Jaffari
Department of Software Engineering, Mehran University of Engineering and Technology, Jamshoro, Pakistan
e-mail: mashal_valliani@hotmail.com

A. S. Asghar
e-mail: aghasabaasghar@yahoo.com

R. Jaffari
e-mail: rabeea.jaffari@faculty.muet.edu.pk

S.-L. Peng et al. (eds.), *Intelligent Computing and Innovation on Data Science*, Lecture Notes in Networks and Systems 118,
https://doi.org/10.1007/978-981-15-3284-9_51

process. Moreover, the physical trial rooms are not safe as the stores might have hidden cameras installed in them. A woman from South China found a hidden camera in H&M store while trying out clothes [2]. In addition to this, as per NBC Washington news, it was suspected by the police that four stores Gap, Old Navy, Forever 21, or H&M in Northern Virginia had hidden cameras installed in their fitting rooms [3].

On the other hand, the Web-based business platform is one of the principle determinants in which the individuals spend their cash and the organizations direct their businesses. As per an e-marketer research report, the Internet business industry will appreciate development from 18 to 25% consistently with the overall exchange to arrive at approximately 4 trillion USD by the year 2020 [4]. The quick development of this industry will profit businesses and online clients, and however, there are disadvantages associated with it. Researchers have recognized that the Web-based apparel retailers need to deal with 30–40% returns because of clients buying wrong-sized products [5, 6] which in turn incurs huge losses to the business. According to a similar study, at least once in a lifetime, 69% percent of Internet apparel clients encounter size issues and the probability of the reports, with the cost for the returned merchandise being taken by the clients, is overwhelmingly 60% [7]. This finding demonstrates that getting an incorrect size is an extreme issue in the Web-based trade. Moreover, a business owner's benefit is greatly affected as removing the article of clothing from the business cycle for more than seven days can cause the seller to lose up to 20% of the value on the piece of clothing, especially in the seasonal and sales time [8]. Considering these issues a few virtual fitting room frameworks are on the rise nowadays. As per reports, Markets and Markets expects the worldwide virtual dressing room market to increase by 2024 up to USD 7.6 billion from USD 2.9 billion in 2019, at a CAGR of 20.9% [9].

According to the survey conducted for VEDZA, 79.5% of the users said that they need virtual fitting rooms. By looking at the demand of the user and to overcome the issues faced during shopping, we propose VEDZA, a system which implements the concept of virtual shopping assistant using augmented reality. The system is feasible to be used both in physical stores as well as the online ones. The aim of the system is to eliminate the insecure and frustrating process of physical fitting rooms and provide the best shopping experiences to the users.

2 Related Work

Numerous amounts of efforts have been done to develop virtual fitting rooms and there are many related platforms available. These are anyway constrained in extension, usefulness, and functionality and are discussed below.

2.1 Zugara: Virtual Dressing Room Product

It has three (3) products: WSS for Web, Kiosks, and instore retail. The product operates with customary webcams as well as Kinect. WSS for Kiosks tablet integration is also available. Zugara limits in utilizing pieces of clothing with long sleeves or if it goes over the elbow and knees. Trial on the body in motion is not possible [10].

2.2 triMirror

A virtual fitting room that allows real clothes to be tried before purchasing on the real-dimensioned avatar body only and not on real-time users [11].

2.3 Awaseba: The Virtual Fitting Room

An e-commerce platform where first-time clients just transfer their photograph onto the Awaseba server, and that photograph works as their virtual self-inside the system after which user can virtually try on clothes and accessories [12].

2.4 Virtual Mirror Technology

The Virtual Mirror Technology is implemented using augmented reality and RFID technology. In the system, the user brings the piece of garment in front of a mirror which scans and embeds it with the image of the person. The output can be seen as a reflected image in the mirror [13].

2.5 SenseMi: Virtual Dressing Room Mirror, Retail Mirror, Smart Mirror

It uses augmented reality in which a 3d model of a cloth or accessory is embedded in the live video of the customer and tracks the movement of the user providing realistic appearance. It uses mirror technology which is quite expensive to implement [14].

2.6 The Virtual Dressing Room Trend

It uses WSS which stands for Webcam Social Shopper software. It is developed for online customers. In this system, a woman holds a garment in front of webcam and can experience on the screen [15].

2.7 Amazon: Virtual Dressing Room

This application helps to make a better idea about what looks good on you. It has several products customized only for online stores [16].

2.8 Fits.me

The system is based on data bank consisting of metrics of human body which is obtained from a person through an Internet framework. Robotic mannequin of an individual physical form is created and used. The user sees its mannequin on the screen which resembles their own body [17].

2.9 Virtual Fitting Room

It is an android application which detects human face only and is just designed for jewelry try on [18].

2.10 Virtual Dressing Room

It is a framework that uses Kinect to track and display the movement of the customer in real time. Users' facial details as well as body size details are captured to develop a realistic mannequin upon which a 3d model is embedded [19].

Figure 1 illustrates the comparison of various features between these platforms. The proposed framework VEDZA's performance is far better as compared to the rest of the frameworks as it operates using depth sense camera using augmented reality technology and is developed for both online and traditional shopping customers.

Platforms / Features	triMirror	Zugera: Virtual Dressing Room Products	Awaseba: The Virtual Fitting Room	SenseMi: Virtual Dressing Room Mirror, Smart Mirror, Retail Mirror	Virtual Dressing Room Trend	Amazon: Virtual Dressing Room	Virtual Mirror Technology	Fits.me	Virtual Dressing Room System based on Kinect Design	Design and Implementation of Virtual Fitting Room based on Image Blending
Based on Augmented reality	X	✓	X	X	X	X	✓	X	✓	✓
Depth sense Camera Used	X	✓ (Kinect)	X	✓ (Kinect)	X	X	X (uses RFID)	X	✓	✓
Virtual try-on (online)	✓	X	✓	X	✓	✓	X	✓	✓	X
Virtual try-on (in-store)	✓ (pc)	✓ (pc)	X	✓ (mirror)	X	X	✓	✓	✓	✓
Ease of searching	X	✓	✓	✓	✓	✓	✓	✓	✓	✓
Output on Computer Screen	✓	✓	✓	X (mirror)	✓	✓	X (mirror)	✓	✓	✓
Try-on real human body	X (avatar)	✓ (	X (photo)	✓	✓	✓	✓	X (mannequin)	X (mannequin)	✓
Body in motion can be tracked	X	X	X	✓	✓	X	✓	X	✓	X
Optimized for ecommerce	X	✓	✓	✓	✓	✓	✓	✓	X	X
Different Categories support (Garments, Accessories)	✓ (Garments)	✓ (limited 3dmodel used only sleeveless and up to knee length)	✓ (Garments, accessories)	✓ (Garments, accessories, make-up items)	✓ (Garments)	✓ (Garments, accessories)	✓ (Garments)	✓ (Garments)	✓ (Garments)	✓ (Accessories i.e. only jewels)

Fig. 1 Comparison of features of similar platform with proposed system

3 Proposed System

VEDZA is based on augmented reality which lets customers try on clothes and accessories virtually in the real-world environment. The framework is developed using Unity3d and uses a depth sense Kinect camera for body detection, body localization, and relevant motion detection. 3d models of clothes and accessories are used to view the product in different orientations providing realistic experiences. Later, object transformation takes place to superimpose 3d model of the product on the detected user body using Kinect which can be experienced on the screen by the user. Various categories of clothing and accessories such as shirts, pants, shorts, complete suits, wrist watches, bags, and hats are available for selection and try out for both men and women.

3.1 Workflow

Figure 2 highlights the workflow of user interaction with VEDZA. The first activity is the selection of product category followed by the finalization of the desired product for virtual trial. If the product is from the clothing category, user can select the size and orientation (front or back) after which Kinect detects the human body and in last the product is displayed superimposed on the user.

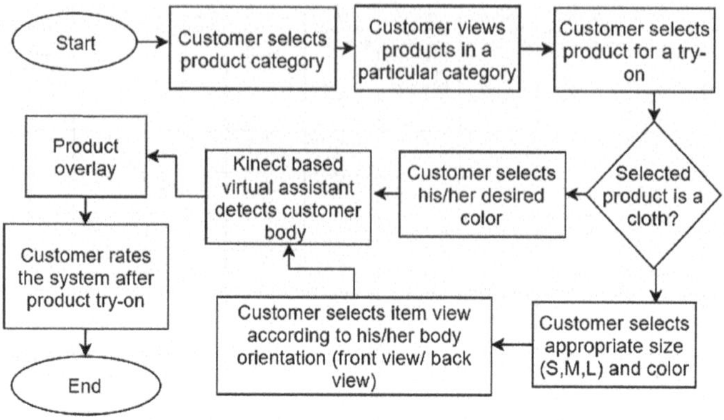

Fig. 2 System's conditional workflow

4 Implementation

The proposed system VEDZA is implemented as follows:

4.1 Autodesk 3d Studio Max

3d models of clothes and accessories are created using 3d studio max by applying the process of meshing, rigging, and texturing.

4.2 Unity3d

In the development of the system, Unity3d is used. The framework is programmed using C#. Unity is used for real-time creations providing built-in system for creating the user interface and custom tools. To reach maximum possible target unity offers "build once and situate anywhere" which is a multiplatform support. The system is universally compatible for different platforms as it has been developed using Unity3d [20].

4.3 Kinect

Microsoft Kinect is being used in the system as the input device to detect live human body joints and movement. As the user selects the product for trial, Kinect device is triggered to be turned on. The user stands in front of the device for the body detection.

In a single skeleton, Kinect can track up to twenty body joints. It can also track at a time six skeletons, and out of six tracks, two human skeletons are fully tracked [21].

5 Results and Evaluations

This section depicts the output result of the proposed system in Fig. 3 and the usability characteristics in Fig. 4. A user survey conducted to check the system feasibility depicts that around 42.2% of the users usually use fitting rooms before buying while 50.6% of the end users find it tiring to such rooms and an overall 79.5% of the users find the need for virtual fitting rooms important.

The user's ratings from a survey conducted for various characteristics regarding the application are depicted in Fig. 4. In the bar chart, *y*-axis represents linear scale consisting of 1–5 points where 1 is the minimum and 5 is the maximum scale, while the *x*-axis indicates the percent of user's response with different colors showing various parameters that are rated by the users.

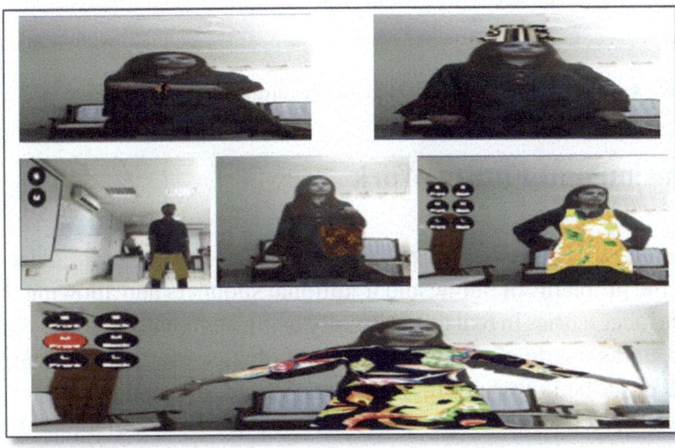

Fig. 3 Final view of the proposed system showing the 3d models embedded on the users in real time

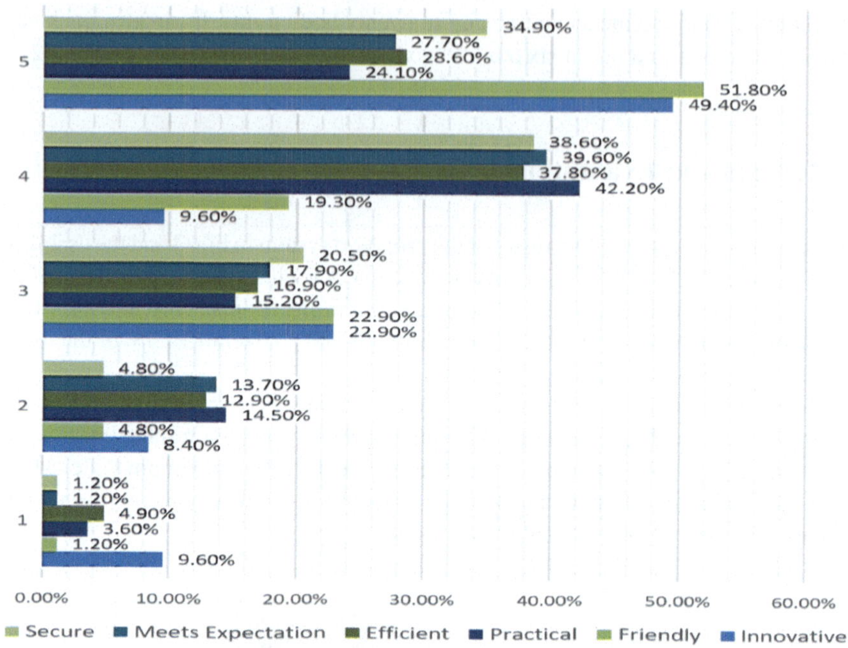

Fig. 4 End user's ratings for various characteristics regarding the application

6 Conclusion and Future Work

In this research, VEDZA was proposed as a virtual fitting room addressing and overcoming the problems of online and traditional shoppers and allowing them to try on clothes and accessories in real time virtually using augmented reality. The research also gave an understanding of the market technologies available in comparison to VEDZA and described in what different ways VEDZA has outperformed each one of them. The research also includes the results of a survey highlighting the need for virtual shopping assistant and experience of the users for the proposed system. The findings depict a positive user support for VEDZA in particular and physical and Web-based virtual fitting rooms in general.

In future, VEDZA can be extended to operate using gestures and voice. Another feature that can be included is a full 3d rotation of the product model with the user body movement to provide a high-class real-time experience and guarantee user satisfaction. Lastly, if user wants to buy any product, they can add it to the cart using the application during the virtual trial and all the shopping details will be transferred to the counter automatically saving users time while billing.

Acknowledgements This research has been completed with the tireless efforts and contributions of all the individuals involved. The authors would like to express greater gratitude to Mehran University of Engineering and Technology, Jamshoro, Pakistan, for providing the necessary resources and environment needed to complete this project. We would also like to acknowledge the effort and time of all those reviewers who participated in the survey and provided reviews and feedback for the need and usability of the research.

References

1. Singh G (2017) Fast fashion has changed the industry and the economy. [online] Available at:https://talkmarkets.com/content/services/fast-fashion-has-changed-the-industry-andtheeconomy?post=141276. Accessed 9 Aug 2019. TalkMarkets
2. Mengxiao C (2017) Hidden camera found in H&M store's changing room. [online] News.cgtn.com. Available at: https://news.cgtn.com/news/3d416a4e3159444e/share_p.html. Accessed 9 Aug 2019
3. Swalec A (2019) Fitting room filmings possible at 4 Va. Shopping centers. [online] NBC4 Washington. Available at: https://www.nbcwashington.com/news/local/Shoppers-MayHave-Been-Filmed-in-Fitting-Rooms-at-4-Northern-Virginia-Shopping-Centers-Police-Say504046181.html. Accessed 9 Aug 2019
4. Worldwide Retail Ecommerce Sales Will Reach $1. 915 Trillion This Year, 2016
5. Brooks AL, Brooks E (2014) Towards an inclusive virtual dressing room for wheelchair-bound customers. In: 2014 international conference on collaboration technologies and systems CTS, pp 582–589
6. Jonstromer H, Rentzhog M, Aner E (2012) E-commerce—new opportunities new barriers" Kommerskollegium The National Board of Trade, [online] Available at: http://www.kommers.se/In-English/Publications/2012/E-commerce-NewOpportunitiesNew-Barriers/
7. Noordin S, Sahari N, Tengku Wook TSM (2016) Fitting in Malaysia online clothing industry. PG symposium
8. Kramer A (2011) The virtual fitting room. [online] strategy + business. Available at: https://www.strategy-business.com/article/00073. Accessed 10 Aug 2019
9. Marketsandmarkets.com (2019) Virtual fitting room market by software and services—2024 | MarketsandMarkets. [online] Available at: https://www.marketsandmarkets.com/Market-Reports/virtual-fitting-room-market-132071646.html. Accessed 10 Aug 2019
10. Feiner SK (2002) Augmented reality: a new way of seeing. Sci Am 286(4):48–55
11. Zugara. (n.d.). Virtual dressing room technology. [online] Available at: http://zugara.com/virtual-dressing-room-technology Accessed 10 Aug 2019. Anon, (n.d.). Home. [online] Available at: https://www.trimirror.com/. Accessed 10 Aug 2019
12. Trendland Online Magazine Curating the Web since 2006. (n.d.). Awaseba: The Virtual Fitting Room. [online] Available at: https://trendland.com/awaseba-the-virtual-fitting-room/. Accessed 10 Aug 2019
13. Quytech Blog. (n.d.). Virtual mirror technology—it will change the way you shop. [online] Available at: http://www.quytech.com/blog/how-virtual-mirror-technologywillchange-the-way-you-shop/. Accessed 10 Aug 2019
14. SenseMi. (n.d.). Virtual dressing room mirror, smart mirror, retail mirror | SenseMi. [online] Available at: http://sensemi.com/. Accessed 10 Aug 2019
15. Klarna Knowledge (2018) The virtual dressing room trend: what retailers need to know—Klarna Knowledge. [online] Available at: https://www.klarna.com/knowledge/articles/thevirtualdressing-room-trend-what-retailers-need-to-know/Accessed 10 Aug 2019
16. Amazon.com. (n.d.). [online] Available at: https://www.amazon.com/MyCoolLookcomVirtual-Dressing-Room/dp/B00QJSPI9U. Accessed 10 Aug 2019
17. Rakuten Fits Me (n.d.). Rakuten fits me fit recommendation technology—Rakuten Fits Me. [online] Available at: https://fits.me/. Accessed 10 Aug 2019
18. Patel B (2016) Design and implementation of virtual fitting room based on image blending. Int J Adv Eng Res Dev 3(4, April-2016):452–458

19. Mok K, Wong C, Choi S, Zhang L (2018) Design and development of virtual dressing room system based on kinect. IJ Inf Technol Comput Sci 39–46
20. Unity. (n.d.) Products—Unity. [online] Available at: https://unity3d.com/unity. Accessed 11 Aug 2019
21. Jana A (2012) Kinect for windows SDK programming guide. Packt Publishing Ltd.

The Impact of Organizational Innovation on Financial Performance: A Perspective of Employees Within Dubai Ports World

Ali Ameen, Mohammed Rahmah, Osama Isaac, D. Balaganesh, Midhunchakkaravarthy, and Divya Midhunchakkravarthy

Abstract Firms' performance in terms of financial performance can be considered as an organizational innovation issue. For the reason being firms' performance is considered as a result of product innovation, process innovation and administrative innovation in developing and implementing effective ideas, the performance is crucially depending on effective organizational innovation. This research aims to examine the impact of organizational innovation (product, process and administrative) on the financial performance of organizations within DP World in the UAE. A personally administered questionnaire being distributed by the researcher will be used to collect data from respondents within DP World in the UAE. Random sampling method was adopted to select the employees who use smart government services, and only 403 out of 700 respondents were achieved a response rate of 58%, which is considered as a healthier survey response rate, while 372 were analysed after removing missing data, outliers and suspicious responses. Partial least squares (PLS) structural equation modelling-variance based (SEM-VB) method was employed to assess the research model by utilizing the software SmartPLS 3.0. The proposed research model explained 58% of the organizational innovation (OI). Organizational innovation had a positive direct effect on the OI within DP world in the United Arab Emirates (UAE). The results of the current research have the potential to give further insights into the innovation of organizations strategies.

Keywords Organizational innovation · Financial performance · Dubai ports world · UAE

The original version of this chapter was revised: The author's name has been corrected from "Midhunchakkravarthy" to "Midhunchakkaravarthy". The correction to this chapter is available at https://doi.org/10.1007/978-981-15-3284-9_86

A. Ameen (✉) · M. Rahmah · O. Isaac · D. Balaganesh · Midhunchakkaravarthy · D. Midhunchakkravarthy
Lincoln University College, Kota Bharu, Selangor, Malaysia

S.-L. Peng et al. (eds.), *Intelligent Computing and Innovation on Data Science*, Lecture Notes in Networks and Systems 118,
https://doi.org/10.1007/978-981-15-3284-9_52

1 Introduction

Firms' performance in terms of financial performance can be considered as an organizational innovation issue [1]. For the reason being firms' performance is considered as a result of the process, product and administrative innovation in developing and implementing effective ideas, the performance is crucially depending on effective organizational innovation.

There are many expanding and successful business organizations in the United Arab Emirates nowadays. A business success does not only depend on the intelligence and hard work of the founder alone but the whole workforce, starting from the lowest to the highest position in an organization. Interest and research on organizational innovation and organizational performance have notably increased among the managements and academics.

The UAE performance overview shows that it is doing well among the world's countries as it ranks 27th overall performance, comparing to the USA as the world's No. 1. However, in the innovation ecosystem, it shows that the UAE ranks 35th in the world in terms of innovation capability. It is clear that the UAE is trying to become a leading technology centre based on the innovation strategy of the Fourth Industrial Revolution [2, 3]. This means that there is a space for innovation improvement which will be reflected on the overall performance of the organization and thus the country. Various global indicators have created a clear image that helps in understanding the position of country level according to a set of measures that are recognized internationally [4–7].

2 Literature Review

2.1 Organizational Innovation (OI)

As suggested by Amabile [8], innovation is not the same as creativity. Innovation is the only thing that defines the generation of new ideas. Innovation leads to the creation and implementation of new processes, products and ideas [9]. Hence, creativity can be regarded as a component of innovation [10].

The work environment for most organizations is turbulent with fast transformations in market conditions and increasing market uncertainties, advancement in information technology, shortened product life cycles and growing competition [11, 12]. Therefore, innovation is an essential aspect for the sustenance and growth of organizations under such adverse environments [13]. Organizations regard innovation to be a crucial variable for their survival [14]. Besides, according to the views of [15], the ambitious goals of organizations can be achieved only by the means of innovation. This is one of the key resources for achieving economic growth and sustainability in the twenty-first century [16, 17].

Even though innovation has played an important role in sustaining the growth of the services sector as well as the manufacturing sector, the focus of innovation studies has primarily been on manufacturing, while only a few studies have explored the role of innovation in the services sector, specifically in the banking sector [18, 19]. The financial services have undergone substantial innovation. On the contrary, the services sector has several distinguishing features that are unlike the characteristics of the products (goods) manufacturing sector. Most of the researches analyse four distinct characteristics: heterogeneity, inseparability, intangibility and perishability. In most contemporary organizations, adopting technology not only uses ICT to fill up some forms and records but rather it is also a tool that performs the process of identification, accumulation, analysis, measurement, preparation, interpretation and communication of the information used by management to plan [2, 20–22]. It is used in evaluating and controlling within an organization and to assure appropriate use and accountability for their resources [20–22].

Numerous studies have investigated the relationship between innovation and performance. Some evidence was found that showed that there is a positive influence of innovation on business performance. In Brazil, the results indicated that efforts put in innovation will likely lead to impacts, and these impacts could possibly imply improved organizational learning performance. A hypothesis is therefore suggested:

***H1**: Organizational innovation has a positive effect on financial performance.*

2.2 Financial Performance (FP)

Financial performance identifies the way in which we hold our shareholders [23]. As per [24], the financial perspective is generally used as a measurement tool in administrative accounting. There is an absence of a typical set of these measurements that can be applied to different firms and environments. Previous investigations measure this standpoint using various metrics, for instance, profit for every employee, return on equity (ROE), earnings per share (EPS), net operational income, profit margin, revenue growth, economic value added (EVA), return on investment (ROI), revenue for every employee and growth in common equity [25].

Financial terms such as value at risk, profitability and market value can define performance. However, performance is frequently used in other environments, such as marketing (that includes number of customers retained over a certain period, customer satisfaction, etc.), operations (that includes throughput time, product or service quality, efficiency, number of outputs, effectiveness, etc.) and others [26].

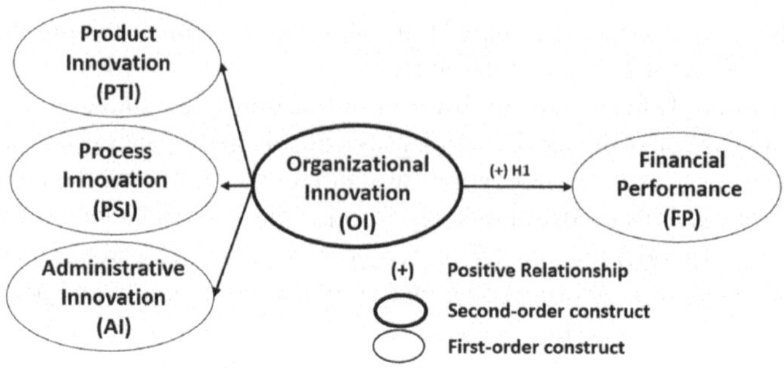

Fig. 1 Conceptual framework

3 Research Method

3.1 Overview of the Proposed Conceptual Framework

This research proposes a research model based on balance scorecard which include aspect finance [27] and organizational innovation postulated in the literature which examined the relationship between organizational innovation (product, process and administrative) and financial performance. Based on the above, the research model for this research is depicted in Fig. 1.

3.2 Instrument Development and Collecting Data

The respondents are employees from DP World in the UAE. A personally administered questionnaire being distributed by the researcher will be used to collect data from respondents within the sample populations in the current research. Random sampling method was adopted to select the employees who uses smart government services, and only 403 out of 700 respondents were achieved a response rate of 57.57%, which is considered as a healthier survey response rate, while 372 were analysed after removing missing data, outliers and suspicious responses. Partial least squares (PLS) structural equation modelling-variance based (SEM-VB) method was employed to assess the research model by utilizing the software SmartPLS 3.0.

4 Analysing Data and Findings

Partial least squares (PLS) structural equation modelling-variance based (SEM-VB) method was employed to assess the research model by utilizing the software Smart-PLS 3.0 [28]. Analysing data through the second-generation multivariate data analysis technique which is SEM offers a simultaneous analysis which leads to more accurate estimates [29, 30].

4.1 Measurement Model Assessment

The individual Cronbach's alpha, the composite reliability (CR), the average variance extracted (AVE) and the factor loadings exceeded the suggested value [31, 32] as illustrated in Table 1.

The degree to which the articles distinguish among concepts or measure different constructs is demonstrated by discriminant validity. Fornell–Larcker was employed to analyse the measurement model's discriminant validity. Table 2 shows the outcomes for discriminant validity by employing the Fornell–Larcker condition. It was discovered that the AVEs' square root on the diagonals (displayed in bold) is bigger than the correlations among constructs (corresponding row as well as column

Table 1 Measurement model assessment

Constructs	Item	Loading (>0.7)	M	SD	α (>0.7)	CR (>0.7)	AVE (>0.5)
Product innovation (PTI)	PTI1	0.950	3.99	1.01	0.939	0.961	0.891
	PTI2	0.946					
	PTI3	0.935					
Process innovation (PSI)	PSI1	0.942	3.78	1.02	0.940	0.962	0.893
	PSI2	0.940					
	PSI3	0.952					
Administrative innovation (AI)	AI1	0.853	3.73	1.05	0.847	0.896	0.683
	AI2	Deleted					
	AI3	0.837					
	AI4	0.785					
	AI5	0.831					
Financial performance (FP)	FP1	0.896	3.68	1.11	0.911	0.937	0.789
	FP2	0.910					
	FP3	0.859					
	FP4	0.888					
	FP5	Deleted					

Table 2 Fornell–Larcker criterion

	AI	FP	PSI	PTI
AI	**0.827**			
FP	0.680	**0.888**		
PSI	0.680	0.674	**0.945**	
PTI	0.613	0.660	0.691	**0.944**

Note Diagonals represent the square root of the average variance extracted while the other entries represent the correlations

values), suggesting a strong association between the concepts and their respective markers in comparison to the other concepts in the model [33, 34]. According to [36], this indicates good discriminant validity. Furthermore, exogenous constructs have a correlation of less than 0.85 [35]. Therefore, all constructs had their discriminant validity fulfilled satisfactorily.

4.2 *Structural Model Assessment*

The structural model can be tested by computing beta (β), R^2 and the corresponding t-values via a bootstrapping procedure with a resample of 5000 [36].

Figure 2 and Table 3 depict the structural model assessment, showing the results of the hypothesis tests. Organizational innovation positively influences financial performance. Hence, H1 is accepted with ($\beta = 0.764, t = 29.217, p < 0.001$). Organizational innovation explains fifty-eight per cent of the variance in financial performance. The

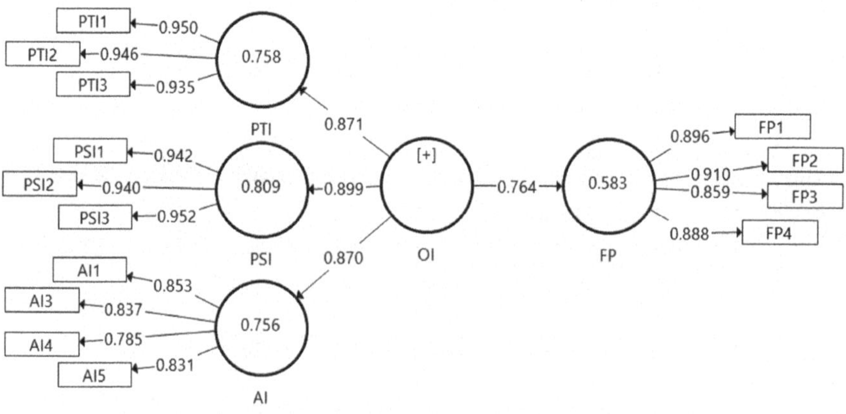

Key: OI: Organizational Innovation; PTI: Product Innovation; PSI: Process Innovation; AI: Administrative Innovation; FP: Financial Performance

Fig. 2 PLS algorithm results

Table 3 Result of direct effect hypotheses

Hypothesis	Relationship	Std Beta	Std Error	*t*-value	*p*-value	Decision	R^2
H1	OI → FP	0.764	0.026	29.217	0.000	Supported	0.58

Key: *OI* Organizational innovation; *FP* Financial performance

values of R^2 have an acceptable level of explanatory power, indicating a substantial model [37, 38].

5 Discussion and Implications

The main objective of this research is to investigate the impact of organizational innovation in term of (process, product and administrative) on the financial performance within DP world in the UAE. This objective has one hypothesis that needs to be tested which is: organizational innovation has a positive impact on the organizational performance.

This hypothesis was supported with ($\beta = 0.764$, $t = 29.217$, $p < 0.001$) which indicates significant effect of product innovation on organizational performance. The findings imply that the more DP world develop new product and services, introduce and diversify product to suit customer needs and try applying a new idea/technology at DP world organization. There will be no problem in financing the work of the Organization and various programs, and to assess the financial side of our programs have a role in future funding and reflected on the performance of the Organization.

This suggests that DP world may want to pay attention to their organizational innovation in term of product, process and administrative to improve their financial performance of the organization. The more innovative the organization is the better and higher financial performance of organization will be. Thus, the objective of this research is achieved. The research contributes critical theoretical value by highlighting those components of organizational innovation that contribute significantly to organizational performances.

6 Conclusion

This research aims to increase knowledge in the field of organizational innovation and organizational performance regarding the UAE. Through examinations of the effects of product innovation, process innovation and administrative innovations on the performances of public-sector organizations, the research adds valuable knowledge to the field of public sector as well as academic studies regarding the UAE. This article has shed some light on the organization performance in the public sector in the UAE and the importance of organizational innovation in that regard and proved that organizational innovation plays a role helping the organizations to improve their financial performance and compete to stay alive.

References

1. Mostafa M (2005) Factors affecting organisational creativity and innovativeness in Egyptian business organisations: an empirical investigation. J Manag Dev 24(1):7–33
2. Ameen A, Almari H, Isaac O (2019) Determining underlying factors that influence online social network usage among public sector employees in the UAE. In: Faisal Saeed FM, Gazem N (eds) Recent trends in data science and soft computing. IRICT 2018. Advances in Intelligent Systems and Computing, Recent Tre, vol 843. Springer International Publishing, Springer Nature Switzerland AG, pp 945–954
3. Alkhateri A, Abuelhassan AS, Khalifa AE, Nusari GSA, Ameen M (2018) The impact of perceived supervisor support on employees turnover intention : the mediating role of job satisfaction and affective organizational commitment. Int Bus Manag 12(7):477–492
4. Al-Ali W, Ameen A, Issac O, Nusari M, Alrajawi I (2018) Investigate the influence of underlying happiness factors on the job performance on the oil and gas industry in UAE. Int J Manag Hum Sci 2(4):32
5. AlShamsi R, Ameen A, Isaac O, Al-Shibami AH, Sayed Khalifa G (2018) The impact of innovation and smart government on happiness: proposing conceptual framework. Int J Manag Hum Sci 2(2):10–26
6. Al-Obthani F, Ameen A, Nusari M, Alrajawy I (2018) Proposing SMART-Government model: theoretical framework. Int J Manag Hum Sci 2(2):27–38
7. Haddad A, Ameen A, Mukred M (2018) The impact of intention of use on the success of big data adoption via organization readiness factor. Int J Manag Hum. Sci 2(1):43–51
8. Amabile T (1983) The social psychology of creativity: a componential conceptualization. J Pers Soc Psychol 45:357–376
9. Trott P (2005) Innovation management and new product development. Pearson Education Limited
10. West MA, Farr JL (1990) Innovation and creativity at work: psychological and Organizational strategies. Wiley
11. Dinopoulos E, Syropoulos C (2007) Rent protection as a barrier to innovation and growth. Econ Theory 32(2):309–332
12. Madrid-Guijarro A, Garcia D, Van Auken H (2009) Barriers to Innovation among Spanish Manufacturing SMEs. J Small Bus Manag 47(4):465–488
13. Bohlmann JD, Spanjol J, Qualls WJ, Rosa JA (2012) The interplay of customer and product innovation dynamics: an exploratory study. J Prod Innov Manag 30(2):228–244
14. Govindarajan V, Trimble C (2005) Ten rules for strategic innovators: from ideas to execution. Harvard University Press
15. Cooper RG (2011) Perspective: the innovation dilemma: how to innovate when the market is mature. J Prod Innov Manag 28(s1):2–27
16. Gumusluoğlu L, Ilsev A (2009) Transformational leadership and organizational innovation: the roles of internal and external support for innovation. J Prod Innov Manag 26(3):264–277
17. Atalay M, Anafarta N et al (2011) Enhancing innovation through intellectual capital: a theoretical overview. J Mod Account Audit 7(2):202
18. de Vries EJ (2006) Innovation in services in networks of organizations and in the distribution of services. Res Policy 35(7):1037–1051
19. Droege H, Hildebrand D, Forcada MAH (2009) Innovation in services: present findings, and future pathways. J Serv Manag 20(2):131–155
20. Ameen A, Ahmad K (2012) Towards harnessing financial information systems in reducing corruption: a review of strategies. Aust J Basic Appl Sci 6(8):500–509

21. Ameen A, Ahmad K (2011) The role of finance information systems in anti financial corruptions: a theoretical review. In: 11 international conference on research and innovation in information systems (ICRIIS'11), pp 267–272
22. Ameen A, Ahmad K (2013) A conceptual framework of financial information systems to reduce corruption. J. Theor. Appl. Inf. Technol. 54(1):59–72
23. Kaplan RS, Norton DP (2005) The balanced scorecard: measures that drive performance. Harv Bus Rev July-August
24. Al Mseden NA, Nassar MA (2015) The effect of balanced scorecard (BSC) implementation on the financial performance of the jordanian companies. In: The international business and social science research conference (2015)
25. Rao MP (2000) A simple method to link productivity to profitability. Manag Account Q 1(4):12–17
26. Verweire K, van den Berghe L (2004) Integrated performance management: a guide to strategy implementation. SAGE Publications
27. Abu-Qouod G (2006) The role of strategic management towards improving institutional performance in public organizations in the Hashemite Kingdom of Jordan. Cairo University-Eygpt
28. Ringle CM, Wende S, Becker J-M (2015) SmartPLS 3. Bonningstedt: SmartPLS
29. Isaac O, Abdullah Z, Ramayah T, Mutahar AM, Alrajawy I (2018) Integrating user Satisfaction and Performance impact with technology acceptance model (TAM) to examine the internet usage within organizations in Yemen. Asian J Inf Technol 17(1):60–78
30. Isaac O, Abdullah Z, Ramayah T, Mutahar AM (2018) Factors determining user satisfaction of internet usage among public sector employees in Yemen. Int J Technol Learn Innov Dev 10(1):37
31. Kline RB (2010) Principles and practice of structural equation modeling, 3rd edn. The Guilford Press, New York
32. Hair JF, Black WC, Babin BJ, Anderson RE (2010) Multivariate data analysis. New Jersey
33. Fornell C, Larcker DF (1981) Evaluating structural equation models with unobservable variables and measurement error. J Mark Res 18(1):39–50
34. Chin WW (1998) The partial least squares approach to structural equation modeling. In: Marcoulides GA (ed) Modern methods for business research. Lawrence Erlbaum Associates, New Jersey. Lawrence Erlbaum, Mahwah, NJ, pp 295–358
35. Awang Z (2014) Structural equation modeling using AMOS. University Teknologi MARA Publication Center, Shah Alam. Malaysia
36. Hair JF, Hult GTM, Ringle C, Sarstedt M (2017) A primer on partial least squares structural equation modeling (PLS-SEM), 2nd ed. Sage, London, Thousand Oaks
37. Cohen J (1988) Statistical power analysis for the behavioral sciences, 2nd ed. LawreAssociatesnce Erlbaum
38. Chin WW (1998) Issues and opinion on structural equation modeling. MIS Q 22(1):7–16

Document Content Analysis Based on Random Forest Algorithm

Wan M. U. Noormanshah, Puteri N. E. Nohuddin, and Zuraini Zainol

Abstract The aim of this study is to develop a classification model with capabilities of performing text analysis, ID labeling, or tagging to an unstructured and uncategorized dataset, and perform supervised classification with trained datasets as input to predict the output of classification. The proposed technique classifies the dataset into four categories (i.e., crime, education, marriage, and sports) fittingly using random forest technique. The framework of text analysis document classification consists of five stages which are (i) collecting news dataset, (ii) data pre-processing (iii) document term matrix and weighting term, (iv) classification using random forest technique, and (v) text analytics and visualization results. This study presents a classification model which is able to perform text analysis during search for terms variable that appears frequently across the dataset.

Keywords Classification · Random forest · Document term matrix · Term frequency-inversed document frequency

1 Introduction

Knowledge Discovery in Database (KDD) is an analytic process of extracting interesting patterns and knowledge in a huge amount dataset [1] by applying a specific algorithm or technique. In general, there are seven stages in KDD process: data cleaning, data integration, data selection, data conversion, data mining, pattern interpretation, and knowledge representation [2]. Data mining (DM) comprises many different techniques and algorithms that are attempted to fit a model to the data [3]. Some common examples of DM techniques can be found in [4]. One of the most

W. M. U. Noormanshah · P. N. E. Nohuddin (✉)
Institute of Visual Informatics, National University of Malaysia, 43600 Bangi, Malaysia
e-mail: puteri.ivi@ukm.edu.my

Z. Zainol
Department of Computer Science, Faculty of Science and Defence Technology, National Defense University of Malaysia, Sungai Besi Camp, 57000 Kuala Lumpur, Malaysia
e-mail: zuraini@upnm.edu.my

S.-L. Peng et al. (eds.), *Intelligent Computing and Innovation on Data Science*, Lecture Notes in Networks and Systems 118,
https://doi.org/10.1007/978-981-15-3284-9_53

popular techniques applied in DM is classification. Classification is an unsupervised learning that is used for predicting group membership for data instance. Classification technique is used in many domains such as predicting customer behavior [5], medical diagnosis [6], education data [7], and transportation [8].

According to [9], it is forecasted approximately 90% the growing data is going to be in the usage of unstructured text databases. These unstructured texts theoretically encompass useful knowledge and information. As such, Google handled almost 20 petabytes of data per day and yet it still increases proportionally year by year and until 2018, it has increased until 2.5 quintillion bytes of data [10]. Data can be found in diverse forms such as structured (e.g., databases), semi-structured (e.g., open standard JSON, NoSQL, etc.), and unstructured (e.g., text files, email, social media data, websites, etc.). This paper describes and categorizes set of documents into group using random forest classification technique.

This paper continues with the next four sections as follows: Sect. 2 discusses on some relevant sub topics of document classification, document term matrix, TF-IDF and random forest technique. Section 3 portrays the framework of the Text Analysis Document Classification. Section 4 discussed the results of the experiment. Finally, we conclude this paper with a conclusion and future direction in Sect. 5.

2 Background and Related Work

2.1 Document Classification

The rapid growth of unstructured documents in digital world has attracted many researchers to explore on document classification. One of a major issue that needs to be looked at in many domains is an automated method of classifying documents. As an example, it is needed by an information retrieval system and Web search engine to sort text bases into sets of semantic categories. Document classification is one of techniques in machine learning (ML) and also in natural language processing (NLP) [5]. This method is beneficial for editors, news, websites, blogs, and also individuals who work with bunch of documents. The aim of performing document classification is to manage and sort the documents in order. Good document classification is significance for an organization from small to large entities that deals with mountains of data as it may involve various processes such as organized, classify, analyses, knowledge sharing, and process storing [6]. Respectively, maintaining and classifying a large amount of information manually from variety types of paper-based document or electronic document will be time—consuming and cost matter in term of labor. In this study, we will focus on the supervised classification technique, which is capable of: (i) labeling and training dataset, (ii) find terms variable of a trained dataset, and (iii) perform document analysis and visualize it.

2.2 Document Term Matrix (DTM)

DTM is a two-dimensional matrix table, where every row in the matrix represents a document vector with one column for an entire term of a corpus. The purpose of DTM is, to represent the existence of terms in a corpus, where the presence of term will be recorded as "1" while absence term will be represented as "0" in the DTM table [11].

Consider a corpus of documents and a dictionary of terms containing a bag of words that appear inside the document. The rows of the matrix *i* represent the term to be analyzed, while the columns *j* represent the document used in the analysis. Each entry of *(i, j)* represents either the term *i,* appears 1 or not 0 in document *j.* Let say if the term appears double in the document, in the matrix table the term will be recorded as 2. By performing DTM, terms similarities between one document to another document in the same corpus can be count. Also, a significant term of a document can be recognized by analyzed sparse matrix or also called as sparsity. Sparse matrix is an efficient way of representation of the information contained in DTM, and it is necessary to be used when encountering many words or cases [12]. DTM is most likely tends to be recorded as 0's. For this case, the sparsity result will show the absence of terms.

2.3 Term Frequency-Inversed Document Frequency (TF-IDF)

One of the methods to compute a term weight across corpus is TF-IDF. The term weight is calculated based on a formula that produced statistical value to estimate the importance of a term in corpus [13]. TF is a frequency of word count for numbers of words appear in a document, while IDF is to measure how significant of a term for the whole documents or corpus. Most search engines apply the TF-IDF for ranking words. Each word/term is assigned, respectively, with its TF and IDF scores. The score is defined as TF*IDF weight for each term. The algorithm of TF-IDF working with a formula of term *t*, document *d* and weight *Wt, d* of term *t* and document *d* in a corpus is given by with equation [14] where:

$$Wt, d = \mathrm{TF}t, d \log\left(N/\mathrm{DF}t\right).$$

- TF*t, d* is total of incidences for *t* in a document *d.*
- DF*t* is the quantity of documents has the term *t.*
- *N* represents the total amount of documents in a corpus.

For example, 1000 words document contains around 140 times term "cut" and calculation for TF of "cut" is TFcut = 140/1000 = 0.14 while IDF calculation throughout the corpus that have 1000 documents for term "cut" is IDFcut = log(1000/*x*), *x* is a number of terms "cut" appears over all documents. Let us assume $x = 140$, the IDFcut will be 0.85. The TF-IDFcut will be the results of TFcut*IDFcut and it will be 0.119. The calculated value of the selected term is increasing consistently to the number showed up in the documents. It is used to check how relevant the term is throughout documents contained in a corpus [15]. Based on the above review, in this study, we incorporated TF-IDF as an attribute pruning technique as a prominent technique to measure the significance of terms.

2.4 Random Forest Algorithm for Classifying Document Subject Domain

Random forest (RF) is a subclass of decision tree classification procedure. It works like a big set of a decorrelated decision tree. The more the number of trees, the robust it can be. The rationale of selecting RF classification is for building trees, and the input variables (i.e., text) are chosen randomly. These algorithm properties allow RF to generate an accuracy evaluation called an "out of bag" error by using the withheld training data [16]. The measurement of variable significance is based on mean reduction calculation whether a variable is significant or not to generate the decision tree. Bagging is used to generalize unbiased and noisy model to produce another model with small classification variance. RF corrects for DT technique habit of overfitting to their training sample. The RF algorithm's primary concept is to get a forecast of each tree and select the best alternative by means of vote [17]. Class prediction using the RF is done by executing the new element that needs to be classified into each tree in the RF. Each tree will give a result on which class the new element belongs to. If the new element mostly labeled as class number 2, then it will fall into group class number 2. The decision tree made up in RF based on the randomness of two possibilities: (i) start with each tree consisting of a random sample from the initial data set, and (ii) randomly select a subset of features to produce the next split.

3 Framework of Text Analysis Document Classification

Figure 1 presents the framework for Text Analysis Document Classification (TADC). It comprises of four modules which are (i) collecting news dataset, (ii) data pre-processing, (ii) DTM and weighting terms, (iii) RF classification, and (iv) visualization. After a set of documents are converted into a corpus, the dataset goes through a data pre-processing segment. Data pre-processing embraces activities of collecting sets of online new pages, reformatting, and cleaning the raw data. The next stage in the second module is terms extraction. The process of ranking and extracting these keywords is based on the TF-IDF method. The keywords are trained and classified

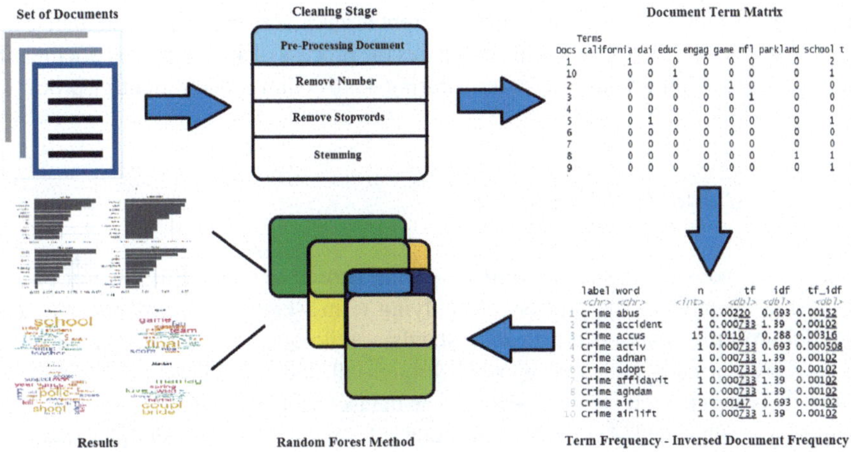

Fig. 1 Framework of TADC

using the RF method and classified according to their group. Finally, the results of the classification are visualized.

3.1 The Dataset

Dataset used in this experiment consists of a selected document from BBC online news. The selected online news pages contain a mixture of four different categories of topics that we aim to classify them into four categories of documents: education, sport, crime, and marriage. In this experiment, 801 pages of news dataset are collected and will be treated as 801 distinct documents. The dataset collection is manually downloaded from Kaggle Web portal [18].

3.2 Data Pre-processing

Data pre-processing removes meaningless data such as stop words, numbers, and symbols. At this stage, all documents will be converted into a corpus. After that, standardization of text is done by transforming all term into lowercase format, remove symbols, and numbers. Next, all stop words and white space are stripped from the documents. It is significance to remove stop words in a corpus to enhance the quality of the dataset. Once completed, the stemming process takes place. Stemming is a process where a term will be trimmed into its root term. The objective is to standardize each term into its root term. For example, the terms "bullying" and "bullied" are pruned and simplified as "bully." Each term that has the same root can be grouped.

This process minimizes the number of different terms across the corpus and increases the frequency of term appears. This facilitates the process of analyzing text document and measuring the significance of a term during word counting and weighting process.

3.3 DTM and Text Weighting

During the indexing phase, the term weighting is used to obtain the value of every term in a document. The process of applying term weighting in this paper is the TF-IDF. After the datasets are cleaned, the dataset is sorted into DTM. The purpose is to represent each term that appears through all document in form of a matrix table. This step is vital for us to analyze the dataset using the summary results. The summary result of DTM provides important information (i.e., sparsity percentage). Removing sparse terms is another approach to reduce the complexity of the dataset. In this experiment, we set sparse equal to 0.98, it would take effect to remove terms that missing over 98% of the corpus and terms that appear at least 2% are retained in the model.

We re-run multiple experiments to obtain the exact required quantity of terms that produce less complexity to the model. So that, setting sparse in DTM is a significance process because the results can influence a future model. In constructing DTM, first, an individual term that appears in each document is gathered in the TF-IDF list to create a feature set of each document. Second, each individual term is associated with respective TF-IDF score and sorted according to their score. Next is to construct features set for representation. The feature set for the entire document collection is created by uniting each document's preserved distinct phrases. Finally, using the built feature set, the term-document variable is produced for each document in the DTM. The aim of this process is to measure the importance of a term in reference to a topic in a set of documents.

3.4 Classification Using RF

Basically, the more frequent certain terms present over documents the more significant the terms as a variable factor to a certain topic. It is important to find the terms variable so that the classification model is able to use the variable as reference for the future classification. The set of terms variable is the output for the further stage. We construct a chain of terms that represents each group that needs to classify any future dataset that has the same attribute content. For example, sport news dataset that we have analysed over 30% of dataset, showing chain of terms that covered the dataset is "tournament," "final," "game," "fans," "score," "player," "team," "win," "match," and "NBA" (Fig. 2).

The output of the weighted terms from TF-IDF results are placed into the RF model. Based on the chain of terms variable, the model starts to create a random

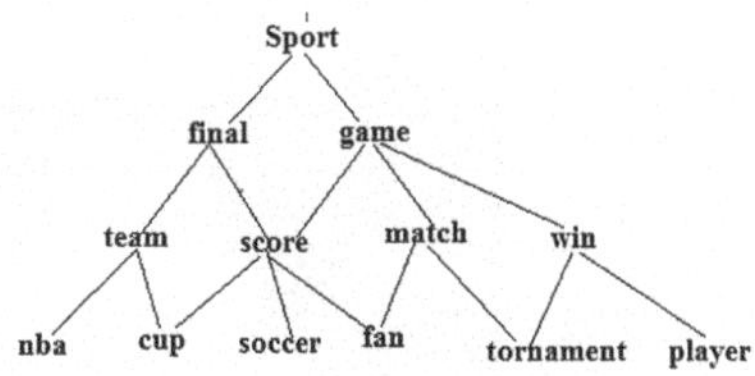

Fig. 2 Chain of terms for sport news dataset

subset of training sample and random value. Each random subset does not have the same value as the other random subset. Random subset generates multiple decision trees. In this experiment, we set it to 200 of decision tree to be generated. Next stage, the news documents are tested into the model and classified into decision trees in the RF model. The labeling result of classification tree may be different from one another as each decision tree comes from a different random subset. However, the final classification of the document is finalized based on most voted group the decision tree labeled.

4 Experimental Results

In this section, the outcomes of the experiment are presented and discussed. As shown in Fig. 3, the out of bag (OOB) prediction error rate representing how much the terms will be missed during the classification of dataset. The lower the OOB error rate, the better and more accurate the dataset classify by the model. In this study, the OOB prediction rate is 9.24%. By applying the RF, the model generates 200 decision trees by using the set of chains produced by 73 terms variables, *mtry* is 37 trial, and variable of importance mode is Gini impurity. The dataset that consists of 801 documents is classified into four different categories for evaluation. All the four categories have their own ID and can be easily differentiated. The evaluation part has two stages which are (i) training and (ii) testing. During the training stage, 30% (i.e., 241 documents) of dataset will be used. The 30% of the dataset are extracted randomly to build the classifier. The other 540 documents are used as the testing dataset to test the classifier.

In Fig. 4, we visualize the results of classification from the testing stage. The top 10 highest terms for each group of the testing dataset will be selected for graph plot. The group will be labelled into four main categories (i) "Crime," (ii) "Education,"

Fig. 3 Final model classification details

```
Call:
 ranger::ranger(dependent.variable.name = ".outcome",
haracter(param$splitrule), write.forest = TRUE,

Type:                             Classification
Number of trees:                  200
Sample size:                      801
Number of independent variables:  73
Mtry:                             37
Target node size:                 1
Variable importance mode:         impurity
Splitrule:                        gini
OOB prediction error:             9.24 %
```

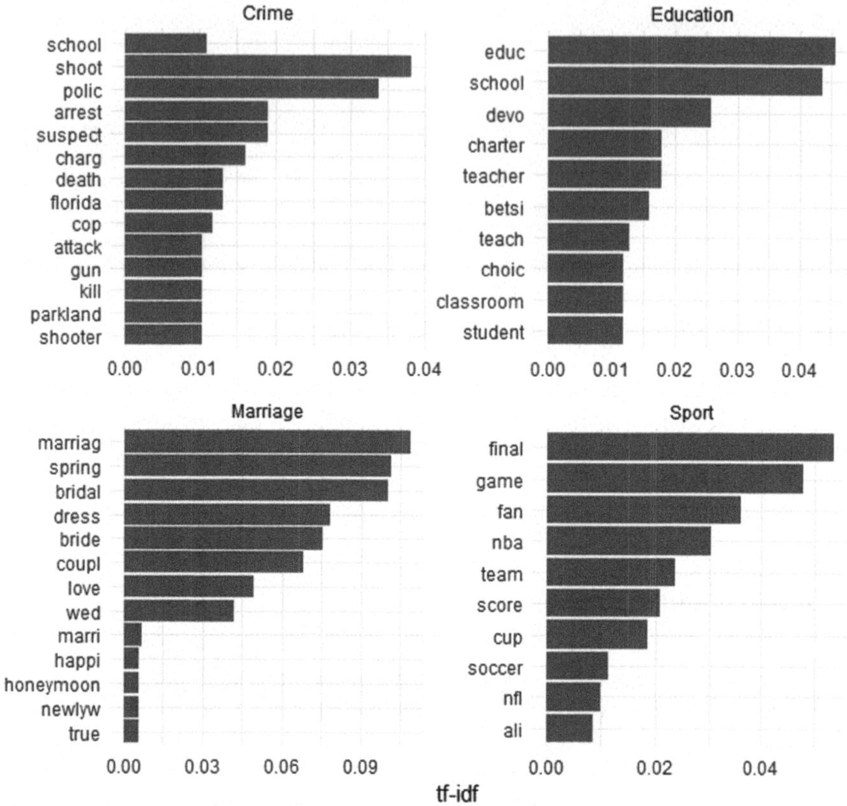

Fig. 4 Graph plot of terms classified into four major group

(iii) "Marriage," and (iv) "Sport." For example, in "Crime" group, the most frequent terms are "shoot," "police," and "arrest." This shows that these terms are frequently present for each news that related to crime section. By setting these terms in the classification model as terms variable, the new set of dataset related to crime section will be automatically classified into "Crime" major group.

This model of classification can be applied to many types of dataset and not limited to news dataset only. Once the model has been trained, new terms variable will be stored in the model classification as classifier. Automatically the model able to do classification for future document that has approximately similar content with the trained dataset. Lastly, we visualize the word cloud of each categorized major group dataset that has been classified in Fig. 5. The word cloud purposely show the document content of the news dataset of each group.

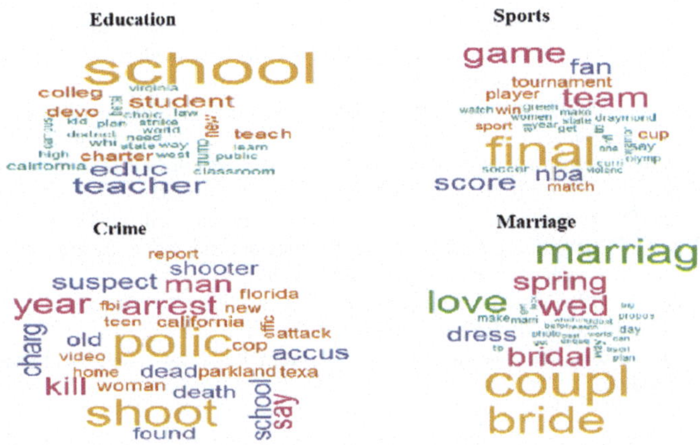

Fig. 5 Word cloud for grouped dataset after classified into four major group

5 Conclusion and Future Work

The significance of this research is the development of a classification model to effectively classify a document. The findings of this research can be used as an enhancement in company day-to-day operation, organizing database, and classifying data. Alongside the experiment, the model capable of doing ID labelling or tagging the dataset by recognizing the title of a document. This research will be extended the TADC into an application for classifying document by using the model template.

References

1. Osmar RZ (1999) Chapter i: introduction to data mining
2. Han J et al (2011) Data mining: concepts and techniques. Elsevier
3. Dunham MH (2006) Data mining: introductory and advanced topics. Pearson Education India
4. Nohuddin PNE et al (2018) A case study in knowledge acquisition for logistic cargo distribution data mining framework. Int J Adv Appl Sci 5(1):8–14
5. De Caigny A et al (2018) A new hybrid classification algorithm for customer churn prediction based on logistic regression and decision trees. Eur J Oper Res 269(2):760–772
6. Priyadarshini MG (2018) Decision tree algorithms for diagnosis of cardiac disease treatment
7. Rahayu SB et al (2018) Case study of UPNM students performance classification algorithms. J Eng Technol 7(4.31):285–289 (2018)
8. Kamarudin ND et al (2018) Performance comparison of machine learning classifiers on aircraft databases. Def S & T Tech Bull (1985–6571), 11(2):154–169
9. Zainol Z et al (2018) VisuaUrText: a text analytic tool for unstructured textual data. J Physics Conf Series 1018(1):012011
10. Dean J et al (2008) MapReduce: simplified data processing on large clusters. Commun ACM 51(1):107–113
11. Zhou M et al (2016) Priors for random count matrices derived from a family of negative binomial processes. J Am Stat Assoc 111(515):1144–1156

12. Zainol Z et al (2017) Text analytic of unstructured textual data: a study on military peacekeeping document using R text mining package. In: International conference on computing and informatics. School of Computing, UUM, pp 1–7
13. Fortuna B et al (2005) Visualization of text document corpus. Informatica 29(4)
14. Ramos J (2003) Using TF-IDF to determine word relevance in document queries. In: Proceedings of the first instructional conference on machine learning, vol 242. Piscataway, NJ, pp 133–142
15. Kalra S et al (2019) Automatic classification of pathology reports using TF-IDF Features
16. Felton BR et al (2019) Using random forest classification and nationally available geospatial data to screen for wetlands over large geographic regions
17. Makel J et al (2019) Performance of random forest machine learning algorithms in binary supernova classification. High Energy Astrophys Phenom 1–18
18. Misra R (2018) News category dataset. Retrieved from: http://www.kaggle.com/rmisra/news-catgory-dataset

Microblogging Hashtag Recommendation Considering Additional Metadata

Anitha Anandhan, Liyana Shuib, and Maizatul Akmar Ismail

Abstract Microblogging is used to broadcast short messages in form of text, pictures, links, and videos to the followers or subscribers on the Internet. The hashtag is the keyword or metadata that is used to mark messages, which allow users to classify or find the related posts easily. However, posts in microblogging environment are not properly tagged due to data sparsity and finding the popularly used relevant hashtag for the tweets. In this paper, hash tag recommender (HTR) is the proposed method using matrix factorization with tweet's tags for user input text. To eliminate data sparsity, hashtag recommendations are generated from similar tweets. Hashtags based on the current trends, time, and location are recommended for the short messages of input by the user. To achieve this, the proposed method calculates the score for each tags, which are identified for similar tweets, and recommendations are generated. Results of this study show that hashtag recommendation outperforms the previous methods. The significance of the proposed HTR for twitter data set is more accurate on considering various metadata such as time and location.

Keywords Hashtag · Location · Twitter · Matrix factorization · Recommendation · Tagging · Microblog

1 Introduction

Microblog is a blogging service in which users generally post information in a different form of sources such as text, pictures, and videos on the Internet. Unlike the regular blogs, the information is shared in small pieces, and the posts are called as

A. Anandhan (✉) · L. Shuib · M. A. Ismail
Faculty of Computer Science and Information Technology, University of Malaya, Kuala Lumpur 50603, Malaysia

L. Shuib
e-mail: liyanashuib@um.edu.my

M. A. Ismail
e-mail: maizatul@um.edu.my

S.-L. Peng et al. (eds.), *Intelligent Computing and Innovation on Data Science*, Lecture Notes in Networks and Systems 118,
https://doi.org/10.1007/978-981-15-3284-9_54

microposts. The messages are possessed in real-time with other users in the platform expressing ideas, comments, and events. There as many microblog platforms available on the Internet. Twitter is a popular microblog, and the messages posted in twitter are called tweets. Twitter supports all rich social media types and has wide user base [1]. Millions of tweets are posted in twitter, and uniquely 1 billion visits to the site every day.

Twitter allows the users to insert the relevant keyword with hash (#), named as hashtag, which brings to get the core idea of the relevant topics. This gives more strength of the information dissemination. The hashtag for microblogs is annotated by authorized users who posted tags. Therefore, recommending the microblog hashtags is attracted and makes attention to researchers' in recent years. Twitter helped great success in social services targeted for marketing, event detection, promoting policies, and alerting about natural disasters predictions, etc. Nevertheless, many consistent information is sidelined in microblogs because of overloading information; the performance of the practical execution of this platform becomes ineffective [2]. Existing researchers directly used the crawled microblogs with hashtags, which are labelled by user. But sometimes users may post with irrelevant hashtags; however, there are lots of noisy and instance hashtags which are real crawled data from Twitter [3]. The hashtag may be grouped as partial and partially relevant which leads more weak in relevancy. Temporal hashtags are the correct one, which are relevant, partial is unintentional, which are weak, and one more is irrelevant.

Many researchers proposed for hashtag recommendation with different approaches. Based on the method, they converted the suggestions to the translation process from content to hashtags [4], based on topics [5]. Accordingly, as the information received is quite larger in content, implicit and explicit feature quantification guides to create the user profile. Hence, influx of recommender system helps to solve this information overload which is considered as one of the major problems [6]. Generally, social media resources, such as Twitter, Facebook, and YouTube are classified and categorized using Hashtags, and it has become a popular way to classify information.

Trending topics on social media sites can be searched and followed using hashtags. The sign # (hash) is prefixed in front of the word of phrase without any whitespace characters or in the form of punctuation. Users in Twitter can create a new tag or follow existing hashtags. Figure 1, shows the association of tweets and hashtags.

Fig. 1 Tweets and Hashtags

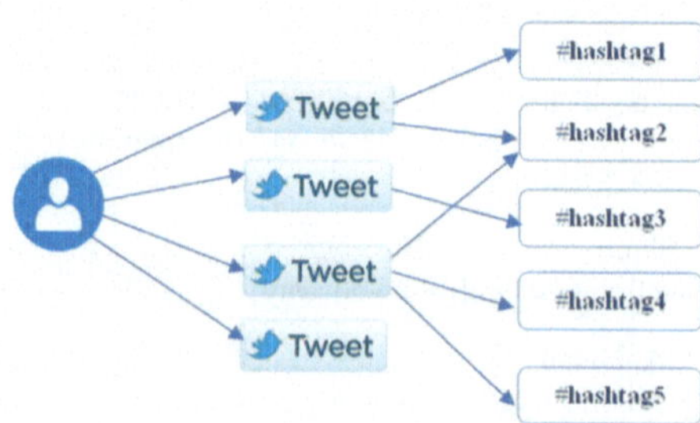

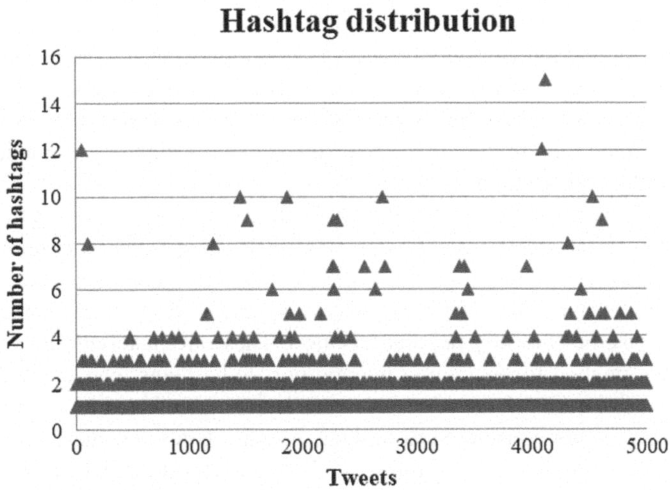

Fig. 2 Hashtag Distribution

When hashtag used in posts, twitter immediately turns it into a hyperlink. The user can follow interesting hashtags to get recent updates. When the hyperlink is clicked, it leads to the Web page containing the relevant and recent tweets marked with hashtag [7]. Since the hashtag is important for classifying the information, it is important to use correct or popular hashtags in the tweets. Therefore, using topic modelling feature of the microblog posts is to be analysed and then improve the accuracy of a hashtag recommendation. This study proposes topic modelling for the tweets containing one or more hashtags as shown in Fig. 2, which is very important is used as topics and all the words in the topics are words that describe the topic.

2 Related Work

Hashtag recommendation problem has been attracted, and considerable research has focused on the field of social network. The short messages used in hashtags are having short life span, which is very challenging for researchers to understand [8]. Twitter can post a short text called a tweet. Hashtag recommendation mainly helps the users to find similar users, which provide references for short text evaluation for hashtag recommendations especially considered microblogs. Self-attentive neural network is adopted to compute the weights of each word and combine the representations of words. As a result, the representation of entire microblog posts is obtained and then the coupling relationship between words for hashtags in a microblog post is analysed. Accuracy of hash tag recommendation is computed after coupling relationship process. The hashtags of a microblog are related to not only the text but also to the publication time and authors' occupation (check sentence). Researcher obtained the hashtag recommendation accuracy through attention model, each key and its unique value are the inputs, and the query for each word is the output. The attention weights, the similarity between the key and the query are computed and then normalized using a softmax function to obtain the weight of the value that corresponds to each key. Finally, statistical representation is obtained through a weighted summation of the value for finding the accuracy [9].

Recommendation for hashtag method in Twitter assists the users to avail their relevant hashtag of the tweets. An approach to tweet data that remarkably make a better performance for recommending the proper hashtags in recommendation systems. Three salient attributes are—first one is tweets, second is the user characteristics, and final one is the currently popular hashtags. These are materially accounted for the proposed method for extracting the appropriate hashtags for recommendation. Stack of crawled twitter data and noisy words is to be eliminated by pre-processing method, which is not considered as hashtag. The attributes considered for the proposed method are hashtags, hyperlinks, symbols, and emoticons which are removed using the appropriate techniques from the content of tweets. Using published tweets, proposed system builds user characteristics containing data about content, hashtags, and social interactions. As a first step, tweets text contents are compared to discover the similar users, which help to find and improve the certainty in tweets [10]. Second, the similarities of messages are discovered using vector of real numbers and computing cosine similarity. Finally, the system refines the user characteristics using the similar tweets. Decisively, the hashtag is derived from the identical tweets and prioritized based on their similarity score. Tran et al [11] experimented the tweets which contained the exclusive information of the users to recognize their relevant hashtags. It showed the significant performance improvement in the hashtag recommendation [11]. Brief experiments on Twitter about the hashtags that are in the form of very short texts in microblogs. The topical representations generated by Hashtag-LDA and TOMOHA are still having to a confined hashtag recommendation performance. Hence, it is proposed to have a new hashtag recommendation on multi-features (HRMF) on microblogs, extracting the set of hashtag for suitable candidate and ranking mechanism. HRMF is to explore the User-Hashtag Topic Model by using short text Expansion (UHTME), which relieves the deficiency of data by widening the short to long text. UHTME modelled with multi-various microblogs considered hashtags of tweets, users together with texts for generating the topical representation by this main feature as vector form, and this topical method mainly collects candidate hashtags with rank of the hashtags [12]. The most pertinent candidate set is to be collected for hashtag recommendation, which help implicitly and explicitly for similar users. Extensive experiments confirm the performance of hashtag recommendation in HRMF and the semantic portrayal capability of representation UHTME [12].

Even though many experiments performed on image tags, which including suggestions of hashtag in the form of text. State-of-the-art methods combining the visual and textual data are the primary step in finding the issues about the suggestion of the hashtag, which mainly avoids the difficulties in categorization of information, by multimodal. Topical translation model was proposed for generating the process of translation using topic-specific word triggers. Translation process has made a collaboration of gaps within visual words of images and corresponding hashtags. Methodology exposes the hashtag recommendation for multimodal microblog posts, which are the primary base work on this task. Generally, experiments are based on textual information, but here the researcher experimented topical translation model combined the text with visual information [13]. Collection of the textual microblogs contains content of text, images, and appropriate hashtags, which are identified by the authors which give aid for researchers reinvestigating the similar task or relevant topics using multimodal in social medium data.

The proposed approach finds important hashtags for grouping the information using correct or popular hashtags in the tweets. Though hashtags are important, large portion of hashtags does contain hashtags, and hence, the automatically recommending hashtags for tweets have received considerable attention. Previous research observations on real Twitter data highlight that this proposed method works better than the related methods. Review results explored the geolocation prediction methods on Twitter data [14].

3 Research Methodology

In this section, the research methodology for recommending hashtags is based on the input tweets. Adding proper hashtags to tweets ensures the tweets reach wide audiences and are easily searchable. So, it is important to select hashtags that are popular in the social network sites. Many tweets in the community are posted without proper hashtags and sometimes with no hashtags. These tweets are not easily searchable and might be unnoticed in the community. Our method recommends hashtags using topic modelling and semantic analysis. Figure 3, **Research methodology,** shows our proposed method that recommends hashtags for user's tweets.

3.1 Pre-Processing

To retrieve the recommended hashtags for the tweet, it is essential to remove all the noise words. The noise words are any words that are not useful in search and can be ignored. In the pre-processing step, all the noise are identified and removed as follows: (1) remove short words (tweets), (2) tweets are split into individual tokens, and each token is compared with list of noise words in database and removed. Our data set includes a set of stop words commonly used. There are around 200 stop words in the data set. Our analysis shows around 5902 tweets containing stop words in the selected 8680 tweets, which are around 67% of tweets. The detailed statistics is provided in Table 1, stop words statistics. The input tweets are processed to remove all the stop words, so the result tokens t are used to match the tweets in the database and all similar tweets are retrieved for processing.

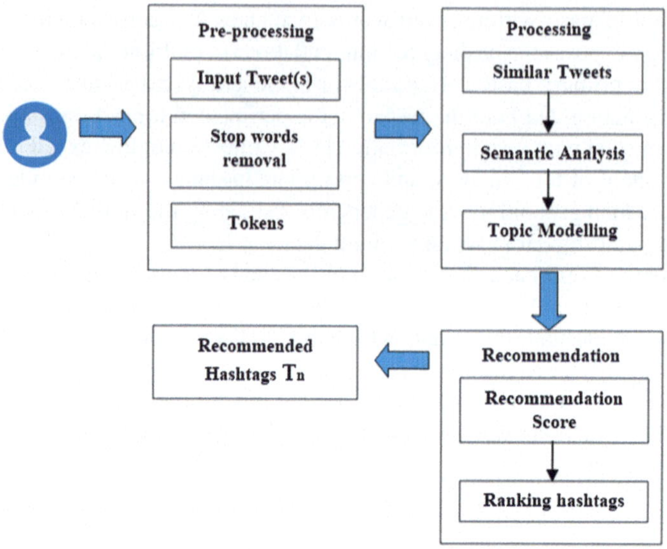

Fig. 3 Research methodology

Table 1 Stop words statistics

Attribute	Value
Stop words	200
Tweets contain stop words	5902
Tweets in training set	8680

Algorithm 1 Searching similar tweets

***Parameters**: TS – Tokens, N - Noise words*
***Output**: SE - Search Expression, Tw - Tweet result*
Begin
T(i) = SET {Tokens TS(i) From I}
N(i) = SET {Noise Words N(i)}
for all token t = 1,.....T(i) do
for all n N(i) do

if T(i) = n
TS(i) = remove n from T(i)
end if
end for
SE = {Search Expr using t} + SE
end for
for all s SE(i) do
Tw(i) = {search s in archive from SE(i)} + Tw
End for
End

3.2 *Users and Tweets Metadata*

Proposed method uses metadata of user and tweet for recommending hashtags. User posts tweets from various locations, and when the topics are modelled in database, the location, date/time stamp are stored in the database. The location of the author is compared with hashtag location and its date/time of recent posting tweets while hashtag recommendation for the input tweet (Table 2).

Figure 4, **hashtag distribution,** shows hashtag distribution posted across users in various geolocations. Our analysis reveals Australia has more number of hashtags with count 72 and UK with count 60. The detailed hashtag distribution for country is given in Fig. 5, **topic distribution and tokens**. The least hashtags for country are found for countries Spain and Scotland. Our proposed method compares the location of the tweet author and recommends hashtags accordingly.

Table 2 Hashtag distribution

Country	Hashtag count
Australia	72
UK	60
Philippines	53
USA	31
California	29
Malaysia	21

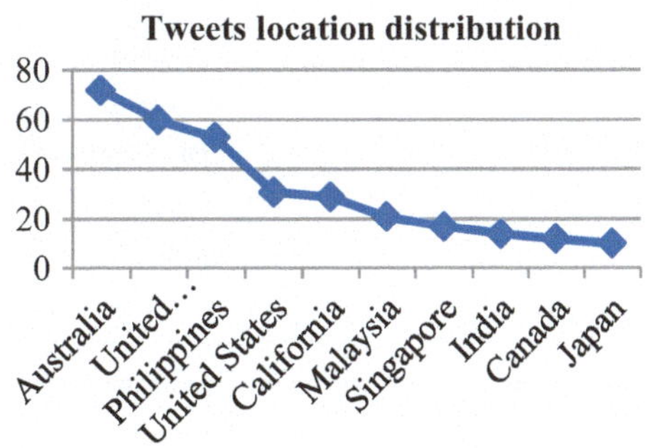

Fig. 4 Hashtag distribution

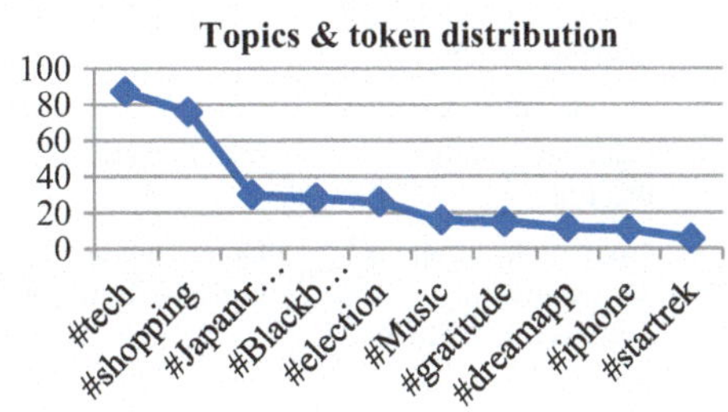

Fig. 5 Topic distribution and tokens

3.3 Retrieve Similar Tweets

Pre-processing step ensures all the stop words are removed from the tweets. The tokens in the tweets are used to retrieve all similar tweets for processing. Our system generates search expression that contains the similar tokens with hashtags. Algorithm 1, searching similar tweets, shows the steps for generating the search expression using tokens.

3.4 Topic Modelling

The tweets in the training data set are processed, and all the hashtags are retrieved. Tweets contain one or more hashtags, and it is used as topics, and all the words in the topics are words that describe the topic. Semantic analysis on the topics and words is performed to ensure that the only relevant words are there in the topic Table 3, topic distribution and tokens, and Table 4, input tweets and recommended hashtags using metadata, provides the list of topics in the training data set. The tokens in the input tweets are compared with the tokens of hashtags. All the hashtags matching are retrieved for processing. Table 4,input tweets and recommended hashtags using metadata, shows list of tweets and all the topics retrieved from the database.

Table 3 Topic distribution and tokens

Hashtags	Words count
#Music	16
#gratitude	15
#election	26
#Blackberry	28
#Japantravel	30
#tech	87

Table 4 Input tweets and recommended hashtags using metadata

#	Tweet	Tokens	Location	Hashtags
1	Buy or sell any products using online portal with amazing offers	Buy, sell, products, online, portal, offers	UK	#shopping, #amazon, #tag, #openhacklondon
2	Election day! Vote the right party in this election	Election, vote, right, party, election	India	#election, #indiavotes, #votes, #pisay
3	Great dude … oh yes! Movie lives in hollywood … Great friend!	Dude, movie, hollywood, friend	USA	#followfriday, #hollywood, #awesome, #designers, #movies

Table 5 Score and ranking of recommendation

Hashtag	Score
#shopping	9.2
#amazon	7.2
#tag	2.1
#openhacklondon	0.8

3.5 Hashtag Ranking and Recommendation

Our method retrieves multiple hashtags using topic modelling search. Since multiple hashtags are retrieved to recommend hashtags, it is essential to calculate score or ranking for these hashtags. Using score and ranking, hashtags are recommended to the tweet. Location, date/time attributes are considered for calculating scores, so accurate recommendations can be made. The score for the selected hashtag for the tweet 1 is shown in Table 5, score and ranking of recommendation above. The score is calculated using metadata location and date/time, and based on the calculation, the most suitable tag for tagging the tweet is #shopping. And the least suitable tag is #openhacklondon. Since the hashtags are recommended based on the recent date/time, these hashtags are suitable for tagging the input tweet.

4 Discussion

Proposed method has been evaluated using multiple tweets, and recommended hashtags are more accurate. Also, the additional metadata used for retrieving similar tweets enables more accurate hashtags that are recommended for the input tweet. Proposed method performed multiple experiments with different ranges around more than 1200 tweets for the input tweet for Twitter hashtag recommendation. Based on the input tweet tokens, the hashtag is to be recommended using geo-location. The final result is the set of hashtags that resemble the related hashtag for the input tweet. The experimental details and results for the same are discussed in the above sections. For all our experiments, different input tweets were conducted using crawled data from twitter to verify the effect of the proposed method.

First, collect the group of user's data on Twitter to construct the topics modelling for various hashtags in the tweets. Noise words are removed from the tweets, and using semantic analysis, various topics are revealed. Second, the input tweets are split into tokens, and all stop words removed. The tokens are then searched in database, and appropriate matching topics are retrieved. There could be more than one topic that can be retrieved from the database. Third, the metadata of user and tweets are used for filtering the topics. Location of the user and date/time of posting tweets are used for retrieving more accurate tweets. Based on the parameter, the scores are calculated by the number of topics in the result set.

Fourth, the hashtags are ranked using the score, and top-T hashtags are recommended for the input tweet. Different experiments were conducted using crawled data from Twitter to verify the effect of proposed method. Metadata of user and tweet are used for recommendation besides the method that uses the explicit or implicit similarities of tweets. The selected hashtags and ranking by score increased the accuracy of hashtag recommendation for tweets than baseline methods.

5 Conclusion

Proposed hashtag recommendation is mainly based on location and date/time metadata. Our method retrieved all the hashtags and tokens from the tweets in the data set. Using topic modelling, the system generates hashtags with appropriate token from the database. Metadata of tweets and users are stored in the database and used for rankings hashtags to be recommended for input tweets. To overcome the sparsity of data problem, the hashtag tweets from identical users are used. The results show that our proposed method is able to recommend hashtag semantically relevant to the tweets. The proposed system does not utilize the important metadata for recommendation, and the proposed method produces better results due to the additional metadata. In future, we will consider improving the performance of the proposed method by considering social links of the author and tweets.

References

1. Anandhan A, Shuib L, Ismail MA, Mujtaba G (2018) Social media recommender systems: review and open research issues. IEEE Access 6:15608–15628
2. Kumar A, Ahuja H, Singh NK, Gupta D, Khanna A, Rodrigues JJ (2018) Supported matrix factorization using distributed representations for personalized recommendations on twitter. Comput Electr Eng 71:569–577
3. Burke R (2002) Hybrid recommender systems: survey and experiments. User Model User-Adap Inter 12(4):331–370
4. Sedhai S, Sun A (2014) Hashtag recommendation for hyperlinked tweets. In: Proceedings of the 37th international ACM SIGIR conference on Research & development in information retrieval. ACM, pp 831–834
5. Ding Z, Qiu X, Zhang Q, Huang X (2013) Learning topical translation model for microblog hashtag suggestion. In: Twenty-third international joint conference on artificial intelligence
6. Katarya R, Arora Y (2018) A survey of recommendation systems in twitter. In: 2018 4th International conference on computational intelligence & communication technology (CICT). IEEE, pp 1–5
7. Ben-Lhachemi N, Nfaoui EH (2018) Using tweets embeddings for hashtag recommendation in Twitter. Procedia Comput Sci 127:7–15
8. Zangerle E, Gassler W, Specht G (2011) Recommending#-tags in twitter. In: Proceedings of the 2nd international workshop on semantic adaptive fs social web (SASWeb 2011), Girona, Spain, pp 67–78
9. Yang D, Zhu R, Li Y (2019) Self-attentive neural network for hashtag recommendation. J Eng Sci Technol Rev 12(2):104–110

10. Godin F, Slavkovikj V, De Neve W, Schrauwen B, Van de Walle R (2013) Using topic models for twitter hashtag recommendation. In: Proceedings of the 22nd international conference on world wide web. ACM, pp 593–596
11. Tran VC, Hwang D, Nguyen NT (2018) Hashtag recommendation approach based on content and user characteristics. Cybern Syst 49(5–6):368–383
12. Kou F-F, Du J-P, Yang C-X, Shi Y-S, Cui W-Q, Liang M-Y, Geng Y (2018) Hashtag recommendation based on multi-features of microblogs. J Comput Sci Technol 33(4): 711–726
13. Gong Y, Zhang Q, Huang X (2018) Hashtag recommendation for multimodal microblog posts. Neurocomputing 272:170–177
14. Zheng X, Han J, Sun A (2018) A survey of location prediction on twitter. IEEE Trans Knowl Data Eng 30(9):1652–1671. https://doi.org/10.1109/tkde.2018.2807840

Analysis and Forecast of Heart Syndrome by Intelligent Retrieval Approach

Noor Basha, K. Manjunath, Mohan Kumar Naik, P. S. Ashok Kumar, P. Venkatesh, and M. Kempanna

Abstract At present scenarios in the world, heart disease analysis and prediction are two demanding factors to be faced by the doctors that are very ridiculous, and in this regard, health industries will generate enormous amount of data. To reduce huge range of deaths from diseases like heart disease, cancer, tumour and Alzheimer's disease, doctors must find the rapid and effectual analysis and detection techniques to be used, where various algorithms are used to learning the machines and create very important responsibilities in study and prediction of various diseases in humans. The key intension of this article is characterized on forecasting and analysis of various heart-related syndromes in patients with wide range of age by means of machine learning algorithms and techniques. In this case study, many parameters are considered to do analysis and predict heart disease of patients, where KNN, logistic regression and decision tree algorithm are used to calculate accuracy and performance.

Keywords Heart syndrome · K-nearest neighbour · Logistic regression · Decision tree

N. Basha
Department of CSE, VIT, Bengaluru, India

K. Manjunath
Department of CSE, Govt. Polytechnic, Chennasandra, Bengaluru, India

M. K. Naik
Department of ECE, NHCE, Bengaluru, India

P. S. A. Kumar (✉)
Department of CSE, DBIT, Bengaluru, India

P. Venkatesh
Department of TCE, DBIT, Bengaluru, India

M. Kempanna
Department of CSE, BIT, Bengaluru, India

S.-L. Peng et al. (eds.), *Intelligent Computing and Innovation on Data Science*, Lecture Notes in Networks and Systems 118,
https://doi.org/10.1007/978-981-15-3284-9_55

1 Introduction

Due to busy schedule as well as routine assignments, peoples are facing severe stress and anxiety. Moreover, some other peoples are addicted with chronic habitual behaviour, like consumption of cigars and gutka, and those peoples are suffering from chronic diseases like heart diseases, cancer, liver problems, kidney failures, etc. To cure such patients with chronic diseases, is a very big hurdle to creative medical practitioners and medical researchers to solve the current world issue and objectives. Regarding this new challenge, IT professionals are provided hand-to-hand support to predict such disease early and cure as well as recover the patients from the chronic disease.

1.1 Heart Syndrome—Case Study

In this world, each person is unique in his attributes and his behaviour, out of which each person may have dissimilar readings of pulse rate and blood pressure. In general, a healthy human pulse rate must be in the range of 60–100 bpm and BP with a range 120/80–140/90 (mm Hg), and these benchmarks are proved.

Nowadays in throughout the world, for accidental or abrupt death, heart syndrome is one key basis, i.e., many peoples are affected by heart disease which is regardless of age in both men as well as women. This is because of improper dieting and consumption of alcoholic contents, cigars, etc., on irrespective of attributes like gender, diabetes, age, BMI, etc., which also added up this disease to humans rapidly. In this paper, we tried to do analysis and predict the heart disease in view of various factors like age, gender, blood pressure, heart rate, diabetes, etc., even though prediction of heart disease is one of the tricky jobs to researchers as well as doctors.

At present, various tools and techniques are available in the market to predicting the diseases, but still we expected some flaws in the analysis and predicting algorithms. Nowadays, based on big data approach, machine learning algorithm plays very important responsibility to explore and develop concealed knowledge and information about the chronic diseases.

2 About Literature

Coronary heart disease narrows down the coronary arteries. Basically, coronary arteries will supply both oxygen as well as blood to heart, and if the heart functioning is not proper, then it causes to malfunctioning of heart which leads to ill or death to a person.

Dewan et al. [1] discussed in detail the cardiovascular disease and different symptoms of heart attack. The different types of classification and clustering algorithms and tools were used.

Thomas and Theresa Princy [2] made use of many classification algorithms to predict the severe heart syndromes based on risk rate, where the author specifically used data mining approach.

Amin et al. [3] made use of an artificial neural network and genetic algorithm to predict health diseases. In this reference, author collaborates the data mining approach with association rules and classification techniques. In this regard, the model developed by the author is so efficient on predicting the heart syndrome. Shilaskar and Ghatol [4], evolution based feature selection is one of the effective method to select critical feature in a data set. Purusothaman and Krishnakumari [5], with critical factors and effective model an experienced medical practitioner can predict heart disease. Miao et al. [6] made use of various classification algorithms to create effective data analysis model on prediction of severe heart syndromes. Usually, dataset may contain noise features, and it abruptly corrupts the valid data, so they tried to reduce the noise by cleaning and pre-processing the dataset and also tried to reduce the dimensionality of the dataset. They found that good accuracy can be achieved with neural networks.

In some other literature, we referred an analysis using data mining. The analysis showed that using different techniques and taking different number of attributes gives different accuracies for predicting heart diseases. Even though the heart disease data set may contains many features and syntactically related duplicate information, in this regard we must refine the data set efficiently. This has to be pre-processed. Also, they say that feature selection has to be done on the dataset for achieving better results.

3 Methodology and Data Analysis

Generally in diabetic patients, high glucose content in the blood may cause damage in blood vessels as well as nerves in the body. If a person is suffering from diabetes on long period, then in future that person has higher probability to get heart disease, i.e., imagine if the person is diabetic, and addicted with alcoholic contents and chain smokers will naturally raise and develop the riskier heart disease.

Figure 1 symbolizes the graph of heart syndrome on people with age and count, if a person is directed by stressful life, he can easily damage his arteries and is accepting very big chance of coronary heart disease.

With the symptoms of high blood pressure, it formulates the person's heart to work very harder to pump the blood, it causes to strain the heart, and moreover, it relates to damage many blood vessels. Abnormal cholesterol levels in the body promote heart diseases and corpulence (obesity). Along with this, improper dieting as well as family history also causes chronic heart disease to the individual. In general, senior citizens and age-old peoples will easily hit by the heart diseases, due to many factors like age, gender, abnormal or unhealthy diet and stress, etc. In this regard, men are big victims or big risk of heart disease prone.

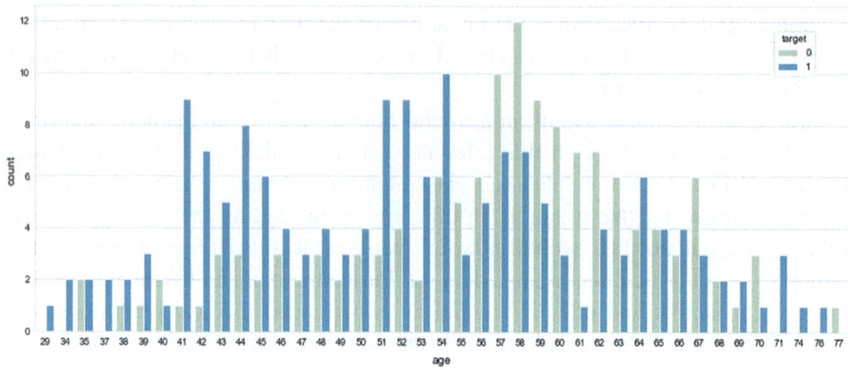

Fig. 1 Number of people who have heart disease based on age

At present, huge amount of research work is related towards heart diseases analysis and prediction system, where many researchers used various techniques and algorithms of machine learning and deep learning. The aim of ML and DL techniques is to achieve better accuracy and efficiency, so that doctors and patients can easily analyze and predict the heart attack chances.

3.1 Data Sources

The dataset used in this article is from Kaggle Web. Basically, Kaggle supports various dataset publication contents, where datasets are open source, very easily accessible data formats and supported to all platforms and work with any tools.

Table 1 specifies the features and represents the various conceptions used to create an effective system model to classify and validate the severity of heart syndrome in critical patients.

Table 1 Various characteristics used in system model

Sl. no.	Features and conceptions
1	Age (referred in months and years)
2	Sex (male = 1, female = 0)
3	CP (chest pain) in patients, with category, like (normal angina = 1, atypical angina = 2, non-angina = 3, asymptomatic = 4)
4	Chol (checked cholesterol content in serum)
5	FBS (fastening blood sugar)
6	Thalach (it referred with respect to maximum heart rate)
7	Exang (exercise-induced angina)

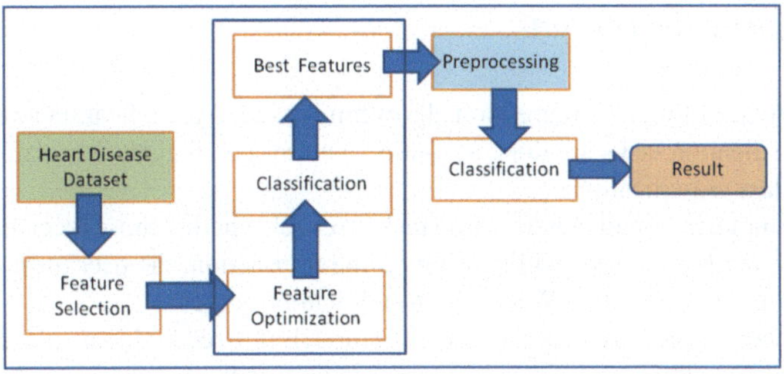

Fig. 2 Proposed architecture of the system

Some of the general symptoms of heart syndrome in severe and critical patients are like severe chest pain, shortness of breath, indigestion, burning sensation in chest, severe pain in stomach, sweating and fatigue, vomiting sensation, dizziness with anxiety and variations in heartbeat.

Figure 2 represents the proposed architecture of the system used to apply various machine learning algorithms to examine and forecast the syndrome of severe and critical coronary heart patients. In this model, heart disease data is considered to be an input data, and then, data is pre-processed by replacing non-available values with column means.

3.2 K-Nearest Neighbours (KNN)

In KNN algorithm, data is classified and regressed, where algorithm learns to evaluate the outcome from specific dataset. It performs well even if the training data is large and contains noisy values. In KNN algorithm, data is divided into two sets, i.e., training data and test sets, where experimental result set is mint for model building and training, where *k*-value is decided. Now, test data to be predicted on the model is built. There are different distance measures.

Pseudocode of KNN algorithm

```
Classify (A, B, C)
        A: training data,
        B: class lables of A,
        C: unknown samples
for i=1 to m
        do
compute distance d(A, c)
        end for
compute set I containing indices for the K smallest distance d (A, C)
        return majority label for {Y, where i € I}
```

3.3 Logistic Regression

The importance of logistic regression algorithm is used to classification tasks, where many classification tasks are done at routine pattern, i.e., logistic regression is used for multiclass classification.

For example, in e-mail classification task, to check whether those received e-mails are spam e-mails or not, or while doing online transaction, the user must be aware about whether the classified Website is fraudulent or not, etc.

Logistic regression is one of the statistical models utilized for binary classification, where it predicts the type (this *or that*, *yes or no*, *A or B*, etc.).

Logistic regression is a classification algorithm 1 that works by trying to learn a function that approximates $P(Y \mid \mathbf{X})$. It makes the central assumption that $P(Y \mid \mathbf{X})$ can be approximated as a sigmoid function applied to a linear combination of input features. Logistic regression is the building block for artificial neural networks.

Mathematically, for a single training data point (x, y) as,

$$P(Y = 1|\mathbf{X} = \mathbf{x}) = \sigma(z) \text{ where } z = \theta_0 + \sum_{i=1}^{m} \theta_i x_i$$

Equivalent forms of above equation can be written as,

$$\begin{aligned} P(Y = 1|\mathbf{X} = \mathbf{x}) &= \sigma\left(\theta^{\mathrm{T}}\mathbf{x}\right) && \text{where we always set } x_0 \text{ to be 1} \\ P(Y = 0|\mathbf{X} = \mathbf{x}) &= 1 - \sigma\left(\theta^{\mathrm{T}}\mathbf{x}\right) && \text{by total law of probability} \end{aligned}$$

In the probability of data, probability of $Y \mid \mathbf{X}$ algorithm is used to create and select the maximized theta value. State log probability function and partial derivatives with respect to theta can be written as,

(a) An algorithm that can choose optimal values of theta.
(b) How the equations is derived.

Finally, logistic regression algorithm totally depends on its θ value.

3.4 Decision Tree

Decision tree is one of the intelligent retrieval techniques for regression and categorization of datasets, where algorithm perform very effectively on continuous and categorical attributes, i.e., decision tree algorithm divides the population into two or more similar sets based on the most significant predictors. In this algorithm, entropy will calculate each and every attribute, and then, it split the dataset with the help of other predictors with maximum information gain or minimum entropy.

Decision Tree Algorithm Pseudocode

Step 1: select the root node in tree
Step 2: find the best attribute in a set
Step 3: split the set into subsets
Step 4: generate the subset with unique value
Step 5: redo from step 1
Step 6: generate subset
Step 7: check leaf node in tree.

4 Result Analysis

The above-mentioned machine learning algorithms are used in this dataset implementations, where logistic regression algorithm has very high accuracy compared to other two algorithms, given in Table 2.

Below result set represents the various key performance indices of the patient's dataset like precision, recall, $f1$-score and support of all three individual algorithms KNN, decision tree and logistic regression algorithm to determine the accuracy score.

Accuracy score of KNN algorithm is: 72.527%

Accuracy	0.73	0.73		91
Macro_avg	0.73	0.73	0.73	91
Weighted_avg	0.74	0.74	0.73	91

Accuracy score of decision tree algorithm is: 73.27%

Accuracy	0.73			91
Macro_avg	0.73	0.73	0.73	91
Weighted_avg	0.74	0.73	0.73	91

Accuracy score of logistic regression is: 82.52%

Accuracy	0.82			91
Macro_avg	0.82	0.82	0.82	91
Weighted_avg	0.82	0.82	0.82	91

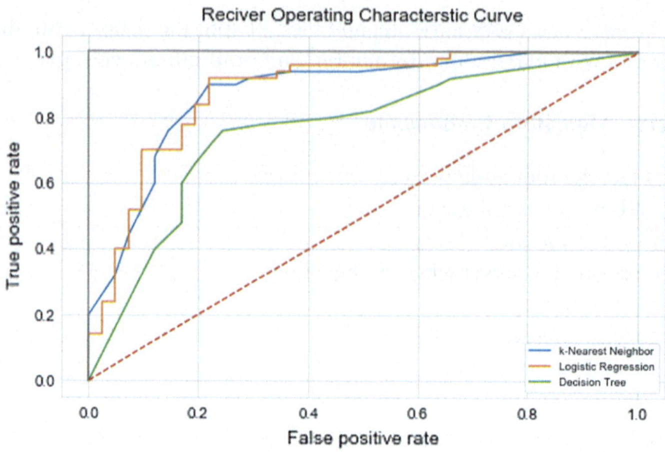

Fig. 3 ROC of KNN, logistic regression and decision tree

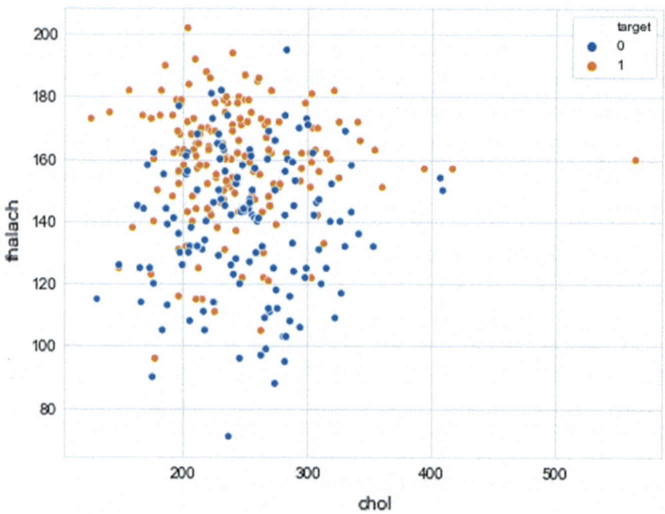

Fig. 4 Scatter plot for thalach versus chol

Table 2 Accuracy of algorithm

Approach	Accuracy
KNN	72.52
Decision tree	73.27
Logistic regression	82.52

The machine learning models are evaluated using the ROC metric. This can be used to understand the model performance, and it is shown in Fig. 3.

Figure 4 represents the scattered plot of heart syndrome with respect to thalach versus chol.

5 Conclusion and Future Work

Parental history or hereditary symptoms will lead to many chronic diseases to peoples, out of which heart disease is one among. If we identify the chronic diseases in early stage, it can be cured, so medical or hospital dataset is collected from Kaggle Web to analysis with different algorithm to check the accuracy score on key attribute with heart disuse patients. While implemented on this system model for heart disease patients by KNN, decision tree and logistic regression algorithm, with verification of key attributes, we found that logistic regression algorithm performs very effective and efficient performance on accuracy score for heart disease prediction. With inference of this customized model, machine learning algorithm provides very valuable knowledge on analysis and prediction of many chronic diseases. In this regard, researchers are helpful to the needy persons, doctors and society.

References

1. Dewan A, Sharma M (2015) Prediction of heart disease using a hybrid technique in data mining classification. 978-9-3805-441 6-8/15/$31.00 c 2015 IEEE
2. Thomas J, Theresa Princy R (2016) Human heart disease prediction system using data mining techniques. In: ICCPCT
3. Amin SU, Agarwal K, Beg R (2013) Genetic neural network based data mining in prediction of heart disease using risk factors. In: IEEE conference in ICT, pp 1227–1231
4. Shilaskar S, Ghatol A (2013) Feature selection for medical diagnosis: evaluation for cardiovascular diseases. Expert Syst Appl 40(10):4146–4153
5. Purusothaman G, Krishnakumari P (2015) A survey of data mining techniques on risk prediction: heart disease. Indian J Sci Technol 8(12):1
6. Miao KH, Miao JH, Miao GJ (2016) Diagnosing coronary heart disease using ensemble machine learning. IJACSA 7(10):30–39

Statistical Analysis of Literacy Rates Using Indian Census Data

Suresh Vishnu Bharadwaj and S. Vigneshwari

Abstract There exist many prevalent issues in today's world, many of which seem to have no feasible solution. One such issue is the disparity in the education provided between males and females. Gender equality is an important topic for discussion and reform in the modern world, and yet a full, thorough understanding of the intensity of the issue is yet to be elicited. Using a combination of modern computational statistical analysis and the vast, extensive trough of census data, this paper provides a framework for gaining a better viewpoint on the issues plaguing us currently, thus taking a step closer to eradicating them.

Keywords Statistical analysis · Data science · Indian census 2011 · Literacy rates · Education levels

1 Introduction

Providing education for all citizens of a nation is one of the pillars of a well-functioning country, which is why literacy rates of a country play a vital role in assessing the overall strength and growth of the economy of that country. Countries with higher literacy rates are usually marked as a more developed nation compared to countries with lower literacy rates, as they can provide good education for all its inhabitants. But there still seems to be a disparity in the education levels and literacy rates between males and females in certain countries. Despite many measures to counter this, the problem cannot be fully addressed until the exact nature and extent of this difference is examined and analysed properly.

There are many possible methods to calculate the exact magnitude of the difference in literacy rates, one of the most efficient being statistical analysis. Statistical analysis methods are statistical functions applied to real-world applications, ranging from

S. Vishnu Bharadwaj (✉) · S. Vigneshwari
Department of Computer Science and Engineering, Sathyabama Institute of Science and Technology, Chennai, India

S. Vigneshwari
e-mail: vigneshwari.cse@sathyabama.ac.in

S.-L. Peng et al. (eds.), *Intelligent Computing and Innovation on Data Science*,
Lecture Notes in Networks and Systems 118,
https://doi.org/10.1007/978-981-15-3284-9_56

fields such as biology to the social sciences. Statistical analysis is generally applied to data related to the census when the entire population cannot be reached for that particular survey, in which case advanced statistical tools such as random sampling are used. But the proposed system uses similar statistical tools to extract information from a census where the entire population has participated.

2 Literature Review

Many advances have been made to field of statistical analysis, which form the basis for the processes which are used in the proposed framework.

Diez et al. [1] have put forth a definitive book on different statistical methods which can be used for different types of data using the R programming language. It delves into many examples for each type of function used, with examples ranging from sports to gender discrimination. It also explains in detail the concepts of confidence intervals and hypothesis testing, which are used in the proposed framework.

Freedman [2] has conducted some seminal work in the field of statistical analysis, creating a definitive guide on the different functions that are used in both a theoretical and practical manner. He has applied such methods on different types of data, including comparing education levels, in a non-computational, purely mathematical manner.

Lomax and Hahs-Vaughn [3] have presented various statistical analysis tools which can be used, mainly dealing with Regression, One-factor ANOVA and Two-factor ANOVA. The last two methods mentioned helps to compare the means of different groups in data, and are an important foundation for the analysis done in the proposed framework.

Using the methods presented in these works and many more, the census data has been used as a knowledge base for extracting useful information computationally.

Chawda et al. [4] successfully used the census data to perform data mining for the purpose of demographic progress analysis, to support the municipal corporations in their decision-making processes. Techniques such as linear regression, decision tree and ARIMA were used for prediction and forecasting. Monte-Carlo simulation was also used for target-based progress tracking.

Robert et al. [5] published an influential paper on the application of statistical methods to census data. Their research deals with the estimation of income levels in areas with very small population levels, less than 1000. The method used to carry out the estimation was the James-Stein estimator, and includes a modified form of linear regression.

Lasse et al. [6] implemented modelling and prediction functions to predict individual salaries. The dataset used was from Finland, collected as part of the Finnish pension reform package. After separating the data into quartiles on the basis of wage, and then modelling the data using a combination of some individual factors, such as age and duration of career, and some general factors, such as the GDP at that time, they were able to predict with high accuracy the salaries of different individuals.

Sharath et al. [7] utilized the census data from the USA to predict income and economic hierarchy using data analytics methods. They were able to successfully predict the demographics based on different parameters, to an acceptable level of accuracy. Different classifiers such as k-means clustering were used.

Sheng and Gengxin [8] proposed a system to perform data mining on census data using R programming. The package used was classification and regression trees (CART), which is very popular as well as essential in machine learning functions. They achieved a high level of accuracy as well.

Priyanka et al. [9] presented a study for the analysis of suicide data in India, and subsequent prediction using the given data. Methods such as Pearson correlation were used to compute which factors were highly correlated, and using those factors, a linear regression model was built to predict future instances. The results that have been noted have around 99% prediction accuracy.

Ross et al. [10] compared the relation between the mortality rates and income inequality in both the Canada and the USA using their respective census data of the early 1990s. Multiple linear regression methods were used to check the correlation between the two factors. The results show that income inequality was a significant explanatory variable in the USA, but not in Canada.

3 Materials and Methods

For statistical analysis, the software that is used here is R. R is an open-source statistical programming language, supported by the R Foundation for Statistical Computing. Although there are plenty of debates today about which is the better programming language for analytics, R or Python, since it is primarily a statistical language, R and its IDE, RStudio are used here. There are many packages that have to be imported, the most important being StatsR [11].

The dataset that is used is the Indian census data from the year 2011, which is available online. [12] Since the data is extremely big, only the essential columns are retained and used. This can be done using R, or any other tool such as Excel.

The research questions for which the analysis is performed are:

1. Is there a statistically significant difference in the literacy rates of males and females nationwide? In other words, does gender play a role in the literacy rate?
2. Is there a statistically significant difference in the literacy rate of people in rural and urban areas? In other words, does the region in which a person lives play a role in the literacy rate?
3. Is there a statistically significant difference in the male and female literacy rates, in terms of the region? In other words, does the region affect the difference in female and male literacy rates?

All the functions in the statistical analysis automatically create a null and alternate hypothesis, and the default $p = 0.05$ is used here as the p-value. Confidence intervals are also created using those functions.

3.1 Algorithm

1. **Setup**

1.1 Import the required packages, dplyr, ggplot2 and statsr, using the library function.

1.2 Import the dataset using the read.csv function.

2. **Data Pre-processing**

2.1 Using the mutate function, convert the different literacy values, which contains number of literate people, into proportions or literacy rates. The literacy rates of the entire population, females and males are obtained.

2.2 With the filter function, separate the data in terms of the region, urban, rural and the entire nation.

3. **Exploratory Data Analysis**

3.1 Using the summary function, the literacy rates are analysed and summary statistics are created based on the different variables created based on gender and region.

3.2 The ggplot function is used to create various plots, such as scatterplots, line graph and boxplots, for the different variables created.

4. **Statistical Analysis**

4.1 The t.test function is used to compare the means of the male and female literacy rates.

4.2 The inference function in the statsr package is used to compare the difference in literacy rates of rural and urban areas.

4.3 The inference function is used again to check if region is a significant factor in the difference between male and female literacy rates.

3.2 Proposed System

Figure 1 represents the above algorithm using a flowchart. Although it is not necessary to follow the same linear order of steps given, it is advised as it will result in a more streamlined and efficient result.

4 Results and Discussion

Although the centre of Fig. 2 is not clear due to the overlapping of text, it is evident which states have a smaller difference in the literacy rates of males and females (Meghalaya, Nagaland, Mizoram, Kerala) and which states have a larger difference in the literacy rates of males and females (Bihar, Rajasthan, Dadra & Nagar Haveli, Daman & Diu).

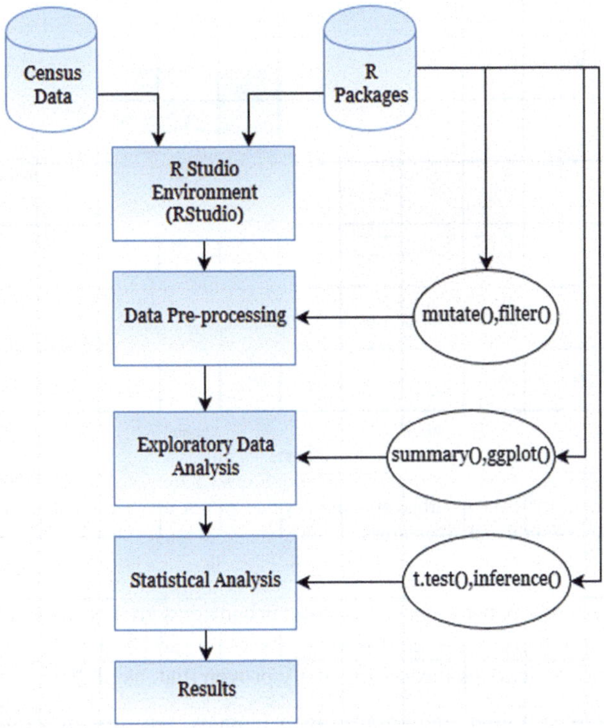

Fig. 1 The proposed framework, based on the algorithmic approach above

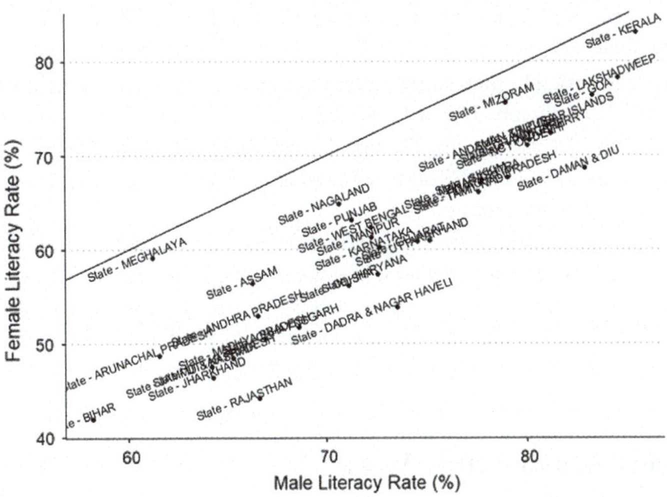

Fig. 2 Male literacy rate (%) versus female literacy rate (%). Even if the central parts of the plot are not completely legible due to an overlap in text, they can be neglected because the bottom and top parts of the plot prove to be far more important, to gauge an understanding

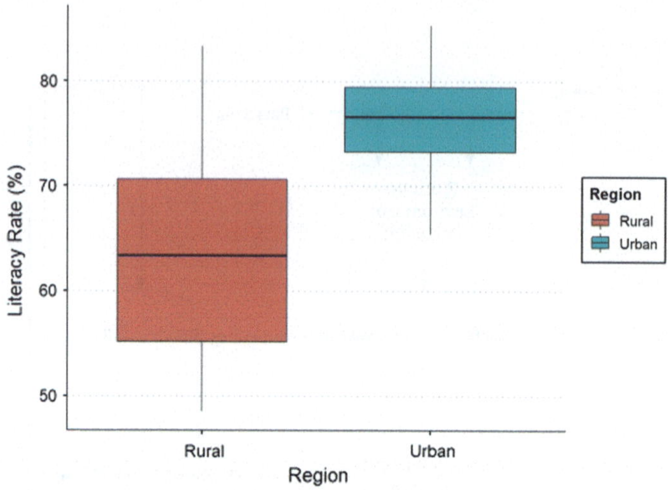

Fig. 3 Boxplot of rural versus urban area literacy rate (%). A very clear and obvious difference in the literacy rates of rural and urban areas

The literacy rate percentage of rural and urban areas are represented in Fig. 3, and it is evident that there is a huge difference between rural and urban areas.

The statistical analysis has conclusively proven that, as of 2011:

1. Males have a higher literacy rate than females, with the difference being statistically significant. It can be stated with 95% confidence that males have a 7.283–15.965% higher literacy rate than females. This can also be interpreted as: for every 1000 males that are literate, the number of females that are literate are only 840–927.
2. Urban areas have a higher literacy rate than rural areas, with the difference being statistically significant. It can be stated with 95% confidence that urban areas have a 9.3914–16.5704% higher literacy rate than rural areas. This can also be interpreted as: for every 1000 urban people that are literate, there are 844–906 rural people that are literate.
3. Rural areas have a higher difference of male and female literacy rates, compared to urban areas, with the difference being statistically significant. It can be stated with 95% confidence that rural areas have 3.2307% to 7.5271% higher male literacy rate than female literacy rate, compared to urban areas.

5 Conclusion and Future Work

The proposed framework proved to be very successful in properly analysing and extracting the necessary outputs from a large dataset. This framework can be implemented on many different datasets, such as other census or survey data, stock market data, sports analytics data or any other significant dataset in any type of field. These methods can also be used to compare the data between different years and extract any useful information from that.

References

1. Diez DM, Barr CD, Cetinkaya-Rundel M (2012) OpenIntro statistics. CreateSpace
2. Freedman DA (2009) Statistical models: theory and practice. Cambridge University Press, pp 26
3. Richard GL, Hahs-Vaughn DL (2007) Statistical concepts: a second course, pp 10. ISBN 0-8058-5850-4
4. Chawda M, Rane R, Giri S (2018) Demographic progress analysis of census data using data mining. In: 2018 Second International Conference on Inventive Communication and Computational Technologies (ICICCT)
5. Fay III RE, Herriot RA (1979) Estimates of income for small places: an application of James-Stein procedures to census data. J Am Stat Assoc 74(366a)
6. Koskinen L, Nummi T, Salonen J (2005) Modelling and predicting individual salaries: a study of finland's unique dataset. Finnish Centre for Pensions
7. Sharath R, Chaitanya SK, Nirupam KN, Sowmya BJ, Srinivasa KG (2016) Data analytics to predict the income and economic hierarchy on census data. In: 2016 international conference on computational systems and information systems for sustainable solutions in IEEE 2016 Journal, pp 249–254
8. Sheng B, Gengxin S (2010) Data mining in Census data with CART. In: 2010 3rd International Conference on Advanced Computer Theory and Engineering (ICACTE)
9. Priyanka SS, Galgali S, Priya SS, Shashank BR, Srinivasa KG (2016) Analysis of suicide victim data for the prediction of number of suicides in India. In: 2016 International Conference on Circuits, Controls, Communications and Computing (I4C)
10. Ross NA, Wolfson MC, Dunn JR, Berthelot J-M, Kaplan GA, Lynch JW (2000) Relation between income inequality and mortality in Canada and in the United States: cross sectional assessment using census data and vital statistics. BMJ 320(7239):898–902
11. StatsR package for R [online] https://github.com/StatsWithR/statsr
12. Data Source, Census Data 2011, [online] Available: www.censusindia.gov.in

Superlative Uprising of Smart Farming to Discovering the Magnitude and Superiority of the Agri-Data in Hybrid Techniques

K. Tharani and D. Ponniselvi

Abstract This exploration concentrated on shrewd cultivating in agribusiness. The recent innovations increase the quality and quantity of agro-products. Based on the field and the soil moisture, the cultivation brings a profit. The plant can be affected by fungi, bacteria, and viruses. It affects the plants shortly. The maladies at the beginning time on the plants are exceptionally hard to discover. Earth's perception will be founded on a Decision Support System (DSS). This methodology will apply in a proposed system to improve the soil continuum. Information mining procedures are connected here to improve the surplus and vitality framework. Be that as it may, in a current framework, they were utilizing a SAR procedure for the topographical debacle. Grouping is used to isolate the information for horticulture, and pre-preparing is used to identify the commotion and evacuate the unimportant information. For finding the ideal outcome, the K-means, fuzzy, KNN, and ANFIS are utilized for the finished structure. On account of these sicknesses, horticulture will elevate the ranchers to misfortune and influence the generation. Shrewd cultivating by applying information mining systems will expand profitability and benefit, just as it expands contamination security and the nature of the items.

Keywords Decision support system (DSS) · Fuzzy · K-means · KNN—K-means nearest neighborhood · ANFIS—artificial neural fuzzy inference system

1 Introduction

In the agriculture fragment, farmers and agribusinesses needed to choose incalculable decisions reliably, and flighty complexities incorporate the various factors affecting them. An essential issue for cultivating orchestrating desire is the definite yield estimation for the different harvests connected with the organization. Data mining strategies are a fundamental system for accomplishing sensible and convincing responses to this issue. Agribusiness has been an unquestionable goal for tremendous data. Biological conditions, vacillation in soil, input levels, mixes,

K. Tharani (✉) · D. Ponniselvi
Vivekanadha College of Arts and Science, Tiruchngode, Namakkal, India

S.-L. Peng et al. (eds.), *Intelligent Computing and Innovation on Data Science*, Lecture Notes in Networks and Systems 118,
https://doi.org/10.1007/978-981-15-3284-9_57

and product expenses have made it much increasingly huge for farmers to use the information and get help to choose essential developing decisions. Today, plant affiliations work with a lot of information. Dealing with and recuperation of basic data in this abundance of cultivating information are significant. Utilization of information and correspondences development enables robotization of expelling imperative data with an ultimate objective to get learning and an example, which engages the part of the arrangement and more straightforward data extraction really from electronic sources, and to move to confirm electronic course of action of documentation which will enable creation cost decline, higher yield, and higher market cost. Data mining despite information about yields engages agrarian undertakings to anticipate floats about customer's conditions or they are direct, which is practiced by researching data from interchange perspectives and finding affiliations and associations in evidently superfluous data. Crude information about rural endeavors is sufficient and different. It is important to gather and store them in a composed structure, and their reconciliation empowers the making of an agrarian data framework. Information mining in farming gives numerous chances to investigating shrouded designs in these accumulations of information. These examples can be utilized to decide the state of clients in farming associations. The data mining techniques involve predicting the quality and quantity of the data resistant properly. And fuzzy helps to provide the possible outcome which helps to denote the binary values. And it is needed to detect the nearest node on the Agri-Data which helps to predict the KNN data mining techniques. Finally, we make a framework like the performance and profit ratio for the given data.

2 Related Works

Shilpa Ankalaki et al. said the viability of the diverse quality measurements and grouping strategies advancing the proper number of bunches is exhibited tentatively for leaf informational collection with the number of groups shifting from five to fifteen. When the proper number of bunches is resolved, the exhibitions of all bunching methods are assessed for fitting the gathering of the information into the number of groups [1]. Jharna Majumdar Email author, Sneha Naraseeyappa et al. proposed that horticulture has been a conspicuous objective for huge information. Natural conditions, changeability in soil, input levels, blends, and product costs have made it even more pertinent for ranchers to utilize the data and get help to settle on basic cultivating choices [2]. MCS Geetha et al. conveyed to incorporate crafted by different creators in a single spot, so it is helpful for analysts to get data about the present situation of information mining strategies and applications in setting to the farming field. It gives an overview of different information mining methods utilized in horticulture which incorporates artificial neural networks, K-closest neighbor, decision tree, and Bayesian system, fuzzy set, support vector machine, and K-implies [3]. Niketa Gandhi et al. determined this audit and recommended that further examinations are expected to see how these procedures can be utilized with complex rural datasets for harvest yield forecast coordinating regular and spatial factors by utilizing GIS advances [4]. Guilherme M. Sanchesab and Henrique C. Junqueira Francob et al. proposed the target of the present examination was to explore the connection between the physical and concoction properties of soils and sugarcane yield, in this way recognizing the dirt parameters that decide the last efficiency of the field [5]. Abhishek B Mankar and Mayur S Burange et al. depict the yield expectation issue that can be comprehended by utilizing data mining methods. This work expects to discover reasonable information models that accomplish high precision and high consensus as far as yield forecast abilities [6]. In the event that observational models are required, they will depend on utilizing propelled procedures; however, they will require appropriate calculation tuning and highlight building to separate the vast majority of the data from datasets. In view of the outcomes, we prescribe following the displayed work process for the improvement of yield models [7]. Ramesh A Medar and Vijay S Rajpurohit depict data mining is a creating investigation field in cultivating harvest yield examination. Data mining is the route toward perceiving the hidden models from a great deal of data. Yield desire is a critical agrarian issue that outstanding parts to be disentangled subject to available data [8]. Hetal Patel and Dharmendra Pate et al. portrayed, as can be seen, that the fittingness of information mining procedures is to a limited degree dictated by the various kinds of rural information or the issues being tended to [9]. Thayn, J. B. and Price, K. P et al. said that the root means square slip-ups between these datasets kept running from 9.4 to 10.9 days, significantly greater than is commendable for most phenology looks at. The reason is that vegetation phenology concentrates must use exact brief data to depict changes in vegetation consistency. Veenadhari Suraparaju et al. declare that the information mining in the application in agribusiness is a moderately new methodology for anticipating/foreseeing agrarian harvest/creature the board [10]. Jharna Majumdar Email author Sneha Naraseeyappa Shilpa Ankalaki et al. proposed that information mining strategies are an essential methodology for achieving pragmatic and compelling answers to this issue. Farming has been an undeniable objective for enormous information. Ecological conditions, changeability in soil, input levels, blends, and item costs have made it even more pertinent for ranchers to utilize the data and get help to settle on basic cultivating choices [11]. This examination shows that the ANN-based forecast model is an appropriate method for anticipating oil yield at another site and to enhance the yield of turmeric oil at a specific site by changing the alterable parameters of the expectation model and hence is of enough business centrality by Abdul Akbara and Sanghamitra Nayaka et al. [12].

3 Methodology

For the most part, the information mining structures fuse to foreseeing the qualities as the ideal one which serves to the consolidated attributes for the covariance arrange. So it considers because of building up the procedure in fluffy and finds the closest ideal in the fluffy KNN methods. However, before that how about we need to discover

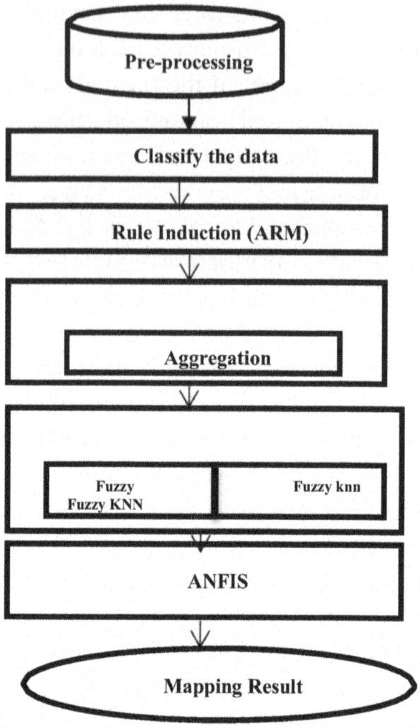

Fig. 1 Overall framework about agriculture working progress

the isolation of Agri-information from bunching by utilizing the information mining characterization techniques individually. Subsequently, the datasets have been put away in the database appropriately. At that point, it goes to control the procedure of pre-handling systems. It is required to expel the commotion of the information and change of the given informational collection control which suits for the accumulation and smoothing capacities (Fig. 1).

At that point, the likelihood must be found by Bayesian procedures that include getting to the K-implies preparing methods. Here the classification of information is controlled by the

Objective Function:

$$J = \sum_{i=0} \sum_{k=1} W_{ik}\left(x^1 - \mu_k\right)^2 \tag{1}$$

where

The variable interims between two hubs for $i = 0$ till m and $k = 1$ till k. However, the target capacity point portrays the consistent worth k. in the event that the worth has been delivered 1 for $k = \text{argmin}\ (x1 - \mu\ k)\ 2$. Be that as it may if the worth is return by 0 to venture out the procedure of the target capacity point in the Agri-information.

Euclidean Distance:

$$J(v) = \sum_{i=1} \sum_{J=1} W_{ik}(x_1 - V_J)^2 \tag{2}$$

where the Euclidean distance between the matrix variable is I, j. And the intervals have to be determined as $i = 1$ till C (cluster center) between $j = 1$ till c in the ith cluster of the given origin.

K-means Clustering:

$$V_i = (1/C_i) \sum_{j=1} X_i \tag{3}$$

Cluster Variation:

$$1\backslash mk \sum_{i=1} (X_i \cdot \mu_{ck})^2 \tag{4}$$

These are all the control which has been done at this point to anticipate the ideal qualities in the keen cultivating strategies which include in the information mining methods. At that point, it goes on the change in the given framework to draw out the transpose of tonal qualities in the Agri-information to the database framework. How about we discover the closest neighbors during the time spent supporting the k-implies closest neighborhood in the information mining concentrate to recognize the probability information. Fuzzy maxims have been met to deliver the best outcomes, for example, TP, TN, FP, and FN in the given double codes. The given information was controlled to proceed onward the finish system between the transformative aggregate informational collection in the reliable information. The ANFIS methods were to find the set hypothesis of information things coverts to the fuzzification to de-fluffy the ideas of Takagi–Sugeno model seems to get the exactness to bring about the information mining procedures. It distinguishes and locates the best arrangement in the aggregate control information preparing which aides of basic leadership controller in the framework.

4 Experimental Results and Discussion

Likewise, the hybrid techniques used to detect the smart farming system produced the variation of results used to detect the best solution to develop the agriculture and implement the profit supports to the farmer. After the refreshing of the ensuing parameters by cross-breed calculation and refreshing of the reason parameters by the back proliferation angle drop calculation, the last upgraded ANFIS model of the warm power plant is obtained. The control of information results has indicated where the anticipated yields of the ANFIS models are execution versus test numbers for the informational indexes of the savvy cultivating. What's more, the fluffy principles and the parameters of the models for the agribusiness informational indexes are additionally given (Figs. 2, 3 and 4).

The mix of given lattice esteems has been controlled to give an enormous measure of transformative qualities for the five parts. It demonstrates the plot variety in the inserting device to perceive the populace potential outcomes in shrewd cultivating procedures. Finally, it lets out the correct procedures and advances the keen method for getting to the information mining strategies (Fig. 5).

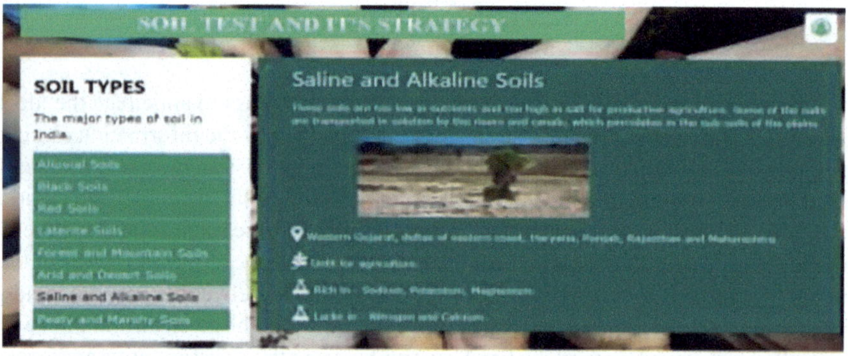

Fig. 2 Soil types

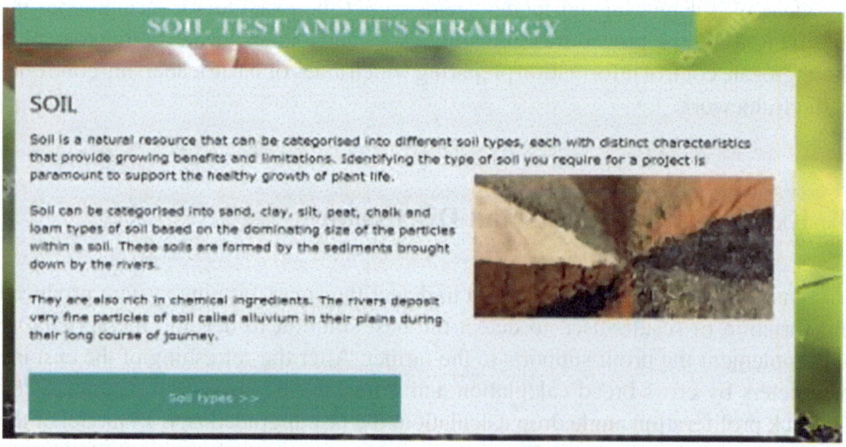

Fig. 3 Description of soil

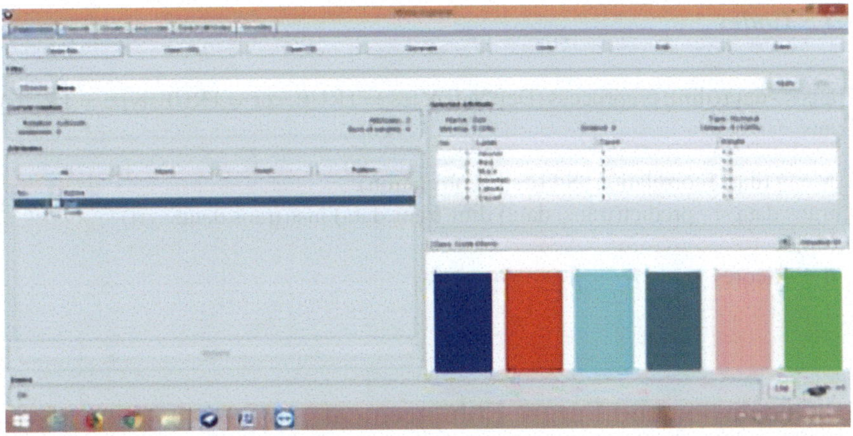

Fig. 4 Variances of manipulating variables report

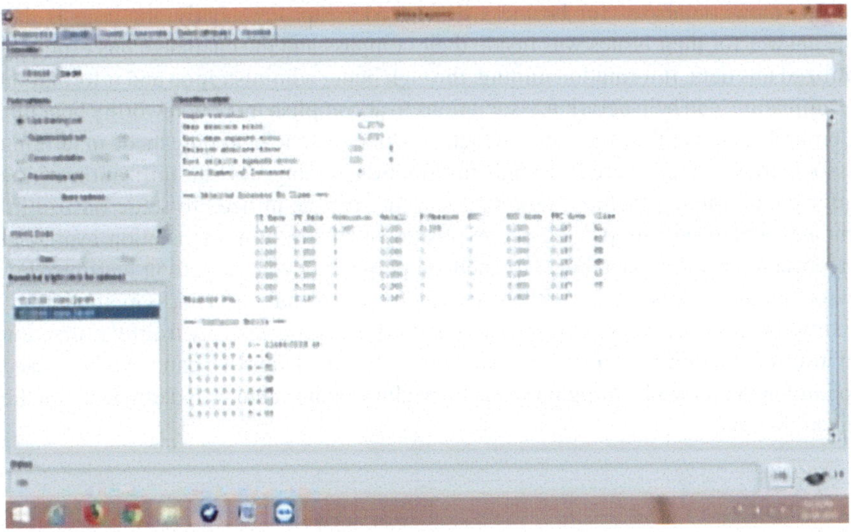

Fig. 5 Cross-validation result

4.1 Codes

```
Trans < - data (dim) preprocess (Dependencies = TRUE, cor = TRUE, type = TRUE)
Trans

  Mean (data$correlation) Sd (data$correlation)
  trans.data < - predict(trans, data) dim(trans.data) hist(trans.data$agri)
  biplot
  d < - Scores
```

5 Conclusion and Future Enhancement

Farming associations and their administration attempt each day to discover data in enormous databases for business basic leadership. Frequently, the case is that the answer for their issues was inside their compass and the challenge has officially utilized this data. Information mining, through better administration and information examination, can help agrarian associations to accomplish more noteworthy benefits. Comprehension of the procedures which are done and choices being made in farming associations is empowered through information mining. By the utilization of the information mining method, procured learning can be utilized to settle on fruitful choices which will propel the accomplishment of the rural association available. Information mining once began presents a perpetual cycle of getting learning. For associations, it speaks to one of the keys which focus to make a business methodology. Extraordinary endeavors put resources into finding a progressively fruitful utilization of information mining in agrarian associations. Further feature research has been used to implement the embedding methods to produce high accuracy and efficiency in the given domain.

References

1. Liu S, Guo L, Webb H, Ya X, Chang X (2019) Internet of Things monitoring system of modern eco-agriculture based on cloud computing. IEEE Access (99):1–1
2. Wu T, Luo J, Dong W, Sun Y, Xia L, Zhang X (2019) Geo-object-based soil organic matter mapping using machine learning algorithms with multi-source geo-spatial data. IEEE J Sel Top Appl Earth Observations Remote Sens 12(4)
3. Del Frate F, Schiavon G, Solimini D, Borgeaud M, Hoekman DH, Vissers MAM (2003) Crop classification using multiconfiguration C-band SAR data. IEEE Trans Geosci Remote Sens Adv Search 41(7):1611–1619
4. Miyaoka K, Maki M, Susaki J, Homma K, Noda K, Oki K (2013) Rice-planted area mapping using small sets of multi-temporal SAR data. IEEE Geosci Remote Sens Lett 10(6)
5. Chou J-S, Nguyen T-K (2018) Forward forecast of stock price using sliding-window metaheuristic- optimized machine-learning regression. IEEE Trans Ind Inf 14(7)

6. de Roo RD, Du Y, Ulaby FT, Dobson MC (2001) A semi-empirical backscattering model at L-band and C-band for a soybean canopy with soil moisture inversion. IEEE Trans Geosci Remote Sens 39(4)
7. Cristofani E, Becquaert M, Lambot S, Vandewal M, Stiens JH, Deligiannis N (2018) Random subsampling and data preconditioning for ground penetrating radars. IEEE Access 6
8. Du H, Mao F, Li X, Zhou G, Xu X, Han N, Sun S, Gao G, Cui L, Li Y, Zhu D, Liu Y, Chen L, Fan W, Li P, Shi Y, Zhou Y (2018) Mapping global bamboo forest distribution using multisource remote sensing data. IEEE J Sel Top Appl Earth Observations Remote Sens 11(5)
9. Jackson TJ, O'Neill PE (1990) Attenuation of soil microwave emission by corn and soybeans at 1.4 and 5 GHz. IEEE Trans Geosci Remote Sens 28(5)
10. Tomičić I, Schatten M (2016) An agent-based framework for modeling and simulation of resources in self-sustainable human settlements: a case study on water management in an eco-village community in Croatia. Int J Sustain Dev World Ecol 23(6)
11. Qin L, Feng S, Zhu H (2018) Research on the technological architectural design of geological hazard monitoring and rescue-after-disaster system based on cloud computing and Internet of things. Int J Syst Assur Eng Manag 9(3):684–695
12. Ishitsuka N, Saito G, Murakami T, Ogawa S, K Okamoto (2003) Methodology development for area determination of rice planted paddy using RADARSAT Data. J Remote Sens Soc Jpn 23(5):458–472

A Survey on Load Balancing in Cloud Computing

Ashish Mishra, Saurabh Sharma, and Divya Tiwari

Abstract Cloud processing is a cutting-edge worldview to give benefits throughout the Internet. Burden adjusting is a core part of cloud computing and keeps away from circumstance in which a few hubs become overburden while the others are inactive or then again have modest effort to do. Load adjusting is able to get better the quality of service (QoS) measurements, including reaction throughput, cost, time, execution and asset usage. In this survey, we revise the writing on the undertaking booking and load-adjusting calculations and present another grouping of such calculations. The development of distributed computing dependent on virtualization innovations carries immense chances to have virtual asset requiring little to no effort with a requirement of owning any framework. Virtualization innovations empower clients to gain, arrange and be charged on compensation per-use premise. In any case, cloud server farms for the most part contain heterogeneous product servers facilitating diverse virtual machines (VMs) with possible different particulars and swinging asset uses, which possibly will reason imbalanced resource usage inside servers that may prompt execution corruption and administration-level understanding (SLAs) infringement. An arrangement focusing on burden adjusting calculations for VM position in cloud server farms is explored, and the overviewed calculations are characterized by the order. The objective of this paper is to give an exhaustive and similar comprehension of existing writing also and help specialists by giving an understanding of potential future improvements.

Keywords Cloud computing · VM · Load-balancing algorithms · SLA · Load balancing

---

A. Mishra (✉)
Faculty of Computer Science and Engineering, Gyan Ganga Institute of Technology and Sciences, Jabalpur 482003, India

S. Sharma
Amity University Gwalior, Gwalior 474005, India

D. Tiwari
Gyan Ganga Institute of Technology and Sciences, Jabalpur 482003, India

S.-L. Peng et al. (eds.), *Intelligent Computing and Innovation on Data Science*, Lecture Notes in Networks and Systems 118,
https://doi.org/10.1007/978-981-15-3284-9_58

1 Introduction

Nowadays, cloud computing is a cutting-edge innovation in the computer services system's field to give administrations and services to customers whenever they want. In a cloud computing framework, assets are circulated all around the globe for quicker adjusting to customers. The customers can effectively get to data through different gadgets, for example, PCs, cell telephones, PDAs and tablets. Distributed computing has confronted numerous difficulties, counting security, productive burden adjusting, asset booking, scaling, QoS the executives, server farm vitality utilization, information locking furthermore, administration accessibility and execution checking. Burden adjusting is one of the fundamental difficulties and worries in cloud conditions; it is the way toward allocating and reassigning the heap among accessible assets so as to expand throughput, even as limiting the expense, reaction time, resources usage and improving execution just as vitality sparing. The service-level agreement (SLA) and client fulfillment could be given by magnificent burden adjusting methods. In this manner, giving the proficient burden adjusting calculations and components is a key to the achievement of distributed computing situations. A few looks have been done in the field of burden adjusting and assignment planning for cloud situations. Notwithstanding, our examinations demonstrated that regardless of the key job of load-adjusting calculations in distributed computing, particularly in the coming of huge information, there are a couple of far-reaching audits of these calculations.

In ordinary server farms, applications are associated with specific bodily servers that are frequently over-provisioned to control the upper-certain exceptional burden. Such layout makes server farms high-priced to hold up with squandered energy and ground place, low asset usage and vital the executives overhead. With virtualization innovation, cloud server farms grow to be step-by-step adaptable and supply better assist to on-request distribution. It conceals server's heterogeneity, empowers server aggregate and gets better utilization of servers. Cloud server farms are profoundly powerful and flighty due to:

(1) Irregular usage of resource styles of customers constantly soliciting for VMs,
(2) Variable resource usages of VMs,
(3) The overall performance of hosts,
(4) Uneven prices of incoming and outgoing of record center clients when taking care of various burden levels may change extraordinarily. These circumstances are anything but difficult to trigger uneven stacks in cloud server farm, and they may likewise prompt execution corruption and administration-level understanding infringement, which requires a heap adjusting component to moderate this issue. Burden adjusting in mists is a component that disseminates the overabundance dynamic nearby remaining task at hand in a perfect world adjusted over every one of the hubs [1]. It is connected to accomplish both better client fulfillment and higher asset use, guaranteeing that no single hub is overpowered, in this way improving the framework generally speaking execution. For VM planning with load-adjusting intention in cloud computing, it expects to appoint

VMs to be appropriate and balance the resources use inside the majority of the hosts.

Appropriate load-adjusting calculations can assist in using to be had assets preferably, consequently restricting the asset utilization. In this architecture all the host are located in the bottom layer which contains all the hardware resources. In the second layer all the hosts are allocated to virtual machine by a cutting edge technology which is further controlled by Virtual Manager. Figure 1 demonstrates the relationship among VMs, software and the hosts in cloud server farms. In this architecture all the host is located in the bottom layer which contains all the hardware resources [2]. In the second layer all the hosts are allocated to virtual machine by a cutting edge technology which is further controlled by Virtual Manager. In the Upper layer Application Environment is created as per the request created by End User [3, 4]. The VM's finished applications possibly will have predefined conditions among them. All hosts might be apportioned with server VMs and VMs are introduced with simply certainly one kind of application. Load adjusting calculations are related each at the software and the VM Machines.

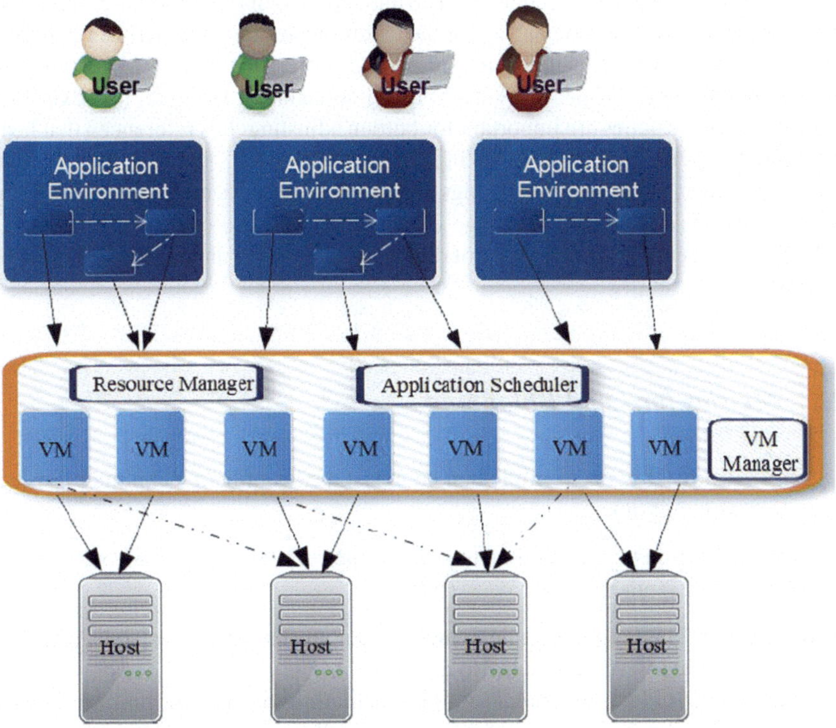

Fig. 1 Application, host and VM relationship in data center

2 The Load-Balancing Policies, Metrics and Model

The version of balancing of load seems to determine 2 in which we are capable to see the heap balancer gets clients' solicitations and runs load-adjusting calculations to disseminate the solicitations some of the VMs. The load balancer chooses which VM wants to be doled out to the subsequent solicitation. The server farm controller is liable for project the executives. Errands are submitted to the load balancer, which plays load-adjusting calculation to allocate errands to a cheaper VM. VM administrator is liable for VMs. The virtualization is a prevailing innovation in allotted computing. The number one motive of virtualization is sharing pricey tool among VMs. VM is a product execution of a pc that going for walks frameworks and applications can hold strolling on. VM Machine are placed in Data Centers and It can be accessed by User from anywhere else by an Open Network. Clients are placed everywhere in the globe, and their solicitations are submitted arbitrarily. Solicitations need to be appointed to VMs for making geared up. In this manner, the venture mission is a critical problem in disbursed computing. On the off risk that a few VMs are overburden while others are inactive or have piece paintings to do, QoS will decline. With the diminishing of QoS, clients end up unhappy and may additionally, moreover, depart the framework and stay away for the indefinite destiny. A hypervisor or digital machine display (VMM) is finished to make and address the VMs. VMM gives four duties: multiplexing, suspension (stockpiling), association (resume) and existence relocation. Those duties are important for burden adjusting. It is referenced that heap adjusting wants to consider undertakings: asset designation and task-making plans. The final effects of those undertakings are the excessive accessibility of property, energy sparing, developing the usage of property, decrease of charge of utilizing belongings, defensive the capability of disbursed computing and reduction in carbon outflow.

2.1 Metrics for Load Balancing

Throughput: This size is implemented to test the quantity of techniques finished in line with unit time.

Time of Response: It is all time that the framework takes to provide service to a submitted assignment.

Makespan: This length is implemented to compute the satisfactory finishing time or even as the assets are specific to a patron.

Scalability: It is the functionality of a calculation to carry out uniform burden adjusting in the framework as ordinary with the situations upon growing the quantity of hubs. The favoured calculation is profoundly adaptable.

Fault Tolerance: It entails a desire for the functionality of the calculation to carry out burden adjusting in case of nice disappointments in excessive first-class hubs or as a substitute interfaces.

Time of Migration: The quantity of time is required to transfer a task from an overloaded node to a low-loaded one.

Imbalance Degree of VM: This size estimates the awkwardness among VMs.

Performance: It records and analyzes machine performance after acting a load-balancing set of rules.

Consumption of Energy: It computes the degree of energy expended with the aid of all hubs. Burden adjusting abstains from overheating and thusly diminishing energy utilization with the energy required to run a data center can be broken down broadly over all of the hubs.

Emission of Carbon: It analyzed the quantity of carbon produced through way of all belongings. Load balancing has a key function that decreases this metric though into power consumed by computing resources and power consumed by Infra [5, 6].

2.2 *Load-Balancing Algorithm's Taxonomy*

In this part, we present the cutting-edge grouping of burden adjusting calculations. In certain examinations, load-adjusting calculations have been organized in view of factors: the circumstance of the framework and individual who commenced the machine. Calculations depending on the condition of the framework are grouping as dynamic and static. Some static methods are min–min, spherical Robin, opportunistic load balancing (OLB) and max–min algorithms. A part of the dynamic calculations include fashions, as an instance, ant colony optimization (ACO).

Monitoring of Loads: In this evolution, the heap and the condition of the assets are observed.

Synchronization of Loads: In this, the state and load in order are switched over.

Criteria of Rebalancing: It is essential to compute a new workload distribution and then take decisions for load balancing based on the new calculation.

Migration of Tasks: In this stage, the actual data movement occurs. When system decides to transfer process/task, this particular step will be executed.

Static algorithm's qualities are:

1. They pick dependent on a set popular, for instance, input load.
2. They are now not adaptable.
3. They need earlier studying approximately the framework.

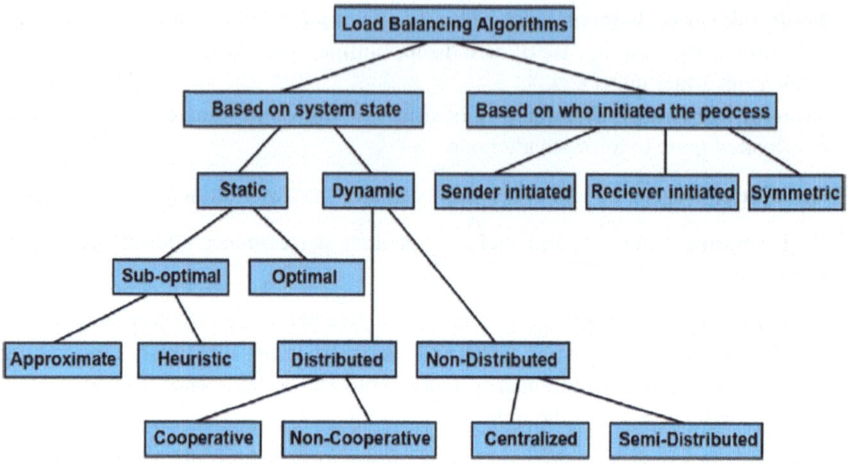

Fig. 2 Classifications of load-balancing algorithms

The functions of dynamic calculations are:

1. They pick relying on the prevailing situation of the framework.
2. They are adaptable.
3. They beautify the exhibition of the framework.

Dynamic calculations are remote to training: conveyed and non-dispersed. Within the conveyed technique, all hubs accomplish the active and dynamic load-balancing computation in the framework and the undertaking of burden adjusting is shared among them. The connections of the framework hubs take structures: agreeable and non-beneficial (Fig. 2).

In the agreeable structure, the hubs cooperate to accomplish a classical objective, for example, to diminish the time of reaction everything being equal. In the non-helpful structure, every node works freely to accomplish a nearby task, for instance, to diminish the time of reaction for a neighborhood task. Non-dispersed calculations are isolated into classes of set of two: incorporated what's more and semi-disseminated. In the incorporated structure, a solitary node called the focal node executes the load-adjusting computations and it is totally in charge of burden adjusting.

3 Cloud-Based Load-Balancing Challenges

Load balancing in cloud processing has confronted a few difficulties. In spite of the fact that the theme of burden adjusting has been extensively examined, in light of the heap adjusting measurements, the present circumstance is a long way from a perfect one. In this segment, we survey the difficulties in burden offsetting with the point of structuring run of the mill burden adjusting methodologies later on. A few investigations have referenced difficulties for the cloud-based burden adjusting.

3.1 Virtual Machine's Migration (Security and Time)

The control on-request nature of cloud computing shows that on the factor while there may be an administration demand, the assets have to accept. Now and then, assets (frequently VMs) need to be relocated from one Data center to other Data Center [7, 8]. So the Data Center architect need to think for Load Balancing algorithm those will be more Energy Efficiency while moving VM image from one location to another location [9, 10].

3.2 Cloud's Spatially Separated Nodes

Spatially allocated nodes in a cloud hub in allotted computing are circulated geologically. The undertaking for this case is that the heap adjusting calculations ought to be planned with the intention that they do not forget parameters, for example, the Load balancing algorithms having capabilities to switch data and virtual machines among the multiple nodes. It also separate user data and program.

4 Conclusion

Load balancing of jobs on VMs is a pinnacle assignment in cloud computing that has informed noteworthy consideration from analysts. This text speaks to a reducing aspect survey of problems and difficulties of burden adjusting. As indicated via this research, a massive assessment has been completed on several load-adjusting structures considering diverse measurements. In mild of our statistics, we have characterized those heap adjusting techniques into some classifications: widespread load-balancing techniques, normal phenomenon-based load balancing, project-based complete load balancing and load balancing of agent-based type. From every type, we have got mentioned focal factors, weaknesses, mind and challenges with those systems. Several load-adjusting calculations are available that consolidate maximum load-adjusting measurements and supply better asset to utilize and much less time of response. But, there may be a need to enhance the techniques for growing execution of the framework in some time. Alongside the improvement in framework execution and asset use, load-adjusting strategies are likewise that specialize in inexperienced processing, strength sparing and mission load the board, which requires developing new calculations. We surveyed multiple algorithms for load balancing for Cloud Computing. The vital part of this paper is comparison of different algorithms considering the characteristics like fairness, throughput, fault tolerance, over head, performance, and response time and resource utilization. The limitation of existing work is that each cloud computing algorithm does not address the related issues like fairness, high throughput and equality. Future work is to mitigate the above problem.

References

1. Randles M, Lamb D, Taleb-Bendiab A (2010) A comparative study into distributed load balancing algorithms for cloud computing. In: 2010 IEEE 24th international conference on advanced information networking and applications workshops (WAINA), IEEE, pp 551–556
2. Jiang Y (2016) A survey of task allocation and load balancing in distributed systems. IEEE Trans Parallel Distrib Syst 27(2):585–599
3. Mann ZA (2015) Allocation of virtual machines in cloud data centers: a survey of problem models and optimization algorithms. ACM Compu Surv (CSUR) 48(1):11
4. Milani AS, Navimipour NJ (2016) Load balancing mechanisms and techniques in the cloud environments: systematic literature review and future trends. J Netw Comput Appl
5. Kansal NJ, Chana I (2012) Cloud load balancing techniques: a step towards green computing. IJCSI Int J Comput Sci Issues 9(1):238–246
6. Coffman Jr EG, Garey MR, Johnson DS (1996) Approximation algorithms for bin packing: a survey. In: Approximation algorithms for NP-hard problems, PWS Publishing, pp 46–93
7. Voorsluys W, Broberg J, Venugopal S, Buyya R (2009) Cost of virtual machine live migration in clouds: a performance evaluation. In: IEEE international conference on cloud computing, Springer, pp 254–265
8. Hu J, Gu J, Sun G, Zhao T (2010) A scheduling strategy on load balancing of virtual machine resources in cloud computing environment. In: 3rd international symposium on parallel architectures, algorithms and programming, IEEE, pp 89–96
9. Speitkamp B, Bichler M (2010) A mathematical programming approach for server consolidation problems in virtualized data centers. IEEE Trans Serv Comput 3(4):266–278
10. Ye K, Jiang X, Huang D, Chen J, Wang B (2011) Live migration of multiple virtual machines with resource reservation in cloud computing environments. Cloud Computing (CLOUD), 2011 IEEE International Conference on, IEEE, 2011; 267–274

Detection of Hard Exudate from Diabetic Retinopathy Image Using Fuzzy Logic

S. Jeyalaksshmi, D. Padmapriya, Divya Midhunchakkravarthy, and Ali Ameen

Abstract Diabetic retinopathy, otherwise called diabetic eye illness, is a therapeutic condition in which damage occurs to the retina because of diabetes and is a main source of visual deficiency. As exudates (exudates are mass of cells and fluid that has seeped out of blood vessels or an organ) are among early clinical indications of diabetic retinopathy, their location would be a basic resource for the mass screening errand and fill in as an essential. A procedure is proposed which depends on morphological image processing and fuzzy logic to recognize hard exudates from diabetic retinopathy retinal image in this dissertation. At the underlying stage, the exudates are distinguished utilizing mathematical morphology that incorporates image preprocessing utilizing HSV colour model and elimination of optic disc. The hard exudates are separated utilizing an adaptive fuzzy logic algorithm that utilizations values in the RGB colour space of retinal image to form fuzzy sets and membership function. The fuzzy output for all the pixels in every exudate is calculated for a given input set corresponding to red, green and blue channels of a pixel in exudates. Since, digital image is formed from combination of pixels, during image acquisition process, the quality of the image diminishes from the point they are captured. To get a quality image, image quality metrics are applied on the proposed algorithm. Then, fuzzy output is computed for hard exudates according to the proportion of the hard exudates detected. By comparing the results with hand-drawn ground truths, it has been obtained that the sensitivity and specificity of detecting hard exudates are 81.75% and 99.99%, respectively.

Keywords Diabetic retinopathy · Fuzzy logic · MSE · PSNR

S. Jeyalaksshmi (✉) · D. Padmapriya
Department of BCA & IT, VISTAS, Chennai, India
e-mail: pravija.lakshmi@gmail.com

D. Padmapriya
e-mail: padmapriya.scs@velsuniv.ac.in

D. Midhunchakkravarthy
Centre of Postgraduate Studies, Lincoln University College, Kota Bharu, Selangor, Malaysia

A. Ameen
Faculty of Computer Science and Multimedia, Lincoln University College, Kota Bharu, Malaysia

S.-L. Peng et al. (eds.), *Intelligent Computing and Innovation on Data Science*,
Lecture Notes in Networks and Systems 118,
https://doi.org/10.1007/978-981-15-3284-9_59

1 Introduction

Diabetes can lead to damage in the eye called diabetic retinopathy. The retina is your seeing part in the back of the eye. High blood sugar levels cause damage to your retinal blood vessels, which in turn causes bleeding and leakage of exudates in your retina and in advanced disease. This leads to the formation of abnormal new blood vessels. High blood pressure and high blood fat increase the damage on your retina. Diabetic retinopathy is the most common cause of blindness in the working population in the western world. The early stage of diabetic retinopathy is named as micro-aneurysm which looks tiny, dark red spots which form clusters in circular shape.

The aim of this research is to develop a system to detect hard exudates in diabetic retinopathy using non-dilated diabetic retinopathy images. The exudates are identified using morphological method and categorized into hard exudates and non-hard exudates using adaptive fuzzy algorithm. Therefore, early detection of diabetic retinopathy is important because ophthalmologist can be able to treat patients using advanced laser treatment.

2 Reviews on Related Work

Akara et al. [1] have proposed a fuzzy c-means (FCM) clustering method to detect exudates. Contrast enhancement is applied for four image-based features, namely intensity, standard deviation on intensity, hue and a number of edge pixels, which are provided as input parameters to a coarse segmentation routine using FCM clustering method.

Basha and Prasad [2] have proposed an approach using fuzzy logic for hard exudates detection from digital fundus images. Initially, morphological segmentation is used to detect abnormal regions in fundus images. Fuzzy sets for different values in the colour space are used to form fuzzy rules. The output is calculated as the average of the values in the different colour spaces.

Fraz et al. [3] proposed the methodologies in two-dimensional retinal images taken from a retinal camera, and a survey of techniques is presented. They reviewed, analysed and categorized the retinal vessel extraction algorithms and techniques and methodologies. The performance is analysed on two available databases, Drive and Stare.

Many techniques have been employed to the exudate detection.

Gardner et al. [4] proposed an automatic detection of diabetic retinopathy using artificial neural network. The exudates are identified from grey-level images. The fundus image was analysed using a backpropagation neural network. The technique did not work well on low contrast images.

Reza et al. [5] described and approached for the detection of exudates using k-means clustering algorithm. They preprocessed retinal images by contrast limited adaptive histogram equalization. Then, they segmented the preprocessed colour retinal images using k-means clustering technique [6, 13].

3 Methodology

For testing the proposed algorithm, we have chosen nearly fifty images from the available DR database, DIARETDB and MESSIDOR.

Firstly, the fundus images are preprocessed by implementing the colour space conversion, removing the noise from the fundus image and by filtering the image [7, 8, 12]. Then, the segmentation and morphological operation are performed on the fundus image to obtain mask image. The image is segmented to extract the blood vessel from the retina because the exudate and the blood vessels exhibit the same colour.

In this work, three linguistic variables x_r, x_g and x_b are taken as input and are denoted by Gaussian combination membership functions (MFs). Each exudate is calculated to verify whether it is hard exudates or not by the values of RGB colour space. When hard exudates are extracted, the values are computed to determine the presence of DR hard exudates according to the proportion of area of a hard exudate using the crisp value of each hard exudate [9, 10]. Determination of image quality is an important part of digital image processing as many different types of noise degrades the quality of image. There are many different techniques to evaluate the quality of image [11]. The most commonly used technique is pixel-based difference measures which include peak signal-to-noise ratio (PSNR), mean squared error (MSE). Hereby working on real-time images and later adding noise(speckle, salt and pepper, Gaussian) to images and then calculating and comparing the PSNR and MSE value for different images. Sensitivity and specificity parameters are used to test performance of the proposed technique. These measurements are calculated using four parameters, namely the true positive (TP) rate (the number of exudate pixels correctly detected), the false positive (FP) rate (the number of non-exudate pixels erroneously detected as pixels of an exudate), the false negative (FN) rate (the number of exudate pixels not detected) and the true negative (TN) rate (the number of non-exudate pixels correctly identified as non-exudate pixels) (Fig. 1).

The four parameters used are the true positive (TP) rate (the number of exudate pixels correctly detected), the false positive (FP) rate (the number of non-exudate pixels erroneously detected as pixels of an exudate), the false negative (FN) rate (the number of exudate pixels not detected) and the true negative (TN) rate (the number of non-exudate pixels correctly identified as non-exudate pixels) (Table 1).

It is shown that average sensitivity and specificity obtained are 81.75% and 100%, respectively, in detecting exudates. It is clear that the proposed technique detects hard exudates using DR with accuracy of almost 99% in all images.

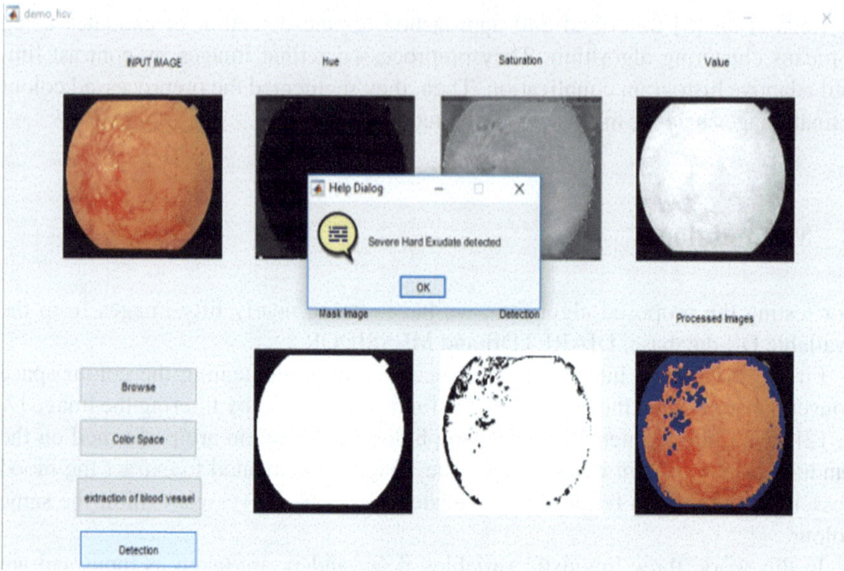

Fig. 1 Final processed image

Table 1 Performance in detecting exudates

Image	TP	FP	FN	TN	Sensitivity	Specificity	Precision	Accuracy
1	321	0	49	431 630	86.76	100	100	99.99
2	51	0	36	431 913	58.62	100	100	99.99
3	3680	0	1012	427 308	78.43	100	100	99.99
4	599	0	46	431 355	92.87	100	100	99.99
5	542	37	42	431 379	92.81	99.99	93.61	99.99
6	270	318	30	431 382	90	99.93	45.92	99.92
7	3304	787	391	427 518	89.42	99.82	80.76	99.73
8	2463	87	227	431 686	91.56	99.98	96.59	99.93
9	2585	0	1339	428 076	65.88	100	100	99.69
10	4090	255	452	427 203	90.05	99.94	94.14	99.84

4 Result Analysis

In this dissertation, there are totally 10 images which have been tested to detect the hard exudate. The quality of the image is also tested on the 10 images. The performance of the image quality assurance is also tested for the 10 fundus images. The average performance of the four fundus image is shown in below.

Figure 2 describes the comparison between the Gaussian and speckle noise for the fundus image for PSNR. The PSNR value is used to detect the noise ratio with respect to the image. By comparing both the algorithms, it is easily predicted that speckle noise has the highest dB (decibel) than the Gaussian noise. Although a higher PSNR generally indicates that the reconstruction is of higher quality, in some cases, it may not. One has to be extremely careful with the range of validity of this metric. Typical values for the PSNR in lossy image and video compression are between 20 and 50 dB, provided the bit depth is 8 bits, where higher is better. For 16-bit data, typical values for the PSNR are between 60 and 80 dB. Acceptable values for wireless transmission quality loss are considered to be about 20–25 dB.

Figure 3 describes the comparison between the Gaussian and speckle noise for the fundus image for MSE. Both the algorithms have minimum value with respect to error, but there is difference in points between the algorithms. The smaller value of the MSE represents the better results (Fig. 4).

SSIM is used for measuring the similarity between two images. The SSIM index is a full reference metric; in other words, the measurement or prediction of image quality is based on an initial uncompressed or distortion-free image as reference. SSIM is designed to improve on traditional methods such as peak signal-to-noise

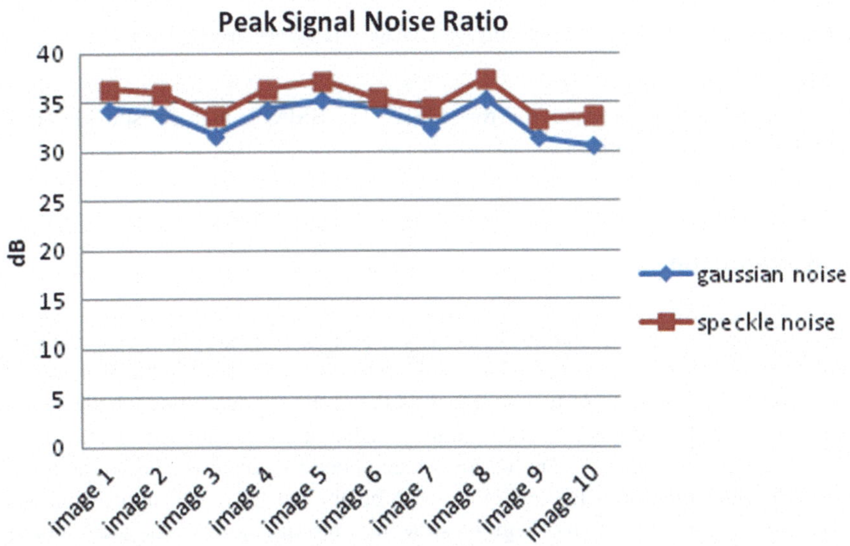

Fig. 2 Comparison between Gaussian and speckle noise for PSNR

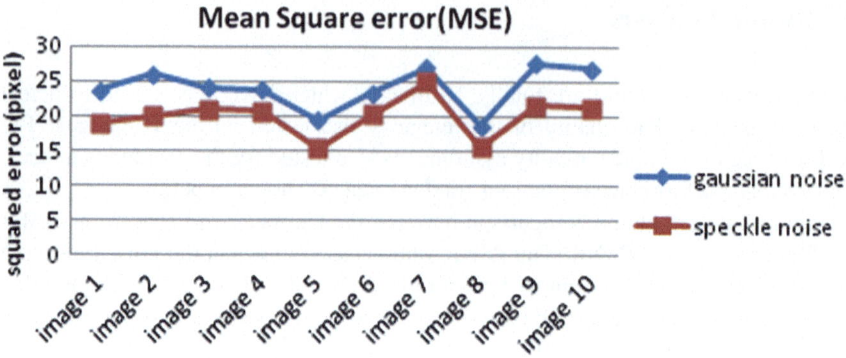

Fig. 3 Comparison of MSE value

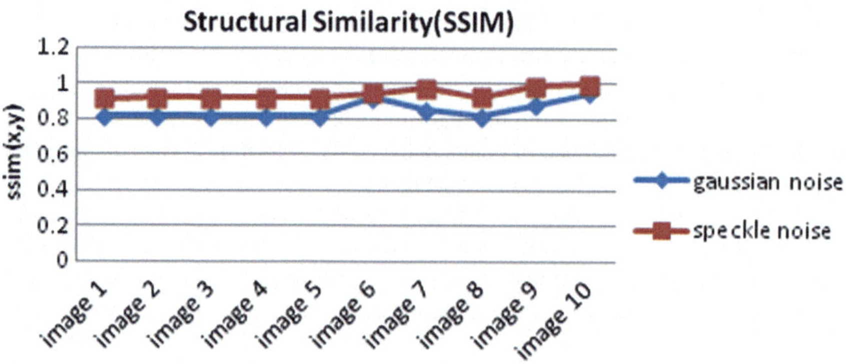

Fig. 4 Comparison for structural similarity between the original and dissorted image

ratio (PSNR) and mean squared error (MSE). The SSIM is high for speckle noise than that of Gaussian noise.

5 Conclusion

This research proposes a technique to identify exudates using morphological methods and categorize these exudates into hard and non-hard exudates using the principle of fuzzy logic. The strength of this approach is the ability to determine whether each exudates is hard exudates or not, separately. It has used the intensity band of the HSV image at this stage. As the fundus image generally contains a high amount of noise, different preprocessing techniques can be applied for noise suppression and enhancing features to equalize regions showing uneven contrast. The technique is fully automated and can be applied to a database of retinal images without changing any parameters during execution of the algorithm.

In the field of image processing, image quality assessment is a fundamental and challenging problem with many interests in a variety of applications, such as dynamic monitoring and adjusting image quality, optimizing algorithms and parameter settings of image processing systems and benchmarking image processing system and algorithms. The estimation of different image quality metrics (PSNR, SNR, MSE) is an essential task in digital image processing as it provides a better way for image quality assessment and its improvement. It can be observed that higher the value of PSNR and lower value of MSE are desired results. SSIM is a human visual system-based metric which uses the luminance, structural and contrast information present in the given image as like in HVS model. These validation results show the robustness, feasibility of the SSIM, and it can perform better than PSNR.

In future, in work to expand the detection system to recognize micro-aneurysms and haemorrhages, there may be a problem in separating the pathologies from small vessels. If the small vessels are missed during this step and are confused with micro-aneurysms or haemorrhages, it may be possible to combine more than one detection technique to make a final decision on the detected area being either a vessel or micro-aneurysm or a haemorrhage.

Images are taken from a public available database domain for the experiments presented in this dissertation. It is prudent, however, to verify this technique using other databases containing DR images. Also, the work needs to be improved to detect exudates pixels having very low intensity values.

References

1. Akara M (2009) Automatic exudate detection from non-dilated diabetic retinopathy retinal images using fuzzy C-means clustering. J Sens 9(3):2148–2161
2. Basha SS, Prasad KS (2008) Automatic detection of hard exudates in diabetic retinopathy using morphological segmentation and fuzzy logic. Int J Comput Sci Netw Secur 8(12)
3. Fraz M, Remagnino P, Hoppe A, Uddanonvara B, Rudnicka A, Owen C, Barman S (2012) Blood vessel segmentation metholodilogies in retinal image—a survey. Comput Methods Programs Biomed 108:407–433
4. Gardner GGI, Keeting D, Williamson TH, Elliott AT (1996) Automatic detection of diabetic retinopathy using an artificial neural network: a screening tool. Br J Ophthalmology 80:940–944
5. Reza W, Eswaran A, Hati S (2008) A quadtree based blood vessel algorithm using RGB components in fundus images. J Med Syst 32:147–155
6. Iqbal MI, Gubbal NS, Aibinu AM, Khan A (2006, October) Automatic diagnosis of diabetic retinopathy using fundus images, Masters Thesis, Blekinge Institute of Technology
7. Kavitha S, Duraiswami K (2011) Automation detection of hard and soft exudate in fundus image using colour histogram thresholding. Eur J Sci Res 493–504
8. Kumari VV, Narayanan NS (2010) Diabetic retinopathy—early detection using image processing techniques. Int J Comput Sci Eng 02(02):357–361
9. Mehdi A, Hamidreza M (2003) Locating the optic nerve in a retinal image using the fuzzy convergence of the blood vessels. IEEE Trans Med Imaging 22(8):951–958
10. Niemeijer M, van Ginneken B, Stall J, Suttorp-Schulten M, Abrámoff M (2005) Automatic detection of red lesions in digital color fundus photographs. IEEE Trans Med Imag 24:584–592
11. Jeyalaksshmi S, Prasanna S (2018) Measuring distinct regions of grayscale image using pixel values. Int J Eng Technol 7(1.1):121–124

12. Suseendran G, Chandrasekaran E(2017) A study on inventory strategies and optimum profit with python programming. J Adv Res Dyn Control Syst 15:592–596
13. Nathiya T, Suseendran G (2019) An improved HNIDS in cloud real time prediction using fuzzy decision making combination rule. Int J Recent Technol Eng 7(5S3):261–267

Evaluating External Public Space's Performance in the Cisadane Riverfront, Tangerang, Indonesia

Rahmi and A. H. Fuad

Abstract The Cisadane Riverfront is an external public space, where many people of different ages come to walk and sit around. Their visitation can impact its nature. This paper examined public space's performance in Kota Lama Tangerang toward outdoor activities, whether of locals or visitors. One of the important considerations notes that this area is located in a national heritage site of Indonesia. It is crucial for the government to improve this historical area because it was a zero point of development in Tangerang city. The research involved drifting, site observation, and literature review. Using variables of the Good Public Space Index, these data were measured. The results indicated that Green Promenade had higher performance than the other two destinations. This paper conducted the causal aspect of external public space's performance in the Cisadane Riverfront as a consideration for its future revitalization.

Keywords External public space · Performance · Riverfront · Outdoor activity · Sustainability

1 Introduction

Humans need a facility to accommodate their movements while performing activities. Revitalizing Cisadane Riverfront is an effort to maintain public space in the Kota Lama (Old City) of Tangerang. The Cisadane Riverfront is one of the places for migrating birds every year that fly from the Thousand Islands (North Jakarta) to Depok (North–South) and Bekasi to Tangerang (East–West) [1]. This phenomenon is one of the attractions of the Cisadane Riverfront. The presence of green open spaces along the river can also improve river water quality [2]. The river, located ten kilometers from the mouth of the Java Sea, has its own ecosystem, and many residents also maintain the environment as their livelihood. Hence, there is an interrelationship that occurs between humans, animals, plants, and water in these public spaces. The

Rahmi · A. H. Fuad (✉)
Universitas Indonesia, Depok, Indonesia

S.-L. Peng et al. (eds.), *Intelligent Computing and Innovation on Data Science*, Lecture Notes in Networks and Systems 118,
https://doi.org/10.1007/978-981-15-3284-9_60

Table 1 External public space at Cisadane Riverfront and its character

External public spaces	Facilities	User	Surrounded building
Gajah Tunggal Park	Children playground, nature, food court, plaza	Kid, teenager, adult	Shopping mall, school, university, mosque
Cisadane Flying Deck	Viewing deck, charging spot, nature	Teenager, adult	Kampung Bekelir
Green Promenade	Children playground, nature, local culinary, fishing spot	Kid, teenager, adult	Benteng Herritage Museum, historical Kalipasir Mosque, Boen Tek Bio Temple, mini-river port

users' outdoor activities and public space should be managed to maintain sustainability. Public space's performance is evaluated based on its associated activities as a destination [3]. In this paper, the activities are connected to their characteristics to identify their correlation with the public space's performance. This paper aims to evaluate the public space's performance toward outdoor activities in the Cisadane Riverfront.

2 Literature Review

2.1 *Public Space*

The Cisadane Riverfront is an external public space [4] because all people can enter the open space free of charge, and the place itself is not managed by a private company. There were three destinations in Cisadane Riverfront that were examined (Table 1), namely Gajah Tunggal Park, the Cisadane Flying Deck, and Green Promenade. Each destination has different characteristics, which emerge from the facilities that are provided there. Based on its function, public space is categorized into positive, negative, ambiguous, and private space [5].

2.2 *Outdoor Activities*

There are three types of outdoor activities: necessary, optional, and social [6]. They depend on the intention of the user. In a riverfront, possible activities are limited by regulations. The user cannot perform activities that leave a negative impact toward nature and its habitat. The outdoor activities in public space are also classified into three types: process, physical contact, and transition [7].

3 Method

The Cisadane Riverfront at Kota Lama Tangerang has three destinations (Fig. 1), which are Gajah Tunggal Park, the Cisadane Flying Deck, and Green Promenade, which is a green belt along Kalipasir Street. Each destination has different characteristics. Gajah Tunggal Park consists of several active elements, such as a playing ground, food corner, mosque, and two-meter pedestrian paths in the riverside. The Cisadane Flying Deck consists of passive elements, as it only has decks along the riverside and a charging booth. In comparison, the Green Promenade along Kalipasir Street has both elements. It has pedestrian pathways, fishing spots, water contact spots, and local street food vendors.

These parks were examined for one month by observing and asking some locals and visitors. Many activities were seen to happen during the research. Tangerang is about 18 meters above sea level [8]. Meanwhile, when the rainwater flows from the higher city to the Cisadane River, it will affect some cities along the river, as their activities are implemented in public spaces. Some variables [3] were used to measure the public space's performance: intensity of use, intensity of activity, duration of activity, variation of use, and variation of users. Using a mix of methods, the data was analyzed. Scoring was used to measure the public space activities, while open questions were asked regarding the kinds of common user activities around those public spaces.

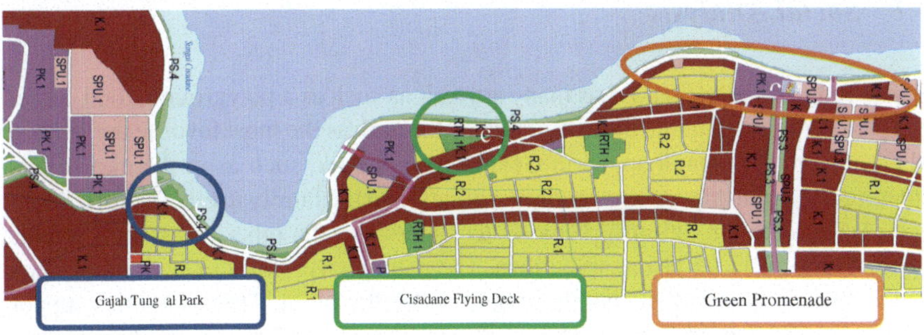

Fig. 1 Location of Gajah Tunggal Park, Green Promenade, and Cisadane Flying Deck. *Source* Local Government of Kota Tangerang

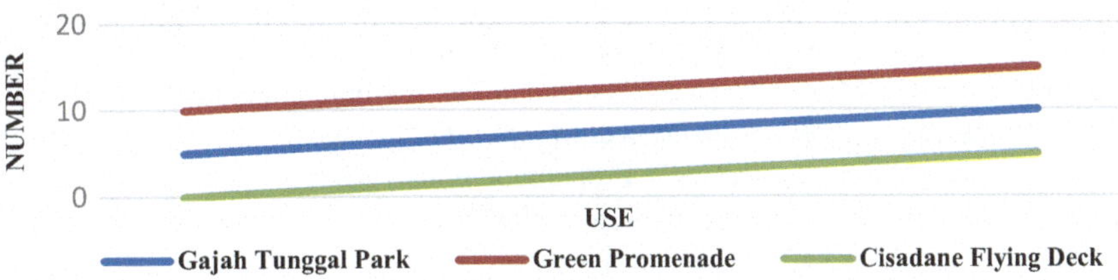

Fig. 2 Variation of use of public space in Cisadane Riverfront, Kota Lama Tangerang

The data was evaluated using social and spatial analysis [9]. Social analysis considered how people act in this public space and its impact on the socioeconomic condition. Spatial analysis considered how this public space contributes to fulfilling people's needs and making an impact toward spatial growth.

The object of observation was the users of open public space, including locals and visitors. Locals were limited to users who stay within a radius of one kilometer from the open public space, and visitors constituted users who originate from outside a radius of one kilometer from the public space.

4 Results and Discussion

According to this classification (Table 2), the results showed that Green Promenade had the highest performance. Gajah Tunggal Park had the second highest performance, and the Cisadane Flying Deck had the lowest.

There was uncommon condition appeared. The all-day usability in Green Promenade described Green Promenade in high intensity of using, while the fishing activity had made it. Fishing spot in Green Promenade attracted user in long duration. While, with the higher number of user, voice could disturb the fishing activity.

4.1 Social Analysis

Gajah Tunggal Park (Fig. 3) has many attractions such as a playground, food corner, and praying room, but this did not guarantee the park as the most favorite destination. Many interactions may happen in Gajah Tunggal Park, such as in the playground, creating a relationship among children, parents, and the family itself. The food corner and plaza create a relationship among teenagers. On the other hand, the Cisadane Flying Deck provides visual attractions. The natural river flows in front of the deck, and another spot here is a charging booth and selfie point. These facilities support adolescent and couple visitors. The results saw that Green Promenade (Fig. 4) had a higher performance than the other two destinations. More activities take place in the promenade (Fig. 2), so it attracts people to come throughout the day. Given its total of 10 h of peak visitation, Green Promenade achieved the best public space performance in the Cisadane Riverfront. This might be the result of certain activities, such as fishing, sitting, and eating. Physical contact activities can increase the performance of public spaces [7]. Moreover, Green Promenade serves many local Tangerang foods, which attracts many people from noon until night. They come to enjoy the Tangerang culinary spots in the riverside. Green Promenade is also greener than other places, so visitors and locals can enjoy the area under the shadows of the surrounding foliage.

Table 2 Public space measurement toward users' activities

Performance	Destination	Activity					
		Gehl			Zhang and Lawson		
		2011			2009		
		Necessary	Optional	Social	Process	Physical contact	Transition
H	Green Promenade	L	V	M	M	M	
A	Gajah Tunggal Park	L	M	M	M	M	M
L	Cisadane Flying Deck		M	M	M		

Performance	Destination	Public space								
		Mehta					Carmona			
		2007					2008			
		Intensity of use	Intensity of activity	Duration of activity	Variation of use	Variation of user	Positive	Negative	Ambiguous	Private
H	Green Promenade	All day	25–50 users in 1 h	10 h	>10	Yes	Yes	no	Yes	No
A	Gajah Tunggal Park	Afternoon	25–50 users in 1 h	3 h	5–10	Yes	No	No	Yes	No
L	Cisadane Flying Deck	Afternoon	<25 users in 1 h	3 h	<5	No	No	Yes	Yes	No

Note L = Local, V = Visitor, M = Mixed, H = High, A = Average, L = Low

Fig. 3 Location of Gajah Tunggal Park. *Source* https://www.topibambu.com/2018/06/wisata-taman-gajah-cisadane-tangerang.html

Fig. 4 Location of Green Promenade. *Source* Rahmi

4.2 Spatial Analysis

The Kota Lama Riverfront had different user characteristics, and those three areas had their own types of visitors. This is caused by the surrounding environment. For example, Gajah Tunggal Park is placed near an education center and mall. Most of its visitors comprise teenagers and families who come to enjoy the park attractions. Next, the Cisadane Flying Deck (Fig. 5) is situated near the home waste recycling

Fig. 5 Location of Cisadane Flying Deck. *Source* Rahmi

industry. This industry generates a negative impact toward the flying deck, which offers only visual attractions, as it is partly disturbed by the presence of the home waste recycling industry. Lastly, the promenade is surrounded by several heritage buildings, including the Kalipasir Mosque (since the 1400s), Boen Tek Bio Temple (since the 1600s), and Walet Ranch (since the 1800s). These heritage buildings were part of the reason for Kota Lama's nomination as one of Indonesia's heritage sites [10]. Spatially, Green Promenade is also adjacent to the Tangerang rail station, and this is the reason why visitors to the promenade come from many places, whether local or international. Green Promenade also provides spots for direct contact with water. A mini-port is available for the annual traditional Chinese ceremony. The mini-port was the zero point of development when the Chinese came to Tangerang in the 1600s. They arrived and built several buildings, bringing their culture to the area. This historical background creates a specific spatial characteristic in the promenade environment. Thus, the promenade also reached the highest performance because of some aspects such as water contact spots, nearby rail station, and nearby heritage buildings. Physical setting, imagery, and activities should be considered in creating identity and giving a sense of public space [11].

Overall, Green Promenade was highly visited because it offers physical contact with nature and traditional or local cuisine. It means that the physical nature and local cuisine create a better impact toward the users' intention in the Cisadane Riverfront. Moreover, Green Promenade is shaded by many trees, providing users with more comfort at noon. Green Promenade also had inclusiveness, meaningfulness, safety, comfort, and pleasure, which are important to the quality of public space [12]. Comfort is more important for the riverfront area because of factors such as humidity, sunlight, and wind. Under the shadow of a tree in Green Promenade, users can enjoy nature anytime. Additionally, urban open spaces should also provide a place for strangers to meet [13].

Green Promenade is a culinary spot in the city of Tangerang. Various local foods of Tangerang are served, such as durian soup, *es doger*, cilok, and others. The local food becomes a potential aspect of the community, so the economy runs positively due to users coming to the area. Green Promenade has some sitting spots that are provided by street vendors. Unfortunately, the street vendors themselves have not been well organized. There is a tendency for them to use parts of the street, so this often causes traffic congestion, especially after working hours. The culinary time operates from 11:00 to 21:00. The 10-h duration (Fig. 6) operates by involving 20 street vendors along Green Promenade. This will continue to grow and become crowded if not managed as soon as possible, especially considering that the people of Tangerang celebrate the Cisadane River Festival and Dragon Boat Competition each year at that location. Street vendors may receive benefits from users who come to the celebration. Social interactions among community members in this area will strengthen their sense of solidarity [9].

It should also be noted that not all users in public spaces intentionally seek recreation and sports, where these activities are optional for them [4]. However, public space planning must also attract visitors who have a need to go there, such as street

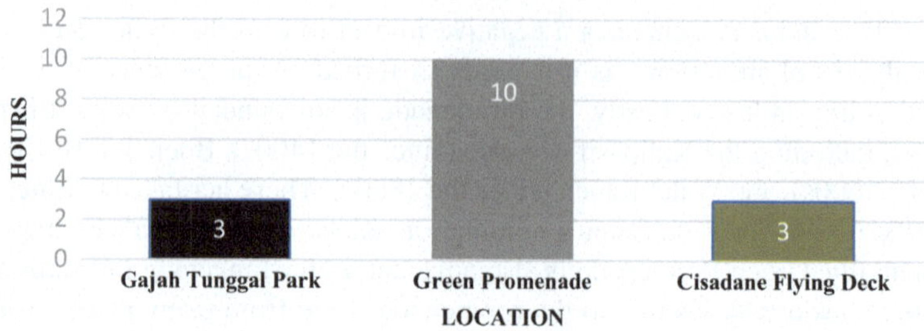

Fig. 6 Duration of activities

vendors, parking attendants, and garden cleaners. There also needs to be a utility zone for those who visit the area for more than six hours.

The existence of a public space's composition has an impact on its environment, both negative and positive. With the human invasion and disturbance of nature, the theory of urban ecology is needed [14]. To prevent city development from becoming too complicated to control, the government, academics, and practitioners must make regulations in urban planning by implementing urban ecological theory. Thus, the movement of people in urban areas can maintain the existing ecosystem as much as possible. Development will be carried out as a form of fulfillment of needs, and it must be in accordance with environmental ethics and in consideration of the cycle that occurs in the built environment. Damaged ecosystems will inevitably exist, but the damage should be minimized. If humans want to preserve the natural ecosystem, it is necessary to consider existing ecosystems. It is also possible to form one, although an artificial ecosystem is very reliant on certain conditions because it depends on the nature where the ecosystem is recognized, so it cannot be generalized. Thus, it is necessary to look further into the history of ecosystems that are formed in certain zones [15]. For instance, the riverfront is a site of river ecosystems.

5 Conclusion

The public space's performance in this paper is related to many aspects. Green Promenade, the space with the highest level of performance, was frequented by visitors, even though it only has optional activities and no transition facilities. This shows that Green Promenade has succeeded in strongly attracting visitors and distinguishing itself by enabling physical contact with nature. By nature, the range of visitor's ages becomes wider.

The public space's performance is intended to give a point of consideration for the Local Government of the city of Tangerang in managing public spaces. Social movement inside these public spaces is affected by floods. Based on the causal aspects that were analyzed above, the government or other planners should examine people's preferences regarding public spaces in Kota Lama Tangerang.

This study is limited only to the riverfront public space. For further research, it is essential that the surrounding environment of the public space is also examined and another relevant variable should be involved to improve the measurement in certain condition like fishing activity in this research.

Acknowledgements The authors wish to give thanks to the Hibah PITTA 9 of Universitas Indonesia for financial support and the Local Government of the city of Tangerang for data collection.

References

1. https://metro.tempo.co/read/410628/jakarta-bangun-koridor-hijau-migrasi-burung
2. Rottle N, Yocom K (2011) Basics landscape architecture: ecological design. AVA Publishing, Switzerland
3. Mehta V (2007) Performance measures of public space. In: 43rd Isocarp Congress
4. Carmona M, Heath T, Oc T, Tiesdell S (2003) Public space—the dimensions of urban design. Architectural Press, USA, p 111
5. Carmona M, de Magalhães C, Hammond L (2008) Public space: the management dimension. Routledge, London, New York, p 62
6. Gehl J (2011) Life between buildings: using public space, vol 6. Van Nostrand Reinhold Company, New York
7. Zhang W, Lawson GM (2011) Meeting and greeting: activities in public outdoor spaces outside high-density urban residential communities. Urban Des Int 14:207–214
8. https://allaboutangerang.blogspot.com/2016/03/geografis-kota-tangerang.html
9. Bonenberg W (2015) Public space in the residential areas: the method of social-spatial analysis. Procedia Manuf 3:1720–1727
10. Irfan (2015) https://banten.antaranews.com/berita/24171/kota-tangerang-wakili-banten-masuk-kota-pusaka
11. Shawket MI (2018) Identity in urban spaces of residential compounds: contributing to a better environment. Hous Build Nat Res Center J 14:235–241
12. Mehta V (2004) Evaluating public space. J Urban Des 19:53–88
13. Thompson CW (2002) Urban open space in the 21st century. J Landsc Urban Plann 60:59–72
14. Niemelä J (1999) Is there a need for a theory of urban ecology? Urban Ecosyst 3:57–65
15. Grose M (2016) Constructed ecologies. Routledge, Melbourne

A Comparative Study of Cryptographic Algorithms in Cloud Environment

M. Revathi and R. Priya

Abstract Cloud is a medium for providing virtual services and resources that can be accessed and utilized easily. Along mechanism in Internet storage content is saved online, retrieved from various distributed and interconnected resources which form a cloud across the network. The prime concern faced by the Internet content system securing, as a result the owner encrypts and utilizes an search technique for accessing the data. When using a cloud storage mechanism, the third party gains the control to import your highly secured and private data to put up in the cloud. The data owners have to deal with various security issues because of entailing intense safety and confidentiality. The present paper outlines the cloud types based on cloud usage and services offered. The focus of the research being to provide secured data sharing by comparing and assessing the functionality of different symmetric encryption algorithms like parallel homomorphic encryption algorithm (PHEA), KUNodes, triple data encryption algorithm (3DEA), Also by implementing PHEA there is flexibility in data sharing, fine-grained access control, improvised average execution time and memory and data size. The research work concludes that PHEA is highly suitable for data sharing compared to other techniques of symmetric encryption.

Keywords Security · Cryptography · Symmetric · Asymmetric · Encryption · Decryption · Fine-grained access control

1 Introduction

One of the demanding and popular Internet technologies prevailing today is the cloud computing. It is a huge zone holding data and resources that can be accessed and shared anytime from anywhere. Cloud offers the benefit of sharing computer

M. Revathi (✉)
PhD Research Scholar, Department of Computer Application, Vels University, Chennai, India
e-mail: vm_revathi@yahoo.co.in

R. Priya
Associate Professor, Department of Computer Application, Vels University, Chennai, India
e-mail: priyaa.research@gmail.com

S.-L. Peng et al. (eds.), *Intelligent Computing and Innovation on Data Science*,
Lecture Notes in Networks and Systems 118,
https://doi.org/10.1007/978-981-15-3284-9_61

resources. Since the user is mainly concerned with security, providing utmost safety while sharing data in a cloud environment is of prime concern. To share safe and secured data among sender and receiver, various symmetric encryption algorithms are imbibed. The cloud is primarily responsible for managing and situation and theft of data and that is safely transmitted to other users. The paper assesses various existing researches associated with methodologies and resolutions concerning secure data sharing. The focus of the research being to provide secured data sharing by comparing and assessing the functionality of different symmetric encryption algorithms like parallel homomorphic encryption algorithm (PHEA), KUNodes, triple data encryption algorithm (3DEA). For the purpose of data sharing, the existing paper considers and discusses parallel homomorphic encryption algorithm (PHEA), making the cloud data authorization easy, yielding in high data security and efficiency. These methodologies guarantee that if by any chance an unauthorized user retrieves the cipher text, the plaintext remains strictly inaccessible. The KUNodes algorithm deals with security along with identifying entire set of participants in the entire system. During process of data sharing, there is participation of multiple nodes. For enhanced and secure data sharing, the re-encryption technique is imbibed in a cloud environment. The AES encryption algorithm involves the whole data block utilizing one square matrix through every round by utilizing permutation and substitutions. The 3DES-triple data encryption algorithm exists as a replacement against DES because of its improvised key length The name '3DES' as implemented. Lastly, ECC-elliptic-curve cryptography algorithm is utilized for building instantiate public key cryptography conventions for the purpose of digital signatures and instance executing keys. All the symmetric algorithms mentioned and discussed are being implemented in public cloud.

2 Related Work

Arockia Panimalar et al. for service-based architecture the approach of cloud computing is employed. To make best use of cloud computing, the responsible individuals must incorporate appropriate parameters to guarantee security. ECC-elliptic-curve cryptograph is imbibed in almost every communication domain, be it mobile computing, server-based encryption, remote sensor systems and image encryption. Cloud computing alongside elliptic-curve cryptograph is a new approach and possesses massive scope of research. The sole target is offering data security with ECC to assure secrecy and data confirmation amidst clouds [1].

B. Thiyagarajan et al. put forths that when an organization/business is uploaded on cloud it portrays a genuine concern regarding the security and integrity of client data and organization officials too. The prime concern remains safeguard processes. For that outsourced algorithms like AES, assures to be efficient. These algorithms are beneficial as they permit the user to restrict any unauthorized personal to access or retrieve user data [2].

BurhanUl Islam Khan et al. put forwards the fact that the process of data sharing within cloud environment is vulnerable to various attacks and threats. Using the traditional technique of cryptography is inefficient in delivering desired and effective authentication requirement. To resolve this issue of data security, SSM—secure-split-merge scheme is imbibed. This scheme employs a unique strategy utilizing an encryption key (AES 128 bit) to split the data. These blocks of splits are then managed on different server racks among various cloud zones [3].

Cheng-Kang proposes in paper an algorithm and i.e. KAC—key aggregate cryptosystem. A fixed sized cipher text is generated by utilizing new public key cryptosystem in order to achieve better flexibility, safety and efficiency while data sharing. Key aggregate cryptosystem is imbibed for cryptography using the key aggregate encryption-decryption algorithm. The disadvantage being, high cost value [4].

Kaitai Liang et al. present research work on functional PRE—proxy re-encryption scheme (i.e. DFA-based) for securely public. PRE being encrypted data converted to cipher text that is linked to a different string via, which been given the re-encryption key. In any case, plaintext cannot be retrieved by the proxy. Functional proxy re-encryption (PRE) algorithm which is DFA-based is being imbibed, yielding in high cost [5].

Kamalakanta Sethi et al. propose the technique of homomorphic encryption in which there is need to decrypt the cipher text and one can directly conduct operations on cipher text. For standardized computations to be performed on cipher text, various efficient techniques are being presented and employed by utilizing the 'DGHV scheme-2010' which stands for 'The KGS—Keg Generation Server' being used to perform 'cipher text refresh' procedure. Moreover, for a relative performance study, a parallel implementation of the above algorithms is exhibited [6].

Kanagavalli Rangasami et al. put forwards that data security incorporates network security, control strategies, service access and data storage. In the technique of homomorphic encryption, data security is offered in a way that operations could be performed on encrypted data itself [7].

Mohamed Nabeel et al. propose the scheme of privacy preserving delegated access control (PPDAC) employed by decomposition ACPs within cloud environment. The data is made invisible in the cloud and the owner needs to manage few attribute factors only. Though still it faces the drawback of network connections dependency and high cost. Algorithms being employed are gen graph, policy decomposition, optimization and random cover [8].

N. Hemalatha et al. investigated the significance related to scenario. The technique of cryptography symmetric encryption is being selected as it is efficient in managing massive data along with remarkable speed linked to storing data in the cloud. An overall outlook of cloud technologies is exhibited along with necessary characteristics, delivery models, classifications and many encryption algorithms. To maintain the cloud confidentiality, a comparative study based on various encryption algorithms is imbibed [9].

Nidhi Grover proposes that in a cloud environment various encryption algorithms are being employed to safeguard user data from malicious attacks and threats. Few techniques/algorithms that are presented in the paper include DES, AES, blowfish and RSA. Also the mentioned algorithms are being compared in contrast with multiple attributes that must be an integral part of an encryption algorithm. There are drawbacks, benefits and shortcomings related to every algorithm. The recommended AES algorithm yields in enhanced data security, speed, improved performance, scalability and minimal memory prerequisite when compared with rest other existing algorithms [10].

3 Proposed Work

3.1 Overview

In a cloud computing environment, application platforms and computer resources are widely dispersed across the Internet via demand and pay principal. Unfortunately, this private information when stored and shared online is unprotected by any physical limitations and this leads to security issues. The technique of cryptography proves valuable and efficient in promising secure data sharing of information among diverse entities by filtering and transmitting only the incomprehensible information and strictly permitting data access to authorized users only. For safe communication, selecting the appropriate cryptographic algorithm is mandatory which can offer efficiency, security and integrity.

3.2 Encryption Technique

To safeguard confidential information, encryption techniques are being enforced. These are categorized as following:

Symmetric Key Encryption This employs a single unique process as well as decryption.

Asymmetric Key Encryption Here two utilized which restricts any unauthorized data access. That is, the receiver has the private key that remains unshared and the public key stays public to all. Technique of symmetric encryption is quite fast in the process of encryption and also it consumes very less computing power. The above encryption process works cipher and splits entire data into chunks of blocks. Asymmetric encryption public key algorithm proves to be effective in securely transferring data and encryption keys in situations where in either users does not have chance to decide or approve on a secret key. In public key encryption algorithms, the keys that are being utilized are lengthier, thus the transmitted data is much secure.

Following symmetric encryption algorithms are examined in below section: KUNodes, 3DEA-triple data and PHEA-parallel homomorphic encryption algorithm. PHEA is quiet effective in handling huge volume of data.

KUNodes

In this, multiple nodes are involved during the process of data sharing which are being: key authority, number of users involved, data provider and storage server. In cloud computing, the data needs to be shared securely; hence, every member is well connected/linked or say dependent on one another. Key matching is mandatory for security purpose. A key is generated by the key authority which is issued to both—the data provider and the user for securely sharing data in the cloud.

Advantages:

1. Encryption time is less compared to 3DES.
2. Prevention from data theft.

Disadvantages:

1. Designing is tough.
2. Loss of data since data sharing is from node to node.

Triple Data Encryption Algorithm (3DEA)

3DES—Triple data encryption standard also mentioned as TDEA-triple data encryption algorithm. Initially, when DES algorithm was designed, key length of 56 bits was appropriate and sufficient but it was prone to attacks by mere physical force. 3DES offers a basic technique of incrementing key length rather than creating an entire block cipher, also it safeguards from brute-force attack. 3DES utilizes 168 bit key (which being total of three separate keys), whereas DES utilizes 56 bit key. In 3DES, either all three keys are same or might be first and third are same. Furthermore, the text is split into 64 bit block, utilizing 8 S-boxes and conducting 48 rounds of processing. 3DES is highly complicated and framed to safeguard the data from various attacks. The '3DES' as the data undergoes DES encryption three times.

Advantages:

1. Higher computational speed.
2. Fast and quick encryption speed.

Disadvantages:

1. Extremely high memory usage.
2. Involves 48 encryption rounds.

Elliptic-Curve Cryptography algorithm

Elliptic-curve cryptography algorithm is utilized for building instantiate public key cryptography conventions for the purpose of digital signatures and instance executing keys. The main motto and inspiration in employing ECC being that it utilizes more keys of small size along with set of executions. ECC allows encrypting a piece of information many number of times with entirely separate keys so that the resultant cipher text undergoes decryption in one go and with only single key. ECC supports both cryptography as well as decoding. Ultimately, a cipher text is generated which undergoes decryption using a single key with mere one decoding cycle.

Advantages:

1. Results in high security be employing-164 bit key.
2. Low power consumption and offers enhanced functionalities to the batteries.

Disadvantages:

1. Is quiet complex and size of encrypted message is increased.
2. Implementation is tough.
3. Suffers from high encryption time.

Parallel Homomorphic Encryption Algorithm (PHEA)

PHEA is a form of cryptography scheme where algebraic operations are on the cipher text and plaintext. As a result, a chunk of cipher text is mutually produced by various parties remaining unaware of the plaintext worked upon by others. A different technique of parallel homomorphism encryption is presented in the paper. This encryption algorithm examines the algebraic link among the algebraic operations on the cipher text as well as plaintext. According to the paper, link amidst algebraic operations performed on the cipher text as well as encryption keys are examined. Simultaneously, achieving the fine-grained access control is making cloud data sharing flexible enough.

Advantages:

1. Faster encryption process.
2. Yielding in high efficiency.
3. No loss of data.

3.3 *Data Security*

Data security signifies safeguarding data against unauthenticated access, alteration or damage. Threats linked with data security falls under two categories: internal and external threats. Internal threats arise from within attacks as the CSPs and cloud users are purely responsible for this. Attacks from an external source are defined as external threats since any third party can also access the data by robbing private data of the user (Fig. 1).

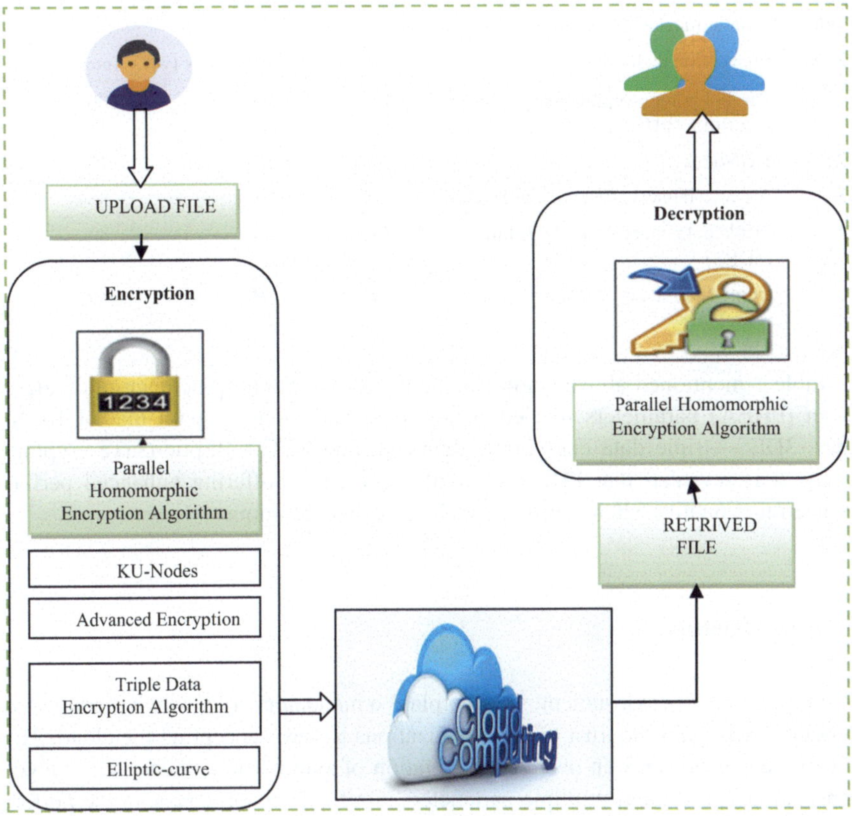

Fig. 1 Design of architecture

4 Result and Discussion

The presented work investigates the security concerns related outsourced being public. To achieve this, PHEA is employed which ascertains security of outsourced data, the request is forwarded to the data owner. The user is checked for authorization by the data owner and the decryption is key then granted to the user. By employing this technique in public cloud, leakage of confidential data from unauthenticated person can be avoided. Both the data user and owner must ascertain that the cloud is protected from any sort of external threats in order to ensure client's safety against data loss and theft. PHEA-parallel homomorphic encryption algorithm outperforms in the realm of authorization, data security and integrity. To obtain data security, data is being added to the third party. For implementing suitable symmetric encryption algorithm, there should be complete awareness regarding performance, strength and shortcomings taking various parameters into consideration. The recommended PHEA technique yields in efficient outcome, considering the following factors: integrity, average execution time, block size, efficiency and memory usage.

Table 1 Comparative analysis of overall performance

S. No	Encryption techniques	Time (ms)	Efficiency (%)	Integrity (%)
1	Parallel homomorphic encryption algorithm (PHEA)	1.13	96	100
2	KUNodes	2.56	85	70
3	Advanced encryption standard (AES)	1.41	91	85
4	Triple data encryption algorithm (3DES)	3.63	88	82
5	Elliptic-curve cryptography (ECC)	3.8	89	89

Table 1 mentioned above compares the parallel homomorphic encryption algorithm (PHEA) parameters like efficiency, time and integrity with that of KUNodes, 3DES—triple data encryption algorithm and ECC—elliptic-curve cryptography. It is revealed that PHEA is highly efficient in offering enhanced performance/functionality when compared with rest other techniques.

5 Conclusion

In Internet system environment, security plays a mandatory role, which if addressed genuinely will gain the trust of the organizations towards accepting the cloud. The existing paper presents an overall investigation of symmetric key encryption algorithm taking in account multiple parameters. On the basis of assessing the performance, the outcome reveals that PHEA , KUNodes, 3DES—triple data encryption elliptic-curve cryptography (ECC) delivering better security by considering availability of resources. The comparison done and the analyses reveals that PHEA ascertains to be more effective, fast, secure and a low storage consumption algorithm with enhanced encryption functionality, with absence of any shortcomings and restrictions compared to rest of the encryption algorithms which faces few drawbacks and issues in storage and performance. Eventually, PHEA is evaluated further taking into account various factors.

References

1. Arockia Panimalar S, Dharani N, Pavithra S, Aiswarya R (2017) Cloud data security using elliptic curve cryptography. Int Res J Eng Technol 4:32–36
2. Thiyagarajan B, Kamalakannan R (2014) Data integrity and security in cloud environment using AES algorithm. IEEE
3. Khan BUI, Olanrewaju RF, Baba AM, Lone SA, Zulkurnain NF (2015) SSM: Secure-Split-Merge data distribution in cloud infrastructure. In: ©IEEE, conference on open systems, 2015, pp 40–45

4. Chu C-K, Chow SSM, Tzeng WG, Zhou J, Deng RH (2014) Key-aggregate cryptosystem for scalable data sharing in cloud storage. IEEE Trans Parallel Distributed Syst 25(2):468
5. Liang K, Au MH, Liu JK, Susilo W, Wong DS (2014) A DFA based functional proxy re-encryption scheme for secure public cloud data sharing. IEEE Trans Inf Forensics Sec 9(10):1667
6. Sethi K, Majumdar A, Bera P (2017) A novel implementation of parallel homomorphic encryption for secure data storage in cloud. In: ©IEEE, international conference on cyber security and protection on digital service, 2017
7. Rangasami K, Vagdevi S (2017) Comparative study of homomorphic encryption methods for secured data operations in cloud computing. In: ©IEEE, 2017 international conference on electrical, electronics, communication, computer and optimization techniques, pp 551–556
8. Nabeel M, Bertino E (2014) Privacy preserving delegated access control in public clouds. IEEE Trans Knowl Date Eng 26(9):1200
9. Hemalatha N, Jenis A, Donald AC, Arockiam L (2014) A comparative analysis of encryption techniques and data security issues in cloud computing. Int J Comput Appl (0975–8887) 96(16):1–6
10. Grover N (2014) A comparative-study of various data encryption techniques in cloud computing. Int J Comput Sci Technol 5:163–168

Discovering the Androgen Transition and Prognostic Cardiovascular Disease by Hybrid Techniques in Data Mining

A. Revathi and P. Sumathi

Abstract In ongoing decades, coronary illness has been recognized as the main source of death in the world. In any case, it is considered the most preventable and controllable illness simultaneously. As demonstrated by the World Health Organization (WHO), the early and helpful finish of coronary sickness accept an astonishing activity in checking its empowering and lessening related treatment costs. Considering the regularly expanding development of coronary illness prompted fatalities, specialists have embraced various information mining methods to analyze it. As indicated by results, the use of similar information mining strategies prompts various outcomes in various datasets. This examination attempts to help human services masters to early analyze coronary illness and evaluate related hazard factors. The cardiovascular have been easily detected by the techniques of Fuzzy, CFS—Co-related feature selection, ZeroR in the hybrid processing. Fuzzy used to detect the processing of facts and produce the nearest possibilities of cardiac disease. CFS has revealed the process of prediction and constructed the variables used to FS—feature selection and VS—variable selection, respectively. ZeroR used to detect negative processing in the cardiovascular.

Keywords Fuzzy · ZeroR · CFS—Co-related feature selection · FS—Feature Selection · VS—Variable Selection

1 Introduction

The social security industry accumulates enormous extents of coronary affliction information which, heartbreakingly, are not "mined" to find camouflaged data for unimaginable fundamental specialists. The diminishing of blood and oxygen supply

---

A. Revathi (✉)
Department of Computer Science and Applications, Vivekanandha Arts and Science College of Women, Sankari, Salem, India

P. Sumathi
PG and Research Department of Computer Science, Government Arts and Science College (Autonomous), Coimbatore, India

S.-L. Peng et al. (eds.), *Intelligent Computing and Innovation on Data Science*, Lecture Notes in Networks and Systems 118,
https://doi.org/10.1007/978-981-15-3284-9_62

to the heart prompts coronary illness. This research has to give a prediction and detection process of current methods of information disclosure in databases utilizing information mining procedures which will be productive for remedial specialists to take a compelling choice. The goal of this appraisal work is to anticipate much more unquestionably the closeness of coronary ailment with a diminished number of characteristics. From the start, thirteen properties were identified with imagining coronary disease. Each datum mining technique fills another need dependent upon the showing objective. Sincere classifiers have worked splendidly in various eccentric veritable conditions. Henceforth, the adages of Bayes control are preparing behind the AI component separately. At that point, it has the procedure of some reasonable constraint to control it. It uncovers the best throughput of the medicinal services' framework inserting ideas. It was determined to give the best answer for social insurance preparing which aids the IoT necessities. Regardless, botches for the presumption in the control handling to the opportunity and the nonappearance of open probability data. Observations exhibit that hypothesis performs reliably when a reduction in different characteristics.

Heart disease (CD) is one of the significant reasons for death. A significant assignment is to distinguish the cardiac malady in all respects minutely and decisively. Generally, restorative symptomatic mix-ups are dangerous and excessive. By and large, they are provoking passing. Information mining procedures are imperative to limit symptomatic blunders just as to improve the patient's security. Information mining methods are extremely successful in structuring a therapeutic emotionally supportive network and advance the capacity to decide the concealed examples and relationship in clinical information.

2 Related Works

Chou et al. proposed a privacy-preserving compressive analysis (PPCA) technique at a low-intricacy system dependent on the subspace-based portrayal. It plays out the encryption and unscrambling on compressive sensing (CS) on the Web and isolating the sign in disconnected [1]. Mortazavi et al. depicted the postoperative intricacies expectation framework to extricate the information from EHR. Accomplice explicit models foresee the disappointment causes and contaminations [2].

Din et al. depicted an unmistakable comprehension of the Internet of medical things (IoMT) with machine learning (ML). This technique used to distinguishing the eHealth care framework to counteract basic infection and analyze it [3]. Amin et al. proposed seven distinctive classification procedures in the coronary illness expectation model. It improves the precision to analyze cardiovascular illness [4]. El-Bialy et al. depicted a fast choice tree and pruned C4.5 tree are used for the near of separated information from various informational collections. It is resolved to apply the mix to the AI calculations for finding the CAD infection [5].

Kavitha et al. proposed principal component analysis (PCA) to accomplish the arrangement precision with the numerical model. It evacuates excess and conflicting information to improve precision and decrease the dimensionality from high to low [6]. Lou et al. proposed an RFRS highlight choice framework and an arrangement framework with an outfit classifier with three phases. In a subsequent framework used, the C4.5 classifier inherits fuzzy system to locate the most extreme precision [7]. Wiharto et al. proposed a novel crossbreed model for diagnosing coronary illness, which gone before by highlight choice and intelligent relapse utilized in the systematic technique. It determines to follow the stage by utilizing the neural system [8]. Köhler et al. proposed a vortex extraction strategy to decide the appropriate vortices for the cardiovascular bloodstream. This technique totally actualizes to GPU [9].

Cardoso et al. depicted a few information mining methods like grouping which include determination, oversampling techniques, and programmed order calculations to recognize the infection. It is an assessment of arbitrary woods arrangement calculation to accomplish the most elevated precision [10].

Dey et al. prescribed the PCA strategy to expand the precision with the least traits for managing different AI calculations. This strategy predicts coronary illness by breaking down-regulated AI calculations [11]. Paul et al. described a fluffy-choice emotionally supportive network (FDSS) in light of hereditary calculation. It forms the whole datasets to extricate the powerful traits and applies the weighted fluffy standards to construct an FDSS for foreseeing coronary illness [12]. Verma et al. proposed a novel half-breed technique for CAD analysis, which incorporates connection-based component subset (CFS) and K-implies bunching calculations. This strategy recognizes CAD [13]. Yogaamrutha et al. portrayed an alternate order calculation like SVM with classes of heart ailments like myocardial localized necrosis and stroke. The expectation model keeps up the enormous informational index with missing traits with the information mining apparatus called WEKA [14].

3 Methodology

World future insights infer that coronary illness is winning more in number. So, it is important to manufacture a proficient smart confided in the computerized framework which predicts the coronary illness precisely depending on the indications as indicated by sexual orientation/age and area learning of specialists in the field at the most reduced expense (Fig. 1).

Data mining frameworks are basic to confine systematic errors similarly to improve the patient's security. Data mining techniques are incredible in organizing a restorative genuinely steady system and improve the ability to choose the unnoticeable models and relationships in clinical data. In this paper, the utilization of request technique, a decision tree for the acknowledgment of coronary disease has been introduced. Gathering tree uses various components including age, glucose, and circulatory strain; it can recognize the probability of patients fallen in CSV by using less systematic tests that put aside time and money.

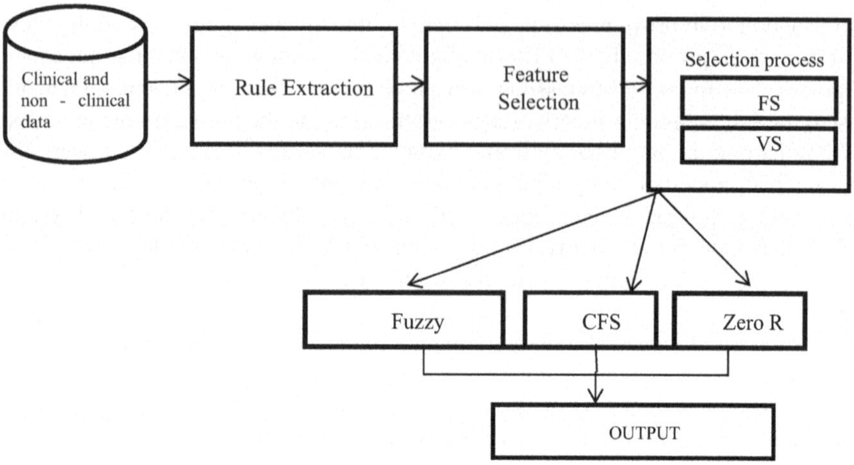

Fig. 1 Overall framework

Therefore, the likelihood has been directed by the embedding which supports the best-first search computation. In like manner, it used to depict the bits of learning concerning cardiovascular agony in the remedial organization structure autonomously. Besides, it confines to find the expressions of making the truth for the proximity and nonappearance which access the twofold characteristics. The clinical data has been shown for a couple of sorts of clinical procedures for coronary sickness.

3.1 *Fuzzy*

The fuzzy utilizing the procedure of including in the Takagi–Sugeno model which creates the closest ideal in the coronary illness. The potential results are used to vanish the rest of the users through the document recovering to pre-preparing in the information mining methods. Here, the control is,

$$(x \text{ and } y) = \min(x,\ y) \tag{1}$$

$$(x \text{ or } y) = \max(x, y) \tag{2}$$

$$\text{Not}\, x = 1 - x \tag{3}$$

$$(x \text{ implies } y) = \max(x,\ 1 - y) \tag{4}$$

These are all the given procedure that has decided the procedure of microcontroller and settled on a choice to the fuzzification and de-fuzzification to the separate arrangement of preparing information. In this, procedure has found the minimum esteem as appeared (1) and maximum estimation of (2) proclaimed too.

At that point, it delivers the negative outcome additionally which has a change in the given procedure in the fluffy framework with backings of set hypothesis capacities. Here,

$$\mu(S\ U\ T)^{(X)} = \left(\mu\ S_{(x)} \text{ or } \mu\ T_{(x)}\right) \tag{5}$$

$$\text{Union} = \max\left(\mu\ S_{(x)} \text{ or } \mu\ T_{(x)}\right). \tag{6}$$

3.2 *Co-related Feature Selection*

Any of the variable or object has been depending on the same data, and the value of manipulation is determined as,

$$X = 2Y \tag{7}$$

$$U = V2 \tag{8}$$

Hence, the part decision has been executed by the parallel overseeing while simultaneously getting to the data affirmation in the coronary illness for looking structure. It passes on the perfect course of action in the embeddings.

3.3 *ZeroR*

- The essentialness of setting up a standard of execution for AI issues.
- How to figure an example execution using the zero-rule methodology on a backslide issue.
- How to learn an example execution using the zero-rule procedure on a game plan issue.

Furthermore, in addition, it contains the SD, Mean, Absolute mistake an incentive on the implanting in default capacity including the procedure of standard in the relapse systems.

Regression:

$$Y = MX + B \tag{9}$$

$$Y1 = A + BX \tag{10}$$

Lose confidence evaluation is used to find conditions that fit data. When it has a fall away from the confidence condition, we can use the model to make needs. One kind of breaking faith appraisal is a prompt assessment. Precisely, when a relationship coefficient displays that information is undoubtedly going to have the choice to anticipate future results and a scatter plot of the information seems to shape a straight line, it can utilize principal direct fall away from the faith to locate an insightful point of confinement. It overviews from basic variable-based math, the condition for a line and what might be appeared differently in relation to direct break faith as appeared as (9, 10) formulae.

4 Result and Discussion

Data mining development gives a customer an arranged approach to managing novel and disguised models in the data. The found information can, in addition, be utilized by the supportive masters to reduce the measure of contradicting drug influence, to propose dynamically sensible restoratively identical decisions. The discovered data can moreover be used by the helpful specialists to diminish the amount of opposing medicine sway, to propose progressively reasonable medicinally equivalent choices.

Anticipating patient's future leaders on the given history is one of the noteworthy employments of data mining techniques that can be used in human administrations on the board. The openness of huge proportions of therapeutic data prompts the necessity for stunning data examination instruments to isolate supportive learning. Subsequently, it will get to the tremendous data by the human administration framework. Additionally, so also, the data has been distinguished to mining the data by the consistently based backings of IoT. Examiners have for a long while been stressed over applying quantifiable and data mining mechanical assemblies to improve data examination on tremendous enlightening records. Disease investigation is one of the applications where data mining gadgets are showing triumphs (Figs 2, 3, 4, 5, and 6).

Thus, the result has shown as,

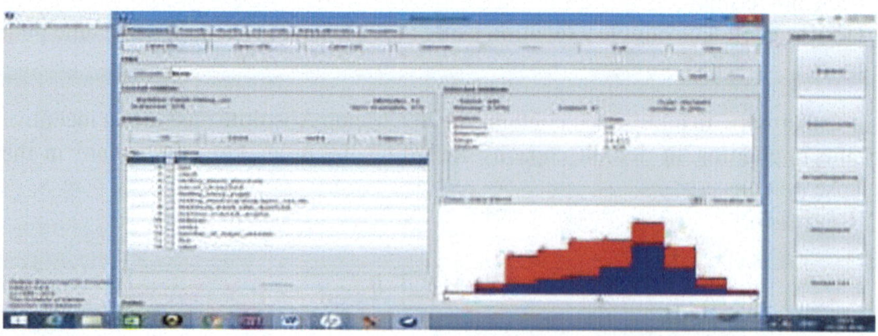

Fig. 2 Accuracy

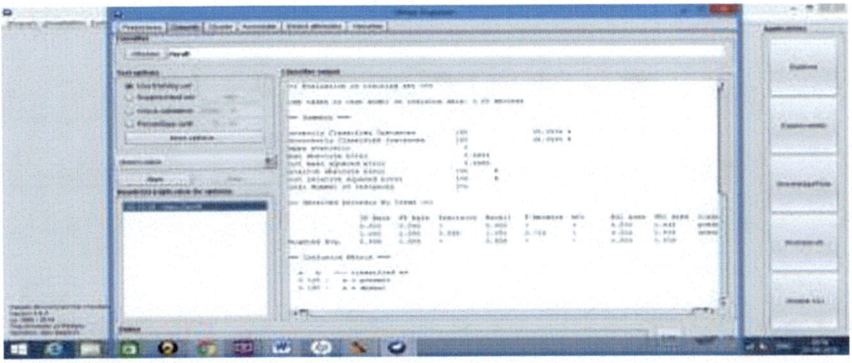

Fig. 3 Fuzzy set

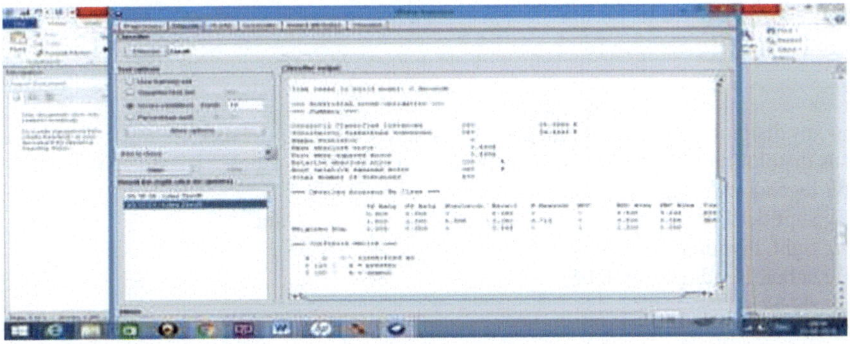

Fig. 4 Cross-validation

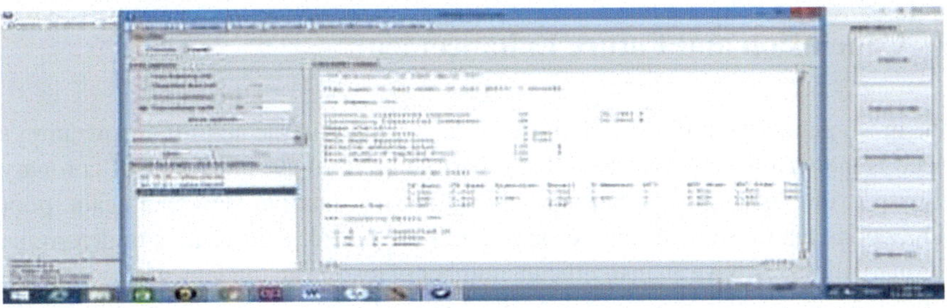

Fig. 5 ZeroR

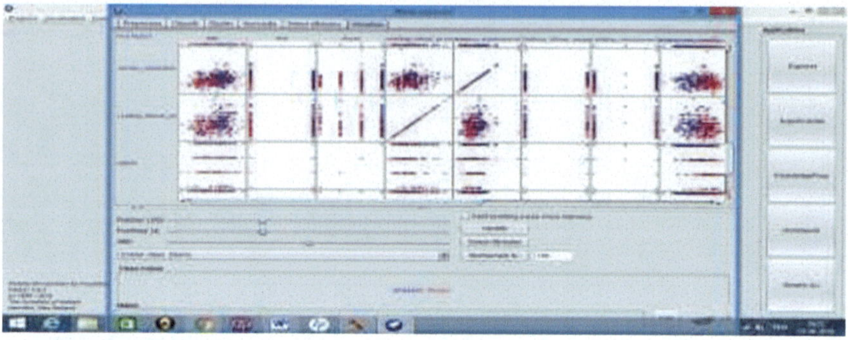

Fig. 6 Plot matrix visualization

4.1 Coding

```
sns.distplot(data[i][result['diagnosis']==0], color='g', label = 'benign')
sns.distplot(data[i][result['diagnosis']==1], color='r', label = 'malignant')
plt.legend(loc='best')
fig.suptitle('Breast Cance Data Analysis')
fig.tight_layout()
fig.subplots_adjust(top=0.95)
plt.show()
```

5 Conclusion and Future Enhancement

The most, generally, utilized technique for data mining in the helpful organization's piece is gathering. The expansive request strategy used for the figure of coronary disease is the decision tree that is used in this investigation. Now and again, poor perceptions lead toward death. All specialists are not all that masters to determine coronary illness to have an insignificant number of tests. The fundamental motivation behind this examination is to analyze the heart patients all the more unequivocally and all the more precisely with a base number of tests (decrease of traits). This exploration assumes an imperative job in the cost decrease of treatment and analyzes illness and extra improvement of the restorative investigations. The purposed research work can further be helped and used for the forecast of different kinds of heart sicknesses. Coronary disease is one of the essential wellsprings of demolition around the world, and it is fundamental to predict the contamination at less than the ideal stage. The PC-supported frameworks help the specialist as a device for anticipating and diagnosing coronary illness. The goal of this survey is to far-reaching about heart-related cardiovascular illness and to brief about existing-choice emotionally supportive networks for the forecast and conclusion of coronary illness bolstered by information mining and half-breed shrewd methods.

References

1. Chou CY, Chang EJ, Li HT, Wu AY (2018) Low-complexity privacy-preserving compressive analysis using subspace-based dictionary for ECG telemonitoring system. IEEE Trans Biomed Circuits Syst 12:801–811
2. Mortazavi BJ, Desai N, Zhang J, Coppi A, Warner F, Krumholz HM, Negahban S (2017) Prediction of adverse events in patients undergoing major cardiovascular procedures. IEEE J Biomed Health Inf 21:1719–1729
3. Din IU, Almogren A, Guizani M, Zuair MA (2019) A Decade of Internet of Things: analysis in the light of healthcare applications. IEEE Access 7:89967–89979
4. Amin MS, Chiam YK, Varathan KD (2019) Identification of significant features and data mining techniques in predicting heart disease. Telematics Inf 36:82–93
5. El-Bialy R, Salamay MA, Karam OH, Khalifa ME (2015) Feature analysis of coronary artery heart disease data sets. Procedia Comput Sci 65:459–468
6. Kavitha R, Kannan E (2016) An efficient framework for heart disease classification using feature extraction and feature selection technique in data mining. In: International Conference on Emerging Trends in Engineering, Technology and Science (ICETETS), pp 1–5
7. Liu X, Wang X, Su Q, Zhang M, Zhu Y, Wang Q, Wang Q (2017) A hybrid classification system for heart disease diagnosis based on the RFRS method. Comput Math Methods Medicine
8. Wiharto W, Kusnanto H, Herianto H (2017) The hybrid system of tiered multivariate analysis and artificial neural network for coronary heart disease diagnosis. Int J Electr Comput Eng 7(2):1023
9. Köhler B, Gasteiger R, Preim U, Theisel H, Gutberlet M, Preim B (2013) Semi-Automatic Vortex Extraction in 4D PC-MRI Cardiac Blood Flow Data using Line Predicates. IEEE Trans Vis Comput Graph 19(12):2773–2782
10. Cardoso A, Silveira T, Dias D, Tuler E, Ferreira R, Rocha L, Gomes C (2018) Combining data mining techniques to enhance cardiac arrhythmia detection. In: International conference on computational science, pp 321–333
11. Dey A, Singh J, Singh N (2016) Analysis of supervised machine learning algorithms for heart disease prediction with a reduced number of attributes using principal component analysis. Int J Comput Appl 140:27–31
12. Paul AK, Shill PC, Rabin MRI, Akhand MAH (2016) Genetic algorithm-based fuzzy decision support system for the diagnosis of heart disease. (ICIEV). In: 5th International Conference on Informatics, Electronics and Vision (ICIEV), pp 145–150
13. Verma L, Srivastava S, Negi PC (2016) A hybrid data mining model to predict coronary artery disease cases using non-invasive clinical data. J Medical Syst 40(7):178
14. Yogaamrutha SC, Cenitta D, Arjunan RV (2019) Forecast of coronary heart disease using data mining classification technique. J Adv Res Dyn Control Syst 11(4):25–36

Crop Prediction Based on Environmental Factors Using Machine Learning Ensemble Algorithms

Tatapudi Ashok and P. Suresh Varma

Abstract India being an agricultural country, the most part of economy depends on yield growth. Agriculture largely depends on rainwater and also depends on diverse soil parameters, namely nitrogen, phosphorus, potassium and weather aspects like temperature and rainfall. The technological growth in agriculture will increase the crop productivity. Remote sensing systems like IoT systems are being more widely used in smart farming systems, and these systems produce generous amount of data. Machine learning is a promising research arena to anticipate the crop based on the data patterns. The proposed system will integrate the IoT sensors like PH sensor, moisture, rainfall, temperature and humidity sensors to observe the data from those sensors and applying machine learning algorithms: Random Forest and GDBoost. A prediction of the most suitable crops according to the current environment is made. This work provides producers a stronger forecast to plant what kind of plants in their farm area depending on the parameters mentioned above to enhance smart farming's productivity.

Keywords Agriculture · IoT · Ensemble algorithms · Machine learning · Data analytics · Prediction

1 Introduction

The proposed system in this paper is a class of crop prediction system for increasing the production based on the key technologies: the Internet of things and machine learning techniques. Sensor technology has been advanced, and various sensors for humidity, temperature, moisture content in soil and acidity of the soil are used to sense the required elements. Machine learning technology predicts the crop based on the sensor data. Uses of these technologies are helpful to the farmer for better production rate in agriculture. The main aim in the agriculture is to improve the crop yield with low operational cost and less pollution in environment. Potential development and yield rely on many distinct characteristics of manufacturing such

T. Ashok (✉) · P. Suresh Varma
Adikavi Nannaya University, Rajamahendravaram, AP, India

S.-L. Peng et al. (eds.), *Intelligent Computing and Innovation on Data Science*,
Lecture Notes in Networks and Systems 118,
https://doi.org/10.1007/978-981-15-3284-9_63

as climate, soil properties and management of irrigation and fertilizer. Farming is the backbone of any economy. In a country like India, which is experiencing ever-increasing food demand due to rising population, advances in the agricultural sector are needed to meet needs. Agriculture has been considered as the main and foremost culture practiced in India since ancient times. Ancient people are growing the crops in their own land and have been adapted to their needs. The natural crops are therefore cultivated and used by many creatures like humans, animals and birds. The sector of agriculture is slowly degrading since the creation of fresh innovative technologies and methods. Because of this, abundant individuals of invention are focused on cultivating artificial products that are hybrid products where life is unhealthy. Nowadays, there is no awareness of contemporary individuals about growing the plants in the correct moment and location. These cultivating methods also change the seasonal climatic circumstances against the basic resources such as land, water and air that contribute to food insecurity. By evaluating all of these issues such as weather, temperature and multiple variables, there is no adequate solution and technology to solve the scenario we face. In India, there are several methods to boost agricultural financial development. There are several ways in which crop yield and crop quality can be increased and improved.

1.1 IoT in Agriculture

The Internet of Things is an interconnection of computer systems both mechanical and digital with the unique ability to move information on a network without human or computer communication. In the field of agriculture, few researchers have suggested IoT-based architectures with computer training to forecast crop form. Machine learning IoT is mature technology, and a ton of agricultural work has been undertaken. This scheme controls plant development form. The proposed scheme helps to make predictive decisions and analyze information detected and gathered from the detectors and placed in the database and analyzed using algorithms for machine learning. Sensors sense the crop yield information for different humidity, temperature, rainfall, pH value parameters. Etc. is placed in memory through IoT systems that are further used to predict crop varieties that have a direct effect on plant development after predictive choices are produced and forwarded to the end user for further interference, resulting in a stronger end user advantage.

1.2 Machine Learning

Machine learning is a technology that is commonly used for issues with agriculture. It is used in the analysis of big information collections and in the information sections to create helpful categories and patterns.

The machine learning method's general objective is to obtain the information from a collection of data and turn it into an understandable framework for further use.

This paper's primary goal is to develop a scheme that can forecast the sort of plant depending on the characteristics of soil and climate. As the population is growing in today's globe, it is expected to be in billions as the years go by and we need to enhance crop output to supply those billion people. The population is growing, and on the other side, the agricultural territory is declining owing to multiple factors such as heavy industrialization, development of commercial markets and erection of residential buildings being created on the agricultural land; therefore, to supply to the ever growing number of consumers, there is a need to boost output which can be attained by applying appropriate The most significant item that is required in everyday lives is smart farming.

Section 2 describes related work, whereas Sect. 3 analyzes the proposed system and architecture, Sect. 4 gives the results and Sect. 5 concludes the paper.

2 Related Work

The statistical method, namely the technique of multi-linear regression, and the method of data mining, namely the clustering technique based on density, were used to predict crop yield analysis [1]. Kalman filter (KF) is used in the suggested method to obtain noise-free performance information with predictive assessment and transmit this data for cluster-based WSNs. Decision tree uses predictive tools for crop output forecasting, plant ranking, land classification, weather assessment and decision-making plant disease forecasting. The system integrates the components of IoT such as null (IoT Gateway) and Mobius (IoT Service Platform) to provide a smart plant development monitoring solution for consumers [2]. Machine learning algorithm was created using logistic regression to handle pure information and forecast results. It provides the outcome but is less precise than other algorithms [3]. The use of spatial data mining in the agricultural domain has been clarified by authors [4]. They used the K-means algorithm along with progressive refinement of the optimization technique for the assessment of geographic associations. Temporal information, temperature and precipitation are provided and analyzed to improve crop yield and decrease plant casualties. In [5], writers recognize the issue of anticipating the median output of a plant form (e.g., soybean) for a region in which concern dependent on a sequence of remotely sensed images collected prior to harvesting and information implemented to estimate the plant sort by convolutional neural networks. This [6] attempts to chart the development of machine-based learning techniques for precise prediction of crop yield and nitrogen structure evaluation over the past 15 years, suggesting that rapid developments in technology detection and therefore ML techniques will offer cost-effective and inclusive solutions for improved plant and environmental assessment and in decision-making. In this system [7], a blend of Internet and wireless modes of communication approach, remote monitoring system (RMS) is suggested. The primary goal is to acquire agricultural production system data in real time that provides simple links to agricultural equipment such as SMS alerts and guidance of climate model, crops.

3 Proposed System

The study work suggested centers on the use of efficient IoT systems and predictive decision-making. In the design of the system, we included communication flow between various system components and input and output for various modules present in the system. Sensed information is contrasted with information collection that is recorded and generated as a consequence of previous knowledge.

The architecture of the system appears in Fig. 1. Main scheme elements are

- IoT equipment
- Predictive Machine Learning.

Based on the results of the assessment, the farmer will decide from this scheme to select the finest plant for that specific soil to boost the crop's production rate.

3.1 IoT Devices

In IoT-based intelligent farming, detectors (sunshine, humidity, heat, ground moisture, etc.) are used to monitor the plant sector and automate the drainage scheme. From anywhere, the peasants can monitor the circumstances of the ground with the sensor.

This study seeks to provide a scheme for monitoring specific plant temperature and humidity characteristics, land and air pH indicator and fertilizer handling, plant humidity detector for sensing plant humidity level. All these detectors are used as well as generating the information to monitor the harvest. IoT apparatus interfaced

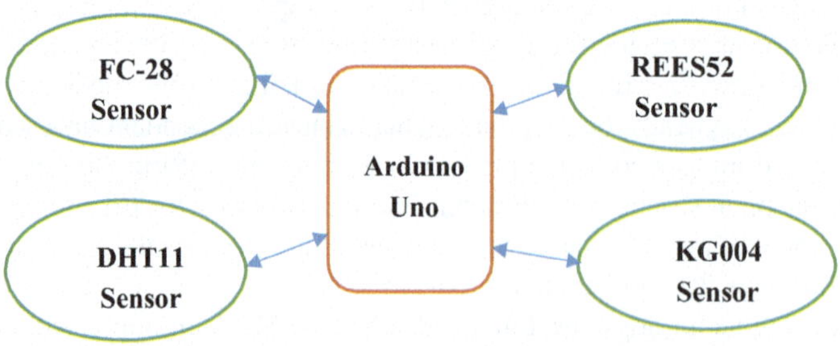

Fig. 1 IoT architecture

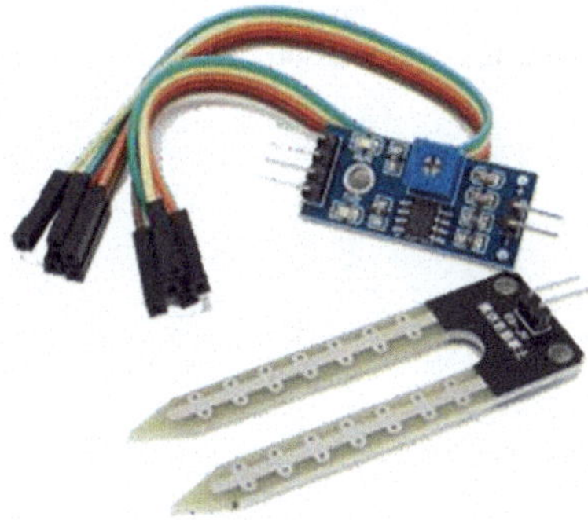

Fig. 2 FC-28 soil moisture sensor

Table 1 FC-28 soil moisture sensor

Input power	Output power	Input current	Output signal
3.3–5 V	0–4.2 V	35 mA	Analog and digital

3.1.1 FC-28 Soil Moisture Sensor

The soil moisture sensor to be used should be quite straight. Sensor samples comprise two large exposed sheets functioning together as a variable resistor. The higher the content of water in soil, the higher will be the conductivity between the buttons, resulting in lowering of resistance and higher SIG production [8] (Fig. 2) (Table 1).

3.1.2 DHT 11 Humidity and Temperature Sensor

DHT 11 temperature and moisture sensor exhibits a mixture of calibrated thermal and humidity sensors through digital signal generation. The utilization of the exclusive digital signal procurement technique and use of detector technology for humidity and temperature assures high effectiveness and improved long term stabilization. This sensor contains an aspect of resistive assessment of moisture and an component of computation of NTC temperature and connections to an 8 bit high-performance microcontroller which provides high efficiency, rapid sensitiveness to response, anti-interference capacity and effective cost wise [9] (Fig. 3) (Table 2).

3.1.3 KG004 Rain Drop Sensor

The rain sensor unit is an easy tool to detect rain. It can be used as a button when the raindrop falls through the packing board and also to evaluate the force of precipitation [10] (Fig. 4) (Table 3).

Fig. 3 DHT 11 humidity and temperature sensor

Table 2 DHT 11 humidity and temperature sensor

Item measurement	Range	Humidity accuracy	Temperature accuracy	Resolution	Package
DHT11	20–90% RH 0–50 °C	±5%RH	±2 °C	1	4 pin single row

Fig. 4 KG004 rain drop sensor

Table 3 KG004 rain drop sensor

Item measurement	Range	Output voltage	Output signal
KG004	−0.3 to + 36 V	+36 V	Digital and analog

3.1.4 REES52 PH Sensor

PH is a metric of a solution's acidity or alkalinity, and the pH scale is between 0 and 14. The pH shows the presence in certain liquids of oxygen[H] + atoms. A sensor that monitors the potential difference between two electrodes can correctly quantify it: a base electrode (silver/silver oxide) and a hydrogen-sensitive glass electrode. That is what the sample is. We also need to use an electronic circuit to properly position the message, and with a microcontroller like Arduino, we can use this sensor [11] (Fig. 5) (Table 4).

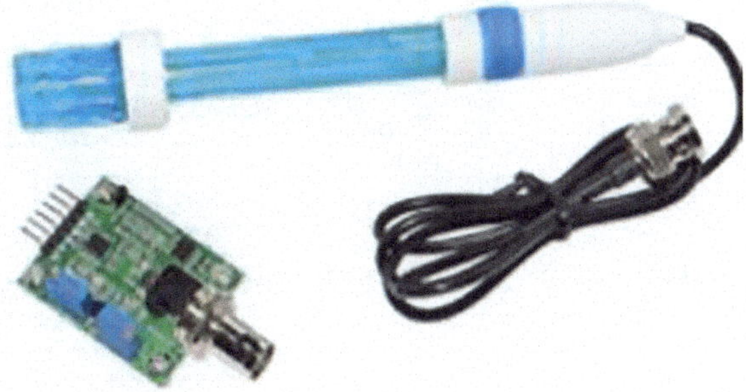

Fig. 5 REES52 PH sensor

Table 4 REES52 PH sensor

Supply power (V)	Current (mA)	Consumption (W)	Working temperature (°C)
5	5–10	≤0.5	10–50

3.2 Machine Learning Algorithm for Prediction

Machine learning is commonly implemented to problems of agriculture. It is used in the analysis of big information collections and in the information sections to create helpful categories and patterns. The machine learning method's general objective is to remove the information from a collection of data and turn it into an understandable framework for further use.

Based on accessible information, this article analyzes the crop yield form. To maximize crop productivity, the machine learning method was used to forecast crop yield. Figure 6 demonstrates the stream of the forecast of suggested crop yield.

As shown in the above situation, devices are implemented on the farm to detect the information linked to moisture, heat, precipitation and pH. Random Forest and GDBoost algorithms are used to identify detected information. The expected outcome demonstrates whatever soil may be appropriate for specific crops and the situation of groundwater [12].

3.2.1 Overview of Data

We obtain information from different locations and arrange datasets in this stage. And for analytics, these datasets are used. Online tools such as Data.gov.in and indiastat.org are also used to generate the accurate information. We contain nearly all the plants that are useful to the farmer in the chart below with the sample information scores. Duration indicates that the plant length indicates the distance temperature needed for the plant in months, minimum and maximum characteristics. N, P, K Values are crop-specific fertilizers, minimum and maximum PH values are related to groundwater standards. Rain fall is that area's scope (Table 5).

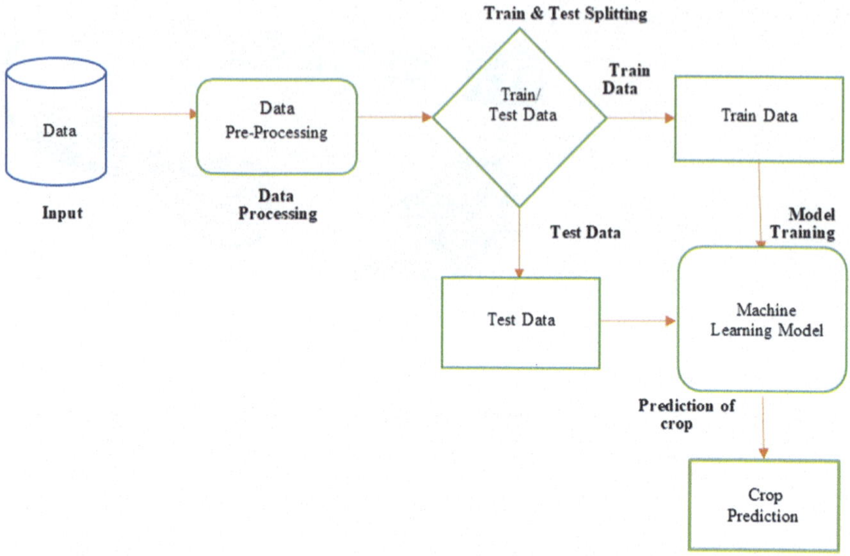

Fig. 6 System architecture

3.2.2 Machine Learning Methods

Machine learning is the method of finding patterns that were previously unidentified and potentially expanding in big sets of data. Using the mined data portrays a model of prediction or model of classification. Datasets gathered from IoT and various sources enter to be much more complicated than traditionally used in the database software practice. Machine learning is primarily classified as descriptive and predictive machine learning. However, in the agricultural sector, for the most part, predictive data mining is used. There are two major classification and clustering methods. Some of the methods below are used to purchase the choice from the information gathered. The methodology suggested includes two stages: the phase of training and the phase of testing. The information was gathered and pre-processed during the training stage. The learning stage uses pre-processed information to train the model. The output quantity is anticipated in the inspection stage depending on the regulations produced. Work begins with phase of pre-processing. The information gathered was pre-processed in this phase. Some information was separated from the information collection during pre-processing. Some of the region was inappropriate for plant manufacturing. So the information will be deleted. Models used in the stage of practice and screening outlined below [13]:

Random Forest:

A combination of tree predictors is termed as Random Forests such that every tree depends on an individually sampled random vector features comprising of equal allocation for all forest crops. The mistake of generalization of forest which converges

Table 5 Data sheet

Crop	Duration	Minimum temperature	Maximum temperature	PH minimum	PH maximum	Rainfall minimum	Rainfall maximum	N	P	K
Bajra	3	18	30	3	8	350	750	L	L	M
Banana	4	15	35	6.5	8.5	450	750	M	VL	VL
Barley	4	12	32	3	8	800	1100	VL	VL	M
Bean	2	14	32	5.5	6.5	300	500	L	VL	M
Black pepper	6	23	33	5.5	6.5	1200	2500	H	VL	M
Blackgram	2	23	35	5	7	500	700	L	H	VL
Bottle gourd	2	24	27	6.5	7.5	400	650	VL	VL	VL

as the quantity of shrubs as the forest becomes large. Usage of a random selection of features to separate each node produces mistake levels that are more stable in terms of noise compared to AdaBoost. Internal measurements assess error, strength and similarity which are used to demonstrate the reaction so as to increase the amount of characteristics utilized in the separation. Internal projections can also be used for variable significance measurement. These concepts can be applied to regression as well [14].

Gradient Boosting:

The primary sources of variation in real and expected scores are noise, variance and bias when we attempt to estimate the destination variable using any machine learning method. Ensemble enables these variables to be reduced. Gradient boosting algorithm gradually, additively and sequentially reduce the bias error of the model with a low variance. By using gradients in the transfer matrix ($y = ax + b + e$), e requires a unique reference as it is the mistake word, and gradient boosting works the same. The loss function is a metric of how the numbers of the healthy model fit the fundamental information. A logical interpretation of the feature of failure would rely on what we try to optimize [15].

3.3 IoT Analytics

IoT analytics is a fully managed service that simplifies the operation of advanced analytics on huge amounts of IoT information without having to worry about the price and complexity typically needed to build an IoT analytics platform. It is the easiest method of using IoT data analytics and gaining thoughts to generate stronger and more accurate IoT apps and use case decisions for machine learning (Figs. 7, 8, 9 and 10).

Fig. 7 Soil moisture data

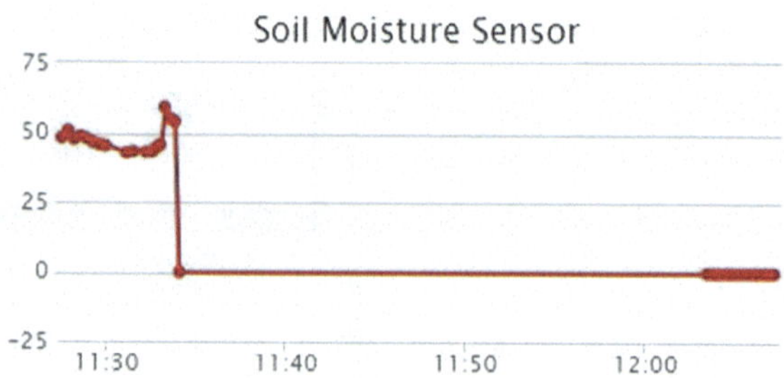

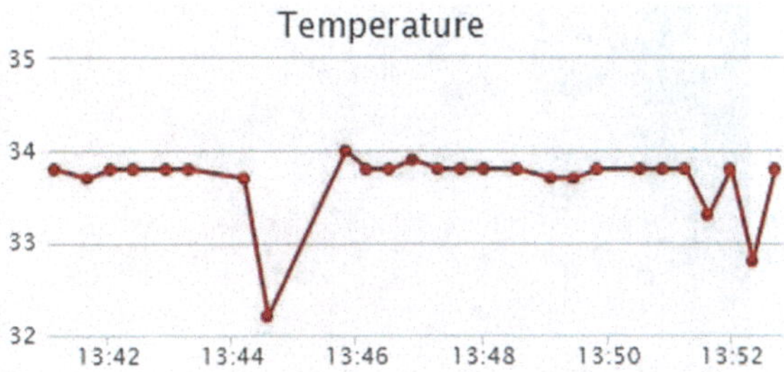

Fig. 8 Temperature data

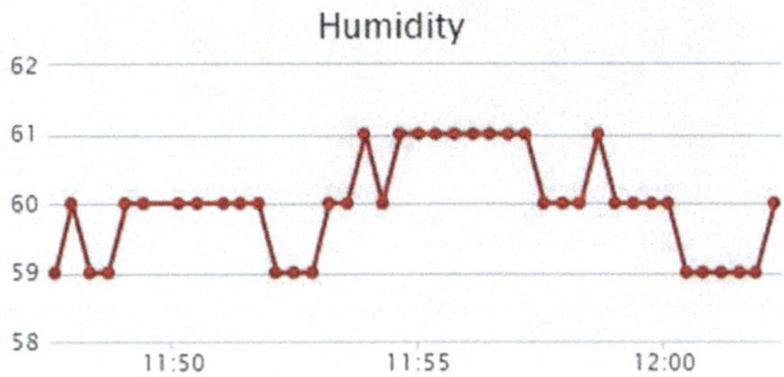

Fig. 9 Humidity data

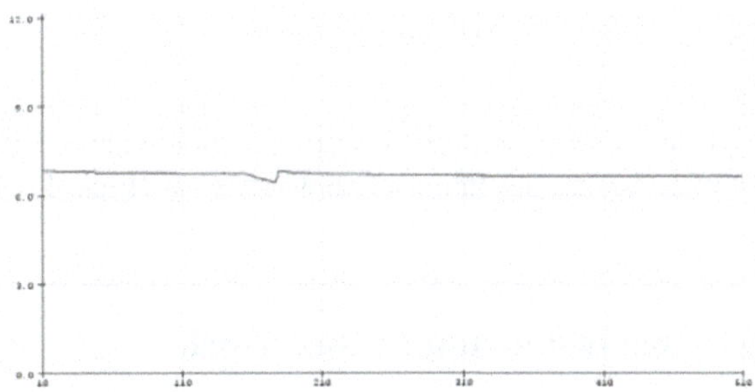

Fig. 10 pH Data

4 Results

After extracting the features from the represented data, we can easily predict on the basis of the crop type on the features we consider. In order to approximately evaluate the effectiveness and efficiency of our experiment, we compare two selective ensemble classification algorithms on the data. Those algorithms are Random Forest and GDBoost. Random Forest is ensemble method to control the bias and variance in data; it improves its accuracy when the depth of the tree is high and lowers when its depth is low. To avoid the over fitting of the model, we put a threshold values to Random Forest Model and that values is 14, with that depth level random forest generated 96.32% accuracy with 3.68% error rate. Gradient boost is also an ensemble method to boost up the model performance, and this method generated

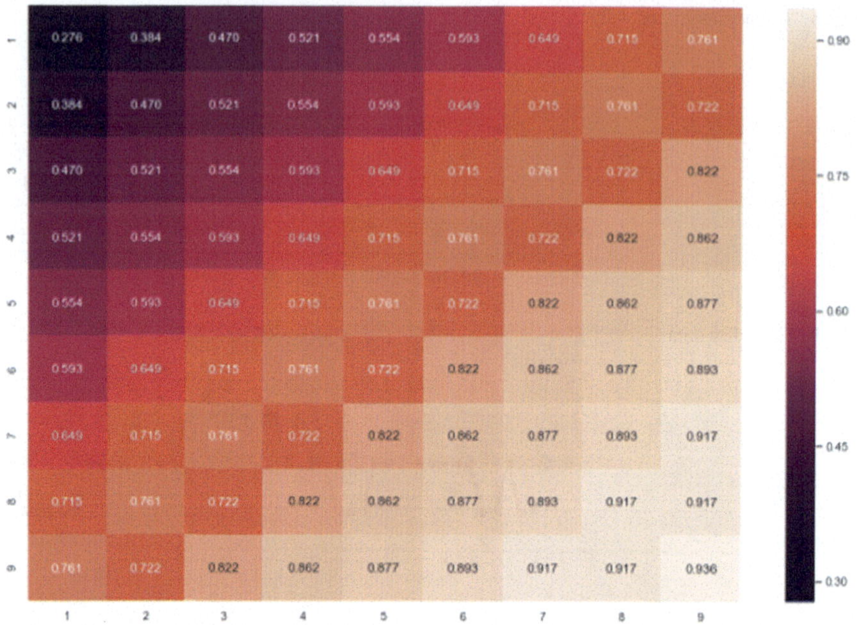

Fig. 11 Random Forest depth generation

highest accuracy with the boosting technique and that is 96.69% and the error rate is 3.61. Accuracies table are showed below (Figs. 11, 12, 13 and 14) (Table 6).

5 Conclusion and Future Work

The suggested scheme lists all possible plants that are viable in a given region, helping the farmer to decide which crop to grow. The scheme has carried out a thorough examination of the soil, climate and pH information and indicates which are the most lucrative plants that can be grown in the appropriate environmental condition. This scheme also examines the previous information output that will assist the farmer gain insight into the market demand and price of different plants. As maximum crop kinds under this scheme will be covered, farmers may learn about the crop that may never have been grown. IoT contributes to the association with the Internet of all farming machines. Different kinds of sensors used in the farm provide real-time farm status information, and the devices can be used to boost the humidity, acidity, etc. Also in view of the currency and inflation ratio, the best lucrative crop can be discovered.

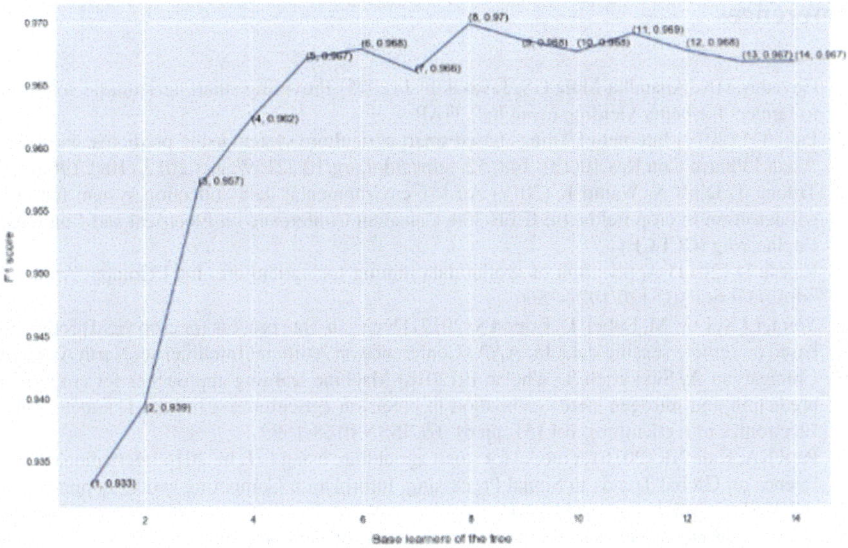

Fig. 12 Scoring values of models on predicted value

Fig. 13 Accuracies of machine learning models

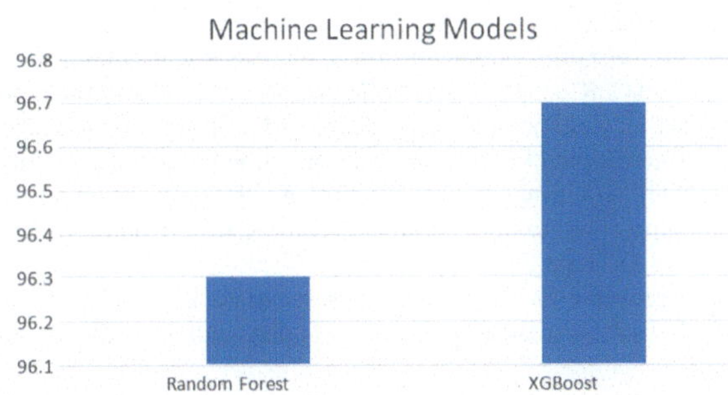

Fig. 14 Error rate of models

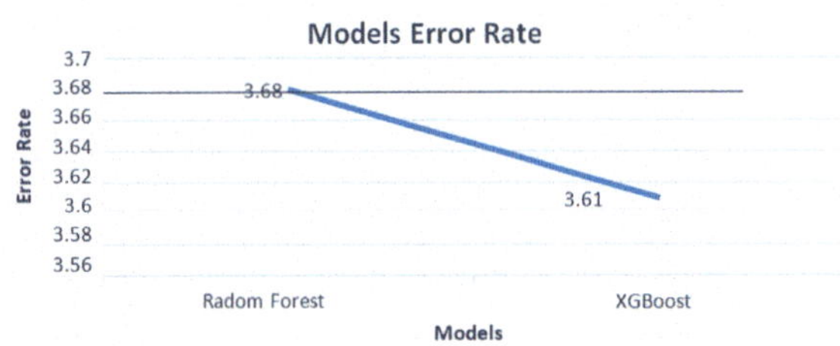

Table 6 Accuracy values of models

Classifier	Error Rate	Accuracy
Random Forest	3.68	96.32
Gradient boost	3.61	96.69

References

1. Gayathri MK, Anandha Mala GS, Jayasakthi J (2015) Providing smart agricultural solutions to farmers for better yielding using IoT. TIAR
2. Patil SM (2017) Internet of Things based smart agriculture system using predictive analytics. Asian J Pharm Clin Res 10(13): 148–52. https://doi.org/10.22159/ajpcr.2017.v10s1.19601
3. Truong T, Dinh A, Wahid K (2017) An IoT environmental data collection system for fungal detection in crop fields. In: IEEE 30th Canadian Conference on Electrical and Computer Engineering (CCECE)
4. Rajesh D (2011) Application of spatial data mining for agriculture. Int J Comput Appl 15. https://doi.org/10.5120/1922-2566
5. You J, Li X, Low M, Lobell D, Ermon S (2017) Deep gaussian process for crop yield prediction based on remote sensing data. In: AAAI Conference on Artificial Intelligence, North America
6. Chlingaryan A, Sukkarieh S, Whelan B (2018) Machine learning approaches for crop yield prediction and nitrogen status estimation in precision agriculture: a review, Computers and Electronics in Agriculture, vol 151, pp 61–69, ISSN 0168-1699
7. Patil KA, Kale NR (2016) A model for smart agriculture using IoT. In: 2016 International Conference on Global Trends in Signal Processing, Information Computing and Communication (ICGTSPICC), Jalgaon, pp 543–545
8. https://www.electronicscomp.com/soil-moisture-sensor-module-india
9. https://www.mouser.com/ds/2/758/DHT11-Technical-Data-Sheet-Translated-Version-1143054.pdf
10. https://www.rhydolabz.com/sensors-weather-sensors-c-137_147/raindrop-sensor-module- p-2336.html
11. https://scidle.com/how-to-use-a-ph-sensor-with-arduino/
12. Ruß G, Kruse R, Schneider M, Wagner M (2008) Estimation of neural network parameters for wheat yield prediction. In: Bramer M (ed) Artificial Intelligence in theory and practice II, IFIP International Federation for Information Processing, Springer,vol 276, pp 109–118
13. Mej´ıa-Guevara I, Kuri-Morales A (2007) Evolutionary feature and parameter selection in support vector regression. In Lecture Notes in Computer Science, Springer, Berlin, Heidelberg, vol 4827, pp 399–408
14. Breiman L (2001) Random Forests. Mach Learn 45:5–32
15. https://towardsdatascience.com/understanding-gradient-boosting-machines-9be756fe76ab

Regression Analysis on Sea Surface Temperature

Manickavasagam Sivasankari and R. Anandan

Abstract This paper is aimed to analyze the sea surface temperature over salinity of regions near the North America Pacific Coast. The usage of different regression analysis techniques has been studied and compared to conclude the best method that will apply in our research on ocean water properties and Potential Fishing Zones. We have considered the systematic reciprocal options among the various analysis methods like linear, decision tree, random forest, and SVR. The optimal transition of the training and testing sets is monitored to arrive at the most responsive and reliable solution.

Keywords Sea surface temperature · Regression analysis · Random forest · SVR

1 Introduction

In the modern world, there is a need to study about the discrepancies of physical versus biological parameters to understand the relationships among them. The need for understanding the ecosystem is the key to both saving it and tapping its potential in an ideal way. The modern-day facilities like the remote sensing technology provide us the ample opportunity and avenues to research the oceanography on a broader scale as it offers a comprehensive view of the oceanographic parameters [1]. In this paper, we have intended to study the sea surface temperature (SST) along with the dependent factor of the sea, i.e., the salinity [2]. This will then be utilized in predicting the fish swarms based on the ecological condition.

Climatic conditions have been the key to human life as well as all other living beings in this universe. In this paper, we look at the effects of the climatic variability of the sea surface and how in turn it affects the marine habitat. Global fishing stocks are

M. Sivasankari (✉) · R. Anandan
Department of CSE, Vels Institute of Science, Technology and Advanced Studies, Chennai, Tamil Nadu, India

M. Sivasankari
Accenture Solutions PVT LTD, Chennai, Tamil Nadu, India

S.-L. Peng et al. (eds.), *Intelligent Computing and Innovation on Data Science*, Lecture Notes in Networks and Systems 118,
https://doi.org/10.1007/978-981-15-3284-9_64

reducing at an alarming rate which has forced the mankind to analyze the behaviors in a smarter way.

1.1 Sea Surface Temperature Analysis

The fishing zone detection is always associated with conditions like thermal facades, salinity, chlorophyll concentration, pH level [3]. This in turn also depends on seasonal variations and external factors [15–17]. We will need to analyze these dependent factors and train the system, so that prediction becomes easier and, in turn, helping us forecast the ecological balance in our ecosystem.

1.2 Salinity Analysis

The salinity of the area of study ranges from 31 to 35, mostly in the range of 33 [4]. This essentially means that 1000 ml of water contains 33 g of salt. Salinity is considered to be the decisive factor which will determine the fish existence as it determines the density of water. The more dense the water is, the particles start sinking in. Deeper oceans are usually more saline than the water near the shore [5].

2 Data Sets and Methods

The California Cooperative Oceanic Fisheries Investigations (CalCOFI) along with the partnership with the California Department of Fish and Wildlife, Scripps Institution of Oceanography and NOAA Fisheries Service provides the longest duration data ranging from 1949. They include temperature, salinity, oxygen, and phosphate observations. The area of study spans across the North America Pacific Coast [6].

In our analysis, we try to understand the thematic relationship between the SST and salinity data. We make use of the regression techniques to analyze and predict the data. Each dataset is unique and would imply the usage of different techniques to draw the conclusion. We have analyzed if there is a relationship between SST and salinity, then we can predict the SST based on the salinity and vice versa. Humidity is also another factor influencing SST and salinity [12–14].

The different regression techniques can be selected based on the nature of the available data. We arrive into a predictive modeling technique which will assimilate the independent and dependent variables. This can be achieved by implementing a series of mathematical and statistical calculations. Here, the independent variable is the temperature and dependent variable is salinity.

In this paper, we also analyze the efficiency of the regression techniques, along with the predictions of SST. The techniques visited upon are linear regression, decision tree regression, random forest regressor, and support vector regressor [4].

3 Methodology

The regression analysis is referred to the process, in which we perform the evaluation of the association between two variables, namely dependent and independent. There are several methods available, but we have studied the behavior of four processes. The classification will be done and will be analyzed based on the training set.

The evaluation of the machine learning algorithms can be done using the below metrics

1. Mean absolute error (MAE)
2. Mean squared error (MSE)
3. r2 Score.

The r2 can be predicted using ground actual target values, estimated target values, and sample weights. The best possible score is 1 and can go to negative also when the model gets worse.

4 Results and Discussion

4.1 Linear Regression

This can be used when there is mapping required for an independent variable against a continuous dependent variable. The missing value has been removed to get non-discrete data. The data given below include the training set and the testing set (Fig. 1).

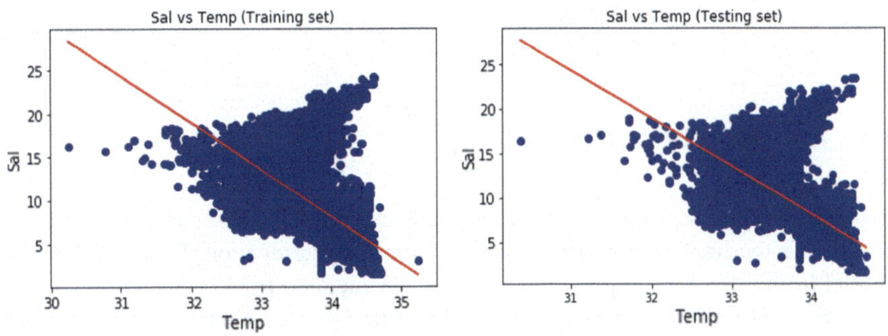

Fig. 1 Linear regression analysis results

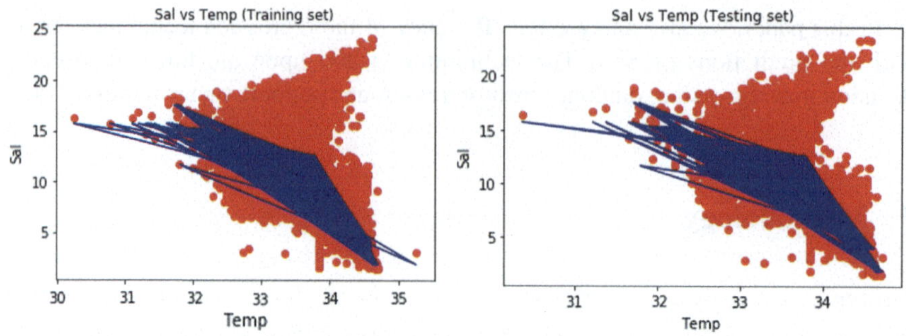

Fig. 2 Decision tree regression analysis results

The measuring values for the technique are given below

Mean absolute error: 2.4985274582997508
Mean squared error: 11.74420964959364
Root mean square error: 3.4269825867070933

r2_score = 0.40211112007952254.

4.2 Decision Tree Regression

The output of this model is a tree structure which will be developed incrementally from smaller subsets. The root node will be the predictor, and it uses the top-down approach. The main algorithm used is ID3 developed by J. R. Quinlan [8–10].

The training set and testing set for this technique as well as the scores are available below (Fig. 2).

Mean absolute error: 2.13981559458496
Mean squared error: 9.110176783512047
Root mean square error: 3.0183069399105267

r2_score = 0.5362077521189327.

4.3 Random Forest Regression

In this technique, we make use of multiple decision trees and combine all of them to generate the final output. This is called bootstrap aggregation. It will collect the dataset from different samples, and each tree is generated from each sample. In this case, the CalCOFI data is analyzed and the testing and training sets are generated along with the scores (Fig. 3).

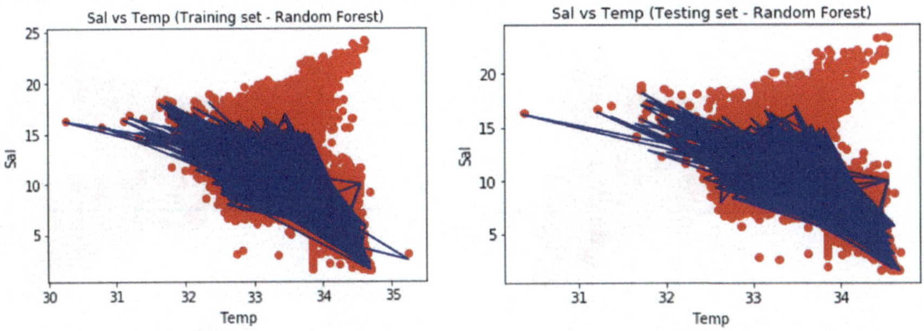

Fig. 3 Random forest regression analysis results

Mean absolute error: 2.182942169093261
Mean squared error: 9.501441697367195
Root mean square error: 3.0824408668078607

r2_score = 0.5162887496422415

This method consists of numerous decision trees and tries to create a group of uncorrelated trees [12–14]. They may not be accurate individually, but prediction by grouping is more accurate.

We use Scikit in Python to create these decision trees and below given is the

```
from sklearn.tree import
DecisionTreeClassifier

tree =
DecisionTreeClassifier(random_state=RSEED)
```

If we notice our results based on the r2 score, we can adapt to random forest over the decision tree as the generalization of testing data seems to be in a much better state.

4.4 Support Vector Regression

This technique is similar to support vector machine. Here the error will be fitted well within the threshold.

Linear SVR is calculated using the below formula

$$y = \sum_{i=1}^{N} (a_i - a_i^*)\langle x_i, x\rangle + b$$

The training and testing training sets give us a linear (Fig. 4)

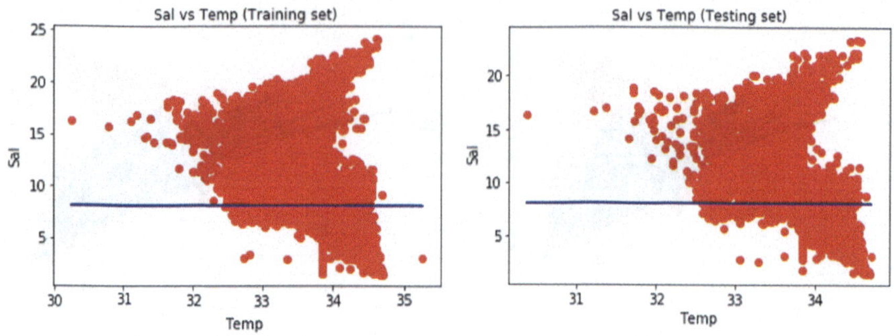

Fig. 4 SVR analysis results

Mean absolute error: 3.623734606961357
Mean squared error: 20.48092238438408
Root mean square error: 4.525585308485973

r2_score:−0.04266835397998281

Relevant preprocessing and processing steps were applied to the CalCOFI data for data corrections mainly focused on enhancing quantitative forest parameters.

5 Conclusion

We have compared the different regression techniques using the CalCOFI data. The mean absolute error at the range if 2–4 and the mean squared error has been between the range of 9 to even 20. The best r2 score using this data during the execution has been in the values ranging 0.51 and some have even displayed negative range.

The data has been taken in the Californian region of the Pacific Ocean as this has been a source of Californian current bringing nutrients to the Pacific region. The average temperatures range between 32F and 34F, whereas the salinity range is at around 10–15 parts per 1000. This is also largely influenced by the depths and evaporation patterns. We have also noticed the highest of even 37 parts. The optimal transition projection shows the relations between salinity and SST, but also factored from anomalous shore-level advection and has low-frequency variability.

References

1. Nayak S, Solanki HU, Dwivedi RM Utilization of IRS P4 ocean colour data for potential fishing zone—a cost benefit analysis
2. Narain A, Dwivedi RM, Kumari B, Chaturvedi N, Solanki H, Mankodi PC, Sudarsan D Relationship between sea-surface temperature (SST) and fish catch data: a feasibility study

3. Condron A, DeConto R, Bradley RS, Juanes, F (2005) Multidecadal North Atlantic climate variability and it effect on North American salmon abundance
4. Schneider N, Di Lorenzo E, Niiler PP (2005) Salinity Variations in the Southern California Current. https://doi.org/10.1175/JPO2759.1
5. Akbari E, Alavipanah SK, Jeihouni M, Hajeb M, Haase D, Alavipanah S 2017. A review of ocean/sea subsurface water temperature studies from remote sensing and non-remote sensing methods. Water (Switzerland) [Internet]. 9:936.
6. Checkley DM. Jr, Lindegren M. Sea surface temperature variability at the Scripps Institution of Oceanography Pier. J Phys Oceanogr 44, 2877–2892 (2014).
7. Mansor S, Tan CK, Ibrahim HM, Shariff ARM (2001) Satellite fish forecasting in South China Sea. Proceeding of The 22nd Asian Conference on Remote Sensing (ACRS), Singapore, OCN-07
8. Zainuddin M, Saitoh S-I, Saitoh K (2004) Detection of potential fishing ground for albacore tuna using synoptic measurements of ocean color and thermal remote sensing in the northwestern North Pacific. Geophys Res Lett 31
9. Tan CK, Mansor S, Ibrahim H, Shariff R (2002) Studies of sea surface temperature and chlorophyll-a variations in east coast of Peninsular Malaysia. Pertanika J Sci Technol
10. Aulicino G, Cotroneo Y, Ansorge I, van den Berg M, Cesarano C, Belmonte Rivas M, Olmedo Casal E (2018) Sea surface salinity and temperature in the Southern Atlantic Ocean from South African icebreakers, 2010–2017, Earth Syst Sci Data 10:1227–1236. https://doi.org/10.5194/essd-10-1227-2018
11. Aulicino G, Sansiviero M, Paul S, Cesarano C, Fusco G, Wadhams P, Budillon G (2018) A new approach for monitoring the Terra Nova Bay polynya through MODIS ice surface temperature imagery and its validation during 2010 and 2011 winter seasons. Remote Sens 10:366. https://doi.org/10.3390/rs10030366
12. Chen J, You X, Xiao Y, Zhang R, Wang G, Bao S (2017) A performance evaluation of remotely sensed sea surface salinity products in combination with other surface measurements in reconstructing three-dimensional salinity fields. Acta Oceanol Sin 36(7):15–31. https://doi.org/10.1007/s13131-017-1079-y
13. Mu Z, Zhang W, Wang P, Wang H, Yang X (2019) Assimilation of SMOS Sea Surface Salinity in the Regional Ocean Model for South China Sea. Remote Sens 11: 919
14. D'Addezio JM, Subrahmanyam B (2016) Sea surface salinity variability in the Agulhas Current region inferred from SMOS and Aquarius. Remote Sens Environ 180:440–452
15. Thombley RL, Goericke R An introduction to CalCOFI surface underway data: Scripps Institution of Oceanography, UC San Diego
16. Scripps Institution of Oceanography (2014) University of California, San Diego, La Jolla, California
17. Rajesh G, Mercilin Raajini X, Vinayagasundaram B (2016) Fuzzy trust-based aggregator sensor node election in internet of things. Int J Internet Protoc Technol 9 (2/3):151

A Multi-objective Routing Optimization Using Swarm Intelligence in IoT Networks

Ganesan Rajesh, X. Mercilin Raajini, R. Ashoka Rajan, M. Gokuldhev, and C. Swetha

Abstract Internet of Things (IoT) network devices, are embedded wireless sensor nodes, constrained with limited battery capacity, storage and processing power. Hence, the available resources have to be utilized efficiently during the process like sensing, computing and communication. This creates a need for an optimized routing algorithm which should reduce the resource consumption like battery power to extend the network lifetime, and also we should consider the other network requirements like delay, throughput and packet delivery ratio. Here, a multi-objective routing optimization algorithm BFOA-R is proposed, based on group foraging behavior of *E. coli* and *M. xanthus* bacteria. The primary objective of the proposed routing is to reduce energy consumption during routing and maximize network life. Existing, particle multi-swarm optimization is used as a benchmarking method to evaluate the performance of BFOA-R.

Keywords Multi-objective optimization · BFOA · Routing · Internet of Things (IoT) · Packet delivery ratio (PDR) · Energy efficiency · Swarm intelligence · Bacterial foraging

---

G. Rajesh (✉) · C. Swetha
Department of Information Technology, MIT Campus, Anna University, Chennai, India
e-mail: raajiimegce@gmail.com

C. Swetha
e-mail: swethacs.96@gmail.com

X. Mercilin Raajini
Department of ECE, Prince Shri Venkateshwara Padmavathy Engineering College, Chennai, India
e-mail: raajii.mercy@gmail.com

R. Ashoka Rajan
Department of Computer Science, SoC, SASTRA, Thanjavur, India
e-mail: ashok.tiruchendur@gmail.com

M. Gokuldhev
Department of CSE, Amritha College of Engineering and Technology, Nagercoil, India
e-mail: ksmdhev@gmail.com

S.-L. Peng et al. (eds.), *Intelligent Computing and Innovation on Data Science*,
Lecture Notes in Networks and Systems 118,
https://doi.org/10.1007/978-981-15-3284-9_65

1 Introduction

IoT network is the internetwork of physical devices and objects with embedded electronics, software, sensing devices, actuators, etc. IoT connects objects to get sensed, monitored or controlled remotely across the existing wired or wireless network infrastructure. The deployment of IoT variants in various application domains, like the Internet of Industrial Things (IoIT), Internet of Medical Things (IoMT), Internet of Secure Things (IoXT), results in improved efficiency, accuracy and also reduces human intervention in the application. Routing of data from source to sink is an integral part of IoT, as it enables the exchange of information between the nodes in the network. Routing over a diverse network is a real challenge as it depends on various parameters. Factors like node failure, connectivity issues, frequently changing network topology and environmental conditions affect routing in IoT.

The malfunctioning of the nodes due to battery drain can make changes in the network topology, and this might lead to a demand in the route rerouting of the packets and sometimes reorganization of the network topology. The scarce energy makes it difficult to decide the next hop node. It is essential to reduce the power utilization of the nodes so that the lifetime of the sensor node as well as that of the network is prolonged. Here, the routing issue in IoT has been addressed and a bio-inspired bacterial foraging optimization algorithm for routing has been proposed for efficient routing in IoT.

Multi-objective optimization, also known as Pareto optimization, is a mathematical optimization concerned problem to optimize two or more objectives simultaneously.

The biological activities or behaviors of various bio-organisms are studied and mapped into an optimization technique. Bacterial foraging optimization algorithm (BFOA) is also one of the swarm-inspired optimization algorithms. Group foraging strategy of a swarm of *E. coli* bacteria is used in multi-optimal function optimization.

This paper is organized as follows. Section 2 of this paper discusses about the research gap by refereeing the literatures. Section 3 proposes BFO-R and its architecture, and Sect. 4 explains the mathematical modeling, algorithm of the proposed work. Section 5 explains the experimental observations and performance evaluation.

2 Related Works

In the past decade, research on optimization has attracted more and more attention in various applications. The most general optimization problem can be simply defined as,

$$\text{Maximize } f(X), \text{ where } X = [x_1, x_2, x_3, \ldots\ldots x_C],$$

where C, is the number of constraints to be optimized by maximization or minimization. Routing is the problem of finding the best path for packets in the network such that the packet delivered with expected quality requirements. IoT network and devices have multi-constraints due to the heterogeneity of devices. Hence, there is a need of optimization in routing the packets. This section discusses existing methods for implementing the bacteria foraging optimization.

In [1–8], various route optimization techniques are proposed like dynamic state routing protocols based, directional transmission-based energy aware routing protocol named (PDORP), improved bacterial foraging optimization (IBFO) algorithm and a hybrid of Optimal Secured Energy Aware Protocol (OSEAP) for energy aware routing approach. In various literatures, the authors explored different biologically inspired approaches for routing in IoT networks. Biologically inspired models for routing and other layers in IoT have been more thriving to put up the wonderful defense in energy-efficient routing in IoT networks.

The literature review [9, 10] shows that till there are some research gaps in the previous approaches in terms of routing overhead, in adequate route selection parameters, not considering dynamic nature of nodes and bit error rate. Hence, here, proposed a multi-objective routing optimization routing (BFOA-R) technique for IoT using bacterial foraging swarm intelligence.

3 Proposed System

Devices in IoT have limited battery capacity, storage and processing power. The available energy has to be utilized properly to achieve more throughput and increase lifetime of the network. Hence, an optimized routing algorithm which utilizes the available energy properly is essential. For this purpose, a bio-inspired optimization algorithm has been proposed.

The bio-inspired optimization algorithms are effective optimization algorithms, where the biological activities of various bio-organisms are observed and are converted into an optimized way. This minimizes the cost and time and maximizes the output. Figure 1 shows the architecture of the proposed system.

3.1 Bacterial Foraging—The Theory

The theory of bacterial foraging optimization can be briefed as follows. *Chemotaxis* simulates the movement of an *E. coli* cell. *Swarming* aggregates the bacteria with high density. *Reproduction* is defined as healthier bacteria asexually split into two bacteria and the least healthy bacteria eventually die. *Elimination and Dispersal*, due to the significant local rise of the temperature, may decay a subset of bacteria group, or a group is moved into a new location [11–13].

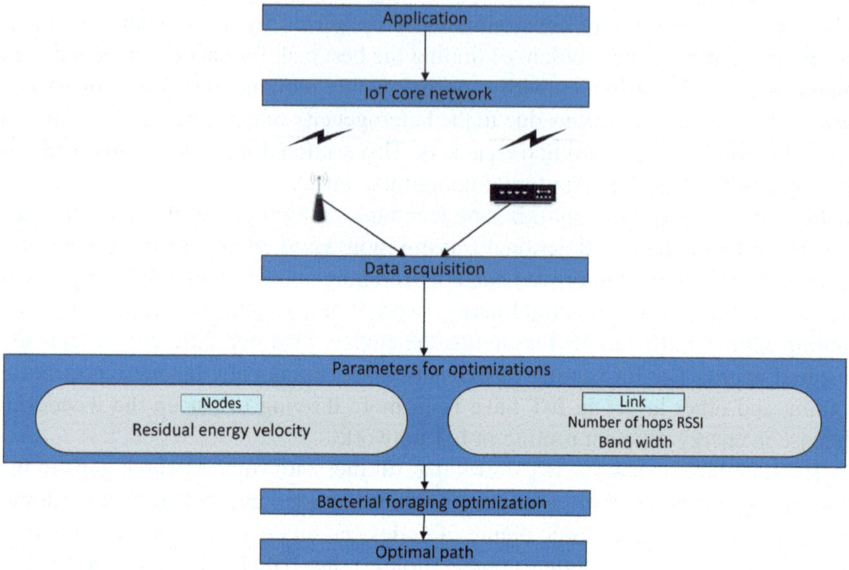

Fig. 1 Architecture diagram for the proposed system (BFOA-R)

3.2 Mapping of Bacterial Foraging for Route Optimization

In this proposed system, the bacterial foraging optimization algorithm is framed in a way such that the chemotaxis process represents the movement of the nodes (bacterium) as the system is a dynamic network. During the swarming process, the mobile nodes arrange themselves and fit into a topology. Reproduction process represents the arrival of new nodes into the system and departure of existing nodes from the system, and finally, the elimination and dispersal process represents the replacement of the weak nodes due to many causes like depletion in the energy.

4 Mathematical Modeling of BFOA-R

The optimization problem is subject to few constraints like node's residual energy, node's transmission power, path length and available channel capacity. The objectives are residual energy (E), number of hops (H), node velocity (V), bandwidth (B) and received signal strength (R).

Consider $S = [E\ H\ V\ B\ R]$
$f(S) = [\max(E, B, R)\&\&\min(H, V)]$

Constraint Function 1:
According to Friis transmission formula,

$$\frac{P_\mathrm{r}}{P_\mathrm{t}} = \frac{A_\mathrm{r}A_\mathrm{t}}{d^2\lambda^2} \tag{1}$$

P_t is the antenna transmitting power, P_r is antenna receiving power, λ is the wavelength of the radio frequency used. A_r is the effective area of the receiving antenna, A_t is the effective area of the transmitting antenna, and d is the distance between antennas.

$$\frac{P_\mathrm{r}}{P_\mathrm{t}} \propto \frac{1}{d^2} \tag{2}$$

$$R - \frac{k_1}{H} \geq 0, \text{ where } \mathrm{k}_1 \text{ is the proportionality constant.} \tag{3}$$

Constraint Function 2:

$$\text{Velocity} \propto \frac{1}{\text{Residual Energy}} \Rightarrow V \propto \frac{1}{E} \tag{4}$$

$$V - \frac{k_2}{E} \leq 0, \quad \text{where } k_2 \text{ is proportionality constant.} \tag{5}$$

Constraint Function 3:
According to Nyquist formula, the bandwidth b and noise in free channel, we can transmit data at a rate up to:

$$C = 2B\ \log_2 M \tag{6}$$

where B is bandwidth and M is the number of signal levels.

$$C \propto B \tag{7}$$

$$C - k_3\ B \geq 0, \text{ where } k_3 \text{ is proportionality constant.} \tag{8}$$

Constraint Function 4:
According to radio energy dissipation theory, energy consumed for transmitting k bit to a distance r is given by

$$E_T(k,r) = \begin{cases} k\left(E_{\mathrm{TX}} + E_{\mathrm{fs}} \times r^2\right) & \text{if}\, r < r_0 \\ k\left(E_{\mathrm{TX}} + E_{\mathrm{mp}} \times r^4\right) & \text{if}\, r \geq r_0 \end{cases} \tag{9}$$

where E_{TX} and E_{RX} represent per bit energy dissipated for transmission and reception, respectively. models, respectively. Energy consumed by receiver for k bit reception is given by

$$E_{\mathrm{R}}(k) = k \times E_{\mathrm{RX}} \tag{10}$$

Considering the free space model the energy consumed is given by the following equations,

$$E \propto k(E_{\mathrm{TX}} + E_{\mathrm{fs}} \times H^2) \tag{11}$$

$$E - k(E_{\mathrm{TX}} + E_{\mathrm{fs}} \times H^2) \geq 0 \tag{12}$$

4.1 *Algorithm: Bacterial Foraging Optimization Algorithm for Routing (BFOA-R)*

See Table 1.

Table 1 Parameters assumed in the algorithm

Symbol	Parameter
n	Number of iterations
p_{size}	Problem size
n_{ed}	Number of elimination dispersal iterations
n_{re}	Number of reproduction iterations
n_{c}	Number of chemotactic iterations
$\mathrm{node}_{\mathrm{life}}$	Lifetime of a node

INPUT: Initialize: n, p_{size}, n_{ed}, n_{re}, n_c, $node_{life}$, $node_{best}$
Calculate: $node_{fitness}$
OUTPUT: Optimized path
PROCESS:

1. Start; Initialize number of bacteria, i.e. nodes.
2. Population InitializePopulation(n, p_{size})
3. For(l=0 to n_{ed})
   For(k=0 to n_{re})
   For(j=0 to n_c)
   ChemotaxisAndSwim(n,p_{size})
4. For (node $\in n$)
   If(fitness(node)$<$fitness($node_{best}$))
   $node_{best}$= node; End; End ; End
   Sortbyfitness(nodes)
5. Selected=selectBynodefitness(nodes)
6. If($node_{life}$$<$ 30)
7. Eliminate(node);End

CHEMOTAXIS AND SWIM function
Chemotaxis and Swim(n,p_{size})
1. For(node$\in$ n)
   Fitness (node)=$node_{life}$ + interaction (node)
2. End

The number of nodes, node's chemotaxis step (n_c), node elimination dispersal (n_{ed}), node reproduction (n_{re}) are initialized. Fitness of nodes is calculated until the condition fails.

5 Performance Evaluation

The proposed system is evaluated using ns-3 simulator. The extensive performance evaluation of the system is tested under two different scenarios of 50 nodes and 100 nodes of network size 200 × 200 m and compared to particle multi-swarm optimization (PMSO) routing algorithm. Table 2 lists the various parameters considered for the simulation.

Figures 2 and 3 plot the average residual energy of the BFOA-R against PMSO at various iterations with 50 and 100 nodes, respectively. It is evident from the graph that the proposed BFOA-R algorithm moderately outperforms the PMSO algorithm in terms of average residual energy.

Figures 4 and 5 show the average packet delivery ratio of the BFOA-R against PMSO at various iterations with 50 and 100 nodes, respectively. Figures 6 and 7 show the number of dead nodes, when 50 nodes and 100 nodes are deployed. The number of dead nodes is reduced by 7% and 14%, respectively, in this case, while using BFOA-R.

Table 2 Simulation parameters

Parameters	Values
Node mobility	Dynamic, 5 m/s
Region	200 × 200
Deployment	Random deployment
Transmission power	3 dBm
Channel model	Free space model
Mobility model	Random way point mobility model
Energy model	Friis loss model

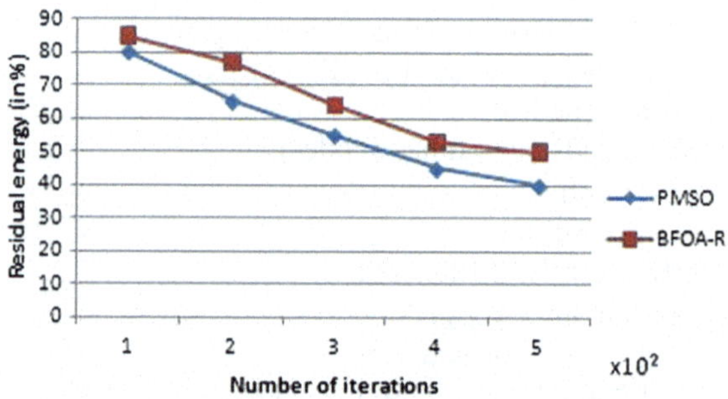

Fig. 2 Analysis of average residual energy for 50 nodes

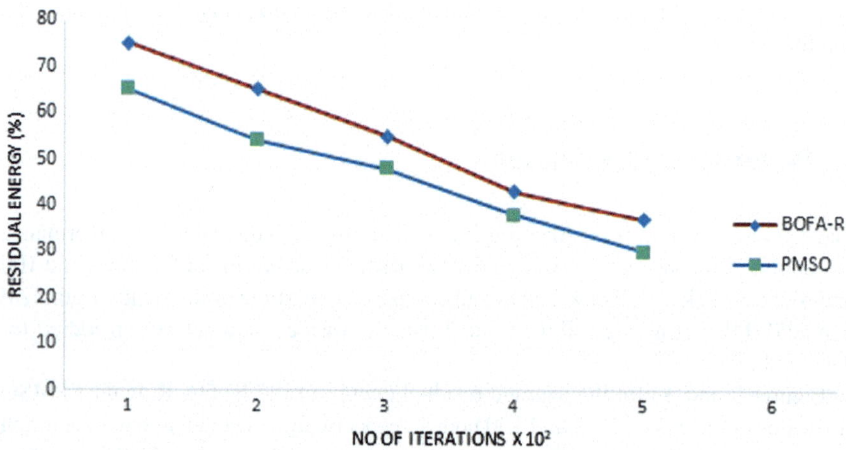

Fig. 3 Analysis of average residual energy for 100 nodes

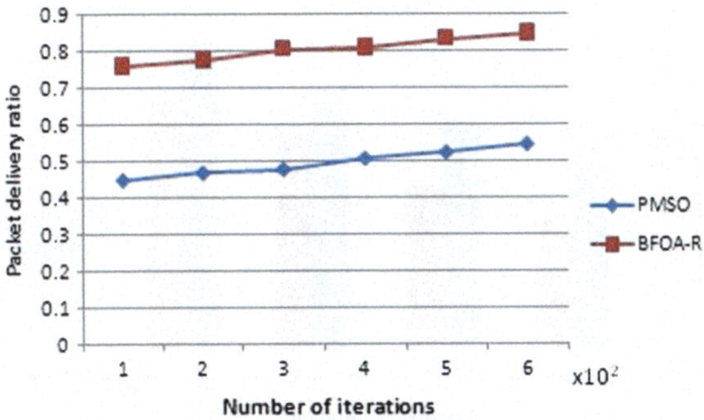

Fig. 4 PDR analysis for 50 nodes

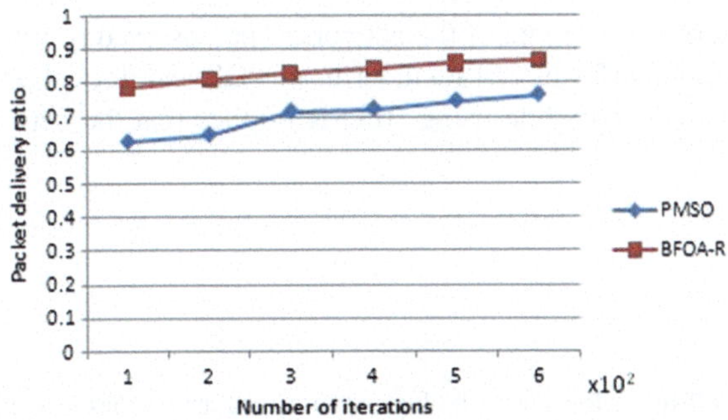

Fig. 5 PDR analysis for 100 nodes

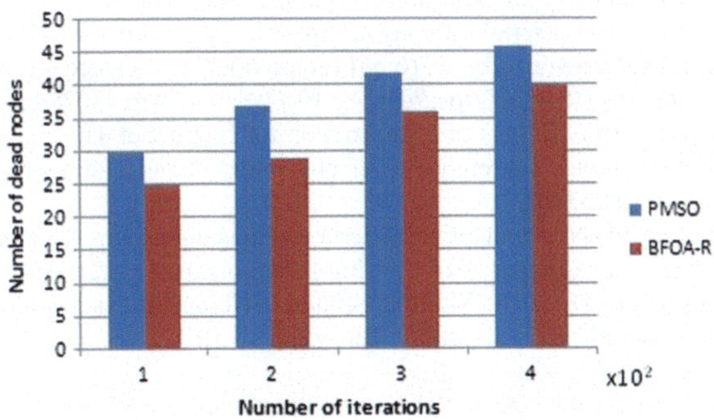

Fig. 6 Analysis of network lifetime for 50 nodes

6 Conclusion

In this paper, we presented a route optimization technique in IoT networks by using bacterial foraging, a swarm intelligence approach. The proposed work is compared

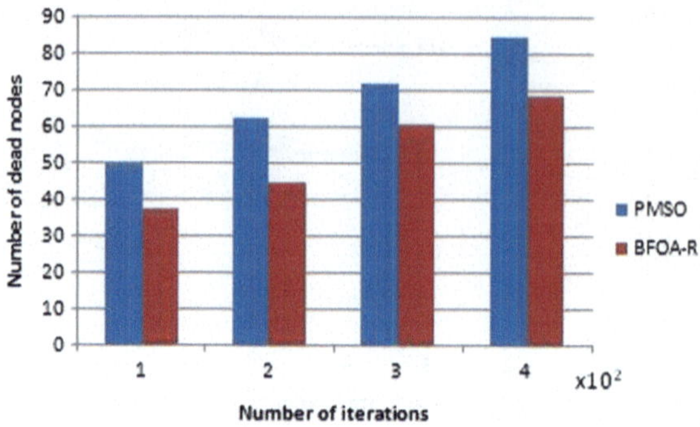

Fig. 7 Analysis of network lifetime for 100 nodes

with PMSO in terms of lifetime of the network. The observed result shows that the proposed BFO-R outperforms PMSO in terms of PDR and throughput. The number of dead nodes is reduced while using BFO-R implies that the network lifetime is extended.

References

1. Banu R et al (2016) A Review on biologically inspired approaches to security for Internet of Things (IoT). In: 2016 International conference on electrical, electronics, and optimization techniques (ICEEOT). https://doi.org/10.1109/iceeot.2016.7754848
2. Brar GS et al (2016) Energy efficient direction-based PDORP routing protocol for WSN. IEEE Access 4:3182–3194. https://doi.org/10.1109/access.2016.2576475
3. Lalwani P, Das S (2016) Bacterial foraging optimization algorithm for CH selection and routing in wireless sensor networks. In: 2016 3rd International conference on recent advances in information technology (RAIT). https://doi.org/10.1109/rait.2016.7507882
4. Mahmud MA et al (2017) Energy efficient routing for Internet of Things (IoT) applications. In: 2017 IEEE international conference on electro information technology (EIT). https://doi.org/10.1109/eit.2017.8053402
5. Nasir ANK et al (2014) Novel adaptive bacteria foraging algorithms for global optimization. Appl Comput Intell Soft Comput 2014:1–7. https://doi.org/10.1155/2014/494271
6. Rani RR, Ramyachitra D (2016) Multiple sequence alignment using multi-objective based bacterial foraging optimization algorithm. Biosystems 15:177–189. https://doi.org/10.1016/j.biosystems.2016.10.005
7. Yusoff M et al (2010) A discrete particle swarm optimization with random selection solution for the shortest path problem. In: 2010 International conference of soft computing and pattern recognition. https://doi.org/10.1109/socpar.2010.5685867
8. Reddy PK, Babu R (2017) An evolutionary secure energy efficient routing protocol in Internet of Things. Int J Intell Eng Syst. https://doi.org/10.22266/ijies2017.0630.38
9. Shao Y, Chen H (2009) Cooperative bacterial foraging optimization. In: FBIE 2009—2009 international conference on future bio medical information engineering. https://doi.org/10.1109/FBIE.2009.5405806
10. Nayyar A, Singh R (2019) IEEMARP—a novel energy efficient multipath routing protocol based on ant colony optimization (ACO) for dynamic sensor networks. Multimedia Tools Appl. https://doi.org/10.1007/s11042-019-7627-z

11. Sahoo SP, Kabat MR (2019) Multi-constrained multicast routing improved by hybrid bacteria foraging/particle swarm optimization. Comput Sci. https://doi.org/10.7494/csci.2019.20.2.3131
12. Sahoo SP, Mahapatra S, Sahu R, Kabat MR (2016) Multi-objective multicast routing based on bacteria foraging optimization. In: Proceedings on 2015 1st international conference on next generation computing technologies, NGCT 2015. https://doi.org/10.1109/NGCT.2015.7375137
13. Sahoo SP et al (2018) A reference-based multiobjective bacteria foraging optimization technique for QoS multicast routing. Arab J Sci Eng 43(12):7457–7472. https://doi.org/10.1007/s13369-018-3090-9
14. Rosário D, Filho JA, Rosário D, Santosy A, Gerla M (2017) A relay placement mechanism based on UAV mobility for satisfactory video transmissions. 2017 16th Annual Mediterranean ad hoc networking workshop, med-hoc-net 2017. https://doi.org/10.1109/MedHocNet.2017.8001638

Prevention of Packet Drop by System Fault in MANET Due to Buffer Overflow

Mohammed Ali Hussain and D. Balaganesh

Abstract Mobile ad hoc networks are a distributed peer to peer multi-hop constrained resources network. Applications of MANET are sensitive, and reliable communication is one of the challenging tasks. Routing protocol goal is to compute the route and send the information in computed route. However, routing in MANET faces the problem of packet drop either by malicious activities or by system fault. Major packet drop due to system fault is buffer overflow. The packet drop due to buffer overflow is due to the problem in either queue length at node buffer or packet forward. Thus, the early congestion control mechanism is needed in MANET to prevent the packet drop due to buffer overflow. This paper addresses the issue by computing the queue length at node buffer during the route computation process. If queue size of the node is more than the predefined threshold in a given time interval, then node does not participate in communication. Performance of the mechanism is validated by network simulator—2. Results are compared with the existing distance vector routing reactive and proactive routing protocols.

Keywords MANET · Routing protocol · Multi-hop · Malicious

1 Introduction

Nowadays, the significance of computing equipment and their connectivity has turned out to be required in our everyday life. Prior, in order to provide communication between computing equipment, the wireless local area stander was utilized. Afterward, there was an extreme interest came for wireless communication, and it accomplished with the help of IEEE 802.11 WLAN specifications. However, current generation requests to build up the wireless communication system with autonomous

M. A. Hussain (✉)
Department of CSE, Lincoln University College, Kota Bharu, Malaysia

D. Balaganesh
Faculty of Computer Science & Multimedia, Lincoln University College, Kota Bharu, Malaysia
e-mail: balaganesh@lincoln.edu.my

S.-L. Peng et al. (eds.), *Intelligent Computing and Innovation on Data Science*,
Lecture Notes in Networks and Systems 118,
https://doi.org/10.1007/978-981-15-3284-9_66

mobile devises. Such system is highly pivotal in crisis administrations, hazard activities, fiasco recuperation and military tasks. The development of the system is achieved by mobile ad hoc network.

The development aim of MANET [1] is to enable wireless communication everywhere at any time. The system is free from infrastructure, autonomous and self-configurable and maintainable. Moreover, the MANET is a wireless communication system with wireless mobile devises distributed in dynamically changing topological network. The system is not controlled by the central administration, and thus, nodes in a network consist of network intelligence to take the decision about the communication. Thus, nodes in a network behave as a router to enable communication in multi-hop manner, and at the same time, node also acts as a host to forward their own information. The characteristics of MANET such as autonomous, self-forming and dynamicity lead to deploy it in critical and sensitive applications such as disaster recovery, military and health care. The communication in these applications must be secured as they are highly sensitive.

Applications of MANET [2] are sensitive, and reliable communication is one of the challenging tasks. Routing protocol goal is to compute the routing paths and send the information in computed routes. However, routing in MANET faces the problem of packet drop either by malicious activities or by system fault. Major packet drop due to system fault is buffer overflow. The packet drop due to buffer overflow is due to the problem in either queue length at node buffer or packet forward. As MANET is very sensitive to delay regarding communication. Thus, the early congestion control mechanism is needed in MANET to prevent the packet drop due to buffer overflow [3–7]. This paper addresses the issue by computing the queue length at node buffer during the route computation process.

2 Proposed Work

MANET consists of wireless mobile nodes, which are constrained with respect to buffer and energy. Moreover, MANET is a peer-peer network, and nodes need to behave as a host and routers. During communication, if node receives the information more than its handling capability, then the information get drop from the node. Then, MANET needs an efficient mechanism to prevent the packet drop from communication path due to insufficient buffer. One can overcome the situation with the help of early congestion control routing.

We design an efficient early congestion prevention technique with the help of queue computation. Every node present in a MANET needs to have the RC filter at transceiver circuit that computes the residual queue length by the technique of exponential average moving average. The weighted constant value is depending on the time constant of RC filter. The queue length is computed by the following equation [8].

$$\text{Average Queue} = (1 - \alpha) * \text{Average queue old} - (\alpha) * \text{Instant Queue} \tag{1}$$

In this equation, is the weighted constant. If this queue is more than the handling capability of the node, then the node drops the packets from the node buffer. Thus, we assign the threshold value for detecting the status of the node regarding packet handling, i.e., Queue handling threshold, and it is computed as follows:

$$\text{Threshold} = 75\,\%\ \text{of buffer size} \tag{2}$$

If computed average queue is greater than the threshold value, then the node does not participate in communication. Otherwise, it participates in communication by computing the buffer packet holding capacity, as follows:

$$\text{Buffer packet holding capacity} = \text{Buffer size} - \text{Average Queue} \tag{3}$$

Proposed routing protocol is a reactive routing protocol with route computation metric as maximum value of Buffer packet holding capacity. Whenever source wants to communicate with destination, it broadcasts the route request packet, same as in the case of AODV routing protocol, instead of the field RREQ packet hop count is replaced by the metric Buffer packet holding capacity. Before computing the Buffer packet holding capacity by node, it checks whether it has the enough buffer to hold the packets by computing average queue and compare to threshold value. If node does not satisfy the threshold status regarding buffer capacity, then the node discards the route RREQ packet. If node satisfies the threshold status regarding buffer packet handling capacity, then it computes the Buffer packet holding capacity value, insets it in the RREQ packet route computing metric field and rebroadcasts it. Finally, destination receives the different RREQ packets, it checks the maximum Buffer packet holding capacity value contains RREQ packet path and creates RREP packet and unicasts it in the path of higher value of Buffer packet holding capacity containing path.

During the communication, every node periodically computes the residual average queue by equation one and compares it with the threshold value of Eq. (2). During the communication, any node finds its residual queue is greater than the threshold capacity, and then, it generates the RERR packet and unicasts it to source node. Then, once again source computes the route based on the above-explained process.

3 Performance Analysis

We have used the Network Simulator 2 to evaluate the performance of the proposed routing protocol and compare it with the distance vector routing protocols in the identical environment. The simulation parameters used for performance analysis are shown in Table 1.

The main aim of the performance evaluation is to check the packet the packet loss due to buffer overflow during the communication in the presence of proposed routing protocol and distance vector routing protocol [9, 10]. Packet loss performance metric

Table 1 Performance analysis parameters

Parameter	Value
Routing	Reactive, proposed
MAC	802.11
Channel	Wireless
Number of nodes	50–100
Mobility	Random way point
Network area	1000 * 1000 m^2

also negatively impacts on the packet delivery. Thus, we also compute the packet delivery ratio of the network along with packet loss. The results are presented in Figs. 1 and 2.

Results (Figs. 1 and 2) are clearly indicating that the packet handling capacity of proposed routing mechanism is good in compared to existing distance vector routing protocols. Current routing mechanism does not handle the packet loss due to buffer overflow as they are based on the distance vector. Moreover, any routing protocol, which is based on the distance vector or resource reservation, causes the higher traffic toward the specific node buffer due its either location or its resource. Thus, our proposed work is based on the packet handling capability y of node with respect to its residual queue size. Thus, performance of the proposed mechanism outperforms in comparison with the existing protocols.

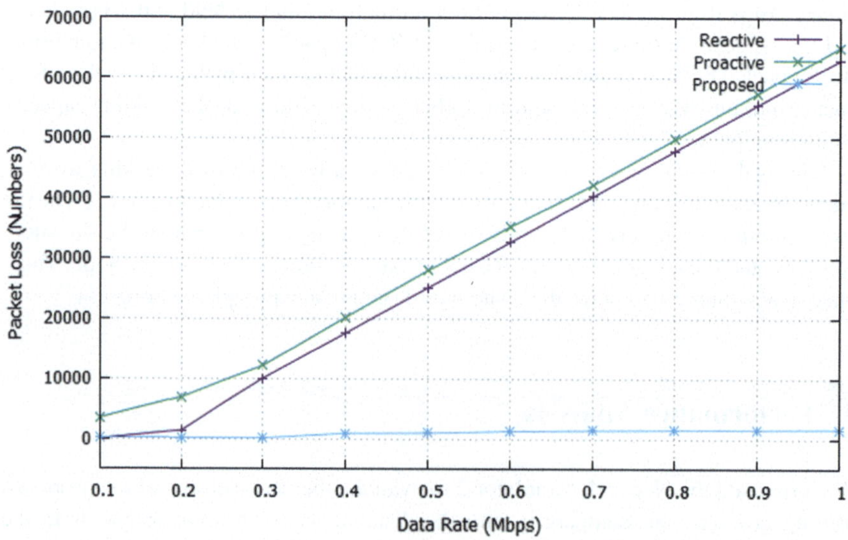

Fig. 1 Packet loss comparison of proposed routing protocol and existing distance vector routing protocols

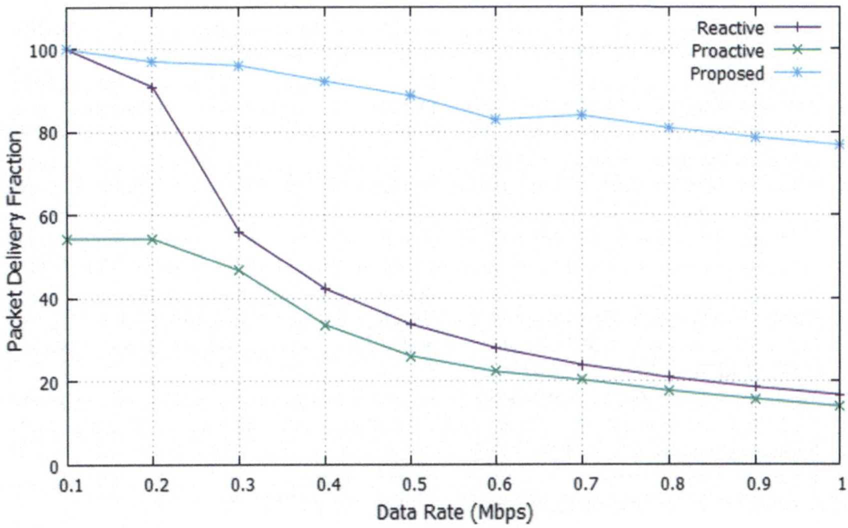

Fig. 2 Packet delivery fraction comparison of proposed routing protocol and existing distance vector routing protocols

4 Conclusion

Mobile ad hoc networks are a distributed peer-peer multi-hop constrained resources network. Applications of MANET are sensitive, and reliable communication is one of the challenging tasks. Routing protocol aim is to compute the route and send the information in computed route. However, routing in MANET faces the problem of packet drop either by malicious activities or by system fault. Major packet drop due to system fault is buffer overflow. The packet drop due to buffer overflow is due to the problem in either queue length at node buffer or packet forward. As MANET is very sensitive to delay regarding communication. Thus, the early congestion control mechanism is needed in MANET to prevent the packet drop due to buffer overflow. This paper addresses the issue by computing the queue length at node buffer during the route computation process. If the queue length of the node is more than the predefined time interval, then node does not participate in communication. Performance of the mechanism is validated by network simulator—2. Results are compared with the existing distance vector routing reactive and proactive routing protocols.

References

1. Hashmi SS (2019) Evaluation of bottleneck intermediate node due to buffer overflow in mobile adhoc networks based on probabilistic model. J Comput Theor Nanosci 16(5–6):2567–2570

2. Sarbhukan VV, Lata R (2019) Impact of mobility and density on performance of MANET. In: Intelligent communication technologies and virtual mobile networks. Springer, Cham, pp 169–178
3. Akhtar N, Khattak MAK, Ullah A, Javed MY (2019) Congestion aware and adaptive routing protocols for MANETs: a survey. In: Recent trends and advances in wireless and IoT-enabled networks. Springer, Cham, pp 159–169
4. Verma A, Khan A (2019) Congestion control using load balancing and multipath routing technique in MANET. Available at SSRN 3372928
5. Vivekananda GN, Reddy PC (2019) Packet loss minimising approach based on traffic prediction for multi-streaming communication over MANET. Int J Wirel Mobile Comput 17(1):1–11
6. Mishra A, Singh S, Tripathi AK (2019) Comparison of MANET routing protocols
7. Kumar J, Kathirvel A (2019) Analysis and ideas for improved routing in MANET
8. Floyd S, Jacobson V (1993) Random early detection gateways for congestion avoidance. IEEE/ACM Trans Netw 4:397–413
9. Wu J, Shi S, Liu Z, Gu X (2019) Optimization of AODV routing protocol in UAV ad hoc network. In: International conference on artificial intelligence for communications and networks. Springer, Cham, pp 472–478
10. Khan MF, Das I (2019). An investigation on existing protocols in MANET. Innovations in computer science and engineering. Springer, Singapore, pp 215–224

Epitome Evolution of Sanctuary to Detect the Interloper in Home Automation

K. Ambika and S. Malliga

Abstract Barrier and assurance become the real aftereffects to the IoT applications and calm, experience some huge hardships. In order to support this rising zone, we, to aggregate things up, review the assessment headway of IoT and spotlight on the resistance. Altogether explore by this strategy for security building and features, the barrier necessities are given. In this point of view, we practice about the appraisal status of a key movement including encryption fragments, the ensuring sensor information cryptographic estimations, and likeness security have quickly laid out the difficulties, Over the top precision, ought to be presented in the significance of gauge status (EEG Longitude Signals) and the way where that sidelong afflictions and prescriptions are of basic centrality in the elucidation of results, ought to dependably be considered. Check of results in single appraisals and self-decision tests are required, thusly, the receptiveness of tests is basic. ATmega328P is used to get the info and yield work which are aid of an Arduino UNO IDE. It is associated with access to the OTA (Over-The-Air) preparing to help the system handling. This system guarantees intruders through the keypad security structure exclusively. It is gotten to by a switch lattice framework. What's more, it ensures enduring without gatecrasher in the framework. Electric power converters are used to create the power supply to the board correspondence between the IDE, at that point, to distinguish and discover the security vitality through the liquid crystal show. It produces the picture relating to prepare for security. At that point, the board has imparted to deliver the key capacity through the presentation which aids in electric power handling. The progress of the order executed by the Door lock for the electric lock will get to the security framework.

Keywords Arduino UNO IDE · OTA—over-the-air · Door lock · Electric power · Liquid crystal monitor · Arduino IDE · Keypad

K. Ambika (✉)
Department of Computer Science, AVS Engineering College, Salem, India

S. Malliga
Department of Computer Science and Engineering, Kongu Engineering College, Perundurai, Erode, India

S.-L. Peng et al. (eds.), *Intelligent Computing and Innovation on Data Science*, Lecture Notes in Networks and Systems 118,
https://doi.org/10.1007/978-981-15-3284-9_67

1 Introduction

The Internet of Things (IoT) is revealed to contain the different structures behind the handling. It produces the shield for insurance to the dependable transmission to start to finish the framework in a particular space. It gathers the required things like a simple way and helps a great deal for the given handling with modest and best to the clients. In addition, it secures them to recognize the weakness as effectively and give information that has spared without mocking of the information in the preparing structure.

Subsequently, this procedure means to actualize for the security reason for the home mechanization. These days everything is behind the procedure of IoT. It contains the sensor, IC, and so on and it joins to control the procedure of insurance given to take the certification to the framework engineering individually. It perceives the way toward decreasing the buffering time to recover the information from the database and its entrance with low control utilization and it is the best bit of leeway of this strategy. The essentialness needs to get to the higher show of the given show. So unmistakably it confines the untouchable getting to and it ensures the consistency of the head concern.

Consequently, this examination advances the execution of security. Whatever the interloper may happen that the data or else dependability is kept up to ensure it as utilized in the IoT-required framework instruments. It upgraded the keypad securing framework because of access by the electric power supply framework to the proprietor between interlopers. It will be connected to the home robotization to put greater security on the solid framework. What's more, in addition, it is utilized to identify the outsider also which aides of the switch lattice framework.

2 Related Works

David Airehrour et al. proposed a period delicate trust-careful RPL directing show (SecTrust-RPL). To give confirmation among Rank and Sybil strikes, the Secure Trust (SecTrust) system is associated [1]. Pericle Perazzo et al. had demonstrated a wireless sensor and actuator networks (WSANs). There are two obligations that are the usage of the RPL security highlights and an assessment of their effect on the WSAN appears by methods for ages [2]. Kumar et al. proposed that an RPL is a coordinating show expected for low control and lossy frameworks. The display of RPL to the extent is a portion of the extra spread information items scattered notice articles created to manage such strikes [3].

Pu et al. proposed a screen-based methodology, called CMD. Every hub screens the sending guardian hub and distinguishes the mischievous activities of the parent hub. The parcel conveyance proportion (PDR) decreases vitality utilization and dormancy [4]. Anthea Mayzaud et al. displayed a scientific classification of the assaults against the convention called RPL. This is the open standard directing convention indicated by IETF [5]. Allot et al. talked about security-related difficulties. Various innovations may take us to trust in IoT applications. The pursue strategies improve security, blockchain, haze registering, edge figuring, and AI [6].

Mario Frustaci et al. has expressed the secrecy in a biological system which is the empowering security issue in IoT. There are three layers called discernment, transportation, and application levels gotten from scientific classification investigation [7]. Dong Wang et al. proposed a multi-get to portable edge processing (MA-MEC). This strategy supports constrained assets, calculation of delicate administrations, and applications. Particularly, the accompanying necessities have researched the protected wiretap coding, asset allotment, signal handling, and multi-hub participation, alongside physical layer key age and confirmation [8]. Florence D. Hudson portrayed a TIPPSS for IoT (Trust, Identity, Privacy, Protection, Safety, and Security) [9]. Granja et al. depicted IP-based correspondence conventions for interchanges stackable to give the required power proficiency, unwavering quality, and Internet availability [10]. Raoof et al. portrayed an RPL strategy for the first-of-its-sort grouping plan. Interruption discovery frameworks (IDSs) feature the order of RPL [11]. Siboni et al. have discussed a security test bed structure that tests a wide range of equipment and programming designs. The general activities are constrained by machine learning (ML) calculations [12]. Tian Donghai et al. proposed simultaneous security observing techniques to dissect the two simultaneous occasions. SIM structure is to send the occasion gatherer into the objective virtual machine and virtualization innovation and multi-center innovation analyzer into confided in execution conditions [13]. Haleh Ayatollahi et al. proposed a positive prescient worth (PPV) of CAD utilizing the fake neural system (ANN) and SVM calculations. The information must be standardized and cleaned before going into SPSS. The SVM strategy predicts high precision and better execution [14]. Nabil Djedjig et al. depicted a Trust Platform Module (TPM) to appraise the various issues of the routing protocol for low-power and lossy networks (RPL). While build topology hub chooses the trust in different hubs [15].

3 Methodology

Gradually improving a security framework with impact by the interloper or criminal with knowing to learn is at higher than the present days. So it considers rousing this exploration for the execution in the security framework which aides if the IoT is significant. IoT is used to give human work as more brilliant than labor separately. Along these lines, it is actualized to the framework security which applies the home robotization by the assistance of the keypad lock preparing in the given framework. It appears to access by the change framework ideas to get to the alphanumeric qualities and ensuring great by the procedure of the electric power supply framework.It is expecting that well-desinged security framework can perform well with less interaction. It contains a shoddy just as best to execute the change of a security framework

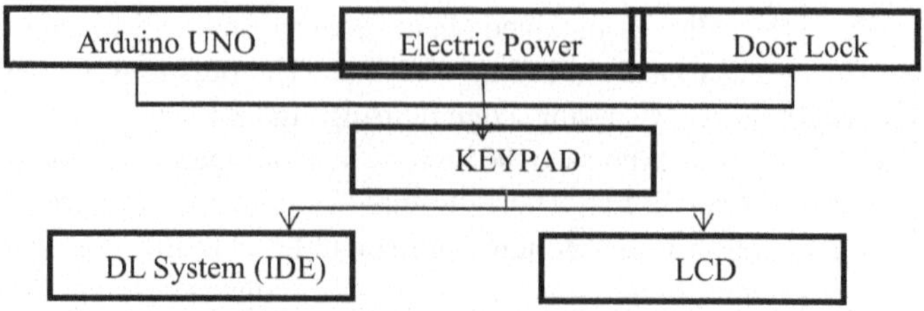

Fig. 1 Framework for security

for actualizing the home robotization which is keeping away from the gatecrasher just as outsider assurance (Fig. 1).

3.1 Arduino UNO

The Arduino board is related to a PC through USB, where it interfaces with the Arduino progression condition (IDE). The customer forms the Arduino code in the IDE, by then moves it to the microcontroller which executes the code, working together with information sources and yields, for instance, sensors, motors, and lights. The Arduino Uno WiFi is an Arduino Uno with a joined WiFi module. The board depends upon the ATmega328P with an ESP8266 WiFi Module encouraged. It contains the electric power supply that has been conveyed by means of the ESP 8266. Since, the procedure has been changed has met by the WiFi underpins with giving the firmware of the framework. It changes over which aides of the OTA with the low power getting to in the specific field.

3.2 Keypad

Arduino with Keypad is dealing with access to pass on the keypad confirmation to the structure. In a standard keypad wired as an *X-Y* switch mastermind, routinely open switches interface a line a section when crushed. On the off chance that a keypad has 12 keys, it is wired as three parts by 4 fragments. It inspects the key encryption strategies which access the switch grid ideas. It is behind the procedure of progress and links the capacity includes in the "*n*" lattice which included when it needs with run-time substance. What's more, it is exceptionally incredible to changes over the key change gatecrasher should identify more learning of this preparing.

The switch matrix has followed by the steps are,

Step 1: Feature selection
Step 2: Select the data set
Step 3: Switch any two rows

Step 4: Multiply a row by a non-zero constant to the matrix
Step 5: Manipulate the result
Step 6: Return the result
Step 7: Stop the process.

And it works on,

$$A.A^1 = I \tag{1}$$

$$\text{So, } A = \begin{pmatrix} 3\,2\,1 \\ 5\,2\,4 \end{pmatrix} \cdot \begin{pmatrix} 2\,4\,1 \\ 3\,6\,3 \end{pmatrix} \tag{2}$$

$$A^1 = \begin{pmatrix} 6\,8\,1 \\ 5\,2\,4 \end{pmatrix} \tag{3}$$

$$I = \begin{pmatrix} 6\,8\,1 \\ 8\,8\,7 \end{pmatrix} \tag{4}$$

where the network needs to change to the inverse framework to the arrangement of switch grid which is utilized to change over,

$$R_1 \leftrightarrow R_2 \tag{5}$$

$$3R_2 \leftrightarrow R_2 \tag{6}$$

Be that as it may, the given framework needs to decide to the procedure of reverse grid for the negative framework develop which recognizes the gatecrasher. Here,

$$A - 1 = 1/\det A \begin{pmatrix} d & -b \\ -c & a \end{pmatrix} \tag{7}$$

If there is a lone key blend that is definitely not hard to find the key advancement anyway taken from the number of plan approaches by the "n" number of key change which raised the hell to the entire grid. Because of the key specifying will be changed by every movement of the dealing within the security structure.

3.3 Electrical Power Consumption

The principle of electric obstruction is a power supply on the electrical device. The essential capability of a power supply proceeds to change the electric stream from a source to the correct voltage, stream, and repeat to control the load. By along these lines, control supplies are immediately and again connoted as electric power converters. A bit of the power accumulations is discrete self-decision bits of apparatus, while others are intertwined with the heap gadgets that they control utilize required. All power supplies have a power input association, which gets a lot of citrus stream from a source, and on any occasion, one power yield affiliations that pass on stream to the store to the structure. That power utilization has been getting to access the double code transformation framework which serves the smaller-scale processor pack of ESP 2866. It decides to distinguish the procedure of intensity on and off framework and it required the insignificant voltage of intensity supply through the handling.

3.4 Liquid Crystal Display

It needs to uncover the preparing message for the administrator else end client. It gives the data to get to its correct or not while running the procedure of home computerization individually. By and large, the screen has been utilized to toss the data like a message to appear by the presentation. It has appeared as an ongoing preparing structure through the change of the bits. It hurls the switch network limit getting ready for protection key to enable it to show up in the instrument separately. In case the pariah has been gotten to the keypad, it hurls the acoustic and message to the contrasting executive with the security structure.

3.5 Door Lock System

An electronic lock (or electric lock) framework has been resolved the procedure of open or bolts to the specific handling by means of the home computerization which assists with sensor recognition. It gets to with backings of paired (0 and 1). Zero looks at the procedure has not begun regardless of whether it is begun the parallel change to 1 The switch network when it turns out badly for the outsider consequently it tosses the message and it looks well. Also, safely it recognizes the gatecrasher and it performs caution or message as the administrator willing separately. This could be the spot security is a fundamental concern. If the structure has revealed to see the interloper with entering the number as upside-down routinely, the door was affected. In addition, it seems to send the acoustic else the message has through sent by procedures for the sign planning which assistants of electric power association.

4 Result and Discussion

The IoT perceives the affliction ID is clear while getting to the installing procedure of the IDE System. By then it shows to the temperature and improvement revelation in like manner in the annoying must be settled as ideal for further arranging. In this context, whatever the arranging has been understood, some security issues occurred while it improving the assurance through this idea for the encryption premise authorities the headway of the switch framework system. As appeared by the IoT, the security system has been picked and gives the best result to the security protection in the home computerization structure. Everything considered, the application dealing with has been obliged by the given transmission framework. By then the yields are (Figs. 2, 3, and 4)

CODING:

```
for(pos=180; pos>=0; pos-=5) //goes from 180 degrees to 0 degrees
{myservo.write(pos); //tell servo to go to position in variable 'pos'
delay(5); //waits 15 ms for the servo to reach the position}
delay(2000);
delay(1000);
```

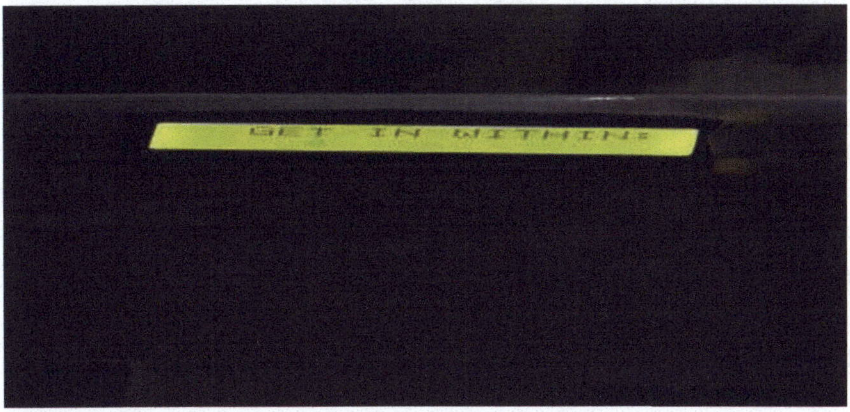

Fig. 2 Stating process

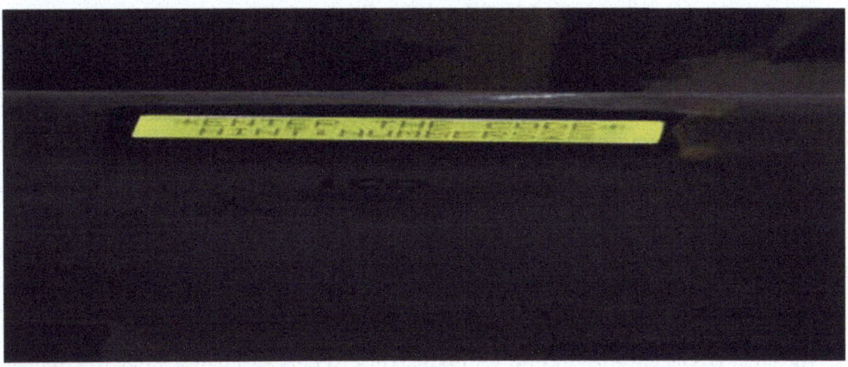

Fig. 3 Enter the process

Fig. 4 Re-locking system

```
counterbeep();
delay(1000);
for(pos=0; pos<=180; pos +=5) //goes from 0 degrees to 180 degrees
{myservo.write(pos); //tell servo to go to position in variable ‘pos’
delay(15);
currentposition=0;
lcd.clear();
displayscreen();}}
```

5 Conclusion and Future Enhancement

This assessment convinces to be executed with the correspondence strategy which supports the Web perspective. It was set up over the world in perspective on the introducing used at a wide level. From now on, the best course is to choose the controlling capacity and security through structure solidly. This moved utilization has been getting the chance to give the exactness, execution, security, uprightness to the trustworthy structure. It is chiefly based on the bad behavior people who need to unauthorize to mocking the data and perceive to impede they’re getting ready while we know the unapproved person. Else, it will work outstandingly encircled and confirmation is extended more encryption dealing with. Further research is required to execute for maintaining a strategic distance from the outsider capacity access in the home computerization framework.

References

1. Airehrour D, Gutierrez JA, Ray SK (2019) SunTrust-RPL: A secure trust-aware RPL routing protocol for Internet of Things. Future Gener Comput Syst 93:860–876

2. An Implementation and Evaluation of the security features of RPL, Pericle Perazzo, Carlo Vallati, Antonio Arena, Giuseppe Anastasi, Gianluca Dini, International Conference on Ad-Hoc Networks and Wireless, 63–76, 2017
3. Kumar A, Matam R, Shukla S (2016) Impact of packet dropping attacks on RPL. In: Fourth international conference on parallel, distributed and grid computing (PDGC), pp 694–698
4. Pu C, Hajjar S (2018) Mitigating forwarding misbehaviors in RPL-based low power and lossy networks. In: 15th IEEE annual consumer communications & networking conference (CCNC), pp 1–6
5. Mayzaud A, Badonnel R, Christ I (2016) A taxonomy of attacks in RPL-based Internet of Things. Int J Net Secur 18(3):459–473
6. Chamola V Saxena V Jain D, Goyal P, Sikdar B (2019) A survey on IoT security: application areas, security threats, and solution architectures. IEEE Access
7. Evaluating Critical Security Issues of the IoT World: Present and Future Challenges, Mario Frustaci, Pasquale Pace, Gianluca Aloi, Giancarlo Fortino, IEEE Internet of Things Journal, 2018
8. Wang D, Bai B, Lei K, Zhao W, Yang Y, Han Z (2019) Enhancing information security via physical layer approaches in heterogeneous IoT with multiple access mobile edge computing in smart city. IEEE Access
9. Hudson FD (2018) Enabling trust and security: TIPPSS for IoT. IEEE Access
10. Granjal J, Monteiro E, Silva JS (2015) Security for the Internet of Things: a survey of existing protocols and open research issues. IEEE Commun Surv Tutor 17(3):1294–1312
11. Raoof A, Matrawy A, Lung C-H (2018) Routing attacks and mitigation methods for RPL-based Internet of Things. IEEE Commun Surv Tutor 21(2):1582–1606
12. Siboni S, Sachidananda V, Meidan Y, Bohadana M, Mathov Y, Bhairav S, Shabtai A, Elovici Y (2019) Security testbed for Internet-of-Things devices. IEEE Trans Reliab
13. Donghai T, Xiaoqi J, Junhua C, Changzhen H (2016) A concurrent security monitoring method for virtualization environments. China Commun
14. Ayatollahi H, Gholamhosseini L, Salehi M (2019) Predicting coronary artery disease: a comparison between two data mining algorithms. BMC Public Health 19(1):448
15. Djedjig N, Tandjaoui D, Medjek F (2015) Trust-based RPL for the Internet of Things. In: IEEE symposium on computers and communication (ISCC), 2015

Endowing Syndrome Empathy to the Epileptic and Cardiovascular Embedding IoT Techniques

S. Sowmiyasree, A. Hariharan, P. Jyothika Shree, and D. S. Gayathri

Abstract Web of things (IoT) improvement has chosen the breaking point of limit and sharp endeavoring to the simple to use to the customers. It is utilized to deliver the best arrangement which aides of the inserting od Arduino IDE with the standards. It has some key highlights of an imperfection in serious issues, for this usage which elevates the defeat because of chronicled issues with access for all system structures. This examination was to pick the purpose of showcases the front line research relating to each area of the model, particularly for considering because of the human services framework and, surveying their characteristics, deficiencies, and overall suitability for an IoT social protection structure. Difficulties that therapeutic organizations IoT appearances including security, protection, wearability, and low-control activity have appeared and recommendations are made for future extraction to this domain. It is executed for the medicinal services framework which decreased the labor work to deliver the exactness and higher execution of this framework Arduino mega used to identify the preparing of the PWM info and yield handling framework. The ESP 2866 Node MCU needs to utilize and distinguish the equipment preparing framework to improve the procedure of association foundations to the framework and IDE. HBS and PS used to distinguish the beat and heartbeat for the ordinary and irregular handling of the installing. At long last, the voltage has isolated to get to the entire preparing by 10K potentiometer.

Keywords Arduino mega · ESP 2866 · Node MCU · HBS—heart beat sensor · PS—pulse sensor · 10K potentiometer

---

S. Sowmiyasree (✉)
Excel College of Arts and Science, Salem, Tamil Nadu, India
e-mail: gnosistechscience@gmail.com

A. Hariharan · P. Jyothika Shree
PSG College of Arts and Science, Coimbatore, India

D. S. Gayathri
Bharathiar University, Coimbatore, Tamil Nadu, India

S.-L. Peng et al. (eds.), *Intelligent Computing and Innovation on Data Science*, Lecture Notes in Networks and Systems 118,
https://doi.org/10.1007/978-981-15-3284-9_68

1 Introduction

It will be exhibited for the enabling shrewd execution of the strategies required to get to the best ideal answer for the clients. Anyway, it has benefits and bad marks for the procedure which has been picked in the field of work. Furthermore, additionally, it was adequate to work for the remote getting to control the intensity of control with the particular IoT necessities. It could be utilized to give better access to human services to those living in country regions or to empower older individuals to live freely at home for more. Basically, it can improve access to medicinal services assets while decreasing strain on social insurance frameworks and can give individuals better command over their very own well-being consistently. The reason for a productive IoT social insurance framework is to give consistent remote seeing of patient flourishing conditions, to check the basic patient conditions and to improve singular satisfaction through an unbelievable IoT condition. New difficulties have been presented with IoT for the security of frameworks and forms and furthermore with the protection issues of an individual's medicinal information. Data security utilizing IoT is exceptionally confused and troublesome; since worldwide availability and openness are the real concerns identified with IoT, security and protection by configuration should be a piece of any IoT use case, undertaking or arrangement.

2 Related Works

Munzel et al. depicted a Social systems administration site (SNSs). To set up the close to home interchanges and cooperation of the individuals' general connectedness of the system [1]. Fan et al. developed a Wuhan Smart Health that contains examination, engineering structure, execution, issues, arrangements, etc. This strategy has a development model, which gives references to different urban areas [2]. Ndiaye et al. portrayed wireless sensor frameworks (WSNs), which contain hundreds and thousands of sensors. Programming Defined Networking (SDN) offers reactions for WSNs by permitting division of the controlling technique for thinking to structure the sensor focus focuses [3]. Pramanik et al. portrayed major information empowered keen medicinal services framework structure (BSHSF). There are five difficulties that are connected to the BSHFS social insurance business situations. The first inside and out is best-in-class huge information and keen human services frameworks in parallel [4]. Venkatesh et al. proposed a measured methodology for IoT applications. This technique utilized general AI to lessen repetition. Nearness, client action, air quality, and area from IoT sensors exhibit a savvy situation [5]. Ali et al. depicted a voice issue decide by voice signal from the discourse through the straight expectation investigation. It recognizes the lower recurrence of 1–1562 Hz. This framework is created by a supported vowel and running discourse [6]. Hossain et al. proposed a voice pathology recognition (VPD). There are two kinds of information conveyed: a voice signal and an EGG signal. Gaussian blend model-based methodology is utilized for the extraction of voice information [7]. Sahoo et al. proposed an electrocardiography (ECG) and multi-channel seismocardiography (SCG) frameworks. The system screens the cardiovascular activities of a patient and exactness [8]. Kim et al. depicted a Gaussian procedure (GP) technique for dynamic strolling designs and inherent wearable sensors. This methodology joins a GP-based state-space demonstrating strategy with a nonlinear dimensionality decrease technique in a one of a kind way [9]. Mehta et al. portrayed a Regional Transport Office (RTO) and Computerized Pollution Check Centers examination of the sensor-based information from the air quality. The client of the portable application serves poisonous quality [10]. Nef et al. proposed a detached infrared (PIR). Exercises of day by day living (ADL) Of their security. The movement of acknowledgment precision is improved by distinguishing covering exercises. PIR-based savvy home innovation improves the working of social orders [11]. Dejian et al. proposed a double center structure for keen urban areas. This drives from the advances arranged on the foundation to resident situated applications. It is firmly identified with the learning situations of urban areas [12].

3 Methodology

The most noteworthy weaknesses incorporate the security chance that accompanies having a lot of touchy information put away in a solitary database, the potential need to normally have the clients have been identified the handling which emotionally supportive network of the sensor. In light of this sensor needs to catch the level of preparing recurrence because of the ailment of patients separately. It will find the apportion of the exactness level in the structure (Fig. 1).

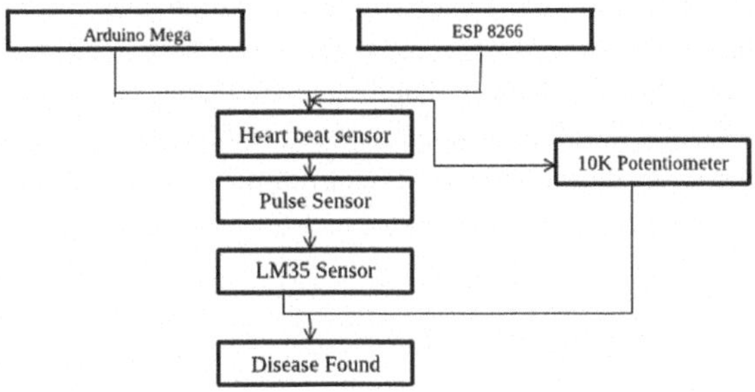

Fig. 1 Framework design for healthcare system

3.1 *Arduino Mega*

The Arduino Mega is a microcontroller named ATmega2560 and it is reliant on the controller of Arduino IDE. It has 54 moved data/yield pins (of which 14 can be used as PWM yields), 16 direct wellsprings of information, 4 UARTs (gear back to backports), a 16 MHz other than it is the fundamental introduction of the oscillator structure pursue to get to the framework. The USB association has been set up for the secured base and association between the framework and the Arduino IDE installing which aides of a power jack, an ICSP header which stacking to get to the buffering information and yield factors, and a reset catch to the capacity empowering and the re-association used to get to the Arduino embeddings. The introducing of A-Mega has been overcome the marked structure which controls the sign by the sign dealing with to the IoT planning. It is the fundamental affiliation establishment for the I/O discoverer which realized by the value control to the therapeutic administration's system. It is specified to check the handling for the essential statement about the which distinguish the header record which stacking by the info and yield variable choice to the preparing of association between start to finish hub.

3.2 *ESP 8266*

NodeMCU is an open-source IoT organize to the chip. It is executed by the 2AA battery work which is controlled through the system for the node. What's more, also, it will be utilized to distinguish the rationale controller which shows the basic leadership framework to mapping just as bypassing to transmit the information inside in the illness discovery framework. The Node MCU approaches by the required low-control utilization which is 3.6 V are executed for the given framework usage. So clearly, it is shabby and the creation is useful for identifying the disease in the utilization of IoT. Every single IoT framework has been controlled as quicker and produces the best exactness for the sensor recognition partook to distinguish it. Regardless of the way that the non-SDK has been presented for the penniless system which is eLua experience to the structure, it is energetically gotten to and produces the best outcome for the center point.

3.3 *Heartbeat Sensor*

Heartbeat sensor is needed to give a computerized yield of warmth beat when a finger is made plans to it. Right when the heartbeat locator is working, the beat LED flashes as one with every heartbeat. This sensor has been perceived by the beat acknowledgment which helpers of the electronic device. It will perceive the beat each second (BPM) and after that executes the data in the correct manner. It calculates the process of,

$$X_K = \sum (W^K)^N . x_n (\text{Interval of } n = 0 \text{ to } n - 1) \tag{1}$$

Thus, the recurrence has been controlled to distinguish the recurrence which is accessed by the Fourier arrangement change with the official of discrete structure. What is more, it dispenses with the multifaceted nature which aides of the number juggling estimation to the given arrangement. Then, the frequency interest calculated to,

$$F = K / N\, f_{\text{sample}} \tag{2}$$

K is constantly a consistent variable that has been resolved to distinguish the example informational collection. It has been executed by the preparation of informational collection. Also, in addition, it will be an improvement to the taken number of factors that are executed by the number-crunching estimation.

The polynomial distribution is declared to,

$$X^K = (((W^{-K} . x_0 + x_1) . W^{-K} + x_2) . W^{-K} + x_3) + X_{N-1}) . W^{-K} \tag{3}$$

It is manipulated to the filtering is,

$$Y = [W^{-K} - Z^{-1}][1/1 - 2\,\text{Re}(W^{-K})Z^{-1} + Z^{-Z}]X \tag{4}$$

It will be used to diminish the tumult to the coefficient elements to the given range and it plays out the switch change had gotten the essential weight taking care of while simultaneously finding the repeat.

3.4 Pulse Sensor

Heartbeat sensor is a fitting and-play pulse sensor for Arduino. It is used to detect the heartbeat from the thump impression and it is required for 4–6 V of the signal detecting system. It works well to detect noise detection which adds on the amplifier for the signal transmission.

The diameter detection is,

$$D = 2r \tag{5}$$

$$C = \phi d \tag{6}$$

Fig. 2 10K potentiometer

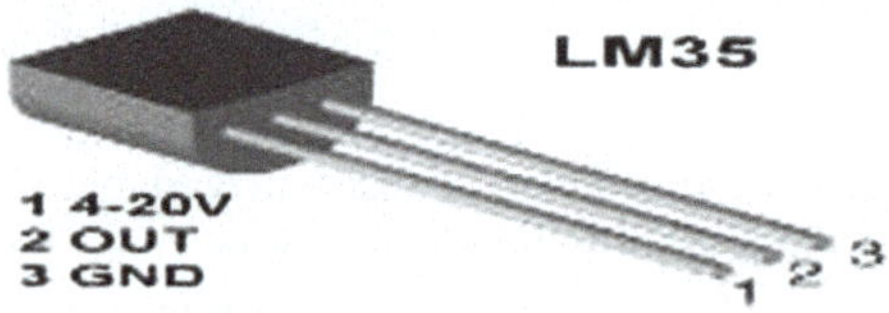

Fig. 3 LM35 sensor detection

The distance across has been estimated by the span of the range and it recognizes the polynomial math capacity that is resolved to the perimeter by the preparing of the transmission capacity scope of the significant change to the framework.

3.5 10K Potentiometer

The picture for a potentiometer is a comparable one as a resistor sets something aside for a jolt in the inside. In a circuit where they are used cautiously as factor resistors or rheostats, only two terminals are wired to various parts. The majority of the three terminals are wired independently when they work as voltage dividers (Figs. 2 and 3).

3.6 LM35 Sensor

LM35 is a consolidated straightforward temperature and recognizes the framework has been uncovered the temperature to the human body identification it decides the information measurements in the precision of the temperature of the precision. It does not require any outer course of action of the inserting or cutting to the time procedure that has been giving conventional accuracy and find the layout of the system. These estimations could be recorded during a couple of activities, for instance, standard walking and recuperation work out. They could be conveyed by means of short-go correspondences to an agreeable, wrist-wearable focal hub, which could then advance data to the cloud through long-ago interchanges. The preparing result has been put away in virtual memory. The administrator or the clients or the individuals who need to realize the status will recover the database to recognize it effectively. Something else, the mystery data (Sensitive) has been covered up to secured unequivocally.

4 Result and Discussion

Subsequently, it has been controlled to play out their activities which aides of the twofold transformation handling for the bits separately. Also, it signifies to give the extra usefulness of the movement caught by the sign handling because of the sensor location too in the IoT system. In a couple of words, IoT human administration systems have been delivered for unequivocal purposes, including recuperation, diabetes the board, helped encompassing living (AAL) for more seasoned individuals, and that is just a hint of something larger. While these structures have been proposed for a wide extent of purposes, they are each unequivocally related through their use of comparable empowering progressions (Figs. 4 and 5).

Thus the above processing used to detect system processing as well for the healthcare system. It was precise to access control mechanisms with different techniques and with energy efficiency. It has different types of protocols for authentication to access with privacy maintain to the admin to doctors. A system is required for the fusion of authentication protocol with an energy-efficient access control mechanism

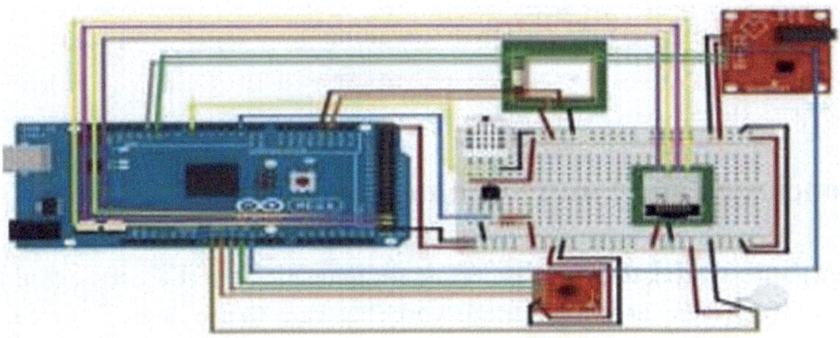

Fig. 4 Structure of the healthcare system

Fig. 5 Accuracy detection

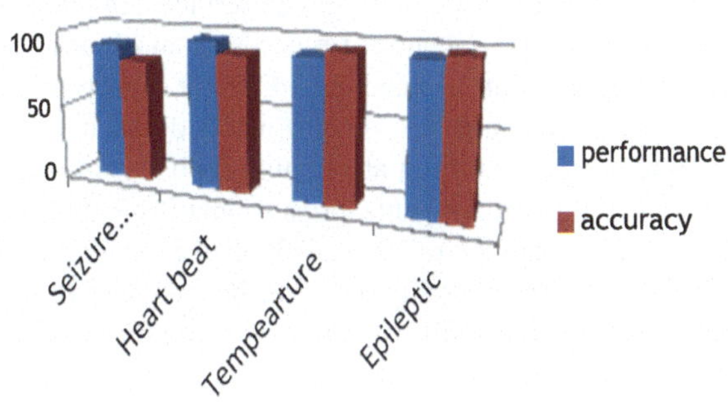

along with the solutions to countermeasure the other attacks in security and privacy of patient healthcare data. After going through the methodology for authentication protocol, for access control, and for energy-efficient access control mechanism, a combined methodology is proposed to be adopted to pool the gap. Barely any techniques have proposed various sorts of conventions for confirmation. A framework is required for the combination of confirmation convention with a vitality productive access control instrument alongside the answers for countermeasure different assaults in security and protection of patient human services information.

4.1 Coding

```
BLYNK_READ(PIN_UPTIME2){
 j = analogRead(A1);
 j = 1024-j;
 j = (33*j)/561;
Blynk.virtualWrite(PIN_UPTIME2,j);}
BLYNK_READ(PIN_UPTIME3)
{k = analogRead(A2);
 if(k < 10) {
  Blynk.virtualWrite(PIN_UPTIME3,70);}
 else if(k < 50) {
  Blynk.virtualWrite(PIN_UPTIME3,"waiting");}
```

5 Conclusion and Future Enhancement

The conventional model we have proposed for managing the advancement of future Internet of things human services frameworks has various use cases. To give setting, this subsection talks about a few of these utilization cases, which incorporate supporting the restoration, helping the administration of constant conditions, checking changes in individuals with degenerative conditions, and observing basic well-being for the arrangement of crisis human services. In the wake of encountering the way of thinking for approval show, for access control, and for essentialness gainful access control instrument, a joined methodology is proposed to be gotten to pool the opening. As the movement keeps being made to lessen the harms, IoT-based structures for remote success enrolling are turning with an obviously reasonable reaction for the approach of helpful organizations sooner instead of later. Each and each framework has been recognized to the malady distinguishing proof which aides of the sensor discovery. Also, in addition, it will be executed to the higher inserting to improve the social insurance framework for making higher precision and execution with less cost.

References

1. Munzel A, Meyer-Waarden L, Galan JP (2018) The social side of sustainability: well-being as a driver and an outcome of social relationships and interactions on social networking sites. Technol Forecast Soc Chang 130:14–27
2. Fan M, Sun J, Zhou B, Chen M (2016) The smart health initiative in China: the case of Wuhan Hubei Province. J Med Syst 40(3):62
3. Ndiaye M, Hancke GP, Abu-Mahfouz AM (2017) Software-defined networking for improved wireless sensor network management: a survey. Sensors 17(5):1031
4. Pramanik MI, Lau RY, Demirkan H, Azad MAK (2017) Smart health: big data-enabled health paradigm within smart cities. Expert Syst Appl 87:370–383
5. Venkatesh J, Aksanli B, Chan CS, Akyurek AS, Rosing TS (2017) Modular and personalized smart health application design in a smart city environment. IEEE Internet Things J 5(2):614–623
6. Ali Z, Muhammad G, Alhamid MF (2017) An automatic health monitoring system for patients suffering from voice complications in smart cities. IEEE Access 5:3900–3908
7. Hossain MS, Muhammad G, Alamri A (2017) Smart healthcare monitoring: a voice pathology detection paradigm for smart cities. IEEE Access 5:3900–3908
8. Sahoo PK, Thakkar HK, Lee MY (2017) Cardiac early warning system with multi-channel SCG and ECG monitoring for mobile health. Sensors 17(4):711
9. Kim T, Park J, Heo S, Sung K, Park J (2017) Characterizing dynamic walking patterns and detecting falls with wearable sensors using Gaussian process methods. Sensors 17(5):1172
10. Mehta Y, Manohara Pai MM, Mallissery S, Singh S (2006) Cloud-enabled air quality detection, analysis, and prediction—a smart city application for smart health. In: 3rd MEC international conference on big data and smart city (ICBDSC), pp 1–7
11. Nef T, Urwyler P, Büchler M, Tarnanas I, Stucki R, Rizzoli D, Müri R, Mosimann U (2015) Evaluation of three state-of-the-art classifiers for recognition of activities of daily living from smart home ambient data. Sensors 15(5):11725–11740 (2015)
12. Liu D, Huang R, Wosinski M (2007) Development of smart cities: educational perspective. In: Smart learning in smart cities, pp 3–14

Blockchain-Based Information Security of Electronic Medical Records (EMR) in a Healthcare Communication System

Rafita Haque, Hasan Sarwar, S. Rayhan Kabir, Rokeya Forhat, Muhammad Jafar Sadeq, Md. Akhtaruzzaman, and Nafisa Haque

Abstract Protecting the digital medical data from cyber-attacks is essential in a healthcare communication system. Blockchain is a distributed as well as shared ledger-based technology which can more readily encourage information security. Blockchain approach is commonly applied for the digital crypto-currency; however, it is awaited to have future conducts in health care. Many healthcare systems today are still subordinated on the back-dated application for keeping healthcare records. Here, the main challenges faced to date are information security and interoperability. This study introduces a blockchain-based model for securing the Electronic Medical Records (EMR). Here, we have used SHA256 secure hash algorithm for generating a unique and identical 256-bit or 32-byte hash value for a particular medical record. In this manuscript, we have focused on five mechanisms (Digital Access Rules,

R. Haque · H. Sarwar
Centre for Higher Studies, Bangladesh University of Professionals, Dhaka, Bangladesh
e-mail: rafitahaque@aub.edu.bd

H. Sarwar
e-mail: hsarwar@cse.uiu.ac.bd

H. Sarwar
Department of CSE, United International University, Dhaka, Bangladesh

R. Haque · S. R. Kabir · M. J. Sadeq · Md. Akhtaruzzaman (✉)
Department of CSE, Asian University of Bangladesh, Dhaka, Bangladesh
e-mail: azaman01@aub.edu.bd

S. R. Kabir
e-mail: rayhan923@aub.edu.bd; rayhan561@diu.edu.bd

M. J. Sadeq
e-mail: jafar@aub.edu.bd

S. R. Kabir · R. Forhat
Department of Software Engineering, Daffodil International University, Dhaka, Bangladesh
e-mail: rokeya35-1028@diu.edu.bd

N. Haque
Department of CSE, Daffodil International University, Dhaka, Bangladesh

S.-L. Peng et al. (eds.), *Intelligent Computing and Innovation on Data Science*,
Lecture Notes in Networks and Systems 118,
https://doi.org/10.1007/978-981-15-3284-9_69

Data Aggregation, Data Immutability, Data Liquidity, and Patient Identity) of data transition for securing the medical records at the proposed blockchain model.

Keywords Blockchain · Secure hash algorithm · Health information technology · Healthcare information security · Interoperability · Cyber-security

1 Introduction

Blockchain innovation has progressed toward becoming nearly as well known for instances including data or information security with respect to its technological potential. Electronic Medical Record (EMR) is the collection of digital health information and stores the information electronically by using a healthcare application [1, 2]. There are a few territories of health care that can be improved by utilizing blockchain technology. This incorporates health informatics, clinical preliminaries, pharmaceutical section, and medical coverage. Ordinary clinical information endeavors are frequently nonetheless which make hindrances to proficient data trade and blocks compelling treatment choice made for patients. This research gives the approach of blockchain technology for ensuring the security of clinical information in the healthcare system.

The 'American Recovery and Reinvestment' law section of 'Health Information Technology for Economic and Clinical Health 2009 (HITECH-2019)' Act conserved approximately 30 billion USD in assets to cooperate EMR assumption by US healthcare suppliers [3]. Because of this exertion, suppliers and healing center utilization of EMR have expanded drastically, while just 9% of non-government intense consideration doctor's facilities had a fundamental EMR in 2008, and 96% had an EMR by 2015 [4]. Cyber-security in various territories is a significant subject all over the world. The healthcare part has quickly turned into a goal for digital assaults [5]. Healthcare divisions are especially sensitive to these sorts of digital assaults. For these perspectives, the security matter is very essential for securing medical data and protecting against cyber-attack at the healthcare system.

The blockchain technology is a recent novel development and innovation that could have conducted in enhancing information security. Blockchain portrayed in detail somewhere else [6] has specific interest to well-being information given its accentuation on sharing, dissemination, and hashing. The objectives of this research paper are to give the ideas about the structure and activity of blockchain in healthcare or medical system and how the utilization of this innovation can be utilized to give security and protection in healthcare communication system. Specifically, more current blockchain endeavors savvy contracts, second-layer frameworks, authorization blockchain facilitate the potential medicinal services utilize cases, and there has been no deficiency of promotion encompassing the capability of the innovation inside human services [7].

In this work, how blockchain innovation may encourage in a healthcare system through five approaches: Digital Access Rules [8], Data Aggregation [9], Data

Immutability [10], Data Liquidity [11], and Patient Identity [12]. Here, we have demonstrated how SHA256 hash algorithm [13] utilized on above five attributes for securing the medical records in a blockchain-based healthcare communication system.

2 Related Works and Research Gaps

In recent research, an idea of healthcare-based blockchain system has been described where how the system may transmit the digital medical data via five approaches [14]. These approaches are Digital Access Rules, Data Aggregation, Data Liquidity, Data Immutability, and Patient Identity. Here, the researchers focused on hindrances to patient-driven interoperability, explicitly clinical information exchange volume, protection and security. Another research has been expressed about data sharing where a blockchain-based method utilizing computerized health identities to confirm members for distance cancer care [15]. In this manuscript, clinical information is put away off the blockchain, and stores encoded information which fills in as pointers to the essential information source. Boulos et al. proposed a geospatial blockchain-based medicinal services technique that uses a crypto-spatial organized framework to include a changeless spatial context [16]. Rahmadika and Rhee also propose a peer-to-peer (P2P) network-based blockchain technology for decentralized personal health information [17]. A recent research article describes the effect of blockchain technology for EMR data in the next-generation healthcare system [18].

Despite the fact that the possibility of an advanced EMR has been portrayed for quite a long time, there has been perceptible footing from the blockchain technology and administrative point of view. Furthermore, above former research works have not expressed any technical approach by using five mechanisms (Patient Identity, Digital Access Rules, Data Immutability, Data Aggregation, and Data Liquidity) in the different steps of the blockchain system for securing the EMR data. According to this research gap, this paper has been committed a research, where a blockchain technology helps the healthcare communication system for securing the EMR information by using the above five attributes in different steps of the blockchain process.

3 Role of Five Mechanisms at Proposed Blockchain Model

The healthcare communication system prescribes interoperability [19] where the efficiency of different digital or software systems to interchanges information, communicate, and authorized data usages. Interoperability is the characteristic that permits the unrestricted sharing of resources between different systems or application [20]. In our experiment, we have set up interoperability-based blockchain system [21] where data transits from one system to another system. In here, we focused Patient Identity, Data Aggregation, Digital Access Rules, Data Immutability, and Data Liquidity and utilized these mechanisms at different stages of blockchain system. The role of the above five mechanisms in our proposed blockchain model are given below.

- **Digital Access Rules**: The doctors and medical staffs can use the features of blockchain system and access the particular modules for the particular EMR.
- **Data Aggregation**: This is a form of data gathering process where information is presented in a report-based or summarized. The summarized EMR texts are encoded at the proposed system for security purpose.
- **Data Immutability**: This mechanism has been used for securing or encoding the medical images or pictures, such as CT scan and X-ray image. At proposed blockchain model, the clinical image or picture data is stored as encoded data.
- **Data Liquidity**: It means to confirm whether the right data is provided to the right user at the right period. It is very vital part of blockchain-based healthcare system where particular encoded data is provided for particular doctor or stakeholder node. Here, the EMR information has been converted to a unique hash code by SHA256 hash algorithm.
- **Patient Identity**: Patient identification is another vital part of the proposed system for setting and acquiring their medical records. In the proposed model, every medical record has been differentiated by a different patient.

4 Blood Test and X-Ray Dataset

In our research, we collect some blood test dataset from a medical center (Life Diagnostic Centre Ltd.). In Table 1, we show some portion of patient blood data where patients are identified by code. Here, the dataset contains bilirubin, SGPT (ALAT), albumin, urea tests, etc. We also used some medical X-ray images which have been collected from this medical center.

5 SHA-256 Hash Algorithm

Secure Hash Algorithm 2 (SHA-2) is a section of cryptographic hash approaches developed by National Security Agency, USA. There are six hash algorithm concepts of SHA-2 algorithmic section that are SHA-224, SHA-256, SHA-384, SHA-512, SHA-512/224, and SHA-512/256. In our experiment, we have used SHA-256 hash algorithm. We used SHA-256 function because of collision resistance. The other benefits [22] of SHA-256 are given bellow.

- The outputs of SHA-256 are shorter, and it saves bandwidth.
- Generally, SHA-512 is not quicker on 32-bit processors, but it is faster on 64-bit processors. On the other hand, SHA-256 is faster on 32-bit processors and also shows well performance on 64-bit processors.

Table 1 A sample portion of medical blood test data

S. no.	Code no.	RBS normal up to 140 m/L	Bilirubin (T & D) normal up to 1.00 mg/dL	SGPT (ALAT) normal up to 41 mg/dl	Urea normal 10–50 mg/dL	Total protein normal 6.7–8.7 g/dL	Albumin normal 3.4–4.8 g/dL	Blood group
1	SA-17-07-1	110.5	0.72	26	27	6.9	3.6	AB+
2	SA-17-07-2	116.9	0.67	30	41	7.7	3.4	A+
3	SA-17-07-3	110.7	0.66	26	27	6.9	3.6	B+
4	SA-17-07-4	116.9	0.72	29	41	7.7	3.4	B+
5	SA-17-07-5	120.7	0.69	32	22	6.7	4.2	O+
6	SA-17-07-6	116.9	0.67	26	23	7.6	3.7	AB+
7	SA-17-07-7	110.7	0.67	26	32	7.9	3.9	B+
8	SA-17-07-8	116.5	0.66	30	21	6.8	3.7	O+
9	SA-17-07-9	110.7	0.66	26	27	6.9	3.6	B+

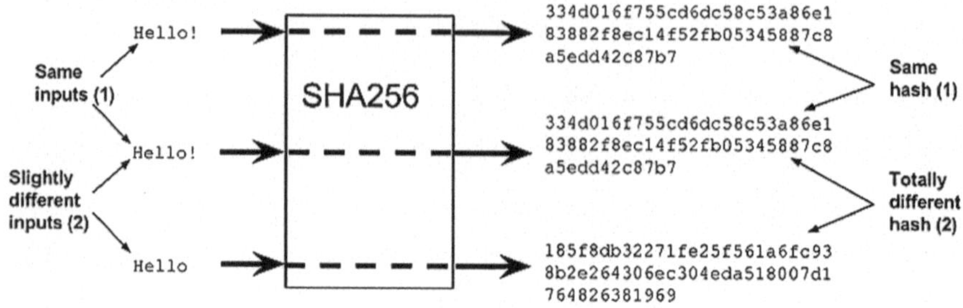

Fig. 1 Different outputs of 'Hello!' and 'Hello' texts at SHA-256 approach

- SHA-256 cannot make collisions, but other types of SHA-2 algorithms are not collision resistance.

In our experiment, SHA-256 produced an almost unique 256-bit (32-byte) signature for a medical data. Figure 1 shows the outputs of 'Hello!' and 'Hello' texts by using SHA-256 algorithm.

6 Proposed Blockchain Model

In the proposed model for securing the medical, the doctors and staffs connected to a blockchain system where they can transfer the medical records. Here, the medical data was converted to a unique hash code by using SHA-256 algorithm. Figure 2 demonstrated the activities of Access Rules, Data Aggregation, Data Immutability, Data Liquidity and Patient Identity at proposed healthcare-based blockchain model with short details.

7 Result and Limitation

The proposed model is under development. Generally, we represented how we can use a hash algorithm in a blockchain system for securing the medical data. In our current work, there are some limitations. We just only used blood test data and get some output of hash values. The uses of X-ray image EMR data are now under experiment. In our extended version of this research, we will show output for image EMR data and also the part of the application.

Table 2 shows the hash data of some blood test results. Here, we identified the patient data by code (see Table 1). In our experiment, we used eleven types of blood test. There are bilirubin, random blood sugar (RBS), serum glutamic pyruvic transaminase (SGPT), serum glutamic oxaloacetic transaminase (SGOT), Gamma-glutamyl transferase (GGT), creatinine, protein, albumin, Hb% (hemoglobin concentration),

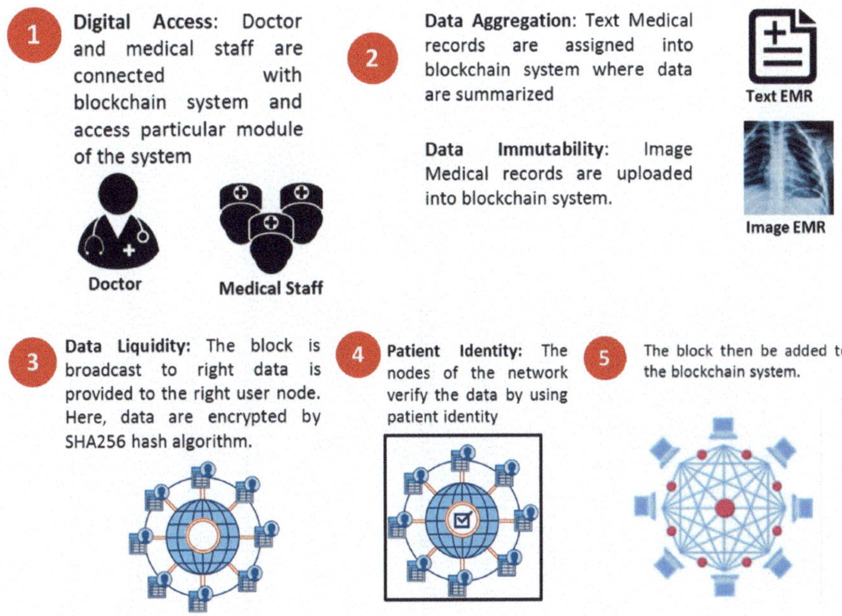

Fig. 2 Steps of proposed blockchain system which focused on Digital Access Rules, Data Aggregation, Data Immutability, Data Liquidity, and Patient Identity

and blood group. Table 2 shows the encoded result of different types of blood test data. Here, the different types of blood test data are converted to 256-bit (32-byte) hash value which passed patient code or identity.

8 Conclusion and Future Works

Cyber-security in different areas is an important topic throughout the world. The healthcare or medical sector has rapidly become a destination for cyber-attacks. Healthcare or medical sectors are particularly delicate to these sorts of cyber or digital assaults. Here, disruption in activities or even divulgence of patient private information can have broad outcomes. For these perspectives, a blockchain-based healthcare communication system is very essential for securing medical data and protecting against cyber-attack. This paper describes a blockchain-based healthcare system for securing the electronic medical data for protecting against cyber-attack.

There has been an increasing move toward five mechanisms for electronic medical data transition at healthcare-based blockchain system. Here, the research revealed the potential part of blockchain system for storing healthcare or medical data as a data security purpose. In this manuscript, we have revealed a dream model to manage and secure the EMR data using blockchain technology.

Table 2 Hash values of different blood test data at proposed healthcare-based blockchain system

Types of blood test	Blood test result	Hash value 256 bits (32 bytes) by using SHA-256 algorithm	Patient code/identity
RBS Normal up to 140 m/L	116.5	83916e81b666e0f554b6a9ed60f48cda71a6bdfe9df4338a75a7b1741ecdee07	
Bilirubin (T & D) Normal up to 1.00 mg/dL	0.66	1778e31697721442d6b9689bc1f79420e64cd03d54c0f3da80105dc6be2c6fdf	
SGPT (ALAT) Normal up to 41 mg/dl	30	624b60c58c9d8bfb6ff1886c2fd605d2adeb6ea4da576068201b6c6958ce93f4	
SGOT (ASAT) Normal up to 37 mg/dl	22	785f3ec7eb32f30b90cd0fcf3657d388b5ff4297f2f9716ff66e9b69c05ddd09	
Urea Normal 10–50 mg/dL	21	6f4b6612125fb3a0daecd2799dfd6c9c299424fd920f9b308110a2c1fbd8f443	
Creatinine Normal Men: 0.70–1.20 mg/dL Women: 0.50–0.90 mg/dL	0.6	b4af4e0d40391d3a00179d935c63b7e15ba8cd3dfa29f218dedc270bd3eb3e79	SA-17-07-83
Total protein Normal 6.7–8.7 g/dL	6.8	ee7e860a857feb60180ac10dcdabc3afc06f9e45f9d761232cb9f8aed33162a6	
Albumin Normal 3.4–4.8 g/dL	3.7	b1133c6354aaff809f802709e53dacce82e609f9dc426c2da77dc68cf6fe64a0	
GGT Normal 11–50 U/L	15	e629fa6598d732768f7c726b4b621285f9c3b85303900aa912017db7617d8bdb	
Hb% Normal 12–14 g/d	11.9	9043cfea1047165f703e0929c7b0d8d649225662ec8e0757a8e476a64edf08c7	
Blood group	O+	4aa8d4a10f941836b0c764d54421edf48c76cafb4b17ffe338b0340194515d89	

In our current work, we just only used blood test data and get some output of hash value. The hashing or encryption of X-ray image data in our proposed system is now under development. This is the key experiment for better output about blockchain-based healthcare communication system in our future works. The proposed system can be moved into a scholarly prototype by various technologies (data mining and decentralized computing). In our future experiment, we want to complete a real-world test for EMR with our own deployed blockchain application.

Acknowledgements This research has been conducted based on MPhil research of our paper's first author Rafita Haque. This research has been proceeded under the supervision of Professor Dr. Hasan Sarwar (second author). Centre for Higher Studies, University of Professionals (BUP) has support for this research. S. Rayhan Kabir (third author) has assisted for collecting the blood test dataset from Life Diagnostic Centre Ltd.

References

1. Halamka JD, Alterovitz G et al (2019) Top 10 blockchain predictions for the (Near) future of healthcare. Blockchain Healthcare Today
2. Mackey TK, Kuo TT et al (2019) 'Fit-for-purpose?'—challenges and opportunities for applications of blockchain technology in the future of healthcare. BMC Medicine 17:68
3. Gnadinger T (2014) New health policy brief: interoperability. Health Affairs
4. Charles D, Gabriel M, Searcy T (2016) Adoption of electronic health record systems among U.S. non-federal acute care hospitals: 2008–2015. ONC Data Brief, paper no. 35
5. Kuo T, Kim H, Ohno-Machado L (2019) The state of research on cyberattacks against hospitals and available best practice recommendations: a scoping review. BMC Med Inform Decis Mak 19:10, PMCID: PMC6330387
6. Catalini C, Gans JS (2016) Some simple economics of the blockchain. Social science research network (SSNR), Rotman school of management working paper no. 2874598; MIT Sloan research paper no. 5191-16
7. Gordon W, Wright A, Landman A (2017) Blockchain in health care: decoding the hype, NEJM Catalyst (2017). Link: https://catalyst.nejm.org/decoding-blockchain-technology-health/. Accessed 28th June 2019
8. Es-Samaali H, Outchakoucht A, Leroy JP (2017) A blockchain-based access control for big data. Int J Comput Netw Commun Secur 5(7):137–147
9. Wang Y, Luo F et al (2019) Distributed meter data aggregation framework based on blockchain and homomorphic encryption. IET Cyber-Phys Syst: Theory & Appl 4(1):30–37
10. Landerreche E, Stevens M (2018) On immutability of blockchains. In: Proceedings of 1st ERCIM blockchain workshop 2018, reports of the european society for socially embedded technologies 2(7)
11. Courtney PK (2011) Data liquidity in health information systems. Cancer J 17(9):219–221
12. Mikula T, Jacobsen RH (2018) Identity and access management with blockchain in electronic healthcare records. In: 2018 21st Euromicro conference on digital system design (DSD), pp 699–706
13. Gowthaman A, Sumathi M (2017) Performance study of enhanced SHA-256 algorithm. Int J Appl Eng Res 10(4):10921–10932
14. Gordon WJ, Catalini C (2018) Blockchain technology for healthcare: facilitating the transition to patient-driven interoperability. Comput Struct Biotechnol 16(2018):224–230
15. Zhang P, White J et al (2018) FHIRChain: applying blockchain to securely and scalably share clinical data. Comput Struct Biotechnol 16:267–278

16. Boulos MNK, Wilson JT, Clauson KA (2018) Geospatial blockchain: promises, challenges, and scenarios in health and healthcare. Int J Health Geogr 17, Article no. 25 (2018)
17. Rahmadika S, Rhee K, Renz J, Wolter D (2018) Blockchain technology for providing an architecture model of decentralized personal health information. Int J Eng Bus Manag 10:1–12
18. Pirtle C, Ehrenfeld J (2018) Blockchain for healthcare: the next generation of medical records? J Med Syst 42:172
19. Shah GH, Leider JP et al (2016) Interoperability of information systems managed and used by the local health departments. J Public Health Manag Pract 22(6 Supp):S34–S43
20. Noura M, Atiquzzaman M, Gaedke M (2019) Interoperability in internet of things: taxonomies and open challenges. Mob Netw Appl 24(3):796–809
21. Kuo T, Kim H, Ohno-Machado L (2017) Blockchain distributed ledger technologies for biomedical and health care application. J Am Med Inform Assoc 24(6):1211–1220
22. Why would I choose SHA-256 over SHA-512 for a SSL/TLS certificate?. Stack Exchange (2017)

Factors Contributing to E-Government Adoption in Indonesia—An Extended of Technology Acceptance Model with Trust: A Conceptual Framework

Wiwit Apit Sulistyowati, Ibrahim Alrajawy, Agung Yulianto, Osama Isaac, and Ali Ameen

Abstract E-Government is an administration framework dependent on correspondence innovation and plans to improve the nature of administration forms from government organizations to the general population through online administrations. The Republic of Indonesia is a part of the ASEAN country that needs improvement of E-Government execution because it is still low and under the country of Singapore, Malaysia, Brunei Darussalam, Thailand, the Philippines, and Vietnam. The low-level of E-Government employment in the Republic of Indonesia needs to be examined by analyzing Internet usage penetration and identifying the cause of low E-Government adoption in the Republic of Indonesia. E-Government becomes an exciting issue in improving good governance, and this study will discuss the conceptual model in supporting E-Government's appropriation in the Republic of Indonesia. This paper explores the technology acceptance model (TAM) and the role of trust in creating the interest of citizens to practice E-Government".

Keywords E-Government adoption · Technology acceptance model (TAM) · Trust

W. A. Sulistyowati · A. Yulianto (✉)
Universitas Swadaya Gunung Jati, Cirebon, Indonesia
e-mail: agung789@gmail.com

W. A. Sulistyowati
e-mail: wiwit.apit@gmail.com

I. Alrajawy · O. Isaac · A. Ameen
Lincoln University College, Petaling Jaya, Malaysia
e-mail: ibrahim2alrajawy@gmail.com

O. Isaac
e-mail: osama4lincoln@gmail.com

A. Ameen
e-mail: ali.ameen@aol.com

S.-L. Peng et al. (eds.), *Intelligent Computing and Innovation on Data Science*,
Lecture Notes in Networks and Systems 118,
https://doi.org/10.1007/978-981-15-3284-9_70

1 Introduction

ICT advancement supports the change of administration given by the legislature to general society, by turning the service manual into an online service, or called E-Government. At first, E-Government became a correspondence choice among the government and the citizens, but with the differences in demographics, economics, society, and global trends, E-Government became the demand and necessity to enter the twenty-first century in order to compete in the world [1]. Therefore, through E-Government, the nature of administration given by the government to residents and organizations can increase and achieve greater efficiency for the parties involved.

According to the review directed by the United Nations in 2018 about E-Government ratings of the Republic of Indonesia, it showed that in 2018, the Republic of Indonesia gained ranked to 107, increasing nine ratings compared to the year 2016. However, the Republic of Indonesia's rankings is still far below ASEAN countries, such as Singapore, Malaysia, Brunei Darussalam, Thailand, the Philippines, and Vietnam. The ranking shows that the Republic of Indonesia should prompt the enforcement of E-Government in all parts of the nation. It is a challenge for the Republic of Indonesia to improve the function of ICT and infrastructure. Governments in different countries manage public finances through the use of ICT in order to create a functional and transparent financial governance [2–6].

For an individual, E-Government can provide a potential benefit, which is more restrained when interacting with the government. Although the citizen has to do activities on specific departments on a predetermined set, the public can perform the activity at the time and place they choose. Therefore, E-Government services provide freedom for the citizen because of the limitations of direct interactions. The advantage of using E-Government administration for individuals is saving time and money [1]. As indicated by the statement [7], the online tax system becomes a worldwide concern through the improvement of ICT that can influence the system of the tax. Also, the electronic services provide tax services in a simpler and faster way, as well as enhance the efficiency of tax administering [8], promoting disclosure, and offering the best quality-service [9, 10]. Besides, E-Government assumes a job in creating transparency, responsiveness, and accountability, but the service is adopted if the citizen considers the service to be reliable [11].

Communication through E-Government occurs in the relationship between governments and governments (G2G), governments with businesses (G2B), and governments with communities (G2P). The citizen has not actualized E-Government involving the government with the citizen (G2P) completely, such as E-Filling, E-KTP, online passport service, and E-Samsat. E-Government service in the Republic of Indonesia related to the payment of vehicle tax is called E-Samsat. E-Samsat has two benefits; for the government, it can present more accurate and up to date data so the government able to predict the realization and potential vehicle tax acceptance in each region. For the citizen, E-Samsat facilitates the public in knowing the amount of vehicle tax issued and making vehicle tax payments through the service of E-Samsat in one province.

According to the information accessed through https://bapenda.jabarprov.go.id that taxpayers often face obstacles in the form of differences in the identification number registered in Service Office Samsat with the identification number of the account holder to be used to pay tax, so that the taxpayer cannot pay the vehicle tax owned via E-Samsat. Another obstacle that arises is a pay code that does not correspond to those listed in the payment system. Some taxpayer complaints to the local revenue body of West Java province related to the use of E-Filling which is quite complicated. The lack of a resident that has access to E-Samsat relates to the high of vehicle taxpayers that conduct transactions manually in the integrated service office. Therefore, receiving and deployment of technology in the utilization of E-Government necessity are to be traced to distinguish the low facility through the E-Government.

Many models used and validated to predict receiving and implement the application of information system (IS), one of which is the technology acceptance model (TAM) [12]. However, TAM ignores the trust factor of the system users. Therefore, it is imperative to explore the role of trust because the low trust owned by the citizen will inhibit the enforcement of E-Government [13]. Trust can impact a resident's plan to use E-Government [14].

2 Conceptual Framework

E-Government adoption by various fields became an essential concern in creating E-Government achievement initiatives [15]. Therefore, E-Government does not guarantee success if the public cannot build the utilization of E-Government administrations [16]. The issue raises questions about how to expand E-Government appropriation in the Republic of Indonesian society. Many earlier studies were reviewing the TAM in the adoption of E-Government [8, 17–20]. However, few studies of E-Government employment conducted in growing nations still require a massive struggle and effort [21, 22].

This study used TAM established based on reasons: (1) The model is convenient to adopt and (2) offers further purity relating to the ties between variables used in the study [8]. TAM was adapted from TRA and was explicitly designed for exhibiting user receiving of information systems. The level where an individual accepts that a specific system's use will be unrestricted of exertion is called as perceived ease of use (PEOU). Ease means freedom from difficulty or hard effort [12]. Acceptance of the application becomes more comfortable to use than others, so the application tends to be acceptable to the user. Some studies explore the prominence of PEOU in predicting technology acceptance [8, 1, 23, 24]. A study was exploring E-Government administration conducted in Jordanian citizens through 75 samples [17] and indicating that PEOU has a positive effect on E-Government adoption. However, other studies [8, 18, 19] were unable to predict the effect of PEOU on the intention to utilize E-Government. Therefore, if the taxpayer believes that the E-Samsat is easy to use, then they will be bound to utilize the system.

The level where people trust that particular system usage will improve its performance is called perceived usefulness (PU), which means that the use of the system can provide benefits [12]. Many studies on various technological applications in different contexts suggest that PU can predict the intention to use technology. For example, studies conducted in Taiwan, Malaysia, Korea, USA, Singapore, Germany, and Iran [23, 25 26–31] suggest that PU is instrumental in technology usage.

Perceived Usefulness is a factor that plays an essential role in influencing people to adopt the technology. Prior researchers considered PU's prominent in the use of E-Government [7, 23, 24, 27]. Studies in the Netherlands analyzed 238 data and the results proved that PU is an essential indicator in the application of E-Government [32]. The results were also reinforced with studies in Malaysia that examined 150 data using regression analysis and the results proved that PU had a positive effect on the intention to use the E-Government service [27]. Studies in Cambodia through 124 respondents showed that PU had a positive effect on the adoption of E-Government [19, 21]. Furthermore, taxpayers feel that E-Samsat is beneficial in the reporting of vehicle taxes and increase their productivity. Therefore, if taxpayers believe that E-Samsat is valuable and comfortable, then they are likely to use the service.

Trust is an essential determinant of Internet technologies adoption [20, 33]. The study of [34] analyzing the achievement of E-Government appropriation in Thailand and the Republic of Indonesia reviewed from portraying the administrations ordinarily offered by governments. Instead, to specify the accomplishment administrations of E-Government, it is compulsory to weigh the factors trust [19–21, 25, 33, 35]. The study E-Government divides the trust into two aspects, namely trust of government (TOG) and trust of the Internet (TOI) [20, 33, 35]. The prior studies proved that trust influenced behavioral intention to use technology [19, 21, 25]. TOG deals with the level of citizen trust in reality and capacity of the government agency to provide E-Government services [11, 33], while TOI is associated with the trust of the citizen that E-Government usage is safe and has no danger to their security [20].

The previous examination has proved that TOG influenced positively on the intention to use E-Government [11, 33, 35]. The intention to use E-Government was explored through 529 respondent answers from 35 provinces in Turkey [36]. The results proved that the TOI has a positive effect on the intention to use E-Government [11, 36], while the TOG does not affect. Adoption E-Government in Mauritius based on 247 data analyzed with structural equation modeling proved that the trustworthiness was the main predictor [33]. E-Government adoption in Africa was analyzed through 282 respondents, and the results show that the intention to use E-Government influenced by the TOI and TOG [37].

The proposed hypotheses are:

H1: PEOU affects positively on the intention to use E-Government among individual taxpayers in the Republic of Indonesia.
H2: PU affects positively on the intention to use E-Government among individual taxpayers in the Republic of Indonesia.
H3a: Trust of the Internet (TOI) affects positively on the intention to use E-Government among individual taxpayers in the Republic of Indonesia.

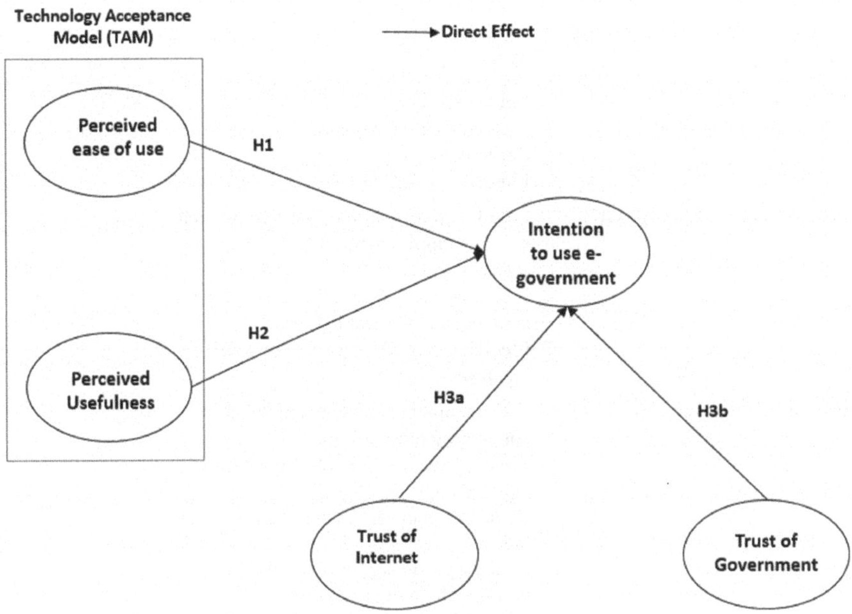

Fig. 1 Conceptual model

H3b: trust of government (TOG) affects positively on the intention to use E-Government among individual taxpayers in the Republic of Indonesia.

Here is a figure of conceptual models in this study (Fig. 1).

3 Operationalization of Construct

The study proposed the theoretical model adoption of E-Government to apply in empirical research. Table 1 shows the operationalization of the variables, which identifies the measurements of each construct and adopted from the previous study.

4 Conclusion

E-Government is a form of commitment and initiatives conducted by the government in enhancing the connection with the citizen and business or organization through increased cost-effectiveness, efficiency in the conveyance of services, description, and knowledge using ICT [34]. The accomplishment of E-Government employment is related to the two parties, namely the Government as the administration

Table 1 Construct measurement

Constructs	Items	Adapted from
Intention to use E-Government	I plan to utilize E-Government later on I intend to use E-Government in regular day to existence I intend to use E-Government later on	[36, 37]
Perceived ease of use (PEOU)	I think that it is simple to get conversant with the employment of E-Government service I feel that the E-Government service is evident and far-reaching I feel that it is effortless for me to obtain the desired service	[20]
Perceived usefulness (PU)	E-Government service helps me in completing more things E-Government service can make my life simpler E-Government service will be useful for me E-Government service can support my productivity and efficiency	[20]
Trust of the Internet (TOI)	The Internet gives enough insurance in making me feel great to get E-Government service I find guaranteed that legitimate and innovative structures enough shield me from issues on the Internet In all purpose, the Internet is strong enough and maintains nature in conducting transactions through E-Government service	[37]
Trust of government (TOG)	I feel that I can believe government offices to make online exchanges sensibly I accept that government agencies can keep my best advantages	[37]

provider and the citizen, both individual and business, as the service user. The challenges in achieving the success of E-Government relate to how public acceptance of the employment of E-Government administrations. Therefore, governments need to identify communities based on demographic characteristics. This is related to uneven Internet penetration in the Republic of Indonesia, as well as the understanding required by the citizen to access E-Government. Perceived ease of use (PEOU) and perceived usefulness (PU) are described in TAM as a factor that significantly affects the interest of people to utilize E-Government administrations. Besides, the trust aspect also becomes an important benchmark to demonstrate the accomplishment of E-Government usage. The citizen who demonstrates the trust of the Internet will give the perception that when accessing E-Government administrations over the

Internet, the security factor of their data privacy will be maintained. Citizens who have trust in the government believe that through e-government, the government agency manages the tax administration with full responsibility.

References

1. Kumar V, Mukerji B, Butt I, Persaud A (2007) Factors for successful E-Government adoption: a conceptual framework. Electron J E-GovMent 5(1):63–76
2. Ameen A, Almari H, Isaac O (2018) Determining underlying factors that influence online social network usage among public sector employees in the UAE. In: Saeed BA, Gazem F, Mohammed NF (ed) 3rd international conference on reliable information and communication technology 2018 (IRICT 2018), Bangi-Putrajaya, Malaysia, 3rd ed. vol 843, Springer, Cham, pp 945–954
3. Ameen A, Ahmad K (2013) A conceptual framework of financial information systems to reduce corruption. J Theor Appl Inf Technol 54(1):59–72
4. Ameen A, Ahmad K (2012) Towards harnessing financial information systems in reducing corruption: a review of strategies. Aust J Basic Appl Sci 6(8):500–509
5. Ameen A, Almari H, Isaac O (2019) Determining underlying factors that influence online social network usage among public sector employees in the UAE. In: Faisal Saeed FM, Gazem N (ed) Recent trends in data science and soft computing. IRICT 2018. Advances in Intelligent Systems and Computing, Recent Tre, vol 843. Springer International Publishing, Springer Nature Switzerland AG, pp 945–954
6. Ameen A, Ahmad K (2011) The Role of finance information systems in anti financial corruptions: a theoretical review. In 11 international conference on research and innovation in information systems (ICRIIS'11), pp 267–272
7. Mustapha B, Obid SNBS (2015) Tax service quality: the mediating effect of perceived ease of use of the online tax system. Procedia—Soc Behav Sci 172:2–9
8. Anuar S, Othman R (2010) Determinants of online tax payment system in Malaysia. Int J Public Inf Syst 1:17–32
9. Lin F, Fofanah SS, Liang D (2011) Assessing citizen adoption of e-Government initiatives in Gambia: a validation of the technology acceptance model in information systems success. Gov Inf Q 28(2):271–279
10. Susanto TD, Goodwin R (2013) User acceptance of SMS-based e-government services: differences between adopters and non-adopters. Gov Inf Q 30(4):486–497
11. Bélanger F, Carter L (2008) Trust and risk in e-government adoption. J Strateg Inf Syst 17(2):165–176
12. Davis FD (1989) Perceived usefulness, perceived ease of use, and user acceptance of information technology. MIS Q 13(3):319–340
13. Veeramootoo N, Nunkoo R, Dwivedi YK (2018) What determines success of an e-government service? Validation of an integrative model of e-filing continuance usage. Gov Inf Q 35(2):161–174
14. Lemuria C, France B (2005) The utilization of e-government services: citizen trust, innovation and acceptance factors. Inf Syst J 15(1):5–25
15. Al-Hujran O, Al-Debei MM, Chatfield A, Migdadi M (2015) The imperative of influencing citizen attitude toward e-government adoption and use. Comput Human Behav 53:189–203
16. Panagiotopoulos PP (2010) Engaging with citizens online: understanding the role of ePetitioning in local government democracy. Internet polit. Policy 2010 an impact assess, no. May, pp 1–15
17. Almahamid S, McAdams AC, Al Kalaldeh T, Al-Sa'eed M (2010) The relationship between perceived usefulness, perceived ease of use, perceived information quality, and intention to use E-government. J Theor Appl Inf Technol 11(1):30–44

18. Carter L, Belanger F (2004) Citizen adoption of electronic government initiatives. Proc Hawaii Int Conf Syst Sci 37(C):1895–1904
19. Sang S, Lee JD, Lee J (2010) E-government adoption in Cambodia: a partial least squares approach. Transform. Gov. People, Process Policy 4(2):138–157
20. Abu-Shanab EA (2017) E-government familiarity influence on Jordanians' perceptions. Telemat. Informatics 34(1):103–113
21. Sang S, Lee JD, Lee J (2009) E-government adoption in ASEAN: the case of Cambodia. Internet Res 19(5):517–534
22. Susanto TD, Aljoza M (2015) Individual acceptance of e-Government services in a developing country: dimensions of perceived usefulness and perceived ease of use and the importance of trust and social influence. Procedia Comput Sci 72:622–629
23. Mustapha B, Sheikh Obid SN (2014) The influence of technology characteristics towards an online tax system usage: the case of nigerian self- employed taxpayer. Int J Comput Appl 105(14):30–36
24. Hamid AA, Razak FZA, Bakar AA, Abdullah WSW (2016) The effects of perceived usefulness and perceived ease of use on continuance intention to use e-government. Procedia Econ Financ 35(October 2015):644–649
25. Gu JC, Lee SC, Suh YH (2009) Determinants of behavioral intention to mobile banking. Expert Syst Appl 36(9):11605–11616
26. Luarn P, Lin HH (2005) Toward an understanding of the behavioral intention to use mobile banking. Comput Human Behav 21(6):873–891
27. Lean OK, Zailani S, Ramayah T, Fernando Y (2009) Factors influencing intention to use e-government services among citizens in Malaysia. Int J Inf Manage 29(6):458–475
28. Greenberg R, Bernad WL, Wing W (2012) The effect of trust in system reliability on the intention to adopt online accounting systems. Int J Account Inf Manag 20(4):363–376
29. Riquelme HE, Rios RE (2010) The moderating effect of gender in the adoption of mobile banking. Int J Bank Mark 28(5):328–341
30. Koenig-Lewis N, Palmer A, Moll A (2010) Predicting young consumers' take up of mobile banking services. Int J Bank Mark 28(5):410–432
31. Hanafizadeh P, Behboudi M, Abedini Koshksaray A, Jalilvand Shirkhani Tabar M (2014) Mobile-banking adoption by Iranian bank clients. Telemat Inform 31(1):62–78
32. Horst M, Kuttschreuter M, Gutteling JM (2007) Perceived usefulness, personal experiences, risk perception and trust as determinants of adoption of e-government services in The Netherlands. Comput Human Behav 23(4):1838–1852
33. Lallmahomed MZI, Lallmahomed N, Lallmahomed GM (2017) Factors influencing the adoption of e-government services in Mauritius. Telemat Inform 34(4):57–72
34. Mirchandani DA, Johnson JH, Joshi K (2008) Perspectives of citizens towards e-government in Thailand and Indonesia: a multigroup analysis. Inf Syst Front 10(4):483–497
35. Zhao F, Khan MS (2013) An empirical study of e-government service adoption: culture and behavioral intention. Int J Public Adm 36(10):710–722
36. Kurfalı M, Arifoğlu A, Tokdemir G, Paçin Y (2017) Adoption of e-government services in Turkey. Comput Human Behav 66(1):168–178
37. Verkijika SF, De Wet L (2018) E-government adoption in sub-Saharan Africa. Electron Commer Res Appl 30(February):83–93

Workload Forecasting Based on Big Data Characteristics in Cloud Systems

R. Kiruthiga and D. Akila

Abstract Resource allocation for big data streams in cloud systems involves selecting the appropriate cloud resources. Big data has certain precise features such as size, speed, veracity, variety, and value. In this paper, a workload forecasting system for resource allocation in big data streams is developed. In this system, the data characteristics such as data type (variety), size (volume), and deviation in data flow rate (velocity) are extracted. Based on these data characteristics, the expected workload of the next time interval is predicted using support vector machine (SVM). Followed by this, the cloud resource manager dynamically allocates the available cloud resources depending on the predicted workload. The presentation outcomes have confirmed that the proposed system has less execution time and achieves better utilization of resources, when compared to the existing tools.

Keywords Big data · Resource allocation · SVM · Workload forecast

1 Introduction

In big data uses, numerous types of enormous information are produced, treated, communicated, and deposited in each instant. Certain big data sellers afford big data as a service (BDaaS) that offers users right to use to serviceable gathered and augmented data, along with their exact modified necessities [1]. Due to the transient nature of cloud servers, management of big data applications becomes complicated [2]. Significant data handling agenda is embellishing more widespread to undertake huge quantities of data in a native or a cloud organized group [3].

Resource allocation for big data streams in cloud systems involves selecting the appropriate cloud resources. Resource allocation based on data characteristics is challenging for big data since the features of information in big data rivulets are strange [4].

R. Kiruthiga (✉) · D. Akila
School of Computing Sciences, Vels Institute of Science, Technology & Advanced Studies (VISTAS), Chennai, Tamil Nadu, India

S.-L. Peng et al. (eds.), *Intelligent Computing and Innovation on Data Science*, Lecture Notes in Networks and Systems 118,
https://doi.org/10.1007/978-981-15-3284-9_71

In static allocation, the user specifies how many doers, interiors, reminiscence, etc., a solicitation can ensure. In dynamic allocation, some idle executors may be released to return certain revenues to the group which may also be returned later on if desired [3]. The issues are

- When a single application is executing using default resource allocation mechanism, it consumes all the resources, thereby preventing other applications from sharing the resources.
- Defaulting source distribution appliances may possibly not exert as any solicitation with a firm target might have to hold in the FIFO line.
- Unfitting source distribution in both stationary and active source distribution methods might disturb the targets.
- A virtual machine (VM) may require more CPU resources, while another VM needs more network bandwidth or memory. Such a dynamic imbalance of resources in individual VM leads to the total inefficiency of cluster resources.
- Cost, performance, and availability are the main concerns of resource allocation.

Some existing examination on cloud data distribution concentrated on only one of these constraints. But, it involves high time complexity. In this paperwork, the data characteristics are extracted based on the type, size, and deviation in data flow rate. Based on these data characteristics, the current data segment is analysed and the expected workload of next time interval is predicted. The cloud resource manager dynamically allocates available cloud resources depending on the predicted workload.

2 Related Works

Kaur et al. [4] have suggested a scheme that foretells the data features regarding size, velocity, diversity, inconsistency, and accuracy. The anticipated values are conveyed in an increase fourfold CoBa. Zhang et al. [5] have suggested a novel temporary load predicting agenda centred on big data skills. Initially, a group investigation is done to categorize regular load decorations for distinct loads by means of keen measure data. Then, a link investigation is made use to regulate acute prominent features. At that time, correct predicting replicas are selected for diverse load designs. In conclusion, the predicted entire structure load is got over an accumulation of a distinct load's predicting outcomes.

Tang et al. [6] have suggested an enhanced LSTM expectation exemplary only if its precise report and a fault back proliferation technique. The relative case studies validate that their suggested amended LSTM expectation exemplary can attain advanced accurateness and actual presentation in extensive computing structures.

Dhamodharavadhani et al. [7] have reviewed the V-characteristics of big data for knowing the evolutionary stages of big data and big data opportunities. This review summarizes that even the basic V's volume, variety, and velocity still make the best part in big data analytics, but none of these stand on their own.

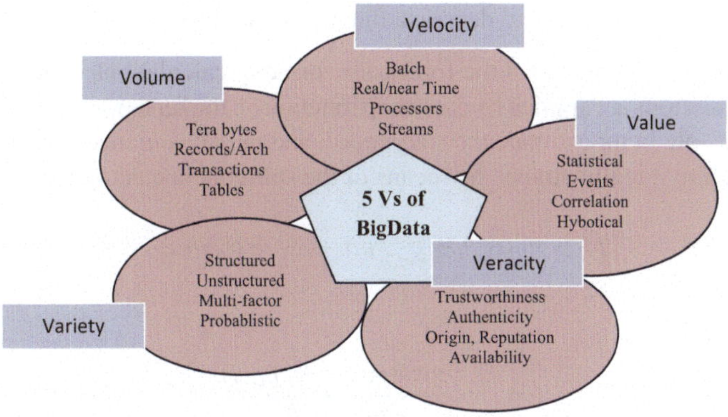

Fig. 1 5V's of big data

2.1 Proposed Solution

2.1.1 The 5V's of Characteristics of Big Data

Big data has been described centred on few of its features. There are five features that have been utilized to describe big data. (similarly known as 5V's) (Fig. 1).

Volume: it denotes the amount of data collected by a group.
Velocity: it denotes the growing rapidity at which this data is made.
Value: it is a significant characteristic of the data described by the added value in which the composed data can take to the anticipated procedure or trade [8, 9].

2.2 Estimating the Volume and Data Flow Rate

The steps involved in this process are explained in this section.

The functions of MapReduce-based framework are illustrated in Fig. 2.

In the beginning, the elementary data rivulet is arbitrarily divided into numerous map functions. The amount of map functions differs along with the data onset rate. Every map function appraises volume and velocity by means of Kalman filter. The assessed values of volume and velocity are directed to the corresponding decrease functions EVol() and EVel(), correspondingly. These two reduce functions combine the obtained values to determine the cumulative volume and velocity.

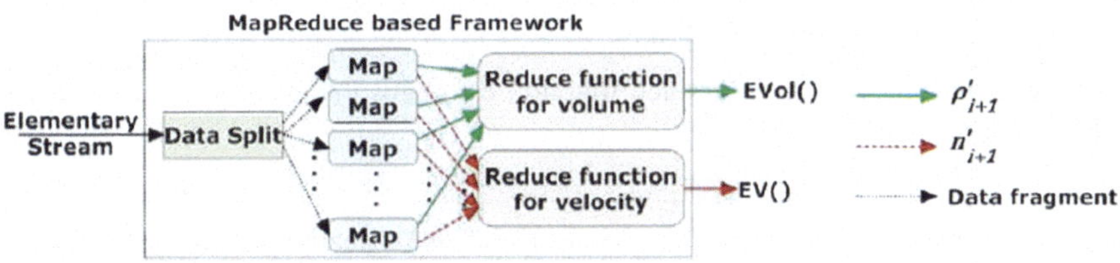

Fig. 2 MapReduce framework

The particulars of map and decrease functions are given below.

Map function: it implements the forecaster corrector calculations of Kalman filter. These calculations are applied by each map function on the fundamental data streams. Initially, the predictor estimate the size, speed, and fault covariance of $(i + 1)$th data segment, afore it really attains, by means of the subsequent calculations

$$\rho'_{i+1} = \alpha_1 \rho_i + \alpha_2 \rho_{i-1} + \cdots + \alpha_q \rho_{1-q+1} \tag{1}$$

where

$$\alpha_j = \frac{\text{covariance}(\rho_i, \rho_{i-j})}{\text{variance}(\rho_j)} \tag{2}$$

$$\eta'_{i+1} = \beta_1 n_i + \beta_2 n_{i-1} + \cdots + \beta_q n_{i-q+1} \tag{3}$$

where

$$\beta_j = \frac{\text{covariance}(n_i, n_{i-j})}{\text{variance}(n_j)} \tag{4}$$

$$\Omega'_{i+1} = \Omega_i + Q \tag{5}$$

where

ρ'_i, ρ_i are the estimated and corrected volumes of ith segment.
η'_i, η_i are the estimated and corrected velocities of ith segment.

Equation (1) evaluates the capacity of $(i + 1)$th data section by putting on unevenness into concern. Akin is the situation for speed evaluation in Eq. 3.

When the $(i + 1)$th data segment is received, the corrector applies the following equations to modify these estimated values.

$$\rho_{i+1} = \rho'_i + K_i(y_i - \rho'_i) \tag{6}$$

$$\eta_{i+1} = \eta'_{i+1} + K_i(z_i - \eta'_i) \tag{7}$$

where

K_i is the Kalman gain at ith prediction step, given by
y_i is the ith measurement of volume
z_i is the ith measurement of velocity.

Reduce function: the EVol() reduce function estimates the average (Avg) of these entire values.

1. If $0 \leq \text{avg}(\rho'_{i+1}) < (\mu(\rho))$, thenEvol() = "Low"
2. If $(\mu(\rho)) - \sigma(\rho)) \leq \text{Avg}(\rho'_{i+1} \leq (\mu(\rho)) + \sigma(\rho))$, thenEVol() = "Medium"

3. If $\text{Avg}(\rho'_{i+1}) > (\mu(\rho)) + \sigma(\rho))$, thenEVol() = "Low"

Similarly, the Avg of all velocities and compare with mean and std. Deviation.

2.3 SVM-Based Total Load Prediction

For diverse load arrangements, diverse support vector machine (SVM) prototypes and factors are advanced to make sure the predicting accurateness inside the essential confines. SVM is an actual method for organization and reversion difficulties. SVMs are overseen learning prototypes with related learning procedures that examine data and identify designs made use for organization and reversion examination. SVM can competently achieve a nonlinear organization by means of the kernel hoax, indirectly plotting the efforts into high-dimensional characteristic places.

The sustenance vector reversion difficulty is exposed underneath:

$$\min_{\omega,b,\xi,\xi^*} \frac{1}{2}\omega^T \omega + C\sum_{i-1}^{n}(\xi_i + \xi_i^*) \tag{8}$$

Subject to:
$y_i(\omega^T \phi(x_i) + b) \leq \varepsilon + \xi_i$,
$(\omega^T \phi(x_i) + b) \leq \varepsilon + \xi_i)$,
$\xi_i, \xi_i^* \geq 0, i = 1, \ldots, n$
where $(x_1; y_1)\ldots(x_n; y_n)$ are a couple of input and output trajectories, n is the numeral of models, ! is mass factor, b is the verge value, and C is fault charge.

Input models are plotted to advanced dimensional area by means of kernel function φ; ϵ_i is the higher working out fault; ξ_i^* is the minor working out fault bound by ϵ-insensitive duct.

The entire structure load is predicted depending on accumulation of a distinct load's predicting outcomes; once the predicting outcome of every consumer's load is got, the predicted whole load L_{total} can be intended by totalling the entire predicted distinct loads, along with route harm L_{loss}.

$$L_{\text{total}} = L_{\text{loss}} + \sum_{i=1}^{n} l_{\text{user(i)}} \tag{7}$$

3 Experimental Results

3.1 Generation of Workloads

The data stream types considered in this work are listed in Tables 1, 2, and 3 which

Table 1 Different stream types used

Stream type	Big data stream
1	Text
2	Image
3	Audio (VoIP)
4	Video

Table 2 Volumes of different workloads

Workload no.	Volume of stream (GB)			
	1	2	3	4
1	10	30	10	60
2	0	0	40	20
3	15	60	0	0
4	0	10	55	0
5	62	0	0	35
6	0	25	30	65
7	25	0	28	0
8	0	35	55	70
9	55	40	0	45
10	0	11	0	50

Table 3 Velocities of different workloads

Workload no.	Velocity of stream (number of)			
	1	2	3	4
1	100	350	120	600
2	0	0	140	220
3	150	700	0	0
4	0	130	580	0
5	450	0	0	350
6	0	225	375	720
7	210	0	200	0
8	0	300	500	700
9	580	410	0	450
10	0	125	0	800

show the volume (size) and velocity (data flow rate) of the workloads with respect to the mixture of given stream types.

3.2 Results

Primarily, the amount of work of 25 GB is served to EC2 work out enhanced c4.huge occurrences at velocity of 50. Every amount of work is served to the structure later

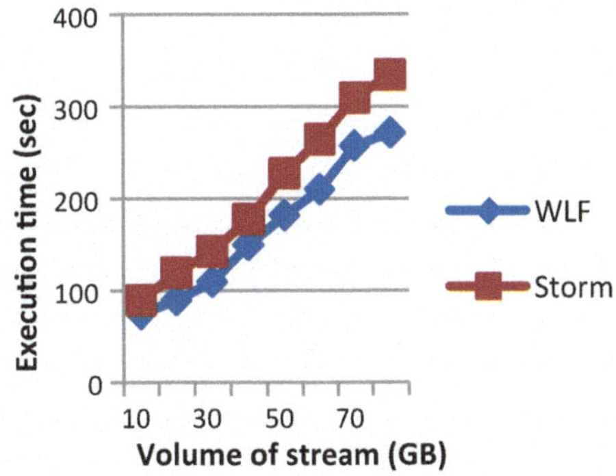

Fig. 3 Execution times for various sizes of workloads

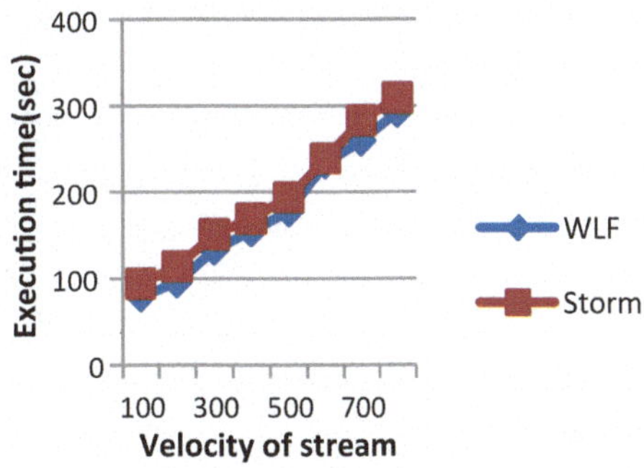

Fig. 4 Execution times for various data generation rates

every 5 min for the period of 1 h. The performance of WLF is compared with the Apache Storm 0.92 tool using the execution time and resource utilization metrics.

A. **Execution time**

In this section, the execution time for different volumes and velocities of workloads is measured and depicted in Figs. 3 and 4, respectively.

B. **Resource utilization**

In this section, the resource utilization (%) for different volumes and velocities of workloads is measured and depicted in Figs. 5 and 6, respectively.

Figures 3 and 4 show the comparison results of both the systems in resource utilization. The outcomes display that use of cloud means is advanced in event of WLF when associated to Storm.

4 Conclusion

In this paper, a work- load forecasting system for resource allocation in big data streams has been developed. In this system, the data characteristics such as the data

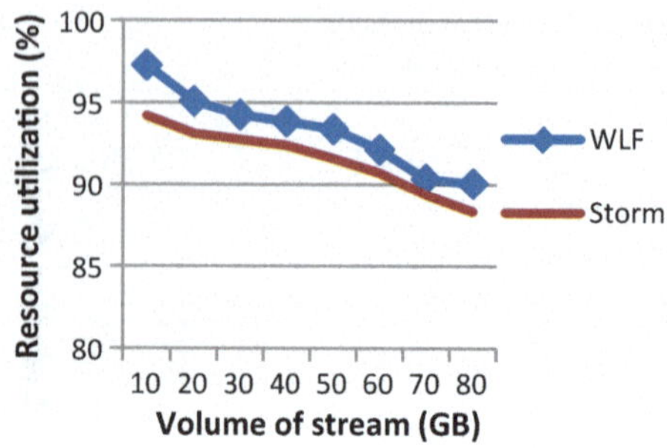

Fig. 5 Resource utilization for various sizes of workloads

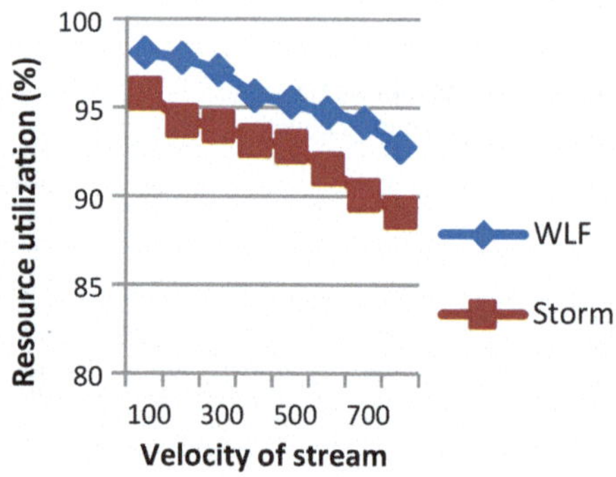

Fig. 6 Resource utilization for various data generation rates

type (variety), size (volume), and deviation in data flow rate (velocity) are extracted. Based on these data characteristics, the expected workload of the next time interval is predicted using support vector machine (SVM). Followed by this, the cloud resource manager dynamically allocates available cloud resources depending on the predicted workload. The suggested scheme is applied in a Java-based solicitation on Amazon EC2 figure improved c4.huge occurrences. The enactment of the WLF scheme is assessed by associating with the Apache Storm 0.92 device using the execution time and resource utilization metrics. Investigational outcomes display that the suggested WLF scheme has fewer execution time and achieves better utilization of resources, when compared to the existing tools.

References

1. Dai W, Qiu L, Wu A, Qiu M (2016) Cloud infrastructure resource allocation for big data applications. IEEE
2. Spicuglia S, Cheny LY, Birkey R, Binder W (2015) Optimizing capacity allocation for big data applications in cloud datacenters. In: IEEE IFIP/IEEE international symposium on integrated network management (IM), Canada

3. Islam MT, Karunasekera S, Buyya R (2017) dSpark: deadline-based resource allocation for big data applications in Apache Spark. In: IEEE IEEE 13th international conference on e-science (e-science), New Zealand
4. Kaur N, Sood SK (2017) Dynamic resource allocation for big data streams based on data characteristics (5Vs). Int J Network Mgmt
5. Zhang P, Wu X, Wang X, Bi S (2015) Short-term load forecasting based on big data technologies. CSEE J Power Energy Syst 1(3)
6. Tang X (2019) Large-scale computing systems workload prediction using parallel improved LSTM neural network. IEEE Acess
7. Dhamodharavadhani S, Gowri R, Rathipriya R (2018) Unlock different V's of big data for analytics. Int J Comput Sci Eng 6(4)
8. Ishwarappa, Anuradha J (2015) A brief introduction on big data 5Vs characteristics and hadoop technology. Preocedia Comput Sci 48:319–324 (Elsevier)
9. Hadi HJ, Shnain AH, Hadishaheed S, Haji Ahmad A (2015) Big data and five V's characteristics. Int J Adv Electron Comput Sci 2(1)

An Empirical Study on Big Data Analytics: Challenges and Directions

Munir Kolapo Yahya-Imam and Felix O. Aranuwa

Abstract The current technology trends have escalated to a stage at which human beings now produce, create and interact with data. Latest technologies all contribute to the production of these data. Henceforth, this huge development of information has prompted the improvement of big data. Many researchers have carried out several studies in this regard. Still, there is a lack of comprehensive literature that addresses the various aspects, technologies, challenges and directions in this area. It is impossible to come to a conclusion without a complete knowledge of the above. Henceforth, this paper examines the difficulties in the appropriation of big data and suggests the best practices to overcome them. Significantly, it analyzed the various areas of big data such as the features, strengths and technological tools for big data process.

Keywords Data · Big data · Analytics · Technology · ETL

1 Introduction

The proliferation of sophisticated technology devices has resulted to the quick development of data. As user increases, so the data they generate. Currently, researches have shown that approximately 5.117 billion people are using mobile phones and about 2.71 billion of them were smart phone users. By implication, 35.13% of the world population now uses smart phones.

Further study has also revealed that smart phone penetration is increasing more than 20% per year. Today, more than 30 million arranged sensor hubs are installed in various industries such as automotive, transportation, manufacturing industries, and utilities which contribute to the huge growth in enterprise data. The sensors increase

M. K. Yahya-Imam (✉)
Faculty of Computer Science and Multimedia, Lincoln University College, Petaling Jaya, Malaysia
e-mail: munir@lincoln.edu.my

F. O. Aranuwa
Department of Computer Science, Adekunle Ajasin University, Akungba, Nigeria
e-mail: felix.aranuwa@aaua.edu.ng

S.-L. Peng et al. (eds.), *Intelligent Computing and Innovation on Data Science*, Lecture Notes in Networks and Systems 118,
https://doi.org/10.1007/978-981-15-3284-9_72

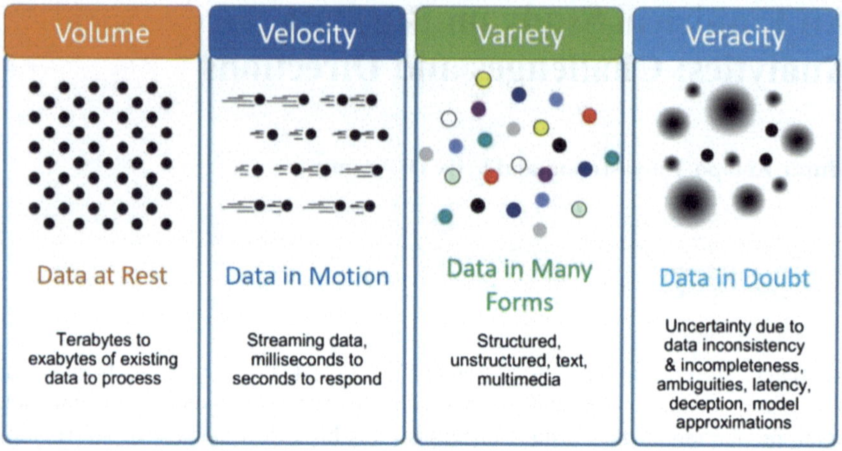

Fig. 1 Features of big data [10]

every year by over 30% [1–3]. It has been predicted from the research from Oracle that as data is growing at a 40% compound annual rate, it will reach 45ZB by 2020.

Essentially, generated data must be captured, organized, stored, analyzed and recovered within an acceptable time frame [4]. According to [5], big data describes data generated at high speeds that require cutting-edge methods and tools to facilitate its storage, management and analysis

Conventional database management systems (DBMS) are currently experiencing great stress in managing such kind of enterprise data. The process of collecting, storing, modeling, analyzing, interpreting and presenting and making intelligent business decisions is more challenging. These difficulties have been marked as the big data issue. Figure 1 exactly portrays the qualities of enormous information dependent on the Vs attributes [6]. In the interim, late investigations have demonstrated that solitary a little bit of enormous information is prepared for examination, while huge segment of the huge information is unstructured, variable, and consequently not prepared for investigation [7–9].

2 Big Data Analytics

The investigations of [11, 12] have portrayed enormous information examination as a technique of taking a gander at gigantic and varied informational indexes to reveal data including shrouded designs that can assist associations with making educated business choices. This thus could prompt increasingly adroit business moves, continuously gainful undertakings, higher advantages and progressively cheerful customers.

Moreover, it empowers huge data investigators and extraordinary assessment specialists to dismember creating volumes of sorted out and unstructured data. For instance, Web snap stream, Web server logs, Web-based systems administration content, content from customer messages, study responses, phone records and machine data gotten by sensors are related with the snare of things.

The significant qualities of enormous information examination are speed and productivity. The capacity to work speedier and remain spry gives affiliations a mighty edge they did not have as of now. Hardly any years back, associations would gather information, run appraisal and uncovered data that could be utilized for future choices. Regardless, today those affiliations can perceive bits of data for quick choices.

2.1 Advantages of Big Data Analytics

The upsides of big data are different. Firstly, statistical reliability is better using bigger population size from high-volume data. Secondly, models can be improved if they include more related factors. Recent studies have shown that proper utilization of big data can possibly invigorate and achieve critical monetary developments.

Additionally, it also promises to improve organizational efficiency and effectiveness, business planning, operations efficiency and novelty [13].

A center in the UK, known as CEBR, has foreseen beneficiaries of big data analytics would be financial institutions, telecommunication industries, retail businesses, private and state-owned companies and trade and engineering enterprises [14–16]. An investigation led by MIT and IBM enquired 3000 administrators and specialists in regards to their huge information-handling capacities. They discovered that firms that utilized big data analytics performed better than those who did not. It shows top-performing firms invariably use data analytics in their business decisions.

Other advantages according to [17–19] include:

- Using big data cuts your costs
- Utilizing big data expands your productivity
- Utilizing big data improves your estimating
- You can contend with enormous organizations
- Enables you to concentrate on nearby inclinations
- Utilizing big data causes you increment deals and unwaveringness
- Utilizing big data guarantees you contract the correct workers
- Better decision-making
- Expanded profitability
- Extortion location
- Greater innovation
- Remain mindful of customer designs.

2.2 Difficulties in Big Data Analytics

Big data faces a few critical difficulties in the business. These include:

Privacy: Privacy is paramount in big data analytics. Privacy must not be compromised as the data passes through the various phases of processing. Lack of privacy has legal implications as well as trust issue.

Access to information: Access to big data must be readily available to users. Access to confidential information, however, must be restricted to authorized users.

Security: Many online applications such as Facebook, Twitter and LinkedIn require users to share private data. While sharing personal data might be beneficial, there is also potential for abuse. So, users must be given the flexibility to expose only selected information. In addition, taking care of immense data, particularly tricky data, can make associations a continuously engaging target for computerized aggressors.

Size: Previously, fast processors could handle the growing size of data. But, today's data grows rapidly than improvements in processor speeds.

Analytics: The primary purpose of data analytics is to discover patterns/features that users need within a certain time frame. Checking the whole data set might not be feasible. It would be easier if the data could be pre-processed before it is analyzed.

New data sources: As the data sources increase, so is the challenge to filter the data, so that only relevant data is investigated.

Unstructured data: Unstructured information, for example, content, diagrams, sound and video, presents extra difficulties. Besides storage, these data require advanced analytics. There are two approaches we can use: (i) Transform unstructured data to structured data and then use classical methods and (ii) develop new methods to handle unstructured data.

Web Semantics: Semantic Web deals with linking documents and data. There are two ways to do this; one way is by relating material that already exists in documents. Another way is by permitting data to be embedded in the Web. Semantic Web technologies were designed to handle a large variety of data sources with outputs ranging from traditional computers to mobile devices to physical sensors and networked embedded systems [20]. The main challenge is to create an ontology which defines contextual relationships and adds meaning to the massive data in the Web, so that it can be divided across spheres of information automatically [21].

Other drawbacks as identified in the literature include:

- Requirement for ability
- Quick adjust
- Equipment requirements
- Expenses
- Trouble incorporating inheritance frameworks.

3 Big Data Processing

Big data can be set up in five stages as showed up in Fig. 2. These stages structure the two standard sub-structures: data management and analytics. Regulating data incorporates methodology and supporting advancements to get and store data and to prepare and recoup it for assessment. Regardless, examination incorporates using methodologies to separate the data to get learning. The target of the whole method is to get bits of learning from the assembled data.

Big data is not just constrained to the way toward breaking down information yet examining information with sharpness. While there are several approaches to handle big data, the best way is to follow a solution-oriented approach. There will be sustainability in the long run if it is theory-driven and not intuition based [23]. Hence, this can be done with the following approaches.

Intuitive Exploration: Suitable for seeing ongoing courses of action from information as they show up.

Direct Batch Reporting: Decent for quickly clarifying information utilizing specially crafted and planned reports appropriate for basic leadership.

Batch ETL (Extract–Transform–Load): Perfect for investigating past patterns and integrating different data sources grounded upon pre-defined queries. Several tools are available in the market to handle these processes which can be collectively called as 'Extraction–Transformation–Loading (ETL)' tools. Figure 3 shows each phase of this process. Data extraction is performed using specialized tools from the various sources and available data connectivity. The extracted data is transformed using a sequence of conversion procedures. The methodology composed chiefly for getting

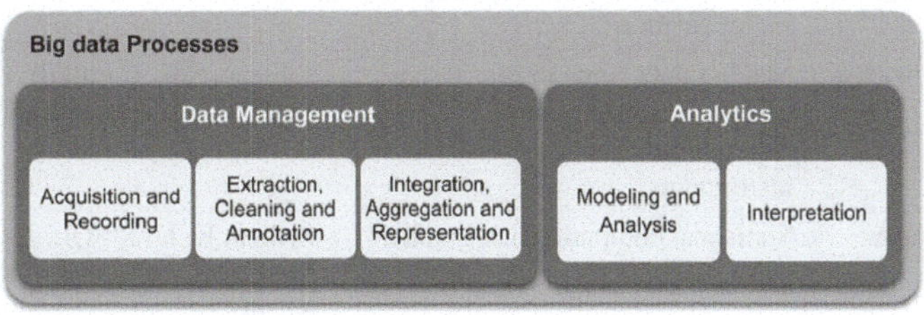

Fig. 2 Stages for extracting insights from big data [22]

Fig. 3 ETL process [24–26]

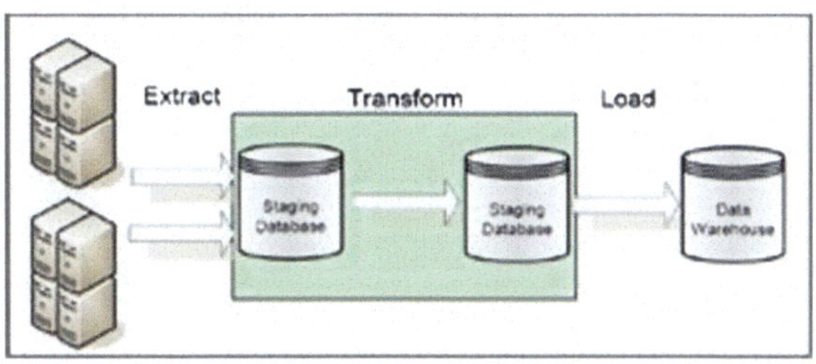

the ideal yield incorporates just the information that is required for the stacking reason.

In addition, there are several open-source tools available to handle the aforementioned stages and techniques. Table 1 gives a summary of some of the popular tools and software.

Table 1 Outline of open-source instruments for big data forms

Phases	Functionalities	Instruments/Software
Data storage	Massive amounts of storage, enormous processing power, limitless concurrent transactions, data security, real-time data	Apache Hadoop, CloudEra, MongoDB, Talend, Microsoft HDInsight, NoSQL, Hive, Sqoop, PolyBase, Presto
Data cleaning	Clean huge unstructured data sets, reshape and redefine the data	Drake, TIBCO Clarity, Trifacta Wrangler, OpenRefine, Winpure, Data Ladder, Data Cleaner Cloudingo, Reifier, IBM Infosphere Quality Stage
Data integration	Integrates between other tools and social media and Web pages	Actian, Attunity, Informatica, SnapLogic, Striim, Syncsort, Talend, Blockspring, Pentaho
Data mining	Discover insights within a database, predictive analysis and decision-making, text analytics, entity analytics	RapidMiner, R, Weka, Orange, KNIME
Information study	Assessing impacts of data patterns. Investigation is posing explicit inquiries and discovering answers in information	NodeXL, Gephi, Qubole, BigML, StatWing
Data visualization	Making data come to life. To pass on complex information experiences, business intelligence instruments	Tableau, Silk, CartoDB, Chartio, Datawrapper, Ideata Analytics
Data languages	Statistical computing and graphics	Python, R, Java, SQL, Julia, Scala, MATLAB, TensorFlow, RegEx, XPath
Data extraction	Extracts data from Website in a structured manner. Transform Web pages into usable data. Acts as a crawler, connector, table extractor, federator and tester	Octoparse, Content Grabber, Import.io, Parsehub, Mozenda, Scraper

4 Big Data Best Practices and Tactics

To gain significant benefits from big data, we must have appropriate methods, techniques and tools for capturing, storing, transferring, sharing, searching, analyzing, visualizing and interpreting the results. However, the approaches and tactics used for handling big data and the benefits gained from them vary from organization to organization. Therefore, below are several best practices that organizations can adopt.

Big data should support business objectives: The basic role of enormous information examination is to acquire arrangements that meet business targets.

Revise policies and rules: Putting these in place will help organizations to control costs and maximize the use of resources. It will also help in knowledge transfer, planning, training people and managing communication.

Data integration: Integration of data from several sources allows the discovery of important relationships and patterns among related factors/variables. Integration must resolve inconsistencies and semantic conflicts in the data.

Capitalize the data: Data can be capitalized only if it is of good quality and has valuable metadata. Hence, data must be collected regularly.

Deploy innovative methods and tools: Conventional methods and tools might not always work well with big data. Organizations must therefore employ innovative tools and cutting-edge technologies to be competitive.

Make the data secure: Growth of data and data analytics might lead to breaches in data. Such concerns must be addressed by means of technology and legal provisions.

Decide on fiscal matters: Some think they have to gather immense measures of information to use sound judgment, not really obvious. Gathering and preparing a lot of informational indexes are costly and past the scope of certain associations. It will be far superior for them to gather the information they really need.

4.1 Current Research and Application Trends in Big Data Analytics

Big data promises to benefit all levels and/or types of audience including government, telecommunication, financial, retail and manufacturing communities. Recent developments in big data are described in [27, 28] might include:

- Novel systems and gadgets for huge information examination and mining
- Analytics for data collected from smart grids and sensors
- Mapping human genome to identify co-occurring gene sequences
- Managing market saturation of individual customers in retail sectors
- Analyzing financial crime data
- Data mining and IOT
- Data mining for counterterrorism.

5 Conclusion

As the world continues with the massive production of data, the same must be managed effectively and efficiently for the benefit of the community at large. These include the Consumers/users, Producers and Developers of tools.

- Consumers of big data include governments, research, financial, retail, healthcare and other businesses. They require relevant and timely information to help them make good decisions.
- Producers of big data include Internet and telecommunication companies as well as governments, research, financial, retail and healthcare institutions.
- They must provide adequate infrastructure to capture, store, organize and transmit data.
- Developers of big data tools must develop and provide tools that are easy to use and affordable. The tools must generate information that the users need and must be easy to understand, interpret and make sense. On the flip side, tools may also be needed to reduce big data to manageable data which requires less computing resources and yet deliver equivalent results.

As Consumers expectation grows, so will the challenges of Producers and Developers. To meet these challenges, big data businesses must provide adequate infrastructure and tools. Specifically, they must:

- Put resources into huge information frameworks to catch, store, arrange and transmit information
- Invest in the development of big data analytic tools
- Link big data analytics to each business strategy and organizational process
- Keep abreast of big data technologies available in the market
- Train, equip and deploy experts in big data analytics.

There is at present an intense deficiency of big data abilities around the world. As the demand grows, universities can collaborate with industries and develop academic curriculum to equip students with the necessary skills set. Similarly, businesses can invest in the development of human capital for this purpose.

References

1. ur Rehman MH, Yaqoob I, Salah K, Imran M, Jayaraman PP, Perera C (2019) The role of big data analytics in industrial Internet of Things. Future Gener Comput Syst 99:247–259 (Elsevier)
2. Dai H-N, Wang H, Xu G, Wan J, Imran M (2019) Big data analytics for manufacturing internet of things: opportunities, challenges and enabling technologies. Enterp Inf Syst. https://doi.org/10.1080/17517575.2019.1633689
3. Manyika J, Chui M, Brown B, Bughin J, Dobbs R, Roxburgh C et al (2011) Big data: the next frontier for innovation, competition, and productivity. McKinsey Global Institute

4. Slack E (2012) What is big data? http://www.storage-switzerland.com/Articles/Entries/2012/8/3_What_is_Big_Data.html
5. TechAmerica Foundation's Federal Big Data Commission (2012) Demystifying big data: a practical guide to transforming the business of Government
6. Ding G, Wu Q, Wang J, Yao YD (2014) Big spectrum data: the new resource for cognitive wireless networking. arXiv preprint arXiv:1404.6508
7. Oracle Enterprise Architecture White Paper (2016) An enterprise architect's guide to big data
8. Oxford Dictionary (2016) Definition of big data in English. Oxford University Press from http://www.oxforddictionaries.com/definition/english/big-data
9. Webopedia (2016) What is big data? Retrieved 29 Aug 2019 from http://www.webopedia.com/TERM/B/big_data.html
10. Big data characteristics https://subscription.packtpub.com/book/application_development/9781787126992/8/ch08lvl1sec42/big-data-characteristics
11. SAS Institute Inc (2019) Big data analytics: what it is and why it matters. Retrieved 29 Aug 2019 from https://www.sas.com/en_us/insights/analytics/big-data-analytics.html
12. Margaret Rouse (2019) Big data analytics, TechTarget. Retrieved 29 Aug 2019 from https://searchbusinessanalytics.techtarget.com/definition/big-data-analytics
13. Lehdonvirta V, Ernkvist M (2011) Converting the virtual economy into development potential: knowledge map of the virtual economy. InfoDev/World Bank White Paper, 1, 5-17
14. Smith D (2012) Big data to add £216 billion to the UK Economy and 58,000 new jobs by 2017. Retrieved on 15 Aug 2019, from http://www.sas.com/offices/europe/uk/press_office/press_releases/BigDataCebr.htm
15. CEBR (2012) Data equity: unlocking the value of big data. Centre for Economics and Business Research White Paper, 4, 7-26
16. King A (2015) 7 benefits to using big data for small businesses. IndustriusCFO. Retrieved 30 Aug 2019 from http://www.industriuscfo.com/7-benefits-using-big-data/
17. Harvey C (2018) Big data pros and cons. Datamation Quinstreet Inc. Retrieved 30 Aug 2019 from https://www.datamation.com/big-data/big-data-pros-and-cons.html
18. van Rijmenam M (2015) The advantages and disadvantages of real-time big data analytics. Datafloq. Retrieved 30 Aug 2019 from https://datafloq.com/read/the-power-of-real-time-big-data/225
19. Labrinidis A, Jagadish HV (2012) Challenges and opportunities with big data. Proc VLDB Endowment 5(12):2032–2033
20. Gandomi A, Haider M (2015) Beyond the hype: big data concepts, methods, and analytics. Int J Inf Manage 35(2):137–144
21. Hayes B (2013) Big data has big implications for customer experience management. IBM Big Data and analytics Hub
22. Golfarelli M, Rizzi S (2009) Data warehouse design: modern principles and methodologies. McGraw-Hill, Columbus
23. Almeida FLF, Calistru C (2013) The main challenges and issues of big data management. Int J Res Stud Comput 2(1)
24. Thirunarayan K, Sheth A (2013) Semantics-empowered approaches to big data processing for physical-cyber-social applications. Semantics for Big Data, AAAI Technical Report FS-13-04
25. Bizer C, Boncz P (2011) The meaningful use of big data: four perspectives—four challenges. 2011 STI Semantic Summit in Riga, Latvia, 6–8 July 2011
26. Gartner (2013) Gartner identifies top technology trends impacting information infrastructure in 2013. Press Release, Retrieved 18 Aug 2019 from http://www.gartner.com/newsroom/id/2359715
27. Amalina F et al (2019) Blending big data analytics: review on challenges and a recent study. In: IEEE Access. https://doi.org/10.1109/access.2019.2923270
28. Singh SK, El-Kassar A-N (2019) Role of big data analytics in developing sustainable capabilities. J Cleaner Prod 213:1264–1273

Task Allocation and Re-allocation for Big Data Applications in Cloud Computing Environments

P. Tamilarasi and D. Akila

Abstract Resource allocation for Big data streams in cloud systems involves selecting the appropriate cloud resources. Since incorrect resource allocation results in either under provisioning or over provisioning, accurate resource allocation becomes challenging in Big data applications. Hence, the objective of this work is to design an optimal solution for resource allocation for minimizing the network bandwidth and response delay. In this paper, a task allocation and re-allocation mechanism for Big data applications is designed. It consists of two important agents: RE-allocation Agent (REA) and Resource Agent (RA). The RA is responsible for mapping the user requirements to the available VMs. The REA monitors the resources and chooses the VMs for resource reconfiguration. Then, it dispatches an allocation or de-allocation request to RA, running in the physical system, based on the varying requirements of virtual machines. Experimental results show that the proposed TARA has less execution time and achieves better utilization of resources, when compared to existing tool.

Keywords Big data · Task allocation · TARA

1 Introduction

Big data consists of large volumes of data generated by various systems in Internet-based applications. Big data processing and analysis are helpful in designing applications for social media, business transactions, etc. Big data denotes the mechanisms involved in extracting, storing, distributing and analyzing huge datasets with various

P. Tamilarasi (✉) · D. Akila
School of Computing Sciences, Vels Institute of Science, Technology & Advanced Studies, Chennai, India
e-mail: tamilmalu@yahoo.com

D. Akila
e-mail: akiindia@yahoo.com

S.-L. Peng et al. (eds.), *Intelligent Computing and Innovation on Data Science*, Lecture Notes in Networks and Systems 118,
https://doi.org/10.1007/978-981-15-3284-9_73

structures and maximum data rate [1]. Cloud environment provides suitable infrastructures for executing huge-sized Big data applications. But, Big data analysis in cloud environment raises various challenges and issues.

Scheduling is defined as a set of policies to manage the flow of work which will be executed by computing resources. But, scheduling more tasks in multi-cloud environment becomes the most challenging issue in the current research [2].

Task scheduling algorithms need to provide high performance and efficient system throughput. Since incorrect resource allocation results in either under provisioning or over provisioning, accurate resource allocation becomes challenging in Big data applications. Since cloud resources are having different characteristics, it is difficult to allocate specific resource for each task [3].

The major challenges of task allocation and scheduling involve:

- Selecting the best cloud resources for each task based on their requirements
- Enhance the task completion time
- Reduce cost of execution (in terms of bandwidth and storage)
- Minimize allocation time
- Utilize the idle resources
- Improve resource utilization
- Increased energy efficiency
- Maintaining fairness.

Our research work aims to design a task allocation and re-allocation model which consists of various servers containing virtual machines (VMs).

In this paper, an optimal resource allocation algorithm to minimize the network bandwidth and allocation latency is designed.

2 Related Works

The system developed by Kaur et al. [4] determines various data characteristics which are expressed as Characteristics of Big data (CoBa). Then, it dynamically clusters the cloud resources by applying self-organized maps (SOM). One among these clusters is assigned to Big data streams depending on its CoBa. Whenever the CoBa of any stream varies, the allocated cloud cluster is also varied.

The task scheduling algorithm proposed by Abed et al. [5] performs Big data processing and storing in cloud environments, depending on the user requirements. To enhance the Big data cloud computing solution, they have used a multi-metric-based solution. The model constitutes multiple control nodes and compute nodes. It also consists of a load-balancing algorithm for task scheduling.

To address the problems of cloud-based Big data applications such as performance, cost and availability, a multi-objective optimization algorithm was developed by Dai et al. [6]. By analyzing the interactions between these metrics, their system has been designed and implemented in experimental test bed.

A joint optimization algorithm proposed by Vakilinia [7] aims to reduce the power consumed by servers and cost of VM migration satisfying the resource and bandwidth conditions. It also handled the server heterogeneity problem. In the re configuration phase, VMs of uncompleted tasks may be migrated and new tasks are allocated to VMs.

Surputheen et al. [8] have proposed a concurrent VM reconfiguration mechanism for Big data tool which is MapReduce on virtualized cloud environments. It adds cores to VMs to execute local tasks temporarily and adjust the computing efficiency of the VMs to contain the scheduled tasks unlike the traditional schemes which can lead to user-friendly configuration methods for cloud resources.

A deadline-aware flexible bandwidth allocation for big-data transfers (DaFBA) algorithm has been developed by Srinivasan et al. [9]. In this approach, the allocated bandwidth can be adaptively adjusted at any time. The scheduling algorithms apply batch processing and dynamic scheduling at each interval for each request. They maximize the acceptance rate and satisfy the deadline constraints.

3 MapReduce Tool

MapReduce is a framework which can be applicable in large parallel data analyses. In MapReduce model, the calculation considers a set of key/value pairs as input and output.

It contains two functions for computation: Map and Reduce. The Map function accepts a pair of input to create a set of temporary key/value pairs, whereas the Reduce function gets a temporary key/values corresponding to that key. It aggregates these value pairs to generate a tiny set of values.

MapReduce with Hadoop implementation employs Hadoop distributed file system (HDFS) which stores data and also the interim results. HDFS maps all the locally stored data to a single file system order enabling the data to be spread at the entire set of nodes [10].

When a task request is received for scheduling, the VMs should have enough processing slots for each task. If no such slot exists, the task is rescheduled to a far away node by transferring the corresponding data. The MapReduce architecture is shown in Fig. 1.

4 Proposed Solution

4.1 Overview

Our proposed scheme consists of two important agents: RE-allocation Agent (REA) and Resource Agent (RA). The REA dispatches de-allocation and allocation requests to the hypervisor based on the physical machine. The RA monitors resource and chooses VM residing at the cluster which demands resource reconfiguration. Then,

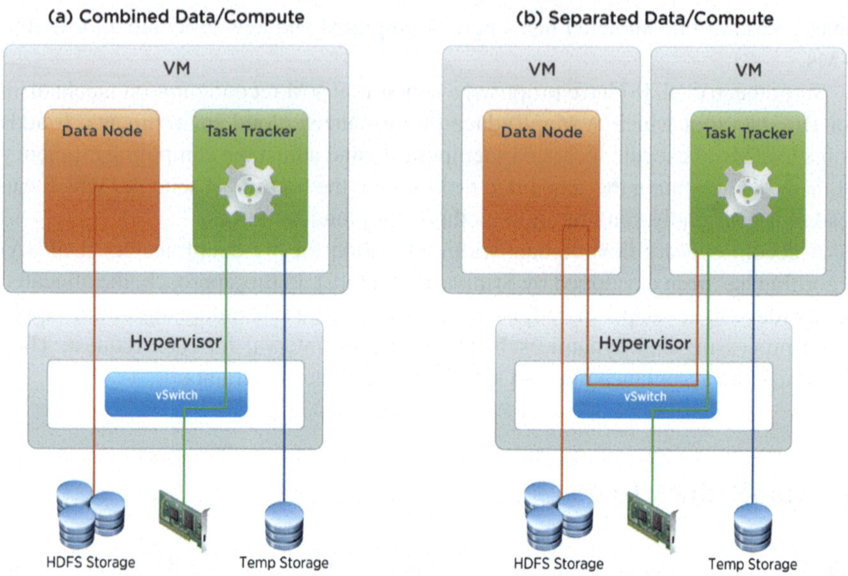

Fig. 1 MapReduce architecture on various VMs

it dispatches an allocation or de-allocation request to RA, running in the physical system.

4.2 Resource Allocation

4.2.1 VM Allocation

A VM allocation $\delta(m_1), \delta(m_2), \ldots, \delta(m_i)$ is defined as mapping of server to set a set of VMs that should satisfy the following two conditions:

1. Each VM should be allocated to at least one server, and no VM is allocated to more than one server

$$\cup_{m_j \in m} \Theta(m_i) = \Theta \quad (1)$$

$$\Theta(m_i) \cap (m_j) = \Phi, \forall\, 1 \leq y, i \neq j \quad (2)$$

2. For each server, the total resource requirements of its hosted VM do not exceed its available resources

$$\sum_{\theta j \in \Theta(mi)} w_i \leq \mathrm{CS}_j, \forall\, 1 \leq i \leq y \quad (3)$$

where W_i is the resource requirement of VM_i and CS_j is the available capacity of server.

4.2.2 Mapping of VMs to Tasks

Let $\{T = T_1, T_2, \ldots T_n\}$ be the set of tasks to be assigned.
For each task T_j, let BWR_j, D_j be the bandwidth requirement and deadline of the task.
Let TS_j be the size of each task T_j
Let $\{S_i = \{VM_{i1}, VM_{i2}, \ldots VM_{im}\}$ be the set of VMs on server S_i.
Let $\{C_{i1}, C_{i2}, \ldots C_{im}\}$ and $\{E_{i1}, E_{i2}, \ldots E_{im}\}$ be the capacities and expected delay of each VMs on S_i, respectively.
Let $Time_{UE}$ be the time required to execute a task of unit size.

Algorithm-1: Resource allocation by RA

For each Task T_j

1. Submit T_j to RA
2. RA extracts BWR_j and D_j from T_j
3. For each VM_{ik} on server S_i, k=12...m
4.   Estimate $E_{ik} = TS_j * Time_{UE}$
5.   If $(BWR_j < C_{ik})$ and $(E_{ik} < D_j)$
6.   Allocate T_j to VM_{ik}
7.   End if
8. End For
9. If(T_j could not find any suitable VM)
10.   Submit Tj to REA
11. End if
12. End For

In Algorithm-1, if a task T_j is submitted to RA with task number, task size, bandwidth required and required deadline), RA compares the task characteristics to each VM characteristics. If a VM with capacity higher than the required bandwidth and expected delay less than the deadline is found, it is selected and task will be assigned to it. If no such VM can be found, then the task request will be forwarded to the REA.

The REA handles task scheduling under load balancing. Hence, it will check the VMs on another server apart from S_i. If another idle VM can be found on any server, it can migrate to that VM such that the traffic load is reduced and number of resources is minimized. If all are engaged, then the task will be put in a waiting queue. These steps are repeated continuously.

4.3 Resource Reconfiguration

In order to process the varying demands on each VM, it may be dynamically reconfigured. When more number of physical resources are found, each VM is allocated with more number of virtual CPUs and memory space during VM execution.

The REA monitors the resources and chooses a VM which demands resource reconfiguration. If the resource requirement W of any VM exceeds the available capacity of the server CS, then it dispatches an allocation request (ALLOC) to RA. On the other hand, if the available capacity *C* of any VM is excess than its demanded resources, it dispatches a de-allocation request (DeALLOC) to the corresponding RA.

Upon obtaining requests from REA, RA reallocates the demanded virtual CPUs to the VM. RA requests the hypervisor of the physical server, to plug or to remove virtual CPUs for VMs present in the system. Algorithm-2 summarizes the steps involved in the resource reconfiguration process.

Algorithm-2 Resource Re-allocation

1. For each VM_{ik} on server S_i, k=12...m
2. REA monitors VM_{ik} at time t
3. If($W_{ik} > CS_i$) ,
4. REA send ALLOC to RA
5. Else if ($C_{ik} > W_{ik}$)
6. REA send DEALLOC to RA
7. End if
8. If(RA receives ALLOC)
9. RA demands hypervisor to hot-plug virtual CPUs
10. Else if (RA receives DEALLOC)
11. RA demands hypervisor to un-plug virtual CPUs
12. End if
13. End For

5 Experimental Results

5.1 Results

The workload with size 25 GB is input to Amazon Elastic Compute Cloud (EC2) at a velocity of 50, after every 5 min for the duration of 1 h. The proposed task allocation and re-allocation (TARA) is implemented on Amazon EC2 compute optimized c4 large instances. The performance of TARA is evaluated by comparing with the Apache Storm 0.92 tool using the execution time and resource utilization metrics.

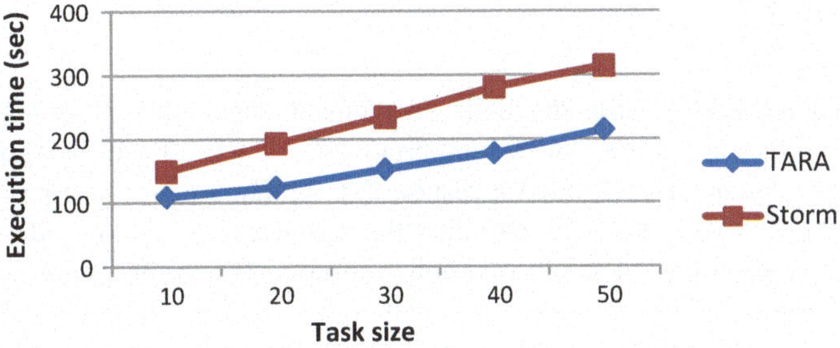

Fig. 2 Execution times for various task sizes

5.1.1 Execution Time

In this section, the execution time for different task sizes is measured and depicted in Fig. 2.

It can be observed from the figure that among all the sizes of tasks, the execution time is lowest when task size is low. Similarly, it becomes highest when the task size is high. The results show execution times for task are less in case of TARA when compared to Storm.

5.1.2 Resource Utilization

The results of resource utilization (%) for different sizes of task are measured and depicted in Fig. 3.

Figure 3 shows the comparison results of both the systems in terms of resource utilization. From the results, it can be observed that TARA achieves higher utilization when compared to Storm.

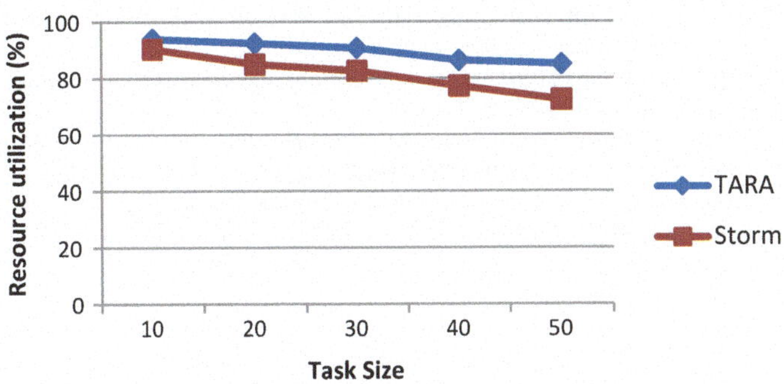

Fig. 3 Resource utilization for various task sizes

6 Conclusion

In this paper, a task allocation and re-allocation mechanism for Big data applications has been designed. It consists of two important agents: RE-allocation Agent (REA) and Resource Agent (RA). The RA is responsible for mapping the user requirements to the available VMs. The REA monitors the resources and chooses the VMs for resource reconfiguration. Then, it dispatches an allocation or de-allocation request to RA, running in the physical system, based on the varying requirements of virtual machines. The performance of TARA is evaluated by comparing with the Apache Storm 0.92 tool using the execution time and resource utilization metrics. Performance results have proved that TARA has less execution time and achieves better utilization of resources, when compared to existing tool.

References

1. Usama M, Liu M, Chen M (2017) Job schedulers for Big data processing in Hadoop environment: testing real-life schedulers using benchmark programs, Digital communications and networks. Elsevier, pp 260–273
2. Mishra SK, Satya Manikyam P, Sahoo B, Obaidat MS, Puthal D, Pratama M (2017) Response-aware scheduling of big data applications in cloud environments. Future Technologies Conference (FTC), Canada
3. Ebrahimi M, Mohan A, Lu S (2018) Scheduling Big data workflows in the cloud under deadline constraints. In: IEEE fourth international conference on Big data computing service and applications
4. Kaur N, Sood SK (2017) Dynamic resource allocation for big data streams based on data characteristics (5Vs). Int J Netw Manage 27(4)
5. Abed S, Shubair DS (2018) Enhancement of task scheduling technique of Big data cloud computing, In: International conference on advances in Big data, computing and data communication systems (icABCD), IEEE, South Africa
6. Dai W, Qiu L, Wu A, Qiu M (2016) Cloud infrastructure resource allocation for Big data applications. IEEE Trans Big Data 4(3):313–324
7. Vakilinia S (2018) Energy efficient temporal load aware resource allocation in cloud computing datacenters. J Cloud Comput Adv Syst Appl 7(2):2–24
8. Surputheen MM, Abdullah M (2017) Dynamic resource allocation scheme for map reduce tool in cloud environment. J Comput Eng 19(4)
9. Srinivasan SM, Truong-Huu T, Gurusamy M (2018) Deadline-aware scheduling and flexible bandwidth allocation for Big-data transfers. IEEE Access 6:74400–74415
10. Tamilarasi P, Akila D (2019) Ground water data analysis using data mining: a literature review. Int J Recent Technol Eng 7:2277–3878

A Systematic Framework for Designing Persuasive Mobile Health Applications Using Behavior Change Wheel

Hasan Sari, Marini Othman, and Hidayah Sulaiman

Abstract Mobile technology holds great potential for designing effective health behavior interventions. Changing individual's behavior and attitude is a growing research topic in the fields of behavioral science and information technology. Persuasive technology (PT), which defined as the technology intended to alter individual's attitude or behavior, has a beneficial influence on changing users' behavior and can lead to a better outcome. Persuasive techniques and models have been utilized to design behavioral change interventions in several contexts, including health care. Among them, the persuasive system model (PSD model) has been widely used in developing persuasive applications. Despite the potential, the PSD model has been criticized for lack of theoretical and evidence basis, which limit its capability in designing effective persuasive applications. Behavior change wheel (BCW) considered a comprehensive systematic framework used behavior techniques to develop behavior change interventions. This study aims to propose an integrated conceptual framework combining PSD and BCW, which could be used to implement successful persuasive mobile health applications.

Keywords Mobile health applications · Persuasive design · Persuasive technology · Behavior change · Behavior change wheel · COM-B model

---

H. Sari (✉) · H. Sulaiman
College of Computing and Informatics, Universiti Tenaga Nasional, 43000 Selangor, Malaysia

H. Sulaiman
e-mail: Hidayah@uniten.edu.my

M. Othman
Institute of Informatics and Computing in Energy, Universiti Tenaga Nasional, 43000 Selangor, Malaysia
e-mail: Marini@uniten.edu.my

S.-L. Peng et al. (eds.), *Intelligent Computing and Innovation on Data Science*, Lecture Notes in Networks and Systems 118,
https://doi.org/10.1007/978-981-15-3284-9_74

1 Introduction

More than 36 million people die annually from non-communicable diseases (NCDs), which equivalent to 63 percent of the total number of global deaths [1]. Example of NCDs includes cardiovascular diseases, cancer, diabetes, and obesity. Several clinical researches declared that these types of diseases, sometimes called "lifestyle diseases," could be prevented and treated through modifying some lifestyle behaviors and habits. Behavior change could be delivered through designing and implementing of behavior change interventions (BCI): "coordinated sets of activities designed to change specified behavior patterns" [2].

Mobile health technology provides a potential opportunity for developing effective behavior change interventions using interactive technology to alter the attitude and/or behavior of the user known as persuasive technology (PT) [3]. The model contains twenty-eight design principles classified under four distinct categories. PSD model had been used to develop support systems for behavior changes [4]. Despite the popularity of PSD in the field of persuasive information systems, it fails to deliver evidence-based interventions [5, 6]. Further, the method in which particular persuasive design features are chosen remains unclear.

The current study, therefore, proposes a framework that uses BCW/COM-B framework, to guide and identify the PSD design principles for a particular setting and environment. We believe that using the BCW/COM-B framework in combination with PSD model will help behavior change, theoretical evidence-based interventions, which could be used in many areas, including healthcare domain.

2 Literature Review

2.1 Persuasive Technology (PT)

Persuasive technology (PT) is defined as "interactive information technology designed for changing users' attitudes, or behavior" [3]. Persuasive technology is a combination of two different terms: persuasion and technology. Persuasion defined as "human communication designed to influence the autonomous judgments and actions of others" [3]. The study of persuasion has received growing attention from researchers in multiple areas [7]. The term technology means the medium that is used for delivering the persuasion such as Web sites, mobile devices, and video games. Persuasive technology applications have been developed in many different areas, such as advertising, online shopping, tourism, and health.

The Functional Trial. Fogs proposed that persuasive technology could be classified into three functional roles: tools, media, and social actors [8]. As tools, interactive technologies can make the target behavior easier or well organized, thereby increasing users' ability to perform the target behavior. As media, interactive technologies have the potential to interact with users, persuading them through rehearsing behavior. As an social actor, technologies can apply social roles to influence users to perform target behavior.

2.2 *Persuasive System Design Model (PSD Model)*

Persuasive systems design model (PSD) is presented by Oinas-Kukkonen and Harjumaa [4]. They proposed a comprehensive model for designing and evaluating a persuasive system that could be used in developing persuasive applications. The principle of PSD is based on social psychology and applied to the domain of human–computer interaction (HCI).

PSD described a set of persuasive principles to serve as a guide for developing and assessing the persuasive systems. The model contains the following components:

- The persuasive system features (seven postulates).
- Context Analysis: comprises the analysis of the intent, event, and strategy.
- The persuasive system features have four categories. Each category contains seven different features.

The main component in PSD is context analysis. Based on this analysis, the principles of the persuasive system could be identified. It includes twenty-eight features divided into four groups: (1) primary task support, (2) computer–human dialog support, (3) credibility support, and (4) social influence.

(1) **Primary Task support**: includes characteristics that help users to perform primary tasks. This stage includes reduction, tunneling, customization, self-monitoring, simulation, and rehearsal.
(2) **Dialogue Support**: includes the features that could be used to help users to keep moving toward the goal. This includes praise, rewards, reminders, suggestion, similarities, liking, and social role.
(3) **Credibility support**: includes features that describe how the system is more credible and thus more persuasive. This includes trustworthiness, knowledge, surface credibility, real-world feeling, power, third-party endorsements, and verifiability.
(4) **Social Support**: includes principles of design to motivate individuals through social influence. This category's design principles are social facilitation, social comparison, normative influence, social learning, collaboration, competition, and appreciation.

2.3 *Behavior Change Theories in Designing Behavior Change Interventions*

Several theoretical approaches have been used in designing interventions. Health Belief Model (HBM) [9], Theory of Planned Behavior (TPB) [10], and Transtheoretical Model are the most widely used in behavioral theories [10]. There are other models focusing on communicating health messages to influence behavior on mass media scale; these include the communication persuasion model and the social marketing theory [11] which could apply to health care.

In addition to the behavior change theories and models, evidence-based frameworks also applied to help the intervention designers to develop and evaluate health interventions such as intervention mapping framework [12] and the medical research council (MRC).

Michie [13] designed a model for behavior change model attached to the COM-B model. This is known as the behavior change wheel (BCW).

2.3.1 Behavior Change Wheel (BCW) framework and the COM-B Model

BCW is an integrated framework based on nineteen (19) behavior change theories and models. The COM-B model is a behavioral model providing a systematic method for analyzing the behavior (s) that needs to change by assessing three main elements: capabilities, opportunities, and motivations.

BCW can assist developers in building comprehensive and evidence-based interventions suitable for both individual and group populations. For each intervention function, different techniques can be used to deliver behavior change techniques (BCTs).

The benefit of BCW/COM-B model is they provide a systematic approach started by analyzing the context where behavior change should occur and end by selecting the behavior change techniques (BCTs) that are appropriate for the particular situation.

The stages required to conduct behavior change using BCW are as follows:

Stage 1: define the behavioral issue: In this stage, the issue which the intervention will solve is defined in behavioral terms (e.g., prolonged sitting). The user who performs the behavior should be identified. Also, the list of all associated behavior that might influence the target behavior should be specified.

Stage 2: select and specify the target behavior: In this stage the behavior that needs to accomplish will be identified (for instance, reduce sitting habit), who requires to do that, what they should bring about change, when and where they need to do it, how many times and with whom?

Stage 3: identify which modifications are essential: the COM-B model in this phase is used to know the habits to alter if the targeted conduct is to be achieved. In this stage, focus group or interview could be used to analyze the situation based on three dimensions in the COM-B model.

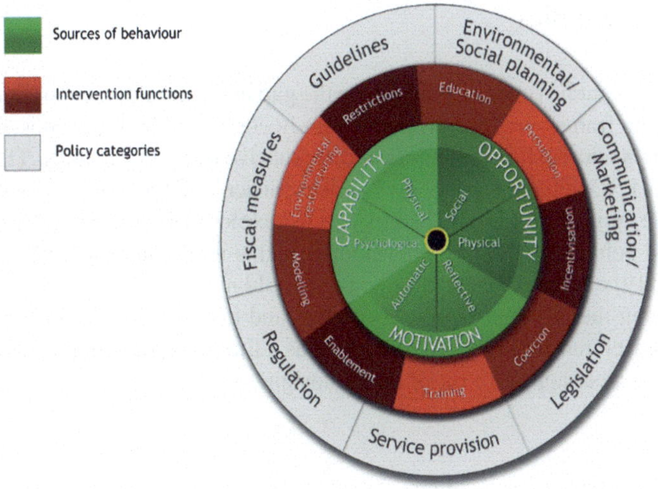

Fig. 1 Behavior change wheel (BCW) [13]

Fig. 2 COM-B mode [14]

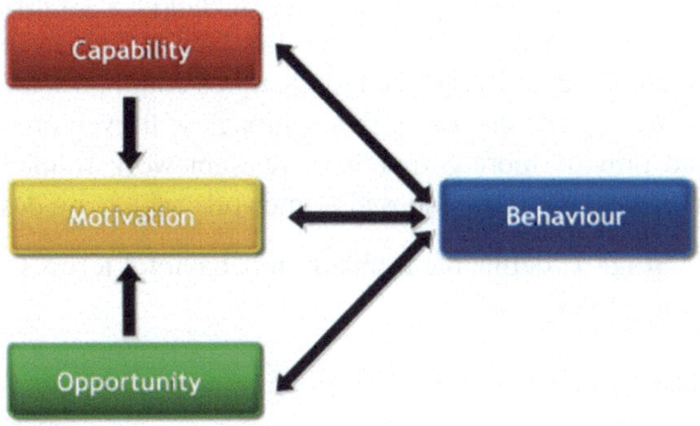

Stage 4: identify the functions and policy categories.

Stage 5: identify behavior change techniques that should be used (there are 93 BCTs taxonomies) (Fig. 2).

2.3.2 Behavior Change Techniques (BCTs)

Behavior change techniques (BCTs) are described as "observable, replicable, and irreducible component of an intervention designed to alter or redirect causal processes that regulate behavior" [15]. Those interventions that combine BCTs have a greater impact on changing individual behavior more than those interventions that are not incorporate BCTs [16, 17].

3 Related Studies

Some previous studies attempted to bridge the gap in the development of a persuasive system design framework. For instance, a study conducted by [6] proposed a model for mapping socio-ecological factors and persuasive design principles. The authors claimed that PSD fails to identify how to select particular PT techniques for designing a persuasive application. Another study by [18] explored how relationships between attitude and behavior guide PSD development, this model named 3D-RAB model. Reference [19] claimed that PSD framework could not provide an explicit approach for addressing the differences in users' needs and how to adapt according to these differences. Reference [20] stated that PSD has a limitation of being too general and not provide a clear approach for practical design.

4 Proposed Framework

To our best knowledge, there is no previous study that links behavior change techniques with persuasive technology techniques. COM-B and BCW framework used as a guideline to identify appropriate behavior change techniques (BCTs), and based on the selected BCTs, the PSD features could be selected.

We believe that using this approach will overcome the limitation in PSD model and provide more comprehensive framework framework. Figure 3 illustrated the main stages in the proposed framework, which are described below:

Stage 1: define the problem in behavioral terms (e.g., prolong sittings, physical inactive).

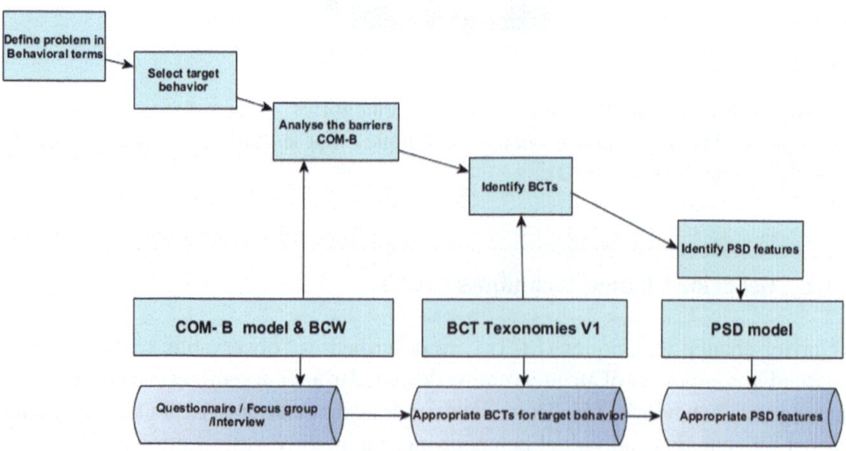

Fig. 3 Proposed framework

Stage 2: select the target behavior (s) (e.g., reduce settings, increase physical activity).
Stage 3: use the COM-B model to evaluate target behavior obstacles and facilitators and list them based on the three dimensions (capability, opportunity, motivation). The analysis could be conducted using focus groups, interview, questionnaire, or a combination of them.
Stage 4: identify BCTs from BCT taxonomies V.1 that address the barriers from stage.
Stage 5: identify the appropriate PSD from four category functions. The suitable PSD features are selected based on the chosen BCTs in stage 4.

5 Discussion

In this paper, we explained how appropriate persuasive design principles could be obtained based on analyzing the context where behavior change should perform using COM-B model and behavior change wheel (BCW). Based on the proposed framework, the PSD features related to the situation could be chosen. Then, through future work with designing persuasive applications, the relevant persuasive design principles could be identified to manipulate COM-B model dimensions and BCW factors.

6 Limitations

This study mainly focused on the description of a systematic approach of how combining persuasive technology principles with behavior change techniques. The framework is required to validate by some experts in behavioral science and information technology. More research on the use of behavioral change methods in developing and evaluating persuasive systems is needed.

7 Conclusion

To design an effective persuasive mobile health intervention, the design stage should be constructed based on behavior change theories. The current study proposed a systematic framework combining behavior change techniques (BCTs) and persuasive design principles. The proposed framework overcomes the limitation in PSD model by integrating behavioral theories in the development of persuasive applications. Future work should be devoted to examining the effectiveness of mobile health applications, developed based on the proposed framework presented in this study.

References

1. World Health Organization (2014) Global status report on noncommunicable diseases 2014
2. Michie S, van Stralen MM, West R (2011) The behaviour change wheel: a new method for characterising and designing behaviour change interventions. Implement Sci 6(1):42
3. Simons HW, Morreale J, Gronbeck BE (2001) Persuasion in society. Communication, p 414
4. Oinas-Kukkonen H, Harjumaa M (2009) Persuasive systems design: key issues, process model, and system features. Commun Assoc Inf Syst 24(28):485–500
5. Wiafe I, Nakata K, Gulliver SR (2011) Designing persuasive third party applications for social networking services based on the 3D-RAB model. In: Communications in computer and information science, vol 185, CCIS, no. PART 2, pp 54–61
6. Mohd Mohadis H, Mohamad Ali N (2015) Using socio-ecological model to inform the design of persuasive applications. In: Proceedings of the 33rd annual ACM conference extended abstracts on human factors in computing systems—CHI EA '15, pp 1905–1910
7. O'Keefe DJ (2015) Message generalizations that support evidence-based persuasive message design: specifying the evidentiary requirements. Health Commun 30(2):106–113
8. Fogg BJ (2003) Persuasive technology: using computers to change what we think and do, pp 1–282
9. Becker MH (1974) The health belief model and personal health behavior. Health Educ Monogr 2:324–473
10. Ajzen I (1991) The theory of planned behavior. Organ Behav Hum Decis Process 50(2):179–211
11. Prochaska JO, DiClemente CC, Norcross JC (1992) In search of how people change: applications to addictive behaviors. Am Psychol 47(9):1102–1114
12. Bartholomew LK et al (2006) Planning health promotion programs: an intervention mapping approach. Planning health promotion programs: an intervention mapping approach
13. Michie S, Atkins L, West R (2014) The behaviour change wheel: a guide to designing interventions
14. Michie S, van Stralen MM, West R (2011) The behaviour change wheel: a new method for characterising and designing behaviour change interventions. Implement Sci 6(1)
15. Michie S et al (2013) The behavior change technique taxonomy (v1) of 93 hierarchically clustered techniques: building an international consensus for the reporting of behavior change interventions. Ann Behav Med 46(1):81–95
16. Harman K, MacRae M, Vallis M, Bassett R (2014) Working with people to make changes: a behavioural change approach used in chronic low back pain rehabilitation. Physiother Canada 66(1):82–90
17. Procter S, Mutrie N, Davis A, Audrey S (2014) Views and experiences of behaviour change techniques to encourage walking to work: a qualitative study. BMC Public Health, vol 14, no 1
18. Wiafe I, Alhammad MM, Nakata K, Gulliver SR (2012) Analyzing the persuasion context of the persuasive systems design model with the 3D-RAB model. Lect Notes Comput Sci (including Subser. Lect. Notes Artif. Intell. Lect. Notes Bioinformatics), vol 7284 LNCS, pp 193–202
19. Wiafe I, Nakata K, Moran S, Gulliver SR (2011) Considering user attitude and behaviour in persuasive systems design: the 3D-Rab model. ECIS 2011 Proc
20. Halttu K (2015) About the persuasion context for BCSSs: analysing the contextual factors, no. BCSS, pp 43–50

Dynamics of Knowledge Management in 4IR Through HR Interventions: Conceptual Framework

Arindam Chakrabarty and Uday Sankar Das

Abstract The world economy has been remaining captive to the exponential growth of knowledge. The concept of knowledge is diversified and multidimensional which essentially includes theoretical constructs, experiential learning, incepts of laboratory results, models and of course its ability to adapt changes. In fact, knowledge economy should be ideally the fusion of indigenous belief and practice and transformation of scientific know-how. The world has witnessed rapid transformation both in society knowledge system and industrial revolution. The twenty-first century has emerged as the torchbearer for fourth industrial revolution which can manifested in designing machines, gadgets that can be embraced with auto-guided instructions, artificially par excellence with human intelligence. The aspiration of fourth industrial revolution (4IR) demands higher order of knowledge, big data analytics and continuous improvement in R&D outcomes. So, it has become emergent to concentrate on the threshold level of knowledge management practices in the transforming economy. This paper has focused on how the interrelations among the level of industrial revolution, knowledge management and transformational HRM practices include KASH protocol using conceptual modelling.

Keywords Knowledge management · 4IR · Human intelligence · Transformational HRM practices

1 Introduction

The progression of knowledge management has been carried away through a long journey. The organization began to understand that human being cannot be compared with machine as a part of neoclassical theory of management. In the beginning of

A. Chakrabarty (✉)
Assistant Professor, Department of Management, Rajiv Gandhi University (Central University), Rono Hills, Doimukh, Arunachal Pradesh 791112, India

U. S. Das
Guest Faculty, Department of Management & Humanities, National Institute of Technology Arunachal Pradesh, Yupia, Arunachal Pradesh 791112, India

S.-L. Peng et al. (eds.), *Intelligent Computing and Innovation on Data Science*, Lecture Notes in Networks and Systems 118,
https://doi.org/10.1007/978-981-15-3284-9_75

twentieth century, the concept of welfare management had been practised by few organizations which paid a special attention on the welfare measures of workers in the factory, but the experience of welfare management practices had not been complacent as it was desired. Prior to the Second World War, the idea of personal management emerged roughly in 1940s which concentrated on measuring performance of labour on various scales, even though this school of thought never recognized the labour as human resource. The importance of training development OD interventions organizational culture climate had not been given due weightage. From 1980, the organization started to implement human resource management over throwing the erstwhile mechanistic and dogmatic view of management. HRM has been evolved as an organic orientation that recognizes and respects labour force as a dynamic resource that can be appreciated over the period of time with the augmentation of knowledge, skill and experiences. The twenty-first century has revolutionized with the advent of superior level of technological advancement. The knowledge-driven economy has been witnessing with a new paradigm, i.e. generation of new idea, product, process, with the succession of high rate of obsolescence. It becomes faster as we proceed towards the present time.

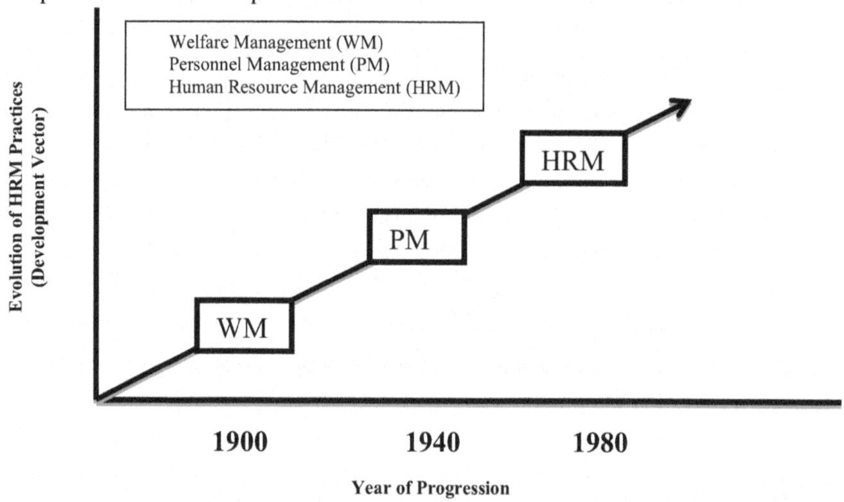

Adapted & Modified from [1, 11, 13]

With the advent, progression and popularity of 4IR, the organizations have explored to recognize the imperative of knowledge management practices at the beginning of twenty-first century. This brings the accumulation of vivid information robust technology and big data compounded with the application of AI, ML and block chain technology, etc. Today, the construct of knowledge management is not confined in accumulating functional super specializations rather it has extended to endless interactions among various dataset from various domains in a multi-varied assortment of knowledge basket with multi-criteria decision-making (MCDM) protocol [14]. This envisages numerous innovative opportunities and new directions that lead to explore knowledge-led dynamic problem-solving mechanism.

1.1 Evolution and Understanding of Knowledge Management

Contemporary business writings have extensively focused on knowledge management and have curated it as a contemporary theoretical discipline and shifted the focus of organizations from tangible products and goods to intangible assets focused on performance and profitability in this competitive environment. Knowledge management has opened up the opportunity to add renewed strategic growth in any business organization [2]. A study 'Emerging Practices in Knowledge Management' conducted by the American Productivity and Quality Center of the USA points out six key strategies of a firm for practice of knowledge management (KM). From a business strategy point of view

1. As a tool to transfer best practices.
2. As a customer-oriented tool.
3. As discipline for personal development.
4. As a tool for intellectual assets management.
5. As a tool for knowledge creation and innovation.

Prominent fortune 500 companies like 'Dow Chemicals' and 'Texas Instruments' were also a part of this survey [6]. KM focuses on gathering of useful knowledge or for the business process so that the employees can readily access knowledge. It also helps to secure specified well-defined set of knowledge practice by preventing from use of inappropriate knowledge. KM is research intensive and involves application of organizational learning capacity over competitive advantage in the long run. Evolution of KM intervention can be categorized into six broad stages which can be further rationalized as depicted below.

Six broad stages	Rationalization of stages
Initiation	KM initiation
Adoption	
Pilot implementation	
Organic growth	KM development
Organizational implementation	
Diffusion	KM maturity

The factors determining the evolution of KM are classified into knowledge self-efficiency, open communication and mutual benefits [4]. The example of companies like Dow Chemicals which is a treasure trove of unorganized intellectual property, whose main business is to earn royalty through licensing of technology and information highlights the importance and needs of knowledge management in order to organize this other wile piled up disorganized knowledge for profit maximization [6].

1.2 Dimensions of Knowledge Management

The knowledge management can be referred in two perspectives, i.e. in terms of capability dimensions and quality ontology. The capability dimension can be broadly categorized into two sub-dimensions, i.e. infrastructure and process. The attributes of infrastructure may include technological led ecosystem, other resources and support facilities structure, culture. The process matrix may comprise acquisition, conversion, application and security. This has been illustrated in the figure mentioned below:

Capability dimensions	Attributes	Meaning
Infrastructure	Technology Structure Culture	Organize fragmented knowledge in an organization Leverage of technological architecture Encouragement of employee interaction
Process	Acquisition Conversion Application Security	KM process of knowledge acquisition Utilization of the existing knowledge Application of knowledge Knowledge protection

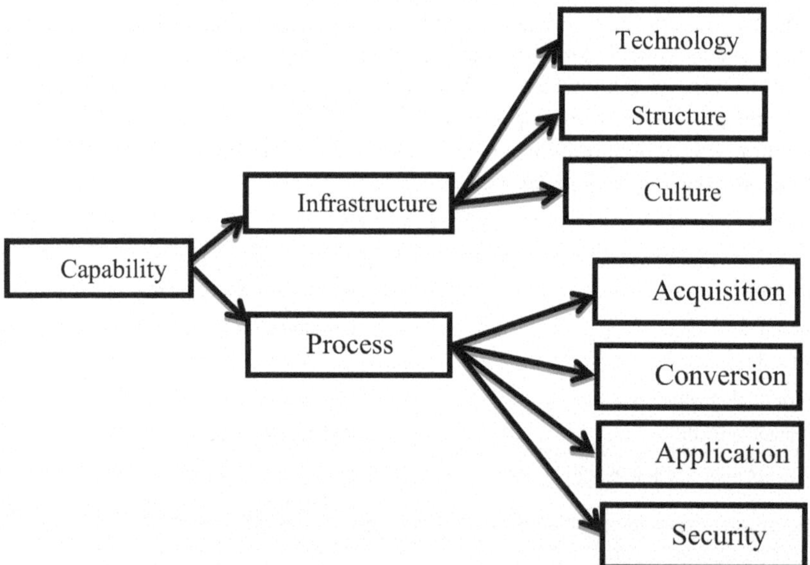

This matrix model helps identify the capability dimensions of knowledge framework and its subsequent branch entities [3]. A conceptual frame work is proposed to manage the quality dimensions of KMS based on the environmental factors and its effects on the same. The resultant framework consists of 36 items grouped into the eight dimensions of KM namely Functionality, Completeness, Reliability, Usability, Access, Serviceability, Flexibility, Security [7, 9].

1.3 Knowledge Management Is an Extension to HRM?

The spectrum of innovation has immensely expanded the ambit of HRM capabilities. The incidence of continuous innovation in every filed of HRM like selection, performance management, training & development etc. has made phenomenal changes to bringforth new directions and domain of thought processes as outcomes that are assimilated in the organizational ecosystem and practiced by the successful mediations and interventions of KM by means of development, dissemination and application of knowledge [8]. Collaborative and holistic practices of KM-induced HRM essentially enhance the uniqueness of organizational competency preferably the knowledge protocol, which positively signifies the association with the extent of innovations not the other way around, i.e. knowledge HRM (KHRM) has no impact on innovation excepting to mediate between collaborative HRM as transformational change agent [5].

1.4 Knowledge Management in the 4IR

There is symbiotic relationship between knowledge management and the progression of 4IR. The fourth industrial revolution has been continuously expanding the knowledge sharing platform so that it can move forward endlessly in consonance with the rapid research and development outcomes. From the beginning of twenty-first century, the world of technological research largely dominated by splendours of electronic gadgets, IoT, machine learning, block chain technology which facilitates to generate record process and interpret the large volume of data which is popularly known as big data analytics which primarily solve the problem by means of various modes of descriptive, predictive and prescriptive data analysis. All these development vectors in the technological framework and high yield application mechanism to solve complex problems have essentially deserved the transformative knowledge management initiatives in the organizational set-up.

2 Objectives of the Study

1. To propose a logical model to understand the interrelationship between progression of industrial revolutions ab initio and individual firms' aspirations for bridging knowledge gaps.
2. To develop a conceptual framework for understanding interrelations and interactions among industrial progression (4IR), knowledge management and transformational HRM practices using HRM competency model.
3. To devise the knowledge-dominated KASH protocol in HR interventions in congruence with the progression of industrial revolution.

3 Research Methodology

This is an exploratory study through which it is attempted to understand the premises and fitness of knowledge management in the emerging 4IR ecosystem. The paper has been designed referring various research papers, reports and suitable application of strategic evaluative protocols widely practised in the academia and the research world.

4 Analysis and Interpretation

4.1 Analysis & Interpretation—I

According to Watson [12] knowledge is regarded as an ability to utilize information in order to add value and influence the decision-making process. It is imperative that the organization should adapt the terminal level of knowledge in a useable form so that there should not be much deviation of standards between industry and firms in terms of creation, transfer and utilization of knowledge [10].

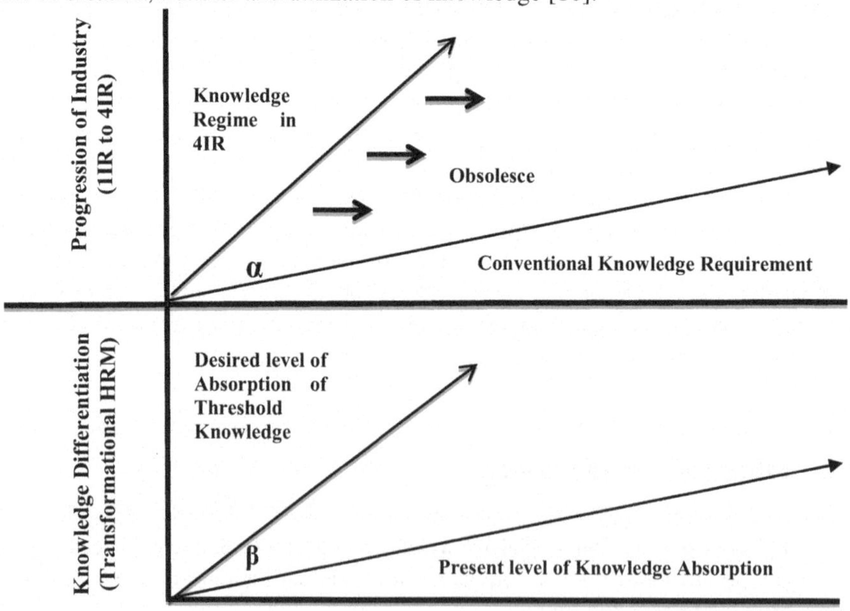

$\alpha \propto \beta$ Model developed by the Authors

The journey of industrial revolution has been largely manifested by the voracity of knowledge which has emerged thorough the continuous process of innovation. In the comparative two-dimensional matrix, the angle (α) between conventional knowledge requirement and knowledge regime in 4IR increases with the fact that 'knowledge regime in 4IR' would tend to incline to Y-axis with the passage of time. Similarly, the angle (β) between the 'present level of knowledge absorption' and the 'desired level of absorption of threshold knowledge' must escalate in proportionate with the time spend and experience gathered. For every organization to survive in the dynamic environment and technological development, the angle α and β must be proportional and highly correlated in order to signify that the organization would remain competitive as it enjoys competency in the incremental knowledge-dominated industrial revolution. If the organization fails to achieve this synergy, it would literary cease to exist. The upsurge of 'knowledge regime in 4IR' tends to incline towards Y-axis which makes the curve stiffer enhancing the value of angle α. As a result of that, it forces to dissociate the previous knowledge set to become obsolete as depicted in the model.

4.2 Analysis & Interpretation—II

The progression of knowledge intends to augment the process of industrial revolution (IR). The set of ongoing innovations essentially land up with a new age and phase of IR; thus, human society moves forward from the primitive era of 1IR to the most advanced knowledge-driven industrial revolution popularly known as Industry 4.0. The industrial environment essentially influences the appropriate inducement of knowledge that can generate higher order of competency uniqueness for the firm. In order to explore these opportunities, the firm needs to invest on high-end resources as well as procurement of superior human resources that can augment and transform the change management initiative at possible encounter. The new era of knowledge management imbibes the HR policies to encourage and promote the best talents to acquire so that the culture of learning organization can perpetuate with higher acceleration as in tune with the expectations of the relevant industry.

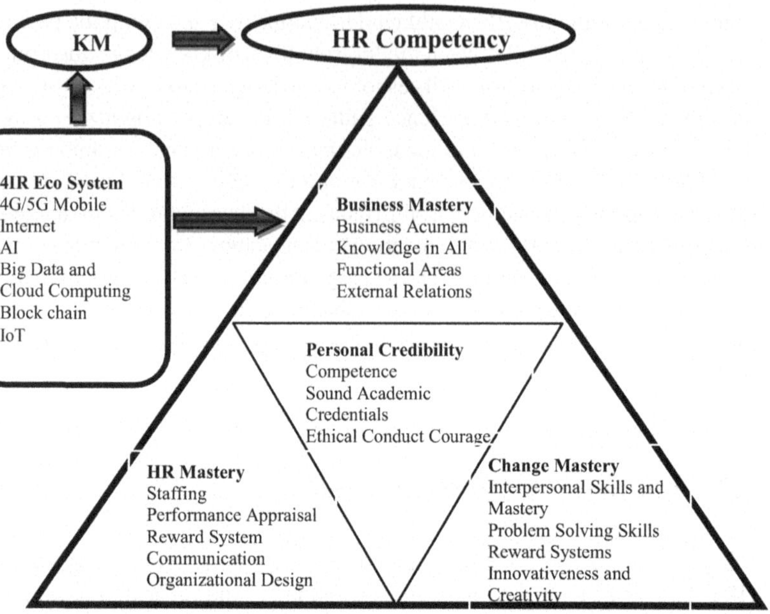

Model developed by the Authors in corporation HR Competency Model adapted from Human Resource management by Christopher Maybe, et. al., Blackwell Business, p.31

The model has been developed to project how the industry environmental factors reign enforces the firm to prioritize knowledge management which can be trickled down even at the bottom of the pyramid in the organizational hierarchy. This holistic development vector has to be inundated by the dynamic HRM practices as depicted above.

4.3 *Analysis & Interpretation—III*

The basic function of HRM revolves on its effective strategies human resource planning, performance management and human resource development which also interacts and correlates each other. One of the important approaches to address the HRM functions may be the successful manifestations of Knowledge, Attitude, Skills, Habits (KASH) protocol.

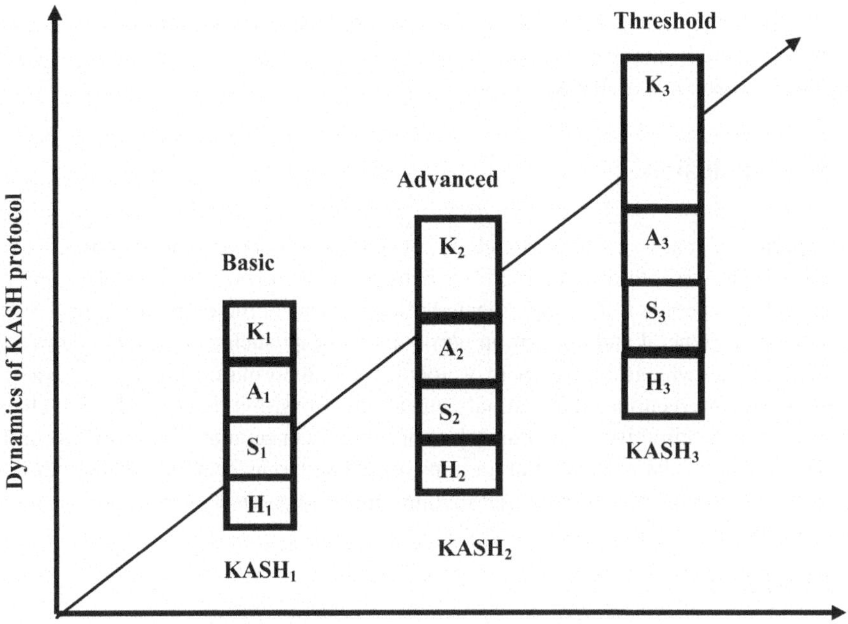

Model developed by the Authors

KASH denotes the assortment of four components: Knowledge (*K*), Attitude (*A*), Skill (*S*) and Habits (*H*) which are primarily required for a certain position of an organization in a mutually exclusive manner. KASH differential matrix examines the differentials of each component with respect to the deviations from the actual level of performance from its desired/expectancy module.

The firm always looks at the KASH differential matrix as illustrated below.

KASH components	Desired KASH set	Actual KASH set	KASH differentials (D~A)
Knowledge (*K*)	K_D	K_A	K_D~K_A
Attitude (*A*)	A_D	A_A	A_D~A_A
Skills (*S*)	S_D	S_A	S_D~S_A
Habits (*H*)	H_D	H_A	H_D~H_A

~ Sign of difference

If ($K_D < K_A$) or ($K_D = K_A$), i.e. the knowledge set desired is either lesser or equal to the knowledge possessed by the existing professional, no training need is identified / required. In general cases, K_D happens to be greater than K_A that means, the desired knowledge is greater than the actual knowledge possessed by the concerned employee that symbolizes the specific requirement of knowledge, i.e. identification of training need on specific knowledge domain. The firm would attempt to minimize the $[K_D - K_A]$ by means of appropriate HR interventions. Similarly, other **KASH** components can also be described. The most feasible '*K*', '*A*', '*S*', '*H*' combinations are generally encouraged for achieving desired HR objectives. With the growing influx of knowledge management, the appropriate '**KASH differential matrix**' needs to be formulated, giving increasing weightage on knowledge components as per the dynamic demands of 4IR and so on.

5 Conclusion

Experiential learning and Research & development generate new idea product process for the welfare of mankind. The benefits of such illustrious development can reach to the people if it is implemented effectively and efficiently. It is a turn for the industry in general and the firm in particular to adapt such changes by augmenting advanced knowledge management protocol. The transformation process needs appropriate HR interventions that can only ensure this transition in an accelerated change management initiative. This paper has presented conceptual framework to understand the interrelations and interventions of KM and transformational HRM through along the progression of industrial revolutions more precisely 4IR ecosystem.

References

1. Berkowitz E, McQuaid K (1978) Businessman and bureaucrat: the evolution of the American social welfare system, 1900–1940. J Econ Hist 38(1):120–142
2. Bernard Nielsen B (2005) Strategic knowledge management research: tracing the co-evolution of strategic management and knowledge management perspectives. Competitiveness Rev Int Bus J 15(1):1–13
3. Fan ZP, Feng B, Sun YH, Ou W (2009) Evaluating knowledge management capability of organizations: a fuzzy linguistic method. Expert Syst Appl 36(2):3346–3354
4. Lin HF (2011) Antecedents of the stage-based knowledge management evolution. J Knowl Manage 15(1):136–155
5. Lopez-Cabrales A, Pérez-Luño A, Cabrera RV (2009) Knowledge as a mediator between HRM practices and innovative activity. Human Resour Manage 48(4):485–503 (Published in Cooperation with the School of Business Administration, The University of Michigan and in alliance with the Society of Human Resources Management)
6. Mullin R (1996) Management: knowledge management: a cultural evolution. J Bus Strategy 17(5):56–59
7. Owlia MS (2010) A framework for quality dimensions of knowledge management systems. Total Qual Manag 21(11):1215–1228
8. Özbağ GK, Esen M, Esen D (2013) The impact of HRM capabilities on innovation mediated by knowledge management capability. Proc-Soc Behav Sci 99:784–793
9. Rao L, Osei-Bryson KM (2007) Towards defining dimensions of knowledge systems quality. Expert Syst Appl 33(2):368–378
10. Teece DJ (2000) Strategies for managing knowledge assets: the role of firm structure and industrial context. Long Range Plan 33(1):35–54
11. Waman AM (2010) The impact of training interventions on the development of competencies of the employees in selected private sector unit in Pune
12. Watson R (1999) Data management: databases and organizations, 2nd edn. John Wiley, New York
13. Wood NJ (1960) Industrial relations policies of American management 1900–1933. Bus Hist Rev 34(4):403–420
14. Wu WW, Lee YT (2007) Selecting knowledge management strategies by using the analytic network process. Expert Syst Appl 32(3):841–847

Clinical Data Classification Using an Ensemble Approach Based on CNN and Bag-of-Words Approach

Bhanu Prakash Battula and D. Balaganesh

Abstract From the past decade, there has been drastic development and deployment of digital data warehoused in electronic health record (EHR). Initially, it is intended for getting patient general info and accomplishment healthcare tasks like billing, but researchers focused on secondary and most important use of these data for innumerable clinical solicitations. In this paper, we addressed the use of deep learning-based clinical note multi-label multi-class approach using ensemble approach based on CNN and bag-of-words approach. And we map those classes for multi-classes. And we perform experiments with Python, and we used libraries of Keras, TensorFlow, NumPy, matplotlib, and we use MIMIC-III data set. And we made comparison with existing works CNN, skip-gram, n-gram and bag of words. The performance results show that proposed framework performed good while classifying the text notes.

Keywords CNN · Electronic Health Care Record · Multi Class Approach

1 Introduction

To distinguish whether a patient [11] experiences a specific illness, a regulated methodology requires the accompanying info: guides to become familiar with the examples for the malady (occasions speaking to patients) and the portrayal well-being status of the patient that is examined. Given that a suitable number of precedents are given (both positive and negative), this basic classification performs well. However, in the therapeutic area, this straightforward case is not experienced that frequently. Surveying the well-being status of a patient does not generally prompt a solitary, singular therapeutic condition. Numerous well-being conditions are corresponded and impact one another, in this way suggesting comorbidity in a patient. There is a level of uncertainty around the phrasing concerning comorbidity. The term characterizes a patient's well-being status when at least two conditions coincide. Various related

B. P. Battula (✉)
Lincoln University College, Kota Bharu, Malaysia

D. Balaganesh
Faculty of Computer Science and Multimedia, Lincoln University College, Kota Bharu, Malaysia

S.-L. Peng et al. (eds.), *Intelligent Computing and Innovation on Data Science*,
Lecture Notes in Networks and Systems 118,
https://doi.org/10.1007/978-981-15-3284-9_76

builds are utilized to allude to comorbidity: multi-disease, dreariness weight, and patient multifaceted nature. Previously, authors propose a top-to-bottom investigation of the comorbidity in patients which incorporates term definition, the nature, significance, and intricacy of this well-being condition. The nearness of various conditions in a patient can be a consequence of possibility, determination or causal affiliations. A typical and straightforward case of cohappening conditions is interminable obstructive aspiratory and liver ailment thinking about that they are brought about by propensities that are connected: smoking and liquor utilization.

In this research, we concentrate on classification of clinical text. It is the problem of assigning documents to different classes. Particularly, medical text or clinical notes plays a key role in patient risk prediction. Clinical note analysis is key and significant importance in the medical field. Because these are free text narratives generated by doctors and nurses, through patient make a diagnosis when he is in clinic. They are regularly joined by a lot of metadata codes from the International Classification of Diseases (ICD) [8, 12], which extant an institutionalized method for demonstrating judgments and systems that were performed amid the experience. These ICD codes have variety of advantages, which vary from admission of hospital to billing and predictive analysis of patient condition. The major problem from traditional approach (manual) is more time consuming and error pruning. Automatic coding will be more accurate and very speedy approach. But mapping of medical text notes with ICD is a difficult task because of two major reasons. The space of the label is high dimensional, due to ICD-9 having 15K medical [4] codes. Second, clinical content incorporates superfluous data, incorrect spellings and non-standard condensing, and a huge restorative vocabulary. These highlights consolidate to make the expectation of ICD codes from clinical notes a particularly troublesome errand, for PCs and human coders alike. In this research work, we are presenting a multi-class labeling model which is mapped to ICD codes.

In this work, we aim to predict the ICD-9 code or its derivatives—from the raw text records. Examples of analyzed EHRs are initial consultation reports, treatment plans, laboratory or other diagnostic results and notes on follow-up meetings. These vary drastically in length and quality and exhibit a plethora of different abbreviations and structures, adding to the existing challenges of developing a global classifier. The patients' documents and codes stand in a many-to-many relationship, joined by a timestamp. However, as the average of codes per document is between 1 and 2, the problem was designed as single label multiple class, thus duplicating all training samples with more than one code. A convolutional neural network was employed for this task. This will be achieved using deep learning-based CNN and bag-of-words approach, and the classification is achieved for mapping for different ICD codes. The rest of the article is organized as follows: Sect. 2 describes the details of existing literature, Sect. 3 presents the proposed framework, Sect. 4 shows experimental setup and results, and finally, Sect. 5 concludes the paper.

2 Literature Work

Ongoing works have illuminated the issues related to straightforwardly applying word embeddings into true applications. Diaz et al. [16, 14] showed that the universally prepared word inserting fail to meet expectations corpus and inquiry explicit embeddings for recovery errands. They proposed locally preparing word embeddings in an inquiry explicit way for the question extension task. Zamani and Croft [2] showed that the hidden presumption in run of the mill word installing techniques is not equivalent to the need of IR errands, and they proposed significance-based models to learn word portrayals dependent on query document importance data, which is the essential target of most IR task.

Tran et al. [15, 10] determine tolerant vectors with their adjusted RBM engineering and at that point train a strategic relapse classifier for suicide hazard stratification. They explored different avenues regarding utilizing the full EHR information versus just utilizing determination codes and found that the classifier utilizing the total EHR information with the eNRBM engineering for idea embedding performed best.

Essentially, deep patient created patient vectors with a 3-layer auto-encoder and at that point utilized these vectors with calculated relapse classifiers to anticipate a wide assortment of ICD-9-based ailment analyses inside an expectation window [9]. Their system indicated upgrades over crude highlights, with prevalent precision at *k*-measurements for all estimations of *k*. In a theoretically comparable style, Liang et al. [3] additionally created patient vectors for use with straight classifiers, however, decided on layer-wise preparing of a deep belief network trailed by help of vector machine for grouping general malady analyses. Since preferably clinical notes related to a patient experience contain rich data about the aggregate of the confirmation, numerous examinations have analyzed result forecast from the content alone. Jacobson et al. [9] looked at the profound unsupervised portrayal of clinical notes for anticipating social insurance related contaminations, using stacked scanty.

Clinical notes regularly incorporate unequivocal individual well-being data (PHI), which makes it hard to openly discharge numerous helpful clinical data sets [1]. As per the rules of the Health Information Portability and Accountability Act, all the clinical notes discharged must be free of delicate data, for example, names of patients and their intermediaries, recognizable proof numbers, emergency clinic names and areas, and geographic areas and dates [9]. Dernoncourt et al. [5, 6] made a framework for the programmed de-recognizable proof of clinical content, which replaces a generally relentless manual de-identification process for sharing confined information. Their structure comprises a bidirectional LSTM arrange also, both character and word-level embeddings. The creators observed their technique to be best in class, with a gathering approach with restrictive irregular field's additionally faring great. In a comparative undertaking, Shweta et al. [15, 13] investigate different RNN models and word installing methods for distinguishing conceivably recognizable named substances in clinical content. The creators show that all RNN variations beat conventional.

3 Proposed Mechanism

Ensemble learning is an outstanding method to join various grouping models into one troupe classifier. A gathering classifier is developed by creating different base classifiers on preparing information and after that consolidating the different expectations utilizing a democratic framework. The goal of troupe learning is to build up a model that can give more exact expectations than every one of its single segment classifiers. Group classifiers have demonstrated to beat the speculation capacity of single classifiers. Principle reasons outfits can improve the exactness of its part classifiers.

There are a few strategies to make classifiers increasingly assorted. Bagging [7] and boosting are two understood procedures that modify the first training information to incorporate a decent variety in the training procedure. The previous strategy utilizes diverse training sets for each segment classifier in parallel, while the last technique prepares numerous classifiers in arrangement on training sets utilizing distinctive weighted perceptions. The two systems can be utilized to determine the inclination change exchange off. This is the issue of all the while limiting two kinds of blunder that utmost directed models to sum up past training information. We can break down the normal mistake of an inconspicuous perception 'o' into the accompanying terms:

- Bias: a blunder brought about by restrictions in the learning strategy
- Variance: a mistake brought about by impediments in the training information
- Irreducible mistake: a blunder coming about because of clamor in the issue itself.

Both bagging and boosting plan to limit the normal mistake, and however, they center on various issues. Bagging tries to decrease change by resampling the training information, while boosting means to limit the predisposition by developing an outfit that has lower inclination than the individual models. Bootstrap conglomerating or bagging [7] is a troupe strategy that prepares various classifiers on various bootstrap tests of the first training information. For each classifier, an arbitrary example with substitution is drawn from training information, containing a similar number of perceptions. Since tests are drawn with substitution, the bootstrap tests can contain copy perceptions. In the wake of training every part classifier on an alternate bootstrapped test, one of the classes can be doled out to any perception 'o' in the characterization stage. In this stage, the packing calculation performs dominant part casting a ballot. All part classifiers $\{p1, p2, \ldots, pN\}$ vote in favor of which of the k classes in C ought to be appointed to unlabeled perception 'o.' The class that gets most votes will be relegated to 'o' as definite forecast. Likely the most notable calculation that applies stowing is Random Forest [13]. The main distinction between Random Forest and the packing calculation utilizing choice trees is the method for highlight determination. On the off chance that a few highlights have an extraordinary impact on the expectation of the class variable, all things considered, they will be chosen in many trees in the gathering. So as to produce an increasingly various troupe of classifiers, Random Forest chooses an irregular subset of highlights at every applicant split, which is additionally called highlight packing. As opposed to the autonomous training of classifiers with sacking, boosting plans to fabricate a grouping of classifiers that rely upon one another. It prepares numerous feeble classifiers in grouping on various training sets $T = \{T1, T2, \ldots, TN\}$. Per capita set holds biased perceptions to create assorted variety in the learning procedure.

Perceptions that were classification error by classifier $pi - 1$ will be allotted higher loads than accurately grouped perceptions to drive the following classifier pi to concentrate more on perceptions that are difficult to arrange.

Here, Fig. 1 explains how the proposed ensemble model was works based on CNN and bagging approach. Initially, data set has been taken and applied the preprocessing and split the data set into number of splits each split contains data. Each data split given to a CNN classifier and train after that assign the train CNN to a bag and give

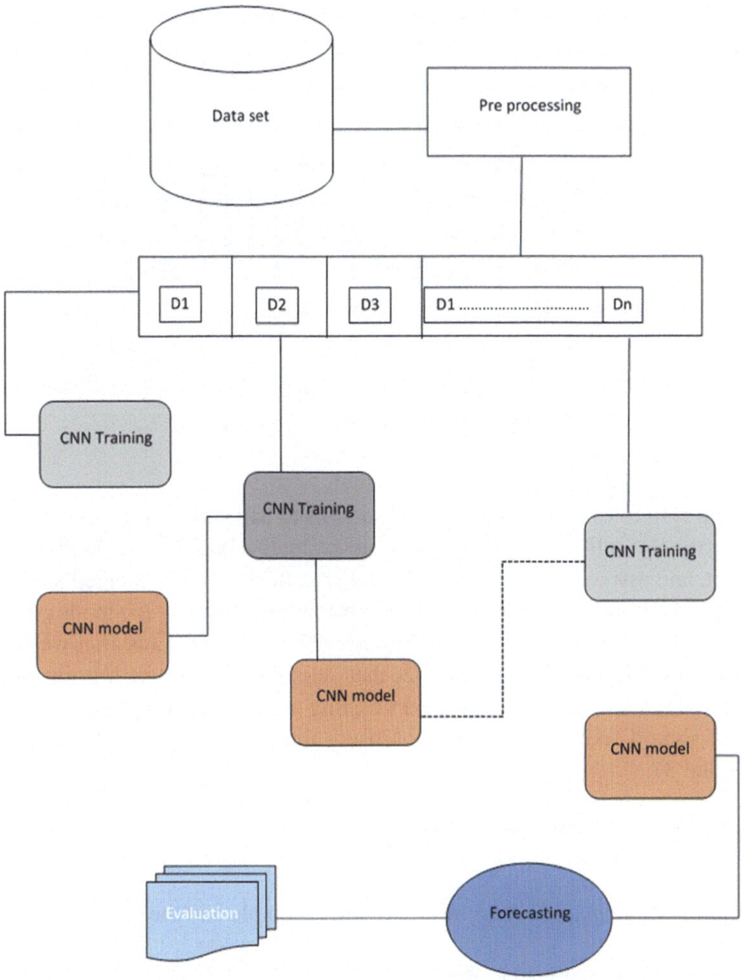

Fig. 1 Ensemble approach based on CNN and bag of words

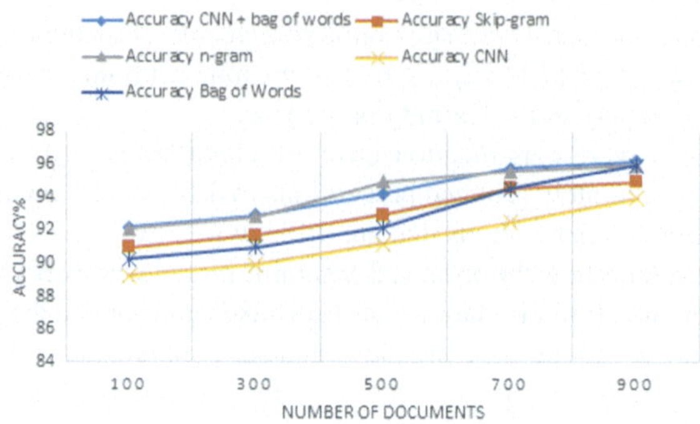

Fig. 2 Accuarcy

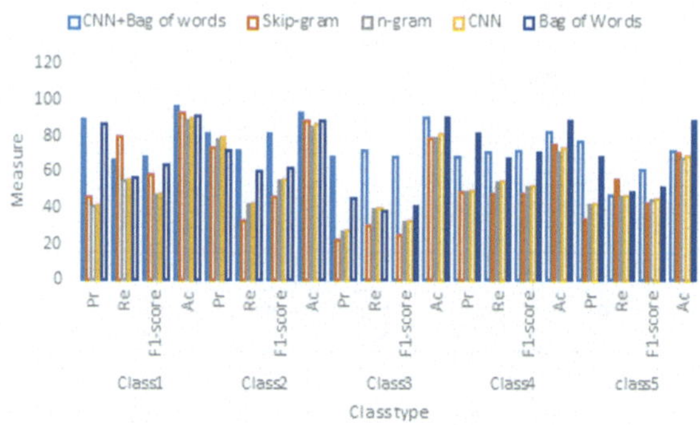

Fig. 3 Recall, *f*-score precession, accuracy of different class labels

the second split to CNN training method add this to bag and so on until the last split. And after the last split, forecast the result and evaluate it based on the highest votes to the object, and assign it to a particular ICD. Error, the loads are changed all the more intensely if the training error is high. Notwithstanding changing the loads everything being equal, a weight *wi* (0 *wi* 1) is relegated to classifier hey. Classifier weight wi is determined by $-\log(i1 - i)$ to guarantee that classifiers with lower training blunders have more impact in the last greater part casting a ballot method. Note that training blunder I should fulfill $0 < I < 0.5$ to ensure that the loads are refreshed correct (Figs. 2 and 3).

Algorithm 1 Bagging Based on CNN

Input: Initial training set *T*, *N* is the size of the ensemble and the algorithm for CNN learning
Output: *P* is the ensemble classifier
Phase-1: Training
1: for i = one to *N* do

2: Si = Bootstrap Sample (T)
3: pi = Learning of CNN (Si)
4: Adding of classifier to P which is ensemble: $P = P\ [pi]$
5: End for
6: Return ensemble $P = \{p1, p2, \ldots, pN\}$
Phase-2: Classification phase
7: Unlabeled observation 'o' classification using all component classifiers $\{p1, p2, \ldots, pN\}$
8: Let vi, k = (1 if hi votes for class Ck
0 otherwise}
9: Attain votes for each and every class
10: Select class with greatest votes: $P(x)$ = arg max $1 \leq k \leq c$ for $k = 1, \ldots, C$
11: Return $P(x)$

In this manner, the calculation ends if the model either gets impeccable characterization or when training blunder is at any rate equivalent to 0.5. This stop condition can be a downside of the calculation when accessible training information is scanty, since the outfit may combine to an ideal training blunder too rapidly. Notwithstanding when a part classifier would most likely arrange a little arrangement of training information accurately, it does not really imply that the group model cannot enhance test information any longer.

4 Experimental Results

To perform experimental analysis, we use Python libraries of Keras, TensorFlow, NumPy, matplotlib, etc. And we used 16 GB RAM and 500 GB HDD with 2 GB graphic card and Intel I5 processor. As a platform, we used Ubuntu 16.04 LTE. And the data set we used here is MIMIC-III. It contains information related to 53,423 particular medical clinic confirmations for grown-up patients (matured 16 years or above) admitted to basic consideration units somewhere in the range of 2001 and 2012. Moreover, it contains information for 7870 neonates conceded somewhere in the range of 2001 and 2008. The information covers 38,597 unmistakable grown-up patients and 49,785 medical clinic confirmations. Mainly here we used three tables' data those are discharge summaries, clinical notes which were written by doctors and nurses who are in ICU and other notes related to ICU.

From that, we performed the experimental evaluation with our proposed model to identify the class of the data which belong to which class of ICD-9 codes. We use accuracy for the classification performance evaluation. The accuracy (ACC) of a classifier is the probability that the classifier classifies a randomly selected sample to the correct class. Here, proposed model CNN–bag-of-words approach gives better accuracy in classification than all other state of artwork.

Here, we saw that the proposed technique has the best execution on practically the majority of the assessed qualities. The n-gram and pack-of-words-based techniques are reliably more fragile than the proposed, validating the discoveries in writing that word installings enhance execution of clinical NLP assignments. We moreover examine in the case of considering longer expressions enhance display execution. Here, we demonstrate the distinction in F1-score between models with expressions length changing from 1st to 5th classes. Whenever the data size increases proposed method does not show much variance, but other models were not shown the considerable performance loss for longer phrases. Proposed model shows significant performance than compared to existing work.

Here, Fig. 4 shows the computation time for training of different mechanisms with respect to number of documents; here, proposed model gives lower computation time for training because it gets features from CNN model, and it is done the features very quickly and accurately. CNN and bag-of-words-based ensemble approach give better than remaining, but other models are not performed up to the mark due to their constraints of not handling medical text. And when the size of the set increases, it does not show much deviation by proposed model, but existing works are considerable shown the decrease in training performance. Here, Fig. 5 shows the computation

Fig. 4 Training time

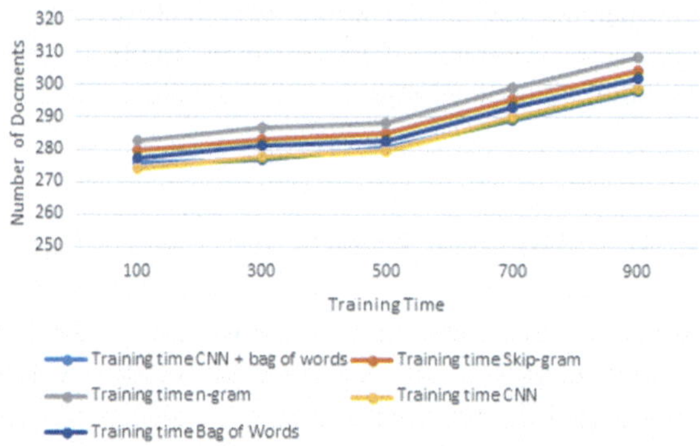

Fig. 5 Testing time

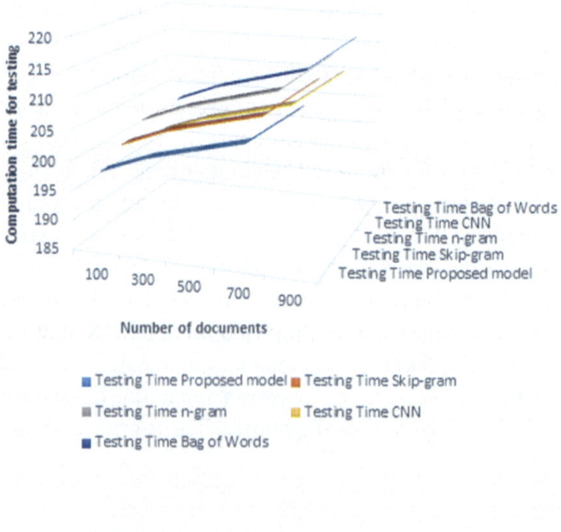

time for testing the different mechanisms with respect to number of documents; here, proposed model gives lower computation time for testing because it gets features from CNN model, and it is done the features very quickly and accurately. CNN based bag of approach gives better than remaining but other models are not performed up to the mark due to their constraints of not handling medical text. The size of the set increases it does not show much deviation by the proposed model but existing works are considerable shown the decrease in training performance.

5 Conclusion

Text classification is a complex task with regular methods, but it gives a significant impact to the society. Medical text classification is a complex task because of its nature. In this paper, we utilized ensemble learning, and it is an outstanding method to join different order models into one group classifier. A gathering classifier is developed by creating various base classifiers on preparing information and after that joining the different expectations utilizing a democratic framework. In this paper, we used deep learning-based clinical note multi-label multi-class approach using CNN model for feature extraction from text notes, the training is based on bag-of-words-based CNN classification, and we map those classes for multi-classes. And we made a comparison with existing works CNN, skip-gram, *n*-gram, and bag of words. The performance results show that the proposed framework performed good while classifying the text notes.

References

1. Angermueller C, Pärnamaa T, Parts L, Stegle O (2019) Deep learning for computational biology. Mol Syst Biol 12(7):878
2. Ayatollahi H, Gholamhosseini L, Salehi M (2019) Predicting coronary artery disease: a comparison between two data mining algorithms. BMC Public Health 19(1):448
3. Che Z, Kale D, Li W, Taha Bahadori M, Liu Y (2015) Deep computational phenotyping. In: Proceedings of the 21st ACM SIGKDD international conference on knowledge discovery and data mining, pp 507–516
4. Choi E, Schuetz A, Stewart WF, Sun J (2016) Medical concept representation learning from electronic health records and its application on heart failure prediction, arXiv, p 45
5. Choi E, Bahadori MT, Searles E, Coffey C, Sun J (2016) Multi-layer representation learning for medical concepts, arXiv, pp 1–20
6. De Sa C, Ratner A, Ré C, Shin J, Wang F, Wu S, Zhang C (2016) Incremental knowledge base construction using deepdive. VLDB J 1–25
7. Dernoncourt F, Lee JY, Uzuner O, Szolovits P (2016) De-identification of patient notes with recurrent neural networks, arXiv
8. Jagannatha A, Yu H (2016) Bidirectional recurrent neural networks for medical event detection in electronic health records, arXiv, pp 473–482
9. Lasko TA, Denny JC, Levy MA (2013) Computational phenotype discovery using unsupervised feature learning over noisy, sparse, and irregular clinical data. PLoS ONE 8(6)

10. Mikolov T, Corrado G, Chen K, Dean J (2013) Efficient estimation of word representations in vector space. In: Proceedings of the International Conference on Learning Representations (ICLR 2013), pp 1–12
11. Miotto R, Li L, Kidd BA, Dudley JT (2016) Deep patient: an unsupervised representation to predict the future of patients from the electronic health records. Sci Rep 6(1):1–10, 26094
12. Nie L, Wang M, Zhang L, Yan S, Zhang B, Chua T-S (2015) Disease inference from health-related questions via sparsely connected deep learning. IEEE Trans Knowl Data Eng 27(8):2107–2119
13. Shweta AE, Saha S, Bhattacharyya P (2016) Deep learning architecture for patient data de-identification in clinical records, ClinicalNLP 2016, p 32
14. Sidey-Gibbons JAM, Sidey-Gibbons CJ (2019) Machine learning in medicine: a practical introduction. BMC Med Res Methodol 19(1):64
15. Tran T, Nguyen TD, Phung D, Venkatesh S (2015) Learning vector representation of medical objects via EMR-driven nonnegative restricted Boltzmann machines (eNRBM). J Biomed Inform 54:96–105
16. Zamani H, Bruce Croft W (2017) Relevance-based word embedding. In: Proceedings of SIGIR, pp 505–514

E-learning in Higher Education in India: Experiences and Challenges—An Exploratory Study

Kiri Taso and Arindam Chakrabarty

Abstract The world community is committed to achieve 17 goals popularly known as United Nation Sustainable Development Goals (UNSDGs) of which education has been given major thrust that has been earmarked in Goal 4. As a member country, India has also attempted to address the issues of education with highest priority that is envisaged by the responses of the government for drafting New Education Policy in 2019. The government is committed to achieve inclusive education that needs the manifestation of e-Learning platform. Since it is difficult to bring the elephantine population under the ambit of conventional education system, this paper has attempted to explore the experiences and challenges of e-Learning mechanism in the higher education system of India.

Keywords E-Learning · UNSDGs · New Education Policy · Inclusive education · Conventional education

1 Introduction

E-Learning can be defined as an online educational learning process. It can simply be understood as 'Internet-Based Learning'. It is an online learning service through which teaching–learning process is carried out. In other words, e-Learning refers to 'the mode of teaching and learning via Internet and website'. E-Learning is adopted by an institution to let the students learn from home and far distance through online mode which would make the teaching–learning process more approachable and convenient to some extent. E-Learning is primarily the network-enabled practices of skills and information transfer between the online learners and resource providers.

K. Taso · A. Chakrabarty (✉)
Rajiv Gandhi University (Central University), Rono Hills, Doimukh, Arunachal Pradesh 791112, India
e-mail: arindam.management@gmail.com

K. Taso
e-mail: kiritaso8@gmail.com

S.-L. Peng et al. (eds.), *Intelligent Computing and Innovation on Data Science*, Lecture Notes in Networks and Systems 118,
https://doi.org/10.1007/978-981-15-3284-9_77

E-Learning refers to the using of electronic application and processes to learn. 'E-Learning can be in other words understood by means of learning through electronic way by adopting modern means of technological learning. E-Learning can be understood as the network-enabled transfer of skills and knowledge to individual as well as masses. It is opposite to offline and non-electronic means of teaching and learning practices (www.economictimes.indiatimes.com/definition/e-learning). Henceforth, the use of computer–desktop and laptop, mobile and tab and other means to deliver teaching–learning process through the use of Internet source can be simply known as e-Learning. To some extent, we can say online learning (e-Learning) is gradually doing away the traditional learning methods.

Concept of ICT:

Information and Communication Technology (ICT) in the field of education is a significant concept to understand. The various curricular developmental projects have been carried out under the assistance of IITs and NITs. The National Mission on Education initiative by the Government of India is expected to boost the Gross Enrolment Ratio (GER) in Higher Education by 5 percentage (%) points during the Eleven Five Year Plan (2007–12). This Mission has two major components which are (i) Content Generation and (ii) Connectivity—along with a provision to provide devices to institution and learner. Besides that the Mission also seeks to provide computer infrastructure and connectivity to over 18,000 colleges and nearly 400 Departments at Universities and Deemed Universities and Institutions in India for a greater national cause. It also provides interactivity and problem-solving approach which will be addressed through the program called 'Talk to a Teacher' Segment.

2 Electronic Learning or Usage of e-Resources Learning

Massive Open Online Course (MOOC):

MOOC is an online course which committed to provide 'Massive' and 'Open' online learning platform via the web. The MOOCs system of learning begins in the year 2001 at the USA. And it became a trendy approach of learning since 2012 onwards (The New York Times, 18 April 2014).

And the below table shows the students admitted in Coursera enrollees:

Sl. No.	Country/region	Percentage (%) Approx
1.	Russia	2.3
2.	Australia	2.4
3.	Canada	3.5
4.	UK	4.5

(continued)

(continued)

Sl. No.	Country/region	Percentage (%) Approx
5.	Spain	4
6.	India	8.7
7.	Brazil	5.2
8.	USA	27.6
9.	Contribution of the different part of the world	42

Source Coursera Enrolees

SWAYAM

The main goal of the SWAYAM is to deliver quality and reachable educational learning prospect to every needed person specially the underprivileged and unreachable section of society. The SWAYAM actually strives to acts as link for those students who are digitally divided and untouched by e-Learning or Digital Revolution process. The indigenously developed IT platform enables the SWAYAM to propagate all the courses that are being thought by the best teacher in the country and are being made universal access to anyone and at anytime at free of cost.

The programmes and features that offered in SWAYAM are giving education from class 9th to postgraduate n which they offer courses like Science, Engineering, Management, Humanities, Mathematics, Arts and Recreation, Commerce, Language, Education and Library courses, etc., given below:

How to access SWAYAM?

SWAYAM can be accessed through two major ways as follows:

Sl.No.	SWAYAM can be accessed through
1	One can access the SWAYAM portal on the web through https://swayam.gov.in
2	One can also access the SWAYAM mobile apps for—***Android and iOS version***

Source SWAYAM, GOI and Swayam learning portal

Review of Literature

Cox [4], (pp. 85–105) explains the necessity of e-Learning which enhanced our understanding on learning and Information Technology (IT) in teaching and learning process in order to have a clarity and consistency of subject and further highlights that although the young generation has wider access to Information Technology (IT) little is known about this impact on their learning process. There is also a need to balance between the formal and informal uses of e-Learning.

Dewan's [5] study reveals that 80% have computer, 80–67% and 20–33% have no computer. Thus, a better infrastructure is required in institution to provide e-Learning curriculum to the e-Learner.

Rana and Lal [7] highlight that there is need of conventional and holistic approach in educational system which will meet the demands of e-Learners at schools, colleges and universities level. The e-Learning institution with the help of World Wide Web (WWW) via Internet tried its best possibilities to promote distance education, virtual and e-Learning approach by delivering and sharing resources, promoting active e-Learning technologies.

Rosenberg [8] says that e-Learning enables us to understand and deals with different web-based contents for teaching–learning process.

Longmire [6] emphasised that 'an e-Learning approach includes a wide range of digital and computer-based learning mechanism'. He further states that e-Learning content is mainly conveyed via Internet, satellite communication, audiotape and videotape, DVD, CD-ROM and TV and still emerging so-called wireless application protocols (WAP).

Agarwal and Nisa [1] focus on the knowledge process outsourcing sector of India. Authors highlight the scenario which had witnessed the rapid change from 'industrial to knowledge-based economy'. Both also highlight the Skyrme [9] and Stiglitz [10] views on 'how the highly skilled labour force is the key to achieved success in the knowledge-based economy and industry'.

According to Tripathi and Jeevan [11], the paradigm shift in teaching–learning process (traditional to e-Learning) is perhaps due to rapid evolvement made in the field of Information and Communication Technology (ICT).

Ali [2] states that the exponential advent in the field of ICT and Internet has greatly influenced and revolutionised the way the knowledge is broadcasted.

3 Objective of the Study

The paper contains the following objectives:

1. To explore various e-Learning portals operating in India.
2. To explain challenges in implementing e-Learning mechanism for effective teaching dissemination process.

4 Analysis I

Due to the emergence of many well-financed institutions which later link with some of the top universities like Udacity, Coursea and edx, etc., at this period, the year 2012 was regarded as the 'The year of MOOCs' as per The New York Times (2 November 2012).

5 Popular e-Learning Firms/Platforms in India: Indicative List

The emergence of cloud computing technology has highly impacted the Online Education Market in India. The cloud technology with its potential capabilities provides a significant amount of data, information and content at single platforms to e-Learning Companies in India. Due to data saving scope, it is easier for the users and providers to procure, manage access and process the information from anywhere and anytime. Another important reason behind the growth of e-Learning markets trends in India is the rising popularity of big data and learning analytics. The technology enables the companies and institutions to provide online courses to the learners. The e-Learning markets due to its significance potentiality and effective results attract many learners to be aware and opt e-Learning courses. This rising awareness on online learning scope has pushed the growth of online education markets in India. The involvement of Information Communications Technology (ICT) in the field of teaching–learning process has led to the increasing demands of alternative educational approach of learning, which provides significant opportunities for growth of the e-Learning companies in India via digital platform. Thus, it is forecasted that Indian e-Learning markets potentiality will be expanded up to US$18 billion by 2022.

E-Learning Institution in India in 2019:

The emergence of cloud computing technology has highly impacted the Online Education Market in India. Due to data saving scope, it is more easier for the users and providers to procure, manage, access and process the information from anywhere and anytime. Another important reason behind the growth of e-Learning markets trends in India is the rising popularity of big data and learning analytics. The technology enables the companies and institutions to provide online courses to the learners. However, it is forecasted that Indian e-Learning markets potentiality will be expanded up to US$18 billion by 2022 (www.technavio.com).

1. **BYJU'S**: BYJU'S is a learning app founded by Byju Raveendran. In 2019, it has a total net worth of $5.4 billion (Rs. 37,000 crore). This firm has efficiently created a K12 learning smartphone app which offers highly effective, adaptive and active engaging learning programmes. This Edetech app not only provides effective tutoring programme at school level but also efficiently delivers a e-Tutoring to various other competitive exams like IIT-JE, UPSC, CAT and GRE, etc.
2. **IGNOU**: IGNOU stands for *Indira Gandhi National Open University* a Central University which is located at Maiden Garhi, New Delhi. It was established in the year 1985. It has a total enrolment of over 4 million students with 67 centres across the country, the reason why it is regarded as world's largest university. The university serves under the motto of—*The People's University*. IGNOU was founded to serve universal and accessible quality higher educational opportunities in India through the means of *Distance and Open Education*. IGNOU offers 226 academic programs like Diploma, Degree and Certificate courses such as

School of Social Science, Sciences, Education, Engineering and Technology, Management Studies, Computer and Information Sciences, Health Sciences, Law, Journalism and New Media Studies, Vocational Education and Training, Foreign Languages and Performing and Visual Arts, etc.

3. **Dexler Education** (2001): It is located in Bangalore (India). The Dexler Education primarily deals with digital education and consultative services in educational sector. The company provides industry-based e-Learning education solution for corporate learning, talent and faculty management and enhances easier mode on e-Learning. Along with its inventive and skilled e-Learning tactics in delivering quality education to the needy students and organisation, the Dexler Education acquired certain position among the highest e-Learning institution in country.
4. **The Educomp Solution (1994)**: It is in Gurgaon and an Indian-based company. Its aims to replace the traditional way of learning with more advance and smarter way of teaching and learning. Educomp Solutions is ranked among the best e-Learning companies in India. As there is saying—*the numbers speaks*, there are 30 million learners across and 65,000 schools in Educomp Solutions in two decades.
5. National Institute of Information Technology (NIIT-1981): It is situated in Gurugram (India). NIIT provides various kinds of e-Learning courses such as managing, self-learning and instruction training, etc. NIIT is specialised in providing knowledge to certain domains such as corporate, skills and career and schools learning groups. NIIT also offers necessary e-Learning facilities to the deserving and socially challenged and deprived students to certain extent.
6. **Edukart** (2017): It is also listed among the top online educational learning companies in India. Edukart is one of Indian higher education enrolment platform for e-Learner. It is an e-Learning entrance coaching site that provides online learning services to the educational seekers. Edukart also offers admission to certain curriculum such as Diploma and Degree Courses along with Entrance and Certificate, etc. Edukart has linked with some well-recognised educational institution in India like Indian School of Business, National Narsee Monjee Institute of Management Studies-School, etc.
7. **Simplilearn**: It is also one of the top e-Learning platform in San Francisco, California (USA) and Bangalore (India). The Simplilearn also delivered various e-Learning programmes such as cloud computing, digital markets and cyber security course, etc., to the online learner. This institution today achieved successful position among the successful online educational institution in India.
8. **Zeus Learning (ZL)**: It *is* also an online learning institution whose headquarters is at Mumbai (India). It occupies top ninth position among the top online learning institution in India. Zeus Learning offers various programmes to the online learner such as software and apps designing, training and solution for mobile and other technological system, etc.
9. **Meritnation**: It provides live online interactive and tutorial classes to the e-Learning seekers. It is an Edu tech start-up, which is a part or division of Applect Learning Systems based in Delhi (India). **Meritnation** is an online learning

providing institution that delivered various types of e-Learning approach to its e-Learner, so that there could be effective online teaching–learning practices.

10. **Excelsoft**: Excelsoft was founded in the year 2000. Excelsoft provides value courses, product and to cater to the needs and demands of all the key educational sectors like K12 learning system, higher education level, corporate learning, etc.

6 Analysis: II

The e-Learning system is vital for rapid teaching, learning and dissemination process but there are inherent challenges as well. A few key indicative challenges are mentioned below.

(a) Lack of uninterrupted power supply is one of the major issues in the online learning process. Since e-Learning system wholly depends on electricity, there has been frequent interruption in power supply that creates disruption in e-Learning process.

(b) Lack of Internet coverage across the country is another key issue to be addressed in order to provide better and quality digital learning capability of the learner of the country.

(c) Technical issues are yet another matter of concern since the entire process of e-Learning revolves around technology, and if there is technical issue that exists in the learning process, it will definitely hamper the e-Learning process.

(d) Lack of professional skills is another issue need to be redressed, as it requires a well-qualified and skilful professional person in the online education system. If the knowledge providers lack the professional skills, then it will again create problems in such teaching–learning process.

(e) Smooth e-Learning process is hindered by the inherent struggle for adaptability of computer skills. In this type of learning pattern, both the teacher and students need to have well versed in the field of computer technology.

(f) Lack of motivation, i.e. self-motivation is another important matter. Since it lacks face-to-face interactive methods, sometimes students remain unmotivated in their learning process.

(g) Reliability of e-Materials is another important concern, as we do not know the reliable source of materials that are being provided to the students.

(h) Most of the e-Platform is unidirectional. In other words, the learning process in the e-Learning process is one-way learning platform to most often. The learner most often did not get time to have face-to-face contact with the resource person [3].

(i) The e-Learning system also suffers from lack of personal or humanistic touch or human factor in the fields of teaching–learning pedagogy. E-Platforms suffer from real-time interactions, since classes are online in nature with time-specific guidance which makes difficult for the learner to attend the exact schedule classes which is another matter of concern for e-Learner.

(j) Lack of adequate e-Materials is another important issue where the learner may face certain problems. The e-Materials are developed as generic not specific to field of inquiry which lacks the flexibility learning capability or interdisciplinary knowledge of the learner.
(k) Huge initial investment for production, preparation and access of materials at the beginning.

7 Limitation of Study

The paper is developed to explore e-Learning practices in the higher education system of India. Since the e-Learning pattern is in the nascent and formative stage in the country, it is difficult to retrieve longitudinal database in terms of number of users, period of usage, qualitative aspect of e-Learning process, etc. So, this paper essentially suffers from adequate relevant information at this moment. However, the paper has attempted to outline the overview of e-Learning process in Indian higher education system.

8 Conclusion

This study can be regarded as a very foundation work in the domain of e-Learning intervention in Indian higher education system. The study indicates that the e-Learning process has gained momentum over a period of time, and it signifies that both the public and private sectors are contributing to this segment that can achieve the inclusive education model up to the extent of higher education level in India.

References

1. Agarwal R, Nisa S (2009) Knowledge process outsourcing: India's emergence as a global leader. Asian Soc Sci 5(1):82–92
2. Ali A (2004). Issues & challenges in implementing e-learning in Malaysia. Retrieved 18 Jan 2008
3. Chakrabarty A, Tagiya M (2018) Physical vis-à-vis virtual resources for excelling career management from a north eastern state of India. In: Bhattacharjee A (ed) Digital impact on human resources practices-text and cases. India: Mittal Publication New Delhi, India, pp. 205–216. ISBN 81-8324-894-2
4. Cox MJ (2013) Formal to informal learning with IT: research challenges and issues for e-learning. J Comput Assist Learn 29(1):85–105
5. Dewan A (2010, July) Scope of technology in higher education in India: A study. In: 2010 international conference on technology for education. IEEE, pp 234–235
6. Longmire W (2001) A primer on learning objects. Learning circuits
7. Rana H, Lal M (2014) E-learning: issues and challenges. Int J Comput Appl 97(5)

8. Rosenberg MJ (2000) The e-learning readiness survey: 20 key strategic questions you and your organization must answer about the sustainability of your e-learning efforts. Retrieved July 25 2005
9. Skyrme DJ (1997) From information to knowledge management: Are you prepared? In: Online information 97, London, 9–11 December 1997, pp. 109–117
10. Stiglitz J (1999) Public policy for a knowledge economy. Remarks Dept Trade Ind Cent Econ Policy Res 27(3):3–6
11. Tripathi M, Jeevan VKJ (2010) E-learning library and information science: a pragmatic view for India. DESIDOC J Lib Inf Technol 30(5):83–90 (www.technavio.com/report/online-education-market-in-india-analysis-share-2018)

A Hybrid Watermarking System for Securing Multi-modal Biometric Using Honey Encryption and Grasshopper Optimization Technique

R. Devi and P. Sujatha

Abstract Digital watermarking is one of the major information hiding techniques for hiding biometrics. The majority of research focuses onto the development of a reliable robust watermarking system. A multi-modal biometrics fingerprint and iris are being used. The biometrics is fused using gradient pyramid technique. The fused template is encrypted using honey encryption. The major intend of this paper is to develop a high secured, robust and reliable watermarking system using some conventional methods fused with an optimization technique namely grasshopper optimization technique (GOA). Many researchers have failed to maintain the two important factors imperceptibility and robustness that define the strength of the watermarking system. The two predominant factors are attained through the proposed algorithm DWTSVDGOA. The imperceptibility and the robustness are measured using some performance measures. The experimental results show that the proposed algorithm DWTSVDGOA gives a NC value as 1, PSNR as 90.75 and SSIM as 0.99. The performance and the evaluation of the technique are found to be more better than existing image watermarking techniques.

Keywords Biometrics · Discrete wavelet transforms (DWT) · Gradient pyramid · Grasshopper optimization algorithm (GOA) · Honey encryption · Image fusion singular value decomposition (SVD)

1 Introduction

Digital watermarking is a standard technology used for broadcast monitoring, security and authentication of digital media. Digital watermarking is a technique used to

R. Devi (✉)
Assistant Professor, Department of Computer Science, VISTAS, Chennai, India
e-mail: newdevi21@gmail.com

P. Sujatha
Professor, Department of Computer Science, VISTAS, Chennai, India
e-mail: sujinagi@gmail.com

S.-L. Peng et al. (eds.), *Intelligent Computing and Innovation on Data Science*, Lecture Notes in Networks and Systems 118,
https://doi.org/10.1007/978-981-15-3284-9_78

implant information into digital data which cannot be easily identified by unauthorized users. The techniques have been used in biometric system in order to secure and authenticate biometric data and also to increase the level of accuracy. A biometric system is an automatic identification of an individual. The digital form of the biometric characteristics is known as a biometric template. Templates are stored in a system database through the sensor devices. The database can either be distributed or centralized which holds only one copy of the template. As per the recent senses, iris and fingerprint authentication growth rate is high when compared to others. Since, these two biometric plays a vital role in biometric authentication, the paper mainly deals with these two predominant features namely fingerprint and iris.

2 Related Work

The author [1] has defined an algorithm using optimized DWT watermarking that provides the quality, imperceptibility and effectiveness. The problem has been optimized with "genetic algorithm (GA)." A pseudo-number sequence was generated. The GA was used to find a specific sub-band to embed the watermark, and a fitness function was also formulated using two factors namely PSNR and correlation factor. The author [2] has proposed a new novel optimization method named "gray wolf optimizer (GWO)" for digital images in wavelet transform. The visibility or scaling factor plays a essential role in positioning the watermark. The author [3] has developed a robust system using "differential evolution (DE)" algorithm fused with SVD and DWT. Venkatram et al. [4] narrated a new optimization technique for digital image watermarking system. The visibility mode of the watermark is the noteworthy constraint that improves the robustness and perceptibility of a watermark. "Gravitational search algorithm (GSA)"-based optimization technique is used to find an optimum position to embed the watermark. Thanki and Borisagar [5] suggested a new watermarking scheme using "artificial bee colony optimization technique (ABC)". Also, the scheme searches for a best optimum position to fit the watermark, to enhance the strength of the system. The author [6] suggested a new scheme on multi-objective bees algorithm (MOBA). Deshmukh and Malviya [7] displayed a multi-modular biometric watermarking strategy for individual distinguishing proof framework dependent on DCT and phase congruency show. Neve et al. [8] have suggested an efficient watermarking scheme using PSO. "Particle swarm optimization (PSO)" was used to manipulate an optimum position to place the watermark on the face image. The author [8] presented a unique finger impression watermarking strategy dependent on SVD and compressive sensing hypothesis.

3 Proposed Work

3.1 Image Acquisition

Image acquisition is the first and foremost step. The paper works mainly on the biometric images especially fingerprints and iris. The fingerprint images are from the dataset FVC2000 consist of four databases namely DB1, DB2, DB3 and DB4. The work has been carried out on the DB2 acquired using a low-cost capacitive with a resolution of 512 dpi of size 256 × 354. The iris images are from the database UBIRIS. V1 composed of 1877 images from 241 persons. In this work, gray-level iris is used for processing.

3.2 Fingerprint Enhancement and Iris Segmentation

Image enhancement and preprocessing are the most important part of image processing to remove noise and edges. It is a process of manipulating an image to enhance in order to get better visual representation of images. This work includes "Otsu thresholding" for binarization, histogram equalization, sharpening and image contrast to increase the intensity of the print. Finally, the fingerprints are filtered using a Wiener filter for de-noising. In this work, the eyes are segmented using Canny edge detection and Hough transform method. It includes cropping, resizing, image conversion, etc. The segmented iris is then polarized using "rubber sheet method" and filtered using "Wiener filter" to remove the noise.

3.3 Image-Level Fusion of Fingerprint and Iris

Image fusion technique is the most critical part of digital image processing Deshmukh and Malviya [7]. The essential characteristics of image fusion are that it should preserve the salient information, avoid artifacts, reliable and robust. In this paper, multi-modal fusion is carried out by combining two different modalities such as fingerprint and iris templates. The enhanced and segmented images fused using three a technique namely gradient pyramid (GP). The reliability of the above technique is measured using some image quality metrics.

3.4 Encryption and Watermark Embedding Procedure

In this proposed work, the fused multi-modal biometric template meant to be the watermark has been encrypted using honey encryption for higher security. The above-fused template is encrypted using the honey encryption scheme to increase the security level. One of the critical and challenging tasks in the watermarking system is to determine the scaling factor. In the system; the scaling factor determines the potentiality of the watermark which determines the robustness and the imperceptibility of the scheme. Therefore, the selection of an optimal scaling factor to embed the watermark enables the digital watermarking system as an optimization problem. **Henceforth, a hybrid watermarking technique (DWTSVDGOA) using nature-inspired optimization technique has been proposed**.

3.5 Watermark Extracting Procedure

The watermarked image undergoes some geometric attacks and signal operations. After applying the geometric attacks and signal operations, the watermark is extracted to identify the strength and the sustainability of the watermark using some major factors. The strength of the watermark yields the strength of the watermarking system.

4 Proposed Algorithm—DWTSVDGOA

The proposed algorithm DWTSVDGOA fuses the two existing techniques DWT and SVD for decomposition and transformation in order to embed the watermark (encrypted fused multi-modal biometric traits) in an optimum scaling factor. The population is initiated using **grasshopper optimization technique (GOA)** to find the best possible position to implant the watermark. The grasshopper optimization techniques used to find the best possible scaling factor for watermarking to achieve the robustness and the imperceptibility of the scheme.

4.1 Mathematical Model of Grasshopper Optimization Algorithm and Its Characteristics

Neve et al. [8] grasshopper optimization algorithm replicates the natural habitat of a grasshopper when it searches for its prey. The same phenomenon is being carried out to locate the best possible solution for an optimization problem. That is why this algorithm is quoted as nature-inspired optimization technique. A mathematical model is formed by replicating the behavior of the grasshopper and is shown below

$$X_i = S_i + G_i + A_i \tag{1}$$

In Eq. 1, where X_i defines the position of the grasshopper, S_i is the social interaction, G_i is the gravity force of the ith grasshopper, [8].

The optimized value is measured based on the fitness function. Three predominant factors namely PSNR, NC and SSIM are combined to define a fitness function. The fitness function is given as

$$\text{Fitness } F = \frac{\text{PSNR}}{100} \times \text{Max}\left[\sum_{i=1}^{n} \text{NC}(w, w')\right] X\text{SSIM}(w, w') \tag{2}$$

4.2 *Pseudo-Code for the Proposed Watermark Embedding Algorithm (DWTSVDGOA)*

STEP 1: Decompose the gray-level cover image into single-level decomposition using Haar wavelet
[LL, LH, HL, HH] = Dwt (Cover_image)
STEP 2: Apply SVD to LL band of the cover image found in STEP 2
[U_imgr1, S_imgr1, V_imgr1] = SVD (LL)
STEP 3: Initialize the grasshoppers population, iteration, lower bound, upper bound and the objective function.
STEP 4: Check the boundaries of the search space, cover image found in STEP 2.
STEP 5: Calculate and identify the fitness function for all the population.

$$\text{Fitness } F = \frac{\text{PSNR}}{100} \times \text{Max}\left[\sum_{i=1}^{n} \text{NC}(w, w')\right] X\text{SSIM}(w, w')$$

STEP 6: Best fitness function is selected to find the optimum scaling factor using GOA.
STEP 7: Read the encrypted fused biometric template as a watermark of size MXN using honey encryption.
STEP 8: Apply single-level Dwt to the watermark (encrypted watermark)
[LL, LH, HL, HH] = Dwt (Watermark_image)
STEP 9: Apply SVD to LL band of the watermark image found in STEP 8
[U_imgr2, S_imgr2, V_imgr2] = SVD (LL]
STEP 10: Embed the modified watermark image found in STEP 12, into the optimum position and is manipulated through the fitness function found in STEP 6.
STEP 11: Apply **inverse SVD** to LL sub-band.
STEP 12: Apply **single-level inverse dwt** get the watermarked image.
Watermarked image = idwt (New LL, New LH, New HL, New HH]
STEP 13: Display the cover image, watermarked image.

The scaling factor is evaluated using grasshopper optimization technique. The upper bound, lower bound and no of iterations and the fitness function are defined and initiated. The algorithm undergoes 100 iterations. A fitness function is generated for each grasshopper. Each grasshopper identifies its next neighbor. The best fitness function is selected based on these parameters where preferably SSIM and NC equal to 1.

4.3 Watermark Extraction Procedure

Pseudo-Code for the Proposed Watermark Extracting Algorithm1 (DWTSVD-GOA)

STEP 1: Read the watermarked image of size MXN.
STEP 2: Apply all the possible geometric attacks and signal operations to the watermarked image to analyze the robustness of the proposed method.
STEP 3: Extract and decrypt the watermark.
STEP 4: Apply idwt to the watermark.
Recovered [LL, LH, HL, HH] = IDWT (Watermark)
STEP 3: Apply inverse SVD to LL sub-band of the watermark found in STEP 4.
[U, S, V] = ISVD (LL)
STEP 5: Calculate the PSNR, NC and SSIM for the watermark.
STEP 6: When SSIM and NC are equal to 1.

5 Experimental Results

5.1 Robustness of the Proposed Algorithm DWTSVDGOA

Robustness is the most important characteristics that provide the strength of the watermarking technique. Robustness means, "ability of a watermarking system to resist change," even after some attacks or modifications. The proposed watermarking technique is evaluated using three quantitative measures, PSNR, SSIM and NCC. All simulation results are performed in MathWork tool MATLAB R2015a under Windows 7 environment. The above Table 1 provides the robustness of the proposed method (DWTSVDGOA). The table consists of the cover image Lena (256 × 256), fused multi-modal template, encrypted fused template (watermark image) of size 256 × 256. To investigate the robustness of the DWTSVDGOA, the watermarked image undergoes some geometric attack and signal processing operations such as Gamma correction, cropping, scaling, rotation, salt and peppernoise, Gaussian attack and so

Table 1 Robustness of the watermarked image using DWTSVDGOA under various signal processing and geometric attacks for the cover image Lena

S. No.	Attack type	Cover image	Fused biometric image (watermark)	Encrypted watermark	Watermarked image with attack	Extracted watermark	PSNR	NCC	SSIM
1	Gamma correction						59.36	1	0.96
2	Cropping						58.31	0.95	0.95
3	Salt and peppernoise						60.54	1	0.96
4	Gaussian attack						75.92	1	0.98
5	Rotation				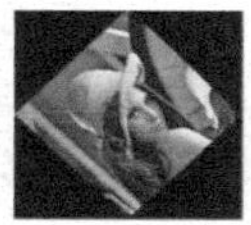		59.38	0.97	0.85

Table 2 Imperceptibility of DWTSVDGOA for the cover images

Cover image (256 × 256)	Watermark (256 × 256)	Watermarked image	PSNR	SSIM	NC
			90.3420	1	1
			89.0682	1	1
			89.7685	0.9876	1

on. The quality of the extracted watermark is ascertained using the metrics PSNR, NC and SSIM. The extracted watermark after Gaussian attack is 75.9299, whereas for other attacks, it may vary from 58 to 60.

5.2 Imperceptibility of the Proposed Algorithm

Imperceptibility is one of the most important characteristics that should retain the fidelity of the watermarked image. It means the quality of the cover and the watermarked image must remain the same for human naked by PSNR, NCC and SSIM.

5.3 Comparing the Imperceptibility of DWTSVDGOA with the Existing Methods in Terms of PSNR and NC

Table 3 gives a comparative study of the proposed methods with other existing methods. The graphical representation of PSNR and NC values are given in the below Figs. 1 and 2.

Table 3 Comparing the PSNR and NC values of DWTSVDGOA with the existing methods to ensure imperceptibility for Lena image

S. No.	Existing techniques	PSNR	NC	References
1	Flirefly + DWT-QR	71.4520	0.98	Yong Guo et al. [9]
2	Bacterial foraging optimization algorithm (BFOA)	55.5	0.96	Nair and Aruna [10]
3	Hybrid DWT-SVD	79.8611	1	Gunjal [11]
4	SVD—Firefly	63.786	1	Vijaya Durga et al. [12]
5	DWT-GWO	61.1463	0.88	Priyanka Panwar et al. [13]

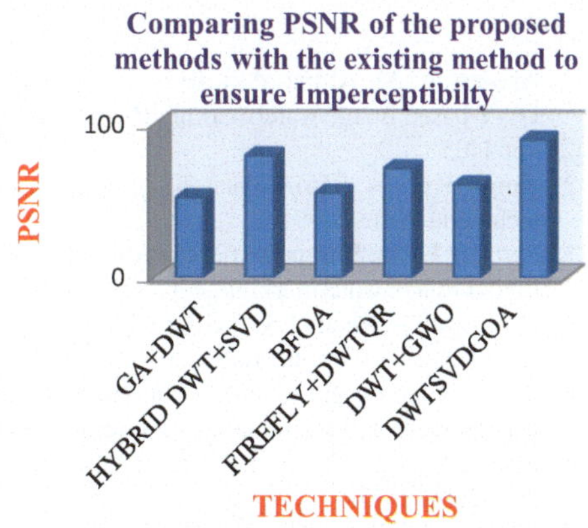

Fig. 1 Performance analysis of the imperceptibility based on PSNR values

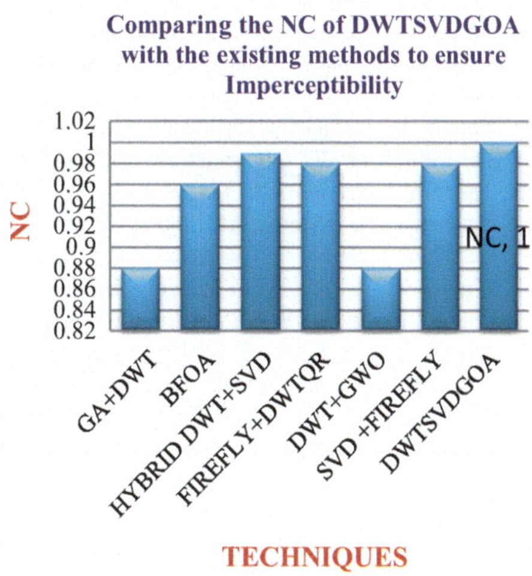

Fig. 2 Performance analysis of the imperceptibility based on the NC values

6 Conclusion

An image watermarking method has been proposed and implemented successfully. The performances of the proposed methods were computed using the metrics PSNR, SSIM and NCC. The results are proving that the proposed method provides good robustness and imperceptibility than the existing methods. The proposed techniques achieved NC as 1 for all cover images and produces PSNR an average of 90.3420.

References

1. Abu-errub A, Al-haj A (2017) Optimized DWT-based image watermarking optimized DWT-based image watermarking (September 2008). https://doi.org/10.1109/ICADIWT.2008.4664405
2. Kumar P et al (2016) Robust digital image watermarking based on evolutionary optimization technique 7(3):92–99
3. Serdang UPM, Ehsan SD (2016) An Overview of multimodal biometric approaches based on digital image watermarking. 13(6):481–494
4. Venkatram N, Reddy LSS, Kishore PVV (2014) DWT-BAT based medical image watermarking for telemedicine applications. 2(1999):18–37
5. Thanki R, Borisagar K (2015) Multibiometric template security using CS theory—SVD based fragile watermarking technique 2 proposed watermarking technique. 12:1–10
6. Roy DA et al (2016) IRIS segmentation using Daughman's method. 978-1-4673-9939-5/16.2668-2675
7. Deshmukh DP, Malviya PAV (2015) Image fusion an application of digital image processing using wavelet transform. 6(11):1247–1255
8. Neve AG, Kakandikar GM, Kulkarni O (2017) Application of grasshopper optimization algorithm for constrained and unconstrained test functions, Int J Swarm Intell Evol Comput 6(3)
9. Guo Y, Li B, Goel N (2017) Optimised blind image watermarking method based on firefly algorithm in DWT-QR transform domain. https://doi.org/10.1049/iet-ipr.2016.0515
10. Nair SAH, Aruna P (2015) Comparison of DCT, SVD and BFOA based multimodal biometric watermarking systems. Alexandria Eng J 54(4):1161–1174. https://doi.org/10.1016/j.aej.2015.07.002
11. Gunjal BL (2015) MEO based secured, robust, high capacity and perceptual quality image watermarking in DWT-SVD domain. https://doi.org/10.1186/s40064-015-0904-z
12. Vijaya Durga K et al (2015) SVD based image watermarking with firefly algorithm. International conference on computer communication and informatics
13. Priyanka Panwar et al (2017) Robust digital image watermarking based on hybrid DWT and GWO optimization technique 7(11):12–17. ISSN:2248-9622

Inadequacy of Li-Fi Disentangles by Laser, Polarizing Beam, Solar, and Formation

D. Balaganesh

Abstract At the present time, the wireless Internet is making a vital role in data communications. Due to the high demand for the requirements in data transmission, people start to use various wireless technologies like Wi-Fi, WiMAX, Li-Fi. It was introduced by Prof. Harald Haas in July 2001 at TED Global talk. Li-Fi is an efficient wireless technology. This paper focuses on existing Li-Fi to compose the enhanced formation to access efficient, secured and better bandwidth. Analyzing the existing Li-Fi technology performance to find the inadequacy which helps to sort out and enhance the technology for further development in Li-Fi. Enhancing technology improves the data transfer speed and improves the efficiency and availability with highly secured and easy formation. The key components of laser, polarizing beam, and solar were used to sort out the inadequacy. This paper discusses a significant contribution to the improvements of Li-Fi data transfer speed, cost-cutting, and uncomplicated formation.

Keywords Li-Fi · Laser · Solar panel · Data transmission

1 Introduction

The first wireless telephone message invented by Alexander Graham Bell in his "photophone" on June 3, 1880, and it is a device that allowed the transmission of sound on a beam of light [1]. After 120 years, Prof. Harald's Li-Fi technology shows that the light transmits the data [2]. This paper discusses what are all the favorable properties possible to use could provide fully efficient wireless technology which has high data transfer speed.

The composition of this paper is as follows: Section 2 discusses the comparison of LED and laser. Section 3 gives the benefit of using solar panel as a receiver. Section 4 discusses the polarize beam. Section 5 examines the performance of the Li-Fi derived

D. Balaganesh (✉)
Lincoln University College, Petaling Jaya, Malaysia
e-mail: Balaganesh@lincoln.edu.my; baga_indian@yahoo.co.in

S.-L. Peng et al. (eds.), *Intelligent Computing and Innovation on Data Science*, Lecture Notes in Networks and Systems 118,
https://doi.org/10.1007/978-981-15-3284-9_79

from the formation. Section 6 evaluates the data transmission rate between laser Li-Fi and LED Li-Fi. Section 7 has given the summarized result which is useful to get more references and knowledge about Li-Fi technology.

2 Comparison of LED and Laser

Li-Fi begins to use LED which is better than radio waves (Wi-Fi) for data transfer. When compared to LED, there are many energy-efficient glowing lights can be used to provide better output by laser diodes.

The spontaneous emission of 850-nm-wavelength junction diode made by semiconductor compound gallium arsenide phosphide is called LED which could transfer data rate up to 100 mbps.

The stimulated emission of monochromatic wavelength radiation produces intense beam which is called as light amplification by the stimulated emission of radiation (LASER) and could transfer data rate up to 100 gbps.

The emissions of laser produce single color beam but LED is having many colors of beams. The direction of laser beam is extremely narrow which helps the receiver to receive 100% output but LED directions are contrary to laser and it covered wide area which is difficult to receive 100% output by the receiver. Overall characteristics of laser beams are coherent, monochromatic, and high intensity but LED is incoherent, different colors, and less intensity beam. Table 1 shows the characteristic of LED and laser.

To identify the efficiency of laser diode and LED, it is helpful to finalize the formation without any inadequacy. Tables 1 and 2, it simplifies to select laser diode due to its efficient characteristics and specifications [3].

3 Solar Panel

Solar panel is used instead of photodiodes for demodulation of data. Solar panels consist of solar cells which work for both storing energy and converting that energy stored in electrical signal. Solar will work more efficient in storing energy and transmitting data for longer hours which will ultimately enhance the overall Li-Fi technology. Solar panel is more beneficial, and it can perform both functions of transmitting data and capturing energy, whereas photodiode requires external device for the requirement of transmitting data. Considering this new technology, some more adulteration can be made to make it more efficient. The solar panel can directly convert the optical signal and modulated light into an electrical signal, without the need for an external power supply. Using solar panel, simultaneous communication and energy harvesting can be realized [4].

Table 1 Comparative characteristic LED and laser

LED	Laser
Current–light output	
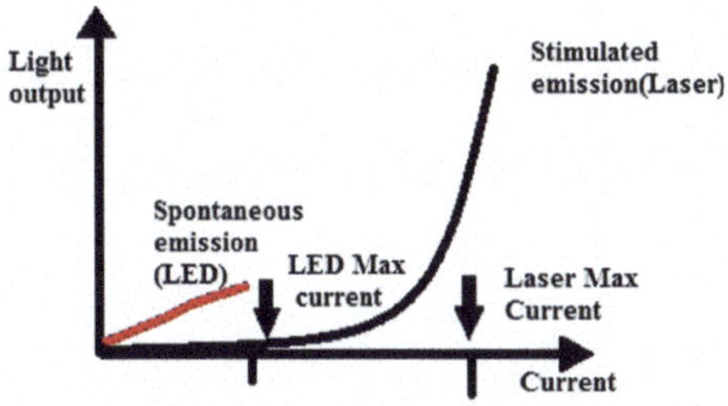	
Wavelength	
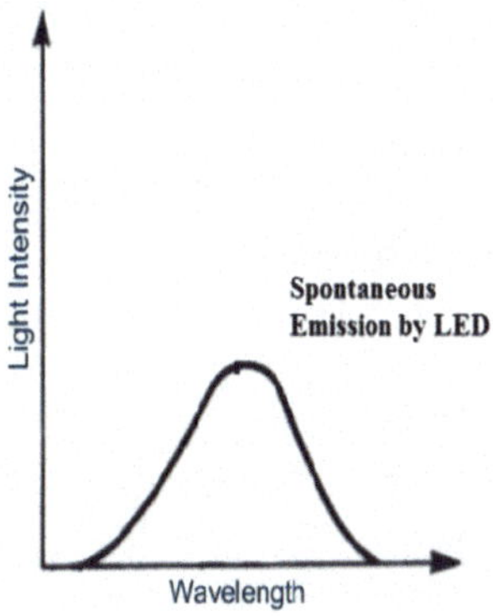	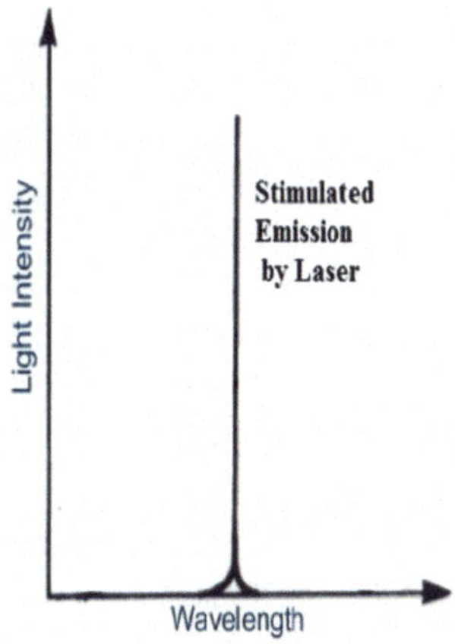
Monochromatic	
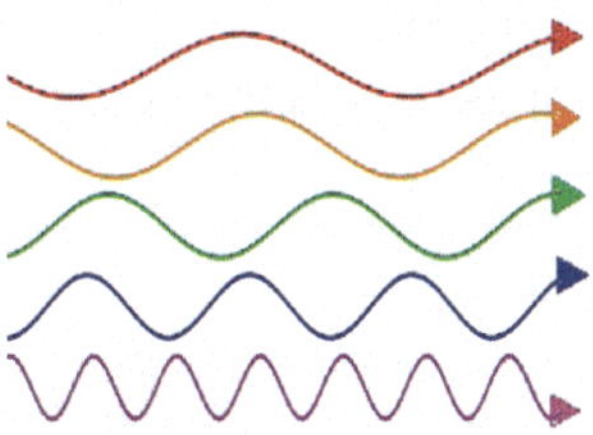	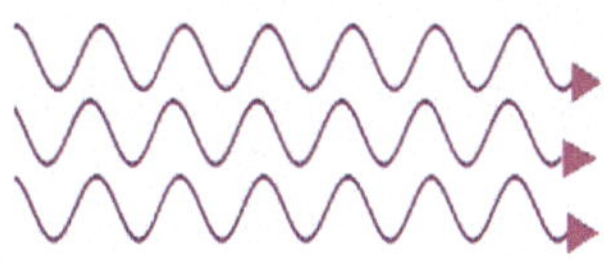

(continued)

By comparing solar cell and photodiode, the solar cells are more efficient to work as a good photosensor. It is optimized to have the maximum conversion efficiency of the incident light to electrical energy. This is achieved by maximizing the photocurrent and the maximum output voltage (Table 3).

Table 1 (continued)

LED	Laser
Coherence	
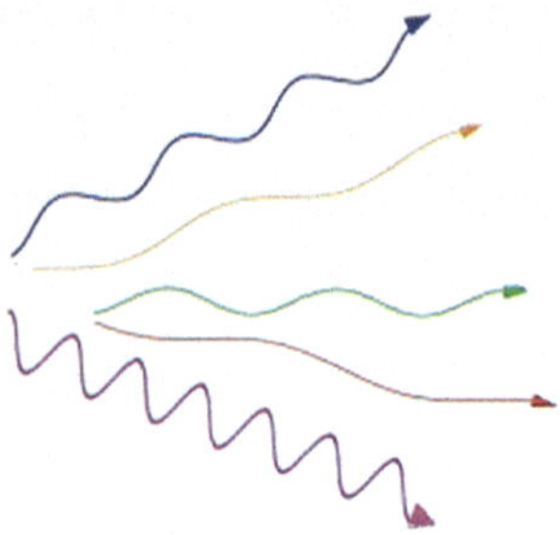	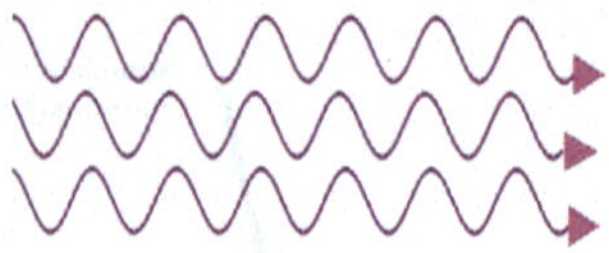

Table 2 LED and laser with respect to various comparison factors/specifications

Specification	Light emitting diode	Laser diode
Output power	Linearly proportional to drive current	Proportional to current above the threshold
Current	Drive Current (50–100 mA) (Peak)	Threshold current (5–40 mA)
Coupled power	Moderate	High
Speed	Slower	Faster
Output pattern	Higher	Lower
Fiber type	Multimode only	Single mode and multimode
Ease of use	Easier	Harder
Lifetime	Longer	Long
Spectral width	Wider, 25–100 nm (10–50 THz)	Narrower, $<10^{-5}$ to 5 nm (<1–2 MHz)
Modulation bandwidth	Moderate, Tens of KHz to tens of MHz	High, Tens of MHz to tens of GHz
Available wavelength	0.66–1.65 mm	0.78–1.65 mm
E/O conversion efficiency	10–20%	30–70%
Eye safety	Generally considered eyesafe	Must be rendered eyesafe, especially for $\lambda < 1400$ nm
Cost	Low	Moderate to high

4 Polarizing Beam Splitters

The laser light beam creates electromagnetic wave which has electric and magnetic elements. The electromagnetic wave has several planes. The 50% of planes are horizontal planes and 50% of planes are vertical planes called unpolarized light while passes through the medium generate single planes or the medium as a splitter which

Table 3 Solar cell versus photodiode

Solar Cell	Photodiode
Area: bigger	Smaller
Wavelength: broad range	Narrow range
Conversion: light to electricity	Light to signal
Designed to operate: forward bias	Reversed bias
Efficiency: power conversion	Generated electron
Quadrant operations: fourth quadrant (+x, −y)	Third quadrant(−x, −y)
Require: no power supply	Needed power supply
Noise: low noise	Large noise
Capacitance: large sensitive	Low not very sensitive

splitting vertical plane separate and horizontal plane separate called polarized light [5].

The medium of splitter used cube splitter which splits laser beam into two beams or more than many beams with the same wavelength and frequency. Using the laser polarization method, data transfer rate increased five times more than normal Li-Fi data transfer rate (Table 4).

The laser polarization data transfer method can possibly transfer 100 gigabits per seconds. From the polarize plane output can possible different wavelengths which can be used as a different data channel. Each beam can be possibly used well by the receiver to get same data rate. The implementation of beam splitter provides conversion output with cost effective in large system (Fig. 1).

Table 4 Polarized beam versus unpolarized beam

Polarized beam	Unpolarized beam
The variation limited to only one plane	The variation happen many plane
It is completely rational in nature	It is completely irrational
It depends on polarized material	It depends on the nature of source
Lightwave can be vertical and horizontal planes	Lightwave known to vibrate multitude directions
Receiver can receive exact position	Scatter wave difficult to position
Appropriate formation can be made without difficulty	Difficulty to arrange proper formations
Same frequency	Random frequency
Constant data rate	Irregular data rate

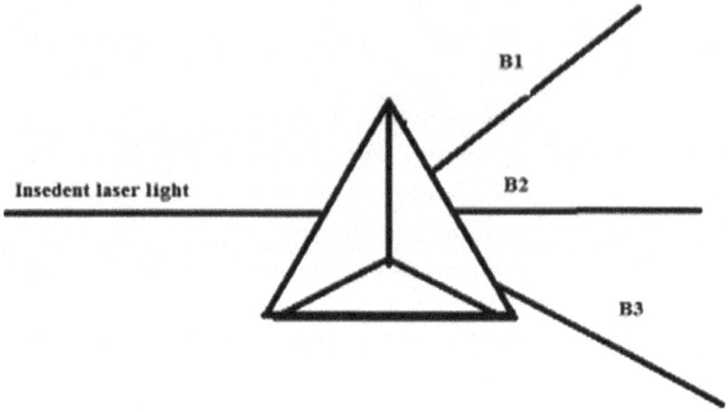

Fig. 1 Principle of beam splitting

Incident laser beam will be split into polarized beam. Characteristics of the splitter beams are equal wavelength, equal diameter, same efficiency, equal power distribution, and uniform data rate.

5 Formation of Li-Fi

Li-Fi is possible to transfer high-speed data which we attested by transmitting audio waves. We have prearranged laser diode as an input supply resource, 40-mm-optical glass prism as a beam splitter, solar panel as a receiver, and speaker as an output. The laser diode received audio signal from the phone which connected by 3.5 mm jack for the conversion of digital-to-analog and including the power supply by 9 V battery. The transmission of light beam by laser diode is separated by many polarized beams by optical glass prism. Each beam can be accessed by solar panel which transfers the signals to the speaker [6].

The formations begin with a phone connected to 3.5-mm-audio jack. It has tricolor wire, namely green, red, and white. The green and red wires were connected to 9 V battery, and white wire is considered as negative. The white wire of the 3.5 mm jack is connected to blue wire of the laser diode which is considered as negative. The red wire of the 9 V battery is considered as a positive line which is connected to 100R resistors, and the other end of resistors is connected to red wire of the laser diode. As we fixed 100R based on Ohm's law, Volts (V) = Current I in amp * R in ohms (Ω) [7, 8] (Figs. 2 and 3).

The novel approach of this paper is the intense beam of light from laser diode properties which are one wavelength and travel in same direction, divided by 40-mm-optical glass prism beam splitter. The one direction single beam is divided into more than two beams. Each beam can be accessed by different solar panels. Each solar panel considers as a receiver which sends the received signal to speakers connected

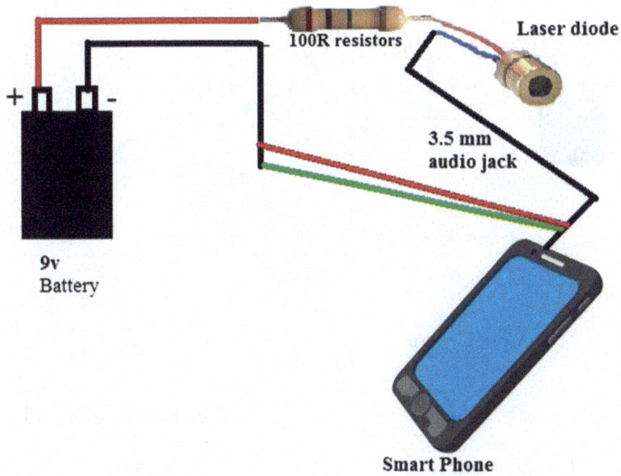

Fig. 2 Source circuit

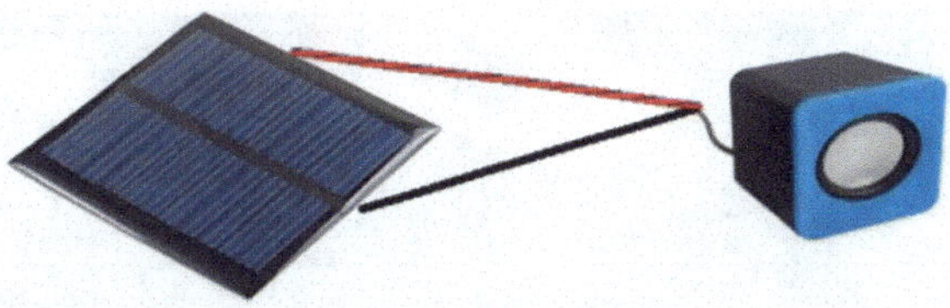

Fig. 3 Output circuit

by 3.5 mm jack. Once the sound played by phone, the wave transfers to laser diode and to beam splitter; it is divided into many beams which are received by different solar panels and transfer to speakers which we can hear the sound wave.

To increase the distance between each part like laser beam-to-beam splitter and beam splitter-to-solar panel, we can hear the same volume of sounds. At the same time, increase and decrease the distance between one solar panel to another solar panel also approximately heard same volume of sounds. By using laser diode, any altering distance is not affected by hearing sound waves. Based on the test proofed, many users can access the same data with same data rate in different places and different angles in a room (Fig. 4).

6 Evaluation of LED Li-Fi Versus Laser Li-Fi

The volume of sounds is evaluated by LED-based audio transmission. Formation of LED-based audio transmission is done as it is laser-based audio transmission [9]. The LED is an input supply resource; 40-mm-optical glass prism is a beam splitter; solar panel as a receiver; and speaker as an output. The LED received audio signal from the phone which is connected by 3.5 mm jack for the conversion of digital-to-analog including power supply by 9 V battery. The transmission of light not shows

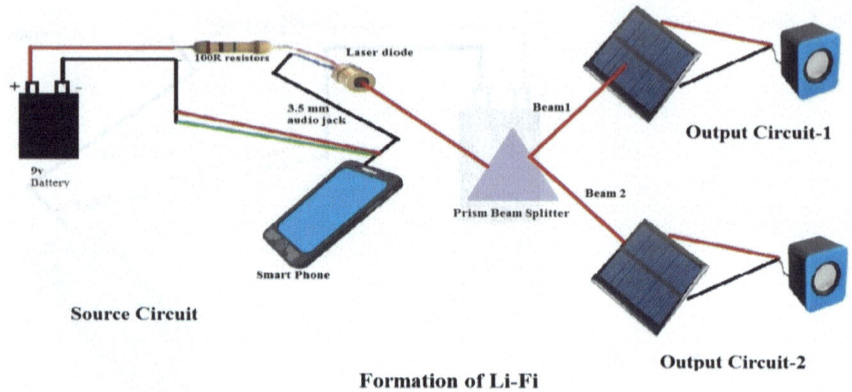

Fig. 4 Observation

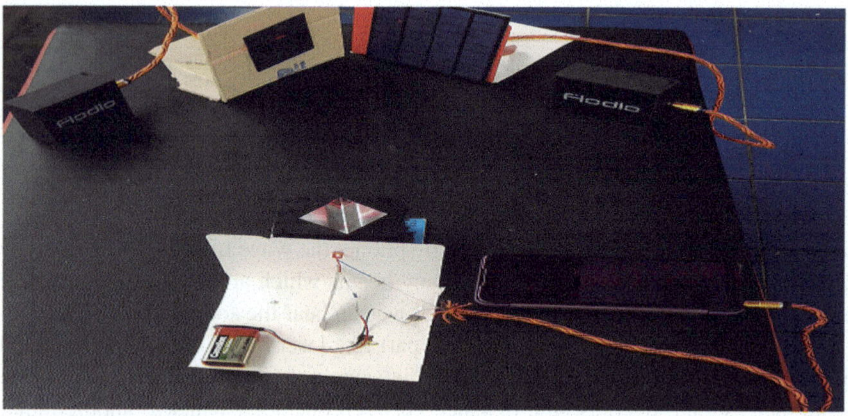

Fig. 5 Practical observation

any reflections by optical glass prism which transfers the signal to the speaker by single solar panel [10] (Fig. 5, Table 5).

7 Conclusion and Future Scope

In this paper the formation of Li-Fi by laser, polarized beam and solar panel given higher data rate, long distance data transmission, single beam divide into too many beam and each beam approximately transfer same data rate which is error free transmission.

The inadequacy of Li-Fi sort out by simple formation can be enhanced, further divide 0 s bits and 1 s bits for more secure data, and join again to access the data.

Table 5 Evaluation of LED Li-Fi technology and laser Li-Fi technology

Sl. No.		Observations	
	Formation changes	LED Li-Fi	Laser Li-Fi
1	Without resistor	Burn out	Little bit heat
2	9 V Battery	Minimum flow maximum drop	Maximum flow minimum drop
3	Resistor values	Less power watts	Maximum power watts
4	Beam splitter	No changes	Many beams
5	Increase the distance between light source to receiver	Sound of the volume decrease	No changes
6	Increase the distance between prism and light sours	Sound of the volume decrease	No changes
7	Minimum size of solar panel	Less capture	100% capture
8	Setup in front of bright light	Produce more noise	No changes
9	Setup in front of sunlight	Difficult to capture	100% capture
10	Change the color of light	No changes	Wavelength increase
11	Between source to receiver placed transparent medium	Volume decrease	No changes
12	Output receiver	Only one receiver	Multiuser receiver
13	Error transmission	i. No sound clarity ii. Signal loss	Error-free transmission
14	Polarized splitter	No reflection	Possible to split 0 bits and 1 bits. To make more secure data transmission

The Li-Fi technology is near to the beginning technology which is assure to provide higher bandwidth based on the demand [11, 12]. Li-Fi technology is more secure, and high-speed data transmission is possible to access wherever you need without affecting the environment. Many users can access the same data rate from a single laser diode beam and is divided by polarized splitter in the Li-Fi technology.

References

1. Pardeshi A, Vyas A (2013) Wireless energy transfer. IOSR J Electr Electron Eng (IOSR-JEEE) 8(1):69–79 (e-ISSN: 2278-1676, p-ISSN: 2320-3331) (Nov–Dec 2013)
2. Haas H, Yin L, Wang Y, Chen C (2016) What is LiFi? IEEE J Light Technol 34(6):1533–1544
3. https://www.rfwireless-world.com/Terminology/LED-vs-Laser.html

4. Agarwal S, Omer Y, Patil TB, Sawant SC (2017) Solar panel cells as power source and Li-Fi data nodes integrated with solar concentrator. Int J Eng Appl Comput Sci. https://doi.org/10.24032/ijeacs/0205/05
5. Haitjema H, Cosijns SJAG, Roset NJJ, Jansen MJ (2003) Improving a commercially available heterodyne laser interferometer to sub-nm uncertainty. Proc. SPIE 5190, Recent developments in traceable dimensional measurements II, (20 Nov 2003). https://doi.org/10.1117/12.508542, https://www.sciencedirect.com/book/9780081020555/smart-sensors-and-mems
6. Vinnarasi A, Aarthy ST (2017) Transmission of data, audio signal and text using Li-Fi. Int J Pure Appl Math 117(17):179–186 (ISSN: 1311-8080 (printed version); ISSN: 1314-3395 (on-line version))
7. Larsen FS (1996) Cerebral circulation in liver failure: Ohm's law in force. Semin Liver Dis 16(3):281–292
8. Henry MP, Ratnayake CK (2005) Electrochemical properties of columns in capillary electro chromatography. I. Ohm's law, resistivity and field strength. J Chromatogram A 1079(1–2):69–76 (24 Jun 2005)
9. Singh G (2015) Li-Fi (Light Fidelity)—an overview to future wireless technology in field of data communication
10. Shetty A (2016) A comparative study and analysis on Li-Fi and Wi-Fi. Int J Compt Appl 150:00975–8887
11. Sharma RR, Sanganal A, Pati S (2014) Implementation of a simple LiFi based system. Int J Comput Technol 1(9), ISSN: 2348 – 6090
12. Tsonev D, Videv S, Haas H (2014) Light fidelity (Li-Fi): towards all-optical networking. Proc. SPIE 9007, Broadband access communication technologies VIII, 900702 (1 Feb 2014)

Phrase Extraction Using Pattern-Based Bootstrapping Approach

R. Hema and T. V. Geetha

Abstract Key phrases convey the important concepts in the text documents. They are useful for document categorization, clustering, indexing, search and summarization and to know about the semantic similarity with other documents. In this research article, a novel method is introduced for extracting key phrases from the chemical literature where the entities and terminologies are not just restricted to chemistry, but they also consist of words from different scientific domains. Six categories of phrases are defined by the chemistry experts such as chemistry, physics, medicine, drugs, cells and toxins. In this work, the extraction of the above six types of phrases is explained using pattern-based bootstrapping approach.

Keywords Domain-specific key phrase extraction · Chemical documents · Features · Seeds · Bootstrapping · Sliding window · Score · Rank

1 Introduction

As it is a challenging task to examine the complete documents to check whether the document would be useful or not, the key phrases would be an alternative for the users to understand the crux of the document. The importance of key phrases inspires the researchers for automatic extraction of key phrases using various approaches. In this work, the focus is on extraction of key phrases from chemical domain. For a specific domain, there is no controlled vocabulary list.

This work explores the idea of automatic identification of phrases from the chemical literature using pattern-based bootstrapping approach. This approach automatically identifies six categories of phrases from the chemical literature. This work starts with a small set of tagged training data. The tagged training data is used to identify the seed word and context features to define a five-window context pattern for each phrase category. The identified patterns are used as seed patterns. The bootstrapping

R. Hema (✉)
University of Madras, Chennai, TamilNadu, India

T. V. Geetha
Anna University, Chennai, TamilNadu, India

S.-L. Peng et al. (eds.), *Intelligent Computing and Innovation on Data Science*, Lecture Notes in Networks and Systems 118,
https://doi.org/10.1007/978-981-15-3284-9_80

algorithm generates new patterns by masking the patterns and extracts new phrases in each iteration, and the process stops when no more new patterns are generated. The contributions of this research article are: (i) The development of a chemical key phrase extraction model is based on pattern-based bootstrapping approach, and (ii) the author introduced a number of domain-specific features for key phrase extraction.

2 Background

This section discusses the related work in the extraction of phrases from text documents. Many researchers have done their research on the key phrase extraction by applying supervised machine learning approaches [1]. Earlier, most of the chemical information extraction systems focused mainly on the named entity recognition systems [2–4]. Our work tries to extract the key phrases from the chemical literature by applying pattern-based bootstrapping method which is a semi-supervised machine learning approach. Shimada et al. [5] used bootstrapping and hierarchical directed acyclic graph structure for sentiment sentence extraction and obtained high accuracy. In the medical domain, Nomura et al. [6] focused on a hybrid approach with bootstrapping and pattern matching method to extract words of complaint/diagnosis. De Benedictis et al. [7] presented GlossBoot, a minimal supervised bootstrapping approach to multilingual glossary learning. In biomedical domain, Movshovitz-Attias et al. [8] used the bootstrapping methodology to extract ontology of categories and seeds. In chemical domain, Hawizy et al. [9] developed a semantic text mining tool called ChemicalTagger to extract information such as units, mixtures, amounts of substances and roles of chemicals as well as action phrases using linguistic context.

In this work, a novel method is introduced for extracting six categories of key phrases from the chemical literature. This work identified the relevant features for the phrases, and the phrases are extracted using context window algorithm which will reduce the noisy phrases. Ranking measure is used to select the phrases. ChemicalTagger tool extracted only the action phrases in chemistry, while this work is aiming to extract phrases related to six fields. The construction of the phrase extractor by using pattern-based bootstrapping approach is described, and a *k*-window algorithm is used for selecting the contexts that are used in the phrase extraction method.

3 Pattern-Based Bootstrapping Phrase Extraction

3.1 Overview of Pattern-Based Bootstrapping Process

Figure 1 shows the overall pattern-based bootstrapping process for phrase extraction.

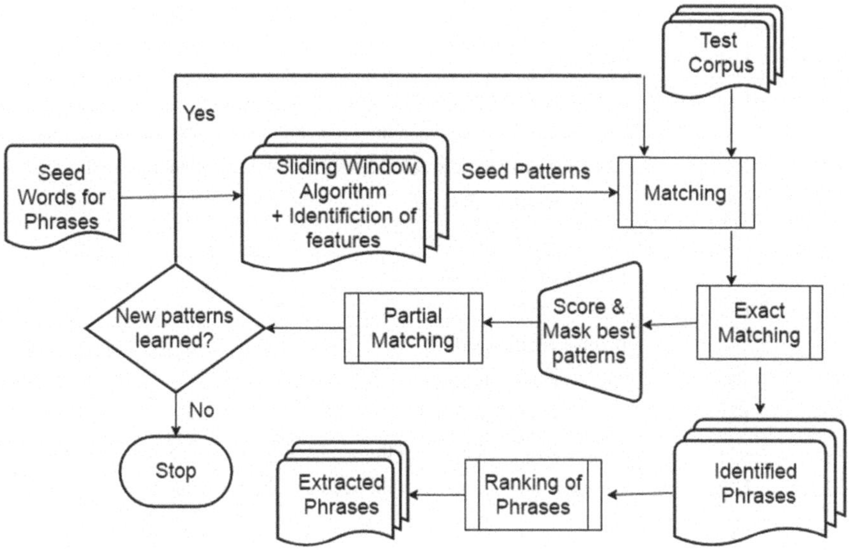

Fig. 1 Architecture for bootstrapping approach phrase extraction

3.2 Initializing with Seeds

For this work, ten thousand chemical research documents from the Chemical Informatics Open Access Journal make the corpus. After surveying the chemical literature in association with domain experts, six categories of key phrases were defined. The process started with a set of seed words for six categories of phrases. The identified seed words are: chemistry-200, physics-45, drug-90, medicine-60, cells-40 and toxins-55.

3.3 Sliding Window Algorithm

In bioinformatics, sliding windows are applied to identify the protein sequence such as the prediction of transmembrane protein segments [10]. Keshtkar et al. [11] used the sliding window to extract the paraphrases of emotion expressions from the texts. In this work, the sliding window is based on the k-window approach to extract the relevant text fragments from chemical documents. The *k*-window algorithm introduced by Bostad [12] is used to identify all the tokens surrounding a seed word in a context window with a size of $\pm k$ words. The context window with five words will define the seed patterns. The pattern captures the sequence of seed word and the tokens surrounding the seed word on both left and right sides. The size of the context window defines the structure of phrases as $w1$, $w2$, $w3$, $w4$, $w5$ where wi ($i = 1, \ldots, 5$) are word tokens in a phrase.

Table 1 List of features for the extraction of six types of phrases

S. no.	Features	Illustration
1.	F1	Sequence of part of speech
2.	F2	Sequence of POS between the seed and the first verb before the seed
3.	F3	Sequence of POS between the seed and the first noun before the seed
4.	F4	First verb before the seed
5.	F5	First noun before the seed
6.	F6	Token before the seed
7.	F7	Seed
8.	F8	Token after the seed
9.	F9	First verb after the seed
10.	F10	First noun after the seed
11.	F11	Sequence of POS between the seed and the first verb after the seed
12.	F12	Sequence of POS between the seed and the first noun after the seed

3.4 Identification of Features

Generally, local and topical features are considered as good measures to identify the word sense disambiguation on contextual analysis. These features include surrounding words and their part-of-speech tags, collocations and keywords in contexts [13]. By analyzing the seed phrases extracted using the above sliding window algorithm, the features that include both lexical and syntactic descriptions of the phrases are considered [14]. Table 1 shows the list of features identified for the phrase extraction.

3.5 Pattern Representation

The design of the pattern plays a major contribution in pattern-based bootstrapping approach. The pattern could be represented in a contextual window [15], and the tuples representing the pattern can be in an order of sequence or can be based on the syntactic structure of a phrase. The pattern consists of five tuples of which one tuple is for seed word, two tuples for a verb and a noun related to the field and two tuples for POS tags.

3.6 Scoring of Phrases

By using each candidate pattern p, a number of chemical phrases can be extracted. At this stage, the extracted chemical phrases are scored according to the matching criteria. The type of chemical phrases we extracted are (i) positive phrases (phrases that match with the patterns), (ii) negative phrases (phrases that match with the patterns but are not relevant to chemical phrases) and (iii) left-out phrases (phrases that are not identified and extracted from the training corpus). With the above types of phrases, the pattern's accuracy and confidence can be calculated with the formulas adopted by Winston Lin et al. [15].

3.7 Ranking of Phrases

Key phrases can be extracted in a relative sense. But, classification measures find it very difficult to take a decision to classify a phrase as key phrase or not. To tackle this problem, the quality of key phrases is evaluated using ranking measures. In ranking method, the top scored phrases are selected as the best ones. Let A_p be the set of patterns in the pattern pool that match any one of the patterns of phrase k. The rank of key phrases can be computed based on the confidence of patterns. In each category, a large number of phrases are extracted from the manually initialized list of seed words. In each iteration, the number of phrases is increased and the extracted phrases are filtered by using ranking measure. This measure is based on the number of phrases that are extracted from less number of patterns. The bootstrapping process starts by selecting a subset of the extraction patterns that are intended to extract the phrases. This subset of patterns is called as "pattern pool." The extracted phrases are selected by ranking measure, and the selected phrases are accumulated in the "phrase pool."

3.8 Masking and New Pattern Generation

The first method of the new pattern generation is the replacement of POS value in the appropriate position p_o. The POS tuple value with the maximum score at position p_o is masked. The new pattern is generated by replacing the tuple value of the original pattern by the new tuple value that occurs most frequently at position k in the test data. The next method of new pattern generation is carried out by shifting the context window to the left or right of the Wi depending on the frequency of occurrence of POS pair of wi, $wi + 1$ or $wi - 1$, wi. Depending on a higher POS pair frequency score, a new pattern is generated. This method of new pattern generation is possible since the chemical nouns and chemical verbs are often associated with POS tags that frequently tend to co-occur together.

There are 2256 different patterns, and about 943 patterns appeared only once. Therefore, patterns appearing more than five times are selected to avoid irrelevant noise. With these patterns, all possible phrases of six types are extracted from the training chemical documents. The output of this step is a list that contains 1,02,735 phrases, and they obey the selected sequence of patterns. Among them, there are 94,432 true instances yielding a precision of 87.7%.

4 Results and Evaluation

In this work, the bootstrapping algorithm learned 1313 patterns and extracted 1,02,735 chemical phrases. The algorithm was evaluated by using two techniques. First, inter-annotator study is conducted to check if the extracted phrases are correct and the degree of agreement between the two annotators is calculated. Second, the performance of the algorithm was evaluated by the standard measures of information retrieval called recall and precision.

4.1 *Evaluating the Inter-annotator Agreement*

The chemical research articles from the test corpus are randomly taken, and the chemical phrases are classified according to the annotation guidelines. The guidelines specify the structure of different categories of phrases that commonly occur in the chemistry literature and contain examples of annotated phrases. For this manual annotation, the 2000 chemical research articles of test data are provided as corpus. Following this annotation process, the proposed bootstrapping algorithm was run over the same test data. The agreement was measured between the two annotators using Cohen's kappa coefficient (Cohen and Jacob 1960), a statistic that measures the agreement between two raters which classify N items into C mutually exclusive categories. The kappa coefficient k is defined by Eq. 1. The equation for k is:

$$k = \frac{P_o - P_e}{1 - P_e} = 1 - \frac{1 - P_o}{1 - P_e} \tag{1}$$

where p_o is the relative observed agreement among raters and p_e is the hypothetical probability of chance agreement. If there is a complete agreement between the raters, then the value of $k = 1$. Otherwise, $k = 0$. The result was obtained by the inter-annotator's agreement for all the extracted phrases as 88.43%. The *kappa* value for all the chemical phrases is 72.76%, which is a significant agreement.

4.2 *Precision and Recall Evaluation*

The experiments were conducted for extracting 6 categories of phrases with the help of the test data of 2000 chemical research articles that are randomly selected from the Journal of Chemical Informatics. To evaluate the automatically generated key phrases, the standard information retrieval measures of precision and recall were used. The precision and recall are calculated using the formulas as follows.

$$\text{Precision} = \frac{N_e \cap N_h}{N_e} \tag{2}$$

Table 2 Precision and recall for each category of phrases from the chemical literature

S. no.	Category of phrases	Precision	Recall
1.	Chemistry	87.47	90.54
2.	Physics	85.29	86.73
3.	Medical	81.34	87.21
4.	Cells	80.68	76.85
5.	Drugs	78.57	79.86
6.	Toxins	72.72	77.28

$$\text{Recall} = \frac{N_e \cap N_h}{N_h} \tag{3}$$

Let N_e be the number of extracted key phrases and N_h be the number of key phrases annotated by the annotators. Table 2 contains the precision and recall values for 6 categories of phrases from the chemical literature.

5 Conclusion

This research shows that six categories of key phrases that commonly occur in the chemical literature can be extracted using a bootstrapping algorithm. This algorithm was developed based on contextual and morphological features that can successfully extract key phrases in six categories. The bootstrapping algorithm achieved good performance results on the data set used by this work. This research compares the related work, human annotators and different data sets with the proposed bootstrapping algorithm. All comparisons suggest that the bootstrapping algorithm for key phrase extraction from the chemical literature is more reliable than the other methods.

This research work highlights several aspects that follow: (1) Bootstrapping approach does not rely on the availability of manually labeled corpora. This approach needs only a small amount of seed patterns to extract the phrases. (2) The success of bootstrapped approach depends on the effectiveness of the seed patterns. Hence, scoring is done for the selection of patterns. (3) Since it needs only a little amount of seed patterns, this approach can be applied to the data sets of high dimensionality. (4) The quality of extracted phrases is measured by a ranking measure which does not need any human intervention. (5) Totally, six categories of phrases are identified from the chemical literature. (6) The comparison is done between the earlier approaches and different data sets. (7) The experiments and evaluation indicate that the bootstrapping algorithm achieved well and produced reasonable outcomes.

References

1. Abulaish M, Anwar T (2012) A supervised learning approach for automatic keyphrase extraction. Int J Innovative Comput Inf Control 8(11):7579–7601
2. Yan S, Spangler WS (2012) Learning to extract chemical names based on random text generation and incomplete dictionary. In: 11th international workshop on data mining in bioinformatics, pp 21—25
3. Akhondi SA, Singh B, van der Host E, van Mulligen E, Hettne KM, Kors JA (2013) A dictionary-and grammar-based chemical named entity recognizer. In: Proceedings of BioCreative Challenge Evaluation Workshop, vol 2, pp 113–120
4. Lowe DM, Sayle RA (2015) LeadMine: a grammar and dictionary driven approach to chemical entity recognition. J Cheminformatics 7(Suppl 1):S5. https://doi.org/10.1186/1758-2946-7-s1-s5
5. Shimada K, Hashimoto D, Endo T (2009) A graph-based approach for sentiment sentence extraction. Published in New Frontiers in Applied Data Mining, Springer, pp 38–48
6. Nomura Y, Suenaga T, Satoh D, Ohki M, Takaki T (2013) Medical information extracting system by bootstrapping of NTTDRDH at NTCIR-10 MedNLP Task. In: Proceedings of the 10th NTCIR conference, pp 732–735
7. De Benedictis F, Faralli S, Navigli R (2013) GlossBoot: bootstrapping multilingual domain glossaries from the web. In: Proceedings of the 51st annual meeting of the association for computational linguistics, 528–538
8. Movshovitz-Attias D, Cohen W (2012) Bootstrapping biomedical ontologies for scientific text using NELL. In: BioNLP. Proceedings of the 2012 workshop on biomedical natural language processing in NAACL, pp 11–19
9. Hawizy L, Jessop DM, Adams N, Murray-Rust P ChemicalTagger: a tool for semantic text-mining in chemistry. J Cheminformatics 3(17). https://doi.org/10.1186/1758-2946-3-17
10. Sipos L, VonHeijne G (1993) Predicting the topology of eukaryotic membrane proteins. Eur J Biochem 213:1333–1340
11. Keshtkar F, InkPen D (2013) A bootstrapping method for extracting paraphrases of emotion expressions from text. Wiley Periodicals 29:417–435
12. Bostad T (2003) Sentence based automatic sentiment classification. Ph.D. Thesis, Computer Speech Text and Internet Technologies, Computer Laboratory, University of Cambridge, Cambridge, UK
13. Mihalcea R (2004) Co-training and self-training for word sense disambiguation. In: Proceedings of natural language learning, pp 33–40
14. Bach N, Badaskar S (2007) A survey on relation extraction. Lit Rev Lang Stat II
15. Lin W, Yangarber R, Grishman R (2003) Bootstrapped learning of semantic classes from positive and negative examples. In: Proceedings of the ICML-2003 workshop on The Continuum from Labeled to Unlabeled Data

UAV's Applications, Architecture, Security Issues and Attack Scenarios: A Survey

Navid Ali Khan, Sarfraz Nawaz Brohi, and NZ Jhanjhi

Abstract Unmanned aerial vehicles (UAVs)/drones have become very popular in recent years as they are widely used in several domains. They are widely used in both military and civilian applications such as aerial photography, entertainment, search and rescue missions, reconnaissance, traffic monitoring, and logistics. Typically, UAVs are operated from a controller or a ground control station (GCS) with the help of different communication protocols such as MAVLink, UranusLink, UAVCAN. These communication protocols are used to exchange messages. The messages contain considerable information about the UAV and certain control commands sent from GCS to UAV or UAV to GCS. Though these protocols provide better communication along with secure aspects, however, mostly there is no subtle mechanism for securing these messages and are prone to many security attacks such as man-in-the-middle (MITM) attack, denial-of-services (DoS) attack, packet data injection attack, and eavesdropping. This can result in serious consequences, for instance, crash land of a military or civilian UAV, steal important data of a military operation, false injection of reports in a reconnaissance or search and rescue operation, and many more. So, there is a need for a secure communication protocol which can ensure the required security standard sets for communication of UAVs. This survey presents the applications, general architecture, attacks on UAVs, and an insight into the security issues of UAV's communication protocols and proposes a new secure communication protocol for the stated issues of UAVs.

Keywords UAVs · Drones · Security · Communication protocols

N. A. Khan (✉) · S. N. Brohi · NZ. Jhanjhi
School of Computer Science and Engineering (SCE), Taylors University, Subang Jaya, Selangor, Malaysia
e-mail: navidalikhan@sd.taylors.edu.my

S. N. Brohi
e-mail: sarfraznawaz.brohi@taylors.edu.my

NZ. Jhanjhi
e-mail: noorzaman.jhanjhi@taylors.edu.my

S.-L. Peng et al. (eds.), *Intelligent Computing and Innovation on Data Science*, Lecture Notes in Networks and Systems 118,
https://doi.org/10.1007/978-981-15-3284-9_81

1 Introduction

Unmanned aerial vehicles, also called drones, have grown tremendously in recent years. They are commonly used in various military and civilian applications [1]. The military market of UAVs is projected to boost by more than 60% by 2020 [2, 3]. Numerous studies [2, 4, 5] have highlighted the fact that soon, the utilization of UAVs for civilian purposes will be higher than the military purposes, and this can eventually overcome the military demands in the future [1].

A UAV is an autonomous or remotely controlled vehicle with no crew onboard [6]. Two different methods can control a UAV: (i) controller or (ii) a ground control station (GCS) [2]. Increasing research and developments in recent years have improved the use of UAVs in various applications. However, UAVs are still in their experimental stages, and a shortage of trained crew members limits its use. Due to the widespread use and security weakness of the UAVs have made them an attractive target for hackers and attackers. Since the technology is new, there are few security solutions for UAVs. Most of these solitons are just proposals, or they are at the beginning of their process of development [7].

This paper presents the various security requirements and threats against UAVs. We believe this research will help academia and practitioners when considering the UAVs. The rest of the paper organized as follow: Sect. 2 comes up with an overview of the various applications, Sect. 3 discusses types of UAVs, their pros and cons, Sect. 4 is about the general architecture of a UAV system, Sect. 5 illustrates the UAVs communication security issues. Section 6 highlights the real attacks on UAVs, Sect. 7, our findings, and in last is the conclusion of the paper.

2 Applications of UAV

UAVs are widely used in both military and civilian applications [8, 9]. They can perform both indoor and outdoor tasks in very challenging environments [10]. According to [11, 12], UAVs can have more than two hundred applications based on their types. The benefit of UAV is that it gives a brisk outline of the objective activity or territory with lesser hazard and threat. The micro-UAVs are most appropriate for reconnaissance missions inside buildings because of their diminished measurement and little size [13]. In addition to military applications, here are some other important UAVs application in civilian use (Table 1).

3 General Architecture and Communication Scenario

The general structure or system of a UAV contains different components such as one or more UAVs, GCS, Global Navigation Satellite System (GNSS), and Air Traffic

Table 1 Applications of the UAVs

Application domain	Studies
Search and rescue mission	[14–17]
Environmental protection	[18, 19]
Shipping delivery	[20]
Space UAVs	[21, 22]
Marine UAVs	[23–25]
Miscellaneous applications	[8]

Control (ATC). All these components communicate with each other through different communication channels or protocols. The GCS is responsible for controlling all the information such as monitoring the UAVs during flight, receiving and processing the data, and upload and update mission instructions. According to [26, 27], the ground control station requires at least three crew members to run a large and complex UAV. The overall UAV system design can be seen in Fig. 1 [28].

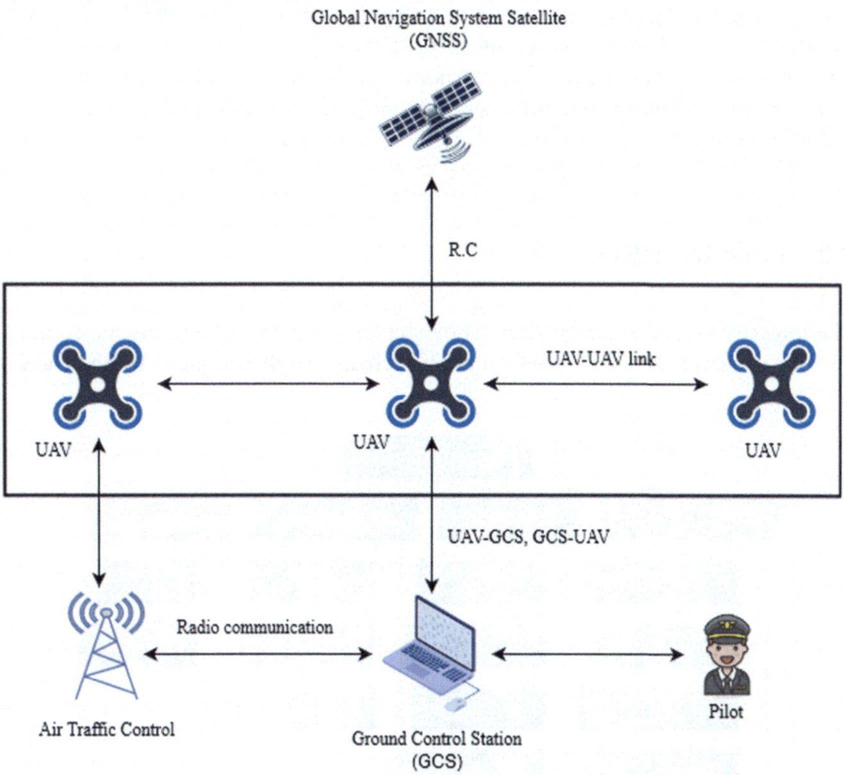

Fig. 1 General structure of the UAV system

4 Security Issues (Threats) Against UAVs

The UAV and GCS normally communicate through a communication protocols [29], such as MAVLink [30], UranusLink [31], UAVCAN [32]. Most of the existing security protocols are either not intended for such an environment [33], and they do not properly utilize the resources [34] or these communication protocols does not provide security measures [29, 35, 36]. Among these protocols, the MAVLink is a conventional and acknowledged lightweight protocol used for communication between GCS and UAVs. However, there is no subtle mechanism for security in this protocol [29, 30, 36]. In what follows, we summarize the UAV's network security in the form of possible attack scenarios, as discussed below, and the taxonomy is given in Fig. 2.

4.1 Confidentiality and Privacy Attacks

An intruder compromises unauthorized access to private and vulnerable information through intercepting data, commands, or instructions passed between UAV and GCS. Confidentiality and privacy requirements restrict the state of communication between UAV and GCS. The confidentiality attacks are further divided into (i) eavesdropping attack [37], (ii) identity spoofing attack [38, 39], (iii) unauthorized access, and (iv) traffic analysis attack [40].

4.2 Integrity Attacks

The integrity of UAVs can be affected by changing the data sent through a communication protocol. The data or information exchanged between the UAV and GCS is

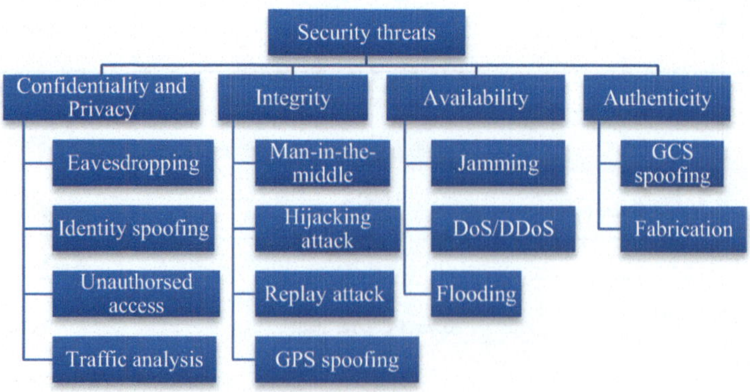

Fig. 2 Security attacks against UAVs [30]

incorrect, and the intruder can use to control these entities for their intended purpose. The integrity violation results in the attacks (i) man-in-the-middle attack [41], (ii) hijacking attack [42], (iii) replay attack [43], and (iv) GPS spoofing attack [30].

4.3 Availability Attacks

Attacks that interrupt the availability can be accomplished by interrupting the connection used to exchange information between the UAV and GCS. The intruder normally either occupies the communication channel or floods a large amount of random traffic to interrupt the communication. This makes the resources unavailable for each other. The availability attack is further divided into (i) jamming attack [41], (ii) denial-of-services (DoS) attack [44], and (iii) flooding attack [36].

4.4 Authenticity Attacks

Attacks on authenticity attempt to create the GCS-UAV think that falsified information is genuine. Authenticity typically ensures that the message, control commands, or other exchange of information are coming from the source it claims to be from. Authenticity is achieved through authentication. Authenticity attack scenarios enable the following attack (i) GCS spoofing attack and (ii) fabrication attack [30].

5 Attacks in a Real Environment on UAVs in Recent Past Based on Literature

The number of known cyber-attacks against the UAV systems was very few until the year 2007 due to their less use and non-popularity. In 2007, numerous incidents of using the satellite to control the transmission of UAVs by some terrorist organizations were reported by different sources [45]. The first case of the mentioned incident was reported back in 2009, when *SkyGrabber* was used to capture the video of UAV from satellite using a satellite antenna by a terrorist group [46–48]. They found out a vulnerability that the video feed was unencrypted [38]. One of the major and serious incidents in the history of UAV's attack is the capturing of US RQ-170 Sentinel UAV military by the Iranian Army [49]. In Sept 2011, the US Air Force Base identified another malware vulnerability in the Predator and Reaper UAV-GCS [50, 51]. In another incident in 2016, the Mexican drug dealers and traffickers deceived the US custom bureau UAV via GPS spoofing attack [52]. In addition, more studies have been undertaken to develop cheap GPS spoofing attacks [53]. In December 2012, Iran launched another attack on ScanEagle, a fixed-wing UAV in the US Air force fleet resulting in capturing it. However, this attack is unknown claim by the Iranian authorities [54].

6 Our Findings

The UAV-GCS communication occurs through communication protocols. The increase in the use of UAVs application demands the communication should be secure. In the literature, it is identified that most of the attacks occur due to the insecure communication in the protocols. There are few protocols available for this purpose, and among them, MAVLink is the most widely used. Though MAVLink provides better communication however, there is no subtle mechanism for securing these messages and are prone to several attacks. Therefore, there is a need for a secure communication protocol which can ensure the required security standard sets for communication between UAVs and ground stations. We will design the protocol in our future work.

7 Conclusion

This paper presents the importance and applications of UAVs in both military and civilian uses. Also, it gives an insight into the general architecture and typical communication scenario. The network security of UAVs is categorized into security requirements and attacks. The security issues are mostly traced back to communication protocols. This paper proposes the need for a secure communication protocol which can address the mentioned attacks and make secure communication possible.

References

1. Saeed AS, Younes AB, Cai C, Cai G (2018) A survey of hybrid unmanned aerial vehicles. Prog Aerosp Sci 98:91–105
2. UAV Roundup 2013—Aerospace America [Internet]. 2013. Available from: https://anyflip.com/efia/wutq
3. Drubin C (2013) UAV market worth $8.3 B by 2018. Horizon House Publications INC 685 Canton ST, Norwood, MA 02062 USA
4. Ping JTK, Ling AE, Quan TJ, Dat CY (2012) Generic unmanned aerial vehicle (UAV) for civilian application-a feasibility assessment and market survey on civilian application for aerial imaging. In: 2012 IEEE conference on sustainable utilization and development in engineering and technology (STUDENT). IEEE, pp 289–294
5. Skrzypietz T (2012) Unmanned aircraft systems for civilian missions. BIGS|Policy Pap. BIGS Policy
6. Watts AC, Ambrosia VG, Hinkley EA (2012) Unmanned aircraft systems in remote sensing and scientific research: classification and considerations of use. Remote Sens 4(6):1671–1692

7. Shoufan A, Alnoon H, Baek J (2015) Secure communication in civil drones. In: International conference on information systems security and privacy. Springer, pp 177–195
8. Hassanalian M, Abdelkefi A (2017) Classifications, applications, and design challenges of drones: a review. Prog Aerosp Sci 91:99–131
9. Saleh M, Jhanjhi NZ, Abdullah A (2020) Proposing a privacy protection model in case of civilian drone. In: 22nd IEEE/ICACT 2020 international conference on advanced communications technology, Korea
10. Saleh M, Jhanjhi NZ, Abdullah A (2019) Proposing a privacy protection model in case of civilian drone. In: 13th international conference on mathematics, actuarial science, computer science and statistics (MACS)
11. Rodríguez RM, Alarcón F, Rubio DS, Ollero A (2013) Autonomous management of an UAV Airfield. In: proceedings of the 3rd international conference on application and theory of automation in command and control systems, Naples, Italy, pp 28–30
12. 20 great UAV applications areas for drones [Internet]. 2014 [cited 2019 Aug 29]. Available from: https://air-vid.com/20-great-uav-applications-areas-drones/
13. Yan R, Pang S, Sun H, Pang Y (2010) Development and missions of unmanned surface vehicle. J Mar Sci Appl 9(4):451–457
14. Stuchlík R, Stachoň Z, Láska K, Kubíček P (2015) Unmanned aerial vehicle-efficient mapping tool available for recent research in polar regions. Czech Polar Rep 5(2):210–221
15. Waharte S, Trigoni N (2010) Supporting search and rescue operations with UAVs. In: 2010 international conference on emerging security technologies. IEEE, pp 142–147
16. TU Delft's ambulance drone drastically increases chances of survival of cardiac arrest patients [Internet]. 2014 [cited 2019 Aug 29]. Available from: https://www.tudelft.nl/en/2014/tu-delft/tu-delfts-ambulance-drone-drastically-increases-chances-of-survival-of-cardiac-arrest-patients/
17. O'Leary A Nora Quoirin: The "dark" unanswered questions after schoolgirl's death in jungle [Internet]. MSN.com. 2019 [cited 2019 Aug 29]. Available from: https://www.msn.com/en-my/news/world/nora-quoirin-the-dark-unanswered-questions-after-schoolgirls-death-in-jungle/ar-AAFU6BS?ocid=spartandhp
18. Restas A (2015) Drone applications for supporting disaster management. World J Eng Technol 3:316–321
19. Yao H, Qin R, Chen X (2019) Unmanned aerial vehicle for remote sensing applications—a review. Remote Sens 11(12):1443
20. Heutger M, Kückelhaus M (2014) Unmanned aerial vehicles in logistics a DHL perspective on implications and use cases for the logistics industry. DHL Cust Solut Innov Troisdorf, Ger
21. Menges P (2006) Artificial neural membrane flapping wing. NIAC phase i study. Final report, Ph D Princ Investig Aerosp Res Syst USA 3:6
22. Sjogren WL, Lorell J, Wong L, Downs W (1975) Mars gravity field based on a short-arc technique. J Geophys Res 80(20):2899–2908
23. Koski WR, Allen T, Ireland D, Buck G, Smith PR, Macrander AM et al (2009) Evaluation of an unmanned airborne system for monitoring marine mammals. Aquat Mamm 35(3):347
24. Fingas M, Brown C (2014) Review of oil spill remote sensing. Mar Pollut Bull 83(1):9–23
25. Reineman BD, Lenain L, Statom NM, Melville WK (2013) Development and testing of instrumentation for UAV-based flux measurements within terrestrial and marine atmospheric boundary layers. J Atmos Ocean Technol 30(7):1295–1319
26. Pratt K, Murphy RR, Stover S, Griffin C (2016) Requirements for semi-autonomous flight in miniature UAVs for structural inspection. AUVSI's Unmanned Syst North Am Orlando, Florida, Assoc Unmanned Veh Syst Int
27. Narayanan RGL, Ibe OC (2015) Joint network for disaster relief and search and rescue network operations. In: Wireless Public Safety Networks 1. Elsevier, pp 163–193
28. Nguyen M-D, Dong N, Roychoudhury A (2017) Security analysis of unmanned aircraft systems
29. Allouch A, Cheikhrouhou O, Koubaa A, Khalgui M, Abbes T (2019) MAVSec: securing the MAVLink protocol for Ardupilot/PX4 unmanned aerial systems. ArXiv Prepr arXiv190500265

30. Koubaa A, Allouch A, Alajlan M, Javed Y, Belghith A, Khalgui M (2019) Micro air vehicle link (MAVLink) in a nutshell: a survey. arXiv Prepr arXiv190610641
31. Kriz V, Gabrlik P (2015) Uranuslink-communication protocol for UAV with small overhead and encryption ability. IFAC-PapersOnLine 48(4):474–479
32. Team U development. UAVCAN communication protocol
33. Kaps J-PE (2006) Cryptography for ultra-low power devices
34. Larrieu N (2014) How can model driven development approaches improve the certification process for UAS? In: 2014 international conference on unmanned aircraft systems (ICUAS). IEEE, pp 253–260
35. Kwon Y-M (2018) Vulnerability analysis of the Mavlink protocol for unmanned aerial vehicles. DGIST
36. Kwon Y-M, Yu J, Cho B-M, Eun Y, Park K-J (2018) Empirical analysis of mavlink protocol vulnerability for attacking unmanned aerial vehicles. IEEE Access 6:43203–43212
37. Li C, Xu Y, Xia J, Zhao J (2018) Protecting secure communication under UAV smart attack with imperfect channel estimation. IEEE Access 6:76395–76401
38. Javaid AY, Sun W, Devabhaktuni VK, Alam M (2012) Cyber security threat analysis and modeling of an unmanned aerial vehicle system. In: 2012 IEEE conference on technologies for homeland security (HST). IEEE, pp 585–590
39. Davidson D, Wu H, Jellinek R, Singh V, Ristenpart T (2016) Controlling UAVs with sensor input spoofing attacks. In: 10th USENIX workshop on offensive technologies (WOOT 16)
40. Trujano F, Chan B, Beams G, Rivera R (2016) Security analysis of DJI phantom 3 standard. Massachusetts Inst Technol
41. Dorri A, Kamel SR, Kheirkhah E (2015) Security challenges in mobile ad hoc networks: a survey. arXiv Prepr arXiv150303233
42. Hartmann K, Giles K (2016) UAV exploitation: a new domain for cyber power. In: 2016 8th international conference on cyber conflict (CyCon). IEEE, pp 205–221
43. Benkraouda H, Barka E, Shuaib K (2018) Cyber-attacks on the data communication of drones monitoring critical infrastructure
44. Hooper M, Tian Y, Zhou R, Cao B, Lauf AP, Watkins L et al (2016) Securing commercial wifi-based UAVs from common security attacks. In: MILCOM 2016–2016 IEEE military communications conference. IEEE, pp 1213–1218
45. Northcutt S (2007) Are satellites vulnerable to hackers? SANS Technol Institute 15
46. Arthur C (2009) SkyGrabber: the $26 software used by insurgents to hack into US drones. Guard 17
47. Rivera E, Baykov R, Gu G (2014) A study on unmanned vehicles and cyber security. Texas, USA
48. Lee YS, Kang YJ, Lee SG, Lee H, Ryu Y (2012) An overview of unmanned aerial vehicle: cyber security perspective. Korea. 12:13
49. Hartmann K, Steup C (2013) The vulnerability of UAVs to cyber attacks-An approach to the risk assessment. In: 2013 5th international conference on cyber conflict (CYCON 2013). IEEE, pp 1–23
50. Javaid AY, Sun W, Alam M (2013) UAVSim: a simulation testbed for unmanned aerial vehicle network cyber security analysis. In: 2013 IEEE globecom workshops (GC Wkshps). IEEE, pp 1432–1436
51. Huang L, Yang Q (2015) Low-cost GPS simulator GPS spoofing by SDR. DEFCON'15
52. Krishna CGL, Murphy RR (2017) A review on cybersecurity vulnerabilities for unmanned aerial vehicles. In: 2017 IEEE international symposium on safety, security and rescue robotics (SSRR). IEEE, pp 194–199
53. Kerns AJ, Shepard DP, Bhatti JA, Humphreys TE (2014) Unmanned aircraft capture and control via GPS spoofing. J F Robot 31(4):617–636
54. He L, Li W, Guo C, Niu R (2014) Civilian unmanned aerial vehicle vulnerability to GPS spoofing attacks. In: 2014 Seventh international symposium on computational intelligence and design. IEEE pp 212–215

A Systematic Survey on Load Balancing in the Cloud

Gutta Sridevi and Midhunchakkaravarthy

Abstract Web growth has been driven by a number of techniques. In massive scale processing, cloud computing is the rising technology. It is a method that consists of varied software and Web-enabled services. As cloud computing grows speedily and purchasers are stringent a lot of services, the traffic on the cloud increases tremendously. Thus, managing load balance is very interesting and important research area nowadays. A decent load balancing algorithm can enhance the performance and resource utilization by dynamically distributing workload among a range of nodes within the system. This paper addresses a comprehensive summary on load balancing algorithms, challenges, and various presently offered load balancing software and its features.

Keywords Cloud computing · Load balancing · Load balancer

1 Introduction

Cloud computing is the emerging and rapidly increasing technology in the current globe. These days, it is of the established and widely used innovation in the domain of IT. It is able to use a variety of Web computer resources through the Web as well as storage devices and apps. The cloud supplier supports the shared pool of resources. The next-generation computing model for their main advantages in self-service on demand, omnipresent network access, risk transfer, and location-independent pooling of resources has been designed for cloud computing [1]. Cloud computing is described as a model that offers useful and interest-based access to a common pool

The original version of this chapter was revised: The author's name has been corrected from "Midhunchakkravarthy" to "Midhunchakkaravarthy". The correction to this chapter is available at https://doi.org/10.1007/978-981-15-3284-9_86

G. Sridevi (✉) · Midhunchakkaravarthy
Department of Computer Science and Multimedia, Lincoln University College, Kota Bharu, Malaysia

Midhunchakkaravarthy
e-mail: sridevi.gutta2012@gmail.com

S.-L. Peng et al. (eds.), *Intelligent Computing and Innovation on Data Science*, Lecture Notes in Networks and Systems 118,
https://doi.org/10.1007/978-981-15-3284-9_82

of resources as well as storage, systems, and services as stated by the National Institute of Standard and Technology [2]. These facilities need stripped-down effort. The main feature of the cloud computing is its versatility, which involves its tendency to develop and decrease computing when necessary [3]. The other quality is adaptable and shows that the enhanced transmission speed and CPU storage, etc., can be adjusted. The high-reliable and secure data storage center is provided to the customers by cloud computing. No data loss, virus attacks, and any other issues need to be concerned by users. When consumers pay to cloud suppliers for what they use, cloud computing avoids costs for hardware, software, and services [4]. The limitations of the cloud are varying the performance, technical issues, security threats in cloud, downtime, Internet connectivity, lower bandwidth, etc.

2 Cloud Infrastructure

2.1 Cloud Deployment

Depends on the user requirements, clouds can be broadly categorized into public, private, hybrid, and community clouds [5].

- Public cloud—Amazon AWS or Microsoft Azure, which are open to the public. Many companies such as Dropbox, Quora, and other startups use Amazon's AWS to provide their services.
- Private cloud—which is solely used by the different teams within a single such as Google and Facebook have their own data centers which are used only by the different teams within the company.
- Hybrid cloud—which uses both the public cloud and a private cloud. It can be used to maintain sensitive data in the private cloud itself and to use the public cloud for less sensitive data.
- Community cloud—given by either within the organization by the external merchant. A cloud is alluded to as community cloud when many businesses share their strategies and the necessities in the cloud. Through many organizations, a community cloud is created and remote access is made to the cloud.

2.2 Architecture for Cloud

The front end and back end are the two components of cloud computing. The front end of the cloud is the client component. It includes interfaces and apps necessary for access to the cloud computing platform. While cloud-related, the background contains the resources needed for the cloud services and controlled by the cloud providers, such as servers, virtual machines, security mechanisms, and data store.

Figure 1 illustrates the deployment models and the services offered by cloud, for example, PaaS, IaaS, and SaaS deploy the cloud as a total framework. Cloud specialist organizations can furnish customers with various sorts of services. They are listed below [5]:

1. SaaS: Software as a service: where the organization providing some software runs it on the cloud and gives users access to that software rather than requiring every person to install that software locally on their system. Facebook, Google, Dropbox, etc., are all SaaS, because they just interact with the user using a Web browser or some other endpoint, while their actual software runs in the cloud.
2. PaaS: Platform as a service: gives a platform to the application engineers, which generally incorporate databases, Web servers, operating systems, and so forth and which makes the activity of the designer a lot simpler as they would prefer not to stress over the hardware and software program stack. The resources gave

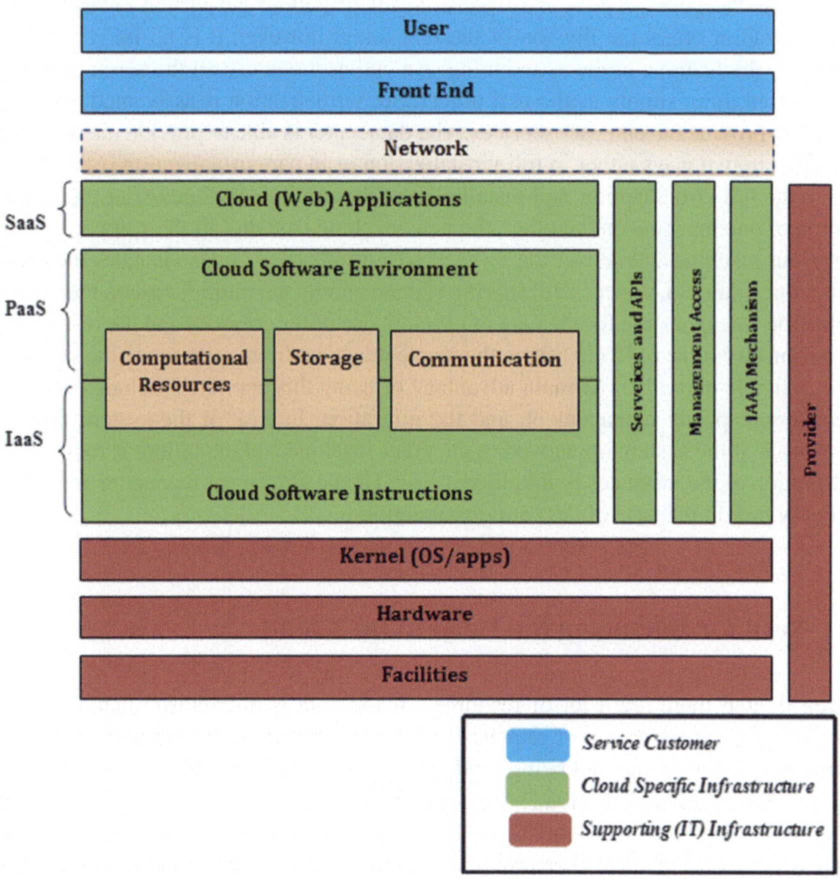

Fig. 1 Cloud computing environment

usually depending on the utilization of the application. Case of such service is Google App Engine.

3. IaaS: Infrastructure as a service: which gives the client direct (remote) access to either a physical machine or a Virtual Machine, storage space, network, etc. The actual details of the infrastructure, such as scaling, security, backup, and location, are abstracted from the client. Amazon AWS's Elastic Cloud Compute is an example of IaaS.

3 Cloud Virtualization

Virtualization in cloud computing is the method of sharing a physical instance of a resource among multiple customers and organizations. Virtual machine (VM) is known as the guest. A host is responsible for providing the virtual environment. Virtualization offers the illusion of the real thing; however, it is no longer real. It gives all the features of the actual thing. An end user can use all the services of the virtualized thing simply as the real object. So, virtualization is associated with the cloud to provide the end user services. The datacenter is able to provide the end user services in two ways either in full virtualization or in paravirtualization [6].

In the full virtualization, full installation of the machine is done. After the installation of one machine on the other, the new machine contains all the software of the previous machine and deliver the services to multiple users. It also isolates the users from one other. So, a successful sharing is done among the multiple users. Paravirtualization is responsible for the care of efficient use of the memory and the resources. It permits a couple of OS to run on the single machine. In this approach, the services are available partially. The main advantage of using this approach includes disaster recovery, capacity management, and the migration. In case of the system failure, instances of the system are moved to the other machine and the failure is recovered. Migration of the instances is simple and easy. Hence, the power management is also an easy task in the case of the paravirtualization.

4 Need for Balancing the Load in the Cloud

Even though there are a lot of resources, if the data is distributed such that it is possible for a large range of requests to hit a small variety of resources, then the remaining resources are not being used at that point. Thus, the efficiency of the pool of resources is affected. By load balancing efficiently, the system will try to distribute the load among as many resources, and as evenly as possible. This results in a massive enhancement in the overall performance of the system. Even if the data is very evenly distributed, a good load balancer should direct traffic so that the load is as even as possible. Even though to manage workload demands by allocating resources among

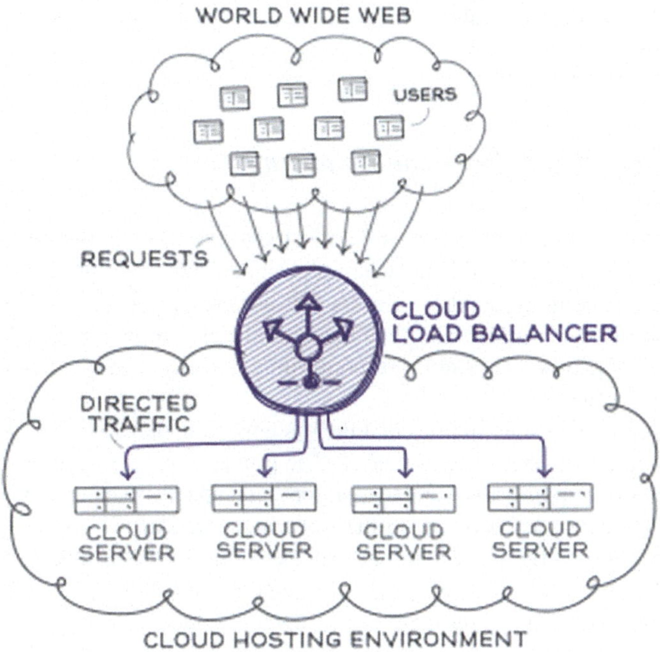

Fig. 2 Load balancing environment in the cloud

multiple systems or servers, a good load balancer is needed to redistribute the data, if possible so that the load is again evened out. Figure 2 illustrates load balancing environment in the cloud.

The primary objective of the cloud load balancing is to achieve full system strength, establish fault tolerance systems, minimize job completion time and waiting time, increase customer loyalty, increase efficiency and performance, and finally improve the resource utilization ratio.

4.1 Categorization of Load Balancing

Load balancing can be done at distinct layers of the cloud stack. At the infrastructure level, load balancing can be done to provision resources, such as scheduling VMs on physical machines, routing traffic on the network based on loads on each route, managing the storage so that a few disks are not overloaded, etc. At the platform level, load balancing can be done in the services that the platform provides. For example, the database or the Web servers provided by the platform can be load balanced. At the software level, balancing the load can be done specific to an application. For example, if the application knows that it will get queries for a particular set of things at different points of time during the day, it can distribute data accordingly at each point of time.

4.2 Factors that Affect Load Balancing

Response time—The architecture is the most significant angle that influences the general efficiency of the load balancing algorithms. Formally, the response time increases with the more number of users that are present in the network. But the hierarchical method works better than the centralized and decentralized strategies and requires less time [7]. Similar response time is showed in the centralized and decentralized strategies.

Server load—Server load implies that the system can manage several requests per second and it usually speaks to as req/sec. The primary goal is to split the workload between the three architectures and to monitor their capacity to handle the workload. The experiment was carried out on the three architectures [7] and the outcomes affirmed that the hierarchical load balancer performs far better due to its design. The load balancing concept in the cloud computing enables users to ensure that the workload is distributed through nodes, thereby reducing the energy consumed. It helps in avoiding the overheating of the cluster. So, the energy-efficient cloud contains the four main elements including the consumer, resource allocator, virtual machine, and the physical machine [8]. These four elements are shown in Fig. 3 and are described below.

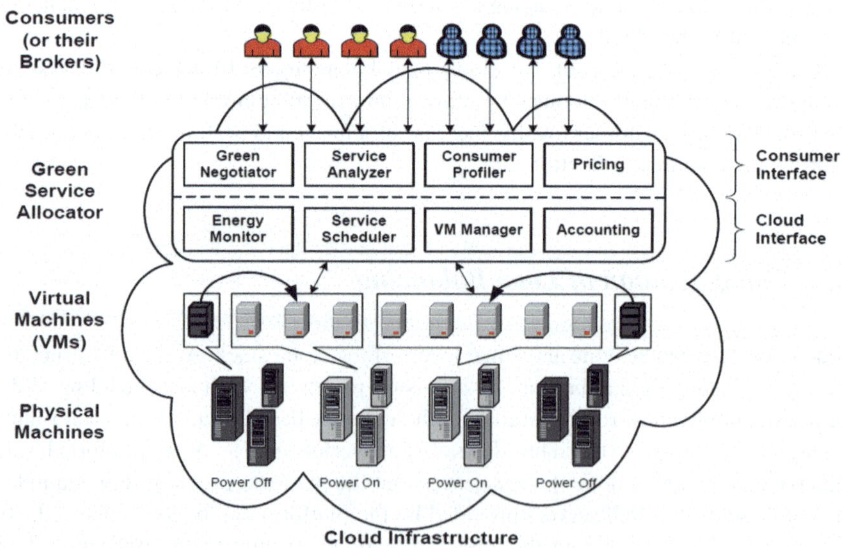

Fig. 3 Cloud computing infrastructure

I. **Consumer**

Consumer may request a service from anywhere globally. For example, in a company, a Web application that runs through the Internet can act as a consumer. This application distributes the workload as per the requirement. Requirement is based on the amount of users who uses the application.

II. **Resource Allocator**

It acts as a consumer–cloud infrastructure interface. It has further many components including green negotiator, Service Analyzer, energy monitor, energy scheduler, etc. Green negotiator makes the cloud provider and customer service-level agreement to preserve quality of service. Service Analyzer checks the requirements specification and decides whether or not to accept the application. To achieve the load balancing strategy, it needs efficient information from the virtual machine and the energy monitor. Virtual machine manger manages the workload across the virtual machines. Service Scheduler helps to schedule the requests across the virtual machines for effective utilization of the resources.

III. **Virtual Machines**

On request, virtual machines (VM) can be launched or stopped. Multiple VM can run at the same time, providing flexibility to update the data resources on a single physical machine. By moving the virtual machine across the physical machine, the nodes can be assembled to save the energy resources.

IV. **Physical Machines**

Physical machine acts as a hardware resource to meet the requirement specification.

5 Algorithms for Load Balancing

Efficiency is increased as the fundamental aim of the load balancing algorithm. The algorithm should ensure scalability so that changes to the system state are made promptly.

A. **Static Load Balancing Algorithms**

For those systems with extremely small load differences, static algorithms are developed. Here, the whole traffic is equally split between the servers. In order to improve the processing efficiency of processors, this algorithm needs prior knowledge of server resources. This strategy is appropriate for Web servers with the same distribution of https requests for the Web traffic across the network. These algorithms are therefore not appropriate for widely differing cloud resources [9]. Examples of such algorithms include round-robin, Min-min, Max-min, etc.

B. **Dynamic Load Balancing Algorithms**

This algorithm first searches for the lightest server on the entire network and assigns load preference to that server. It relies on the execution-time data from the chosen nodes. Based on the computation, this strategy assigns a workload and can be reassigned to the nodes. These algorithms succeed in the balance of the load between heterogeneous resources in the clouds [9]. This method is more precise compared to the static approach. Ant colony optimization and artificial bee colony and throttled load balancing algorithms are the most common in the field of swarm intelligence [10].

6 Some Load Balancing Softwares

In the current industry, there are different kinds of cloud computing service platforms available. They are Amazon Web Services, Kamatera, DigitalOcean, etc. Here in this section, we present a list of load balancing softwares/load balancers that are used by the industrial people based on their requirements. These are illustrated in Table 1.

7 Conclusion

A variety of services were provided by cloud computing across a network to end users. Load balancing is the significant key problem here. Cloud computing has provided end users with a wide variety of services across a network. The main problem here is load balancing, because the overloaded system does not provide efficient services. This paper looked into a methodological investigation on cloud computing architecture, different kinds of load balancing algorithms, and load balancers those are used in the present industry. The major objective of this article is to reinforce the current methodologies in reference to the cloud and discussed leading cloud suppliers for future use. The ultimate goal is to maximizing cloud service providers profit and minimizing cloud user costs. That is the theme of cloud computing.

Table 1 List of presently used load balancing software in the companies

Software/s	Features of load balancer products									
	Authentication	Automatic Configuration	Content Caching	Content routing	Data compression	Health monitoring	Predefined protocols	Redundancy checking	Reverse proxy	SSL offload	Schedulers
jetNEXUS load balancer	Y	Y	Y	Y	Y	Y	Y	Y	Y	Y	N
Avi vantage	Y	Y	Y	Y	Y	Y	Y	Y	Y	Y	Y
Cloud load balancer	Y	Y	Y	Y	Y	Y	Y	Y	Y	Y	N
Loadbalancer.org	Y	Y	Y	Y	Y	Y	Y	Y	Y	Y	N
BalanceNG	N	N	N	Y	N	Y	N	Y	Y	Y	N
ManageEngine OpManager	N	Y	N	Y	N	Y	Y	Y	N	Y	N
LoadMaster	Y	Y	Y	Y	Y	Y	Y	Y	Y	Y	Y
HAProxy	N	N	N	Y	Y	Y	Y	N	Y	Y	N
Snapt balancer	Y	Y	Y	Y	Y	Y	Y	Y	Y	Y	N
Zevenet	N	N	N	Y	N	Y	Y	Y	Y	Y	Y
SimpleNETWORKS	N	Y	N	N	N	N	N	N	N	N	N
LiteSpeed	N	Y	Y	Y	Y	Y	N	N	N	Y	N
Array's ADC	Y	Y	Y	Y	Y	Y	Y	Y	Y	Y	Y
Incapsula	N	N	Y	Y	Y	Y	N	Y	Y	N	N
WebLoad	N	N	N	N	N	N	N	N	N	N	Y
AWS ELB log analyzer	N	N	N	N	N	N	N	N	N	N	N
Noction IRP	Y	Y	N	N	N	Y	Y	Y	N	N	N
NGINX plus	Y	Y	Y	Y	Y	Y	Y	Y	Y	Y	Y
Varnish plus	N	Y	Y	N	N	N	N	N	Y	Y	N
Alteon VA	N	Y	N	N	N	N	N	N	N	N	N

(continued)

Table 1 (continued)

Software/s	Features of load balancer products										
	Authentication	Automatic Configuration	Content Caching	Content routing	Data compression	Health monitoring	Predefined protocols	Redundancy checking	Reverse proxy	SSL offload	Schedulers
NeoLoad	Y	Y	N	N	N	N	N	N	N	N	N
iNetFusion	N	N	N	N	N	N	N	N	N	N	N
ScaleArc	N	N	Y	Y	N	N	N	N	N	N	N
Traffic director	N	Y	N	Y	N	Y	Y	N	N	N	N
TSplus	Y	N	N	N	N	N	N	N	N	N	N
Openmix	N	N	N	Y	N	N	N	N	N	N	N
vThunder	Y	N	N	N	N	Y	N	N	N	Y	N
Pound	Y	Y	Y	N	N	N	Y	Y	Y	N	N
FrugalTest	N	N	N	N	N	N	N	N	N	N	N
Eddie	N	Y	N	N	N	N	N	N	N	N	N
NetScaler	N	N	N	N	N	N	N	N	N	N	N
Google cloud load balancer	N	N	N	N	N	N	N	N	N	N	N

References

1. Hayes B (2008) Cloud computing. Commun ACM 51(7):9–11
2. Mell P, Grance T (2011) The NIST definition of cloud computing. National Institute of Standards and Technology
3. Qaisar EJ (2012) Introduction to cloud computing for developers: key concepts, the players and their offerings. In: Proceedings of IEEE information technology professional conference (TCF Pro IT), pp 1–6
4. Mollah MB (2012) Next generation of computing through cloud computing technology. In: 25th IEEE Canadian conference on electrical and computer engineering (CCECE). 978-1-4673-6/12
5. Goel P (2017) Thesis on load balancing on the cloud. International Institute of Information Technology
6. Avram MG (2013) Advantages and challenges of adopting cloud computing from an enterprise perspective. In: Elsevier proceedings of 7th international conference interdisciplinary in engineering (INTER-ENG), pp 529–534
7. Rayis EA, Kurdi H (2013) Performance analysis of load balancing architectures in cloud computing. In: IEEE proceedings of European modeling symposium (EMS), pp 520–524
8. Buyya R, Beloglazov A, Abawajy J (2010) Energy-efficient management of data center resources for cloud computing: a vision, architectural elements, and open challenges, pp 1–8
9. Gamal M, Rizk R, Mahdi H, Elhady B (2019) Bio-inspired based task scheduling in cloud computing. In: Hassanien A (ed) Machine learning paradigms: theory and application. studies in computational intelligence, vol 801. Springer, Cham
10. Kaur S, Sharma T (2018) Efficient load balancing using improved central load balancing technique. In: 2018 2nd International conference on inventive systems and control (ICISC), Coimbatore, pp 1–5

A Data Tracking and Monitoring Mechanism

Reyner Aranta Lika, Daksha A/P. V. Ramasamy, Danushyaa A/P. Murugiah, and Sarfraz Nawaz Brohi

Abstract There is a rising trend among companies to collect data including personally identifiable information (PII) from service users. An individual's privacy can be violated when his/her PII is accumulated, utilized or divulged. Users of the technology are unable to comprehend what information about them is being utilized, how, when and by whom. With this situation in mind, data transparency is an absolute necessity to combat this issue. Extensive research is being carried out to understand and address the problem. In this research, we propose a security mechanism known as D2TRacker, which can perform data monitoring and data tracking on the processes that are taken on the user's data and ultimately provides data transparency to these service users. Through the development and execution, the results obtained showed that data transparency could be provided to the system users. Overall, this solution serves as a stepping stone in solving the issues encountered. This solution works as intended in a simulated environment, but deployment in real world would need further enhancements.

Keywords Data transparency · Data tracking · PII · Privacy

R. A. Lika (✉) · D. A/P. V. Ramasamy · D. A/P. Murugiah · S. N. Brohi
School of Computing & IT, Taylor's University, Subang Jaya, Malaysia
e-mail: ReynerArantaLika@sd.taylors.edu.my

D. A/P. V. Ramasamy
e-mail: DakshaRamasamy@sd.taylors.edu.my

D. A/P. Murugiah
e-mail: DanushyaaMurugiah@sd.taylors.edu.my

S. N. Brohi
e-mail: SarfrazNawaz.Brohi@taylors.edu.my

S.-L. Peng et al. (eds.), *Intelligent Computing and Innovation on Data Science*,
Lecture Notes in Networks and Systems 118,
https://doi.org/10.1007/978-981-15-3284-9_83

1 Introduction

Technology is changing our concept and understanding of privacy and protection. Data privacy and data protection are firmly interconnected. Data protection is tied in with anchoring data against unapproved access while data privacy is about approved access. This dissimilitude matters since they are woven profoundly into the larger issues of privacy and cybersecurity, the two of which pose a potential threat in businesses, legislative issues and culture [1].

Data protection law has reached a defining moment. The present debate about the subject is lively, and surveying data uncovers that everybody is to a great degree worried about privacy, both on and off the Internet. Personally identifiable information (PII) is a standout amongst the most focal ideas in information privacy regulation. It characterizes the degree and limits of a substantial scope of privacy statutes and regulations [2]. It strikes numerous sound judgments that a person's privacy can be harmed when PII is gathered, utilized or unveiled. The rich tableau of accessible data presents critical concerns for information privacy. The more data about a person that is known, the more probable it turns into information that can be utilized to uniquely recognize or to determine further details about that individual [3].

Data transparency is frequently associated with governments and the information they discharge. The reasoning is that by making information accessible to all for any function, authorities turn out to be more explicable, and citizens are empowered [4]. The world of data is evolving rapidly. However, customer conduct, associated devices and gadgets, data breaches and new plans of action are likewise affecting existing operations. Notwithstanding the powers of progress, organizations still expect access to individual information to work viably. However, customers battle to comprehend what information is being utilized, how, when and by whom [5–7]. While end-users are regarded as the owner of data, they are either oblivious to data collection activities of apps installed or they do not know which data are being collected and for which purpose [8]. Over 50 million records were collected by Cambridge Analytica from Facebook without their explicit permission [9]. A study revealed that if a person decides to read online privacy policies each time that person visits a new website, the time required to read is approximately 201 h a year, which is exceedingly high [10]. Thus, most people tend to ignore privacy policies. The value of privacy is genuinely appreciated mostly after the privacy itself is lost. As the number of companies holding the data owners' information goes up, the chances of potential data breach will also increase. When a data breach occurs, the data subjects' PII might get leaked.

Thus, data is the asset that makes free or close-to-free administrations conceivable. The main reason Skype, Facebook or any online service is free is because of information. Most consumers comprehend this theoretically, however, not in every case particularly. Besides, consumers might be willing to acknowledge the trade-off between an esteem-included administration and revelation of data about them, in a great compensation course of action. The challenge is that this data is covered in the 60-page security policy of which most consumers will never read, and even if they did, they would not even comprehend it. There are several solutions such as

[11–13] reported in the existing literature which track and monitor data. However, there are still myriad security and privacy challenges that currently have not been fully addressed when dealing with the data transparency.

The rest of the paper is organized as follows: Sect. 2 elaborates our proposed solution, Sect. 3 showcases the benchmarking of our solution against other existing solutions, and Sect. 4 ends this paper with a conclusion and future work.

2 Proposed Solution

The proposed solution is a standardized, agent-reliant architecture where the agent serves as a data tracking and monitoring tool. With this agent, it will be possible for users to be updated with the current state of their data. For this tool to function, it needs to be installed on the clients' workstation as well as on the servers of the applications that they are connecting with and transferring data. The agent that resides on the company's server will continuously extract data from its database. This data will then be stored on our database. Following this, if the company decides to modify or delete the data, the change would be detected when the current data extracted is compared with the initial data that is extracted and stored in our database.

As for data transmission, an assumption was made that all company servers will have our agent installed. Privacy policies have often dictated that companies may share data with other parties as needed, but the identity of these third-party companies is almost always unknown. By having the agents installed in the third party's side, the agent will be allowed to check the IP addresses of all connected servers. If a third party connects to the server, this connection will be visible to us. Following this, if the third party contains user details when extracted from their database, we can conclude that data has been transferred to the third party. The architecture diagram of our system is shown in Fig. 1.

With all the data monitoring and data tracking being accomplished, data transparency is a criterion that is to be satisfied. This was done through the implementation of a data transparency report. For easier viewing, the user can click the "Deletion", "Modification" and "Transmission" button in order to view the reports which list the results for the process of their choice. The transparency report user interface can be seen in Fig. 2. Visualization using dashboard was implemented to highlight the results of the data transparency report. The dashboard shown in Fig. 3 details the most dangerous company of which the user is subscribed to. Dangerous in this context means the companies which perform the greatest number of data modification, deletion and transmission. Furthermore, the dashboard also showcases data modification and data transfer information for the users to see the company activity on a daily basis. Finally, a summary of the total number of data modification, deletion and transmission that has happened is also provided.

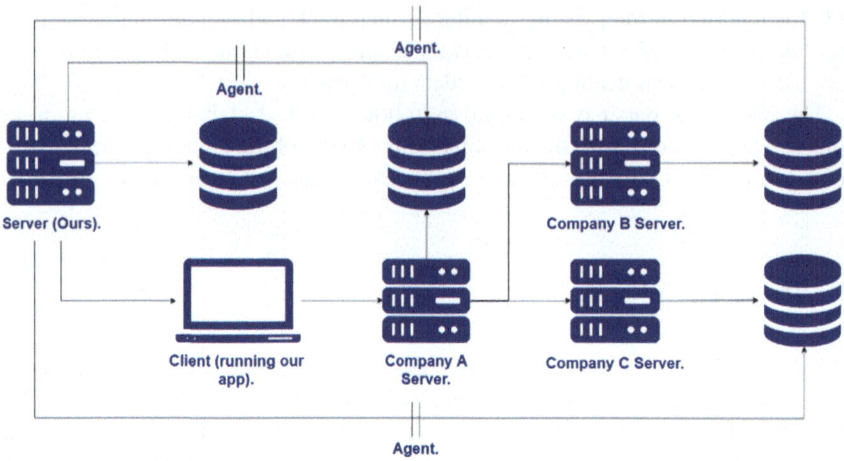

Fig. 1 Architecture diagram of D2TRacker

Fig. 2 Transparency report screen

3 Results and Discussion

The main goal of D2TRacker is to ensure that it provides the users with the ability to know where their data is being transferred and what is happening to their data which is stored with the companies. Based on Tables 1 and 2, data transparency seems to be a very stressed and recurrent goal among most of the other systems.

Other systems mostly had a general theme where companies had visibility over the data that they had collected [2, 5, 11, 12, 15, 17]. In our solution, data tracking

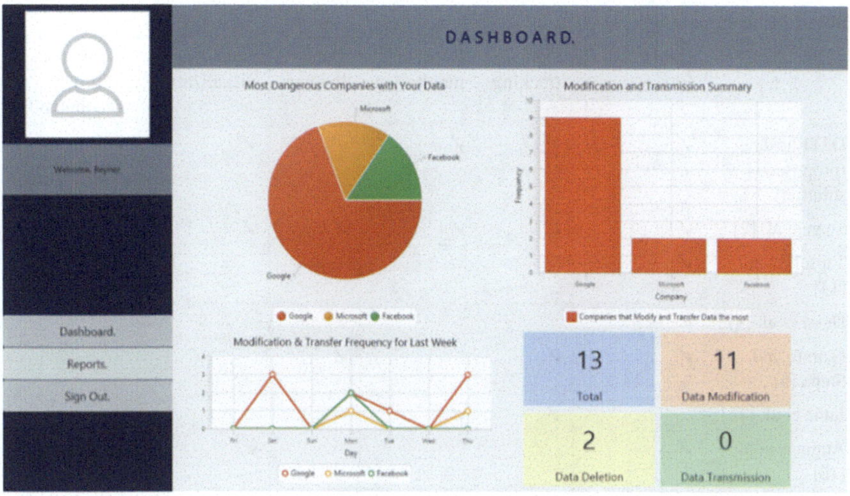

Fig. 3 Dashboard screen

and monitoring work as the main goal to provide transparency to the data owners. By collecting and logging such information, a transparency report can be presented to the end-users of our system.

Data visualization is a characteristic that is present in several solutions that were examined [2, 12, 13, 16, 18, 22, 24]. Visualization in this context means to provide the end-users of the solution the ability to view certain features. In our system, data visualization is showcased through the dashboard. Companies improving accountability is a characteristic that is unique in terms of the theme of the project. The solution involving this characteristic was intended for the companies to improve accountability for their data practices which in turn will improve the trust of the clients [7]. This characteristic is present in our solution because it is taken as an assumption that companies are voluntarily allowing for D2TRacker agents to be deployed at their side.

The user-centric approach in this context means solutions that cater to the end-users of the service for the increased benefit of the data owners. Based on the papers explaining the solution, a user-centric approach is usually met by allowing the data owners visibility or control over their data [4, 7, 13]. In D2TRacker, a user-centric approach is implemented through the usage of a data transparency report.

There are also several points that our solution currently has not satisfied. For example, several solutions [6, 7, 13, 14, 22] contain functions that allow users to take control of their data. Additionally, several studies contain policy verification or policy enforcement which can be used to provide better capacities in securing private information [5, 6, 20, 21]. Another solution utilized machine learning to improve the fairness and transparency of data [16]. D2TRacker currently does not have such functions. There are also several limitations in our solution, such as the

Table 1 Benchmarking Table-I

	Data transparency	Data tracking	Data monitoring	Data logging	Data visualization	Companies improve accountability
D2TRacker (proposed solution)	✓	✓	✓	✓	✓	✓
Suen et al. [2]	✓	✓	✓	✓	✓	
Tapsell et al. [14]	✓					
Dessì et al. [4]	✓					
Fromm and Stepa [5]	✓	✓		✓		
Janic et al. [15]		✓	✓	✓		
Antunes et al. [16]	✓				✓	
Kodeswaran and Viegas [6]						
Trabelsi and Sendor [13]	✓				✓	
Ko et al. [17]	✓	✓	✓	✓		
Gomi [7]		✓		✓		✓
Ko and Will [11]	✓	✓		✓		
Asuncion [18]	✓	✓		✓	✓	
Xie et al. [19]		✓	✓			
Seneviratne and Kagal [20]	✓			✓		
Yu et al. [21]		✓		✓		
Fischer-Hübner et al. [12]	✓	✓			✓	
Tovernić et al. [22]				✓	✓	
Blauw and Solms [23]	✓					
Bier et al. [24]	✓	✓			✓	
De Oliveira et al. [25]	✓	✓		✓		

assumptions taken to justify the functionalities of the solution that prevent real-life implementation. There is also the assumption where companies consent to the deployment of our agent on their infrastructure, which would require complete trust from the companies on the service provided by us.

Table 2 Benchmarking Table-II

	User-centric approach	Policy verification	Policy enforcement	User data control	Assurance of data security	Machine learning utilization
D2TRacker	✓					
Suen et al. [2]						
Tapsell et al. [14]				✓		
Dessì et al. [4]	✓					
Fromm and Stepa [5]		✓				
Janic et al. [15]						
Antunes et al. [16]						✓
Kodeswaran and Viegas [6]			✓	✓		
Trabelsi and Sendor [13]	✓			✓		
Ko et al. [17]						
Gomi [7]	✓			✓		
Ko and Will [11]					✓	
Asuncion [18]						
Xie et al. [19]						
Seneviratne and Kagal [20]		✓	✓			
Yu et al. [21]			✓			
Fischer-Hübner et al. [12]						
Tovernić et al. [22]				✓		
Blauw and Solms [23]						
Bier et al. [24]						
De Oliveira et al. [25]						

4 Conclusion and Future Work

D2TRacker is an agent-reliant architecture where the agent works as a data tracking and monitoring tool, which allows the users to understand the current state of their data. We believe that D2TRacker managed to aid in solving a problem that is highly relevant in our current state. Our system does make several assumptions, but we believe it is a stepping stone towards achieving data transparency. Moreover, there are still some improvements that can be made to our system, such as implementation in a real-life scenario, tracking the movement of images, expanding our scope and more. While there are researches that have been conducted in this subject realm, the specific area focused by them differed from the scope that was envisioned by us. Our system champions for data transparency towards the users whose data have been collected by companies, which is still considered a niche topic in the domain.

References

1. Zhang Q et al (2007) A study on context-aware privacy protection for personal information, pp 1351–1358
2. Suen CH et al (2013) S2Logger: end-to-end data tracking mechanism for cloud data provenance. In: Proceedings—12th IEEE international conference on trust, security and privacy in computing and communications, TrustCom 2013, pp 594–602. https://doi.org/10.1109/trustcom.2013.73
3. Singer S, Sapte DW (no date) Er data privacy : fiction or reality? how much privacy are individuals entitled to under the law, pp 153–170
4. Dessi N et al (2016) Increasing open government data transparency with spatial dimension. Proceedings—25th IEEE international conference on enabling technologies: infrastructure for collaborative enterprises, WETICE 2016, pp 247–249. https://doi.org/10.1109/wetice.2016.61
5. Fromm A, Stepa V (2017) HDFT++ hybrid data flow tracking for SaaS cloud services. In: 2017 IEEE 4th international conference on cyber security and cloud computing (CSCloud), pp 333–338. https://doi.org/10.1109/cscloud.2017.9
6. Kodeswaran P, Viegas E (2010) A policy based infrastructure for social data access with privacy guarantees. In: Proceedings—2010 IEEE international symposium on policies for distributed systems and networks, policy 2010, pp 14–17. https://doi.org/10.1109/POLICY.2010.25
7. Gomi H (2010) A persistent data tracking mechanism for user-centric identity governance. Identity Inf Soc 3(3):639–656. https://doi.org/10.1007/s12394-010-0069-4
8. Gustarini M, Wac K, Dey AK (2015) Anonymous smartphone data collection : factors influencing the users' acceptance in mobile crowd sensing Anonymous smartphone data collection : factors influencing the users' acceptance in mobile crowd sensing. In: Personal and ubiquitous computing. Springer, London. https://doi.org/10.1007/s00779-015-0898-0
9. Revealed: 50 million Facebook profiles harvested for Cambridge Analytica in major data breach (2018). https://www.theguardian.com/news/2018/mar/17/cambridge-analytica-facebook-influence-us-election
10. Mcdonald AM, Cranor LF (2001) The cost of reading privacy policies, p 0389
11. Ko RKL, Will MA (2014) Progger: an efficient, tamper-evident kernel-space logger for cloud data provenance tracking. In: IEEE international conference on cloud computing, CLOUD, pp 881–889. https://doi.org/10.1109/cloud.2014.121
12. Fischer-Hübner S et al (2016) Transparency, privacy and trust—technology for tracking and controlling my data disclosures: does this work? IFIP Adv Inf Commun Technol 473:ix. https://doi.org/10.1007/978-3-319-41354-9
13. Trabelsi S, Sendor J (2012) Sticky policies for data control in the cloud. In: 2012 10th Annual international conference on privacy, security and trust, PST 2012, pp 75–80. https://doi.org/10.1109/pst.2012.6297922
14. Tapsell J, Akram RN, Markantonakis K (2018) Consumer centric data control, tracking and transparency—a position paper. In: Proceedings—17th IEEE international conference on trust, security and privacy in computing and communications and 12th IEEE international conference

on big data science and engineering, Trustcom/BigDataSE 2018. IEEE, pp 1380–1385. https://doi.org/10.1109/trustcom/bigdatase.2018.00191

15. Janic M, Wijbenga JP, Veugen T (2013) Transparency enhancing tools (TETs): an overview. In: workshop on socio-technical aspects in security and trust, STAST, pp 18–25. https://doi.org/10.1109/stast.2013.11
16. Antunes N et al (2018) Fairness and transparency of machine learning for trustworthy cloud services. In: Proceedings—48th annual IEEE/IFIP international conference on dependable systems and networks workshops, DSN-W 2018. IEEE, pp 188–193. https://doi.org/10.1109/dsn-w.2018.00063
17. Ko RKL, Jagadpramana P Lee BS (2011) Flogger: a file-centric logger for monitoring file access and transfers within cloud computing environments. In: Proceedings of 10th IEEE Int. conference on trust, security and privacy in computing and communications, TrustCom 2011, 8th IEEE international conference on embedded software and systems, ICESS 2011, 6th International conference on FCST 2011, pp 765–771. https://doi.org/10.1109/trustcom.2011.100
18. Reddy K, Venter HS (2013) The architecture of a digital forensic readiness management system. Comput Secur 32:73–89. https://doi.org/10.1016/j.cose.2012.09.008 Elsevier Ltd.
19. Xie Y et al (2018) Efficient monitoring and forensic analysis via accurate network-attached provenance collection with minimal storage overhead. Digital Invest 26:19–28. https://doi.org/10.1016/j.diin.2018.05.001 Elsevier Ltd
20. Seneviratne O, Kagal L (2007) Enabling privacy through transparency. Science 317(5842):1188. https://doi.org/10.1126/science.1138728
21. Yu S, Vargas DV, Sakurai K (2018) 'Effectively protect your privacy: enabling flexible privacy control on web tracking. In: Proceedings—2017 5th international symposium on computing and networking, CANDAR 2017, 2018–January, pp 533–536. https://doi.org/10.1109/candar.2017.26
22. Tovernić S et al (2018) Solution for detecting sensitive data inside data lake, pp. 1284–1288. https://doi.org/10.23919/mipro.2018.8400232
23. Blauw FF, Von Solms SH (2018) Towards collecting and linking personal information for complete personal. Springer. https://doi.org/10.1007/978-3-319-91238-7
24. Bier C, Kühne K, Beyerer J (2016) PrivacyInsight: the next generation privacy dashboard. Springer. https://doi.org/10.1007/978-3-319-44760-5_9
25. De Oliveira AS et al (2013) Monitoring personal data transfers in the cloud, In: Proceedings of the international conference on cloud computing technology and science, CloudCom, vol 1, pp 347–354. https://doi.org/10.1109/CloudCom.2013.52

An Efficient Node Priority and Threshold-Based Partitioning Algorithm for Graph Processing

J. Chinna and K. Kavitha

Abstract Partitioning algorithms play vital role while processing large graph data. It is not possible to process such large graphs using a single machine; the graphs are usually partitioned into different clusters. These partitioned clusters can then be used as distributed memory clusters for processing by systems like Pregel. This research work proposed a node priority and threshold-based partition algorithm for graphs. The proposed method clusters the nodes in the graphs into four partitions based on their reachability either as incoming or outgoing. The initial partitions are generated by considering both inward as well as outward strategies. Final partitions are created by combining the merits of both strategies. Incompatible partitions are eliminated, and refined partitions are the results due to this analysis. The entire work is subdivided into four phases. The proposed method is implemented in Java platform. The partitions generated from two real-time datasets are analyzed, and the outcomes are presented in this paper. The datasets considered for the research work is large enough to address the issues of the existing works. The time taken for partitioning is the metric used to assess the proposed algorithm.

Keywords Graph processing · Partitioning · Node priority · Threshold · Clusters

1 Introduction

The tremendous growth in the technological innovations of the Internet has set a new trend in handling data. Graphs are created from many day-to-day applications such as online purchases, blogs, on-road networks and outcomes of scientific simulations [1]. This forms the floors for partitioning in graph processing. Some of the applications of graph processing include parallel processing, to handle complex networks like

J. Chinna (✉)
Research Scholar, Department of Computer Science, Mother Teresa Women's University, Kodaikanal, Tamilnadu, India

K. Kavitha
Assistant Professor, Department of Computer Science, Mother Teresa Women's University, Kodaikanal, Tamilnadu, India

S.-L. Peng et al. (eds.), *Intelligent Computing and Innovation on Data Science*, Lecture Notes in Networks and Systems 118,
https://doi.org/10.1007/978-981-15-3284-9_84

biological networks, geographically embedded networks, etc., road networks, image processing and physical design of VLSI. Systems like Pregel [2] and GraphChi [3] are using distributed memory clusters. Graphs can be partitioned by balancing the two apprehensions [4]. The two concerns are computational load balance and computational overhead. Most of the existing studies elucidated their finding with the help of small graph and failed to extend them for large graphs.

Graph frameworks are assessed in terms of power usage for processing. WattWatcher [5] is a power estimation tool for workloads. It overcomes the drawbacks of the cycle-accurate simulation method which consumes more runtime and is challenging while simulating the software stack. The outcomes of the graph partitions are sometimes analyzed using this kind of tools depending upon the needs. This research work is an initial step toward graph processing framework. The scope of this research work is confined to partitioning of graphs.

2 Review of Literature

The graph partitioning can be optimized either through hardware or software standpoint [6]. Many graph processing frameworks have been proposed earlier and contributed for better processing. Song et al. analyzed the power behavior of different applications using graphs and compared their performances in terms of scalability, performance and energy cost. This research work clearly described the impact of scalability in graph processing. VB-Partitioner [7] was proposed by Lee and Liu. This distributed data partitioning model is developed by considering the efficiency to process queries in clouds as its prime objective. Here, the correlation among the vertices influences the vertex block grouping. The first prototype was built over the Hadoop distributed file system. The GraphChi [3] may be used to improve the iteration speed.

The graphs are usually represented as adjacency matrices. While using adjacency matrices, the partitions are defined by considering two choices. The choice about partitioning the edges and norms for nominating a node as master is vital. The Cartesian vertex-cuts [8] create partition based on 1D block partitioning policy. The masters are created using the hosts of the nodes. The authors of [9] proposed multilevel label propagation (MLP) method for partitioning graphs. They considered both real-time and synthetic graphs for experimental study. The time taken for partitioning is mentioned as several hours in their work. The partitions created are set to be efficient, and its effectiveness is verified through synthetic graphs. Huang and Abadi [10] proposed a light-weight customizable partition framework for Leopard. It is a dynamic graph. The proposed framework considers replication for improving its efficiency.

The framework proposed is fault-tolerant and an accessible locality. In the research work [11], a structure-centric partitioning algorithm is proposed. It reduces the IO resource overhead and cache miss rate and improves scalability as well as effectiveness. They adopted the priority strategy to graph partitions for scheduling them in desired order. Chen et al. [12] proposed a novel framework for graph partitioning. This framework follows vertex-oriented graph partitioning. It is aimed for cloud environment. The experiments were conducted using graphs over 100 GB in every cluster and on the cloud environments such as Amazon and EC2.

The TopoX [13] is a graph partitioning technique based on topology refactorization. This technique is designed to address the prime issues of graph processing, namely reducing the communication cost and balancing the load. The performance of the TopoX is better than that of the existing parallel graph processing prototype PowerLyra by 78%. The minimal overhead due to refactorization and memory consumption is negligible when compared to the improved performance. The PowerLyra [14] proposed by Chen et al. suggested an environment with centralized computation for high-degree and low-degree vertices. The authors also recorded that the existing frameworks such as Pregel [2] results in load imbalance for high-degree vertices where frameworks like GraphX results in more communication cost and huge memory consumption. The PowerLyra makes use of a new hybrid-cut partition algorithm with heuristics. Since it is a combination of edge-cut and vertex-cut, it is noted for its generality and efficiency. GraphA [15] is an adaptive partitioning algorithm proposed by Li et al. This algorithm is implemented using Spark and GraphLab. By contrast to the existing parallel graph processing algorithms, the GraphA performs well in terms of ingress time, memory consumption and time taken for partitioning and computational cost.

3 Proposed Framework

The workflow of the proposed efficient threshold-based partitioning algorithm for graph processing is shown in Fig. 1. The proposed partitioning algorithm can be subdivided into four phases. The individual phases are explained in this section.

3.1 Calculation of Node Priority Calculation

The graph dataset considered for this research work consists of two fields, namely "from node" and "to node". If the node is having more number of inlinks or outlinks or both, the node will be considered are often reachable in the graph. Thus, in-degrees and out-degrees of the nodes play vital role in graph processing. The proposed algorithm calculates the node priority based on the number of inlinks and outlinks. Let us consider $G = (V, E)$ be a directed graph, where $V = \{v_1, v_2, \ldots, v_n\}$ be the set of vertices or nodes and $E = \{e_{ij}, \ldots, e_{mn}\}$ be the set of edges. The ordered pair (v_i, v_j) of vertices is represented as an edge e_{ij}. The in-degree and out-degree of any vertex v_i are denoted as $\deg^+(v_i)$ and $\deg^-(v_i)$, respectively.

The in-degree of a vertex/node $\deg^+$ is the number of edges entering onto a vertex/node $\deg^-$, and the out-degree is the number of edges originating from a

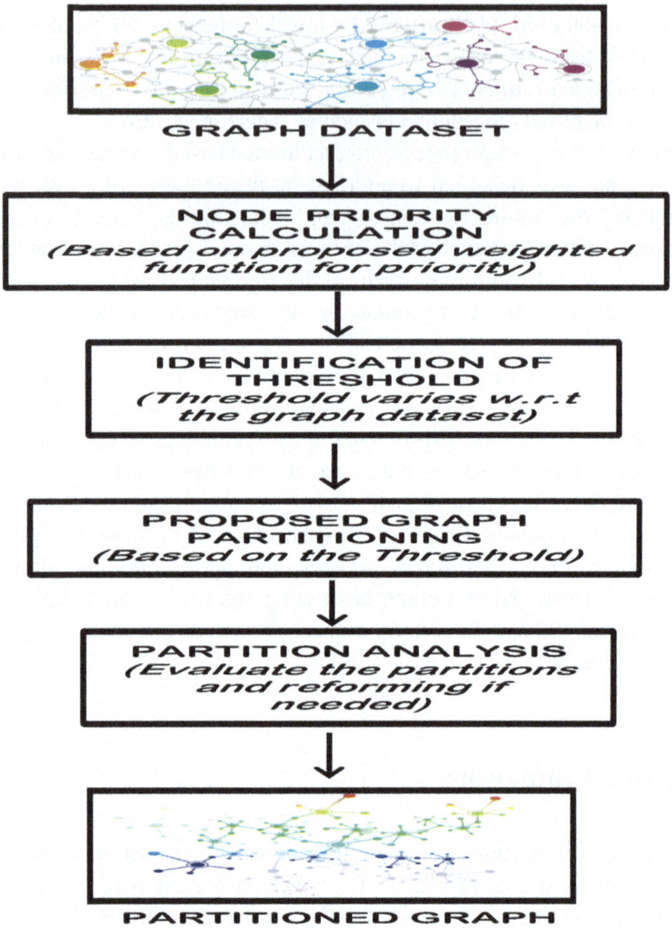

Fig. 1 Workflow of the proposed partitioning algorithm

vertex/node. The in-degree and out-degree are computed as below.

$$\deg^{+}(v_i) = \sum_j e_{ij} \tag{1}$$

$$\deg^{-}(v_i) = \sum_j e_{ji} \tag{2}$$

Here, two different node priorities, namely INP—in-degree-based node priority and ONP—out-degree-based node priority, are computed. Both of them represent the reachability of the node in the graph and are computed using below equations.

$$\mathrm{INP}(v_n) = \alpha(\deg^{+}(v_n)) + \beta(\deg^{-}(v_n)) \tag{3}$$

$$\mathrm{ONP}(v_n) = \beta(\deg^{+}(v_n)) + \alpha(\deg^{-}(v_n)) \tag{4}$$

where

i. α and β are fractions. $0 > \alpha < 1, 0 > \beta < 1$
ii. α should be greater than β. $\alpha > \beta$
iii. Sum of α and β should be equal to 1. $\alpha + \beta = 1$

3.2 *Identification of Threshold Values*

Based on the node priority values, the threshold values will be identified for partitioning the graph dataset. The given graph dataset will be partitioned into four different partitions with respect to in-degree as well as out-degree priorities, namely very high priority (VHP), high priority (HP), medium priority (MP) and low priority (LP). The partitions can be processed in distributed systems. The criteria for forming partitions are decided by the weights assigned to each partition.

$$T_{\mathrm{in}} = w_1 * \max\ \mathrm{INP} + w_2 * \max\ \mathrm{INP} + w_3 * \max\ \mathrm{INP} \tag{5}$$

$$T_{\mathrm{out}} = w_1 * \max\ \mathrm{ONP} + w_2 * \max\ \mathrm{ONP} + w_3 * \max\ \mathrm{ONP} \tag{6}$$

3.3 *Threshold-Based Partitioning*

The upper and lower boundaries for the four partitions are defined based on the corresponding threshold value and weights assigned. The partitions with respect to the in-degree priorities and threshold are listed below.

$$w_1 * T_{\mathrm{in}} > \mathrm{VHP_{in}} \leq T_{\mathrm{in}} \tag{7}$$

$$w_2 * T_{\mathrm{in}} > \mathrm{HP_{in}} \leq w_1 * T_{\mathrm{in}} \tag{8}$$

$$w_3 * T_{\mathrm{in}} > \mathrm{MP_{in}} \leq w_2 * T_{\mathrm{in}} \tag{9}$$

$$0 > \mathrm{LP_{in}} \leq w_3 * T_{\mathrm{in}} \tag{10}$$

The partitions with respect to the out-degree priorities and threshold are given below.

$$w_1 * T_{\text{out}} > \text{VHP}_{\text{out}} \leq T_{\text{out}} \tag{11}$$

$$w_2 * T_{\text{out}} > \text{HP}_{\text{out}} \leq w_1 * T_{\text{out}} \tag{12}$$

$$w_3 * T_{\text{out}} > \text{MP}_{\text{out}} \leq w_2 * T_{\text{out}} \tag{13}$$

$$0 > \text{LP}_{\text{out}} \leq w_3 * T_{\text{out}} \tag{14}$$

Based on the above criteria, the graph dataset will be initially partitioned into eight groups. These groups will be analyzed in the next phase to get the final partitions.

3.4 Threshold-Based Partitioning

The partitions generated with respect to the in-degree priorities and out-degree priorities will be analyzed in this phase. The partitions of in-degree priorities will be compared with that of out-degree priorities. During refining partitions, the vertices in the lesser priority levels may be upgraded to immediate higher priority levels of partitions. As per the threshold-based partitioning, the graph dataset will be partitioned into two grouping cases with eight partitions—four with respect to in-degree priorities (first grouping case) and four with respect to the out-degree priorities (second grouping case). For example, let us consider a vertex v_i. The vertex v_i will be assigned to medium priority partition MP based on in-degree priorities or assigned to higher priority partition HP based on the out-degree priorities. Since the vertex v_i belongs to higher priority level in any one grouping case, it is upgraded to the higher priority level in the partition analysis phase. The eight partitions are thus analyzed, reduced and reformed to get the final set of four partitions in this phase.

3.5 Proposed Graph Partitioning Algorithm

The proposed algorithm for graph partitioning is given below. The input the partitioning algorithm is the graph dataset. The graph dataset considered for the algorithmic design consists of only two attributes, namely "from node" and "to node". The algorithm does not support the graph data in the form of adjacency matrix. The output of the algorithm will be four partitions of nodes.

Algorithm: Node priority and Threshold based Graph Partitioning

Input: Graph dataset G Output: Partitioned Graph

1. Compute the in-degree $deg^{+}(v_n)$ and out-degree $deg^{-}(v_n)$ of all nodes
2. Calculate the corresponding node priorities with respect to in-degrees $INP(v_n)$ and out-degrees $ONP(v_n)$.
3. Optimally find the weights (w_1, w_2, w_3)
4. Identify the thresholds based on INP and ONP
5. Partition the graph data into initial partitions
6. Analyze the initial partitions
   a. Compare the partitions of in-degree and out-degree
   b. Analyze the suitability of refining
   c. If possible upgrade the nodes to next immediate higher partition priority level
   d. Combine the partitions if needed
7. Final partitions **VHP, HP, MP, LP**

4 Results and Discussion

The research work is tested using the datasets collected from Stanford large network dataset collection [16]. From the datasets available, two datasets [17–19] are selected for implementing the proposed partitioning algorithm.

4.1 Dataset Description

The amazon0601.txt.gz is a network dataset which was collected by sneaking the Amazon Website. This network data is based on the similar product purchases made by the customers through Amazon. The Wiki-Talk.txt.gz is a communication network. In Wikipedia, each registered user will have a page named as talk page. Any other registered user can discuss as well as edit the article of Wikipedia. This dataset was created from the page edit history dumps from its inception till January 2008. The details about the dataset are given below (Table 1).

Table 1 Datasets used

Name	Type	Nodes	Edges	Size (MB)
amazon0601	Directed	403,394	3,387,388	45.6
Wiki-Talk	Directed	2,394,385	5,021,410	63.3

Table 2 Number of nodes in the four partitions

Partition ID	Amazon	Wiki
VHP	92,781	646,484
HP	167,026	790,965
MP	131,234	891,939
LP	12,353	64,997
Total	403,394	2,394,385

4.2 Research Outcomes

The configuration of the machine used for implementation is Intel core i7 processor seventh generation with 8M cache and clock speed up to 4.50 GHz. The partitions and the number of nodes in each partitions of the dataset are tabulated below (Table 2).

The percentage of nodes assigned for the partitions of both dataset are illustrated in Fig. 2. The time taken for partitioning the chosen datasets is plotted in Fig. 3. The loading and execution time breakups are clearly depicted in the chart. The execution time is very low than the loading time. It is evident from the figure that the proposed algorithm is having less computational overhead.

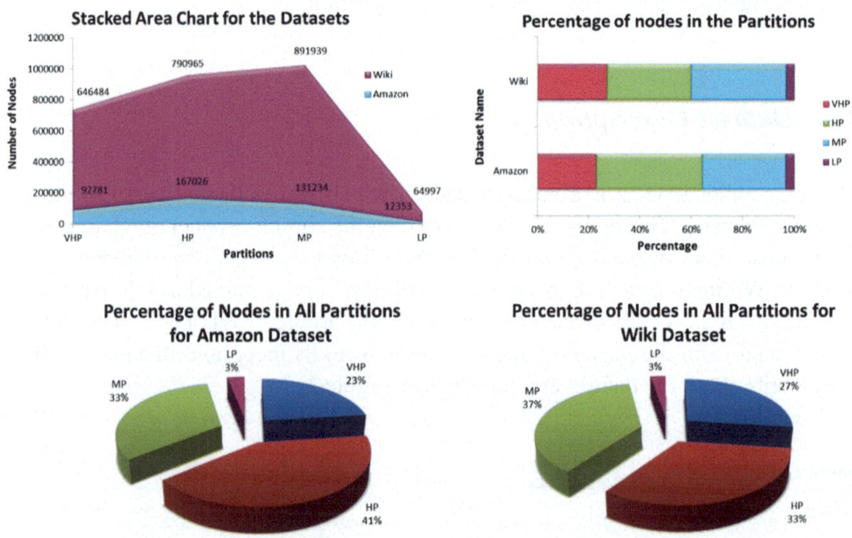

Fig. 2 **a** Stacked area chart for the datasets chosen. **b** Percentage of nodes in the four partitions. **c** Percentage of nodes for Amazon Dataset. **d** Percentage of nodes for Wiki Dataset

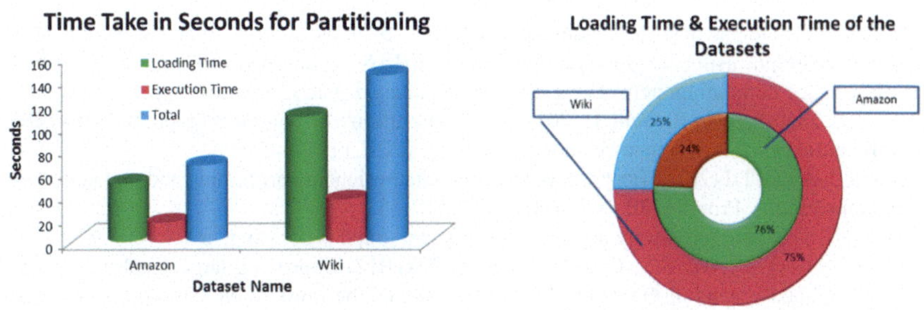

Fig. 3 **a** Time taken (in seconds) for partitioning. **b** Percentage of time taken for partitioning the datasets

5 Conclusion

This research work consists of four phases for partitioning the graph dataset. The priorities of the nodes with respect to their incoming and outgoing link are identified first. Then, the threshold for partitioning is calculated. The initial partitions are analyzed to form the optimal partitions through the analysis phase. The proposed node priority and threshold-based partitioning algorithm effectively group the nodes into four priority classes. Once the graph data is prioritized and grouped accordingly, it will be efficiently processed in distributed systems. This work is a part of energy-efficient graph processing. The outcomes of this work will be further extended to reduce energy consumption while processing graph data.

References

1. Buluç A, Meyerhenke H, Safro I, Sanders P, Schulz C (2013) Recent advances in graph partitioning
2. Malewicz G, Austern MH, Bik AJ, Dehnert JC, Horn I, Leiser N, Czajkowski G (2010) Pregel: a system for large-scale graph processing. In: Proceedings of ACM SIGMOD international conference on management of data, SIGMOD'10, pp 135–146
3. Kyrola GB, Guestrin C (2012) GraphChi: large-scale graph computation on just a PC. In: Proceedings of the 10th USENIX conference on operating systems design and implementation, OSDI'12. USENIX Association, Berkeley, CA, USA, pp 31–46
4. Gill G, Dathathri R, Hoang L, Pingali K (2018) A study of partitioning policies for graph analytics on large-scale distributed platforms. Proc VLDB Endowment 12(4):321–334
5. LeBeane M, Ryoo JH, Panda R, John LK (2015) Watt watcher: fine-grained power estimation for emerging workloads. In: 2015 27th International symposium on computer architecture and high performance computing (SBAC-PAD). IEEE, pp 106–113
6. Song S, Zheng X, Gerstlauer A, John LK (2016) Fine-grained power analysis of emerging graph processing workloads for cloud operations management. In: 2016 IEEE international conference on big data (Big Data). IEEE, pp 2121–2126
7. Lee K, Liu L (2013) Efficient data partitioning model for heterogeneous graphs in the cloud. In: Proceedings of the international conference on high performance computing, networking, storage and analysis. ACM, p 46

8. Boman EG, Devine KD, Rajamanickam S (2013) Scalable matrix computations on large scale-free graphs using 2D graph partitioning. In: 2013 SC—International conference for high performance computing, networking, storage and analysis (SC), Nov 2013, pp 1–12
9. Wang L, Xiao Y, Shao B, Wang H (2014) How to partition a billion-node graph. In: 2014 IEEE 30th international conference on data engineering. IEEE, pp 568–579
10. Huang J, Abadi DJ (2016) Leopard: lightweight edge-oriented partitioning and replication for dynamic graphs. Proc VLDB Endowment 9(7):540–551
11. Si B (2018) A structure-aware approach for efficient graph processing. arXiv:1806.00907
12. Chen R, Yang M, Weng X, Choi B, He B, Li X (2012) Improving large graph processing on partitioned graphs in the cloud. In: Proceedings of the third ACM symposium on cloud computing. ACM, p 3
13. Li D, Zhang Y, Wang J, Tan KL (2019) TopoX: topology refactorization for efficient graph partitioning and processing. Proc VLDB Endowment 12(8):891–905
14. Chen R, Shi J, Chen Y, Zang B, Guan H, Chen H (2019) Powerlyra: differentiated graph computation and partitioning on skewed graphs. ACM Trans Parallel Comput (TOPC) 5(3):13
15. Li D, Zhang C, Wang J, Xu H, Zhang Z, Zhang Y (2017) Grapha: adaptive partitioning for natural graphs. In: 2017 IEEE 37th international conference on distributed computing systems (ICDCS). IEEE
16. https://snap.stanford.edu/data/ Last accessed on 10 Sept 2019
17. Leskovec J, Adamic L, Adamic B (2007) The dynamics of viral marketing. ACM Trans Web (ACM TWEB) 1(1)
18. Leskovec J, Huttenlocher D, Kleinberg J (2010) Signed networks in social media. CHI 2010
19. Leskovec J, Huttenlocher D, Kleinberg J (2010) Predicting positive and negative links in online social networks. WWW2010

Method for Simulating SQL Injection and DOS Attack

K. Rohini, K. Kasturi, and R. Vignesh

Abstract Data is the most important aspect in any computer system. Database powered Web applications are used by the organization to get data from customers and the most used database in SQL. Attacks are created day by day to exploit the database and extract information without authorization. SQL injection in such attack poisons dynamic SQL statements to access the SQL database. GNS3 is an effective tool to model and emulate this attack. Using this setup, we can simulate the setup of particular network systems, execute SQL injection attacks on victim computers and retrieve network logs to analyse them. On the other side, denial-of-service (DoS) attacks are yet another major cybersecurity threats, and more researches have been going on developing DoS attack detection techniques. Modelling DoS attacks is an easy simple job because there are very big contemporary DoS attacks, and in most of the instances, it would be impossible to simulate them. The graphical network simulators can be used (GUI) to detect the attacks. Due to heavy traffic, it is subjected to hammering of trustworthiness over the last few years of uncertainties regarding its reliability. In order to avoid that the filtering based on source IP address is integrated with ports by using an access control management. The most important aim of this work is to simulate an SQL injection attack and denial-of-service (DoS) attack situations in a complex network environment. This model can help researchers to develop the effective countermeasures.

Keywords Graphical network simulators · Network forensics · SQL injection attack · Denial-of-service (DoS) attacks · Internet Control Message Protocol (ICMP)

K. Rohini (✉)
Associate Professor, Dept of Computer Science, VISTAS, Chennai, Tamil Nadu, India
e-mail: rrohini16@gmail.com

K. Kasturi
Assistant Professor, Dept of Computer Science, VISTAS, Chennai, Tamil Nadu, India
e-mail: kasturi.scs@velsuniv.ac.in

R. Vignesh
Tata Consultancy Services, Chennai, Tamil Nadu, India
e-mail: rrvignesh7@gmail.com

S.-L. Peng et al. (eds.), *Intelligent Computing and Innovation on Data Science*,
Lecture Notes in Networks and Systems 118,
https://doi.org/10.1007/978-981-15-3284-9_85

1 Introduction

The computer network technology is rising rapidly, and the growth of Internet usage in the technology is more rapid. The Internet is used for a multitude of reasons by the general individuals, including economic transactions, instructional works and many other activities. Using the Internet to perform significant duties, such as transferring a balance from a bank account, always involves a security risk [1]. Network security is most important issue of computing since several kinds of attacks are growing every day. Of all the attacks in network, DoS and SQL injection are the attacks which are still a major problem for many corporations and enterprises [2]. Even as countermeasures are introduced to prevent them, new types of DoS attacks are used such as the distributed denial-of-service (DDoS) in which numerous compromised computer systems called bonnets attack a destination such as a server, Website or other network resource and cause a rejection of service for users of the intended resource [3]. Moreover, most of the Websites struggle to keep their users' data confidential which is a very difficult and most important task given the day to day discovery of bugs in the codes which are used to build the very Website [4]. The databases behind these Websites store significant data along with sensitive information. Without proper authorization, no user can access the database behind the Website, but with the vulnerabilities, a hacker can access it without authorization. One such attack is SQL injection enabling an interloper to infuse malevolent input into a SQL statement in which the condition will always be true.

1.1 SQL Injection AttackMotivation and Impacts

In SQL injection, malicious SQL statements are executed that attack and control any database server by suppressing the security measures. These vulnerabilities can violate authentication and authorization to retrieve the entire content of the database [5].

SQL injection attacks are unique potential and hazardous vulnerabilities in Web applications. An attacker will find vulnerable user inputs within the Web page or Web application to make an attack with SQL injection. SQL injection directly discloses user input in a SQL query. In order to be executed in the database, the attacker can create malicious input content called payload.

During the exploitation of an SQL injection technique on a susceptible Website, an intruder can do a set of things. Usually, connecting to the database server relies on the user's privileges that the Web application uses. By employing a susceptibility to SQL injection, an intruder can perform:

- Adding, deleting, editing or reading the information of the database
- Reading the source code and writing files of database server
- The exploitation which can still direct towards the total capture
- of the database and Web server.

Attackers can use SQL injections to locate other users' credentials in the database. These users can then be impersonated. The impersonated user can be an administrator having all the privileges of the database [6]. SQL injections allow information selection and output from the database by the attackers. It also enables full access to a database server towards all information. Attackers are able to alter, delete or add records and even drop tables.

1.2 SQL Injection Attacks and Its Types

Data from servers can be exploited through SQL injection in many ways. Popular techniques provide data extracting based on faults, conditional checking (true/false) and timing. The different kinds of SQL injection attacks are as follows:1.2.1 Error-Based Intrusion using SQL Injection.

Using SQL injection based on intrusion error through vulnerable attackers can obtain data from noticeable database faults such as table names and content. In the error-based attacks, the query request returns an error, thereby creating a duplicate entry to the database. For table names and content, the same process operates [7]. Deactivating error notifications on industrial systems helps avoiding the collection of such data by intruders.

1.2.1 Boolean-Based Intrusion Using SQL Injection

Whenever a SQL query fails, there is no noticeable error information on the page. There is a further way to obtain details, as sections of the Webpage quite often vanish or update, or the whole Webpage may refuse to load. Such signs enable attackers to decide if the input variable is susceptible and if it enables data to be extracted. Intruders can test this by inserting conditional check into a SQL request:

```
https://Vqj.com/content.php?id=1+AND+1=1
```

ExtractData: https://Vqj.com/content.php?id=1+AND+IF(version()+LIKE+'5%',true,false)

By this query, if the database version is 5.X, then the Website must load as normal. But if the version is distinct, it will act differently (such as displaying an empty page), showing if it is susceptible.

1.2.2 Time-Based Intrusion Using SQL Injection

In certain ways, while a susceptible SQL query has no tangible effect on the page's output, it could still be possible to obtain details from an inherent database by urging the database to wait/sleep for a specified time period earlier to answering. When the page is not susceptible, loading will take longer than normal, enabling hackers to obtain data, still there are no noticeable deviations upon the page. However, to create 'real' method is altered to the one that requires a few little times to perform, like 'sleep (5)' which gives the database five seconds to sleep.

https://Vqj.com/content.php?id=1+AND+IF(version()+LIKE+'5%
',sleep(5),false).

1.2.3 Out-of-Band Intrusion Using SQL Injection

Always using the out-of-band method is really the only way an intruder can extract data from a database. Such kinds of assaults normally require sending the information straight to a device operated by the intruder from the database server. This may be used by attackers if an injection does not happen immediately after the information has been placed but even at a later date.

The destination makes a DNS application to the domain possessed by the attacker in these needs, with the cause of the query within the sub-domain. This implies that an intruder does not have to see the injection outcome, yet can wait until a request is sent by the database server.

1.3 Impacts of DoS Attack

A denial-of-service (DoS) attack occurs when there is no availability of a service that would normally operate. There may be many reasons for not being available, but it generally relates to infrastructure that is unable to deal owing to overloading ability.

The distributed denial-of-service (DDoS), resulting from a big amount of malicious applications attacks one target. This is often performed via a botnet, where many devices are programmed at precisely the same moment to request a service [8]. Compared to hacking attacks such as phishing or brute-force attacks, DoS generally does not attempt to steal data or lead to a violation of safety, but the loss of reputation for the impacted business may still cost a great deal of time and money.

Public services are still being victimized by DoS attacks, with many sector specialists predicting that this will get much worse gets better. Cybercriminals continue to concentrate their attempts on penetrating critical public infrastructure systems such as energy grids, nuclear facilities, transport networks, economic stability, public health and even the supply of drinking water.

DoS attacks can interfere with the accessibility of vital services that we use as part of our daily lives [9]. The risks of infrastructure attacks have been identified by previous reports such as ransomware attacks. The effects of the DoS operation in an organization can be serious—from economic expenses to an adverse effect on the reputation of a brand. For example, network downtime can have a severe financial effect as it can affect productivity, physical harm and even threaten public safety.

1.4 DOS Attack and Its Types

As quoted by ***Matthew Prince,*** **in a very simpler form,** *"...a Denial of Service attack is when an attacker is trying to generate more traffic than you have resources to handle...".*

The goal of DoS attack demands for information that has the impact of disabling the victim's features. DoS attacks were recorded in 98 Q3 nations, where China was targeted for the biggest amount of attacks (63.30% of all assaults), which is 5.3 pp. higher than the quarter before. The share of South Korea dropped from 14.17 to 8.70%, shifting it to fifth. The USA came in the second despite this country's proportion of assaults dropping from 14.03 to 12.98%. The US Computer Emergency Readiness Team (US-CERT) offers instructions to determine when a DoS attack can take place. Modelling a DoS attack in reality is not an easy task. This can be achieved with the help of an emulator tool Graphical Network Simulator 3. This can demonstrate how a real DoS attack is in real-life situation and at the same time ensuring safety in modelling the attack. The common five types of DDoS/DoS methods or attacks.

1. SYN flood: This type of DoS attacks wherein an intruder delivers a set of requests to a victim's system in a trial for using large volumes of server resources to develop the system mechanism non-responsive to network traffic [10].
2. Teardrop attacks: This intrusion involves providing the victim's device with overlapping, over-sized payloads with broken and unorganized IP fragments. Obviously, the goal is to invade into the server and even bang operating system by the bug during the reassembling of TCP/IP fragmentation. All the operating systems, including UNIX, are at risk for this type of DoS attack.
3. Low-rate denial-of-service attacks: This is almost always a lethal attack on DoS! This type of attack is planned to take benefit of the slow-time dynamics of TCP to have the ability to implement the retransmission time-out (RTO) method to scale back TCP outturn. In brief, an attacker produces a protocol outflow by constantly entering an RTO state by creating high-rate and intense bursts–even at slow time scales for RTO. The outcome of the protocol at the target node will be decreased substantially, even though the invader may have a small typical regular rate, that is hard to detect.
4. ICMP flood: This type of protocol is a connecting protocol which is not frequently used for IP, diagnostic and error activities. AN ICMP flood—transferring any of the excessive numbers of ICMP packets—will dominate a destination server that

tries to process any incoming demand, leading to a DoS dysfunction for the destination server.

5. Peer-to-peer attacks: This network unlike a centralized client–server system could act as a distributive network in which individual network nodes operate as both servers and clients of resources, whenever there is an access request from client–server as well as operating system nodes to the resources supplied by centralized machines.

2 SQL Injection and Dos Attack Simulation Method

2.1 GNS3

Hardly, no widespread network simulator is available that can be used to create many of the simulations. There are benefits and drawbacks for every simulator. In this work, we used Graphical Network Simulator (GNS3) which is an emulator tool and gives results very similar results to real network process [11]. GNS3 enables us to operate a simple topology composed of just a few components on our laptop, to those with various components hosting various servers or enabled even in cloud hosting. One of the main reasons of selecting GNS3 is that it is an open-source software. It supports multiple switching options, allVIRL images and multi-vendor environments.

2.2 Kali Linux

Kali Linux is aimed for distributive, sophisticated, advanced testing and secure auditing platform based on a Debian-based Linux. Kali includes many hundred components for countless information protection functions, such as penetration testing, security scrutinize, computer forensics and reverse engineering. The tool used in this attack is SQL map which injects SQL statements into the space provided for login in the attacker's Website and shows the list of all databases in the vulnerable Website. The tool used in DoS attack is hping3 which is used to flood the victim with the TCP packets endlessly until the service crashes. Figures 1 and 2 show the results in SQL injection attack.

Ping Results towards Web server before attack and during the DoS attack are shown in Figs. 3 and 4.

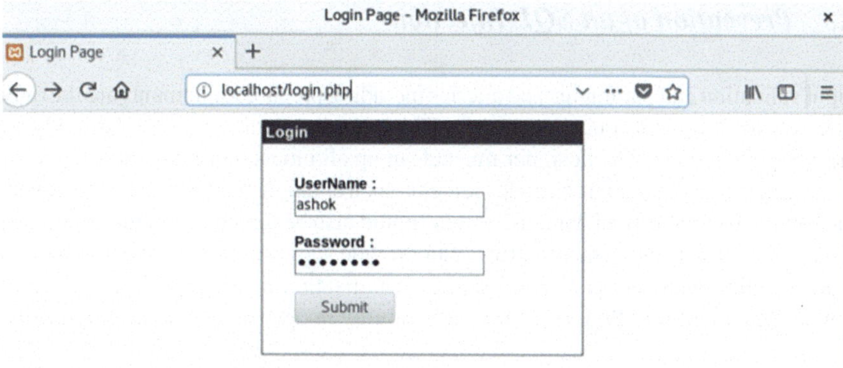

Fig. 1 Password has been taken from the database for user Ashok

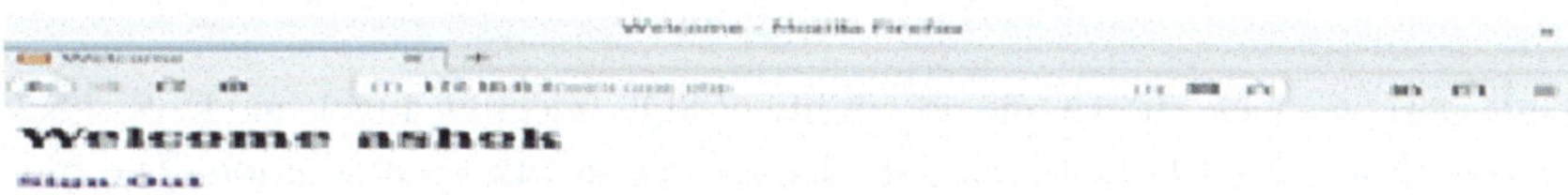

Fig. 2 Content inside the database accessed

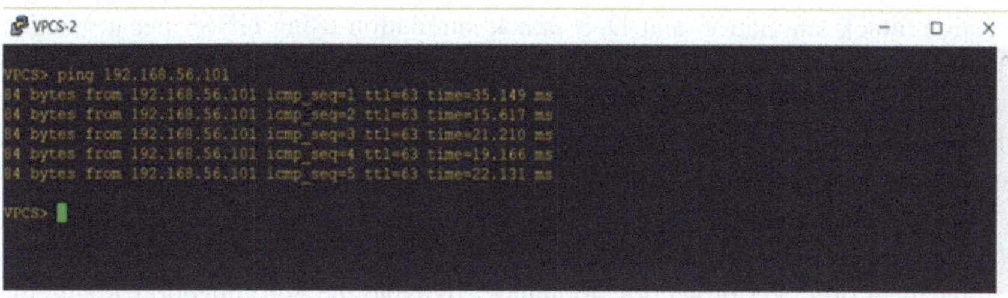

Fig. 3 VPCS-2 TCP ping of Web server before attack

Fig. 4 VPCS-2 TCP ping of Web server during attack

2.3 Prevention of an SQL Injection

Input validation and parameterized queries including prepared statements are the only sure way to avoid SQL injection attacks. The input should never be used directly by the application code. The designer must clean up all inputs, not just inputs from the Web form. Potential malicious code elements such as single quotes must be removed. Moreover, the visibility of database errors should also be turned off on the concerned production sites as the database errors can provide suggestion about the database to SQL injection attackers [12]. Furthermore, sanitization of the input using firewall can be done to prevent SQL injection vulnerability in open source code temporarily.

3 Conclusion

In this work, we have shown one possibility of SQL injection attack and DoS attack. Using a cloud-based Dos security service is a complete ideal solution. The most important thing regarding DoS attacks is to think and implement security measures in advance. There are many types of SQL injection attack through which an attacker could take down a Webserver and its networks. This focused on creating a SQL injection attack simulation and DoS attack simulation using GNS3 because other popular simulators like OPNET, NS3 and others do not have an accurate procedure and result. Furthermore, there are also some drawbacks in using GNS3 especially relative to many other network simulators. Because it uses hardware services to simulate all device's task, and within its topology, scalability is restricted.

This attack can cause major problem to big enterprises and companies. To prevent DoS attacks, we have countermeasures such as using a next-generation firewall, load balancer or a DoS protection appliance. To avoid the SQL injection attack, the user input should be sanitized before using it in SQL database, not displaying error messages, and the needed access privileges should be granted to the accounts used to connect to the database. Simple prevention strategies can save the important data from the databases to be exploited. This simulation of DoS attack method can be more effectively done using the SQL injection method where complex SQL queries are sent to the database that will overload and drain the efficiency of the Web server and the database and exhaust the server resources.

References

1. Harshita N, Ramesh (2013) A survey of different types of security threats and its countermeasures. In: International conference on electrical, electronics and computer engineering, Mysore, ISBN 978-81-927147-3-8, May 2013
2. Vignesh R, Rohini K (2018) Analysis to determine the scope and challenging responsibilities of ethical hacking employed in cyber security. Int J Eng Technol 7(3.27):196–199

3. Mukhopadhyay D, Oh B-J, Shim S-H, Kim Y-C (2010) A study on recent approaches in handling DDoS attacks, 1(7)
4. Mahrouqi A, Tobin P, Abdalla S, Kechadi T (2016) Simulating SQL-injection cyber-attacks using GNS3. Int J Comput Theory Eng 8(3)
5. Patil A, Laturkar A, Athawale SV, Takale R, Tathawade P (2017) A multilevel system to mitigate DDOS, brute force and SQL injection attack for cloud security. In: 2017 International Conference on Information, Communication, Instrumentation and Control (ICICIC), Indore, 2017, pp 1–7, -3–8
6. Sharma C, Jain SC (2014) Analysis and classification of SQL injection vulnerabilities and attacks on web applications. In: International Conference on Advances in Engineering & Technology Research (ICAETR—2014), Aug 2014
7. SejalFarde, Chaudhari S (2016) SQL Injection (SQLI). Int J Adv Res Comput Eng Technol (IJARCET) 5(6)
8. Moore D, Voelker G, Savage S (2001) Inferring internet denial-of service activity. Technical report, DTIC Document
9. Singh S, Bhandari A (2013) Review of PPM, a traceback technique for defending against DDoS attacks. Int J Eng Trends Technol (IJETT), vol 4, ISSN 2231-5381, p 2550, 6 June 2013
10. Balyk A, Karpinski M, Naglik A, Shangytbayeva G, Romanets I (2017) using graphical network simulator 3 for DDOS attack simulation. Int J Comput 16(4):219–225
11. Jelena M (2004) A taxonomy of DDoS attack and DDoS defence mechanisms. 449 Smith Hall, Computer and Information Sciences Department, University of Delaware Newark, DE 19716, Peter Reiher, 3564 Boelter Hall, Computer Science Department UCLA
12. Kaushik M, Gazal O (2004) SQL injection attack detection and prevention methods: a critical review. Int J Innov Res Sci Eng Technol (An ISO 3297: 2007 Certified Organization) 3(4)

Correction to: Intelligent Computing and Innovation on Data Science

Sheng-Lung Peng, Le Hoang Son, G. Suseendran, and D. Balaganesh

Correction to:
S.-L. Peng et al. (eds.), *Intelligent Computing and Innovation on Data Science*, Lecture Notes in Networks and Systems 118, https://doi.org/10.1007/978-981-15-3284-9

In the original version of the book, the author name "Midhunchakkravarthy" has been changed to "Midhunchakkaravarthy" for the below Chapters:

"A Survey on Feature Fatigue Analysis Using Machine Learning Approaches for Online Products"

"A Study on Seismic Big Data Handling at Seismic Exploration Industry"

"The Impact of Organizational Innovation on Financial Performance: A Perspective of Employees Within Dubai Ports World"

"A Systematic Survey on Load Balancing in the Cloud"

The author name "V. Vivekanandam" has been changed to "B. Vivekanandam" for the below Chapters:

"A Survey on Feature Fatigue Analysis Using Machine Learning Approaches for Online Products"

"Green Technology to Assess and Measure Energy Efficiency of Data Center in Cloud Computing"

The chapters and book have been updated with the changes.

The updated versions of these chapters can be found at
https://doi.org/10.1007/978-981-15-3284-9_26
https://doi.org/10.1007/978-981-15-3284-9_37
https://doi.org/10.1007/978-981-15-3284-9_46
https://doi.org/10.1007/978-981-15-3284-9_52
https://doi.org/10.1007/978-981-15-3284-9_82

S.-L. Peng et al. (eds.), *Intelligent Computing and Innovation on Data Science*,
Lecture Notes in Networks and Systems 118,
https://doi.org/10.1007/978-981-15-3284-9_86

The manufacturer's authorised representative in the EU is Springer Nature Customer Service Centre GmbH, Europaplatz 3, 69115 Heidelberg, Germany. If you have any concerns regarding our products, please contact ProductSafety@springernature.com

Printed and bound by CPI Group (UK) Ltd, Croydon, CR0 4YY
25/09/2026
02222853-0019